International Scientific Conference "Investments. Construction. Real Estate: New Technologies and Targeted Development Priorities" (ICRE-2020)

IOP Conference Series: Materials Science and Engineering
Volume 880

Irkutsk, Russia
23 - 24 April 2020

ISBN: 978-1-7138-4744-1
ISSN: 1757-8981

Printed from e-media with permission by:

Curran Associates, Inc.
57 Morehouse Lane
Red Hook, NY 12571

Some format issues inherent in the e-media version may also appear in this print version.

Institute of Physics
Dirac House, Temple Back
Bristol BS1 6BE UK

Phone: 44 1 17 929 7481
Fax: 44 1 17 920 0979

techtracking@iop.org

Additional copies of this publication are available from:

Curran Associates, Inc.
57 Morehouse Lane
Red Hook, NY 12571 USA
Phone: 845-758-0400
Fax: 845-758-2633
Email: curran@proceedings.com
Web: www.proceedings.com

TABLE OF CONTENTS

IOP Conf. Series: Materials Science and Engineering **880** (2020) 011001 doi:10.1088/1757-899X/880/1/011001

PREFACE

The III International Scientific and Practical Conference «Investments. Construction. Real Estate: New Technologies and Targeted Development Priorities 2020» (ICRE-2020) was organized by Irkutsk National Research Technical University (Russia, Irkutsk), Novosibirsk State University of Architecture and Civil Engineering, (Russia, Novosibirsk), Regional Academic Research and Education Center (Russia, Irkutsk), Siberian Federal University, (Russia, Krasnoyarsk), East-Siberian State University of Technology and Management (Russia, Ulan-Ude), Institute of Energy Systems named after L. A. Melentyeva (Russia, Irkutsk), Harbin Polytechnic University (China), Hokkaido University, (Japan), Vinh University, Vinh, (Vietnam), Mongolian State University of Science and Technology (Mongolia, Ulan Bator), The American University of Moldova (Moldova) on 23-24 April, 2020.

The major goal and feature of the ICRE-2020 conference is to bring academic scientists, engineers, industry researchers together to exchange and share their experiences and research results, and discuss the practical challenges encountered and the solutions adopted. The conference program encompassed a wide range of topics and was divided into 10 sections: Building materials, building construction and technologies of a new generation; Energy efficient and green buildings; Housing and municipal services. Urban engineering infrastructure; Architecture and urban planning. The quality of the urban environment; Highways. Transport facilities; Digital economy and information modeling of buildings and structures; Land resources and economic development of the regions of Siberia and the Far East; Economics and organization of construction. Construction Management, real estate, housing and municipal services; Management and innovation in the development of investment and construction industry; Transformation of the real estate market.

For the first time, the conference was held online. In order to ensure the sanitary and epidemiological well-being of the population in the Russian Federation in connection with the spread of a new coronavirus infection (COVID-19), the President of the Russian Federation established a self-isolation regime from April 4 to 30, 2020. Based on this, we are not entitled to hold face-to-face meetings of the conference and hear scientific reports.

The conference organizers decided to hold the conference online. The plenary meeting of the conference was held on April 23, 2020 (Zoom, Bigbluebutton platforms). Sessions on sections of the conference were held on April 24, 2020 (Zoom, Bigbluebutton platforms). All conference participants have registered on the TimePaid platform (we enclose a list of registered conference participants).

More than 150 participants from different regions of the Russian Federation, China, Vietnam, Japan and other countries attended the conference. Conference participants discussed prospective fields of research and important achievements in the industrial and civil construction, architecture and urban planning, design, construction, real estate management and transformation.

On behalf of the Organizing Committee we express our gratitude to the participants for the active work at the conference sections

prof, Peshkov V V

Peer review statement

All papers published in this volume of *IOP Conference Series: Materials Science and Engineering* have been peer reviewed through processes administered by the proceedings Editors. Reviews were conducted by expert referees to the professional and scientific standards expected of a proceedings journal published by IOP Publishing.

IOP Conf. Series: Materials Science and Engineering **880** (2020) 012001 doi:10.1088/1757-899X/880/1/012001

Comparative analysis of the compression strength characteristics of polymer concrete modified by ultra-fine by-products silicon and aluminum industries

V B Balabanov[1] and K N Putsenko[1]

[1]Department of Urban Construction and Management, Irkutsk National Research Technical University St. Lermontov 83, Irkutsk, Russia

E-mail: putsenko_kseniya@mail.ru

Abstract. Presently, one of the actual problems of concrete science is the improvement of concrete formulations by using industrial waste, ultra-fine by-products from the production of aluminum and silicon industry, as well as modification by means of additives. The aim of scientific research is development of new technologies for obtaining concrete with enhanced performance characteristics. The article provides an overview of the causes of concrete destruction and methods for increased physical-mechanical characteristics and of concrete. The results of a literature search on the experience of using silica fume and polymers in concrete are presented. The advantages and disadvantages of using silica fume and polymers in the composition of concrete are identified. A study was carried out to determine the effect of various modifying additives on the physical-mechanical characteristics of concrete. Experimental studies have been conducted on the joint use of silica fume and polymers in concrete mixture. The results of the study are: the optimal amount of silica fume is 10-20 %, the optimal amount of polymer is 1.0-3.5 %. On the basis of tests to determine the strength revealed that the class of concrete increased from B15 to B30 compared with the composition of the control sample.

1. Introduction

Concrete is the only and most widely used building material in the world. According to analytical companies and the assessment of the Ministry of Economic Development of Russia, the construction industry made a key contribution to improving the dynamics of GDP in 2018. In January-December 2018, Russian plants produced a total of 53.8 million tons of cement. Despite the fact, that cement production decreased by 1.6-1.9% (compared with the previous year), demand for it does not decrease [1].

Large volumes of production and consumption of cement are determined by the number of buildings and structures is being built, both in road construction and in civil and industrial.

One of the main reasons for the destruction of concrete is a group of factors integrated on the basis of chemical effects on concrete. These include carbonization, leaching, sulfatization, destruction by chlorides, chemical reaction between cement alkalis and aggregates.

An equally important reason is the physical cause of the destruction of concrete. This means the destruction of concrete resulting from shrinkage, overstretched, mechanical factors (abrasion or impact). This group also includes fire damage and seismic impact.

To prevent most of the factors leading to the loss of technical and operational characteristics, it is possible through the use of a complex of additives. Having analyzed GOST 24211-2008, we can group all concrete additives in the following groups:

• plasticizers,
• mobility regulating additives,
• boosters,
• increase in corrosion resistance,
• additives for frost resistance of concrete,
• mineral additives,
• complex additives [2].

All used concrete additives have both disadvantages and advantages. Most disadvantages can be avoided with proper use of supplements. But, despite this, it should be remembered that the use of additives is inextricably linked with the rise in price of final construction products. In this regard, a problem is the cost of production without loss of technical and operational characteristics. This is possible by replacing any component of the concrete formulation with a material similar in properties. Such materials are waste and by-products of industrial - by-products from the production of aluminum and silicon industry (silica fume), thermal power station waste (fly ash), etc. Silica fume was selected for our scientific research [3-5].

Silica fume is a by-product of ferroalloy and aluminum production. Silica fume (hereinafter SF) easily reacts with calcium hydroxide, which is formed during the hydration of Portland cement. Due to the inherent physical and chemical properties of SF, it is a highly active pozzolan. SF is a dispersed material with a specific surface area in the range of 13,000 - 30,000 m^2/kg [6-9].

Silica fume is available in several basic forms: suspension, compacted, unconsolidated and granular SF.

The main advantages of using SF are to improve the structure of concrete and increase strength. The main disadvantage is that SF is a dispersed material that increases water consumption. By calculating the water-cement ratio, the specific amount of water required for the desired consistency is determined. It should remain liquid, but have a high viscosity [10-14]. Excessive addition of water to the mixture will lead to the following consequences:

• loss of the required consistency,
• increase in setting time,
• loss of volume of the mixture,
• formation of pores and voids,
• decrease in strength and durability.

To regulate the water-cement ratio and give the mixture additional positive characteristics, it is proposed to use a polymer. The use of polymers allows to achieve such properties as:

• high strength and abrasion resistance,
• water holding capacity,
• increase in the duration of setting time,
• high rates of water resistance and waterproofness,
• increased indicators of frost resistance [15-17].

Based on the data obtained as a result of a review of the causes of concrete destruction and methods for improving the physical-mechanical characteristics of concrete, as well as a review of the experience of using silica fume and polymers in concrete, a plan of experimental studies on the joint use of SF and polymer was compiled.

2. Materials and methods

The aim of the research was:

1. Determine the optimal amount of SF in the concrete mixture,

IOP Conf. Series: Materials Science and Engineering **880** (2020) 012001 doi:10.1088/1757-899X/880/1/012001

2. Determine the optimal amount of polymer in the concrete mixture,
3. To identify the optimal combination of SF and polymer,
4. Determine the strength of the resulting concrete.

A series of samples was made with different contents of SF and latex. These samples gained strength in wet conditions. All samples were tested for strength. Samples of compositions that showed the best strength indicators were selected. The data are presented in figures 1-3.

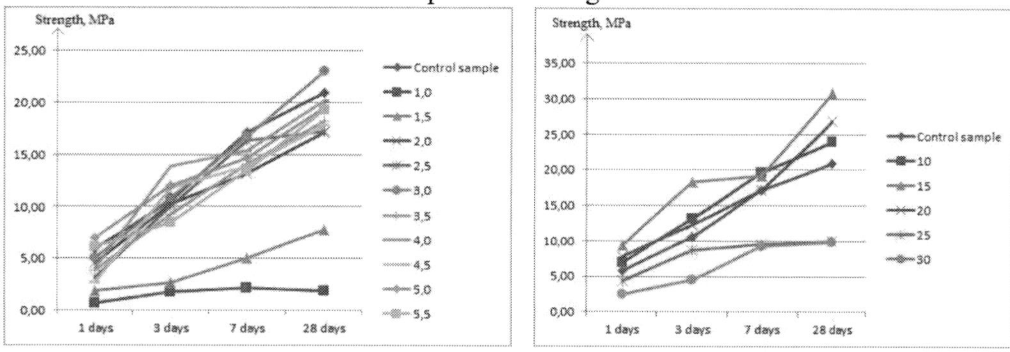

Figure1-2. Graph comparing the dynamics of the strength set of the composition of Control sample + latex (left). Graph comparing the dynamics of the strength set of the composition of Control sample + SF (right).

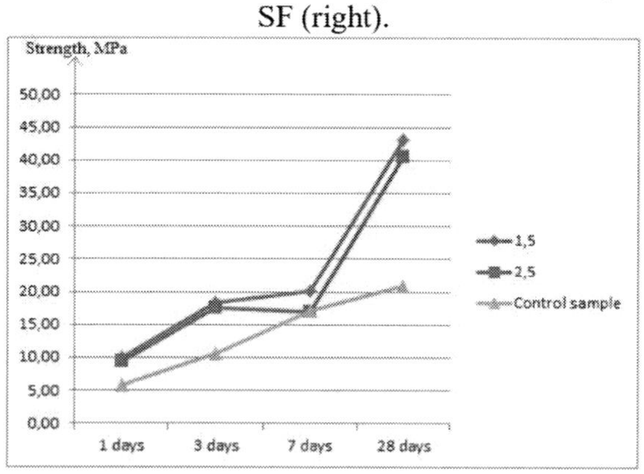

Figure 3. Graph comparing the dynamics of the strength set of the composition of CS + SF + latex (wet conditions of hardening).

After analyzing the graph of the dynamics of the set of concrete strength, we can conclude that the optimal amount of SF is 10-20%, and polymer - 1.0-3.5%.

3. Results

Based on the results of the studies, the following results were achieved:

1. The optimal amount of SF - 10-20%,
2. The optimal amount of polymer is 1.0-3.5%,
3. According to the density analysis, the developed concrete is a heavy type of concrete,
4. Based on tests to determine the strength, it was revealed that the class of the developed formulation increased from B15 to B30 compared to the composition of the control sample.

4. Conclusion

The combined use of silica fume and polymer allows us to achieve a number of positive technical and operational characteristics of concrete. In addition, the use of silica fume allows you to solve a number of current problems:

• import substitution (use of resources of the local raw material base, namely SF of Bratsk),
• environmental friendliness of the final product through the use of silica fume.

IOP Conf. Series: Materials Science and Engineering **880** (2020) 012001 doi:10.1088/1757-899X/880/1/012001

References

[1] Results of 2018. Cement Production in the Russian Federation
[2] GOST 24211-2008. 2008 Additives for concrete and mortar. General specifications (Moscow: Standards Publishing) p 16
[3] Slobodchikova N A, Pluta K V and Dzogii A A 2015 Application prospects of production and consumption waste in building, reconstruction and major repair of motor roads *Irkutsk: Proceedings of Irkutsk State Technical University* **8** pp 126-132
[4] Slobodchikova N A, Pluta K V and Lofler M 2017 Getting the cement on the basis of wastes of industrial production *Irkutsk: Proceedings of Universities. Investments. Construction. Real estate* **7** pp. 62-66
[5] Kujukov S A, Genze D A, Shmatok V V and Zhigajlov A A 2019 A comprehensive approach to improving the durability of soil cement foundation *Permian: Transport. Transport structures. Ecology* **4** pp 80-89
[6] Parker D J 1985 Silica Concrete Current Practice *Sheet Concrete Society* **104** pp 1-6
[7] European Silica Fume Committee, History & manufacturing process
[8] Silica Fume Association *What is silica fume?*
[9] The Silica Fume Association *Silica Silica fume in concrete*
[10] Iler Ralph K 1979 *The Chemistry of Silica: Solubility, Polymerization, Colloid and Surface Properties and Biochemistry of Silica* (New York: Wiley-Interscience) p 896
[11] Electronic Magazine "Construction"
[12] ASTM International, Official website
[13] Holland T C 2005 *Silica Fume User's Manual: Technical Report No FHWA-IF-05-016 for the Federal Highway Administration* (Washington, DC) p 183
[14] Bergna H E 1994 *The Colloid Chemistry of Silica (Advances in Chemistry Series) (USA: American Chemical Society)* p 718
[15] Fedorova G D 2014 An experimental study of the strength of concrete with a complex additive (Moscow: Construction Materials) **4** pp 70-71
[16] Berdov G I 2012 The effect of mineral microfillers on the properties of building materials *Irkutsk: Building Materials* **9** pp 79-83
[17] Falikman V R and Kalashnikov O O 2015 Particularly high-strength quick-hardening concrete *Irkutsk: Concrete and Reinforced Concrete* **1** pp 2-9

IOP Conf. Series: Materials Science and Engineering **880** (2020) 012002 doi:10.1088/1757-899X/880/1/012002

Determination of scale coefficients for samples of autoclave structurally heat-insulating aerated concrete based on fly ash

A Baranova[1] and M Krivyh[1]

[1]Department of Industrial and Civil Engineering, Angarsk State Technical University, Tchaikovsky str., 60, Angarsk, 665835, Russia

E:mail baranova2012aa@mail.ru

Abstract. The article presents the results of determining the scale coefficients for autoclave structurally heat-insulating aerated concrete based on fly ash. In the course of research, the dependences of the compression strength values on the size and area of the working section of samples of autoclave aerated concrete of the grades D500 and D600 are derived. The dependences of mechanical characteristics on its average density in the range from 500 to 600 kg/m^3 are established. For samples of autoclave structurally heat-insulating aerated concrete based on fly ash, the values of scale coefficients were experimentally determined, which in relation to the base sample of 150x150x150 mm are: for prisms of 40x40x160 mm – 1.45, for cubes of 70,7x70,7x70,7 mm and 100x100x100 mm – 1.2 and 0.95, respectively.

Aerated concrete based on fly ash is a type of cellular concrete and is an artificial stone material with evenly distributed pores throughout the volume, obtained as a result of solidification of a rationally selected mixture of a binder, ash, water, additives (if necessary) and a gas-forming agent. Autoclave aerated concrete gains strength under conditions of saturated steam in autoclaves at high temperature (175÷195 °C) and high pressure (0.8÷1.2 MPa) for a certain time (10÷12 hours).

From previously published works [1÷15], it is known that when testing materials for compressive strength, the dimensions of the tested samples have a significant impact.

According to GOST 10180-2012, the compressive strength of concrete that is not determined on basic samples is adjusted using scale coefficients *K*, which are set experimentally for each type and class of concrete, as well as for each test equipment [16].

Since there is no large aggregate in aerated concrete, its strength characteristics can be determined on small samples, for example, on prisms with a size of 40x40x160 mm.

The purpose of the work was to determine the scale coefficients for samples made of autoclave structurally heat-insulating aerated concrete based on fly ash.

The research was carried out on samples of autoclave aerated concrete based on fly ash produced by CJSC «Stroikompleks».

The research methodology was as follows. Samples with dimensions of 40x40x160 mm, 70,7x70,7x70,7 mm, 100x100x100 mm and 150x150x150 mm were cut from blocks of autoclave aerated concrete of grades D500 and D600. Then they were dried to a constant mass at a temperature of 105±5 °C and tested for strength in accordance with GOST 10180-2012 Concretes. Methods for strength determination using reference specimens.

The scale coefficients, standard deviations and variation coefficients were determined using the method set out in application L of GOST 10180.

The test results are presented in tables 1÷3 and shown in figures 1÷4.

Table 1. Mechanical characteristics of autoclave structurally heat-insulating aerated concrete based on fly ash.

Grade of aerated concrete	Size of the sample, mm	Area of the working section of sample, sm²	Compressive strength, R_{com}, MPa	Flexural strength, R_{fl}, MPa
D500	40x40x160	25	2,39	1,29
	70,7x70,7x70,7	50	3,08	-
	100x100x100	100	3,8	-
	150x150x150	225	3,63	-
D600	40x40x160	25	2,95	1,47
	70,7x70,7x70,7	50	3,36	-
	100x100x100	100	4,43	-
	150x150x150	225	4,13	-

The results presented in table 1 show that as the samples size increases from 40x40x160 mm to 100x100x100 mm, the values of the compressive strength of autoclave aerated concrete based on fly ash increase. Further increase in size to 150x150x150 mm leads to a decrease in the values of the compression strength.

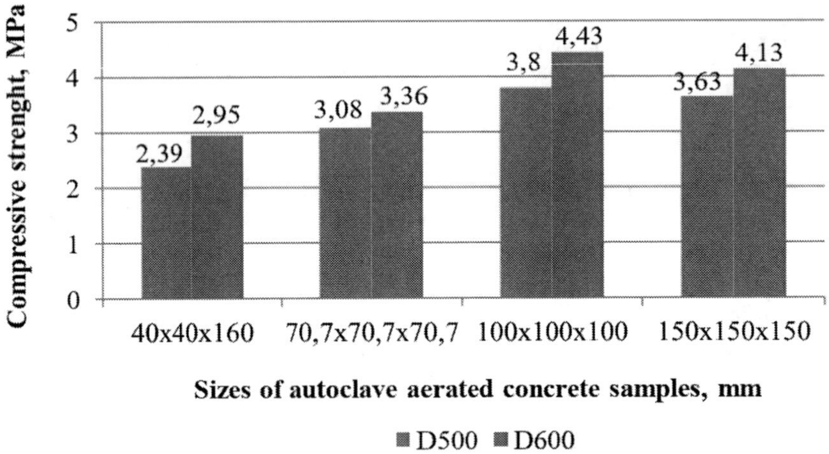

Figure 1. Values of the compressive strength of autoclave aerated concrete based on fly ash, depending on the size of the samples.

The difference in compressive strength between the grades D500 and D600 (figure 1) ranges from 9.1 to 23.4% depending on the size of the samples, while the average value is 15.7 %.

The dependence of the values of the compressive strength of autoclave aerated concrete based on fly ash on the area of the working section of the samples (figure 2) is described by the following equations.

For autoclave aerated concrete of the grade D500, the equation has the form:

$$R_{com} = -0{,}0001 \cdot x^2 + 0{,}0314 \cdot x + 1{,}7035 \qquad (1)$$

For autoclave aerated concrete of the grade D600:

$$R_{fl} = -0{,}0001 \cdot x^2 + 0{,}0336 \cdot x + 2{,}0962, \qquad (2)$$

where x – is the area of the working section of sample of the autoclave aerated concrete based on fly ash, cm².

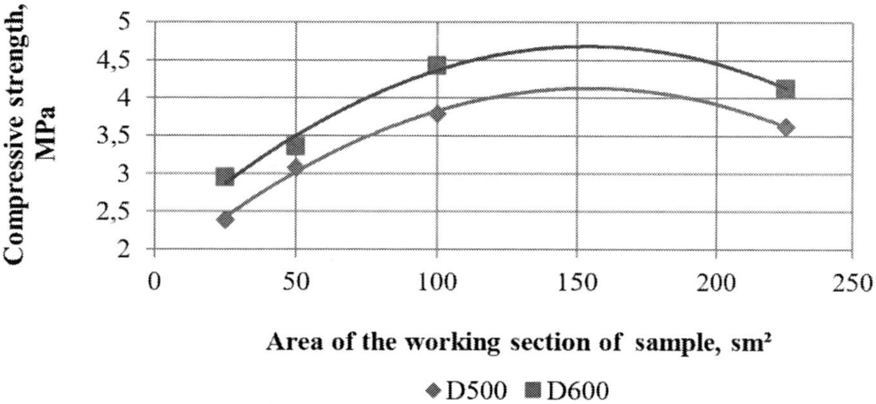

Figure 2. Values of compressive strength of autoclave aerated concrete based on fly ash, depending on the area of the working section of the samples.

According to the graph shown in figure 2, it can be determined that the maximum values of the compression strength of the studied grades of aerated concrete will be obtained on samples with an area of the working section of 150 cm^2.

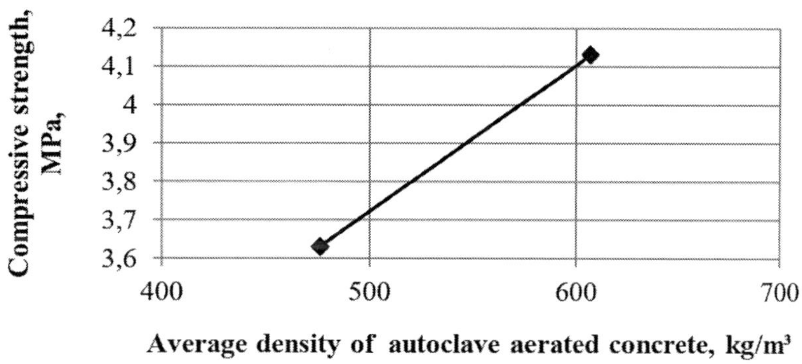

Figure 3. Dependence of the compressive strength of samples of autoclave aerated concrete based on fly ash on the average density.

According to the graph shown in figure 3, it can be seen that the change in the compressive strength of samples of autoclave aerated concrete in the range of average densities from 500 to 600 kg/m^3 for a base sample of 150x150x150 mm is linear and is described by the equation:

$$R_{com} = 0{,}0038 \cdot x + 1{,}8132, \qquad (3)$$

where x – is the average density of autoclave aerated concrete based on fly ash, kg/m^3.

Figure 4 shows that the change in the flexural strength of samples of autoclave aerated concrete based on fly ash in the range of average densities from 500 to 600 kg/m^3 for prisms with a size of 40x40x160 mm is linear and is described by the equation:

$$R_{fl} = 0{,}0014 \cdot x + 0{,}636, \qquad (4)$$

where x – is the average density of autoclave aerated concrete based on fly ash, kg/m^3.

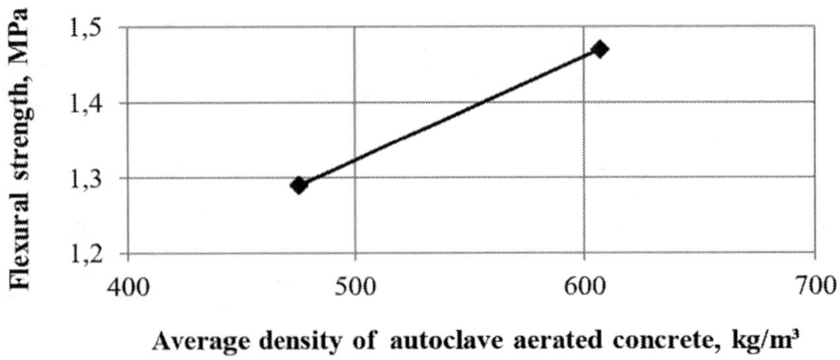

Figure 4. Dependence of the flexural strength of samples of autoclave aerated concrete based on fly ash on the average density.

Values of scale coefficients for samples of autoclave structurally heat-insulating aerated concrete based on fly ash are given in table 2.

Table 2. Experimentally established values of scale coefficients for samples of autoclave structurally heat-insulating aerated concrete based on fly ash.

Grade of aerated concrete	Size of the sample, mm	Scale coefficient, *K*
D500	40x40x160	1,52
	70,7x70,7x70,7	1,18
	100x100x100	0,96
	150x150x150	1
D600	40x40x160	1,4
	70,7x70,7x70,7	1,23
	100x100x100	0,93
	150x150x150	1

For samples of aerated concrete of the studied grades, general values of scale coefficients were derived, as well as standard deviations and variation coefficients were determined (table 3).

Table 3. Scale coefficients, standard deviations and variation coefficients for samples of autoclave aerated concrete based on fly ash.

Grade of aerated concrete	Size of the sample, mm	Scale coefficient, *K*	Standard deviation, S_k	Variation coefficient, *V*, %
D500	40x40x160	1,45	0,0688	4,75
	70,7x70,7x70,7	1,2	0,0214	1,78
	100x100x100	0,95	0,0053	0,56
	150x150x150	1	0	0
D600	40x40x160	1,45	0,05	3,45
	70,7x70,7x70,7	1,2	0,0292	2,43
	100x100x100	0,95	0,0177	1,86
	150x150x150	1	0	0

The variation coefficients (table 3) of the studied grades of aerated concrete do not exceed 15%, accordingly, the obtained values of scale coefficients can be used for calculating the strength characteristics of autoclave structurally heat-insulating aerated concrete based on fly ash in the range of average densities from 500 to 600 kg/m^3.

Thus, in the process of research, the dependence of the values of the compression strength of autoclaved aerated concrete of the grades D500 and D600 on the size and area of the working section of the samples is derived. The dependences of mechanical characteristics on the average density of autoclave aerated concrete based on fly ash in the range from 500 to 600 kg/m^3 have been experimentally established.

For samples of autoclave structurally heat-insulating aerated concrete based on fly ash, the values of scale coefficients were experimentally determined, which in relation to the base sample with size 150x150x150 mm are: for prisms of 40x40x160 mm – 1.45, for cubes of 70,7x70,7x70,7 mm and 100x100x100 mm – 1.2 and 0.95, respectively.

References

[1] Lobastov N A and Chernyavets A N 2009 The effect of the size and geometric shapes of test specimens of structural graphite strength under compression *RTC (OTK)* **11-12** pp 13-16
[2] Eremichev A N 2014 Selection of the relative height of samples for compression tests *VMSTU (VMGTU)* **3** pp 136-140
[3] Kuznetsov N N and Pak A K 2014 Influence of the hard rock specimens size ratio on the results of their strength determination under uniaxial compression *VMSTU (VMGTU)* **17 (2)** pp 246-253
[4] van Vliet M.R.A. and van Mier J.G.M 1998 Experimental and numerical investigation of size scale effects in concrete fracture *MIS* pp 185-206
[5] van Vliet M.R.A. and van Mier J.G.M 2000 Experimental investigation of size effect in concrete and sandstone under uniaxial tension *EFM* **65 (2-3)** pp 165-188
[6] Kadleček V and Špetla Z 1967 Effect of size and shape of test specimens on the direct tensile strength of concrete *RB* **36** pp 175-184
[7] Kvirikadze O L 1974 The effect of size of cube samples for strength properties of concrete (Vliyanie razmerov obrazcov-kubov na prochnostnye harakteristiki betonov) Tbilisi: SabchotaSakartvelo p 53
[8] Tsvetkov S V 2014. On the statistical component of the scale factor when testing concrete *ICE (PGS)* **12** pp 52–54
[9] Sviridov N V and Hirnov V V 2015 Features of rationing of quality indicators of high strength concrete *CT (TB)* **7-8** pp 54-61
[10] Avdeev B A 1972 Technique of determining the mechanical properties of materials (M.: Engineering) p 298
[11] Chait R and Papirno R 1983 Compression testing of homogeneous materials and composites. Philadelphia, Pa: American Society for Testing and Materials p 294
[12] Pangaev V V, Horse M M, Serdyuk V M and Molokov D V 2012 Investigation of the scale factors to determine the strength of cement mortar under compression *PHEI (IVUZS)* **2 (638)** pp 102-108
[13] Tsiskreli G D Lekishvili G L 1966. On the scale effect in concrete CRC (BZ) **10**
[14] Krivyh M V, Baranova A A and Lonshakov I V 2019 The influence of the size of samples and their moisture on the compressive strength of foamed concrete based on microsilica (Vliyanie razmerov obrazcov i ih vlazhnosti na prochnost' pri szhatii penobetona na osnove mikrokremnezyoma) *MTSTP (STNTP)* **1** pp 182-184
[15] Baranova A A, Gass J A and Lonshakov I V 2019 Determination of scale and correction coefficients for fine-grained concrete specimens PUICR (IVISN) **3 (30)** pp 502-509
[16] Gass Y A, Baranova A A and Lonshakov I V 2019 The influence of the size of samples and their moisture on the strength characteristics of fine-grained concrete *MTSTP (STNTP)* **1** pp 178-179

Investigation of water absorption of non-autoclaved foam concretes based on microsilica

A Baranova[1] and V Bygajchuk[1]

[1]Department of Industrial and Civil Engineering, Angarsk State Technical University, Tchaikovsky str., 60, Angarsk, 665835, Russia

E:mail: baranova2012aa@mail.ru

Abstract. The article presents the results of a study of water absorption of non-autoclaved foam concrete based on microsilica. The dependence of water absorption by mass and by volume on the average density of foam concrete in the range of 500÷700 kg/m^3 is derived. It is experimentally established that the water absorption of foam concrete based on microsilica the grade D500 by mass is 100, 9 %, by volume – 53.6 %, for the grade D700 – 64.3 % and 42.4 %, respectively. At the same time, a day after immersion in water, foam concrete the grade D500 gains an average of 69.7% from full water absorption both by mass and by volume, foam concrete the grade D700 – 66.1 %. Full of water absorption of non-autoclaved foam concrete based on microsilica reaches in seven weeks later. The difference in water absorption between the grades D500 and D700 by mass is 56.9 %, by volume – 26.4 %. The closed porosity of foam concrete based on microsilica the grade D500 is 26 %, and for the grade D700 – 32.2 %.

Water absorption is the ability of a material to absorb and retain water in its pores when it comes in direct contact with it, i.e. when it is immersed.

Water absorption of materials is characterized by the ratio of the amount of water that absorbs the dry material when it is kept in water to the mass of the dry material (water absorption by mass W_m) or to the volume of the material (water absorption by volume W_v).

Many Russian and foreign researchers study the water absorption of concretes [1÷18]. It is known that water absorption is affected by the porosity of the material and the nature of its pores (open or closed pores). The greater the open porosity of the material, the higher its water absorption.

The amount of water absorption significantly affects the frost resistance of the material, and subsequently its durability.

Foam concrete is a type of cellular concrete in which porosity is created by introducing pore-forming components, in particular a foaming agent, into the concrete mixture. Therefore, the total porosity of foam concrete, depending on their average density, can reach 90% or more.

The introduction of ultrafine fillers with low bulk density, such as microsilica, into the composition of foam concrete reduces their overall porosity, which has a positive effect on the mechanical characteristics of foam concrete [19].

The aim of the work was to establish the dependence of water absorption of non-autoclaved foam concrete based on microsilica on its average density.

The following materials were used in the research:
- Portland cement of the grade CEM I 42.5 H produced by JSC «Angarskcement»;
- microsilica from filters of dust collectors of JSC «Kremniy»;

- hyperplasticizer based on polycarboxylates «MC-Power-Flow-3100»;
- synthetic foaming agent Penta Pav 430 (grade A).

The research methodology was as follows. The foam concrete mixes for the foam concretes grades D500 and D700 was prepared using the classical technology: the mortar mix (Portland cement (PC)+microsilica (MC)+water+hyperplasticizer) and the foam (water+foaming agent) were prepared separately. Then everything was mixed until a homogeneous mass was obtained. From previously performed work, it was found that the optimal ratio of the binder (PC) to the filler (MC) in foam concrete is 1:1 [20].

Samples of 100x100x100 mm in size were made from foam concrete mixes. After 28 days of normal hardening, they were dried to a constant mass at a temperature of 105±5 °C and tested for water absorption in accordance with GOST 12730.3 Concretes. Method of determination of water absorption.

The test results are presented in table 1 and shown in figures 1÷3.

Table 1. Water absorption of foam concrete samples.

Parameter	D500	D700
The average density, ρ_m, kg/m^3	531	660
Water absorption by mass, W_m, %, after		
1 day	70,3	42,5
7 days	75,8	49,0
14 days	82,3	53,2
21 days	88,0	57,0
28 days	92,6	60,4
35 days	96,5	62,5
42 days	99,5	63,9
49 days	100,9	64,3
Water absorption by volume, W_v, %, after		
1 day	37,3	28,0
7 days	40,3	32,4
14 days	43,7	35,1
21 days	46,7	37,6
28 days	49,2	39,8
35 days	51,3	41,2
42 days	52,8	42,2
49 days	53,6	42,4

The results presented in table 1 show that after a day of water absorption testing, foam concrete the grade D500 gains an average of 69.7 % from full water absorption, both by mass and volume, and foam concrete the grade D700 – 66.1 %. In the future, after seven weeks of testing, the water absorption by mass and volume of foam concrete D500 increases by an average of 43.6 %, while for D700, the increase is 51.4 %. The difference in water absorption between the grades D500 and D700 by mass is 56.9 %, by volume is 26.4 %.

The change in water absorption by mass and by volume of non-autoclaved foam concrete based on microsilica depending on its average density in the range from 500 to 700 kg/m^3 (figure 1) is described by the equations:

$$W_m = 251{,}56 - 0{,}2837 \cdot x, \quad (1)$$

$$W_v = 99{,}702 - 0{,}0868 \cdot x, \quad (2)$$

where x – the average density of non-autoclaved foam concrete based on microsilica, kg/m^3.

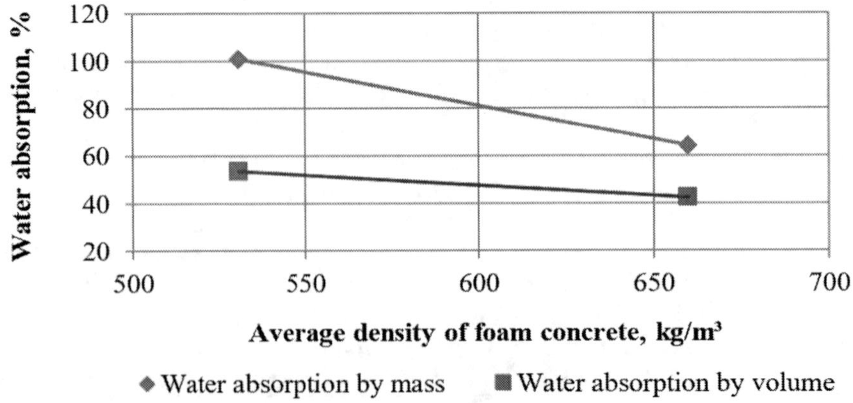

Figure 1. Water absorption of non-autoclaved foam concretes based on microsilica.

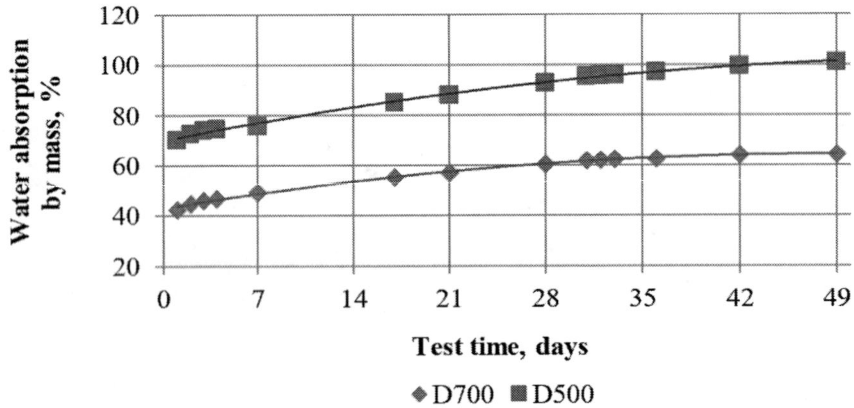

Figure 2. Change in time the water absorption by mass of non-autoclave foam concretes based on microsilica.

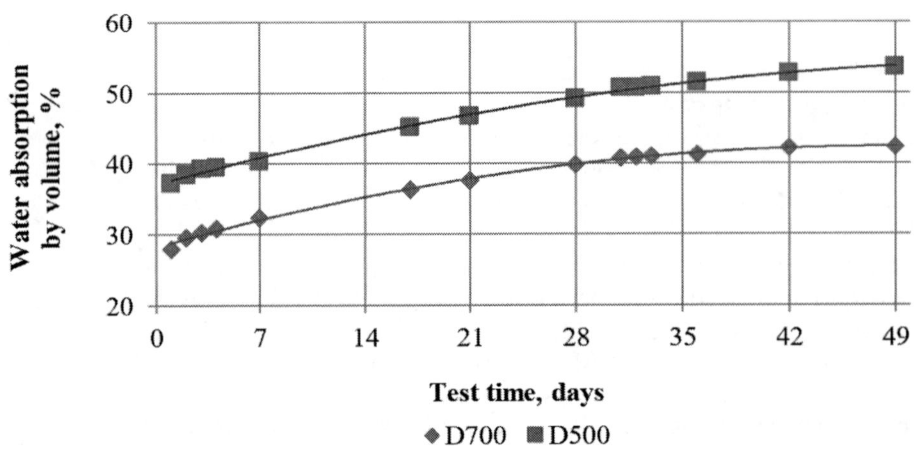

Figure 3. Change in time the water absorption by volume of non-autoclave foam concretes based on microsilica.

The change in time the water absorption by mass and by volume of non-autoclave foam concretes based on microsilica (figures 2, 3) is described by the following equations.

For foam concrete the grade D500:

$$W_m = 69{,}928 - 0{,}0086 \cdot x^2 + 1{,}0623 \cdot x \quad (3)$$

$$W_v = 37{,}114 - 0{,}0046 \cdot x^2 + 0{,}5665 \cdot x \quad (4)$$

For foam concrete the grade D700:

$$W_m = 42{,}836 - 0{,}0092 \cdot x^2 + 0{,}8882 \cdot x \quad (5)$$

$$W_v = 28{,}292 - 0{,}006 \cdot x^2 + 0{,}5832 \cdot x, \quad (6)$$

where x – the test time, days.

The amount of water absorption by volume is numerically equal to the open porosity of the material (P_{op}). Knowing the total and open porosity, you can calculate the closed porosity of the material.

The total porosity of materials in % is calculated by the formula:

$$P_{tot} = \left(1 - \frac{\rho_m}{\rho}\right) \cdot 100\,, \quad (7)$$

where ρ_m and ρ – the average and true densities of the material, respectively, kg/m^3.

The closed porosity of the material in % is determined by the formula:

$$P_{cl} = P_{tot} - P_{op} \quad (8)$$

The values of the total, open and closed porosity of non-autoclaved foam concretes based on microsilica are shown in table 2.

Table 2. Porosity values of foam concrete based on microsilica in depending on its average density.

Parameter	The grade of foam concrete	
	D500	D700
Average density, kg/m^3	531	660
Total porosity, %	79,6	74,6
Open porosity, %	53,6	42,4
Closed porosity, %	26,0	32,2

The results from table 2 shows that the number of closed pores in foam concrete based on microsilica the grade D700 are 23.85 % more than in the foam concrete the grade D500, while the difference in the total porosity between the foam concretes of the studied grades is only 5 %.

Thus, it is experimentally established that:

- water absorption of foam concrete based on microsilica the grade D500 by mass is 100, 9 %, by volume is 53.6 %, for the grade D700 – 64.3 % and 42.4 %, respectively;
- on the first day of the water absorption test, foam concrete the grade D500 gains an average of 69.7 % from full water absorption, both by mass and volume, and foam concrete the grade D700 – 66.1 %. Full of water absorption of non-autoclaved foam concrete based on microsilica reaches in seven weeks later. During this time, water absorption by mass and volume of foam concrete the grade D500 increases by an average of 43.6 %, while for D700 the increase is 51.4 %;
- the difference in water absorption between the grades D500 and D700 by mass is 56.9 %, by volume is 26.4 %;
- closed porosity of non-autoclaved foam concrete based on microsilica the grade D500 is 26 %, the grade D700 is 32.2 %.

Water absorption dependences by mass and by volume of non-autoclaved foam concrete based on microsilica on its average density in the range from 500 to 700 kg/m^3 have been experimentally derived.

References

[1] Andreeva A V and Savvinova M E 2017 Water absorption of fine-grained concrete (Vodopogloshchenie melkozernistogo betona) *XIX ISR in memory of A. A. UKHTOMSKY (XIX MNC pamyati UHTOMSKOGO A.A.)* pp 19-20

[2] Korovkin M O, Eroshkina N A, Teplova M F and Korovchenko I V 2015 Research on the

kinetics of water absorption of fine-grained concrete *YS (MU)* **13** pp 132-135

[3] Hvastunov V L, Kalashnikov V I, Hvastunov AV and Pausk V V 2014 Parameters of water absorption and porosity of powder-activated high-strength concrete with low specific consumption of cement per unit of strength *RAC (RAS)* **4** pp 45-51

[4] Fedyuk R S 2016. Investigation of water absorption of fine-grained fiber concrete on a composite binder *BR (FI)* **2-2** pp 303-307

[5] Merzlyakov A O 2019 Study of the traceability of sodium formate additive and its effect on water absorption of concrete concrete *IS (IN)* **1** pp 289-29

[6] Zhengyu Pan Changsheng and Lijian 2014 The influence of different hydrophobic agent on the water absorptivity of foam concrete *AMSEM* **665** pp 192-195

[7] Liu C., Luo J., Li Q 2019 Water-resistance properties of high-belite sulphoaluminate cement-based ultra-light foamed concrete treated with different water repellents *CBM* **228** UNSP 116798

[8] Nambiar E K Kunhanandan and Ramamurthy K 2007 Sorption characteristics of foam concrete *CCR* **37 (9)** pp 1341-1347

[9] Ioannou I, Hamilton A and Hall C 2008 Capillary absorption of water and n-decane by autoclaved aerated concrete *CCR* **38 (6)** pp 766-771

[10] Kumar E Muthu and Ramamurthy K 2017 Influence of production on the strength, density and water absorption of aerated geopolymer paste and mortar using Class F fly ash *CBM* **156** pp 1137-1149

[11] Namsone E, Sahmenko G and Namsone E 2017 Reduction of the capillary water absorption of foamed concrete by using the porous aggregate *IMST IOP CS* **251**

[12] Lin K-L, Lee T-C and Chang J-C 2013 Water absorption and retention of porous ceramics cosintered from waste diatomite and catalyst *EP&SE* **32 (3)** pp 640-648

[13] Raj A, Sathyan D. and Mini K M 2019.Physical and functional characteristics of foam concrete *CBM* **221** pp 787-799

[14] Abd Elrahman M, Chung S-Y and Stephan D 2019 Effect of different expanded aggregates on the properties of lightweight concrete *MCR* **71 (2)** pp 95-107

[15] Karolina R and Sianipar Y C 2018 The utilization of stone ash on cellular lightweight concrete *TALENTA-CEST IOP CS* **309**

[16] Guimaraes A S, Ribeiro I M and de Freitas V P 2017 A tool to predict water absorption in porous building materials *JPM* **20 (2)** pp 127-141

[17] Dachowski R and Kostrzewa P 2016 The use of waste materials in the construction industry *WMCAUS PE* **161** pp 754-758

[18] Jitchaiyaphum K, Sinsiri T and Jaturapitakkul C 2013 Cellular lightweight concrete containing high-calcium fly ash and natural zeolite *IJMMM* **20 (5)** pp 462-471

[19] Baranova A, Shustov P and Yazina O 2018 Structural and heat-insulating foam concrete of non-autoclaved hardening based on microsilica *MATEC WoC* **212** *N article 01003*

[20] Baranova A and Ryabkov I 2019 Investigation of thermal conductivity of non-autoclaved foam concrete based on microsilica *IOP CS: MSaE V* **667** *N 1*

IOP Conf. Series: Materials Science and Engineering **880** (2020) 012004 doi:10.1088/1757-899X/880/1/012004

Innovations in construction and the practical potential of wood in student projects

Ts B Dagdanova[1] and M B Khadeev[2]

[1] Institute of architecture, construction and design of INRTU, 83 Lermontov street, Irkutsk, 664074, Russia
[2] Architectural Studio, 343/4 Lermontov street, Irkutsk, 664017, Russia

E-mail: m-1@mail.ru

Annotation. The problems of environmental pollution on the planet, the processing of human waste, makes architects and builders look for a solution to the problems by means of architecture and new technologies in construction. The article deals with the construction of wooden and wood structures in the Western countries, in particular, the construction of high-rise buildings as an alternative to the concrete and metal constructions. Wood as a modern high-tech, reliable and environmentally friendly building material in comparison with other materials shows its advantages in many ways. Article presents separate project proposals within the framework of the International Student Competitions "Timber in the city: Urban Habitats" 2018-2019 and 2015-2016 organized by the Association of University Architectural Schools of the USA - ACSA, as well as examples of unique wooden structures designed by famous architects. The XII International Congress on Wooden Construction in Moscow Recommendations, including ones on education and the need to train relevant specialists in Russia, are examined.

The deteriorating situation of ecologies in the world, the ethical implications of: the growing scarcity of freshwater resources; diminution of energy resources and their possible replacements; destruction of pollutants throughout the world; and the processing and recycling of human waste to integrate architects, planners, builders to seek problem resolution through architecture. Today, in the European countries: Finland, Sweden, Italy, Austria, France and in USA and Canada the specialists are building high class and business apartments and public buildings with use of timber construction techniques, multistorey wooden residential blocks are established. The timber construction is a modern, reliable, high-technology and environmentally friendly building material. In Russia, as in many countries, limitations are exist number of storeys of the wooden buildings. For instance, in Finland it is not allowed to build houses with height no more than three floors, in Austria - is no more than five floors, in Canada - is no more than six floors, and in Russia - the height limitation of three floors.

The professionals dealing with the ecological problem really, comparing the pros and cons of the modern industrial timber construction and of traditional reinforced concrete and steel construction, have come to realize that architects and planners may play a leading role in the promotion of the industrial wooden housing concept. In addition, the concrete production, erection and recycling of concrete and other inventive materials causes to intensity carbon emissions and to environmental pollution. The construction of buildings with these materials are estimated to be responsible for half of all greenhouse gas emissions in the world. Wood is an important material for ensuring a sustainable future; it has a neutral carbon footprint, it is renewable, and the manufacture of wood products has a smaller environmental impact than that of competing building materials. Carbon dioxide is remain locked in if the wood was used to make construction materials or processing into woodbased materials. Wood is the unique material and estimated to be the carbon store.

As international practice shows, organizing and holding various competitions, events, festivals helps to receive and propose new ideas. A good examples in the field of training young architects and town-planners in USA and Canada are regular competitions on current themes. For example, for the academic year 2018-2019, Association of Collegiate Schools of Architecture (ACSA) in USA is pleased to announce the International student design ***competition "Timber in the city 3: Urban Habitats"*** in May 2018**.** This was the third student design competition "Timber in the city". Over the year**,** U.S. students had worked on the design project with their teachers. Research and design work on the competition should be structured over the course one or more semesters of the 2018-2019 academic year. An intention of all ACSA competitions is to make students aware that research is a fundamental element of any design solution.

Architecture students (two teams) from the Irkutsk National Research Technical University (INRTU) took part in this competition since the second semester of 4-course (spring of 2019 year). During the project development stage the students found that there are important gaps in skills and knowledge about wooden multistoried structures.

It was noted in the summaries that "Timber in the City: Urban Habitats Competition", organized by the Association of Collegiate Schools of Architecture (ACSA), the Binational Softwood Lumber Council (BSLC) and Parsons School of Design, attracted more than 920 architectural students and faculty. Entrants designed proposals to imagine the transformation of our cities through sustainable buildings from renewable resources, offering efficient affordable construction, innovating with new and traditional timber materials, and designing healthy living and working environments. The winning entrants, with prizes totaling $40,000, were chosen by a panel of leading architects and professors based on the design's ability to integrate wood as the primary structural material while meeting the needs of the local community. The competition focused on a vacant waterfront site in Queens, New York, as a vibrant and vanguard model of healthy, biophilic living for the future of the city. Students were asked to design a mid-rise, mixed-use complex that includes affordable housing, a large community wellness facility, and an early childhood education center, all interlaced with a new exterior public waterfront space. Entrants were challenged to propose construction systems in scenarios that draw optimally on the performance characteristics of not one, but a variety of wood technologies. ACSA is pleased to announce the winners of a student design competition exploring wood as an innovative building material" [1].

"*First Place*: *Aperture*

The University of Maryland's winning submission attracted the jurors with its intelligent use of timber construction techniques. The orientation of the buildings creates a protective plaza linking the buildings to the community with a clear understanding of the urban context.

Second Place: *Re-Gen Growth*

The City College of New York's winning submission stands out for its innovative use of wood, which connects spaces, structure and the user experience all together. The density of the project represents a three-dimensional occupiable city.

Third Place: *Timber Living*

The University of Illinois at Urbana-Champaign's submission has a strong grasp of timber construction and how to assemble it in a clever way. The project shows an advanced integration of sustainable measures" [1].

The programs for this mixed-use development are composed to challenge students and educators to think creatively and critically about the way in which choices about building materials, and the interrelationship of interior space and the exterior environments frame long-term consequences for the health of urban environments. Housing is the largest component of the competition program and presents an opportunity to look closely at the way timber construction can be used effectively in creating buildings based on smaller cellular units (figure 1a). A community wellness and sports facility complements the housing, and offers larger community and collective spaces that will require larger structural spans. An early childhood education center, for children from 6 weeks to 5 years old, calls attention to the critical role these institutions play in the long-term vitality and development of a community.

The materials. The competition challenges participants to interpret, invent, and deploy numerous methods of building systems, with a focus on innovations in wood design on a real site. For thousands of years, solid wood has been used as a building material. Modern timber products and systems have greatly expanded the potential uses of this historic material. Timber is an ideal green building material: it is well suited for a broad range of structural and aesthetic applications, it offers economical construction and high performance characteristics in strength and energy efficiency; and wood is an economic driver to maintain forests and protect jobs in rural communities.

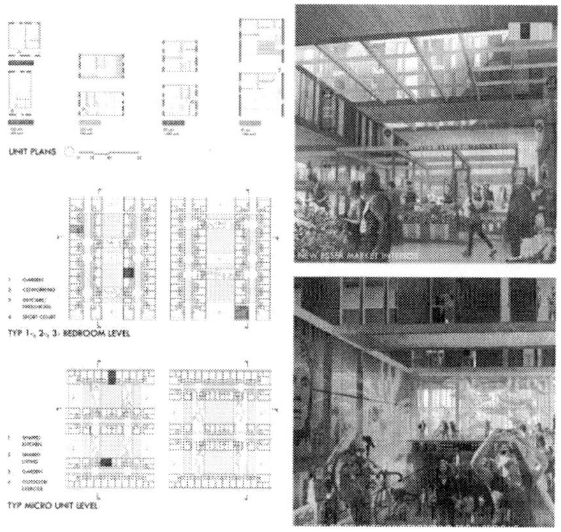

a

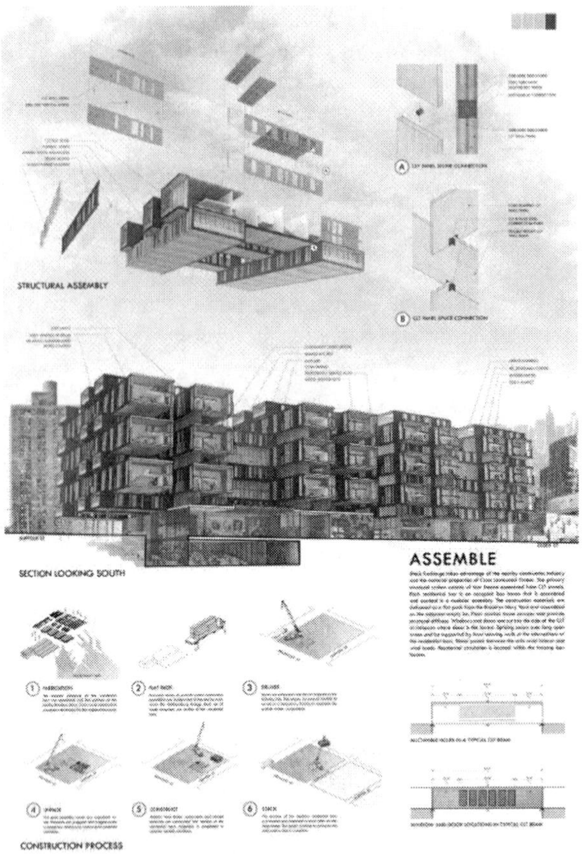

b

Figure 1. Fragments of the project-winner of the competition "Timber in the city 2015-2016": (a) floor Plans based on the idea of stacking bars. Interiors; (b) the process of installing a residential building from prefabricated elements.

Criteria for the judging of submissions will include: timber/wood as the primary structural material, creative and innovative use of timber/wood in the design solution, successful response of the design to its surrounding context, the creative and clear approaches to designing a healthy urban mixed use environment with timber as a central material, and successful response to basic architectural concepts such as human activity needs, structural integrity, and coherence of architectural vocabulary.

The tasks assigned during ***competition "Timber in the city 2015-2016"*** had been like tasks next competitions, especially in term to use wood to construct new settlement units. It was noted in the project summary of the competition winner that "This winning submission is outstanding for its inventive formal strategy and expressive use of timber. The concept of stacking volumes with core atria dropping down is very nice, creating a lot of visual depth and variation not typically found in New York City. The scale of the market and gallery spaces read as great urban rooms with the residential spaces floating above. The contrast of the dark wood exterior and the warm, glowing interior also add to the aesthetic interest of the building. The project takes advantage of mass timber and its natural partnership with modular construction, which pairs well with the project's massing. The story of constructability presented through the diagrams is believable and there is a clarity to the structural assembly and delivery methods (figure 1b). Although the part of stacked volumes is appealing, it would have acoustical issues and thermal implications from the extraneous surface area. Another technical issue is that the CLT is loaded on its weak axis. Lastly, this team was progressive in their approach to the program. There integration of communal residential spaces and areas for urban farming is great, and it is nice to see built-in, systematic environmental strategies" [2].

As for the two competitive projects carried out by the students of Irkutsk National Research Technical University (INRTU), in both cases the tasks were solved: to use natural factors efficiently, such as adjacent river, opening on Manhattan, to form a new favorable social environment that stimulates pedestrian and physical activity, to implement a number of innovations in planning, landscaping, eco-friendly construction and architecture using wooden structures.

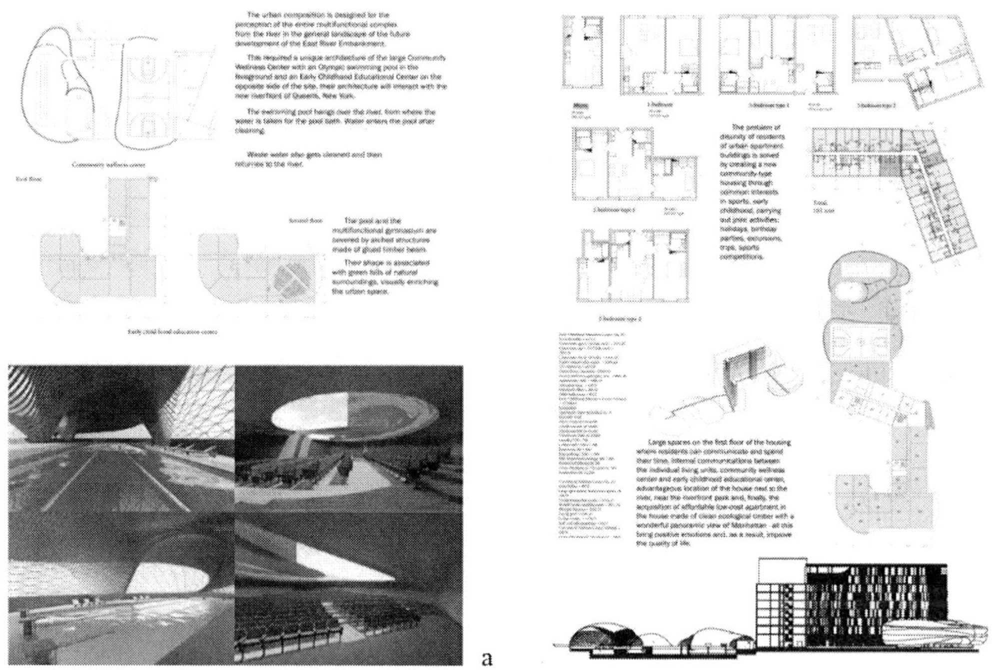

a b

Figure 2. Fragments of the project No.1 students of the INRTU "Timber in the city 2018-2019": (a) floor Plans for the pool and children's center and interiors; (b) apartment Plans of a residential building made of prefabricated elements. Facade.

The town-planning concept of the project No. 1 is to create a coastal boulevard with public buildings inscribed in the landscape of the river and constituting an indissoluble whole with it.

The unique architecture of Community Wellness Center with an Olympic swimming pool on the river and an Early Childhood Educational Center evokes associations with green hills of natural environment, becoming a part of the design of city embankment and of the nearby park (figure 2a). An increasing height of housing units from 7 to 9 floors represents the dynamic development of the area from the embankment to the neighborhood interior. All three objects have internal connections among themselves at the level of the 1st floor. The complex is a new type of communal residential building with shared spaces for residents. It solves the problem of disunity of residents in urban apartment buildings focusing on their common interests in sports, early childhood, carrying out joint activities, thereby improving the quality of life (figure 2b).

Construction elements and Assemble. Residential complex is assembled from prefabricated modules, which are installed as in LEGO. The bearing construction elements (CLT-panels) are connected using studs on the edges of the panels, which are to insert into the grooves of adjacent panels without using metal bolts or nails. Vertical communication stiffness core with staircases and elevator shaft, internal transverse and longitudinal bearing walls are the bearing elements of the residential design.

The minimum of large openings in the inner bearing CLT panels provides structural stiffness to the building. All the construction elements meet the requirements of fire safety, thermal insulation, sound insulation and ecology. External walls are self-bearing, panels are 7-ss TL thick: 280 mm (figure 3). The pool and the multifunctional gymnasium are covered by arched structures made of glued timber beams. The floors are seven-layer CLT panels, reinforced with T-glued beams.

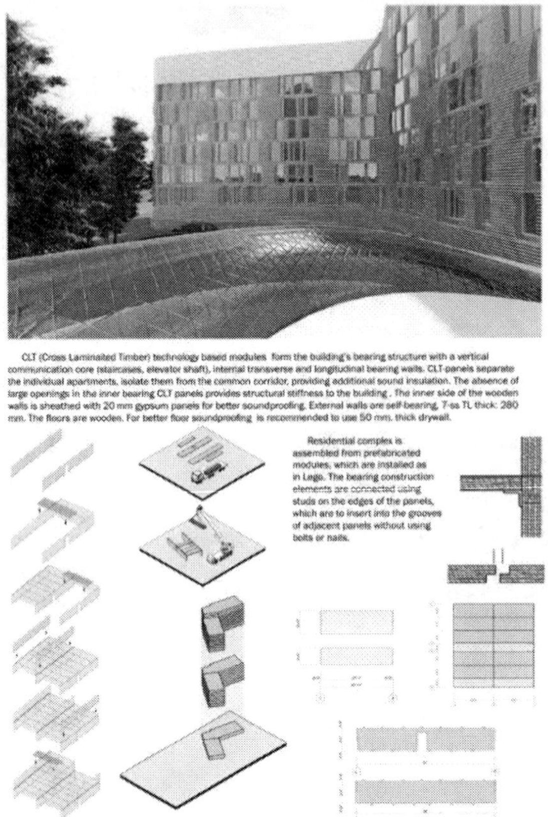

Figure 3. Fragments of the project No. 1 students of the INRTU "Timber in the city 2018-2019". Perspective of a residential building with a pool roof. The process of installing a residential building from prefabricated elements.

Both separate constructional elements and ready-made large building blocks of a high completion level can be delivered to the construction site: blocks are supplied with electricity, communication, water. Compared with the reinforced concrete, they can be mounted two times faster and with a much more smaller number of workers. Thanks to the module's high degree of prefabrication, complete decoration of their internal surfaces, as well as the availability of the necessary fixing elements, the panels are completely ready for installation.

The economic effect is achieved through: quick assemble in the absence of a wet process, the use of environmentally friendly renewable natural material - timber and full prefabrication of the mounted constructional elements.

IOP Conf. Series: Materials Science and Engineering **880** (2020) 012004 doi:10.1088/1757-899X/880/1/012004

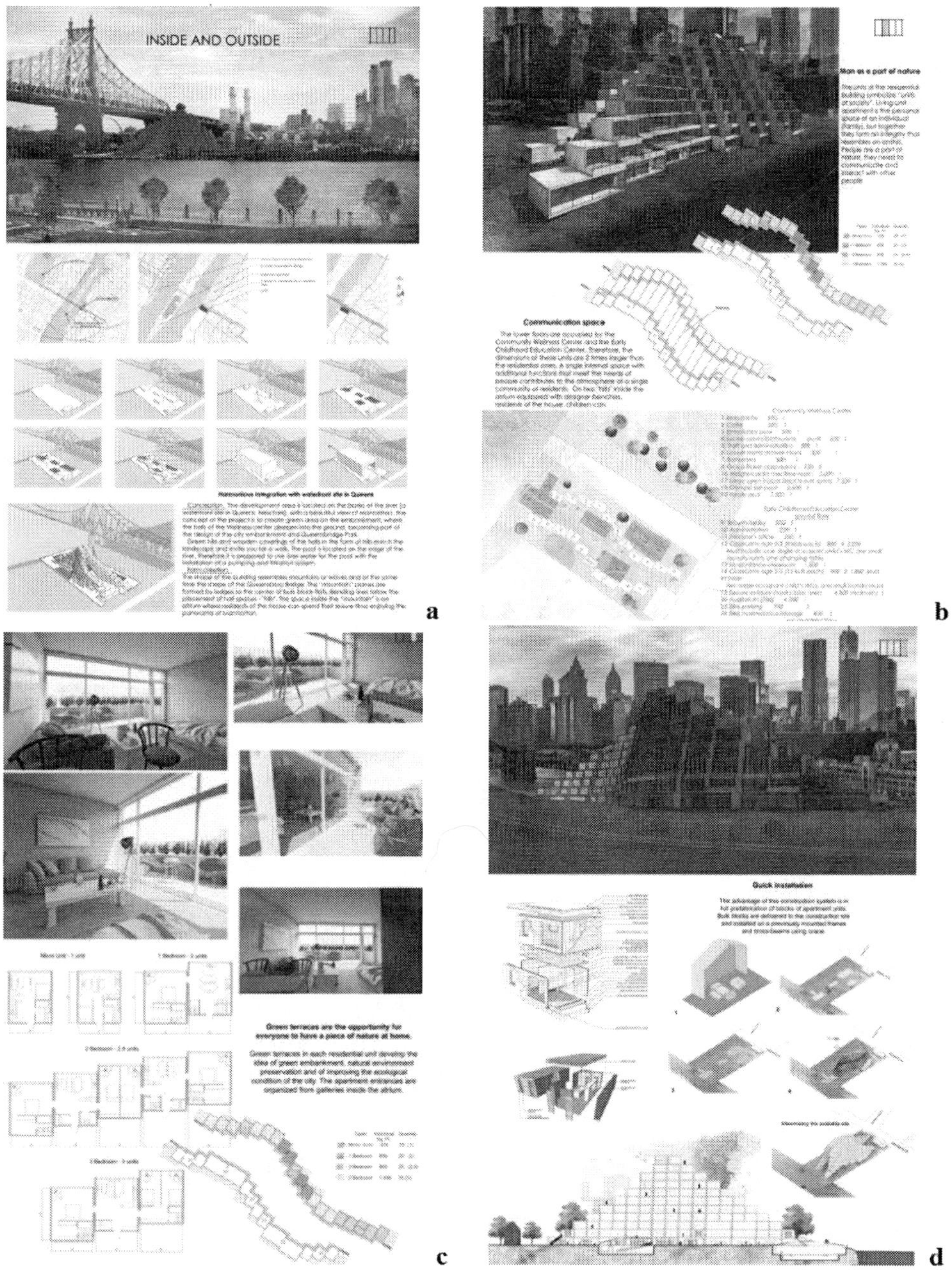

Figure 4. Fragments of the project No. 1 students of the INRTU: (a) idea of the project; (b) plans and perspective of a residential building; (c) the plans of the apartments, interiors; (d) the process of installing a residential building from prefabricated elements.

The second project proposal No. 2 is based on the revival of the idea of building housing from prefabricated three-dimensional blocks of wood, where each represents a separate living room or even a three-dimensional cell for two small bedrooms. The complex composition is a narrowed in upper part multi-luminous atrium space, formed by ledges of put on each other three-dimensional modules, with an exit to the river side. Sports halls and children's facilities occupy the first floors of the complex (figure 4 (a), (b), (c), (d).

Fast construction process. The advantage of this design is that each cell is assembled at the factory. It is delivered ready-made to the site and can be installed by a crane, which speeds up the construction process. Frames and crossbars get installed on the construction site, blocks of residential cells get attached to them. If necessary, it is possible to replace the cells.

I would like to note the original project of the Japanese *architect Shigeru Ban* of the seven-storey building of the ***Tamedia publishing house*** with an area of 1000 m2 in Zurich. The entire internal structure of the new building of the concern "Mediengruppe Tamedia" in Zurich (beams, ceilings, even dowels for fasteners) is made exclusively of wood. This masterpiece, combining traditional materials with the latest technologies and approaches in architecture, was the result of collaboration between Swiss engineers and an outstanding Japanese architect (figure 5 (a), (b), (c), (d).

Figure 5. New building of the publishing concern "Mediengruppe Tamedia" in Zurich, architect Shigeru Ban. Construction fragments and nodes. Example of combining traditional materials with the latest technologies.

Details of the supporting structure of the building from wooden pillars were glued together, and then processed on a milling machine with an accuracy of up to a millimeter. The finished parts were joined on the construction site in a structure which is a five stories high wooden frame. The supporting elements of the structure were connected without the use of bolts and nails. The facade is completely made of glass. This building has a "pleasant working atmosphere". When comparing CO_2 emissions from the wooden building construction and buildings made of concrete and metal, it turned out that in the first case, CO_2 emissions are one-third lower than when they are built from metal. The savings when compared with concrete were about 50%. This example clearly demonstrates that on a global scale, building from wood and wood-containing materials can significantly improve the health of the planet. In addition, wood is the only renewable building material.

In March 2019, Moscow hosted the XII international Congress on wooden construction, the main theme of which was designated as: "a Wooden house is a national priority!" In 2019, the Russian Government approved the strategy for the development of the forest complex until 2030. One of the priority areas reflected in this document is wooden house construction. The regulatory framework improvement and building codes development was launched. The Ministry of industry and trade and the Ministry of construction have already developed standards for new types of building materials. It is possible to begin broad use of wooden structures is possible in the construction of public buildings. Russian companies shared their experience in implementing such projects, including large sports facilities built of wooden structures in St. Petersburg, Kazan, Moscow, Krasnoyarsk, whose architecture is unique (figure 6).

a

b

Figure 6. Major sports wooden facilities in Russia: (a) water park in Saint Petersburg; (b) palace of water sports in Kazan.

Director of Zaha Hadid Architects (Great Britain) architect Jim Severin presented at the Congress a unique project of the world's first fully wooden five-thousand-seat stadium in England. Qatar's 'Al Wakrah' stadium, which is capable of holding 40,000 spectators, the 2022 world cup venue designed

IOP Conf. Series: Materials Science and Engineering **880** (2020) 012004 doi:10.1088/1757-899X/880/1/012004

by the late Zaha Hadid with use of the carbon-neutral of the wooden constructions, was inaugurated on thursday may 16, 2019 (figure 7 (a), (b).

a

b

Figure 7. Stadiums on the design by Zaha Hadid Architects: (a) the world's fully wooden five-thousand-seat stadium in England; (b) the carbon-neutral of the wooden constructions in Qatar.

It turned out that the construction of wooden high - rise buildings is a technically simple process. Almost all the necessary materials for a wooden house are produced in Russia, but its assembly requires high qualification of construction workers. The largest in Europe OSB building slabs production plant - DOC "Kalevala" functions in Karelia, and LVL (Laminated Veneer Lumber) production for frame racks and ceilings is in the city of Torzhok (Tver region). Experts highlight three advanced technologies that are widely used in the West and that are quite applicable in Russia. CLT panels are cross-glued wooden slabs and wall panels (Cross Laminated Timber), whose properties are close to reinforced concrete, but six times lighter than reinforced concrete slabs. CLT panels thermal insulation properties are 3-5 times higher than similar properties of brick or concrete walls. Panels are assembled with bolts and screws without the use of additional equipment. This is a new technical solution. Proper use of wood and adhesive joints can significantly increase the rigidity and strength of the plates.

Houses made of this new material are fast-built and economical. Swedish experts estimate that the construction of wooden multi-storey houses is 5-20% cheaper than the construction of concrete panel houses, since simpler tools are used, and less transportation is required. Such buildings also are energy saving: 1 sq.m energy consumption is only 65 kWh per year. In Russia, in brick houses, heating consumes 130-150 kW per 1 sq. m.

In addition, wooden buildings, even the high-rise ones, built using modern technologies, are eco-friendly and safe. As for earthquake-prone regions like ours - with estimated seismicity of up to 8 and 9 points, prefabricated wooden multi-storey buildings can withstand such earthquakes, since they are also lighter than reinforced concrete and brick. As for fire safety and reliability, wooden structures resist the effects of high temperature for at least 45 minutes from the beginning of the fire. At the high temperature (on different reanalysis data from 400^0 C) metal structures had reduced the bearing capacity. Low energy costs for production, transportation, installation, ease of operation and disposal confirm the advantages of wood as a building material.

Famous canadian architect Jim Heverin, who devoted his life to research and promotion of multi-storey wooden house construction, was surprised that the share of wood construction in Russia – the most forested country in the world – is only a few thousandths of a percent. He was also strunned by the lack of special architectural institutions in Russia and even by the absence of wooden construction training programs.

The Congress adopted a resolution to give priority to education and training of relevant specialists in Russia.

References

[1] https:// www.competitions@ACSA-ARCH.org 2019 Timber in the City-competition winner 10.08.2019
[2] https://www.timberinthecity.com.
[3] https://www.acsa-arch.org
[4] Maevskaya Marianna. Wooden skyscrapers. Traditional materials in innovative technologies of
[5] modern high-rise construction. Contemporary world's architecture, #11. 2018. Available at: http://www.niitiag.ru/pub/pub_cat/sovremennaya_arkhitektura_mira_11. Ed., comp. N. A. Konovalova. - Moscow-SPb.: Nestor-History, 2018.
[6] Rasuleva Yu.V., Kudasheva D.R., Titarenko A.V. (2019). The concept of a multifunctional urban xoskeleton. The Eurasian Scientific Journal, [online] 4(11). Available at: hthttps://esj.today/PDF/18SAVN419.pdf (in Russian)
[7] https://vk.com/@vesti.realty-dzhim-heverin-zaha-hadid-architects-derevo-eto-material-budu
[8] https://sports.inquirer.net/380145/p2fb-zaha-hadid-architects-to-build-groundbreaki
[9] http://proofer.faktor.fi/epaper/LVLHandbook/#2.kogda-zhdat-mnogoetazhnie-zdania-iz-dereva-v-rossii//vk.com/4.12.2019.
[10] https://www.archdaily.com/930422/timber-trends-7-to-watch-for-2020
[11] https://research.thinkwood.com/
[12] Baranova A A and Bobrova A A 2019 Dispersed reinforcement of cellular and fine-grained concrete based on silica fume *Proceedings of Universities. Investment. Construction. Real estate* **9(4)** pp 694–703
[13] Petrov A V and Efimova A K 2019 Nguyen Thanh Tung. Optimisation of technology used for restoring gas-ash concrete outdoor wall panels using modified structural and heat-insulating *concrete Investment. Construction. Real estate* **9(3)** pp 542–549
[14] Selezneva O I, Radaev S S and Baranova A A The impact of thermal treatment on the granulometric and chemical composition of diatomite of the Kamyshlov field when obtaininga liquid-glass binder *Investment. Construction. Real estate* **9(3)** pp 550–555

[15] Peshkov A V 2019 Digitalization of the economy, implementation of national projects and development of the construction industry Investments. Construction. Real Estate: New Technologies and Targeted Development Priorities IOP Conference Series: Materials Science and Engineering **667** 012074
[16] Tyuryukhanov K Yu and Pugin K G 2019 Impact of the surface of particles of moulding sand on the structural formation of asphalt *concrete Proceedings of Universities. Investment. Construction. Real estate* **9(3)** pp 566–577
[17] Dobruskina M A, Petrov A V and Bat-Erdene Z 2019 Improving the technology of retaining walls in the Irkutsk region using gabion baskets *Proceedings of Universities. Investment. Construction. Real estate* **9 (2)** pp 312–323
[18] Moskvitin V A., Emelyanova N A. and Mashovich A Y 2019 Experimental studies of air permeability indicators of composite "Poroplast CF" *Investment. Construction, Real estate* **9 (2)** pp 342–353

IOP Conf. Series: Materials Science and Engineering **880** (2020) 012005 doi:10.1088/1757-899X/880/1/012005

Preparation of highly selective sorbents from metallurgical wastes based on cavitation activation of sorption centers

O G Dubrovskaya[1], A G Bobrik[1], M A Bobrik[1], A I Matushenko[1] and E A Eldarzade[1]

[1] Federal State Autonomous Educational Institution of Higher Education "Siberian Federal University" pr. Svobodny, 82, Krasnoyarsk, 660041, Russian Federation

E-mail: dubrovskayaolga@mail.ru

Abstract. Having analyzed the initial data of RUSAL Krasnoyarsk Aluminium Smelter, we classified 45 types of waste, determined the hazard classes and estimated annual waste accumulation to be stored and disposed of. It has been defined that 14 types of waste can be recycled and reused as constructional materials of various purposes and sorbents for industrial wastewater treatment. In this study, we subjected a lining scrap of various origin, to a complete physical and chemical analysis in order to specify the type of modification and treatment of waste, specifically lining brick scrap, in the Testing laboratory of structural materials and chemical analysis of water of the School of Engineering and Construction, Siberian Federal University, and determined the chemical composition of this waste by the X-ray fluorescence (XRF) method. We also identified the substances suitable for modification of the sorbent – the oxides of aluminium, iron, silicon, and titanium – in the optimal percentage ratio, which resulted in obtaining functional active sorbents with high technical and operational characteristics at low costs for mechanical grinding and activation.

1. Introduction

Taking into account a large amount of metallurgical solid waste, specifically lining brick waste, being more than 14 tons/year, recycling with further reuse as the secondary product in other satellite manufacturing processes is believed to be quite a reasonable technology. One of the promising solutions here is to obtain highly efficient sorbents and use them in industrial wastewater treatment. Both economically and environmentally, it is rational to condition this wastewater for further reuse and closed-loop wastewater management. Process water is essential for many production operations in metallurgical plants, including:

- heat transfer in heating systems, cooling of liquids and solid bodies;
- flushing of gas cleaning filters to remove solid particles;
- surface treatment of heat power equipment;
- washing operations and wet treatment of the territory.

The quality of process water varies depending on numerous production processes; therefore, different technological systems are used in each case of wastewater treatment.

Enhancing wastewater treatment from petroleum products after washing of heat-power equipment requires both upgrading the standard filtration system by implementing new design parameters of the filtering units and utilizing filtration media with high sorption performance.

IOP Conf. Series: Materials Science and Engineering **880** (2020) 012005 doi:10.1088/1757-899X/880/1/012005

The double-layer quartz-carbon filters abundantly used are currently unable to provide the required quality of petroleum-containing effluent cleaning neither in accordance with the Russian hygienic regulations 2015 nor with the process water quality requirements. Besides, these filters are very bulky, difficult in operation and are not cost-effective for regeneration.

The standard water treatment scheme can be upgraded with the following structural and technological elements: filtration unit with replaceable cartridges and filtering bed made of the experimentally obtained sorbent based on a modified lining brick.

2. Materials and methods

We propose to use the experimentally obtained sorption bed based on a clay component as a sorbent. The raw material for this sorbent is waste, namely, lining brick scrap, the composition of which is determined by the XRF method (figure 1).

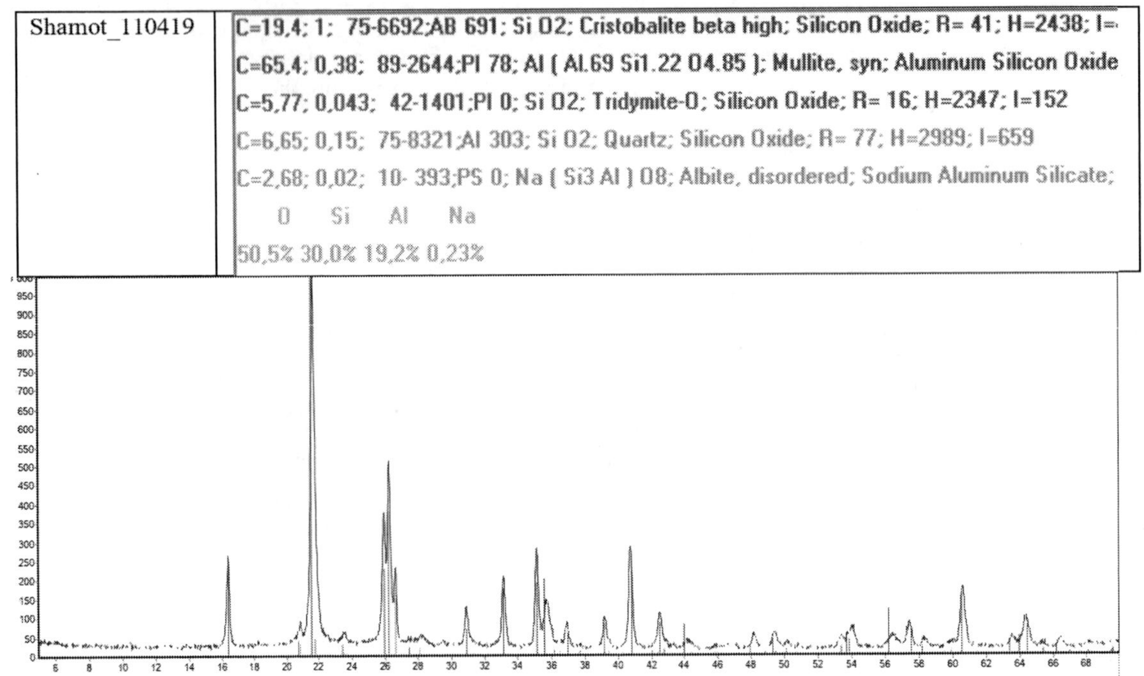

Figure 1. X-ray fluorescence analysis of the raw materials (lining brick waste)water with surfactants; IV – water for SPS

The obtained granular sorbent has a highly-developed active surface with micro-, meso-, and macropores which provide a high selective-sorption capacity to remove petroleum products and heavy metals from wastewater. The laboratory sorbent was tested in the Research Laboratory of the School of Engineering and Construction of SibFU (Testing laboratory of structural materials and chemical analysis of water). The sorbent was exposed to modifications, such as thermo-acid treatment, alkaline treatment, cavitational activation at 5000 rpm for 90 sec.

We determined core physical and chemical parameters of granules of the sorption material for various types of activation and studied the efficiency of removal of petroleum products and a number of heavy metals from wastewater. The sorbent showed stable activity to Fe, Mn, Pb, Zn, and turned out to be less active with Cu and Ni. The sorption properties of this material are significantly reduced in acidic media, but when a pH value was in the range of 8-9.5, sorption occurred in the optimum zone. The qualitative and quantitative composition of pollutants in the untreated wastewater and in the filtrate was analysed with the use of the standard methods of chemical analysis, IR spectrometry and atomic emission spectral analysis.

3. Results and discussion

The prepared sorbent has a high sorption capacity and its degree of extraction of petroleum products and heavy metals is 25% higher than that of the counterpart. At the same time, microscopy of the laboratory sorbent revealed an aluminium-containing surface layer which allows accelerating sorption based on the electrostatic attraction of pollutants. The results of the preliminary study are given in Table 1.

The cavitation-treated modification of the sorbent appeared to be the most effective. The experimental data are given in Figure 2. But a special focus should be placed on the results of studying the sorption capacity of the activated modified sorbent under various temperature conditions. The experimental data showed that, regardless of the temperature of the water fed to treatment, the efficiency of the sorbent activated by hydrothermodynamic cavitation remains consistently high and equals 86.7%.

Table 1. Characteristics of the sorbent

Indicator	Modified sorbent – thermo-acid treatment	Modified sorbent – cavitational treatment at 5000 rpm
Morpho-physical parameters	Granules of free form, with the size of 1.15-1.3 mm; colour varies from grey to brown	Granules of free form, with the size of, 0.03-0.3 mm; colour varies from grey to brown
Sorption capacity (absorptive capacity) for copper, mg/g	17.52 – 25.6	15.79 – 26.7
Temperature of application, °C	+4...+25	+4...+25
Heavy metal extraction degree	81.2%	95.1%
Cu	64%	86.9%
Fe	86.6%	98.9%
Pb	92.9%	99.7%
Maximum portion of the sorbent	5 – 12.8 g/l	5 – 9.4 g/l
Portion of the discharged sorbent	0.7 mg/l	0.68 mg/l
he rated height of the sorption bed in the adsorber	Depending on the diameter (D) of sorption filter 0.80 – 1.50m	Depending on the diameter (D) of sorption filter 0.45 – 0.60m

A comparative analysis of the sorption efficiency is illustrated in the diagram below (figure 2).

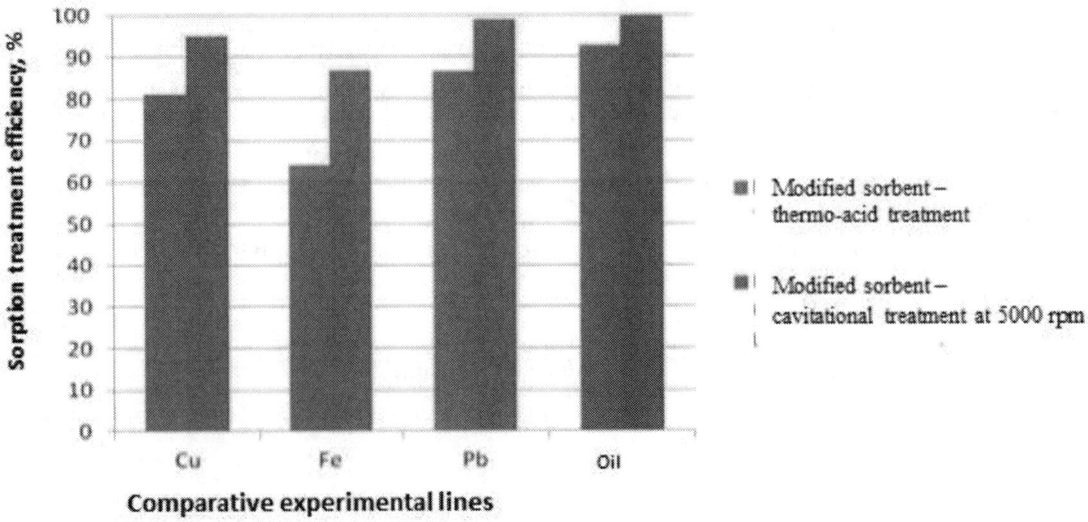

Figure 2. Industrial wastewater treatment efficiency

Table 2. Treatment efficiency

Line	Initial concentration of petroleum products in the water	Filtering conditions	Final concentration (cavitationly modified sorbent)	Cleaning effect (cavitationly modified sorbent)
1	40 mg/dm^3	Standard conditions (+20±2°C)	1.37 mg/dm^3	96.57%
2	40 mg/dm^3	Heating (+40±2°C)	1.02 mg/dm^3	97.45%
3	40 mg/dm^3	Cooling up to +0,1…+0,4°C	5.32 mg/dm^3	86.7%

4. Conclusion

The advantages of the obtained and sorption bed being studies are its high sorption capacity, the efficiency of extraction of petroleum products and heavy metals, the simple design and easy operation of the sorption filter, the maximum use of the filter capacity, the cost-effectiveness, and which is even more important – the high quality of purified water. Thus, the type of sorbent activated in a cavitation unit based on the hydrothermodynamic effects is recommended to be used as a sorption bed. This type of activation allows developing micro-, meso- and macrocracks on the surface of the sorbent, which undoubtedly leads to an increase in the area of the developed surface and results in an increase in the sorption capacity. During the laboratory tests at a pilot filtering unit, the degree of purification from petroleum products was 86.7% (with the final concentration of petroleum products of up to 0.05 mg/l).

These technical solutions allow achieving the required quality of water for reuse. The main advantages of the proposed sorption materials are:

1. the coal industry waste recycling with set variable sorption parameters (thermo-acid, alkaline or cavitation treatment);

2. various types of activation of burnt clay enable setting a selectivity of sorption with a simultaneous increase in sorption capacity – the degree of extraction of petroleum products is 95% at the maximum designated portion of the sorbent.

The introduction of a cluster of similar sorption filters for industrial wastewater treatment and conditioning of process water will make possible to form a closed-loop water use cycle and significantly reduce operating expenses of the enterprise.

This work was funded by RFBR and the Government of Krasnoyarsk Territory in the framework of research project 18-41-242008.

References

[1] Dubrovskaya O G 2014 Resource-saving technologies for neutralizing and disposal of waste from enterprises of the thermal power complex of Krasnoyarsk Territory p 164

[2] Dubrovskaya O G 2011 Conditioning of wastewater from power systems and complexes **6** pp 665-675

[3] Eldarzade E A 2016 Use of purified industrial and storm water as an alternative water source of the Enterprise *Building and Architecture – formation of the environment of vital activity* pp 122-123

[4] Eldarzade E A 2017 Development of closed water treatment systems for oil and gas production enterprises as a basis for environmental security in the region *Current state of science and technology* pp 73-79

[5] Kulikova N I and Nozhevnikova A N 2014 Purification of municipal wastewater with the reuse of water and treated sediments p 400

[6] Gimaeva A R 2011Sorption of heavy metal ions from water by activated carbon adsorbents Sorption and chromatographic processes **11** pp 350-356

[7] Zykova I V 2004 Adsorption of copper ions by ceramic chips from binary and multicomponent solutions *Proceedings of Universities. Chemistry and Chemical Engineering* **47** pp151-167

[8] Tsvetkova A D 2011 Investigation of the process of adsorption of copper ions on modified silicon dioxide *Bulletin of N.A. Nekrasov Kostroma State Universit* **2** pp 27-30

[9] Mukhin V M 2000 Active coals of Russia p 352

[10] Timofeev K L 2015 The kinetics of sorption of indium, iron, and zinc ions by weakly acidic cation exchangers Sorption and chromatographic processes **15** pp 720-729

[11] Wang Ji-Zhong 2005 Removal of heavy metals from wastewater using natural minerals *Bull Mineral Petrol Geochem* **24** pp 159-164

[12] Svilovi'c S 2009 Modeling batch kinetics of copper ions sorption using synthetic zeolite *Journal of Hazardous Materials* **170** pp 941-947

[13] Demirbas Ayhan 2008 Heavy metal adsorption on agrobased waste materials Journal of Hazardous Materials **2-3** pp 220-229

[14] Lokendra S, Thakur S Thakur Lokendra, Parmar 2013 *Mukesh International Journal of Chemical and Physical Sciences* **6** pp 6-19

[15] Tumin Najua D 2008 Adsorption of copper from aqueous solution by Elais Guineensis kernel activated carbon Journal of Engineering Science and Technology **2** pp 180-189

IOP Conf. Series: Materials Science and Engineering **880** (2020) 012006 doi:10.1088/1757-899X/880/1/012006

Cyclic testing of polypropylene fibre reinforced concrete

I G Korneeva[1]

[1]Irkutsk National Research Technical University, Department of Building Production, 83, Lermontov Street, 6640074, Russia

E-mail: inna.k@ex.istu.edu

Annotation. The article studies fatigue limits of polypropylene fibre reinforced concrete under conditions of low-cycle loading of variable intensity. It presents the experimental data based on testing mode under load conditions with constant strain rate. When put to the cyclic test, polypropylene fibre reinforced concrete tends to experience changes in its structural properties. Accordingly, the study has a quantitative assessment of changes in the resistance potential for axial compression of both test concrete samples and prismatic samples subjected to cyclic loads.

Volumetric reinforcement of cement-containing matrices (solutions, concretes) forms [1-5] an internal structure of constructions. The structure is likely to experience external cyclic effects. Similar systems have an increased resistance to fatigue effects due to the ability to redistribute internal stresses between the components of the composite, if distribution is uniform and deformation is compatible.

One of the factors causing fatigue transformations of bearing structural elements in buildings is a cyclic change in their stress-strain state resulted from fluctuating external effects. Structures demonstrate responces to these effects, that are various by nature, and they express this through stress (strain) variation that probably has cumulative consequences. In cement-containing composites, this may result in breaks of internal structure with qualitative and quantitative changes in resistance.

The service life of fibre-containing composites results from targeted design of their structure, including physical and mechanical properties of reinforcing fragments, their size, pore space of matrix, and other factors [1-5].

The previous study has shown that the use of polypropylene fibres with a length of 40 mm allows obtaining optimal structural modification at their volume content of 1.8-2%. The current research is an experimental testing of fatigue endurance of polypropylene fibre reinforced concrete under conditions of low-cycle loading of variable intensity.

Experimental study

For the experimental part of the study, we used prismatic samples (100×100×400 mm) made of fine-grained concrete (cement:fine aggregate:coarse aggregate:water = 1:1,42:3,31:0,55), reinforced with polypropylene fibres with an equivalent diameter of 0.8 mm, a length of 40 mm and a volume fraction of 1.5%. Table1 shows the essential parameters of primary components.

Table 1. Initial indicators of polypropylene properties and fibre concrete.

Material	Strength. MPa	Modulus, MPa	Maximum deformation	Limits R_{ev}		Density, kg / m^3
				R^0_{crc}	R^v_{crc}	
polypropylene	145.54	2767				900
fiber concrete	35.79	22607	$265 \cdot 10^{-5}$	0.32	0.74	2298

concrete	43.57	26580	$247 \cdot 10^{-5}$	0.47	0.88	2325

After a three-month hardening period under the standard storage regulations, fatigue tests began, which eliminates the time factor when analyzing the kinetics of changes in the properties of fibre concrete.

We calibrated the major part of samples (Pulsar 1.2) according to the ultrasound speed, which averaged (4175 m/s) with a variation coefficient of 4%. At the same time, the variability of the prism mass was 2.02%. We divided the samples into two groups: a) test samples, b) samples subjected to cyclic compression with a set voltage drop.

The experimentally approved level of maximum stresses is in accordance with the research objectives – the possibility of fatigue degradation of structure based on a limited number of impact cycles. They are the stresses in the range of the upper and lower limits of microcrack formation Rcr_c (according to O. Ya. Berg [1]) and components of $0.6R_b$. The minimum cyclic loading was 0.2 MPa in the mode of constant deformation rate of 0.04 mm/s, the duration of the load-unload cycle was 1.5 minutes.

In the process of cyclic actions, we collected the following data: diagrams of deformation σ-ε, modulus of elasticity, energy expended during the cycle, as well as deformations of samples in two directions. Automatically recorded data are with an accuracy of 10^{-5}. Control measurements of parameters took place after the 1st and every 10 subsequent cycles of effects with a total duration of 200.

Figure 1 shows the trends of changes in the structural properties of fiber concrete under cyclic effects. At the same time, the indicators recorded after ten cycles are taken as initial indicators. In our opinion, this approach reduces the influence of random factors caused by experimental inaccuracies and the inevitable structural adaptability of the material to external influences. This makes it possible to link observed property changes to controlled cyclic loads on a more reasonable basis.

The analysis of experiment results proved high sensitivity of all parameters to the considered effects. However, they differ in their direction and dynamics. The modulus of elasticity and the magnitude of longitudinal deformations at maximum stresses appear relatively stable with a moderate initial increase (up to 30-40%). There was inaccuracy of kinetic change in the values of transverse deformations. The behavior of particular samples reveals a tendency to stabilize the transverse deformation after 40-50 loading cycles, which indirectly indicates the completion of fatigue transformation with a predominant positive effect.

The changes in longitudinal and transverse deformations affect the kinetics of the Poisson's ratio estimated by their correlation (figure 1D). With a significant variation of experimental data (more than 40%), there is a tendency for its decrease with stabilization by 0.12÷0.14, which is significantly lower than generally accepted values [1-5]. It should be noted that the value of transverse deformations coefficient (figure 1E), estimated by the ratio of increments at individual stages, significantly exceeds the value of 0.5. That is, in accordance with generally accepted ideas [6-9], cyclic loading leads to the accumulation of internal micro-destructions. The fact that the proportion of plastic deformations in total longitudinal deformations at the maximum cyclic loading is constantly increasing confirms this assumption. Although growth dynamics is ambiguous (figure F), there is a clear trend of its increase.

The above experimental data indicate fatigue transformation of polymer fibre concrete under the influence of cyclic effects of the accepted range. The results of ultrasound examination of samples prove this. At the end of the cycle loads, the ultrasound speed was reduced by 11.8%. To quantify changes in the resistance potential, we performed comparative experiments for axial compression of test samples and cyclically loaded prismatic samples. We performed mechanical tests by means of Instron 5989 used in automatic mode with a constant deformation rate of 0.004 mm/s. A sensor of this automatic complex performed force control with an accuracy of 1 kN, and extensometers controlled deformation. Table 2 showns the average values of the monitored parameter. Cyclical effects of the accepted level and their duration did not have any impact on the strength of fine-grained fibre concrete, including the statistical parameters of its distribution. Their variability is less than 4%, which is significantly lower than normalized value.

IOP Conf. Series: Materials Science and Engineering **880** (2020) 012006 doi:10.1088/1757-899X/880/1/012006

The elasticity modulus increases significantly (19.7%) at a high distribution density (about 9%). With comparative strength stability, this indicates an increased sensitivity of deformation properties to fatigue processes. As mentioned earlier, the equipment provided testing with no prior centering accompanied by automatic recording of deformations up to the moment of physical destruction. In this case, differential analysis of the deformation structure on the ascending and descending branches becomes possible. There are initial (ε_0) deformations of "tooth" type [1, 4] with a concavity towards the stress ordinate. Their physical nature is problematic [1, 5] caused mostly by structural defects. It is obvious that they experience residual deformations under cyclic effects, which leads to a nearly two-fold reduction in axial compression tests. Compression of internal structural imperfections helps to reduce the limit of proportionality and longitudinal deformations corresponding to the maximum load. The identity of their changes indirectly indicates the common nature of fatigue processes that cause them.

As for the descending branch of the chart, taking into account the regulatory restrictions on the use of this resistance potential [10-17], the comparative analysis refers to a section corresponding to a 20% stress reduction ($\sigma = 0.8\ \sigma_{max}$). The experimental samples have a longer section length compared to the test samples, which indicates their increased ability to redistribute internal forces.

The transverse deformations of samples subjected to cyclic effects changed significantly. If we take into account that these deformations are associated with the processes of micro-destructions, then their reduction after cycles is due to greater structural adaptability to force influences.

1. Polypropylene fibre reinforced concrete (PFRC) has a high fatigue resistance to a few cycles of moderate intensity forces.
2. Multiple directions and non-monotonicity of changes in significant indicators of PFRC design properties determine the need for a differentiated approach to the analysis and evaluation of fatigue effects from cyclical external influences.
3. Fatigue transformation of PFRC accompanies a decrease in the potential of its ability to internally redistribute efforts.

Table 2. Comparative evaluation of changes in strength and deformation properties.

Controlled parameter	Index	dimension	Experi-mental	Test	Change
Prismatic strength	R_b	MPa	34.57	35.79	0.966
Elasticity modulus	E_b	MPa	27072	22607	1.197
Longitudinal deformation at maximum load	$\varepsilon_{b,\ max} \cdot 10^5$	‰	215	265	0.811
Longitudinal deformation at fracture (0,8 σ_{max})	$\varepsilon_{b,\ ult} \cdot 10^5$	‰	281	318	0.884
Proportionality limit on the ascending branch	$\varepsilon_{b,\ e} \cdot 10^5$	‰	129	159	0.811
Initial deformations (compressions)	$\varepsilon_0 \cdot 10^5$	‰	23	42	0.548
Length of the downward branch of deformations	$(\varepsilon_{b,\ ult} - \varepsilon_{b,max}) \cdot 10^5$	‰	66	53	1.245
Transverse deformation at maximum load	$\varepsilon'_{b,\ max} \cdot 10^5$	‰	88	153	0.575
Transverse deformations at failure	$\varepsilon'_{b,\ ult} \cdot 10^5$	‰	376	474	0.793

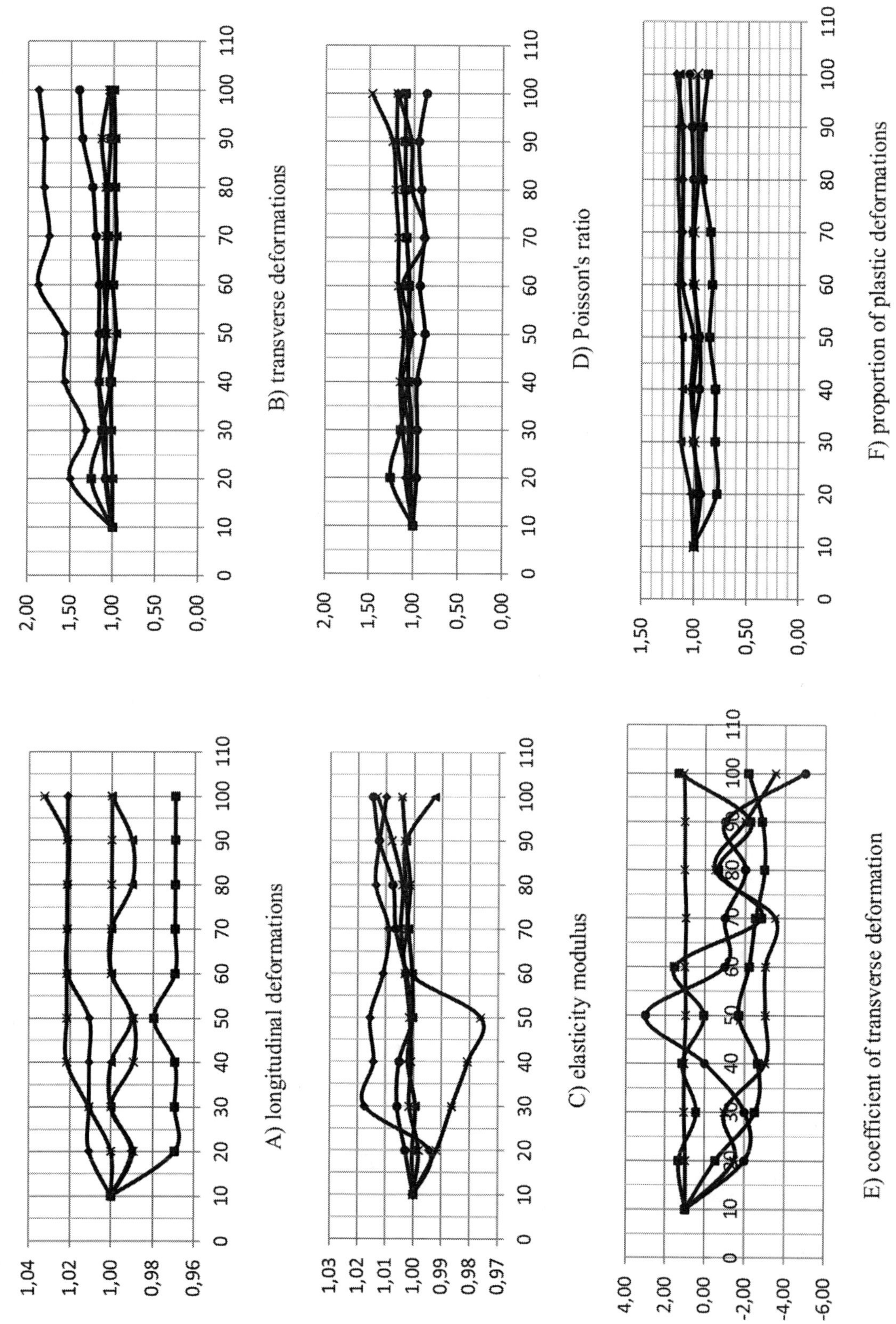

Figure 1. Kinetics of structural properties under cyclic loads.

IOP Conf. Series: Materials Science and Engineering **880** (2020) 012006 doi:10.1088/1757-899X/880/1/012006

References

[1] Berg O Ya *Physical bases of strength of concrete and reinforced concrete* M 1962 p 96
[2] Bazhenov Yu M *Concrete under dynamic loading* M. stroizdat 1970 p 273
[3] Rabinovich F N *Composites based on dispersed reinforced concrete. Questions of theory and design, technology, construction* M. DIA 2004 p 563
[4] Karpenko N I *General models of reinforced concrete mechanics* M stroizdat 1996 p 416
[5] Gvozdev A A (Ed.) 1978 *New in the design of concrete and reinforced concrete structures* (Moscow) p 202
[6] Morris A D and Garrett G G A comparative study of the static and fatigue behaviour of plain and steel fibre reinforced mortar in compression and direct tension. Int *J Cem Compos Lightweight Concr* 1981 **3(2)** pp 73–91
[7] Wang H L and Song Y P Fatigue capacity of plain concrete under fatigue loading with constant confined stress *Mater Struct* 2011 **44** pp 253–62
[8] Cachim P B, Figueiras J A and Pereira P A A Fatigue behavior of fiber-reinforced concrete in compression *Cem Concr Comp* 2002 **24(9)** pp 211–7
[9] Castillo E, Fernandez-Cantelli A and Ruiz-Ripoll M L A general model for fatigue damage due to any stress history. Int *J Fatigue* 2008 **30** pp 150–64
[10] Paskova T and Meyer C Low-cycle fatigue of plain and fiberreinforced concrete. *ACI Mater J* 1997 **94(4)** pp 273–85
[11] Grzybowski M and Meyer C Damage accumulation in concrete with and without fiber reinforcement. *ACI Mater J* 1993 **90(6)** pp 594–604
[12] Zhang J, Stang H and Li V C Experimental study on crack bridging in FRC under uniaxial fatigue tension. *J Mater Civil Eng* 2000 **12** pp 66–73
[13] Yin W and Hsu T C C Fatigue behavior of steel fiber reinforced concrete in uniaxial and biaxial compression. *ACI Mater J* 1995 **92(1)** pp 71–81
[14] Gao L and Hsu T C C. Fatigue of concrete under uniaxial compression cyclic loading *ACI Mater J* 1998 **95(5)** pp 575–81
[15] Lee M K and Barr B I G An overview of the fatigue behaviour of plain and fibre reinforced concrete *Cem Concr Comp* 2004 **26** pp 299–305
[16] Ghosni N, Samali B and Vessalas K Evaluation of structural behaviour of polypropylene fibre reinforced concrete beam under cyclic loading 23rd Australasian Conf. on the Mechanics of Structures and Materials (ACMSM23) 9-12 December 2014 (Byron Bay:Australia) ed. S T Smith
[17] Korneeva I G 2019 Extensibility of fibre reinforced concrete *IOP Conf. series: materials science and engineering* (2019) **667** 012044

IOP Conf. Series: Materials Science and Engineering **880** (2020) 012007 doi:10.1088/1757-899X/880/1/012007

Construction of an axonometric outline of a curved surface of a part

L I Kravcova[1] and I I Kostrubova[1]

[1] Irkutsk National Research Technical University, 83 Lermontov str., Irkutsk, 664074, Russia

E-mail: kravcova@istu.edu

Abstract. The article considers the geometric construction of the orthogonal isometry of a curved surface. To solve this problem, the authors propose to perform additional construction of the contour of the section of the surface in the orthogonal drawing and use the coordinates of the points of this section to construct an axonometric outline of the surface. The mismatch between the axonometric contour of the rotation surface and the corresponding contour of this surface in the projection drawing is illustrated. As a result, the proposed solution allows geometrically accurate construction of support and additional points of the axonometric contours of the revolution surfaces.

1. Introduction

The article discusses solutions to geometric problems associated with the construction of an axonometric drawing of curved surfaces. It is recommended to solve this problem in a certain sequence: first, on the orthogonal drawing to build a line of an axonometric surface contour, and already on its basis to build a outline of a curved surface in an axonometry. Such tasks are encountered in the process of developing architectural and design projects [1-8]. New approaches to their solution are relevant. Despite the introduction of modern computer technology, the a priori value of manual modeling of the architectural form based on axonometric projection as a process ahead of the emergence of a “conceivable concept” remains [4].

It is known that an axonometric drawing, in other words “axonometry", is geometrically constructed according to a projection orthogonal drawing [9, 10]. The basis of the method for constructing axonometric images is that the object together with its coordinate system is projected parallel to the axonometry plane. The basic concepts of these geometric constructions are well known for a long time [7-9, 12]. The shape of the part can be different. It can be multifaceted, curved, or combine both surfaces.

2. Materials and methods

Any axonometric image of an object is performed by constructing its individual points [11]. Building polyhedral surfaces based on their specified geometric shape elements (vertices, faces, edges) does not cause difficulties. Axonometric projections are based on coordinates taken from an orthogonal drawing[9, 10, 14]. We can say that elements of an orthogonal drawing are transformed into elements of an axonometric drawing. In Figure 1, the projections of the pyramid vertices in axonometry are based on coordinates taken from the orthogonal drawing.

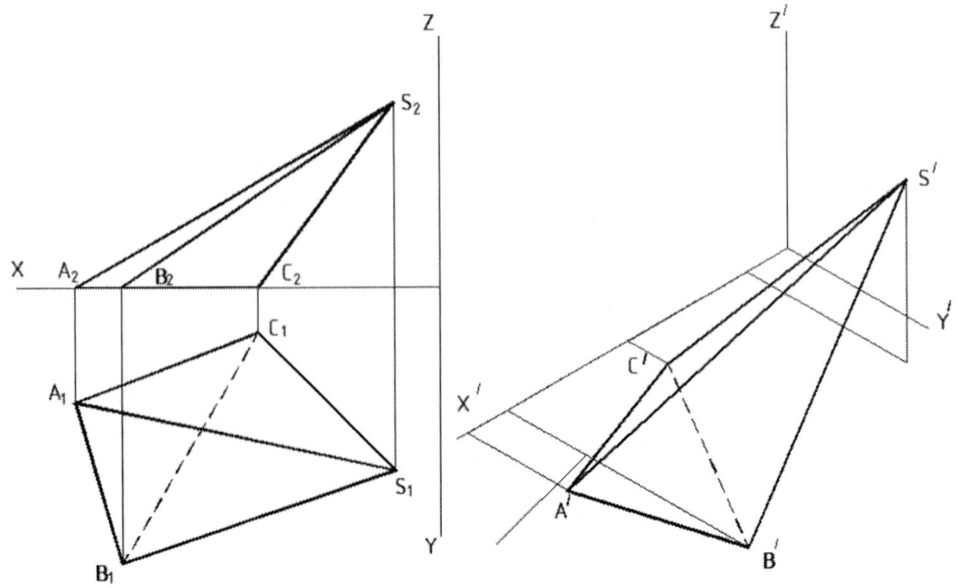

Figure 1. Construction of axonometry of a polyhedral surface

The construction of axonometry of curved surfaces is somewhat more complicated. Before proceeding to the solution of such problems, we first consider some features of the projection process associated with the concept of "surface outline". Obtaining an axonometric drawing is achieved in stages using two wrapping projecting surfaces of different projection directions [12].

It is believed that the projection drawing of the part itself is the result of the intersection of the wrapping, projecting surface with the projection plane. Recall that in orthogonal projection, the wrapping surface is a set of projecting lines that are tangent to the surface of the part and perpendicular to the plane of projections. The direction of projection is equal to the angle of 90 degrees. The line of contact of the wrapping projection surface with the surface of the part is called the visible contour line and its orthogonal projection on the projection plane is called the surface outline, or the visible contour projection. Note that this contour, which belongs to the orthogonal drawing, does not directly participate in the construction of the axonometric outline of the curved surface. The following arguments lead to this conclusion.

The axonometric projection of a curved surface is formed according to the same law as the orthogonal one. That is, the axonometric drawing of the part is the result of the intersection of the newly formed wrapping projection surface with the axonometric plane. To clarify, the newly formed enveloping projection surface is a set of projecting lines that are tangent to the surface of the part and parallel to the direction of axonometric projection. As is known, for orthogonal isometry, the projections of the axonometric projection direction are located at an angle of 45 degrees to the projections of the coordinate axes [9, 12].

It is easy to predict that the expected line of contact of the new wrapping projection surface with the surface of the part, the so-called visible contour line, is not present in the orthogonal drawing. This is due to the fact that there was no need to portray it. But now this complicates or makes it impossible to build its axonometric projection, i.e. the shape of the surface. In other words, there is no contour on an orthogonal drawing that can be used to draw an outline of a curved surface in axonometry. It follows that in order to solve the main problem, it is necessary to add a new visible contour line to the orthogonal drawing in accordance with the chosen direction of axonometric projection.

Consider a possible solution to this problem. As can be seen in Figure 2, the orthogonal drawing of the surface of revolution contains a base - this is a parallel, and the outline is the main meridian. The direction of the axonometric projection S is selected in accordance with the direction of projection of the orthogonal isometry. We believe that the new "wrapping ray surfaces" touches a given surface of revolution along a plane curve. This flat curve is the line of the visible contour. We construct it in the drawing as a line of a section of given revolution surface. First we define the determinant of the section plane. Obviously, it can be defined by three points A, B, C. Each of these points is the touch point of the projecting lines of the wrapping surface with the specified rotation surface. As can be seen on the horizontal projection of the orthogonal surface drawing, points A, B lie at the intersection of the parallel and the meridian does not require additional construction. To set point C in the drawing, an additional geometric solution is required. From Figure 2 we can see that the point of contact of the projecting straight line with the given surface of revolution must lie in the axial horizontal projecting plane. The meridian of the revolution surface and the tangent projection straight line belong to this plane.

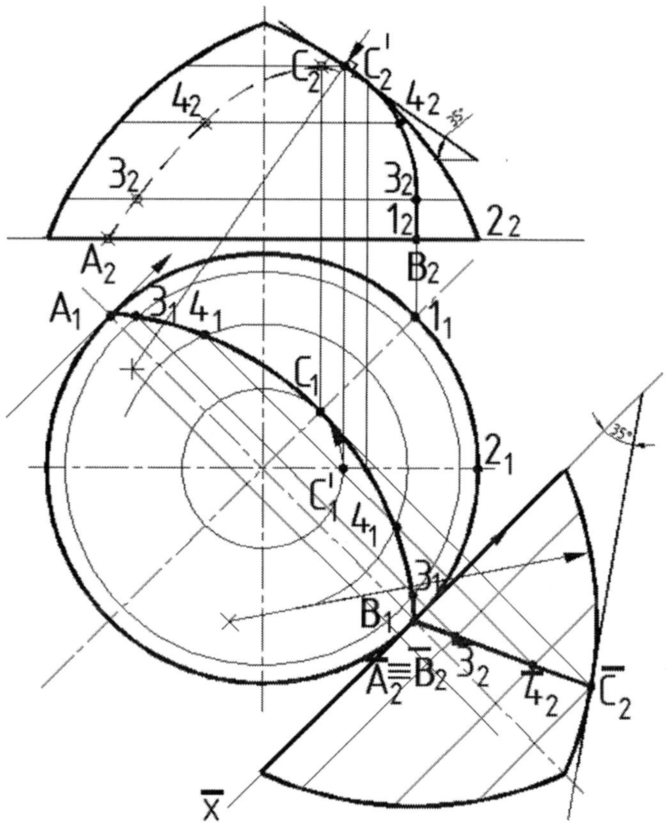

Figure 2. Construction of a line of touch of the enveloping projection surface with the surface of rotation

Next, we show how to find the desired point of tangency C. Note that all constructions associated with finding this point must be solved in the level plane. Therefore, we turn the meridian with the expected point C around the axis of the surface of revolution to the position of the main meridian. On the horizontal projection (Picture 2), point 1 should be turned to point 2 in parallel. Then (this can be seen in the frontal projection of the drawing) in the plane of the main meridian we draw a straight tangent to the meridian and parallel to the direction of the axonometric projection. The angle of this straight line is 35 degrees. The tangent point (C) is constructed as the intersection point of two mutually perpendicular lines. One of which is a direct tangent to the meridian, and the second passes through the center of the arc of the meridian. Now it remains to turn the meridian back to its original position, and with it, the tangent point C should be rotated to the original position of the meridian.

Thus, in the drawing, three points A, B, C are defined. They belong to the secant plane. These points are points of visibility, since they belong to the outline of the surface of revolution during its axonometric projection. That is, these three constructed reference points define the line of touch, or the line of the visible contour in the drawing. Next, you need to determine the intermediate points of this line. Since the construction of axonometry requires the natural value of the coordinates of the points, it is therefore rational to solve this problem by applying methods of transforming the drawing. We use the method of replacing projection planes. An additional vertical projection plane is located perpendicular to the secant plane (A, B, C) and parallel to the axial, horizontally projecting plane (the meridian of the surface of revolution, the tangent projection line and the tangent point C belong to this plane). We use the method of replacing projection planes. An additional vertical projection plane is located perpendicular to the secant plane (A, B, C) and parallel to the axial horizontally projecting plane (the meridian of the revolution surface, the tangent projection line and the tangent point C belong to this plane). To solve the main problem, the meridian is projected in full size on the additional projection plane, and the secant plane is projected by a degenerate projection. A further solution is easy to see in the drawing. The desired line of the visible contour is defined on the frontal and horizontal projection planes by three reference points (A, B, C).

Next, you need to determine the intermediate points of this line. Therefore, on the additional projection plane, we construct a series of lines parallel to the horizontal projection plane. These are degenerate projections of parallels. The next step is to find their intersection points with a degenerate secant plane.

These points (3, 4), like points (A, B, C) belong to the parallels of a given surface of revolution. We complete the missing projections of these points and construct their smooth contour on the frontal and horizontal projection planes. Thus, the resulted drawing is defined by the initial contours of the surface of revolution under consideration and supplemented by a smooth contour of the points (A, B, C, 3, 4). These points are points of visibility and belong to the outline of the surface of revolution during its axonometric projection. This can be seen on the additional projection plane in the direction of axonometric projection.

For greater clarity, the parallels meridians are plotted on an axonometric drawing and a cut is made. Additionally, an isometry of the meridian is constructed and it passes through the top of the surface of revolution and points A, B.

3. Results

Now, according to this drawing, we construct the orthogonal isometry of the surface of revolution. Figure 3 shows an isometry of the base of the revolution surface in the form of an ellipse and the outline of the surface of revolution. The construction of an isometric surface shape is performed according to the coordinates of the smooth contour of the points (A, B, C, 3, 4).

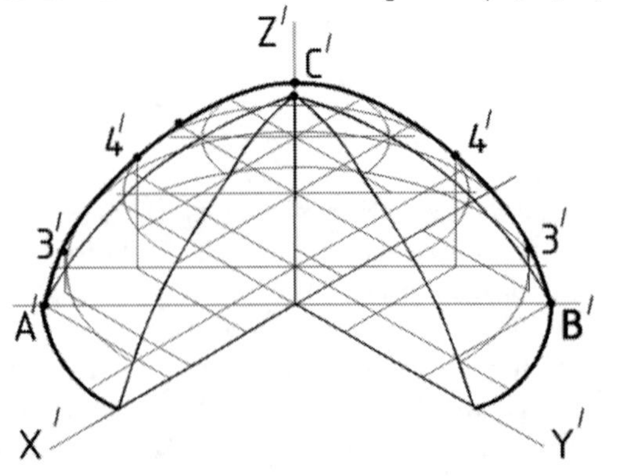

Figure 3. Construction of an axonometric outline of the revolution surface

IOP Conf. Series: Materials Science and Engineering **880** (2020) 012007 doi:10.1088/1757-899X/880/1/012007

4. Conclusion

Analysis of the image of these two contours on an isometry allows us to conclude that the use of an additional circuit specially constructed on an orthogonal drawing makes it possible to most accurately convey the outline shape of a curved surface in axonometry.

References

[1] Bolshakova M S 2018 Axonometric constructions in graphic design *Visual Culture: Design, Advertising,Inf. Technologies: Materials of the XVII Vseros. Scientific Practical conf.* [Omsk: Omsk State Technical University Publishing House] p 15

[2] Eliseev N A 2013 Development of the axonometry in works of scientists of petersburg *XXX Symposium of the Intern. Committee for the History of Technology* [M.-SPb.: ICONTEC] p 9

[3] Efimov D I 2014 *E-learning in continuing education* [Ulyanovsk: Publishing house: Ulyanovsk State Technical University Number] pp 263-264

[4] Malakhov S, Boranov S 2017 Axonometry role as a model motivating the archofunction J.*Innovative Project* B **2** 3 (7) [Samara: State Technical University] p 67

[5] Matveeva M, Kalyuzhnova N and Violin S 2019 IOP Conf. Series: Materials Science and Engineering. 667 doi:10.1088/1757-899X/667/1/012036 6.

[6] Peshkov V V 2019 DOI: 04010 (https://doi.org/10.1051/matecconf/201821204010)

[7] Puras I Y 2016 Modern trends in graphic design Business and Design Review B **1** (2) p 8 3

[8] Rauschenbakh B V 2001 *Geometry of a Picture and Visual Perception* (SPb.: ABC-Classic) p 320

[9] Chetverukhin N F 1963 *Descriptive Geometry* (M.: Higher School) pp 255 – 372

[10] Glazunov EA, Chetverukhin N.F. 1953 *Axonometry* (Moscow: State Publishing House of Technical and Theoretical Literature) p. 292

[11] Gordon V O, Sementsov-Ogievsky M A 2009 *Course Descriptive Geometry* (Moscow: High School) p 272 p.

[12] Klimukhin A G 1963 *Descriptive Geometry* (Moscow: Stroyizdat) p 368

[13] Klimukhin A G 2007 *Descriptive Geometry (*M: Architecture) p 333

[14] Koroev Y I 2015 *Descriptive Geometry* (Moscow: KNORUS) p 423

[15] Kravcova L I, Kostrubova I I and Smolkova E F 2014 Technical drawing (Irkutsk: ISTU) p 88

IOP Conf. Series: Materials Science and Engineering **880** (2020) 012008 doi:10.1088/1757-899X/880/1/012008

Gas-hydrodynamic characteristics of aerotanks

V N Kul'kov[1,3] and E Yu Solopanov[2]

[1]Institute of Architecture, Construction and Design, Irkutsk National Research Technical University, 83 Lermontov Street, Irkutsk 664074, Russian Federation
[2]Institute of Information Technology and Data Analysis, Irkutsk National Research Technical University, 83 Lermontov Street, Irkutsk 664074, Russian Federation

E-mail kulkof.viktor@yandex.ru

Abstract. The purpose of the research was to determine the hydrodynamic indicators of aerotanks determining the depth and effectiveness of wastewater treatment. The study of technological parameters of the aeration tank was carried out on a physical model representing a transverse vertical section of the aerotank. The hydrodynamic similarity in the model and industrial aerotank determined by the identical values of the Reynolds, Froude, and Euler criteria. The study of similar indicators on an industrial aeration tank is very difficult and costly. The model and experimental setup make it possible to vary in a wide range the intensity of air aeration, the design and location of the aerator, control the speed of hydrodynamic flows and the concentration of freely floating sludge in the facility with aeration. To study the gas-hydrodynamic indicators in the aerotank model, original methods based on the laws of physical chemistry have been developed and used. The contact surface of the phases "liquid - gas" from the intensity of air aeration was determined using a chemical method for aerated biological wastewater treatment facilities. The dependences of the specific contact surface of the phases "liquid - gas", referred to the unit of the working area of the aerator, on the intensity of aeration and the depth of immersion of the aerator are obtained. The oxidizing ability of a fine bubble aerator was determined for various aeration intensities. The dependences of the oxidizing ability on the contact surface of the phases 'liquid - gas" in aerotanks were obtained for the first time. It is shown that the oxidation ability of the aerotank is determined by the phase contact surface and, to a large extent, by the hydrodynamic situation in the aerotank, which forms the fields of freely floating activated sludge.

1. Introduction

Wastewater treatment in aerotanks, using a free-floating active biocenosis, allows to achieve the required quality of wastewater treatment, reducing the anthropogenic load on surface water sources. For the stable operation of the aeration tank, which is part of the biological treatment facilities, it is necessary to ensure a uniform distribution of oxygen and silt throughout the volume of the aerated structure [1–3].

The gas-hydrodynamic situation in the aerotank to a large extent determines the effectiveness of wastewater treatment. The direction and speed of hydrodynamic flows and activated sludge suspended in it, as well as the concentration of dissolved oxygen, are determined by the type and intensity of aeration and the location of the aerators in the aerotank. The aeration system provides the supply and distribution of air oxygen to the aerotank, which is necessary to maintain the oxidation of pollutants entering the biological treatment, maintaining activated sludge in a state of suspension, and also blowing off the gases formed as a result of metabolism, the excess of which inhibits the process of biochemical wastewater treatment [4–6].

The possibility of physical modeling of gas-hydrodynamic processes in aerated wastewater treatment plants as a first approximation was determined by the ratio between four dimensionless complexes [7–9]:

$$E_u = X \cdot Re^{-s} \cdot Fr^{-r}, \quad (1)$$

where E_u, Fr, Re – are the Euler, Frood, and Reynolds criteria, respectively; X – constant.

The Euler criterion reflects the effect of the difference in hydrostatic pressure on fluid motion, and its value characterizes the ratio of the change in hydrostatic pressure to the inertia force at similar points. The Froude criterion reflects the influence of gravity or dead weight on fluid motion and is a measure of the ratio of inertia to gravity at similar points. The Reynolds criterion reflects the influence of the friction force on the fluid motion, and its value characterizes the ratio of inertial forces to the friction forces in such flows [7, 8].

Providing a developed contact surface of the "liquid - air" during aeration makes it possible to increase the oxidation power of the aerotank, stabilize the mineralization of wastewater, increase the depth of biological treatment and provide floating activated sludge with aerobic conditions [8–12].

2. Experimental part

The purpose of the research was to determine the hydrodynamic indicators of aerotanks that determine the depth and effectiveness of wastewater treatment using a fine-bubble aerator installed in the corner of the aerated structure. To study the gas-hydrodynamic processes and the distribution of activated sludge in aerated structures, an experimental setup was created with a model cell (figure 1), which is a vertical transverse section of the aerotank (figure 2).

Figure 1. Model cell.

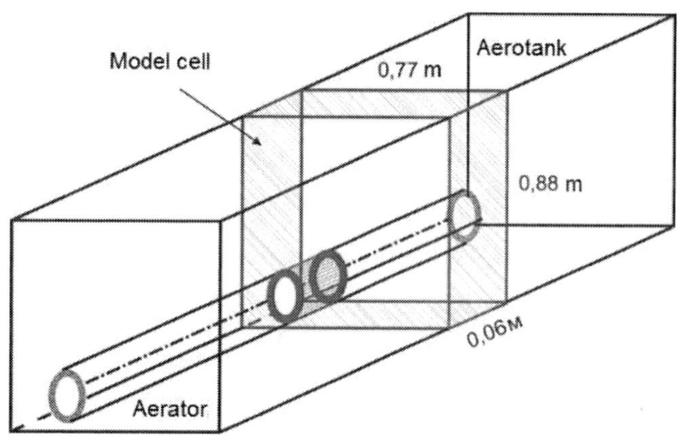

Figure 2. A model cell representing a vertical cross section of an aerotank.

The cell was made of polished silicate glass and had dimensions of 0,06×0,77×0,88 meters. The amount of air supplied by the compressor to the cell was controlled by valves, and the instantaneous air flow was controlled by a rotameter of the RM 0,63 type GUZ [8, 9, 12]. AKVA-LINE aerator consisted of a perforated hull with radial holes. Its dispersant is made in the form of a cylindrical shell of a composition of two grades of porous polyethylene, which ensures contact of air with liquid directly in the upper coating layer. The technology of creating a dispersing coating provides fine-bubble aeration. The coating forms air bubbles, the average size of which remains unchanged or even decreases with increasing air flow [8].

The interfacial surface of the gas-liquid system is an important calculated value, which depends on the geometry of the structure, the conditions of its functioning and the physical properties of the liquid.

The mass transfer rate of oxygen from air of liquid a can be easily changed by changing the interphase surface [5, 8, 13, 14].

The chemical method [8–10, 13, 14] for determining the surface of the contact of phases is based on the known chemical reaction rate. It allows you to determine the integral surface of the contact of phases regardless of the structure of the two-phase layer moving in the structure. To determine the contact surfaceof the phases "liquid - gas", the absorption rate of carbon dioxide by an aqueous alkaline solution of sodium hydroxide was studied. The change in alkali concentration was determined by titration of an alkali solution with decinormal hydrochloric acid [8, 10, 14].

Studies to determine the oxidative ability of pneumatic aeration using the AKVA -LINE fine bubble aerator were carried out on a model of aeration tank filled with tap water, the aerator was installed in the working position, at the bottom of the model in the corner. Sodium sulfite and a catalyst cobalt chloride were used to remove oxygen from water [8, 11]. To determine the concentration of oxygen in water, an ATT–3010 oxygen meter was used. The oxidizing ability of the aerator (gO_2/h) was determined by the equation [4, 5]:

$$OC = \frac{K_q a \cdot C_\infty \cdot V}{1000}, \qquad (2)$$

where $K_q a$ – volumetric mass transfer coefficient, 1/h; C_∞ – limiting concentration of oxygen in water at an infinite aeration time at a given atmospheric pressure, water temperature and its salinity, mg/dm^3; V – volume of water in the cell, dm^3.

To study the hydrodynamics in the physical model of the aeration tank, with the intensity of fine-bubble aeration J = 5.41 m^3/(m^2·h) with a lateral aerator, we used the tracer method based on the introduction of solid particles with the same density in the water as the density of water [6 , 8, 15]. Particles moved along with the fluid flow, which, when using video in a planar two-dimensional model, made it possible to analyze the velocity of any fluid element with its decomposition into components along the coordinate axes.

To determine the scalar and vector values of the wastewater circulation velocity at the i-th space point of the model of the vertical transverse section of the aeration tank (see figure 1), a control square of 80÷80 mm was selected, the coordinates of the mark x_1 and z_1 at the input and x_2 and z_2 were fixed the exit from this square, as well as the time t of finding the mark in this control square.

Speed components were calculated as:

$$\overrightarrow{V_{x,\,i}} = \frac{(x_2 - x_1)}{t}, \quad \text{cm/s}; \qquad (3)$$

$$\overrightarrow{V_{z,\,i}} = \frac{(z_2 - z_1)}{t}, \quad \text{cm/s}. \qquad (4)$$

Since the true velocity is a pulsating quantity, and $\overrightarrow{V_{x,\,i}}$ and $\overrightarrow{V_{z,\,i}}$ are averaged quasistatic components, to determine the latter, we controlled the movement of 10÷15 label particles at each studied point of the model. The distribution of the components of the circulation velocity at a given point was close to the normal law. The medians of the distribution curves coincided with an accuracy of 5% with the arithmetic mean values, which were used in the future to construct a quasi-static vector field.

Velocity modulus values were calculated as

$$|V_i| = \sqrt{\left(\overrightarrow{V_{x,\,i}}\right)^2 + \left(\overrightarrow{V_{z,\,i}}\right)^2}. \qquad (5)$$

When constructing vector and scalar velocity fields, the area of the vertical cross-sectional model of the aeration tank was divided into 10 horizontal (I – X) and nine vertical (1 – 9) sections. Plots of the velocity components V_x, V_z and of velocity modules |V| water were obtained for each of the processed sections in MS Excel, and vector and scalar fields of the fluid circulation velocity were constructed using the CorelDraw graphical editor [6, 8].

The distribution of free-floating sludge in the cross section of the aeration tank was studied by constructing the sludge distribution fields in the aeration tank model with a lateral arrangement of a finely bubble aerator. We used the method of light absorption, measuring the intensity of the light flux in 320 control squares measuring 40÷40 mm. From the difference in the readings of the intensity of the stream passing through the sludge mixture and clean water, the integral value of the loss of light flux intensity attributable to activated sludge was found [6, 8]. To translate the difference in light flux intensities into a dose of activated sludge, a calibration plot of the dependence was constructed.

For the experiments, sludge taken from the secondary settling tanks at the left-bank sewage treatment facilities of the city of Irkutsk was used. The silt index, which characterizes the sedimentation ability of sludge, was ~150 cm^3/g.

3. Discussion of the results

The dependences of the specific contact surface of the phases, referred to the unit of the working surface of the aerator, on the aeration intensity of the lateral fine bubble aerator for various liquid levels in the aeration tank model, are presented in figure 3.

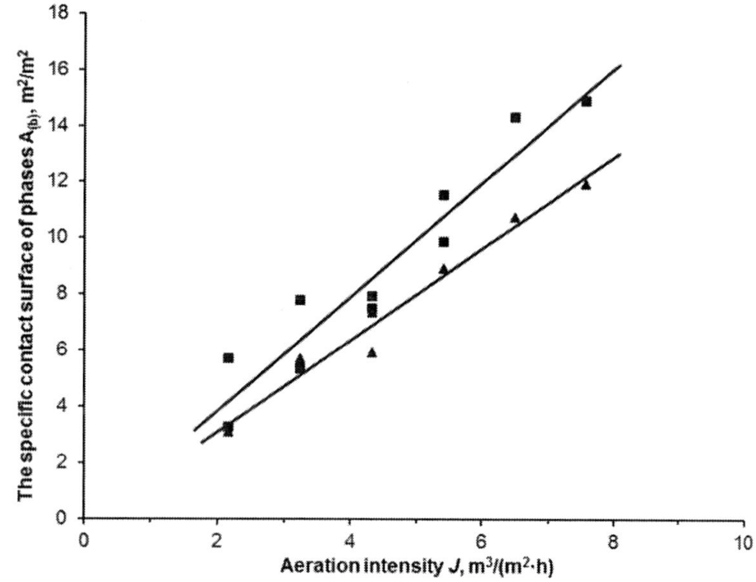

Figure 3. The dependence of the specific contact surface of the phases, referred to the unit of the working surface of the aerator, on the intensity of aeration, with a lateral arrangement of a fine-bubble aerator, for the depth of the aerotank model: ■ – 0,8 m; ▲ – 0,4 m.

With an increase in aeration intensity from 2 to 7 $m^3/(m^2 \cdot h)$, the specific contact surface of the phases increases, approximately three times, for the given depths of the aeration tank model. These dependencies are linear and are described by the following equations:

$$A_{(0,8)} = 2.0315 \cdot J - 0.2156, \quad (6)$$

$$A_{(0,4)} = 1.6357 \cdot J - 0.1854. \quad (7)$$

The use of the aeration tank model with a depth of 0.4 m leads to a 25% reduction in the value of the specific contact surface of phases, referred to the unit of the working surface of the aerator, at any aeration intensity.

The dependence of the oxidizing ability (OC) of a fine bubble aerator on the specific intensity of air aeration is linear and is described by the equation: $OC = a \cdot J$, and the tangent of the angle of inclination of the straight line to the abscissa axis is determined by the depth of the aeration tank model (figure 4).

With a depth of the aeration tank model of 0.8 m, the angular dependence coefficient is $a = 0.5121$, and at a depth of 0.4 m, the dependence of the oxidizing ability on the aeration intensity is smoother and $a = 0.3082$.

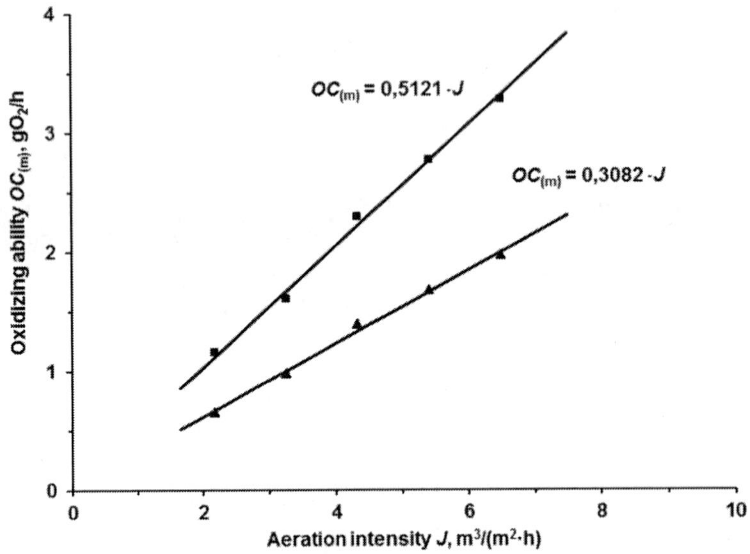

Figure 4. The dependence of the oxidizing ability of a fine bubble aerator AQUA-LINE on the intensity of aeration with a lateral arrangement of the aerator for the depth of the aeration tank model: ■ – 0,8 м; ▲ – 0,4 м.

From the analysis of the dependence, it can be concluded that it is advisable to increase the depth of immersion of the aerator in the aerotank in order to increase the oxidizing ability. The use of a square cross-section of the aerotank also contributes to an increase in oxidizing ability compared to a rectangular cross-section. This is due to the twisting of gas bubbles in the circulating flow of water and the increase in the time of mass transfer of oxygen.

The vector (figure 5) and scalar fields (figure 6) constructed according to the results of research at a fine-bubble aeration intensity $J = 5.41$ m^3/(m^2·h), with the lateral arrangement of the aerator, show field inhomogeneity over the area of the cross-sectional section of the aerotank.

The range of water velocity values varied from 0.6 to 0.05 m/s and less. The highest numerical values of the velocity are realized along the perimeter of the circulation circuit, decreasing towards its central zone. In the center of the contour, the speed is 10÷12 times less than along its perimeter and the center of the contour coincides with the geometric center of the model of the vertical transverse section of the aerotank. The velocities in the corner zones of the cell are ~2 times smaller than the maximum water velocity, which is explained by the chaotic motion of the water flows during rotation in the circulation circuit.

The numerical values of the flow rates of sewage in the vertical cross section of the aerotank obtained by us are in good agreement with the only published data [4]. According to these data, taking into account the heterogeneity of the field, the flow velocity varies from 10 cm/s to 70 cm/s with a lateral arrangement of the aerator in the aeration tank. It should be noted that there is no information on the intensity of aeration and the design of the aerotank in the literature.

An extensive "stagnant zone" in the central region of the aerotank cross-section model, in which speeds of 5÷10 cm/s of water flows are realized, promotes rapid coagulation of activated sludge into large agglomerates. The low flow rates do not contribute to the delivery of dissolved oxygen to the

IOP Conf. Series: Materials Science and Engineering **880** (2020) 012008 doi:10.1088/1757-899X/880/1/012008

stagnant zone, however, they are sufficient to maintain activated sludge in state of suspension. The oxidizing ability of this area of the aerotank, which depends on the surface of the "liquid – sludge" contact, decreases. This area is ~20÷25% of the entire cross-sectional area of the aerotank.

Thus, a fourth of the volume of the aeration tank is derived from the process of active oxidative ability of organic substances.

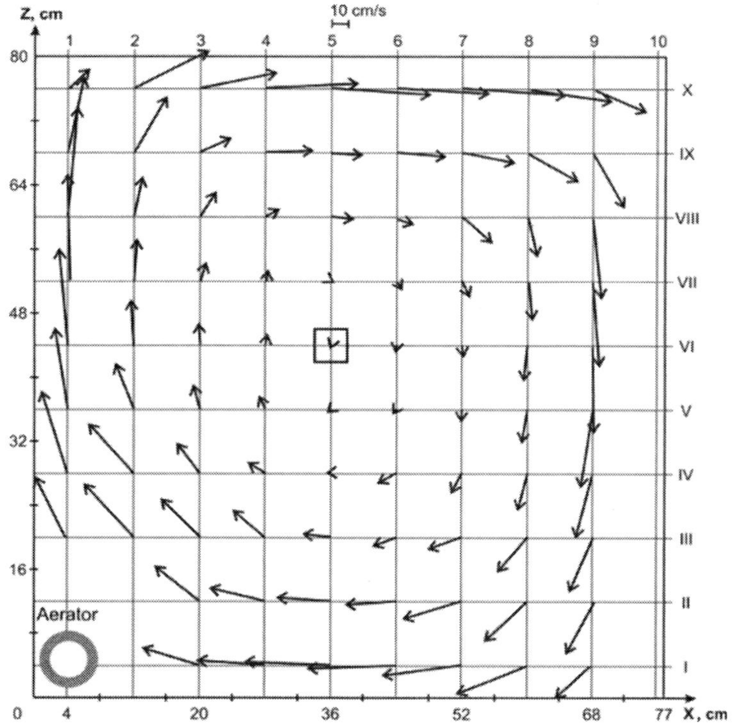

Figure 5. Vector field of water flow velocity in the aeration tank model.

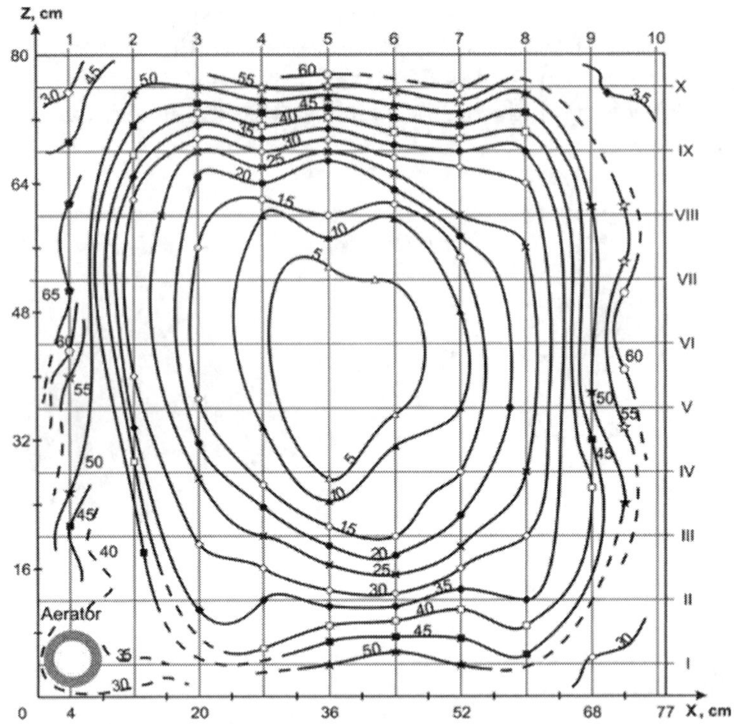

Figure 6. Scalar field of water flow velocity in the aeration tank model: Δ – 5 cm/s; ▲ – 10 cm/s; ○ – 15 cm/s; ● – 20 cm/s; × – 25 cm/s; ◊ – 30 cm/s; ♦ – 35 cm/s; □ – 40 cm/s; ■ – 45 cm/s; ★ – 50 cm/s; ☆ – 55 cm/s; ⬠ – 60 cm/s; ⬟ – 65 cm/s.

A comparison of the numerical values of the water velocities obtained during the simulation and measured on aerotanks with a distributed water supply of the Left-bank sewage treatment facilities in Irkutsk allows us to use physical modeling to study the gas-hydrodynamics of the aerotank.

The distribution field of floating sludge along the vertical cross section of the aerotank, with a lateral arrangement of a fine bubble aerator at a dose of activated sludge $a_i = 3$ g/dm^3, is characterized by an uneven distribution of sludge over the section of the aerotank. The area of maximum sludge concentration (interval 3÷3.1 g/dm^3) has an irregular annular shape and occupies the central peripheral part of the aerotank [6, 8].

In the areas of the model adjacent to the walls, bottom and surface of the wastewater, the concentration of sludge is 2.9÷3.0 g/dm^3. This region is characterized by maximum fluid velocities of 0.45÷0.60 m/s (see figure 5).

Inside the ring-shaped form, a concentration drop is observed from 2.9 to 2.4 g/dm^3, and the realized velocities have a minimum value of ~0.1 m/s and less. A decrease in the rate of fluid circulation in the center of the ring promotes coagulation of activated sludge with the formation of large flakes, crushed in water jets with a high speed (figure 7).

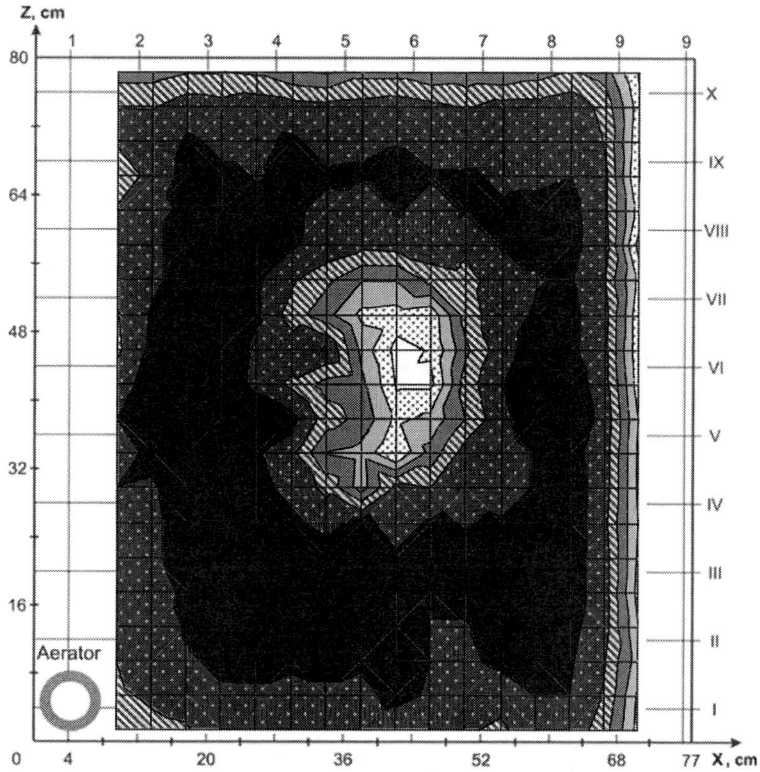

Figure 7. Plane field of activated sludge distribution in a vertical cross-sectional model of the aerotank at a fine-bubble aeration intensity $J = 5.41$ m^3/(m^2·h).

A three-dimensional image of the obtained field of distribution of activated sludge clearly shows the uneven distribution of sludge over the vertical cross section of the aerotank (figure 8).

The sludge-water mixture in the aeration tank is a heterogeneous (multiphase) system, in which the biologically purified wastewater serves as the dispersion medium, and the activated sludge flakes are formed as a complex three-level cellular structure surrounded by an exocellular substance of biopolymer composition. The precipitation of flocs of activated sludge (when its concentration in the sludge mixture is more than 0.5÷1 g/dm^3) occurs with the formation of a visible phase boundary between the clarified water and sludge.

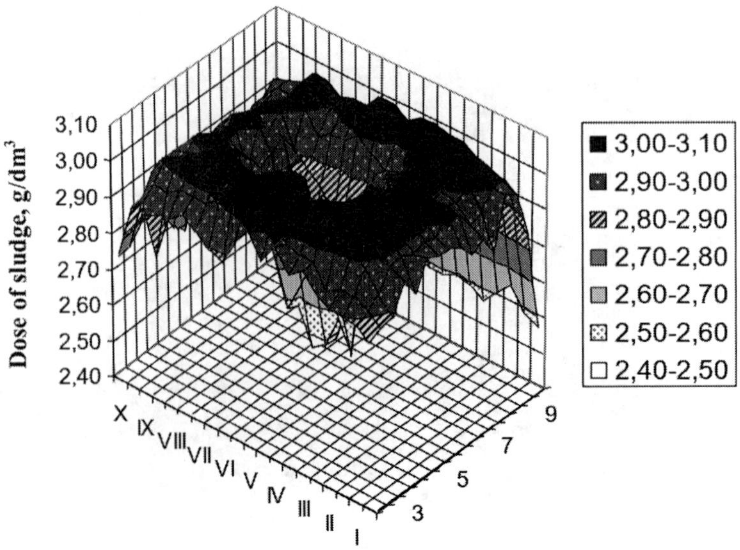

Figure 8. Three-dimensional and planar distribution fields of activated sludge in a vertical cross-sectional model of the aerotank at a fine-bubble aeration intensity $J = 5.41$ m³/(m²·h). The concentration of activated sludge is given in g/dm³.

For the sludge index values $I_i = 150$ cm³/g and a dose of activated sludge $a_i = 3$ g/dm³, the rate of decrease in the interface between the clarified water and sludge is 7.23 mm/s, which is much lower than the flow rates along the perimeter of the circulation circuit and, to a first approximation, coincides with its center.

Our analysis of the molecular kinetic properties of the colloidal system made it possible to prove that, for a given aeration intensity, $J = 5.41$ m³/(m²·h) and $a_i = 3$ g/dm³, the sludge flakes move in the flows of the circulation circuit, providing the oxidation power of the aerotank.

4. Conclusion

An experimental setup with a model cell representing a vertical cross section of an aerotank is proposed for studying gas-hydrodynamic processes and the distribution of activated sludge in aerated structures. The possibility of physical modeling of gas-hydrodynamic processes on the aerating plants of wastewater treatment as a first approximation is shown.

The contact surface of the phases "liquid - gas" was determined using a chemical method for on the aerating plants of wastewater treatment from intensity of air aeration. The dependences of the specific contact surface of the phases, referred to the unit of the working surface of the aerator, on the intensity of aeration and the depth of immersion of the aerators are obtained. Dependencies are linear. Mathematical expressions obtained.

The oxidative capacity of a fine-bubble aerator for different aeration intensities of a water-silt mixture was determined. The dependences are linear, the tangent of the angle of inclination of the lines to the abscissa axis is determined by the depth of the aerotank model.

For the first time, the dependences of the oxidizing ability on the contact surface of the phases "liquid - gas" were obtained. It was shown that the phase contact surface does not completely determine the oxidation ability of the aerotank. The hydrodynamic situation in the aeration tank, which forms the fields of activated sludge, is very important for the oxidative ability of the aerotank.

The obtained vector and scalar fields of fluid velocities at the lateral position of a small-bubble aerator show the in homogeneity of the velocity field in the model of the vertical cross section of the aerotank. The maximum velocity of fluid flows is realized in the main circulation circuit and reaches ~60 cm/s. There are "stagnant zones" in the center of the aerotank model, with fluid flow rates of 10 cm/s or less. These speeds, however, are sufficient to maintain activated sludge in state suspension. In the stagnant zone, sludge flakes coagulate, which reduces the intensity of the oxidative process of organic pollution in this zone, reaching ~25% of the model area of the vertical section of the aerotank.

An analysis of the molecular kinetic properties of the colloidal system showed that for a given aeration intensity $J = 5.41$ $m^3/(m^2 \cdot h)$ and a dose of activated sludge $a = 3$ g/dm^3, the sludge flakes move in the flows of the central circulation circuit providing the oxidative power of the aerotank.

The distribution fields of activated sludge are obtained in the vertical cross-sectional model of the aerotank with a lateral arrangement of a fine-bubble aerator. The non-uniform distribution of silt over the cross section of the aerotank model is shown. The area of maximum sludge concentration has an irregular annular shape and occupies the central peripheral part of the aeration tank model. The central stagnant zone is characterized by active coagulation of sludge into large agglomerates, which reduces the "liquid - sludge" surface, which determines the oxidation of organic substances in the aerotank.

References

[1] Ermolin Yu A and Alekseev M I 2017 *Water and Ecology* **2** 18–27.

[2] Zhmur N S 2003 *Technological and biochemical processes for wastewater treatment at facilities with aeration tanks* (Moscow: AKVAROS) p 512

[3] Voronov V Yu and Yakovlev S V 2006 *Sewage and wastewater treatment* (Moscow: Association of Construction Universities) p 704

[4] Popkovich G S and Repin B N 1986 *Wastewater aeration systems* (Moscow: Stroyizdat) p 136

[5] Karelin Y A, Zhukov D D, Zhurov V N and Repin B N 1973 *Treatment of industrial wastewater in aerotanks* (Moscow: Stroyizdat) p 223

[6] Kul'kov V N, Solopanov E Yu, Evteeva I V and Razum A S 2008 *Bulletin of Irkutsk State Technical University* **4** 48-52

[7] Kutateladze S S and Styrikovich M A 1976 *Hydrodynamics of gas-liquid systems* (Moscow: Energy) p 296

[8] Kul'kov V N and Solopanov E Yu 2009 *The phase contact surface in aerobic wastewater treatment* (Irkutsk: Irkutsk State Technical University Publ.) p 144

[9] Solopanov E Yu and Kul'kov V N 2006 *Materials of Int. Scientific-Practical Conf. "Investments. Construction. Real estate"* (Irkutsk) (Irkutsk: Irkutsk State Technical University Publ.) pp 114–117

[10] Meshengisser Yu M 2002 *Water and Ecology* **2** pp 68–71

[11] Karelin Y A, Repin B N, Afanasyev A F and Ponomarev V V 1981 *Water supply and sanitary equipment* **5** pp 7–9

[12] Kul'kov V N. and Solopanov E Yu 2007 *Collection of reports of the All-Russian Scientific and Practical Conference "Water: problems and solutions"* (Tyumen) issue 5 (Tyumen: Tyumen State University of Architecture and Civil Engineering Publ.) pp 51–57

[13] Krenn Walfgang 1974 Stereametrischen Methoden zur Bestimmung von Volument und Oberfluche von Blasen oder tropfen ous deren Abbildungen *Chem. Ing. Techn.* **23** 46–48

[14] Rodionov A I, Kashnikov A M, Ulyanov B L, Shpagin N S and Strogonov E F 1967 *Chemical Industry* **3** pp 209–212

[15] Kul'kov V N 1995 *Fields of circulation of electrolyte and gas in aluminum electrolysis cells* (Moscow: Deposited in VINITI, № 2510 B95) p 110

IOP Conf. Series: Materials Science and Engineering **880** (2020) 012009 doi:10.1088/1757-899X/880/1/012009

The study of the phase composition and microstructure of composite binders using industrial waste Transbaikalia

S Lkhasaranov[1], L Urkhanova[1] and D Danzanov[1]

[1]Department "Construction Materials, Roads and Woodworking", East Siberia State University of Technology and Management, Klyuchevskaya street, 40 V, building 1, Ulan-Ude, 670013, Russian Federation

E-mail: solbon23187@mail.ru

Abstract. The further development of industry is decelerated by lack of energy resources. This is especially significant for the construction industry. Considerable energy savings can be achieved by using specially prepared raw materials. The modern complex approach to solving problems in civil engineering requires additional development of resource-saving technologies, which imply the widespread use of associated industrial products, designed to significantly reduce the energy intensity of cement and concrete production without a fundamental change in technology. The article presents the results of studies on the possibility of obtaining a composite binder using technogenic waste - fly ash and polycarboxylate type superplasticizers. It was investigated that natural pearlite and fly ash can be used as part of a multi-component fine cement. With the introduction of fly ash and superplasticizers, the technological and mechanical properties of cement change. High mechanical properties of the composite binder were obtained when the content of fly ash in the amount of 30 wt. % This is due to the structure-forming role of fly ash during Portland cement hydration. Using the method of differential thermal analysis and IR spectroscopy, it is shown that the combined use of fly ash and superplasticizer leads to a change in hydration processes, the binding of portlandite with the formation of an additional amount of calcium hydrosilicates.

1. Introduction

Recently, environmental aspects of the development of engineering and technology have been given closer attention, and therefore, work related to the use of technogenic materials in the construction industry has been actively resumed [1-8].

It is known that cement is the most expensive component of concrete mix, the cost of which depends on the cost of concrete itself. The issue of saving cement for the production of concrete and reinforced concrete products and structures is one of the important issues of modern construction. Improving the quality of concrete requires the use of new composite binders instead of conventional Portland cement, with improved physical and mechanical properties [1, 3, 6, 7].

At present, more than 1.5 billion tons of ash and slag waste have accumulated in ash dumps in the Russian Federation, the amount of which increases by 25 million tons annually. At the same time, no more than 14% of the generated waste is disposed of. At the same time, ash and slag waste is a valuable anthropogenic resource: in the Russian market there is a demand in the cement industry, civil and road construction for certified secondary products from ash and slag waste.

2. Materials

In current research to obtain composite binders were used Portland cement CEM I 32,5, fly ash from Gusinoozersk thermal power plant, super plasticizers Neolite 303 and Sika Viscocrete®. Composite binders were obtained with consistent milling of fly ash (0-50 weight %) with OPC. Specific surface area of composite binders were 450-480 m^2/kg. From these binders were prepared samples with dimensions 40*40*40 mm. Samples were stored in molds at t = 20-22 °C, relative humidity of 90-95%, and without the molds above the water for a period of 28 days.

3. Results and Discussion

Studies have shown that the use of fly ash as part of composite binders leads to a change in normal consistency, setting time and physical and mechanical properties (Table 1, Fig. 1).

Table 1. Characteristics of composite binders using fly ash

Parameter	with Neolite when ash content, wt.%			with Sika Viscocrete when ash content, wt.%			OPC
	10	30	50	10	30	50	
Normal consistency, %	27	25	22	24	23	20	28
Setting time, min:							
Start of setting	130	140	150	140	150	160	90
End of setting	290	300	310	300	320	330	245

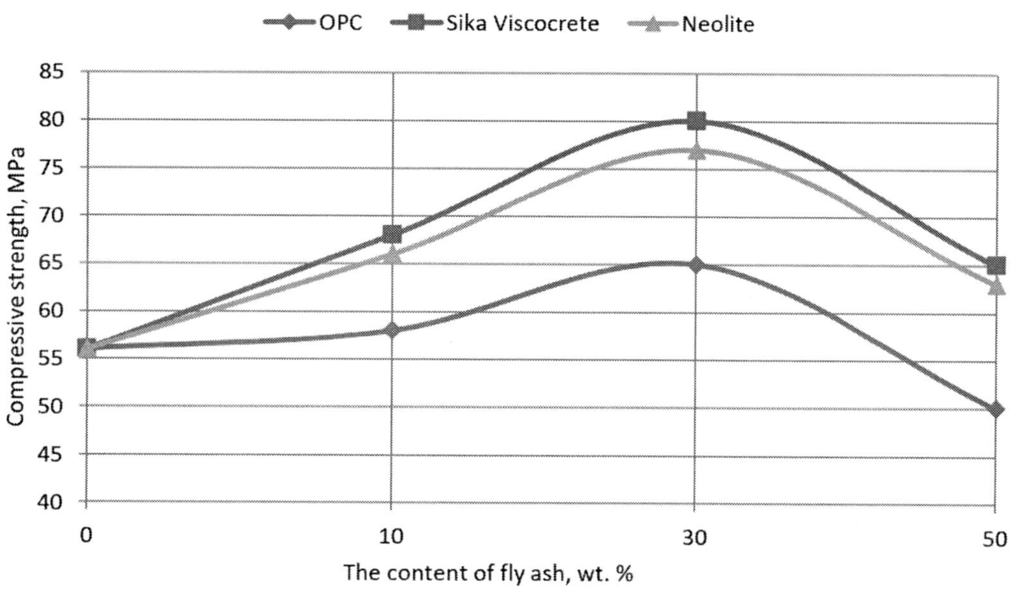

Figure 1. Compression strength of Portland cement with fly ash at 28 days

The introduction of Neolite 303 and SikaViscocrete additives increases the compressive strength by 10-15%. In addition to the water-reducing effect, which helps to increase physical and mechanical characteristics, superplasticizers significantly affect the processes occurring in the cement system.

A change in the mechanical properties of composite binders is associated with a change in the structure of cement stone. Fly ash when mixed with Portland cement acts as an active mineral additive, is involved in the structure formation of cement.

To assess the effect of fly ash on Portland cement, a set of research methods was carried out, including differential thermal and IR spectral analyzes, and proving a change in the phase composition of the cement stone.

IOP Conf. Series: Materials Science and Engineering **880** (2020) 012009 doi:10.1088/1757-899X/880/1/012009

The results of the differential thermal analysis (Fig. 2) of the cement matrix indicate a change in the endoeffect in the temperature range 515-520 ° C in ordinary Portland cement to an exoeffect in Portland cement with fly ash.

a

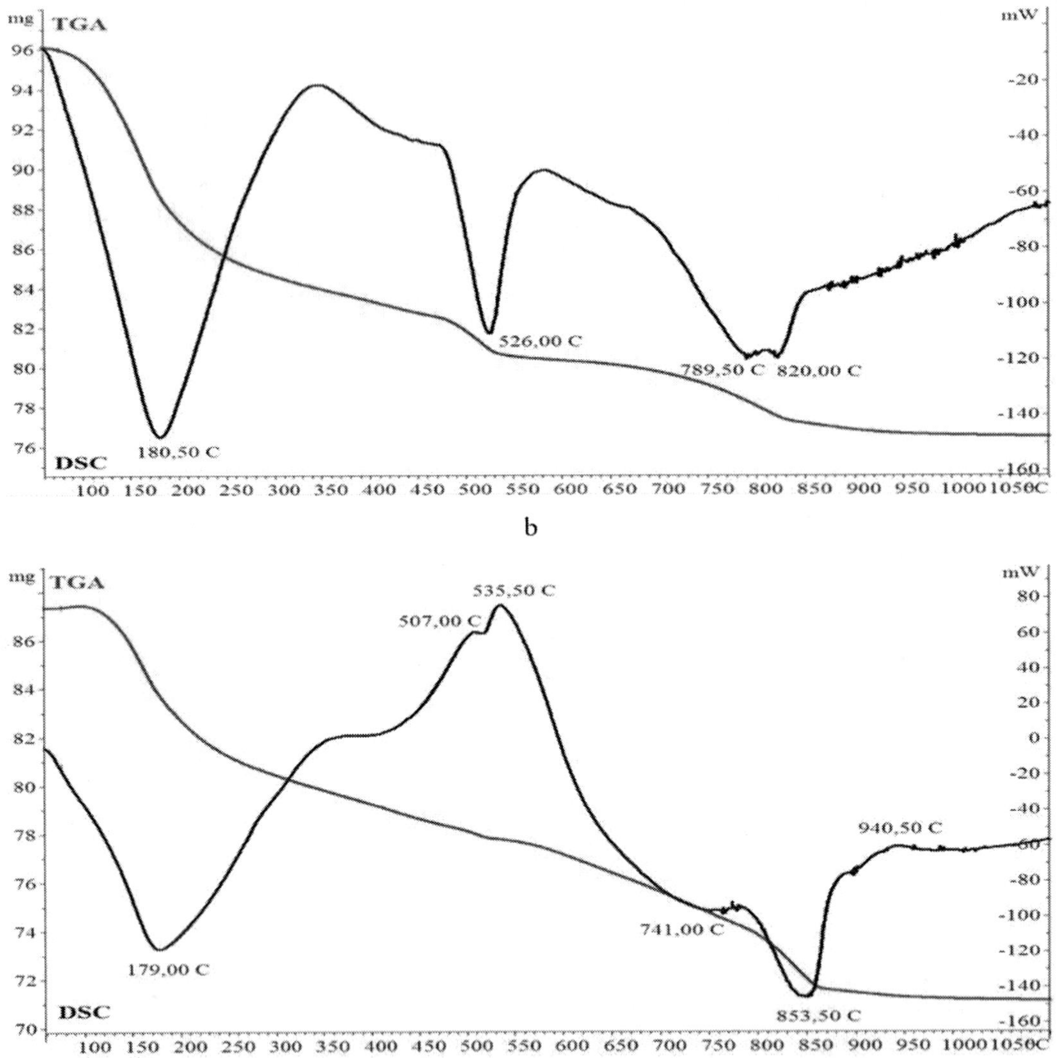

Figure 2. Curves of differential scanning calorimetry of Portland cement (a) and Portland cement with fly ash - 30% (b)

This indicates the binding of calcium hydroxide to the active components contained in fly ash. When comparing the DTA curves, a change in the basicity of the formed calcium hydrosilicates is observed: the endothermic effect in the temperature range of 815–820 ° C shifts to the right, towards the temperature increase to 853 ° C. A wide temperature range is associated with the fact that the introduction of fly ash leads to the formation of an additional amount of calcium hydrosilicates that differ from traditional ones. In Portland cement with fly ash, additional low-basic calcium hydrosilicates are formed, which are structured along the surface of the solid phase, with the formation of a denser composite with improved physical and mechanical properties.

The change in the phase composition of the cement stone is confirmed by IR spectral analysis on an IRAffinity-1 IR Fourier spectrometer (Fig. 3).

a

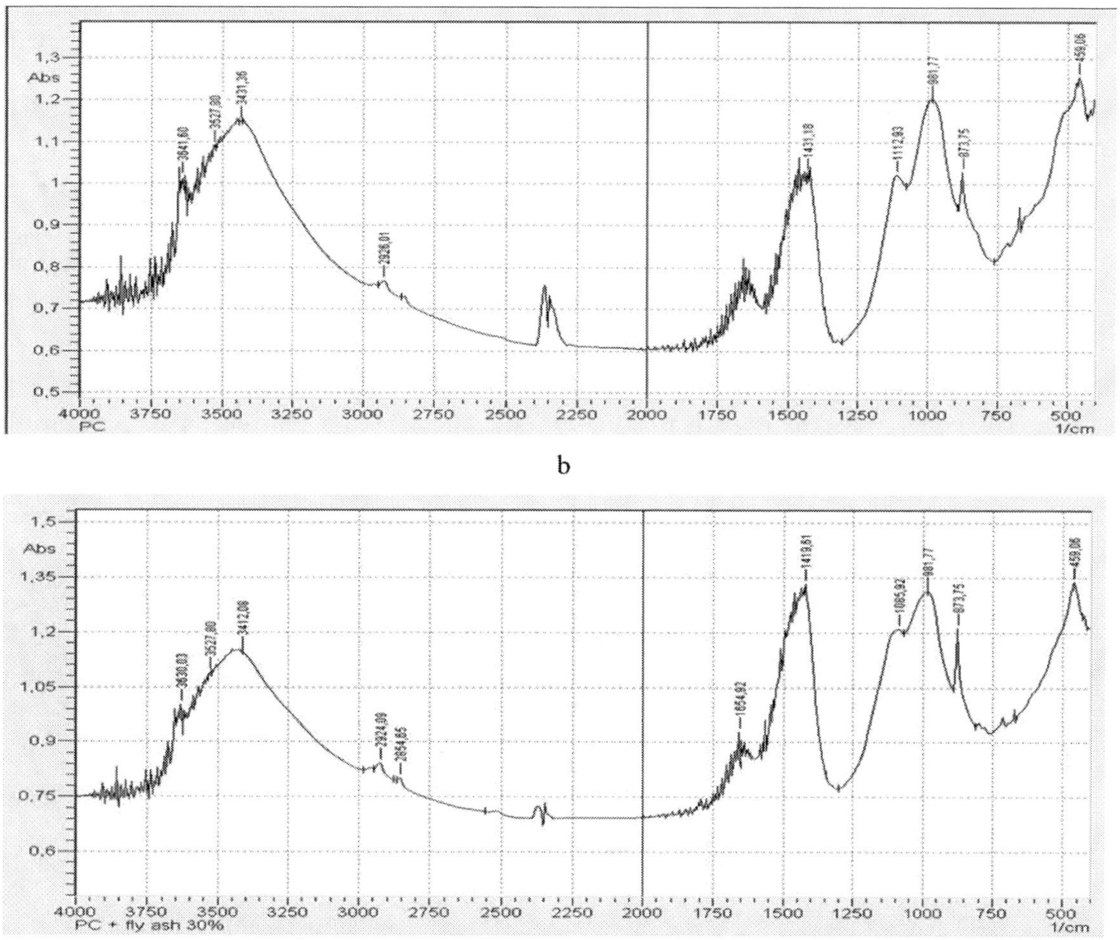

Figure 3. IR spectra of Portland cement (a) and Portland cement with 30% fly ash (b)

An analysis of the IR spectra showed that with the introduction of fly ash there is a change in the intensity of the absorption band in the frequency range 1000-1100 cm-1, which correspond to vibrations of Si-O bonds, which is associated with the formation of calcium hydrosilicates. This indicates a change in the process of hydration of Portland cement with the introduction of fly ash and the formation of an additional amount of calcium hydrosilicates. The shift in the frequencies of the absorption bands corresponding to calcium hydrosilicates also suggests that the structures formed differ from traditional ones.

4. Conclusion

The results showed that the introduction of fly ash in an amount of 10-50% does not lead to a decrease in the strength of the composite binder compared to conventional Portland cement. The greatest result was shown by a composition with 30% fly ash. Nevertheless, an increase in fly ash content up to 50% can significantly save material costs for obtaining a binder and concrete based on it, and may be appropriate for a feasibility study.

With the introduction of superplasticizers in the composition of composite binders, the W / C ratio decreases, the rheological properties of the binders improve, and their strength increases. The addition of SikaViscoCrete and Neolite to the binder in an amount of 0.6% leads to an increase in the strength of the composite binder.

The analysis (DTA and IR spectroscopy) allows us to conclude that the introduction of fly ash contributes to the directional formation of a high-strength stone structure from low-basic calcium hydrosilicates and a decrease in the content of calcium hydroxide.

References

[1] Alfimova N, Sheychenko M, Karatsupa S, Yakovlev E, Kolomatskiy A and Shapovalov N 2014 Features of application of high-mg technogenic raw materials as a component of composite binders *Research Journal of Applied Sciences* **9 (11)** pp 779-783

[2] Garg M and Pundir A 2014 Investigation of properties of fluorogypsum-slag composite binders–Hydration, strength and microstructure *Cement and Concrete Composites* **45** pp 227-233

[3] Fediuk R and Yushin A 2016 Composite binders for concrete with reduced *permeability IOP Conference Series: Materials Science and Engineering* **1** 012021

[4] Martinez-Lopez R and Escalante-Garcia J I 2016 Alkali activated composite binders of waste silica soda lime glass and blast furnace slag: Strength as a function of the composition *Construction and Building Materials* **119** pp 119-129

[5] Peshkov V V and Gamayunova E V 2018 Specific features of investment support for the repair and reconstruction of buildings and facilities *Proceedings of Universities. Investment. Construction. Real estate* **8** pp 56-72 DOI: 10.21285/2227-2917-2018-2-56-72

[6] Zhang Z, Wang Q and Yang J 2017 Hydration mechanisms of composite binders containing phosphorus slag at different temperatures *Construction and Building Materials* **147** pp 720-732

[7] Lesovik R, Klyuyev S, Klyuyev A, Netrebenko A and Durachenko A 2014 High-strength fiber-reinforced concrete containing technogenic raw materials and composite binders with use of nanodispersed powder *Research Journal of Applied Sciences* **9(12)** pp 1153-1157

[8] Sun J, Wang Z and Chen Z 2018 Hydration mechanism of composite binders containing blast furnace ferronickel slag at different curing temperatures *Journal of Thermal Analysis and Calorimetry* **131(3)** pp 2291-2301

[9] Kirca O 2006 Temperature effect on calcium aluminate cement based composite binders (Middle East Technical University)

[10] Selezneva O I, Radaev S S and Baranova A A The impact of thermal treatment on the granulometric and chemical composition of diatomite of the Kamyshlov field when obtaininga liquid-glass binder *Investment. Construction. Real estate* **9(3)** pp 550–555

[11] Tyuryukhanov K Yu and Pugin K G 2019 Impact of the surface of particles of moulding sand on the structural formation of asphalt *concrete Proceedings of Universities. Investment. Construction. Real estate* **9(3)** pp 566–577

[12] *Proceedings of Universities. Investment. Construction. Real estate* **10(1)** pp 108–115https://doi.org/10.21285/2227-2917-2020-1-108-115

[13] Komarov A K, Ivanov I A and Lundenbazar B 2019 Theory and practice of the use of gabions for forming protective structures *Proceedings of Universities. Investment. Construction. Real estate* **9(1)** pp 78–89 DOI: 10.21285/2227-2917-2019-1-78-89

[14] Kazimirov I A and Peshkov V V 2019 Determination of price behaviour in the secondary residential real estate market using a multidimensional regression model *Proceedings of Universities. Investment. Construction. Real estate* **9(3)** pp 476–487 DOI: 10.21285/2227-2917-2019-3-476-487

Composite low-water demand cements – one of the most promising and available technologies in terms of "Ecology" national project

S V Makarenko[1], B M Lozovskij[1], O V Khohryakov[2], V G Khozin[2] and D I Baishev[3]

[1] Irkutsk National Research Technical University, 83 Lermontov str., Irkutsk, 664074, Russia
[2] Kazan State University of Architecture and Engineering, Department of technology of construction materials, units and constructions, 1 Zelenaya str., Kazan 420043, Russia
[3] Ølen Betong, Avtomobilistov, site №6, Kola city, 183052 p.b. 3350, Russia

E-mail: makarenko_83_07@mail.ru

Abstract. In current article results of research of composite low-water demand cement (LWDC) based on Angarsk plant's cement CEMII/A-F 32,5R, dump ash from TPP-10, marble powder from Sludjanskoj quarry and micro silica from LLC "Bratsk ferritic alloys plant" (Irkutsk region) are performed. Goal of the study is development of efficient ecology-friendly technology for high-quality cements with raw materials local to Irkutsk region. As a result of the study, technological, physical and mechanical properties of cement paste in fresh and hardened state were determined. Standard consistency W/B of LWDC-based cement pastes' could be changed in range between 19 and 26%, and compressive strength of hardened cement paste can be variated from 22 to 60 MP. 2-stage 3-factor regressive analysis for compressive strength values of hardened cement paste at age 3,7,28 and after thermal curing was done, regression equations were obtained.

1 Introduction

In Russian Federation ecology problem is escalated in status of national project, which main goal is efficient by-products and wastes usage. It will be realized in terms of 11 federal projects, e.g."Implementation the best available technology", which is agreed with Irkutsk region aims of realization of strategy of construction materials industry' development strategy, initiated by government of the region. As foundation of this strategy, development of construction materials based on by-products and local raw materials was chosen.

It is absolutely obvious that industry of construction materials could have one of the biggest capacity of large by-products and wastes range, e.g. fly ashes of thermal power plant (TPP). Basing on this fact, industry of construction materials could be considered as implementation of this project in near future.

In European countries raw material mining, goods production, its consumption and waste generation are considered as chain link [1-5]. It is believed that such an approach will allow to preserve environment from deterioration. Due to this principle laws, which drive construction materials' producers to implement wide usage of by-products and wastes, especially in technology of mineral binders and ceramics, are adopted.

Technology of LWDC has quite good capacity of large-tonnage production wastes utilization [6-8]. This technology is safe for the environment, has lesser energy consumption and easier to implement in comparison with OPC production' one. Using an LWDC technology makes it possible to decrease significantly CO_2 emission (up to 3-4 times), and utilize up to 70% of wastes such as fly ash per 1 ton of binder, also it could be built in "classic" technology of OPC without colossal investments and readjustments. LWDC technology could be rightfully assigned as one of the best available technologies for mineral binders' production.

2. Materials and methods

In current work evaluation of composite LWDC was made with usage of three types of mineral fillers. First one is a dump ash obtained from TPP-10 (total amount of ash dumps of TPP of JSC "Irkutskenergo" is about 80 million ton and annual increase is about 1,7 million ton). According to data obtained by x-ray spectral analysis, this ash includes silica (20%), mullite (13%), albite (4%), aluminosilicate mineral (about 1%) and the rest is aluminosilicate glass phase.

Table 1. Properties of dump ash from TPP-10

Property	True density (g/cm^3)	Bulk density (g/cm^3)	Voids content (%)	Specific surface area (cm^2/g)	pH of 10%-water solution
Value	2,03	0,76	64%	1334	9,6

Second type of filler – marble from Sludjanskoj quarry (Irkutsk region), evaluated source is about 120 million ton.

Third type – micro silica MK-85 from LLC "Bratsk ferritic alloys plant", produced according to TU-5743-048-02495332-96.

As a clinker containing part of LWDC, CEMII/A-F 32,5R SC "Angarsk cement and mining plant".

Table 2. Properties of cement

Property	Standard consistency W/B (%)	Setting time (min)		Activity at 28-day age (MPa)
		Beginning	End	
Value	2,03	75	180	35

Choice of current binder was made due to its low consumption in Irkutsk region in order of search of possibility of its modification and properties' enhancement.

Choice of superplasticizer was made from most popular types of such ones in current region:

1. Polyplast Ligno – admixture based on modified lignosulfonates.
2. Polyplast SP-1 (naphthalene formaldehyde).
3. Polyplast PK (PCE).

Table 3. Properties of superplasticizers by comparison of standard consistency W/B and price values

Type of admixture	Dosage (%)	Standard consistency W/B (%)	Price incl. VAT (RUR/kg)	Remarks
Polyplast Ligno	1	25,5	56	Significant retardation at 1-day age
Polyplast SP-1	1	24,5	80	---
Polyplast PK	1	21,5	360	---

Basing on obtained values of standard consistency W/B and prices, it was decided to make an evaluation of composite LWDC with SP-1 admixture. Apparently, low water-reducing effect of Polyplast PK was obtained due to addition of lignosulfonate agent as intensifier of grinding, which has ceased PCE-based admixture's efficiency.

Since the fillers used differ in hardness, the LWDC was obtained in a sequentially separate manner using a vibration-rod mill. This method consisted of preliminary joint grinding of cement, ash and superplasticizer and their subsequent grinding with marble with the addition of micro silica.

For a comprehensive assessment of the effect of fillers on the properties of composite LWDC and a reduction in the amount of experimental work, method of mathematical planning were used. It was a 2-level 3-factor experiment (2^3). Factors and range of variation are given in table. 4, the planning matrix is tab. 5. In this case, the specific surface area of the LWDC ranged between 7000 and 7500 cm^2/g, depending on the content of silica fume in accordance with the conditions of regression analysis. The output parameter of the experiment is the compressive strength of hardened cement paste at the age of 3, 7 and 28 days of normal hardening and after thermal curing under standard isothermal conditions at a temperature of 80 °C (method GOST 310.4).

Table 4. Factors and range of variation

Factors	Level of factor			Range of variation
	-1	0	+1	
X_1 (cement content in LWDC)	30	50	70	20
X_2 ratio micro silica: ash : marble, %	20/60/20	10/60/30	0/60/40	10
X_3 superplasticizer content by LWDC mass, %	0,5	1	1,5	0,5

Table 5. Planning matrix

Number of composition	X_0	X_1	X_2	X_3
1	+	+	-	+
2	+	-	-	+
3	+	+	+	+
4	+	-	+	+
5	+	+	-	-
6	+	-	-	-
7	+	+	+	-
8	+	-	+	-

Initially, the specific surface area and normal density of the LWDC were determined, adopted according to the planning matrix: and the compressive strength of cement stone at the age of 3, 7, 28 days of normal hardening (Table 6).

3. Results

In table 6 physical and mechanical properties of LWDC, determined by laboratory tests, are performed. Specific surface area of LWDC compositions were determined with PSH-12 in every case, according to variation of content during regression analysis, so as compressive strength of LWDC at age 3, 7, and 28 days of normal hardening and after thermal curing. Regression equations are obtained, which display influence of basic factors on output parameter – compressive strength at different ages. At table 7 values of basic criteria of evaluation of obtained model adequacy – Fischer and Student coefficients and dispersion of regression quotients.

Table 6. Technological, physical and mechanical properties of LWDC-based cement paste

№ composition	Specific surface area (cm^2/g)	Standard consistency W/B (%)	Compressive strength at age, MPa			
			3 days	7 days	28 days	Thermal curing
1	7120	20,5	31,57	44,83	58,5	43,95
2	7573	19,5	12,83	27,7	33,1	24,6
3	7558	21	16,73	24	34	28,95
4	7550	22	6,33,	9,33	20,75	9,3

5	6980	24,5	23,1	33,6	44,9	38,15
6	7488	25	15	20,77	30,85	24,2
7	7128	26,5	15,	21,67	30,05	28,9
8	7121	25,5	6,7	10,77	22,35	11

Table 7. Values of Fisher coefficients and variance of model adequacy

Type of index	3 days	7 days	28 days	Thermal curing
S^2_{ad}	10	12	26,4	4,5
F_{obs}	7,23	3,57	8,25	0,61
F_{table}	9,12	9,12	9,12	9,12
Model adequacy check	the model is adequate, because $F_{exp} < F_{tab}$			
Regression coefficient dispersion	0,17	0,42	0,4	0,92
Confidence interval	0,41	1	0,94	2,17
Student coefficient	2,36	2,36	2,36	2,36

Based on the results of the work, the regression equations for composite LWDCs are obtained:
– 3 days of normal conditions hardening:

$$R_3 = 15,99+5,76x_1-4,65x_2+0,875x_3 \quad (1)$$

– 7 days of normal conditions hardening:

$$R=24,08+6,94x_1-7,64x_2+2,38x_3 \quad (2)$$

– 28 days of normal conditions hardening:

$$R28 = 34,31+7,55x_1-7,53x_2+2,28x_3 \quad (3)$$

– after thermal curing:

$$R_{npon} = R=26,13+8,86x_1-6,59x_2+0,575x_3 \quad (4)$$

4. Discussion

As follows from the table. 3, LWDC has a standard consistency W/B from 19.5 to 26.5% compared with cement CEM II A / Z 32.5B, whose' one is 28%.

Analyzing the data obtained from the regression equations, it can concluded that a significant contribution is made by micro silica on the output parameter (strength of the LWDC stone) in the quantitative corresponding contribution of X_1 (type of binder) for all hardening ages (3, 7, 28 days). Moreover, the significance of the contribution increases (pozzolanic activity) by 7 days and remains almost unchanged till 28 days. An analysis of the kinetics of hardening of the studied composite LWDC indicates 50% of the set of strength at the age of 3 days, 80% of the set of strength at the age of 7 days and after thermal curing. Having considered the regression equation at the age of 28 days of normal hardening, it can be said about the sufficient strength of the LWDC-30, given in addition that the actual content of the clinker component in terms of the LWDC is no more than 124%.

Reliability of the presented results was justified by calculating the confidence interval, checking the model for adequacy taking into account the Fisher coefficients (Table 7).

5. Conclusion

Basing on the foregoing, it can be concluded that the introduction of the technology for the production of LWDC can undoubtedly make a significant contribution to the implementation of the national project "Ecology", by implementing the federal program "Iplementation of the best available technologies", at least, on the scale of the Irkutsk region and regions where solid fuel is a fundamental type. Thus, LWDC, no doubt, predicts a reliable future and widespread distribution.

IOP Conf. Series: Materials Science and Engineering **880** (2020) 012010 doi:10.1088/1757-899X/880/1/012010

Reference

[1] Skorohod M A 2016 State and improvement of the competitiveness of the cement market of the Eurasian Economic Union (EAEU) *9tn International Cement Conference. Non-Commercial organization Soyuzcement*

[2] Chomaeva M N 2016 Ecological problems of the environmental impact of the chemical industry (on the example of cement production) *National Security and Strategic Planning* **2-1** pp 141-143

[3] Kunn Connenhall 2013 CEMBUREAU - cement and energy market in Europe and the world *Cement and its application* **3** pp 22-33

[4] Rickert J & Müller K 2011 Efficient composite cements - a contribution to reducing CO_2 emissions. (Alitinform International analytical review *Cement. Concrete. Dry mixes* **2** pp 28-49

[5] Karpenko N I and Yarmakovsky V N 1997 The main directions of resource conservation in the construction and operation of buildings *Building materials* 7 pp 12-21

[6] Yudovich B E, Dmitriev A M and Zubekhin S A 1997 Cements of low water demand - binders of a new generation *Cement and its application* **4** pp 15-18

[7] Khozin V G, Khokhryakov O V, Sibgatullin I R, Gizzatulin A R and Kharchenko I Ya 2014 Carbonate cements of low water demand - a green alternative to the cement industry in Russia Building materials **5** pp 76-82

[8] Tyuryukhanov K Yu and Pugin K G 2019 Impact of the surface of particles of moulding sand on the structural formation of asphalt *concrete Proceedings of Universities. Investment. Construction. Real estate* **9(3)** pp 566–577

[9] Dobruskina M A, Petrov A V and Bat-Erdene Z 2019 Improving the technology of retaining walls in the Irkutsk region using gabion baskets *Proceedings of Universities. Investment. Construction. Real estate* **9 (2)** pp 312–323

[10] Moskvitin V A., Emelyanova N A. and Mashovich A Y 2019 Experimental studies of air permeability indicators of composite "Poroplast CF" *Investment. Construction, Real estate* **9 (2)** pp 342–353

[11] tenders for industrial and civil construction Investments. Construction. Real Estate: New Technologies and Targeted Development Priorities IOP Conference Series: Materials Science and Engineering **667** 012063

[12] Baranova A A and Bobrova A A 2019 Dispersed reinforcement of cellular and fine-grained concrete based on silica fume Proceedings of Universities. Investment. Construction. Real estate **9(4)** pp 694–703

[13] Petrov A V and Efimova A K 2019 Nguyen Thanh Tung. Optimisation of technology used for restoring gas-ash concrete outdoor wall panels using modified structural and heat-insulating concrete Investment. Construction. Real estate **9(3)** pp 542–549

The using of liquid stone for the restoration and creation of copies of archaeological objects

A Ya Mashovich[1], V L Ruposov[1] and E M Ineshin[2]

[1]Irkutsk National Research Technical University, 83, Lermontov Street, Irkutsk, 664074 Russia
[2]Irkutsk State University, 9, Sukha Bator Street, Irkutsk, 664011 Russia

E-mail: ruposov@istu.edu

Abstract: It is an actual problem to investigate archaeological objects outside their native locations. New building materials make it possible to make the most realistic copy of archaeological objects, without destroying the object and extracting it from the natural environment. Liquid stone allows you to create objects with the imitation of color and texture of the original. This material can withstand high loads and it is used for prosthetics of the elements of archaeological structures. The technology of repetition of such objects allows making mass circulations of the most interesting archaeological objects. The technique makes it possible to restore objects from the remaining photos and description.

1. Introduction

The use of new materials in various fields of science and technology allowed to develop the directions of use that had not previously come to the attention of researchers. the methods for creating replicas of complex objects have proliferated with the advent of 3D modeling and reverse engineering. New materials make it possible to simulate the shape, color and texture.

The purpose of the article is to show how the new building materials can be used to recreate the cultural heritage.

The main objective of the study is to determine the possibilities and methods for creating models of archaeological and cultural objects using liquid stone technology.

2. New building material liquid stone

New materials with the properties which were not available to the construction and restoration technologies have appeared recently. Such materials have expanded the possibilities of the construction and restoration industries and have improved aesthetics and cost-effective creation and reconstruction of various objects. Aesthetics of materials makes the technology attractive for residential use. The economy is caused by the ability to use recycled materials. A similar material is a liquid stone. This material appeared on the market at the beginning of the 21st century was used to decorate the interior. The elements for bathrooms and kitchens, such as sinks, cutting tables, etc. were made of such a material. This material has attracted attention by the fact that it simulated a natural stone with the method of making compositions of mineral fillers and pigments cemented by polymeric resin.

Our method consists of several stages. On the first stage the material is a composition in the liquid state. The second stage is applied to a substrate by casting or spraying. On the last stage it is cured by adding the curing agent causing the chemical reaction of hardening. If required, the surface is ground. The experiments showed that the material has good adhesion and it does not peel off from concrete, wood, as well as iron, masonry, ceramics, plastic, glass, etc. The features of this material do not only create a surface with different color and textural characteristics, but also it may be a structural member itself, it can manufacture the volume elements in use, it can be exposed outdoors at various temperature and humidity conditions [1]. Besides the fact that the production of a liquid stone is environmentally safe, its composition can include production wastes [2]. Industrial waste can give a product both a natural and an artificial look. The researchers note the possibility of creating a diversity of textures and shapes, functionality, beauty, hygiene, maintainability, durability and excellent appearance, allowing a liquid stone to find its place at the market [3]. Initially, the manufacturer of a liquid stone deals only with major manufacturers such as: Corian - DuPont (US), Montelli - DuPont (China), Staron - Samsung (Korea), Hi-Macs - LG (Korea), Akrilika - Acrylic (China), Safas - Granicoat (USA) [4]. Now this technology has been mastered by small businesses and it can be implemented on any industrial site [5].

3. Technology Implementation Examples

As a result of the research with this material, various products have been created: kitchen worktops and sinks, large and medium sized chess. This material was used as covering for paving tiles of various sizes. A.Ya. Mashovich has developed the patterns of paving slabs, street chess; chess for a dynamic game (see figure 1). Our experience allows creating almost any shape and color, or a complex repeat structure.

Figure 1. Samples of developed products

4. Archaeological aspects of the use of technology

Archaeological and cultural objects of a great value, in some cases, can not be removed from the location of their original residence and be placed in museums, often located in remote and inaccessible areas. It is not possible to fully involve them in active cultural and scientific revolution, making the subject of aesthetic observation. For example, petrogliphs elements masonry structures ancient buildings, stone sculptures, and structural elements burial al. The process of cultural familiarization of certain types of rock art is highly toxic to the objects themselves, it is conjugated with such changes for a safe and affordable way to review, which lead to the loss of the natural appearance and the surrounding landscape [11]. Something similar happened with the works of rock art in the caves of France and Spain, but the harmful effects of microflora imparted by tourists influenced on the works of primitive art.

Modern 3D scanning methods allow you to create not only accurate digital models of the surviving archaeological objects, which can serve as the basis for creating museum copies for exhibiting rock art, but also to recreate the once lost elements of archaeological heritage objects [7]. For example, the cleaved portion of sculptures or statues, the lost elements of decoration plants, the components of structures that are in a fragmented state. It is also possible to male prosthetic appliance of the structural parts of ancient structures, for the reconstruction of the lost initial appearance (column element, pillars, arch element, etc.) of a building or a structure. In such museum expositions, such as in open-air museums, it is possible to create archaeological expositions as close to the natural environment as possible using liquid stone technology: copies of rock paintings [8], constructions of burial mounds, megalithic structures and other objects. Often, the transfer of these archaeological heritage objects to the museum for a number of natural or legislative reasons is not possible.

The creation of such replicas is possible according to the preserved measurements, drawings, drawings and photographs. Prosthetics of fragments of ancient structures or sculptures, with the reconstruction of a more complete appearance of an archaeological site through 3D scanning, can provide the most accurate articulation of the preserved elements with prosthetic areas and a tight fit of these parts to each other is unattainable with other restoration methods [9]. In this case, it is possible to recreate the bearing abilities of ancient structures, which is important in the reconstruction of a more complete appearance of complex objects.

Figure 2. Samples of figures made of liquid stone and imitation of various types of materials

5. Description of technological stages

As a result of this study, the stages of creating replicas of archaeological objects were formed:

1. Creating an exact replica of a real object. The replica can be digital as a result of laser or optical scanning [1] or stereo shooting. An analogue replica may be made of gypsum or rubber.
2. Digital processing of the object, creating a form for filling.
3. Selection of the recipe liquid stone for repetition of color and texture and texture of the material cultural heritage.
4. Mold filling, hardening, extraction from the mold.
5. Finalization of the element of a liquid stone.
6. Transfer of the object to the customer (museums, theaters, cultural entities, educational sciences).

6. Research findings

The work done on the basis of these steps makes it possible to create different objects. For example, figure 2 shows chess pieces with an imitation of the surface of marble and cast iron, as well as the samples for creating surfaces of various colors and textures. The main conclusion of these studies is the ability to use this technology to create a replica of archaeological sites, with increasing or decreasing scale. A liquid stone is also a very durable material, so it can be used to replace the missing parts of archaeological sites and it can have sufficient bearing capacity. As a result, a great need for the research in this approach to the creation of replicas of real objects has been found.

References

[1] Wang H, Luo Y, An C., Huang L. and Zhang D 2019 Application of imaging polarimeters to enhanced detection of stone carving *Journal of Cultural Heritage* **40** pp 92-98

[2] Ruposov V L, Alexejenko V V and Mashovich A Y 2019 On the technology of utilizing drilling sludge for the purpose of building material production *IOP Conference Series: Earth and Environmental Science* **229(1)** 012024

[3] Khater H M, Ezzat M 2018 Preparation and characterization of engineered stones based geopolymer composites *Journal of Building Engineering* **20** pp 493-500

[4] Krasilnikov D A, Kartashov T A 2014 Influence of mineral additive on the physical and mechanical properties of the ready liquid stone composition. *Actual problems of the humanities and natural sciences* **4-3** pp 79-81

[5] Mashovich A, Ruposov V and Moskvitin V 2018 Development of technology of production of powlines for conducting works with the use of local types of raw materials in low-populated and district areas *Izvestiya vuzov. Investments. Building. The property* **3 (26)** pp 112-121

[6] Åhfeldt L K 2010 3D-scanning at the Archaeological Research Laboratory 2006–2009 URL: http://goo.gl/jwJlT

[7] Martin B Sweatman, Alistair Coombs 2018 Decoding European Palaeolithic Art: Extremely Ancient knowledge of Precession of the Equinoxes Wayback Machine *Athens Journal of History*

[8] Roussot A.Breuil et Lascaux 1990 Lascaux, premier chef d'œuvre de l'humanité. Les Dossiers d'Archéologie **152**

[9] Pires H, Rubio J M and Arana A E 2015 Techniques for revealing 3d hidden archeological features: morphological residual models as virtual-polynomial texture maps *International Archives of the Photogrammetry, Remote Sensing and Spatial Information Sciences* **6** pp 415-421

[10] Iakushkin O, Budlov E, Uteshev A and Grishkin V 2019 Automated creation of unique editable textures for three-dimensional models of archaeological artefacts *Lecture Notes in Computer Science* pp 752-760

[11] *Draft Medium-term Strategy* 2014-2021 / UNESCO. General Conference, 37th, 2014-2020 y.

IOP Conf. Series: Materials Science and Engineering **880** (2020) 012011 doi:10.1088/1757-899X/880/1/012011

pp 36-48. URL: http://unesdoc.unesco.org/images/0022/002200/220031R.pdf

IOP Conf. Series: Materials Science and Engineering **880** (2020) 012012 doi:10.1088/1757-899X/880/1/012012

Investment support for the implementation of scientific developments as a basis for the production of construction materials

A V Peshkov

[1] Irkutsk National Research Technical University, 83 Lermontov str., Irkutsk, 664074, Russia

E-mail: expertiza@istu.edu

Abstract. The introduction of new types of materials is one of the conditions of economic growth, as well as the means to meet public needs and realize the interests of society, to constantly expand the market space of competing subjects of the economy. The development and introduction of new types of materials into production is of great importance for the State as a means of increasing competitiveness and eliminating the dependence of the State on the mismatch of life cycles of produced materials. Many interesting technologies and materials are emerging today, but there are many barriers to innovation and to modernization in the construction industry. The construction industry is the main opportunity for the growth of competitiveness and prosperity of the national economy. Therefore, its high-tech development is so necessary in modern realities. Moreover, the knowledge and application of new technologies in construction is extremely important for citizens, which will lead to savings in the construction and maintenance of residential premises. In general, the construction industry needs to get state stimulation of the development and introduction of innovative technologies in construction, which will allow to dramatically increase the volume of construction in the country and stimulate the introduction of innovations, otherwise against the background of the growing progress in developed countries construction revolution risks a final loss of competitiveness in this sphere.

Ensuring the quality of durability of building materials and structures is one of the current tasks of modern construction. The priority areas of development of the industry of construction materials are ensuring high operational properties and competitiveness of products, reducing the energy intensity of production processes through the use of local raw materials, as well as the use of new innovative technologies [1,2]. For this purpose new approaches and techniques of design and synthesis of construction materials are required, consisting in rational selection of raw materials and complex of modern modifiers, which allow to achieve specified physical and mechanical characteristics of construction materials and products from them.

The production of construction and finishing materials is a huge industry with a wide range of products. However, it should be noted that the quality of construction materials used by the manufacturer does not always meet the requirements of the quality management system. Many unscrupulous manufacturers do not specify the composition, properties of the materials or indicate the composition that is not true, together with the conduct of poor quality construction works, an

IOP Conf. Series: Materials Science and Engineering **880** (2020) 012012 doi:10.1088/1757-899X/880/1/012012

unsatisfactory result can be obtained. Thus, the quality of construction materials and production technologies are today a huge problem for the consumer.

A number of factors can be identified, both at the level of legislation and at the level of the industry, which influence the use of innovative materials in construction. At the level of legislation, in connection with the adoption of 261 Federal Law "On Energy Saving and Increase of Energy Efficiency and on Amendments to Certain Legislative Acts of the Russian Federation," there was a need for the use of energy-efficient materials in the construction of new buildings and structures, as well as in their reconstruction. In this case, innovative materials are the basis for the development of technical solutions [3].

Building energy-efficient buildings is profitable. For the developer it is an opportunity to sell at an increased price, for the buyer - to invest at the stage of purchase of real estate, and further save in its operation.

At the level of the construction industry, one of the motivations for introducing modern materials and technologies into the sphere of construction is to reduce the cost of construction at the stage of production of structural elements and construction of buildings directly. For example, the use of composite materials in the manufacture of Reinforced Concrete products, such as reinforcement, fibre or various aggregates - foam glass, Penetron "Admixa," can not only improve the physical and technical characteristics of construction materials, but, accordingly, increase the service life, but also will allow significant savings. Thus, the use of composite reinforcement from glass-plastic in non-bearing elements of buildings allows to achieve savings from 15 to 28% at the stage of manufacture of Reinforced concrete products [4,5].

Another important motivation for the developer to promote the introduction of modern materials and technologies is the significant expansion of the operational capabilities of buildings and structures. For example, thanks to the use of "penetrating waterproofing" at the stage of construction of buildings and structures, the developer has the opportunity to use spaces under buildings for commercial sale: it is possible to build underground parking lots or to use these areas for commercial hiring. At the same time, the developer can derive maximum profit from the sale of real space with minimal investment, as well as exclude risks related to warranty repairs and compensation for damage. With high land costs, it makes sense to make the most of it for commercial purposes. It is modern technologies that allow such results to be achieved.

The developer often faces various tasks, which cannot be solved without the application of non-standard technical approaches. It can be a task to build multi-storey buildings on bad soil, or to build complex monolithic structures without a large number of bearing elements or any other task that can be solved by using modern construction materials in the development of technical solutions. For example, due to composite materials, it is possible to strengthen structures without increasing the weight of structures, as well as without loss of space and area.

It is worth pointing out that the construction industry as a whole is immune to new technologies, namely the low costs of Research & Development (R & D), conservatism of designers, contractors and consumers themselves are the main characteristics of this industry. Of all industrial spheres, innovation in construction comes almost least [6]. However, with the passage of time and the massive penetration of innovative developments into all spheres of activity, the conservative construction industry is forced to change its established traditions and introduce processes of radical change.

The main obstacle to the development and introduction of innovative technologies in the construction sector is the absence of a clear State policy in the field of construction. There is practically no system of economic stimulation of innovation in construction.

The system of laws and regulations in the field of energy saving in construction is still pending. This does not allow the formation of technological corridors in construction. In Russia, the authorities have appointed a road map for construction "Improvement of the entrepreneurial climate in the sphere of construction". However, there are doubts that this road map will fundamentally change the situation in the industry. Major problems for housing construction, for example, are the lack of clear urban planning policies, the inconsistency of urban planning plans and plans of natural monopolies, the insufficient investments of the State in engineering and social infrastructure, and the insufficiency of existing mechanisms for financing construction.

In analysing innovations in the sphere of state regulation of the construction sphere, it can be argued that the general state strategy provides for further liberalization of the sphere of construction and reduction of the role of the state. It is difficult to imagine that such a strategy, not supplemented by modern tools of state stimulation of development and introduction of innovative technologies in construction, will allow to dramatically increase the volume of construction in the country and stimulate the introduction of innovations [7].

New technologies play a huge role in any production, their introduction reduces costs and speed of production, improves the quality of products. New patents determine the emergence of new products that can make a healthy competition with the old ones. In terms of promoting the use of innovative materials, the most effective is the initial introduction of innovative materials and technologies at the company 's business process level. Innovative processes should permeate all scientific, technical, production, marketing activities and ultimately focus on meeting market needs and achieving commercial results, as does any business process [8]. The innovation process is considered as a set of sequential work - from obtaining theoretical knowledge to consumer use of goods created on the basis of new knowledge. Innovative process does not come to an end with introduction (emergence in the market of a new product, service or bringing to design capacity of new technology), is not interrupted after introduction as in process of distribution the innovation is improved, gains new consumer properties. At the same time, the enterprise needs to introduce the concept of "innovation cycle," which involves establishing a feedback between the consumer of a new product, service, method, technology and scientific component of innovation [9,10].

The development and production of a new type of construction material requires the presentation of an innovative idea in the form of a business plan. At the same time, it is important to justify the prospects of innovation in the market, to assess the possibilities of its production implementation, to determine the efficiency of technological investments, as well as to analyse the impact of innovation on the economic growth and development of the enterprise economy as a whole [11]. In each construction organization, the process of introducing and developing new innovative materials should consist of several defined stages. The main steps are shown in Picture 1.

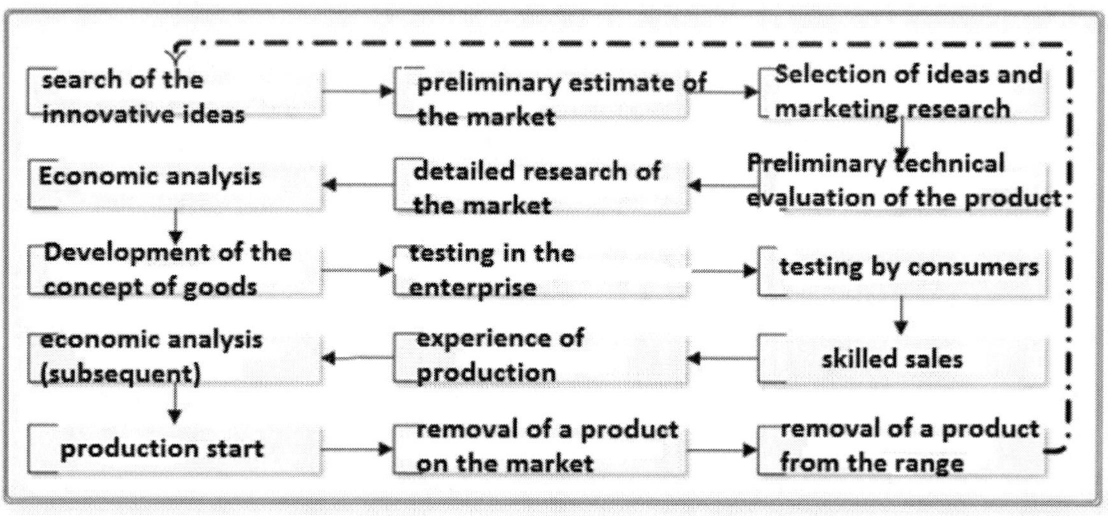

Picture 1. Stages of implementation of a new type of construction material in the company

Each step reflected in the scheme implies a need for a certain level of staff with specific skills, knowledge. Therefore, it will be necessary to resort to training of personnel, which will be carried out in two main directions: training of theory and practice of carrying out works on the implementation of this project, as well as retraining and advanced training of personnel in those areas in which there will be a need during the implementation of the program. In case of promotion of innovations in construction production, even lower-level personnel - builders, construction workers, finishing masters and so on - will have to undergo training in correct handling of materials of the new generation. Compliance with the stages of the scheme will allow to minimize risks of introduction of innovative materials and prepare for potential difficulties in their implementation.

Russia now sees a marked trend towards innovation in construction and production. Large developers, such as PIC, Gals, Donstroi and others, are actively working on innovative directions in the field of construction materials, design, implementation of BIM-technologies, in the process of management of facilities and utilities. Now the industry is already entering a new round of development, when a lot of statistical data have been accumulated, experience of using various materials and technologies has been analysed [12].

A striking example of an innovative construction company in Russia is the largest company Technonikol. For example, last year one of the significant innovations was the ELVATOP polymer membrane, based on the Elvaloy® polymer plasticizer developed by DuPont ™. Elvaloy® plasticizer membranes are in high demand in the EU and the USA: they are durable and have high chemical resistance. But for Russia this product is new. Not so long ago the innovative material for facades lining - tiles TECHNONIKOL HAUBERK, made on the basis of glass-canvas, improved bitumen and granulate from natural basalt, has been released [13]. The facade tile has a patent, produced at one of the factories of the corporation in Ryazani. The material is highly reliable and easy to install, and is also affordable.

"Among our other innovative products recently introduced to the market, the aluminized steam-insulating membrane PAROBARIERE, which is characterized by increased strength, reliability and the highest steam resistance. Basalt thermal insulation of GreenGuard without exaggeration can be called a breakthrough in the segment of stone wool. In its production, a safe biopolymer binder of organic origin is used, which is harmless and environmentally friendly. The technology has been developed by our specialists, "- continues Eugene Voilov [14,15]. Moreover, in April 2019, the company launched the production of construction films in Ryazan to provide reliable steam, hydro- and wind protection. The plant has equipment, which has no analogues in Russia. Deliveries of products are planned both to various regions of the Russian Federation and abroad - to the CIS and Eastern Europe.

The structure of Technonikol Company includes a scientific center, the main directions of which are the creation of new roofing, hydro- and heat-insulating materials, modernization of existing technology. The Scientific Centre has an extensive research base. Laboratories occupy an area of more than 250 square meters and are equipped with modern equipment, with the help of which it is possible to study physical and mechanical characteristics of materials in a wide range of temperatures, to accurately determine the structure and composition of raw components, to test the durability of materials.

The research activity of the Scientific Center is aimed at studying and preventing the causes of aging materials, expanding the working ranges of temperatures of roofing materials, increasing the range of their use. Scientific research and the introduction of advanced technologies will allow TechnoNICOLE to market several new products each year [16].

The key to successful development of production tomorrow is continuous work in the field of research and development. The development of new technologies is the main goal of the Scientific Center.

Main directions of activity of the Scientific Center:

– Improvement of techniques of the analysis of raw materials, materials and finished goods, introduction of express techniques.

- General and methodological support for thermal insulation laboratories of TechnoNICOLE.
- Development of new promising products and directions.
- Basic studies on finished products and raw materials; Creation of a library of samples of raw materials, products of the company 's plants.

- Work on applications of enterprises and divisions: carrying out chemical, thermal, microscopic and other types of analysis, power in solving technological problems.
- Raw material selection tests and recommendations.
– Patent work.

The above-mentioned directions are investments of TechnoNICOLE Corporation in scientific development - it is support of the Russian industry of construction materials and preservation of competitiveness of domestic products on the world market [17,18].

As noted earlier, the introduction of new types of materials is directly linked to economic progress and, although it may lead to rapid economic growth, there are some contradictions, including economic ones. The introduction of new construction material creates new needs, which could lead to an increase in unmet demand. There may be negative consequences associated with unpredictable results of introducing certain achievements into production (pollution, accidents, disasters); Underestimating the meaning of the human factor. Important problems of global scale were increasing distance from sources of raw materials and energy, exhaustion of natural sources of raw materials both in quantity and in terms of its physical properties [19]. As a result of the study of the economic problem of introducing new types of materials, the following conclusions can be drawn. First, the State must act actively to improve the quality and structure of aggregate demand. Secondly, the State has an obligation to actively stimulate the supply of scarce factors of production. Third, it is necessary to identify and support those technologies whose development will indicate the interests of domestic enterprises and provide them with competitive advantages in the world market. Fourth, the state should improve the mechanism of regulation of legal relations in the market of scientific technical products in order to ensure consistency of interests in carrying out technological developments, their patenting and sale. Fifth, it is necessary to define a system of tax incentives for the realization of the interests of innovative enterprises and to develop and implement programmes to increase the level of education of the population and professional training of personnel in the scientific and technical sphere.

References

[1] Akinina V P and Abasov R G 2013 Public-private partnership as a catalyst for innovation processes *Finance and credit* **9** p 51

[2] Artemenko A A 2015 Actual issues of the innovative development of construction *Young scientist* **11** pp 742-744

[3] Matveeva M 2018 Synchronization of the activity of participants during the implementation of housing construction projects *Proceedings of Universities. Investment. Construction. Real estate* **8** pp 31-41 DOI: 10.21285/2227-2917-2018-2-31-41.

[4] Matveeva M V, Doroshenko T G and Sukhanova I A 2019 On the issue of cost and timing of contract tenders for industrial and civil construction *Investments. Construction. Real Estate: New Technologies and Targeted Development Priorities IOP Conference Series: Materials Science and Engineering* **667** 012063

[5] Kalyuzhnova N, Violin S and Matveeva M 2019 Smart Specialization of Russian Regions: Prospects and Limitations *Investments. Construction. Real Estate: New Technologies and Targeted Development Priorities IOP Conference Series: Materials Science and Engineering* **667** 012036

[6] Skolubovich A Yu and Matveeva M V 2019 Development of a methodology for assessing the readiness of enterprises for public-private partnership in the field of public water supply *Proceedings of Universities. Investment. Construction. Real estate* **9** pp 274–283 DOI: 10.21285/2227-2917-2019-2-274-283

[7] Chupin V R and Maizel D I 2020 Aggregation and decomposition in routing oil and gas pipeline systems *Proceedings of Universities. Investment. Construction. Real estate* **10(1)** pp 108–115https://doi.org/10.21285/2227-2917-2020-1-108-115

[8] Biglova A A 2018 State of the public procurement system in Russia *Science, education,*

IOP Conf. Series: Materials Science and Engineering **880** (2020) 012012 doi:10.1088/1757-899X/880/1/012012

innovation: approbation of research results Materials of the International (correspondence) scientific and practical conference pp 153-156
[9] Gaskova Yu.I. "Integrative approach to the formation of a mechanism for the coordinated development of industry and infrastructure in the region" // Proceedings of the II International scientific and practical conference "Economics and modern management: theory, methodology and practice". - Penza: 2018. - p.358-362.
[10] Komarov A K, Ivanov I A and Lundenbazar B 2019 Theory and practice of the use of gabions for forming protective structures *Proceedings of Universities. Investment. Construction. Real estate* **9(1)** pp 78–89 DOI: 10.21285/2227-2917-2019-1-78-89
[11] Peshkov V V and Gamayunova E V 2018 Specific features of investment support for the repair and reconstruction of buildings and facilities *Proceedings of Universities. Investment. Construction. Real estate* **8** pp 56-72 DOI: 10.21285/2227-2917-2018-2-56-72
[12] Yaskova N Yu and Sarchenko V I 2018Development of strategic approaches to the creation of aqua-territorial industrial complexes in the Arctic zone *Proceedings of Universities. Investment. Construction. Real estate* **8** pp. 84-93 DOI: 10.21285/2227-2917-2018-2-84-93
[13] Gertsekovich D, Gorbachevskaya L, Grigorova L and Peshkov V 2019 Return on investment in REIT real estate funds *Investments. Construction. Real Estate: New Technologies and Targeted Development Priorities IOP Conference Series: Materials Science and Engineering* **667** 012063
[14] Matveeva M, Kalyuzhnova N, Skripka S 2019 Intellectual specialization of Russian regions: prospects and limitations *IOP Conf. Series: materials Science and engineering* **667**
[15] Sayfutdinova R V 2016 *Modern Fundamental and Applied Researchers* **1(20)** pp 167-173
[16] Vovseenko E A 2018 *Financial Instruments and Hedging: New Rules of IFRS* European Proceedings of Social and Behavioural Sciences **L** DOI: https://dx.doi.org /10.15405/epsbs.2018.12.167
[17] *The project "INNOVATIVE RUSSIA - 2020"* 2011 (Strategy for innovative development of the Russian Federation for the period until 2020)
[18] Trifilova A A 2009 Assessment of the effectiveness of innovative development of the enterprise. *Finance and Statistics* p 511
[19] Safronov I V 2008 The concepts of "innovation" and "innovation activity": the essence and content *Questions of modern science and practice* **4 (14)** p 224

IOP Conf. Series: Materials Science and Engineering **880** (2020) 012013 doi:10.1088/1757-899X/880/1/012013

Technology for repairing and improving energy efficiency of enclosing structures of large-panel buildings of the 1-335 series

A V Petrov[1], A G Petunin[1] and E V Samarkina[1]

[1] Irkutsk National Research Technical University, 83 Lermontov str., Irkutsk, 664074, Russia

E-mail: termina@mail.ru

Abstract. The article deals with technological and design solutions for repairs with increasing the thermal properties of wall panels of large-panel buildings of the first series of mass developments. A technology for restoring wall panels using modified structural and heat-insulating concrete is proposed. The technology of repair of gas-concrete panels increases the thermal characteristics of building envelope structures to the level close to the standard, reduces their physical wear. Analysis of repair and restoration works of gas-concrete panels with modified material showed the rationality and financial attractiveness in comparison with the method of shotcrete and subsequent insulation and installation of a ventilated facade.

1. Introduction

Starting from the 60s, there was a mass development of Soviet cities with residential buildings of the 1-335S and 1-335A series (AC, CB) - large-panel buildings with incomplete and full precast concrete frames, respectively.

Buildings of the 335 series, operated for 30-40 years or more without repair and protection measures, have typical defects. These are mainly foci of frost destruction in areas where condensation accumulates, resulting in local defects and damage to the cellular concrete of load-bearing wall panels (zones of window and door openings, mounting and intersectional seams). The destruction of the solution of external panel seams activates the physical aging of materials, and, as a result, the formation of cold bridges, violation of steam and waterproofing with regular moisture saturation of the seam cavity. In this case, along with operational inconveniences and violations of the appearance and facade, irreversible processes of metal corrosion of welded embedded parts occur – and this is one of the issues of ensuring the structural reliability of the building.

To date the standard operating time and technical resource of wall bearing and enclosing structures of large panel buildings of the 1-335 series have been exhausted.

2. Materials and methods

In the Irkutsk region, numerous surveys confirm the presence of defects and damages in the external wall panels made of gas-reinforced concrete, indicating a decrease in the operational reliability of buildings of efficiency. For seismically active areas, the regulatory requirements have been significantly increased, which the buildings under consideration objectively cannot fully comply with.

A very acute problem is the continuous stratification of the north-west orientation load-bearing wall panels, which are subject to intense effects of atmospheric precipitation, with the threat of their sudden collapse. The technical condition of the enclosing structures is significantly aggravated for houses of the 1-335S series with an incomplete frame, when all the load from the floor structures is transferred to the stratified emergency panels, resulting in a drawdown of the crossbars and floor slabs with a threat to human life.

The reasons for the appearance of damage to load-bearing wall panels, first of all, lie in the properties of the material itself – gas-concrete:

- increased hygroscopicity of cellular concrete, which includes gas-reinforced concrete;
- low protective properties of gas-reinforced concrete in relation to the reinforcement and, as a result, corrosion of the external reinforcement grids, which leads to the stratification of load-bearing wall panels. This is due, primarily, to the significant decline in the alkaline environment of concrete due to carbonation of lime, directed-porous structure of gas-concrete and the presence of sulfur in the ash-the ash wastes CHP production.

The second main reason for the stratification of gas-reinforced concrete load-bearing wall panels is a gross violation of the production technology of concrete products, namely, uncontrolled "immersion" of the outer grid to a depth of 150-200 mm from the outer surface instead of 40 mm, provided by the project. Compliance with the design requirements for the reinforcement of load-bearing wall panels would preserve the structural integrity of the main part of the material with the inevitable corrosion of the reinforcing mesh, without creating an emergency condition of the external panels of the building and the possibility of restoring the texture-protective layer of concrete at minimal cost.

Moscow has adopted a law on housing renovation, which provides for the demolition of more than 4,500 "khrushchevok" and relocation of their residents to new homes. The program is designed for 20 years and about 2 trillion rubles will be allocated from the city budget.

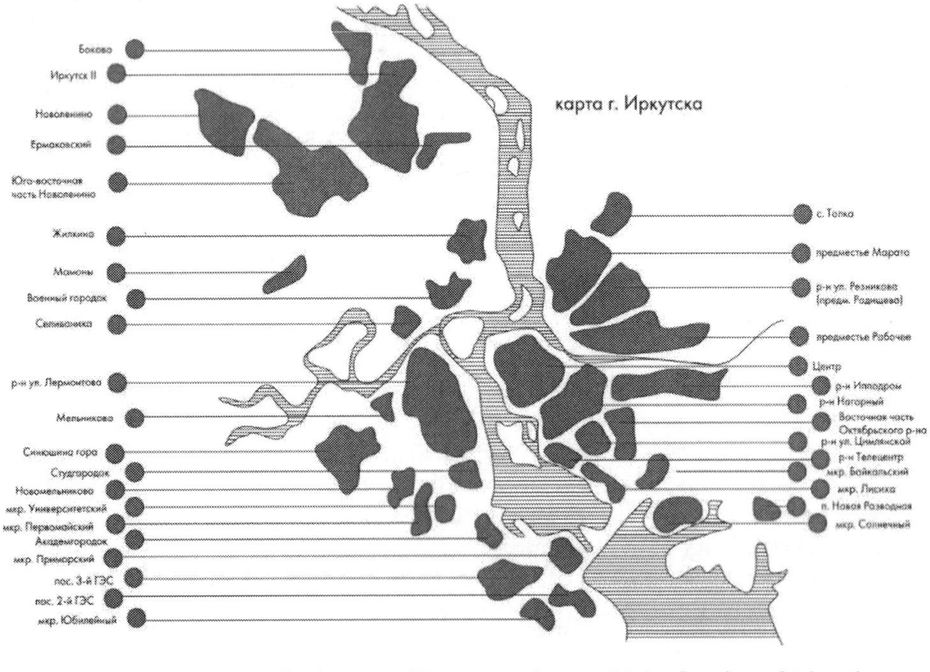

Figure 1. The layout of houses series 1-335 in the city of Irkutsk.

In the Irkutsk region, there are more than 1,800 houses of the 335 series, and the demolition and construction of new housing will require more than 600 billion rubles. Only in Irkutsk, the minimum necessary volumes of repair, modernization and reconstruction of these series are more than 700 thousand square meters of total area. Figure 1 shows the layout of residential buildings of series 1-335 in Irkutsk.

It should be noted that in June 2017, Vladimir Putin explained at a large press conference that the regions of Russia do not have such financial opportunities, so the repair of large-panel buildings of the 1-335 series is the only solution to the problem.

3. Results

A technology for restoring and improving energy-saving characteristics of gas-reinforced concrete wall panels with modified structural and heat-insulating concrete has been developed. This technical solution was tested on the emergency panels at the end of the building (figure 2), located at the address: Angarsk, 10th microdistrict, 36.

Figure 2. Spontaneous collapse of gas-reinforced concrete wall panels in the north-west orientation in Angarsk.

The technological process of restoring wall panels is divided into several types of work:

1st - fixing of eaves blocks and dismantling of drainpipes.

2nd – collapse of the emergency layer of concrete with the help of an aerial platform. Special attention is paid to safety in the event of a collapse with the site fencing and the calculation of the danger zone.

3rd – preparatory work. These include the development of the site: installation of scaffolding; preparation of the site for the preparation of modified light concrete; organization of sites for the installation of household and storage facilities; organization of temporary water and power supply, as well as manual electrified tools, etc.

4th - reinforcement works. A reinforcing mesh is installed with attachment to wall panels and columns using chemical anchors.

5th - shuttering works. Before installing the formwork in the lower part of the building, an equilateral corner is attached to the anchors for supporting the formwork. As an experiment, we used wood-plywood formwork, which was installed in tiers.

6th – place concrete work. Production of concrete works is carried out in tiers over the entire width of the building's end face with careful compaction with a deep vibrator.

7th – finishing thermal insulation works. The facade is insulated with a finishing heat-insulating paint coating in several layers, which together with light concrete increase the thermal characteristics of the repaired panels. Figure 5 shows the technological stages of repair and restoration work on the end of the building of the 1-335 series with a full frame.

4. Discussion

Over the past few years, there has been a growing need for emergency repair of large-panel buildings of series 1-335 of various modifications in Russian cities.

In large-panel houses of the 1-335S series, the crossbars are supported directly on the wall panels. One of the main technical defects of such houses is the construction of support units for interfacing runs of inter-floor floors with metal consoles, made up of two channels № 12, embedded in the supporting ribs of wall panels and being the supports of the runs. As a result of physical wear and corrosion of embedded parts, in March 2005, a five-story large-panel building collapsed in the city of Temirtau in the Republic of Kazakhstan, and in 2006, a part of the house collapsed in Arkhangelsk.

To strengthen the support nodes, Omsk specialists developed and patented an anchor device for unloading the support consoles (figure 3).

For this problem, TbilZRIED recommends that the reinforcement of the crossbar support units on the outer two-layer wall panels, without the device of wall columns (figure 4).

Figure 3. Strengthening of support units of houses of the 1-335 series with an incomplete frame in Omsk.

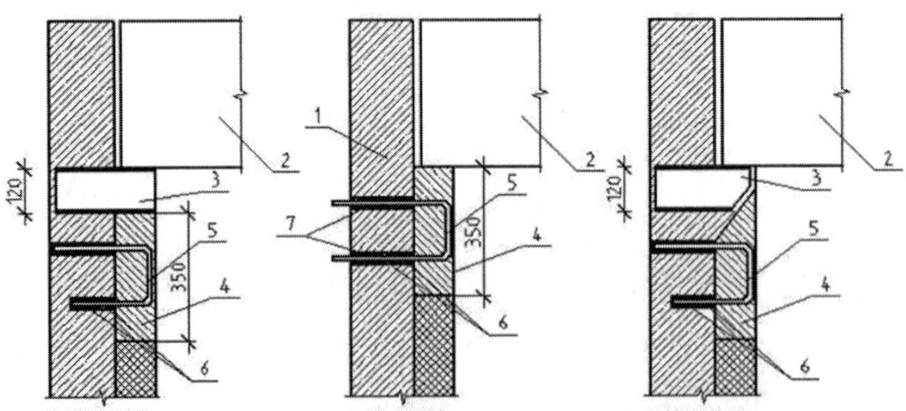

Figure 4. Ways to enhance existing consoles depending on their state.
1 – panel edge; 2 – crossbar; 3 – channel bar №12; 4 – gain console; 5 – armature; 6 – polymer solution; 7 – pit

Reinforcement of the crossbar support units is performed by gluing pre-made reinforced concrete elements-consoles, with rebar releases anchored by epoxy polymer solution in pits pre-drilled in the

IOP Conf. Series: Materials Science and Engineering **880** (2020) 012013 doi:10.1088/1757-899X/880/1/012013

panel ribs. According to the recommendations, consoles can be made separate for each edge of the panel or paired into two mating edges. The use of paired consoles also allows you to increase the spatial rigidity of the frame.

Currently, since 2004, in Irkutsk, two main methods of restoring the external walls of residential buildings of the 335 series (full frame) have been tested in practice:

1. Dismantling of failed wall panels and replacing them with new products in full;
2. Restoration of the load-bearing capacity of wall panels with reinforced shotcrete with subsequent installation of the insulated ventilated facade system.

The new technology for restoring the operational reliability of wall panels using modified lightweight concrete gives significant advantages in terms of labor costs, and economic efficiency, compared with shotcrete, increases by up to 30 percent.

5. Conclusion

Thus, the proposed technology for repairing gas-reinforced concrete wall panels with modified structural and heat-insulating concrete can be implemented on residential buildings of the 1-335 series of various modifications in the Irkutsk region. The use of this technology will not only increase the operational reliability of gas-concrete panels and reduce their physical wear, but will also solve the current problems of wall panel layering with less financial costs, without resorting to replacing wall panels and temporary relocation of residents.

a) b)

c) d)

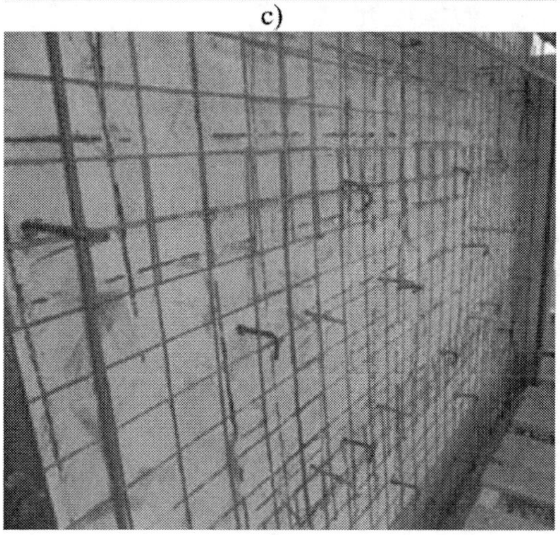

e) f)

Figure 5. Technological stages of repair and restoration works of the end of the building series 1-335 with a full frame.
a – securing cornice blocks; b – collapse of the emergency layer; c – installation of armature; d – installing formwork; e – production of concrete works; f – general view of the restored end of the building

References

[1] Petrov A V, Peshkov V V and Petunin A G 2015 Large-panel buildings of the 1-335 series with external wall panels made of gas-reinforced concrete: repair, reconstruct or demolish *Bulletin of ISTU* **11 (106)** pp 85-91

[2] Petrov AV and Petunin A G Reconstruction of large-panel buildings of the 1-335S series with an incomplete frame in the Irkutsk *region News of universities. Investment. Construction. Realty* **2 (3)** pp 105-110

[3] Pinus B I and Gagarski V V 2009 Enhancement of seismic resistance of large-panel buildings *Proceedings of the International scientific and practical conference, Irkutsk* pp 167-173

[4] Dmitriev A N, Zhdanovsky B V, Koloskov V N and Sukhachev V P 1987 Problems of reconstruction and demolition of five-story panel houses in Moscow *Industrial and Civil Engineering* **4** pp 28-30

[5] Bulgakov S N 2005 Building materials, equipment, technologies of the 21st century, Moscow pp 60-61

[6] Bulgakov S N 1995 The concept of reconstruction of a 5-storey building of the 60s-70s. *Industrial and civil construction* **6** pp17-19

[7] Bulgakov S N 1999 Reconstruction of residential buildings pp 47-63

[8] Matveev E P 1998 Technology of reconstruction of low-rise residential buildings by the method of superstructure of floors from volume blocks *Housing construction* pp 10

[9] Mikhalko V R 1986 *Repair of structures of large-panel buildings*

[10] *Recommendations for inspection and assessment of the technical condition of large-panel and stone buildings* 1988 p 57

[11] *Reconstruction of buildings and structures* 1991 p 351

IOP Conf. Series: Materials Science and Engineering **880** (2020) 012014 doi:10.1088/1757-899X/880/1/012014

Changes properties of concrete with cyclic low-temperature impacts

B I Pinus[1]

[1]Irkutsk National Research Technical University, 664074, Lermontov St. 83, Russia

E-mail: pinus@istu.edu

Abstract. The analysis of concrete fracture in the process of exhaustion of its frost resistance. Statistical regularities and kinetics of ultimate strength upper limit of microdestruction and deformative values are established. A steady decline in the fracture toughness was found during cyclical freezing and thawing (FTC). Fatigue transformation of concrete long-term operation in severe climatic conditions is characterized with an increase in the danger of brittle fracture.

1. Introduction

Long-term operation of reinforced concrete structures is objectively associated with a certain probability of stress-strain state fluctuations due to changes in the magnitude and conditions of application of loads, background manifestation of seismic activity, alternating temperature differences and other factors [1-5].

The most important factors in the interaction of construction with the environment [1-5] is the change in damping properties of concrete caused by cyclic degradation and hysteretic energy dissipation. Fatigue fractures occur at fluctuating impact levels significantly lower than the calculated values.

In a comparative cyclical manifestation, such impacts can be classified as few-repeated and the structural-mechanical changes by them as low-cycle fatigue. Their distinctive feature is the ambiguous, cumulative nature of the consequences, as well as the interdependence of the impact magnitude (number of cycles) and the functional response of object.

2. Research methodology

The fracture toughness was estimated by the ratio of energy absorbed during compression of concrete samples at the stages of conditionally elastic (W_e) and plastic (W_{pl}) deformation. In this case, the level of stresses corresponding to the upper boundary R^{v}_{crc} and plastic, taken until stress $\sigma_b = 0.8\sigma_{ult}$ on the descending branch of the σ-ε diagram, crack formation is assumed to be the limit of conditionally elastic resistance. The equipment used in the study (Instron 5989) allowed for automatic loading with a constant deformation rate (0.05 mm/s) and simultaneous recording of σ-ε and energy of absorption up to the moment of complete destruction of test samples. The influence of severe climatic conditions was simulated by daily cycles of freezing (- 40 ± 2 °C) and thawing in water at a temperature of 20 ± 2 °C. The frequency of mechanical tests was taken in accordance with the previously established [10, 11] kinetics at FTC of significant structural properties of concrete. Its specificity is governed by the ambiguity of consequences of frosty transformation of concrete and is due to the simultaneous development of opposite phenomena determining the extreme nature of changes in strength and deformability. This determines the feasibility of intermediate (in the range of normalized frost resistance

F) valuations at the stages of ascending and descending vectors reflecting the dynamics of changes in the controlled parameters.

Experimental studies were carried out on prismatic samples (100×100×400) made of concrete composition (Cement:Sand:Gravel:Water=1:1.75:2.9:0.45) under normal conditions of monthly hardening and long-term (more than 2 months) storage until temperature and humidity tests. Standard tests of cubic samples established frost resistance of concrete, which is 35 cycles of the adopted level of cooling. Selection of samples is performed using calibration rejection by their density, estimated by the speed of ultrasound. The total number of samples tested in the main mode was 84 prisms. The minimum number of samples at the stage of T-W impact was 12, thereby allowing to perform their statistically acceptable generalization. The agreement of the theoretical and experimental distribution of the indicators of the controlled parameters was evaluated according to the Pearson criterion [12] as ensuring the minimum error in accepting the hypothesis and evaluating the required quantile of concrete properties.

3. Results and their discussion

Basic statistical distributions of significant parameters of strength and deformability at various levels of exhaustion of the standard resource of frost resistance (n = N/F) are shown in Table 1.

Presented values correspond to:

R^{v}_{crc} - stress level of a sharp (more than 2-fold) increase in lateral deformations at the ascending stage of loading;

$\varepsilon_{b,ult}$ - deformation (‰) at the peak of loading;

$\varepsilon_{b,\max}$ - deformation (‰) at the time of 15% load reduction at the stage of pseudoplastic deformation;

Significant differences in the dynamics of changes in all controlled parameters are evident in both comparison of the average values and the distribution density of their random realizations. Kinetics of strength and the upper boundary of crack formation with increasing negative distribution asymmetry is characterized by comparative statistical homogeneity. The latter indicates increased variability of fatigue strength and, as a consequence, a more significant decrease of the normalized 5% quantile.

Monotonous acceleration with a tendency toward stabilization is observed in the indicators of ultimate compressibility. The asymmetry of distribution and spread of realization are characterized by moderate growth at stages of T-W impacts corresponding to the frost resistance of concrete.

The most significant and statistically significant are the changes in deformation of concrete at the descending stages of loading. A moderate (up to 18%) increase $\varepsilon_{b,\max}$ occurs until the exhaustion (50-60%) of normalized frost resistance resource. Subsequently, a significant decrease in the potential of pseudoplastic deformation is observed with a change in the asymmetry of the distribution and shift of the corresponding quantiles.

Fatigue changes in concrete resistance are clearly supported by diagrams σ_b-ε_b reconstructed based on average values of experimental data at different stages of T-W impacts (Figure 1). There is a steady tendency in increasing the slope of the curves to the axis of deformations, which is an indirect confirmation of the decrease in the modulus of elasticity. Apparent is the increase in deformations on the descending branches with a simultaneous more intense drop in the load.

The observed differences in directivity, kinetics, and distribution of concrete properties indicators confirm the ambiguity of the fatigue effects of cyclic low-temperature and humidity impacts. From that follows the need for a differentiated approach to the selection of criteria of durability for specific design situations. At the same time, the feasibility of a generalized (integral) assessment of structural modification is confirmed. Fracture toughness, as such an indicator, is considered below. It is characterized by the ratio of the energy aspects of concrete resistance to compression at stages of FTC.

Full energy consumption (W), absorbed during testing of samples were determined in an automatic mode as the square of diagrams σ-ε. Fractional components of energy at the stages of conditionally elastic (W_e) and plastic (W_{pl}) deformation are determined from the corresponding sections of the reconstructed diagrams (Figure 1). Where in, functioning of W_e corresponded to the area of the diagram limited by the level $R^{v}_{b,crc}$ and ε^{v}_{b}; and W_{pb} = W-W_e. The assessment of fracture toughness (W_{pl}/W)

obtained in this approach seems to the most optimistic, as objective inaccuracies in the model analysis lead to an overestimation of the plastic resistance potential.

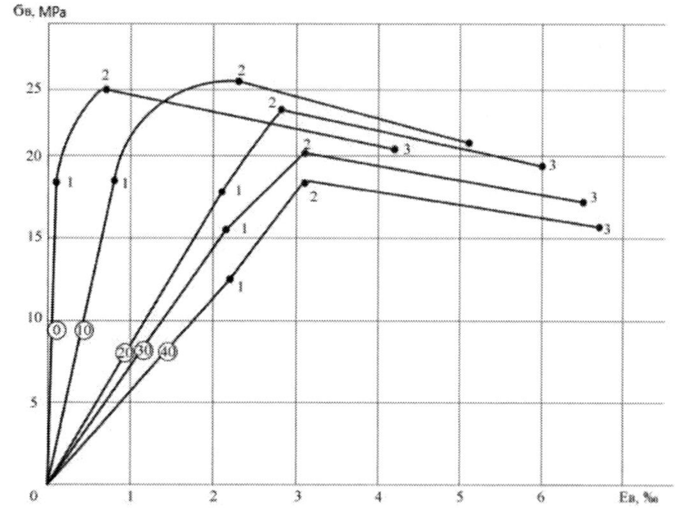

Figure 1. Compression diagram model reconstructed from experimental data (numbers on the curves are numbers FTC).

Figure 2 shows the curves characterizing the change in energy parameters during T-W impacts. A steady tendency towards an increase in the energy dissipated during compression is observed at the initial stages of FTC. Moreover, it is mainly due to the predominant increase (up to 66%) of the elastic resistance potential. Functioning of plastic deformation increases more moderately and does not exceed 14-18%. The maximum energy increase is timed to the moment of exhaustion of 50-60% of the standardized resource (grade) of concrete frost resistance. After the cycles corresponding to "F", functioning of elastic resistance exceeds the initial value, while plastic deformation is less than the initial by more than 10-12%. Based on the tests (N/F=1.29), the components of the energy of fracture are almost identical (Figure 3).

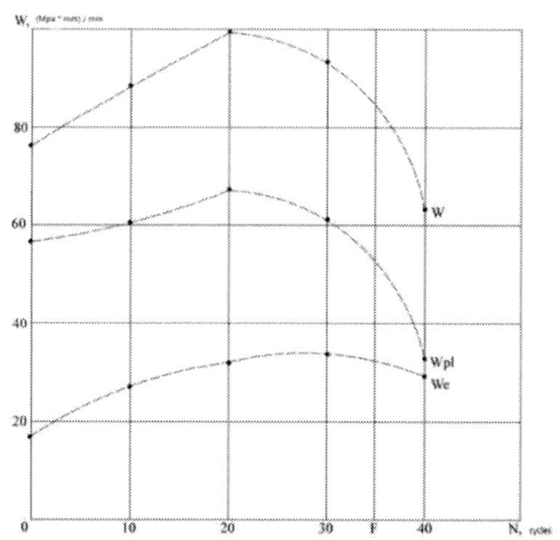

Figure 2. Change in energy of fracture during T-W impacts.

Such kinetics of energy parameters is a reflection (consequence) of the structural modification of concrete at FTC, leading to a change in the physical laws of its resistance to force impact. Moreover, objective data suggest an increase in the likelihood of a brittle nature of fracture as the frost resistance resource is exhausted. This is confirmed by a steady decrease in fracture toughness at FTC, numerically equal to the ratio $W_{pl}(N)/W(N)$, (Figure 3). If normative methods of calculation using nonlinear deformation models are considered and redistribution of forces in the sections of statically indeterminate systems, the decrease in the fracture toughness in reinforced concrete elements of the "northern" execution should be regulated. The correction provided for in clause 6.1.27 [14] is clearly insufficient. This is confirmed (Figure 3) by a significant difference in the kinetics strength and fracture toughness.

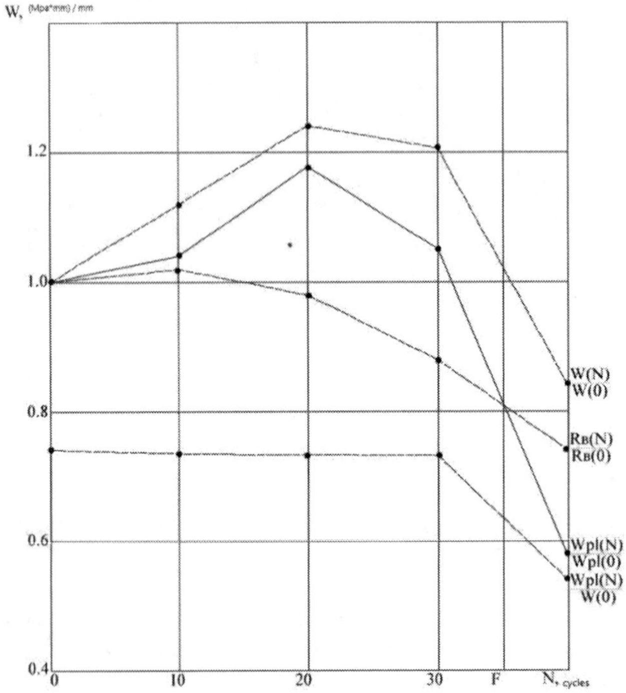

Figure 3. Dynamics at FTC of strength, energy and fracture toughness.

4. Conclusions

1. Fatigue transformation of the structure of concrete subjected to cyclic temperature and humidity impacts is characterized by a steady decrease in fracture toughness, evaluated by the ratio of energy consumption of plastic and total deformation.
2. Long-term operation of reinforced concrete elements in severe climatic conditions is associated with an increase in the danger (risk) of brittle fracture.
3. When designing the fracture toughness, it is advisable to perform calculations in a probabilistic formulation by using differentiated coefficients of reliability depending on a particular design situation.

Table1 Statistics of distribution of concrete properties indicators at stages FTC

Magnitude of frost impact	Prismatic strengh, R_b			Ultimate deformations, $\varepsilon_{b,ult}$			Max deformations, $\varepsilon_{b,max}$			Upper boundary of microdestruction, $R^{v}_{b,crc}$		
	Average, MPa	Dispersion, MPa^2	Coeff. of asymmetry	Average, ‰	Dispersion	Coeff. of asymmetry	Average, ‰	Dispersion	Coeff. of asymmetry	Averag, MPa	Dispersion, MPa^2	Coeff. of asymmetry
0	24.0	9.12	-0.38	1.64	0.10	-0.58	2.55	0.24	-0.57	18.0	9.1	-0.50
0.29	24.5	12.1	-0.43	2.34	0.18	-0.54	2.67	0.21	-0.51	19.4	13.2	-0.56
0.57	22.8	13.8	-0.49	2.88	0.32	-0.59	3.02	0.25	0.50	17.5	16.1	-0.59
0.89	20.2	29.7	-0.81	3.15	0.78	-0.84	2.78	0.45	0.72	15.0	14.0	-0.75
1.29	18.5	26.1	-0.84	3.18	1.04	-0.96	1.91	0.23	0.75	12.0	13.2	-0.90

IOP Conf. Series: Materials Science and Engineering **880** (2020) 012014 doi:10.1088/1757-899X/880/1/012014

References

[1] Yarema S Ya 1973 Staging of fatigue fracture and its consequences. *Physical and chemical mechanics of materials* **6** (Moscow) pp 66-74

[2] GOST R ISO 2394-2016 *Construction structures. The basic principles of reliability* (Moscow: Standartinform) p 72

[3] Zaitsev Yu V 1982 *Modeling of deformations and concrete strength by methods of fracture mechanics* (Moscow: Stroyizdat) p 196

[4] Karpenko N I 1996 *General models of reinforced concrete mechanics* (Moscow: Stroyzdat) p 416

[5] Aktuganov I Z and Pushkin A A 1982 Experimental verification of the principle of linearity of summation of concrete damage under the impact of non-stationary alternating cycles. *Problems of the reliability of reinforced concrete structures* (Kuibyshev) pp 5-8

[6] GOST 29167-91 1992 *Concrete. Methods for determining the characteristics of crack resistance (fracture toughness) under static loading* (Moscow: Goststandart) 24 p

[7] Yudin A P and Tkachenko T A 1976 Deformations and resistance of uniaxially compressed concrete under periodic T-W environmental impacts. *Works of RISI* **15** (Rostov-on-Don) pp 16-25

[8] Moskvin V M, Kapkin B M, Mazur B M and Podvalniy A M 1967 *Resistance of concrete and reinforced concrete at a negative temperature* (Moscow: Stroyizdat) p 131

[9] Vlasov G M and Bokarev S A 1984 Accounting for changes in the strength and deformation characteristics of concrete after alternate freezing and thawing. *Universities News "Construction and Architecture"* **11** pp 113-114

[10] Pinus B I, Pinus J N and Khomyakova I V 2015 Changing of structural properties of concrete during cooling and freezing. *Bulletin of ISTU* **2** (Irkutsk) pp 111-115

[11] Pinus B I 1985 *Ensuring the durability of reinforced concrete structures at low temperature impacts.* Dissertation Doctor of Technical Sciences (Moscow) p 367

[12] Druzhinin G V and Voronkova O V 1983 Comparison of methods for constructing an empirical distribution function based on a small number of observations. *Reliability and quality control* (Moscow: Standard) pp 12-14

[13] Pinus B I 2018 Consideration of cumulative effects in concrete structures at low temperatures impact. *MATEC Web of Conferences* **212** (France: EDP Sciences)

[14] SP 63.13330.2012 *Concrete and reinforced concrete structures. The main provisions* (Moscow: Minregion RF) p 154

IOP Conf. Series: Materials Science and Engineering **880** (2020) 012015 doi:10.1088/1757-899X/880/1/012015

On the issue of ensuring reliability of reinforced concrete structures of "northern execution"

B I Pinus[1]

[1]Irkutsk National Research Technical University, 664074, Irkutsk, Russia, Lermontov St, 83

E-mail: pinus@istu.edu

Abstract. The work is devoted to the design justification of the required resource of bearing capacity of bent reinforced concrete elements of "northern execution" based on ensuring the required reliability at the time of exhaustion of the frost resistance of concrete. The dynamic model of functioning of structure in severe climatic conditions is considered as a set of static tests at individual stages of cyclic temperature and humidity effects (FTC) formed on the basis of corresponding experimental samples of strength and deformability of concrete. For the criterion for reliability of structure, it is taken the value of strength and deformability of concrete, at which its bearing capacity over a normal section has a probability of at least 99.7%. Using graphic-analytical modeling, limit (failure) values were established and correction of the normative dependency by introducing a differentiating coefficient of reliability, accounting for the level of reinforcement and design negative air temperature, was proposed. The technical and economic unreasonableness of excessive, more than 80% of the boundary, reinforcement of normal sections of bent reinforced concrete elements of "northern execution" is established.

1. Introduction

Modern design methods allow us to consider reliability as a response of a building structure or a result of its interaction with the environment. In such a formulation, it can be identified with the probability of the occurrence of functional parametric failures in the form of undesirable limit states [1].

In the absence of statically representative data on structural failures in adverse operating conditions, the required design service life can be justified by the correction of calculated values of materials' resistance, taking into account the physical and statistical laws governing their change. At the same time, adjustable and controlled parameters should have increased sensitivity to the probable failure, kinetics of change, and acceptable capabilities of ensuring information-statistical representation.

Taking into account the existing practice of design, operation, and control, it is advisable to select controlled parameters according to the criteria of the standard and recommendations [2-3] employing calculated dynamic models.

The fundamental difference between the latter from the corresponding normative functional dependences is the presence of time (time, cycles) and statistical distribution factors of significant response parameters of structures a priori accepted as random variables.

The experimental and analytical substantiation of such an approach to designing for reliability in the bearing capacity of bent reinforced concrete elements of "northern execution" is the goal of this work.

2. Materials and methods

A dynamic model of operation of structures in severe climatic conditions is considered as a set of static tests after a certain number of cycles (*N*) of temperature-humidity (T-W) impacts. Its acceptability follows from the Bailey principle [4], the adequacy of which for the conditions under consideration is confirmed experimentally [5-7]. The preferred test frequency is determined by physical laws of frost destruction of concrete, as a non-stationary process of an extreme form [8-10].

Taking into account the specifics of control and standardized indicators of quality as well as the possibility of summarizing numerous experimental data, the time factor is analyzed in the form of relative number of cycles of T-W impacts ($n = N/F$) [11].

Controlled parameters of reliability are established by numerical modeling using criteria-based normative equations of bearing capacity and standardized indicators of their variability (tolerances). Comparative dynamics of FTC of strength and deformation properties was established experimentally [8, 14].

Prismatic strength of concrete $R_b(n)$ and its ultimate deformability $\varepsilon_{bu}(n)$, evaluated by the totality of random implementations at individual test stages were taken as experimental controlled indicators of reliability of bent elements according to the criteria of the standard [3]. Moreover, the durability of structures corresponds to the number of cycles, after which the probability of a decrease in $R_b(n)$ is within the established tolerance (SP, SNiP, GOST) and is 99.7%.

The necessary correction of the calculated resistance of concrete based on ensuring reliability in the bearing capacity of a bent element for the duration of service life, corresponding to the frost resistance resource, is established by the graphic-analytical method in the following sequence.

The functional model of the strength of a bent element along the normal section in the parameters of random realizations is converted to:

$$\tilde{R}_b = \frac{\tilde{\sigma}_e^2 A_s^2}{2\tilde{b}(\tilde{\sigma}_e \tilde{A}_s \tilde{h}_0 - \tilde{\mathrm{M}}_0)} \qquad (1)$$

$\tilde{\mathrm{M}}_0$ – a set of random values of the bearing capacity of an element under the assumption of a standardized variability of all parameters of the normative dependence.

Numerical modeling determines the failure value of concrete strength R_{ba}, corresponding to the probability of a parametric failure of an element based on the bearing capacity of not more than 0.3%. Wherein, options for different levels of reinforcement (from minimum to maximum expected) and variability (from standard to predicted) are considered. Based on the numerical analysis, graphs of the dependence of the probability of parametric failure (α = 0.3%) on the adopted reinforcement and dispersion of concrete strength are constructed. At the same time, various options are considered for correcting the reduction in the estimated value when evaluating the design value of the bearing capacity MO.

Below, as an example, presented are the curves of changes in the failure value (under reliability condition) of concrete's strength for two reinforcement options and various coefficient of variability (C_v) of the concrete strength index (Figure 1). The significant influence of the level of reinforcement ($\mu' = \mu/\mu_R$) and statistical distribution indicators on the magnitude of failure value of concrete strength R_{bn}, which indirectly characterizes the reliability of structure in bearing capacity, is clearly confirmed. The graphs of the risk of failure for the two most common options of correction of the calculated value γ_{bF} = 0.85 and 0.7 are also presented here. Ambiguity in the effect of corrective procedures on the relative (to the boundary) level of reinforcement and the density distribution of concrete strength is observed.

The obtained databank of graphical dependences of failure values R_{ba} in the scattering fields μ' - C_v was adopted as the basis for an indirect assessment of reliability of reinforced concrete structures of "northern execution" according to the results of standard compression tests of prismatic samples after cyclic freezing and thawing of various intensities. About one hundred samples of normally hardened concrete (Cement : Sand : Gravel : Water = 1: 1.75: 2.9: 0.45) were exposed to temperature-humidity impacts (minus 42° C and thawing of water 20 ± 2°C). Periodically, the samples were tested for compression with a constant deformation rate (0.05 mm/s) with automatic control of deformations (σ-ε diagram). The number of simultaneously tested prisms (12-18 items) made it possible to obtain statistically reliable data on the main distribution indices [8-11] of concrete strength at the stages of exhaustion of the normalized resource of its frost resistance.

Using experimental values of statistics and the databank of failure strength values under the conditions of ensuring the required reliability of structures (not less than 0.997), average values of R_{ba} were established for bending elements with different relative levels of reinforcement μ' (Table 1).

As it has been expected, the failure value of concrete strength, considered a criterion for ensuring the necessary reliability for bearing capacity of a normal section, depends on the level of reinforcement and is a sensitive indicator to the impact of external conditions. Its monotonic increase is observed with increasing reinforcement and practical stabilization at $\mu' = 0.85\mu_R$. Similar tendencies are preserved in the time aspect, up to the moment of exhaustion of the resource of normalized frost resistance of concrete F.

A comparative assessment of the distribution statistics R_b(n) and R_b allows us to state that at the initial stages of cyclic T-W impacts (n $\leq$ 0.3), the calculated functional model of norms ($\gamma_F = 1.0$) provides the required level of reliability in the strength of the normal section for the entire range of reinforcement under analysis. A simple interpolation of the presented data shows that in prospect (n > 0.3) the risk of parametric failure may exceed the permissible level (0.3 %) in elements with $\mu' \geq (0.4\text{-}0.45)\mu_R$.

In our opinion, there is a fundamental difference in the basic premises of the normative (deterministic) method for calculating bearing capacity and the probability-statistical forecast of durability. The first assumes the presence of a boundary height of the compressed zone ε_R, which for a given reinforcement a priori excludes the possibility of a dual nature [12, 13] of parametric failure. The second postulates the probability of a different type of destruction, due to the statistical laws of the joint distribution of significant indicators of the structural properties of reinforcement and concrete. That is, parametric failures due to the destruction of the compressed zone until the tensile yield of reinforcement is quite likely when reinforcement is significantly lower than the boundary (μ_R). As μ increases, its probability increases, which is confirmed by the dynamics of the failure value of R_{ba}. The need is confirmed for differentiated correction of the design resistance taking into account the established laws and assurance of the necessary level of the design reliability of structure for the period of operation in severe climatic conditions identical to its frost resistance.

The data presented in the table indicate the possibility of achieving parity of R_b(n) and R_{ba} by acceptable reduction of the calculated value by 15-30% (γ_F = 0.85-0.7). Wherein the level of reinforcement of bent elements should not exceed 0.8 μ_R. It is not difficult to verify that ensuring the reliability of more reinforced sections is associated with an economically problematic decrease in the design resistance and calls for an appropriate rationale.

3. Conclusions

Taking into account the physical and statistical laws of frost destruction of concrete and reinforced concrete [5, 7, 14], the temperature of the operating environment (T) and the structural features of elements (μ'), the correction of the criterion equation of bearing capacity by introducing the coefficient of reliability of concrete is proposed in a first approximation.

$$\gamma_{bF} = (1 - 0.0075\text{T})(1.36 - 0.48\mu') \leq 1 \qquad (2)$$

The value of γ_{bF} = 1-0.5 at a design temperature of up to minus 60° C and $\mu' \leq 1.2\mu_R$.

Table1. Changes of failure value R_{ba} in FTC.

n	Distributions statistics of concrete strength		γ_{bF}	Failure value R_{ba} for bent elements with μ' equal to				
	Average R_b(n), MPa	Coefficient of variability C_{vRb}, %		0.181	0.398	0.833	1.087	1.449
0	24.0	12.5	1	7.5	11.9	15.4	15.6	16.0

0.29	24.5	14.2	1	8.1	12.8	16.6	16.8	17.1
			1	11.8	18.1	23.6	23.8	24.3
0.86	20.2	20.7	0.85	-	-	21.0	21.3	22.2
			0.70	-	-	17.0	17.4	19.8
			1	14.5	22.1	28.8	29.6	29.9
1.0	19.7	23.4	0.85	-	20.4	24.8	25.1	27.3
			0.70	-	18.2	20.8	21.7	24.9
			1	20.0	38.6	38.6	39.2	41.6
1.29	19.1	27.0	0.85	17.4	33.4	33.3	35.3	38.2
			0.70	-	29.0	29.0	30.5	34.8

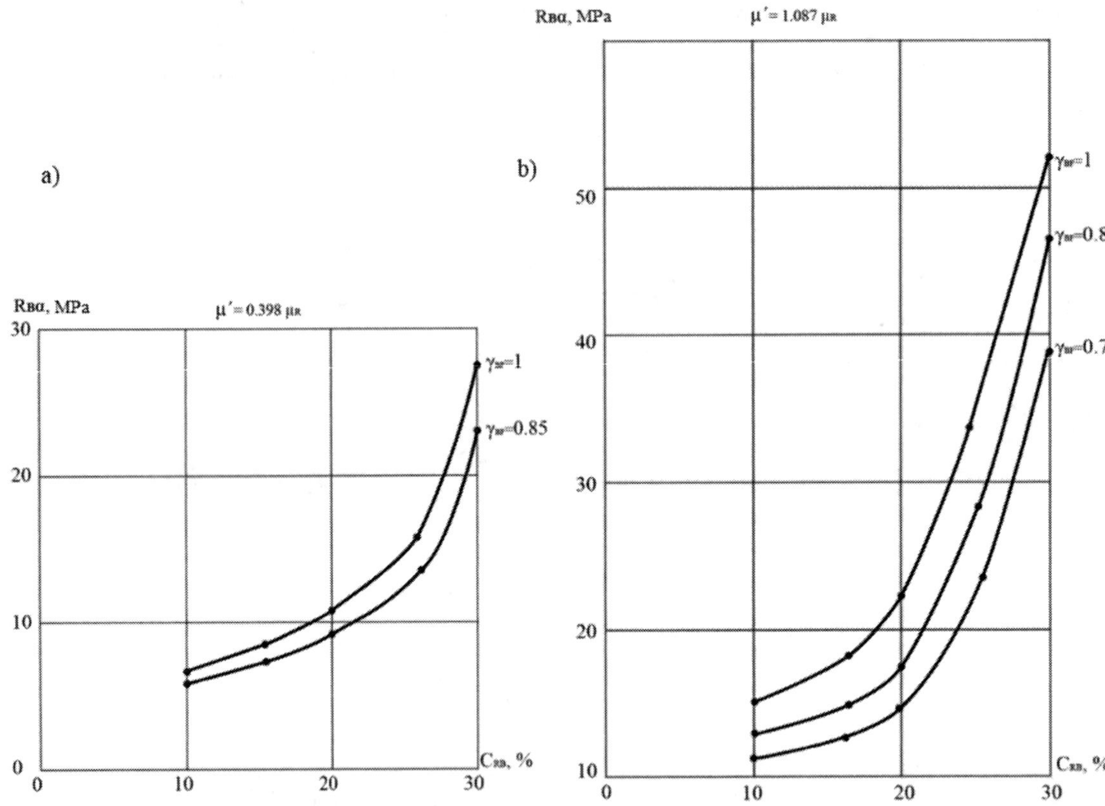

Figure 1. Change $R_{b\alpha}$ in the field of dispersion of coefficient of variability of concrete strength with relative reinforcement a) $\mu' = 0.398\mu_R$; b) $\mu' = 1.087\mu_R$

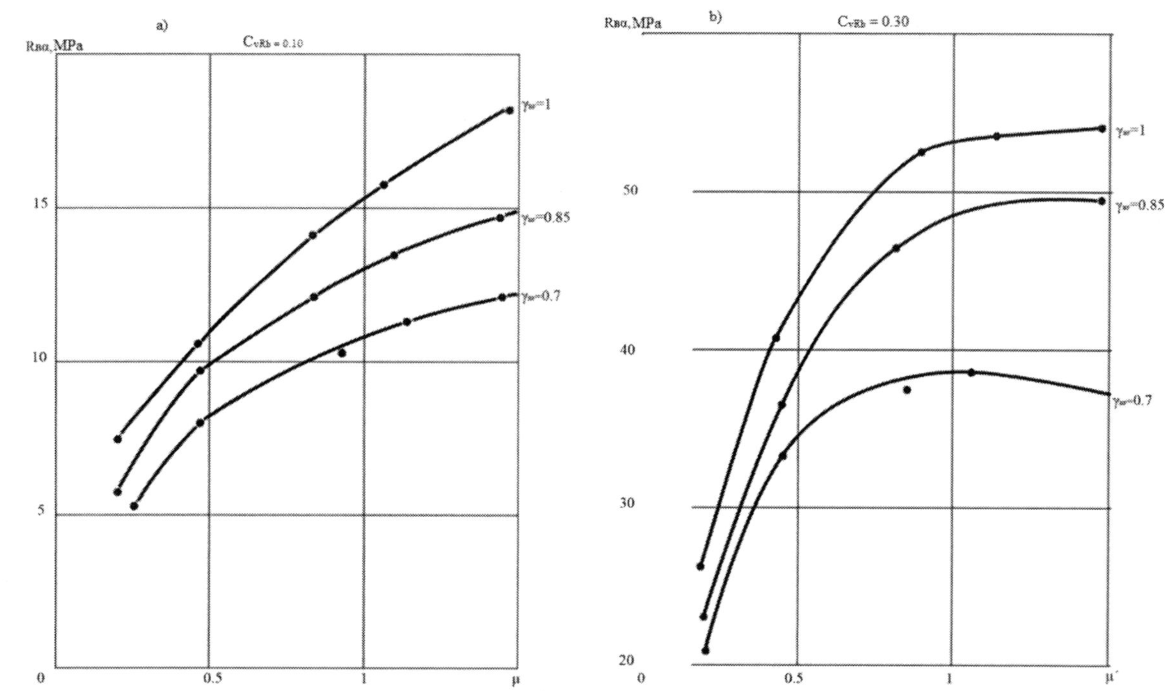

Figure 2. Change in R_{ba} for bent elements of different levels of reinforcement with a coefficient of variability of concrete strength a) $C_{vRb} = 10\%$; b) $C_{vRb} = 30\%$

References

[1] GOST R ISO 2394-2016 *Construction structures. The basic principles of reliability* (Moscow: Standartinform) p 72

[2] GOST 27.201.81 1982 *Reliability in technology. Assessment of reliability indicators with a small number of observations using additional information* (Moscow: Gosstroyizdat) p 27

[3] Recommendations on using additional information for reduction of time of reliability test. 1974 (Gorky: Gorky branch of VNIIMASH) p 141

[4] Karpenko N I 1996 *General models of reinforced concrete mechanics* (Moscow: Stroyzdat) p 416

[5] Aktuganov I Z and Pushkin A A 1982 Experimental verification of the principle of linearity of summation of concrete damage under the impact of non-stationary alternating cycles. *Problems of the reliability of reinforced concrete structures* (Kuibyshev) pp 5-8

[6] Vazhenin V V 1966 On the issue of frost resistance of concrete and reinforced concrete. *Problems of the reliability of reinforced concrete structures* (Chelyabinsk) pp 197-199

[7] Gladkov V S 1972 On the destruction of hydraulic concrete under repeated freezing and thawing in non-stationary mode. *Works of coordination meetings on hydraulic engineering* **73** (Leningrad: Energia) pp 133-142

[8] Pinus B I, Pinus J N and Khomyakova I V 2015 Changing of structural properties of concrete during cooling and freezing. *Bulletin of ISTU* **2** (Irkutsk) pp 111-115

[9] Pinus B I 2018 Consideration of cumulative effects in concrete structures at low temperatures impact. MATEC Web of Conferences **212**.01025 (France: EDP Sciences)

[10] Trofimov B Ya and Pogorelov S N 1986 The study of concrete deformation during cyclic freezing. InterVUZ Bulletin *Performance of building materials under the impact of various operational factors* (Kazan) pp 79-82

[11] Pinus B I and Pinus J N 2018 Probabilistic analysis of the strength and reliability of reinforced concrete structures under low-temperature impact. MATEC Web of conferences **212**.01026. ICRE (France: EDP Sciences)

[12] Mikhailov K V (Ed.) 1977 *New on the strength of reinforced concrete* (Moscow: Stroyizdat) p 272
[13] Gvozdev A A (Ed.) 1978 *New in the design of concrete and reinforced concrete structures* (Moscow) p 202
[14] Pinus B I 1985 *Ensuring the durability of reinforced concrete structures at low temperature impacts*. Dissertation Doctor of Technical Sciences (Moscow) p 367

IOP Conf. Series: Materials Science and Engineering **880** (2020) 012016 doi:10.1088/1757-899X/880/1/012016

Main achievements and directions of scientific research in the field of foundation construction on sites composed of permafrost in the Russian Federation

V E Rozina[1] and A K Komarov[1]

[1]Federal State Budgetary Educational Institution of higher Professional Education INRTU, department "Construction production", 664074, Irkutsk, Lermontov str., 83, Russia

E-mail: vikt.rozina@yandex.ru

Abstract. Construction in the northern construction-climatic zone associated with the solution of complex engineering problems. This is primarily the monitoring of the temperature state of the cryolithozone, which depends on a combination of climatic, engineering-geological, hydrogeological, environmental and technogenic factors. This is even more relevant in connection with the fact of degradation of permafrost. Generalized and classified technical solutions for the construction of foundations of buildings and structures built on the I and II principles. At the same time, technical audit of design solutions and scientific support during the entire life cycle of objects is required. In order to improve and implement the progressive direction in foundation construction, variants of design solutions for self-cooling freestanding columnar foundations of factory production are proposed and the work schedule is detailed.

Permafrost soils in natural conditions are located in the northern regions of the Russian Federation, the United States of America, Canada, partially Scandinavian countries, Greenland, and for high-altitude permafrost island character is located in other countries of the world.

On the territory of Russia, permafrost soils (cryolithozone) occupy 65% of the country's territory.

The temperature of permafrost soils is known to range from -0.5°C to -10°C and depending on this, soils are divided into: low-temperature and high-temperature; hard-frozen (firmly cemented with ice, characterized by brittle fracture and almost incompressible) and plastic-frozen (cemented with ice, but having viscous properties and compressibility under external load).

The temperature state of the cryolithozone is not stable.

Factors that influence the natural environment include: instability of arctic ecosystems, solar radiation, surface albedo, and polar night.

The consequences of human life are influenced on warm-balance characteristics of ground coat layer surface in winter and summer.

The consequences of natural and climatic phenomena characterized by global warming and man-made processes are manifested in the form of degradation of the earth's cryolithozone.

As a result of a comprehensive analysis of data from weather stations and geocryological hospitals, it was found that within a few decades, the border of permafrost can move north by 200-500 km. [1,2]

At the same time, the northern lands (in particular in Russia) contain the main natural resources, the development of which is necessary to ensure the viability of the country's economy.

In all countries, including Russia, two basic principles of the use of permafrost soils as the foundations of buildings and structures are used (in other countries-in a different version and interpretation). [3,4]

I principle-permafrost soils of the base are used in a frozen state, preserved during construction and during the entire period of operation of the structure;

II principle-permafrost soils of the base are used in thawed or thawing condition.

To implement the I principle of construction on permafrost soils (solid-frozen low-temperature soils), the following basic engineering solutions are provided and developed: arrangement of ventilated subfields or cold ground floors of buildings with natural and incentive ventilation; laying in the base of the construction of pipes, boxes with forced ventilation; application of ventilated foundation structures with forced ventilation; the installation of the seasonally operating cooling devices mainly deep action (even in the plastic-frozen soils); the device of heat shields reducing the thermal impact of buildings, structures on the frozen ground.

It is mandatory to take into account the depth of seasonal freezing-thawing of the soil (merging and non-merging permafrost).

The use of the II principle of construction on permafrost soils (and this is usually plastic-frozen high-temperature soils), as well as a fairly complex engineering task and it is connected with indepth termophysical calculation and with observing exact rules of works production.

Technical solutions used in construction practice: preliminary (before the construction of the building) artificial thawing of base soils; replacement of icy soils with thawed non-sedimentary soil (sandy or large-block); increasing the depth of foundations with bear against the rock or other low-compressible soils. [5,6]

Intensive development of the northern regions of Russia has been going on for several decades, and in the last 10-15 years, the state and enterprise investors allocate very large funds for this.

One of the main directions of improving the quality of work in the design, construction and operation of buildings, structures, highways are technical audit of design solutions and scientific support during the entire life cycle of objects. [7,8]

The purpose of the technical audit is to assess the effectiveness of the use of new technologies and equipment in specific conditions for a given production task, competitive evaluation of options for a set of heterogeneous criteria, structural and parametric selection of the best options, assessment of the payback period of investments and risk prediction.

Scientific support includes systematization of scientific research, observations and results of geocryological and geotechnical monitoring, methods of mathematical modeling of thermodynamic and thermal interactions of objects and geotechnical systems with permafrost soils.

A progressive direction for improving construction on permafrost soils is the use of self-cooling support systems-foundations. [9,10]

The thermocouple partially rises above the ground, and there are holes in the structure. Cold air due to convection falls down to the base, transfers the cold to the ground.

The use of self cooling support systems has a number of advantages over other systems:

- combining cooling and supporting structure into a single unit;
- reducing the cost of cooling systems-the absence of special self-cooling devices (SSCD) which are installed separately from the foundations;
- cooling of deep soil layers as opposed to surface cooling;
- higher durability, reliability and aesthetics;
- ecological safety;

In the development of this direction of foundation construction on permafrost soils, the construction of a columnar self-cooling foundation is proposed and is currently being tested.

On Figure 1 the variants of the design solution of self-cooling supports for buildings and structures are given. In this case, it is a free-standing columnar reinforced concrete foundation of factory manufacture. Channels and cavities in the foundation are located in such a way as to provide natural (and if necessary forced) ventilation of the internal surfaces of the structure and mainly the ground surface at the level of the upper boundary of the permafrost layer of the base

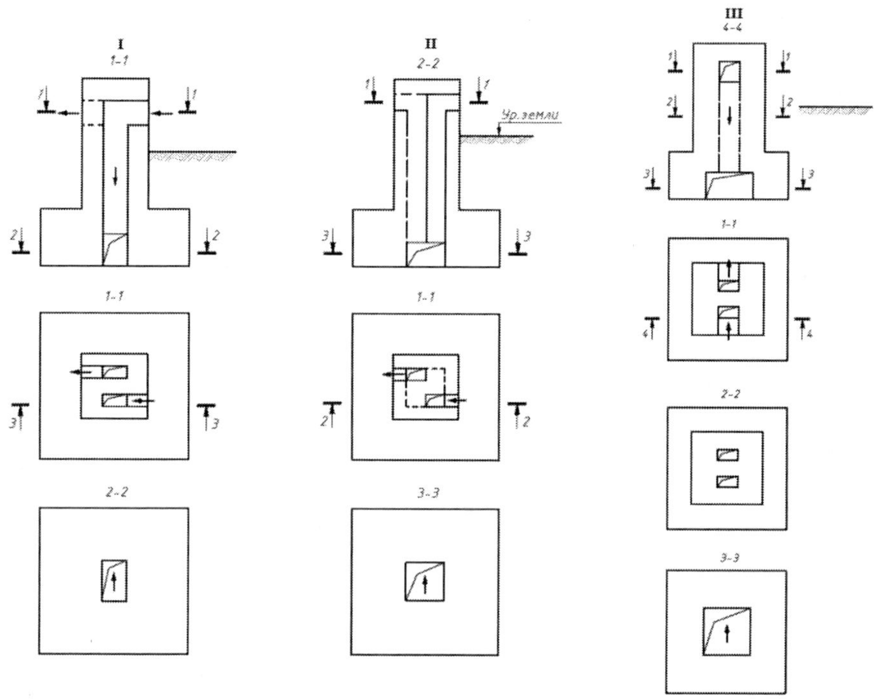

Figure 1. Design options for a self-cooling columnar foundation. (I, II, III)

Installation of such foundations in the design position must be performed at certain times of the year and in strict compliance with the rules of work.

Production time is limited. The start time of the excavation works (i.e. development of pits for free-standing foundations) must match the beginning of the period reduction in average daily outdoor temperature below 0°C. During this period, the soil thawed over the entire depth of the active layer, and the top "crust" is already partially frozen soil (if this is the case) is insignificant and does not preclude the development of soil conventional mechanisms. The pit is developed to the full depth to the level of the foundation sole. The base is prepared in the traditional way. The prefabricated foundation with a factory-applied layer of waterproofing is mounted in the design position.

The channel openings are located above ground level. The foundation is installed in such a way that the inlet and outlet openings of the ventilation channels are located taking into account the direction of the prevailing winds. Further, without interruptions in the production of works, it is necessary to perform backfilling of the pit with soil with a layer-by-layer seal.

The work must be performed as soon as possible, providing in advance the availability of the necessary machines, mechanisms, materials, structures and labor resources.

This technology will avoid excessive thermal effects on the permafrost soil and after completion of work to ensure the penetration of cold air inside the foundation to cool the base.

During the construction of both the I and II principles there is a task to improve the system of temperature monitoring of permafrost soils of the base of buildings and structures under construction.

It is necessary to control, and sometime to operate a temperature regime of soils in the course of operation of objects.

A modern measuring system allows monitoring temperature conditions continuously during the operation of objects. The set of equipment includes: thermocos installed in thermometric wells; controllers; loggers; personal computers for recording and processing data. The equipment works continuously for several years.

The measured temperature values from the thermocos sensors are recorded on a MicroSD memory card. Data collection is carried out by copying a data file using a communication line to a personal computer in the form of an archive over a radio signal.

Thus, as a result of many years of fundamental research in the Russian Federation, almost any engineering problems in construction in the northern construction and climate zone are solved. Self-cooling free-standing columnar foundations of factory manufacture (Pic.1) designed for construction on the principle of I – one of the most effective areas of foundation construction

References

[1] OJSC "Fundamentproekt", Moscow, Comprehensive studies of the properties of frozen soils in the study of large infrastructure projects International Association of Foundation Builders *Collection of reports of the international scientific and technical conference "Technologies for the design and construction of foundations on permafrost soils"* October 16-17, 2014, Moscow, All-Russian Exhibition Center.

[2] Simmakers LLC, Moscow, Forecasting Permafrost Thawing in the Frost 3D Universal Program// International Association of Foundation Builders *Collection of reports of the international scientific and technical conference "Technologies for the design and construction of foundations on permafrost soils"* October 16-17, 2014, Moscow, All-Russian Exhibition Center

[3] SP 25.13330.2012 Foundations and foundations on permafrost soils. Updated edition of SNiP 2.02.04-88

[4] GOST 20276-2012 Soils. Field methods for characterizing strength and deformability

[5] LLC Scientific-Production Firm "Dorcentr", Tyumen, Analysis of trends in the development of innovative technologies, structures and materials used in the design, construction and operation of facilities in the permafrost zone International Association of Foundation Builders *Collection of reports of the international scientific and technical conference "Technologies for the design and construction of foundations on permafrost soils"* October 16-17, 2014, Moscow, All-Russian Exhibition Center

[6] Makarov A V, Zhuravlev A V, Tian V Yu 2019 Features of the construction of foundations in permafrost soils *Engineering Bulletin of the Don* **1**

[7] OJSC NPP Etalon, Omsk, System of temperature monitoring of permafrost soils with data transmission over the air International Association of Foundation Builders *Collection of reports of the international scientific and technical conference "Technologies for the design and construction of foundations on permafrost soils"* October 16-17, 2014, Moscow, All-Russian Exhibition Center

[8] Simmakers LLC, Moscow, Innovations in the mathematical modeling of thermal regimes of permafrost Soils nternational Association of Foundation Builders *Collection of reports of the international scientific and technical conference "Technologies for the design and construction of foundations on permafrost soils"* October 16-17, 2014, Moscow, All-Russian Exhibition Center

[9] Scientific Research Institute of Transport Construction (TsNIIS) OJSC, Moscow, Self-cooling permafrost support systems International Association of Foundation Builders *Collection of reports of the international scientific and technical conference "Technologies for the design and construction of foundations on permafrost soils"* October 16-17, 2014, Moscow, All-Russian Exhibition Center

[10] Andreev V S, Nabokov A V, Passek V V 2017Development of a complex of self-cooling support systems as a fundamentally new direction in the construction of bridges and roads on permafrost *Prospects for Science* **4 (91)** p 18

IOP Conf. Series: Materials Science and Engineering **880** (2020) 012017 doi:10.1088/1757-899X/880/1/012017

Polymerasphaltconcrete together with the use of modified effective components

V E Rozina[1] N I Shestakov[2] and A K Komarov[1]

[1] Irkutsk National Research Technical University, 83 Lermontov str., Irkutsk, 664074, Russia
[2] National research Moscow state University of civil engineering, department of building materials and materials science, 129337, Moscow, Yaroslavl highway, 26, Russia

E-mail: vikt.rozina@yandex.ru

Abstract. In this paper, we consider the possibility of joint use of polystyrene, its block copolymers and transformer oil to produce polymerasphaltconcrete. It is established that the properties of bitumen change when different polymers and their quantity are introduced into its composition. It was found that the introduction of polystyrene together with a plasticizer expands the temperature range of operation, which leads to an improvement in the physical-mechanical and hydrophysical properties of polymerasphaltconcrete. The regularities of the influence of the composition and quantity of polymers on the physical, mechanical and operational properties of asphalt concrete based on modified binders are revealed. It was found that the optimal amount of polymers and plasticizer is 3.5% and 10% of the mass of bitumen.

The road surface must provide maximum resistance to fatigue damage and be resistant to changes in temperature of daily and seasonal cycles. One of the promising areas that allow us to solve these problems is the use of bitumens together with plasticizing polymers (PBB). The cost of PBB is higher than ordinary bitumen by an average of 60-65 %, but roads are not built only from bitumen, the share of bitumen in asphalt concrete is only 6 %, calculations show that the increase in the cost of building 1 km of road is about one percent [1-4]. Given the increase in the service life of the road in two or three times, the use of PBB is certainly economically justified. Thus, an urgent task is to develop a technology for the production of polymerasphaltconcrete based on a polymer-bitumen binder obtained from ground polystyrene and transformer oil-processing.

The most important advantage of the polymer-bitumen binder (hereinafter referred to as PBB) is the ability to regulate their properties due to the ratio of the polymer content and the plasticizer, which creates a quality material suitable in any climatic conditions for different driving conditions. The most popular modifiers are thermoplastic elastomers of the SBS type: Kraton D (company KratonPolimers), Calprene (Dynasol company) and others. Along with them, synthetic latexes of the Butonal NS series (BASF company) and thermopolymers of the Elvaloy series (du Pont company) are widely used. These polymers reduce the sensitivity of bitumen to temperature changes, increase the cohesive strength and heat resistance of binders, give them elasticity, and improve their behavior at low temperatures. The current trend in Russia, which is to reduce the consumption of expensive polymers for modifying road bitumen, necessitates an in-depth study of bitumen-polymer binders. At the same time, a significant scientific and practical interest is represented by BPB prepared on bitumens of different brands. When developing PBB compositions, it is necessary to carefully choose the combination of the "bitumen — polymer — plasticizer" system, since there is a possibility of their incompatibility, or use special techniques aimed at reducing their competing ability. [5-7]

There are certain requirements for polymers that determine the effectiveness of their use for modifying bitumens. Thus, the macromolecules of the polymer must have a tendency to association and the ability to immobilize the largest possible volume of the dispersion medium. Also, the polymer must be quickly and well distributed in the dispersion medium of bitumen without destruction, forming a structural grid in the bitumen that retains strength at high and elasticity at low temperatures. The availability and cheapness of the polymer is also important. In this regard, it is very important to study the use of polystyrene as a polymer and oil-processing of transformer plants as a plasticizer of the medium.

The structure of polymers is shown in pictures 1-3.

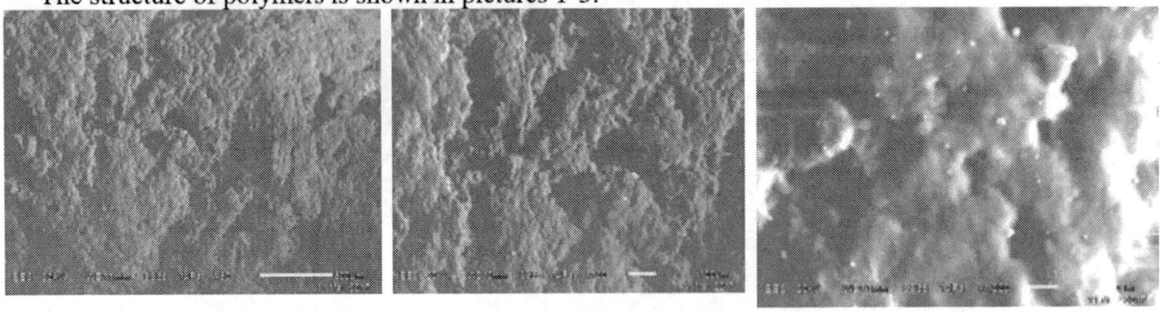

A) 50x magnification *B) 100x magnification* *C) 1000x magnification*

Picture 1. Surface structure of styrene-butadiene-styrene SBS

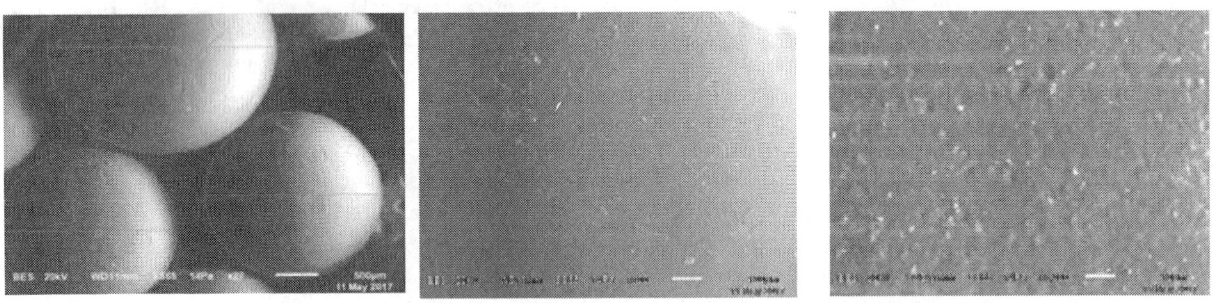

A) 50x magnification *B) 100x magnification* *C) 1000x magnification*

Picture 2. Surface structure of bead styrene before grinding

The surface of beaded styrene is expressed by clear spherical and low-porosity structures. Such surfaces are most often inactive to sorption processes with organic and organic substances, and adhesive forces can only be formed from the conditions of chemical activity and chemical-sorption capabilities of the surface. Therefore, to ensure the conditions for the formation of a homogeneous binder composite, it is necessary to change the surface type of the structure or its dispersion. In this case, when working with the bitumen melt, the polymer will be more active, which will lead to an increase in the quality characteristics of the resulting PBB. For further work with polymer, it produced its disintegration in vibroerase. The surface structure is shown in picture 3.

A) 50x magnification *B) 100x magnification* *C) 1000x magnification*

Picture 3. Surface structure of ground styrene

From numerous foreign data, it is known that when polymers are introduced into bitumen without plasticizers, at least 5-6% of the polymer by weight is required to obtain PBB with optimal properties. In this case, the viscosity of the resulting binder is significantly higher than the viscosity of bitumen, which can lead to technological difficulties in the preparation of asphalt concrete mixtures at asphalt plants. Introduction of the plasticizer allows to provide the required temperature regime (no higher than 160°C) and significantly increase the efficiency of the polymer being introduced, i.e. obtain a PBB with a developed spatial structural grid with a minimum polymer content of 2-2.5%, and exclude a colloidal mill from the necessary equipment set.

The selection of prototypes of polymer-asphalt-concrete mixtures for the I-th road-climatic zones carried out in a continuous grading of the grain composition according to the requirements of GOST 9128-2013 "asphalt concrete mixture, polymerasphaltconcrete, polymerasphaltconcrete for roads and airfields. Technical conditions". Preparation and testing of polymerasphaltconcrete samples was performed in accordance with GOST 12801-98 "Materials based on organic binders for road and airfield construction".

PBB is a three-component system, so changing the quality of one of the components inevitably leads to a change in the quality of the PBB itself. Since all components are manufactured on an industrial scale, changes in their performance within the specified brands, of course, take place.

To determine changes in the properties of PBB with different polymer and plasticizer compositions, the penetration and softening temperature of the initial bitumen and the resulting PBB of different compositions were determined at the first stage. Variants of compositions are presented in table 1.

Table 1. Component variants of PBB compositions based on BND 90/130 bitumen

№ composition's	1	2	3	4	5	6	7	8	9	10	11	12	13	14	15	16	17	18
Polystyrene, %	2	3,5	5	-	-	-	2	3,5	5	-	-	-	2	3,5	5	-	-	-
Styrene-butadiene-styrene, %	-	-	-	2	3,5	5	-	-	-	2	3,5	5	-	-	-	2	3,5	5
Transformer oil, %	5	5	5	5	5	5	10	10	10	10	10	10	15	15	15	15	15	15

The depth of penetration of the needle at 25°C characterizes the plasticity and viscosity of the binder, its technological properties, and therefore, indirectly, the workability of polymerasphaltconcrete mixtures. Plasticity increases with an increase in the plasticizer content, and with an increase in the polymer content in PBB, this effect is noticeably less pronounced. It is important to note that if the oil content in the PBB is less than 5%, and even more so without it, the plasticity of the PBB is significantly reduced. Therefore, with a low content of plasticizer in PBB, you should be afraid of a sharp deterioration in the technological properties of polymerasphaltconcrete mixtures, namely, workability and compaction, as well as a decrease in labor productivity at asphalt plants in their manufacture due to the high viscosity of the binder.

The softening temperature is the most important operational indicator and characterizes the heat resistance of bitumens and their transition from the elastic-plastic rheological state to the viscous one, which is characterized by the absence of a spatial structural grid in the binder. The results of determining the softening temperature of the obtained PBB are presented in table 2.

Table 2. The softening temperature of the PBB

№ composition's	Softening temperature, °C	Needle penetration depth, 0.1 mm, at a temperature of 25°C	№ composition's	Softening temperature, °C	Needle penetration depth, 0.1 mm, at a temperature of 25°C
1	44,3	111	10	34,8	147
2	46,9	88	11	36,4	112
3	48,3	69	12	39,2	93
4	43,1	116	13	33,5	247
5	45,8	91	14	35,3	131
6	48,1	72	15	37,0	118
7	36,1	141	16	32,8	258
8	38,9	110	17	34,3	137
9	40,1	88	18	36,8	122

According to the results obtained, it is obvious that the softening temperature changes with an increase in the plasticizer content when polystyrene is introduced into the bitumen in an amount of 2% in the same way as when styrene-butadiene-styrene is introduced into the bitumen. As the results of the study show, the introduction of polymers and plasticizers changes the technological properties and composition of bitumen, which should lead to changes in the physical, mechanical and operational properties of polymerasphaltconcrete. The results of determining the basic and hydrophysical properties are presented in table 3.

Table 3. Basic and hydrophysical properties of polymerasphaltconcrete based on various polymers and 10% plasticizer

Composition of the binder / Indicators	BND 90/130 Source	Polystyrene, mass.%			Styrene-butadiene-styrene, mass.%		
		2	3,5	5	2	3,5	5
The porosity of the mineral skeleton, %	19,8	19,6	19,0	18,7	19,3	19,1	18,8
Residual porosity, %	4,01	3,69	3,57	3,40	3,66	3,61	3,38
Water saturation, %	2,79	2,11	1,97	1,92	2,08	2,01	1,98
Swelling, %	3,56	3,41	3,40	3,11	3,50	3,21	3,07
The coefficient of water resistance	0,95	0,96	0,98	1,01	0,96	0,96	0,99

According to the obtained data, it was found that the introduction of the polymer and the plasticizer has a positive effect on hydrophysical properties resulting polymerisation. The main requirement for polymerasphaltconcrete: they must have high structural and mechanical characteristics that determine their durability and reliable performance in engineering design. This can be achieved by bringing their strength as close as possible to that of a defect-free body.

Table 4. Strength characteristics of the obtained polymerasphaltconcretes

Composition of the binder / The limit of compressive strength	BND 90/130 Source	Polystyrene, mass.%			Polystyrene, mass.%		
		2	3,5	5	2	3,5	5
At t=50°C, MPa	1,0	1,1	1,3	1,3	1,1	1,2	1,3
At t=20°C, MPa	2,4	2,6	2,8	3,0	2,8	2,9	3,2
At t=0°C, MPa	7,4	7,6	7,7	7,9	7,8	7,8	7,6

Analyzing the physical and mechanical properties of polymerasphaltconcrete, it should be noted that the mechanical properties of polymerasphaltconcretes with the content of plasticizer in PBB are slightly higher than the standard values.

The dependence of the strength limit of asphalt concrete under compression at O, 20 and 50°C on the content of the plasticizer in the PBB of the studied polymerasphaltconcrete mixtures meet the requirements of GOST 9128-2013.

Thus, the use of polymer and plasticizer significantly alter the physical and mechanical properties of the PBB by increasing the temperature range of health binder, the work of adhesion of bitumen to mineral materials. The polymer and plasticizer content of 3.5 and 10%, respectively, should be considered the optimal amount for obtaining PBB. The developed compositions of polymerasphaltconcrete using different polymers and different amounts of them are very effective, and the introduction of ground polystyrene together with a plasticizer, leads to an improvement in the strength of asphalt concrete at 20° and 50° C, increasing water resistance and shear resistance.

References

[1] Pokrovsky A V 2014 A brief review of the experience of using cast polymerasphaltconcrete on artificial structures in the North-Western region of the Russian Federation *Online journal of science* **5 (24)**

[2] Gokhman L M 2017 Substantiation of regulatory requirements for polymerasphaltconcrete according to GOST 9128-2013 *Bulletin of the KNARU* **79**

[3] Balabanov V B, Romanov A V and Klimentyev I M 2013 Polymer-bitumen binders, plasticized waste automotive oils *Bulletin of the INRTU* **6 (77)**

[4] Akhunova R R, Biglova R Z, Zadkin M A, Talipov R F, Mustafin A G 2012 Butadiene-α-methyl styrene copolymer-modifying additive in the composition of road bitumen *Bulletin of the Bashkir University* **№3**

[5] Lebedeva K Yu 2013 Research of physical and mechanical characteristics of bituminous-rubber composite binders *JSRP* **2**

[6] Mullahmetov N R, Kemalov A F, Kemalov R A and Kostromin R N 2010 Modification of road bitumen with rubber *Bulletin of the Kazan technological University* **7**

[7] Doshlov O I and Speshilov E G 2013 Polymer-bituminous binder – a high-tech basis for a new generation of asphalt *Bulletin of the INRTU* **6 (77)**

[8] Kazimirov I A and Peshkov V V 2019 Determination of price behaviour in the secondary

residential real estate market using a multidimensional regression model *Proceedings of Universities. Investment. Construction. Real estate* **9(3)** pp 476–487 DOI: 10.21285/2227-2917-2019-3-476-487

[9] Matveeva M 2018 Synchronization of the activity of participants during the implementation of housing construction projects *Proceedings of Universities. Investment. Construction. Real estate* **8** pp 31-41 DOI: 10.21285/2227-2917-2018-2-31-41.

[10] Larina O P and Khalabuda E I 2019 Accident prevention during construction and installation works at height *Proceedings of Universities. Investment. Construction. Real estate* **9(4)** pp 734–741 https://doi.org/10.21285/2227-2917-2019-4-734-741

[11] Selezneva O I, Radaev S S and Baranova A A The impact of thermal treatment on the granulometric and chemical composition of diatomite of the Kamyshlov field when obtaininga liquid-glass binder *Investment. Construction. Real estate* **9(3)** pp 550–555

[12] Tyuryukhanov K Yu and Pugin K G 2019 Impact of the surface of particles of moulding sand on the structural formation of asphalt *concrete Proceedings of Universities. Investment. Construction. Real estate* **9(3)** pp 566–577

[13] Dobruskina M A, Petrov A V and Bat-Erdene Z 2019 Improving the technology of retaining walls in the Irkutsk region using gabion baskets *Proceedings of Universities. Investment. Construction. Real estate* **9 (2)** pp 312–323

IOP Conf. Series: Materials Science and Engineering **880** (2020) 012018 doi:10.1088/1757-899X/880/1/012018

Construction and technical examination of wooden structures (on the example of a wooden country house)

E V Samoilenko[1] **and V V Peshkov**[2]

[1]Federal budget institution Irkutsk Laboratory of forensic examinations, Krasnokazachya, 131, Irkutsk, 664081
[2]Irkutsk National Research Technical University, 83 Lermontov str., Irkutsk, 664074, Russia

E-mail: catherine.samojlencko@yandex.ru

Abstract. In practice, the construction of wooden buildings of low-rise residential construction is carried out based on the experience and knowledge of the builders, and does not agree with the legal framework of the construction industry. Therefore, many low-rise and garden houses are typical, unequal spacing of the beams, overestimated the distance between the bearing walls, the construction of attic floors without regard to bearing loads of the lower floors, uneven box joints timber, use of faulty material or materials from different batches with different parameters and characteristics. All these violations in the construction and operation of objects made of wooden structures lead to civil disputes in the framework of which it is necessary to conduct construction and technical expertise.

1. Introduction

Wood is the oldest material used by man in a variety of industries. Its strength, hardness, wear resistance and at the same time good deformability, elasticity, water absorption and water repulsion, hygroscopicity, external properties (texture, texture, gloss, color), dielectric qualities, anisotropy make wood a unique and versatile material.

Russia is the richest country in the world in terms of timber reserves, and the lands of the forest fund occupy the largest area in comparison with lands of other categories, which favorably affects the popularity of the material during construction. According to **BR 55.13330.2016** «Residential single-family houses»: «it is possible to carry out the construction and reconstruction of residential single-family houses with the number of ground floors of no more than three» [15]. Therefore, wooden structures are widespread in the construction of cottages, private and country houses.

The legal side has a particular impact on the construction industry, namely obtaining a building permit. The new edition of the **Urban Planning Code of the Russian Federation (GRK RF)** establishes part 3. Article 48, the following: «Implementation of the preparation of project documentation is not required during the construction, reconstruction of an individual housing project, a garden house» [18]. In the event that the construction of townhouses or houses with a storey of more than three floors is carried out, the owner's passport, a document establishing ownership of the land and an application for permission to build, must add project documentation, in accordance with Decree of the **Government of the Russian Federation of 16.02.2008 No. 87** «On Composition sections of project documentation and requirements for their content" [19].

2. Main defects and damages of wooden structures

All defects of wooden structures arise due to: the use of low-quality materials, improper construction and operation

2.1. Vice of wood

Taken into account during construction and technical examination, stated in **GOST 2140-81**, divided into groups types and varieties:

- Knots branches (open and overgrown are divided according to the shape of the cuts, by position and relative position, by condition, by exit to the surface);
- Cracks (by type, position, depth and width);
- Defects in the shape of the tree trunk (stamina, rounded or ribbed lumps, ovality, growth, curvature simple or complex);
- Wood structure defects (fiber inclination, roll, traction wood, sagging, curl, eyes, pockets, core, germination, pseudo-nucleus, spotting, tar, etc.);
- Chemical color spots (by type and color intensity);
- Mushroom lesions (mushroom spots, mold, saponified mushroom stains, browning, rot, hollow);
- Biological damage (wormhole, parasitic plants, from birds);
- Foreign inclusions, mechanical damage (foreign inclusions, carbonization, peeling of the bark, cut bevels, pickle, risks, waviness, hairiness, notch, washed down, chips, burr, dents, etc.);
- Warp (along the seam and along the edge, transverse, winged) [4].

2.2. Defects and damages of wooden structures

The most common practice, obtained during the operation and construction, in the course of lectures **E.G. Zhuravleva** shows: rot, wall roll, lower crown crowns, bulging, crevices on horizontal joints, fixing crowns with nails, hanging upper crowns on window frames, weathering of the sealant from the joints, etc. [7].

3. Subject of civil disputes

The most frustrated in practice during the construction and technical examination comes down to establishing:

- compliance of the project data with the requirements of regulatory and technical documentation, and with the provisions of the contract (part 1 of article 754, part 3 of article 744, part 4 of article 744, part 1 of article 761 of the **Civil Code of the RF** [22]);
- causes of destruction: complete or partial loss of the functional, operational and other properties of the object (part 1, part 2 of article 741; part 2 of article 745; part 2 of article 748; part 2 of article 351; part 3 , part 4 of article 744; part 3 of article 753; part 1 of page 761; part 4 of article 743; part 2 of article 1065; part 1 of article 1079; article 1100 of the **Civil Code of the RF** [22] and etc.);
- compliance of types, volumes, quality and cost of work, requirements of normative and technical documentation, project provisions, contract terms (part 1 of article 754; part 2 of article 748; part 3 of article 753 of the **Civil Code of the RF** [22]);
- the possibility of using residential buildings for living (articles 166-170, part 6 of article 753, part 1 of article 222 of the **Civil Code of the RF** [22]);
- the presence (absence) of defects in the supporting structures, if any, the causes of the appearance and development of defects (part 2 of article 9.4 of the **Code of Administrative Offenses** [1]; part 1 of article 741 of the **Civil Code of the RF** [22]), their belonging to the category of hidden or explicit (166-170 of the **Civil Code of the RF**) defects, the possibility and cost of eliminating them (part 6 of article 753 of the **Civil Code**

of the RF [22], Article 5 of the **Law of the Russian Federation** «On Protection of Consumer Rights» [21]), etc.

4. The regulatory framework

In the production of construction and technical examinations of wooden low-rise buildings, it consists of the requirements:

- **GOST 11047-90** «Details and wooden products for low-rise residential and public buildings. Technical conditions» [3];
- **GOST 2140-81** (ST SEV 2017-79, ST SEV 2018-79, ST SEV 2019-79, ST SEV 320-76, ST SEV 321-76, ST SEV 391-76, ST SEV 3286-81, ST SEV 3287-81, ...) «Visible defects of wood. Classification, terms and definitions » [4];
- **GOST 30974-2002** «Corner joints of wooden block and log low-rise buildings. Classification, design, dimensions» [5];
- **BR 131.13330.2018** «Construction climatology. Updated edition of SNiP 23-01-99 *» [8];
- **BR 14.13330.2018** «Construction in seismic areas. Updated edition of SNiP II-7-81 *» [10];
- **BR 17.13330.2017** «Roofing. Updated edition of SNiP II-26-76» [11];
- **BR 20.13330.2016** «Loads and impacts. Updated edition of SNiP 2.01.07-85 *»[12];
- **BR 30-102-99** «Planning and development of territories of low-rise housing construction» [13];
- **BR 42.13330.2016** «Urban planning. Planning and development of urban and rural settlements. Updated edition of SniP 2.07.01-89 *»[14];
- **BR 55.13330.2016** «Residential single-family houses. Updated edition of SniP 31-02-2001» [15];
- **BR 64.13330.2017** «Wooden structures. Updated edition of SNiP II-25-80» [16];
- **BR 70.13330.2012** «Bearing and enclosing structures. Updated edition of SNiP 03.03.01-87» [17];
- **Federal Law of July 22, 2008 N 123-ФЗ** (as amended of December 27, 2018) «Technical Regulation on Fire Safety Requirements» **[21] and others.**

5. Production of construction and technical expertise of wooden structures using the example of a country house

Federal budget institution Irkutsk Laboratory of forensic examinations, in the last two years, has carried out several construction and technical examinations of wooden structures for the purpose of ships and investigating authorities. Violations of the erection of log cabins, country houses, wooden roofs are detected during the operation of the facility and lead the facility to an emergency state.

In the course of the construction and technical examination in civil matters, a number of discrepancies were identified: volumetric planning and design decisions specified in the contract and the builder's estimate. The construction of the object was carried out according to the "Design project" developed by the construction company, without the accompanying justification of space-planning and structural solutions.

Inspection and inspection of the facility was carried out in accordance with **GOST 31937-2011** «Buildings and structures. Rules for the inspection and monitoring of technical condition» [6] and **BR 13-102-2003** «Rules for the inspection of load-bearing building structures of buildings and structures» [9].

Actual volumetric - planning and design solutions identified during the inspection: 1 storey building (7.0x8.0m and 2.7m high) with an attic floor of the same area, 2.58m high. and an attached garage (3.9x6m). The beam of the common wall of the garage and the house with a section of 100x180mm differs from the dimensions of the main load-bearing walls of the house and the garage 180x180mm. The internal walls in the house are also made of 100x180mm timber. Interventional insulation - jute. Ceilings - planks on wooden beams of timber. The roof is gable (Broken ramp), rafter legs from a wooden board. Roofing - Profiled steel sheets. Drainage - organized from the northwest slope of the roof along the suspended gutters made of PVC; unorganized from the southeast slope. Door and window openings - windows in wooden frames with single glazing. Door and window openings are made of 40x180mm boards. Foundation - Screw piles, Structural diagram: 1 floor: frameless with longitudinal and transverse load-bearing walls made of timber; Attic floor: frame of wooden boards.

The erected property does not comply with the requirements of building codes and the mandatory requirements in the field of construction specified in the «Regulatory Base» section for the following violations:

- Deviations of building structures exceed the maximum deviations: - Deviation of the edges of the walls of the walls from the vertical by 1 m. Length and walls of the partitions from the vertical by 1 m. Height to 17 mm - non-compliance with clause 8.1.7. 1. **BR 70.13330.2012** «Bearing and enclosing structures. Updated edition of SNiP 3.03.01-87 » [17].
- Longitudinal deep cracks were recorded on the surface of the beam with a disclosure width of up to 9 mm., Blunt obzol and violation of the integrity of the beam - extrusion, cleavage. - non-compliance with table No. 1 **GOST 11047-90** «Details and wooden products for low-rise residential and public buildings. Technical conditions» [3]. Along the perimeter of the property there are numerous areas between the bars with cracks..
- From the attic side, natural ventilation in the roof is not provided through openings. - non-compliance with clause 4.4. **BR 17.13330.2017** «Roofs. Updated edition of SNiP II-26-76» [11].
- Snow retention devices are not provided in the object under study - does not comply with clause 6.4.4.5 and clause 6.4.4.7. 3.**BR 17.13330.2017** «Roofs. Updated edition of SNiP II-26-76» [11].
- The partition in axes 2-3 / AB is attached to the load-bearing wall along the 3 axis end-to-end with fastening with wooden dowels, which does not exclude the possibility of transferring to them horizontal loads acting in their plane - does not correspond - clause 6.5.2. **BR 14.13330.2018** «Construction in seismic areas. Updated edition of SNiP II-7-81 *» [10].
- The spatial rigidity of the rafter system is not provided, since there are no horizontal connections between the vertical frames in the direction parallel to the ridge - does not comply with clause 9.6. **BR 64.13330.2017** «Wooden structures. Updated edition of SNiP II-25-80» [16].

Identified inconsistencies of the constructed property, it is necessary not only to indicate how the actual violations, but also to justify them with the necessary calculation methods developed by construction and technical mechanics.

A.Yu. Butyrin highlighted the most common: «methods for determining internal stresses in parts of structures from external loads, temperature effects, etc., methods for determining deformations, studying stability conditions, studying various changes in deformations during long-term operation of structures» [2,91].

During the construction and technical examination of a country house located in the Irkutsk region, calculations were made in accordance with the requirement of **BR 64.13330.2017**, clause 4.3. «According to the bearing capacity (1st group of limiting states) and deformations that do not interfere with normal operation (2nd group of limiting states), taking into account the nature and duration of the load» [11]. During the calculation, the formulas and coefficients of SP 20.13330.2016 «Loads and impacts» [12] and **BR 64.13330.2017** «Wooden structures" were used» [11]. The calculations were carried out according to the measured drawings built in the AutoCAD software package. During the calculation, the loads from all the beams of the first and attic floors were determined and it was revealed that:

- The load-bearing capacity of the first and second groups of limiting states of ordinary beams of the interfloor overlapping with a span of 4.24 m, taking into account the load from the roof, is not provided - does not meet the requirements of **BR 20.13330.2016** «Loads and impacts» [12].

- The load-bearing capacity of the main beam of the interfloor overlap in the first group of limit states is not provided - does not meet the requirements of **BR 20.13330.2016** «Loads and impact» [12].

All identified defects and construction defects of the wooden structure of the country house were summarized in table. 1:

Table 1. Description of materiality and disposability, as well as ways to eliminate the shortcomings of a country house

Construction flaws	Material / non-material Disposable / unrecoverable
The deviation of the faces of the wall crowns from the vertical by 1 m. Of length and the walls of the partitions from the vertical by 1 m. Of height to 17 mm.	A significant drawback. Disposable flaw. Dismantling / installation of walls and partitions with jacks, winches, and a truck (as an emphasis), with the partial addition of new material. The fastening of the timber is recommended to be performed with the help of pins. With a survey, the identified areas should be cleaned.
Longitudinal deep cracks on the surface of the beam with an opening width of up to 9 mm., Blunt obzol and violation of the integrity of the beam - extrusion, cleavage.	Unacceptable wood defects according to **GOST 11047-90** [3]. We will eliminate the means of dismantling / installation of walls and partitions with the partial addition of new material.
Along the perimeter of the property there are numerous areas between the bars with cracks.	A significant drawback. Disposable flaw. To eliminate this drawback, it is necessary to pierce the gaps and gaps.
From the attic side, natural ventilation is not provided through openings in the roof.	A significant drawback. Disposable flaw. The correct technological device of ventilation grilles in the gables of the attic.
The object under study does not provide snow-holding devices.	Not a significant drawback. Disposable flaw. The device of snow-holding devices on the roof
The partition in axes 2-3 / AB is attached to the load-bearing wall along the 3 axis end-to-end with fastening with wooden dowels, which does not exclude the possibility of transferring to them horizontal loads acting in their plane - does not correspond - clause 6.5.2. **BR 14.13330.2018** «Construction in seismic areas. Updated edition of SNiP II-7-81 *» [10].	A significant drawback. Disposable flaw. Dismantling / installation of the partition, followed by its fastening with flexible connections, on the basis of **GOST 30974-2002** «Corner joints of wooden block and timbered low-rise buildings. Classification, design, dimensions» [5].
The spatial rigidity of the rafter system is not provided, since there are no horizontal	A significant drawback. Disposable flaw.

connections between the vertical frames in the direction parallel to the ridge.	Installation of wooden girders in the nodes of vertical frames, as well as the installation of wooden braces between the uprights of the frames in the longitudinal direction.
The load-bearing capacity in the first and second group of limiting states of the ordinary beams of the floor beams with a span of 4.24 m, taking into account the load from the roof, is not provided.	A significant drawback. Disposable flaw. To exclude the transfer of load from the roof to the floor beams by dismantling the central pillars, by installing struts with the vertical frame (tightening) resting on the upper belt. The upper belt should be reinforced. Dismantling racks above the beams of the garage, installing a strut with support on a load-bearing wall.
The bearing capacity of the main beam of the interfloor overlap in the first group of limit states is not provided.	A significant drawback. Disposable flaw. Strengthening the supporting section of the main beam by installing steel corners on three sides that distribute the load on the walls.

The causes of all the shortcomings, except for deviations of the faces of the wall crowns from the vertical, are a violation of the technological process during construction. It is impossible to determine the reason for the deviations of building structures exceeding the maximum deviations, because there can be many reasons, for example: non-compliance with the design, design errors, violation of the process during construction, weather and climatic conditions (when one side of the house warms up better than the other, it sits faster).

6. Conclusion

On the example of the construction and technical examination of a country house located in the Irkutsk Region, a mismatch of the contract, technical specifications and estimates with the actual work performed was revealed, which is a common feature of low-rise construction as a whole.

The defects and damages considered within the framework of the requirements of regulatory documents revealed significant violations of the technological process during construction. All defects identified as a result of an expert diagnostic examination must be eliminated in accordance with the regulatory and technical requirements in force in the Russian Federation, but it is impossible to rectify the identified defects of the property without dismantling the main load-bearing structures of the house, followed by determining the suitability of the building material for further use - i.e. . definition of quality.

In the process of conducting an examination, a shortcoming of the lack of design documentation was found, as a justification of the space-planning and structural solutions of the construction site. Violation of the technological process during construction, justified by the lack of accounting for design loads, led to the emergency state of the facility.

References

[1] «Code of Administrative Offenses of the Russian Federation» dated 30.12.2001 N 195 -FL (ред. от 06.02.2019) URL: http://www.consultant.ru/document/cons_doc_LAW_34661/ (as amended on 21.03.2019).

[2] Butyrin A.Yu. Theory and practice of judicial construction and technical expertise. - M .: Gorodets, 2006 .—544p.

[3] GOST 11047-90 «Details and wooden products for low-rise residential and public buildings. Technical conditions Wooden parts and wood products for construction. Part 2. Gates, parts and products, floor and cover panels, floor beams, parquet products, glued structures,

fiberboard and chipboard: Sat. GOSTs». - M .: IPK Standards Publishing House, 2002. URL: http://docs.cntd.ru/document/9054219 (amended on: 03.04.2020).
[4] GOST 2140-81 (ST SEV 2017-79, ST SEV 2018-79, ST SEV 2019-79, ST SEV 320-76, ST SEV 321-76, ST SEV 391-76, ST SEV 3286-81, ST SEV 3287-81, ...) «Visible wood defects. Classification, terms and definitions, measurement methods» (edition January 01, 2020) URL: http://docs.cntd.ru/document/1200004894 (amended on: 03.04.2020).
[5] GOST 30974-2002 «Corner joints of wooden block and log low-rise buildings. Classification, design, dimensions». M .: Gosstroy of Russia, GUP TsPP, 2003. URL: http://docs.cntd.ru/document/1200031045 (amended on: 03.04.2020).
[6] GOST R 31937-2011 «Buildings and constructions. Rules for inspection and monitoring of technical condition» (edition 01.01.2014) URL: http://docs.cntd.ru/document/1200100941 (amended on: 21.03.2019).
[7] Zhuravlev E.G. Examination of the investment and construction process. - Irkutsk: Publishing House ISTU, 2016. - 73 p..
[8] BR 131.13330.2018 «Construction climatology. Updated edition of SNiP 23-01-99 *» M: Standartinform, 2019. URL: http://docs.cntd.ru/document/554402860 (amended on: 03.04.2020).
[9] BR 13-102-2003 «Rules for the inspection of load-bearing building structures of buildings and structures» (edition 21 августа 2003) URL: http://docs.cntd.ru/document/1200034118 (amended on: 21.03.2019).
[10] BR 14.13330.2018 «Construction in seismic areas. Updated edition of SNiP II-7-81 *». M .: Standartinform, 2018. URL: http://docs.cntd.ru/document/550565571 (amended on: 03.04.2020).
[11] Melekhov E S and Shlepnev O K 2018 Development and reconstruction of district water supply systems *Proceedings of Universities. Investment. Construction. Real estate* **8(4)** pp 114–127 DOI: 10.21285/2227-2917-2018-4-114-127
[12] BR 17.13330.2017 «Roofing. Updated edition of SNiP II-26-76» (edition February 18, 2019) URL: http://docs.cntd.ru/document/456081632 (amended on: 03.04.2020).
[13] BR П 20.13330.2016 Нагрузки и воздействия. Актуализированная редакция СНиП 2.01.07-85* (ред. 28 января 2019) URL: http://docs.cntd.ru/document/456044318/ (amended on: 03.04.2020).
[14] BR 30-102-99 «Planning and development of territories of low-rise housing construction» M .: Gosstroy of Russia, State Unitary Enterprise TsPP, 2000. URL: http://docs.cntd.ru/document/1200004849 (amended on: 03.04.2020).
[15] Yaskova N Yu 2018 Construction in the context of new views on the goals of economic development. *Proceedings of Universities. Investment. Construction. Real estate* **8** pp 133-142 DOI: 10.21285/2227-2917- 2018-1-133-142
[16] BR 42.13330.2016 «Urban planning. Planning and development of urban and rural settlements. Updated edition of SNiP 2.07.01-89*» (edition December 19, 2019) URL: http://docs.cntd.ru/document/456054209 (amended on: 03.04.2020).
[17] BR 55.13330.2016 «Residential single-family houses. Updated edition of SNiP 31-02-2001» M .: Standartinform, 2019 (edition July 10, 2018) URL: http://docs.cntd.ru/document/456039916 (amended on: 03.04.2020).
[18] BR 64.13330.2017 «Wooden structures. Updated edition of SNiP II-25-80». M .: Standartinform, 2017 (edition January 30, 2019) URL: http://docs.cntd.ru/document/456082589 (amended on: 03.04.2020).
[19] BR 70.13330.2012 «Bearing and enclosing structures. Updated edition of SNiP 3.03.01-87» (edition December 26, 2017) URL: http://docs.cntd.ru/document/1200097510 (amended on: 03.04.2020).
[20] Chupin V R and Maizel D I 2020 Aggregation and decomposition in routing oil and gas pipeline systems *Proceedings of Universities. Investment. Construction. Real estate* **10(1)** pp

108–115https://doi.org/10.21285/2227-2917-2020-1-108-115
[21] «Town Planning Code of the Russian Federation» dated December 29, 2004 N 190-FL (edition from 04.24.2020) URL: http://www.consultant.ru/document/cons_doc_LAW_51040/ (amended on: 04.05.2020).
[22] Decree of the Government of the Russian Federation of February 16, 2008 N 87 (as amended on April 28, 2020) "On the composition of sections of project documentation and requirements for their content" URL: http://www.consultant.ru/document/cons_doc_LAW_75048/ (amended on: 30.04.2020).
[23] Peshkov A V 2019 Remote mortgage lending as a resource for housing development Investments. Construction. Real Estate: New Technologies and Targeted Development Priorities IOP Conference Series: Materials Science and Engineering **667** 012073
[24] Law of the Russian Federation of 07.02.1992 N 2300-1 (as amended on 07/29/2018) "On Protection of Consumer Rights" of 12.30.2001 N 195-FL (edition from 06.02.2019) URL: http://www.consultant.ru/document/cons_doc_LAW_305/ (amended on: 21.03.2019).
[25] Federal Law of July 22, 2008 N 123-ФЗ (as amended on December 27, 2018) URL: http://docs.cntd.ru/document/902111644 (amended on: 03.04.2020).
[26] «Civil Code of the Russian Federation (Part One) "dated November 30, 1994 N 51-ФЗ (as amended on August 3, 2018) (as amended and supplemented, entered into force on 01/01/2019). URL: http://www.consultant.ru/search/?q=%D0%93%D0%9A+%D0%A0%D0%A4 (amended on: 20.03.2019).

IOP Conf. Series: Materials Science and Engineering **880** (2020) 012019 doi:10.1088/1757-899X/880/1/012019

Oxidized thermally expanded graphite as a raw material for the production of cement composites

O Selezneva[1], V Orlov[1] and P Shustov[2,3]

[1]Industrial University of Tyumen, Volodarsky str., 38, Tyumen, 625000, Russia
[2]Angarsk State Technical University, Tchaikovsky str., 60, Angarsk, 665835, Russia
[3]Irkutsk National Research Technical University, Lermontov str., 83, Irkutsk, 664074, Russia

E-mail: sel__olga@mail.ru

Abstract. One of the current tasks is to improve the performance of cement composites. Effective ways to solve a problem is to use a new generation of carbon - containing modifiers. The regulation of the physical and mechanical and operational properties of cement composites must be carried out at the micro level, by regulating the structure formation of the cement matrix One of the promising directions is the use of graphene as a carbon-containing modifier. The article represents research data on oxidized thermally expanding graphite. The results of microscopic and laser diffraction granulometric analyzes are represented. The performed studies allowed to characterize the properties of the research material. The analysis of particle size distribution was carried out on a Fritsch NanoTec laser particle size measuring device with a dispersion unit in a liquid space with a total measurement range from 0.01 to 2100 μm and showed a decrease in size from 574 μm to 22.2 μm. Microscopic research was carried out by using a scanning (raster) electron microscope TESCAN Mira 3 with a magnification range of x1 – x1 000,000. The experimental results presented in the present research prove the effectiveness of using oxidized thermally expanding graphite as a modifier of cement composites.

1. Introduction

Graphene is a two-dimensional allotropic form of carbon (figure 1), in which atoms combined into a hexagonal crystal lattice form a layer with a thickness of one atom [1,2]. Graphene has unique properties, which allows to use it in various fields. According to scientific research [3,4,5], the material is very durable, flexible and elastic, able to restore the crystalline structure in case of damage, has a high electrical conductivity of about 100 m / s, high electric capacity, which approaches to 65 kW · h / kg, very light, absorbs radioactive wastes [6,7].

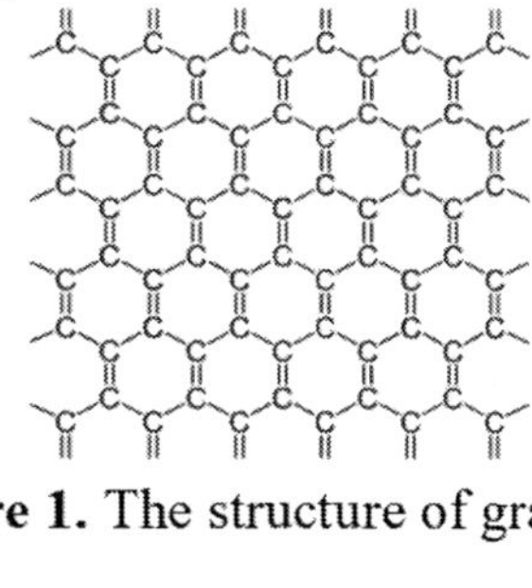

Figure 1. The structure of graphene

According to the literature sources [8, 9], pure graphene is extremely rare. There are various methods for producing graphene [9, 10, 11, 12] from different types of graphite and its oxides with oxygen-containing functional groups and molecules attached at the edges or inside the carbon network.

An important direction in the production of heavy concrete is the transition to cement composites with improved physical, mechanical and operational properties [13,14,15]. To achieve this goal, an integrated approach to solving the problem is required. First of all, the optimization of the composition should be controlled at the micro level with the possibility of regulating the structure formation of the cement matrix. Promising are the studies on the use of carbon-containing components in the form of: carbon tubes, fibers and fullerenes in the production of high-strength concrete [16, 17]. The authors of the following papers [18, 19] consider it possible to obtain durable materials using graphene due to the micro-reinforcing effect. The introduction of carbon-containing modifiers into the composition makes it possible to increase compressive and bending strength, crack resistance, fire resistance, reduce water absorption and, as a result, increase frost resistance. The question still remains unanswered about the effective method of introducing and distributing carbon-containing additives and fibres in a cement composite in connection with low concentrations of the substance. Currently, the most actual task is to develop a technology for producing pure graphene in industrial scales and to reduce the production cost.

2. Methods

The analysis of particle size distribution was carried out on a laser device for measuring particle size with a dispersion unit in a liquid space and with a total measurement range of 0.01 - 2100 μm Fritsch NanoTec.

The structure of the obtained carbon-containing additive was studied using a TESCAN Mira 3 scanning electron microscope with a magnification range of up to 1 000 000 times.

3. Results and Comments

A powder of oxidized thermally expanding graphite was used as a sample material for producing a carbon-containing additive. This material is a grey crystalline powder with a characteristic metallic shine. Layering (scaly structure) due to the laying of the plates in the radial (flat) direction. A photograph of the source material is demonstrated in figure 2 (a, b, c).

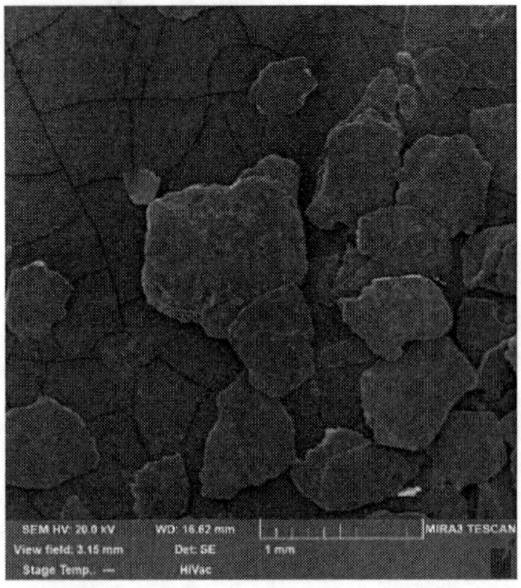

a

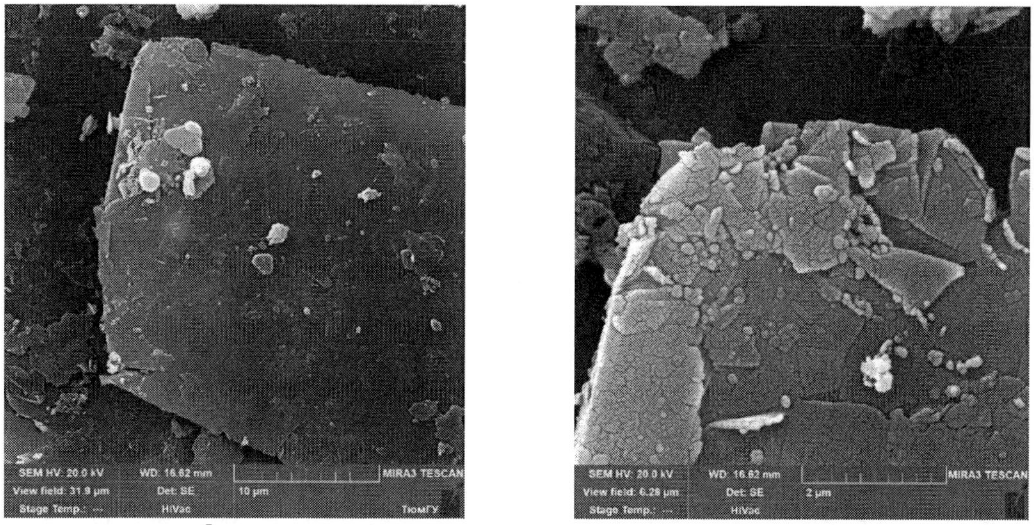

b c

Figure 2. A microphotograph of the sample graphite

The particle sizes of the initial powder vary from 4 to 1350 µm and are represented in figure 3.

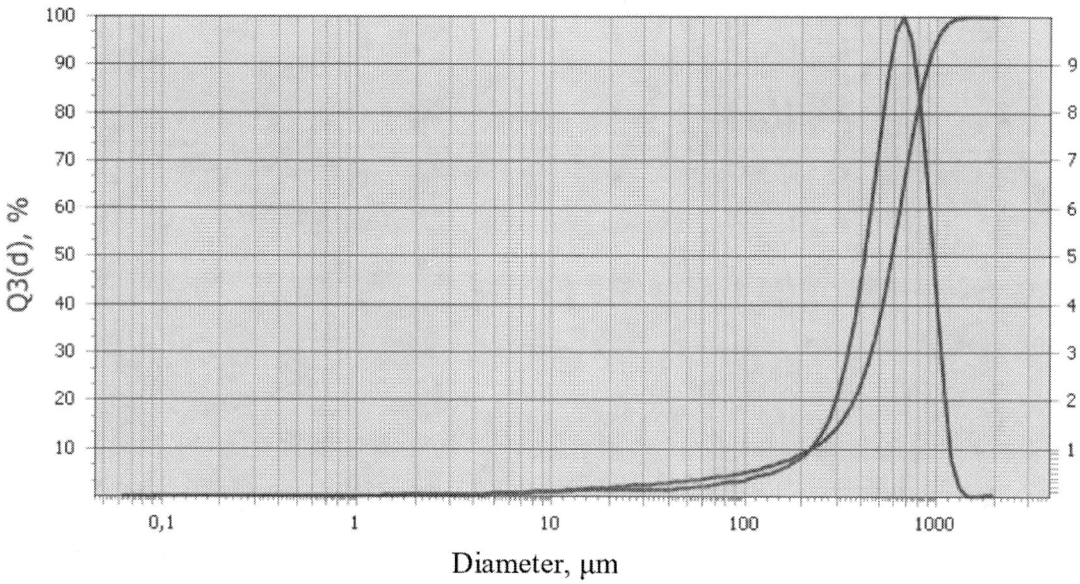

Figure 3. Particle size distribution diagram of sample graphite

The particle-weighted average particle size D [4.3] is 574 µm, which is comparable with the fine sand fraction in the cement stone, therefore, in the initial state it cannot carry out the function of a modifier.

The carbon-containing additive was synthesized according to the method proposed by scientists of Rutgers University (USA) [21], by short-term exposure, not exceeding 10 seconds, of electromagnetic (microwave) radiation (frequency 2450 MHz), power 1000W.

As a result of processing procedure, there is a sharp thermal activation of the raw material with a hundred-fold volume increase of the material due to the removal of oxygen-containing groups. The rapid increase in pressure in the interlayer space leads to the splitting of the initial matrix, with the formation of elongated highly porous flakes (worm-shaped), characterized by a length of up to 20 mm and having a defective structure with a high content of voids. The morphology of the additive consists of "crosslinked" graphene packs and contains up to 99.9% carbon without side liquids. A photograph of the microstructure is shown in figure 4 (a, b, c, d, e, f).

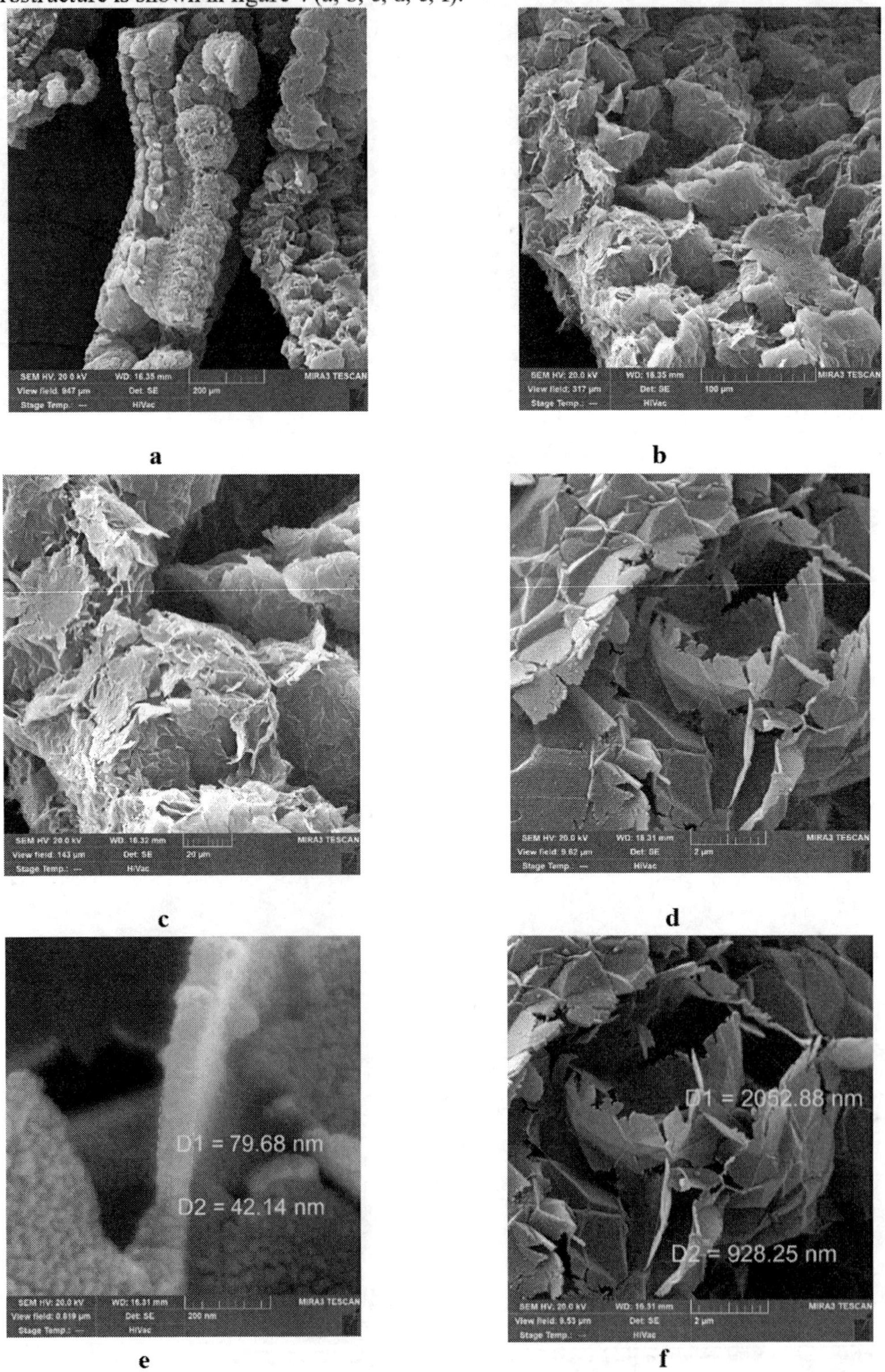

Figure 4. Microstructure of the synthesis of thermally expanded graphite

To assess the particle size distribution, the obtained material is dispersed in distilled water using ultrasonic treatment. The particle sizes of the obtained carbon-containing additives vary from 0.01 to 74 microns and are represented in figure 5.

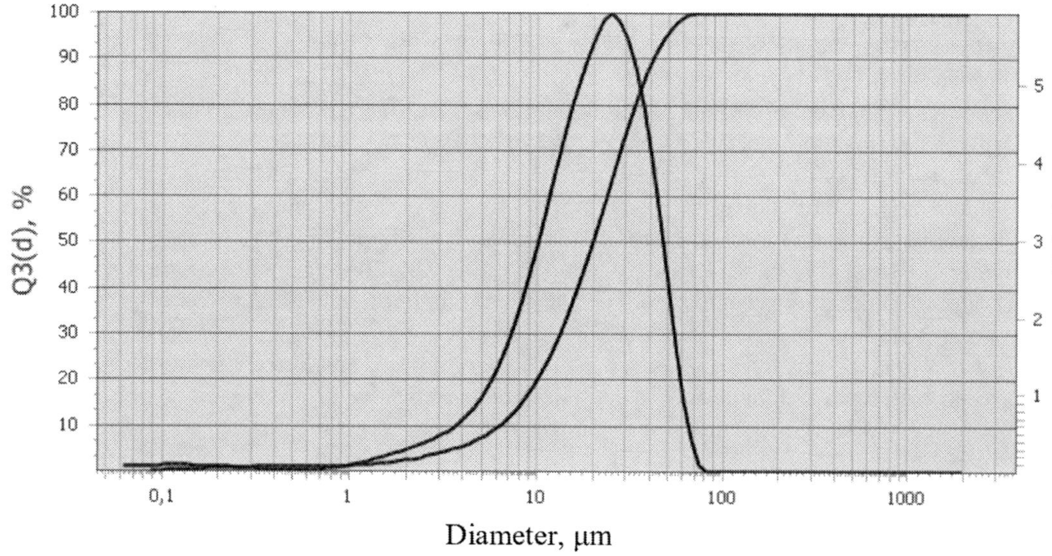

Figure 5. Particle size distribution diagram of carbon-containing additive

The volume-weighted average particle size D [4.3] is 22.2 µm.

The granulometric composition of the additive corresponds to the cement dimension, the content of the additive fraction of up to 10 µm is 21%, which proves the effectiveness of the applied method and allows the additive to be used as a microfiller and modifier of cement composites.

4. Conclusions

As a result of studying the synthesized carbon-containing additive, an improvement in the characteristics of the material was shown compared with the initial oxidized thermally expanding graphite. Having examined the microstructure photographs, the presence of graphene packs with a dimension of 0.01 to 74 µm was diagnosed. Analyzing the data obtained during the particle size analysis, we can conclude that when exposed to electromagnetic radiation, the particle sizes are reduced from 574 microns to 22.2 microns, showing a decrease of about 26 times. A thine fraction of about 10 microns appears. This shows the possibility of introducing a graphite-containing additive into the composition of heavy concrete with a cement binder by optimizing the microstructure due to a denser packing of grains and the creation of additional centres for the growth of new materials.

References

[1] Chuah S, Pan Z, Sanjaan J G, Wang C M and Duan W H 2014 Nano reinforced cement and concrete composites and new perspective from grapheme oxide *Construction and Building Materials* **73** pp 113–124

[2] Pan Z, He L, Qiu L, Korayem A H, Li G, Zu J W, Hu, Collins F, Li D, Duan W H and Wang M C 2015 Mechanical properties and microstructure of a grapheme oxide – cement composit *Cement & Concrete Composites* **58** pp 140–147

[3] Moxmmed A, Sanjayn J G, Duan W H and Nazan A Incorporating grapheme oxide in cement composites: A study of transport properties *Construction and Building Materials* **84** pp 341–347

[4] Ahmadreza Sedaghat, Manoj K Ram, Zayed A, Rajeev Kamal, Natadia Shanahan 2014 Investigation of Physical Properties of Graphene-Cement Composite for Structural Applications *Open Journal of Composite Materials* **4** pp 12–21

[5] Muhit B A, Nam B H, Zhai Lei and Zuyus J 2015 Effects of microstructure on the compressive strength of graphene oxide-cement composites *Nanotecnology in Construction* https://pantherfile.uwm.edu/sobolev/www/ NICOM-5/13_Nam.pdf (date of access 23.11.2015).

[6] Federova G D, Aleksandrov G N and Smagulova S A 2015 Investigation of the stability of a liquid suspension of graphene oxide *Construction Materials* **2** pp 15–21

[7] Federova G D, Aleksandrov G N and Smagulova S A 2016 On the use of graphene oxide in cement systems *Construction Materials* **1-2** pp 21–26

[8] Horszczaruk E, Mijowska E, Kalenczuk R J, Aleksandrzak M and Mijowska S 2015 Nanocomposite of cement/graphene oxide – Impact on hydration kinetics and Young's modulus *Construction and Building Materials* **78** pp 234–242

[9] Wang Q, Wang J, Lu C-x., Lie Bo-w, Jang R and Li C-z 2015 Influence of grapheme oxide additions on the microstructure and mechanical strength of cement *New Carbon Materials* **4** pp 349–359

[10] Bonaccorso F, Lombardo A, Hasan T, Sun Zh, Colombo L and Ferrari A C 2012 Production and processing of graphene and 2d crystals *Materials Today* **12** pp 563–589

[11] Graphene oxide reinforced cement. http://www.monash. edu.au/assets/pdf/industry/graphene-oxide.pdf.

[12] Patent WO 2013096990 A1. Graphene oxide reinforced cement and concrete. Pan Z, Duan W H, Li D and Collins F Declared 21.12.2012. Published 04.07.2013.

[13] Selezneva O I, Radaev S S and Ryasnaya N Z Regulation of the properties of composites based on local raw materials 2011 *Collection of materials of the All-Russian scientific-practical conference* "Actual problems of construction, ecology and energy conservation in Western Siberia." (Tyumen) pp 81-85

[14] Milyaev I V 2009 Optimization of the properties of modified cement stone *Scientific Bulletin of VGASU* **5** pp 102-104

[15] Kharkhardin A N 2011 The structure of the topology of dispersed systems of interacting micro- and nanoparticles *Bulletin of higher education.* Construction **5** pp 119-125

[16] Raki L; Beaudoin J J; Alizadeh R; Makar J M; Sato T Cement and concrete nanoscience and nanotechnology *Materials* **3** pp 918-942.

[17] Cwirzen A; Habermehl-Cwirzen K and Penttala V 2008 Surface decoration of carbon nanotubes and mechanical properties of cement/carbon nanotube composites *Adv. Cem. Res* **20** pp 65-73

[18] Bazhenov Yu M, Alimov L A and Voronin V V 1996 Development of the theory of the formation of the structure and properties of concrete with industrial waste *By universities. Construction* **4** pp 55-58

[19] Falikman V R 2011 Nanomaterials and nanotechnologies in modern concrete *ALITinform: Cement. Concrete. Dry mixes* **5-6** p 34-48

[20] Kozhukhova N I, Bondarenko A I and Strokova V V 2012 The dependence of the structure formation mechanism on the chemical composition as a key factor in the cementing system Construction complex of Russia *The science. Education. Practice: materials of an international scientific-practical conference dedicated to the 50th anniversary of the East Siberian State University of Technology and Management and the Faculty of Civil Engineering* pp 162-164

[21] Voiry D, Yang J, Kupferberg J, Fullon R, Lee C, Jeong H Y, Shin H S and Chhowalla M 2016 High-quality graphene via microwave reduction of solution-exfoliated graphene oxide *Science* **353** pp 1413-16

IOP Conf. Series: Materials Science and Engineering **880** (2020) 012020 doi:10.1088/1757-899X/880/1/012020

Use of secondary raw material for foam concrete production

S A Sherbin[1] **and P S Gorbach**[1]

[1] Federal state-funded educational institution of the higher education "Angarsk state technical university", 665835 Angarsk, Russia

E-mail: pgs@angtu.ru

Abstract. The article discusses the problems of using secondary raw materials for the production of mixed binder compositions and foam concrete. The effect of ash and microsilica on the properties of low-linker binder compositions and large-scale foam concrete was studied.

Foam concrete currently holds an increasingly strong position in construction. This material is characterized by fire resistance, high heat, sound and steam insulation properties, reduced average density, which makes it possible to successfully use it both in monolithic house construction and for manufacture of a large range of construction products: Wall blocks, heat insulation materials, etc. Comparative analysis of existing technologies of foam concrete production indicates the preference of autoclave-free foam concrete in terms of processability, but with the need to solve the problem of their deformability, environmental friendliness, stable high strength indicators, with simultaneous solution of the problem of cement consumption reduction, increase of hardening rate and achievement of uniform porous structure [1,2].

The difficult environmental environment, combined with rising energy prices, encourages interest in the secondary use of concrete in construction. Such attempts are being made in many countries of the world, including Russia. The use of concrete scrap resulting from the disassembly of dilapidated housing in the production of foam concrete is widespread [3-6].

It has been found that the use of wastes generated during the production of cellular concrete (cutting, non-standard products, autoclave treatment wastes) has a positive effect on the forming process - contributes to the stabilization of the forming mixture, creates favorable conditions for the formation of low density articles (300 – 350 kg/m^3) [7-9]. Non-autoclave foam concrete with density of 600 – 800 kg/m^3, strength of 3.0 – 6.0 MPa [10,11] is obtained on the basis of glass.

In the Republic of Buryatia and in the Irkutsk region every year about 130 thousand tons of ash is formed, which is not used, and in huge quantities accumulates in gold mines, which are stationary centers of environmental pollution. A large number of studies are aimed at finding industrial versions of the by-product of thermal power plants - ash and fine by-product of ferrosilicon smelting and its alloys - microsilicon in the production of effective construction materials, including foam concrete [12-14]. Therefore, the use of industrial waste in the manufacture of low-link binder compositions and foam concrete is of interest. For receiving mixed knitting and foam concrete of foam concrete curing were used: Portlandtsement of Siberian Cement holding of the PTs 400 D20, PTs 500 D0 brands, ashes of hydroremoval of the Ulan-Ude and Angarsk combined heat and power plant-1, New and Elansky sand, Mdt =1.35; Microsilica - departure of crystalline silicon shop of Bratsk Aluminium Plant and ZAO "Silicon" (Shelekhov, Irkutsk region); Foaming agents ПБ-2000, Arekom-4, Penta surfactant 430 A, Biopor, Centripor SK 120, protein-based foaming agent. The Bratsk Aluminium Plant microsilica is a very fine additive (Sud = 850–900 m^2/kg), while the ZAO Silicon microsilica has Sud = 180–220 m^2/kg.

In the studies carried out, low-linker binders or mixed cements (MC) were prepared by thoroughly mixing Portland cement, blast-furnace granulated slag, fly ash and fine microsilica in the ratios indicated in table 1. To modify the properties of the MC, complex additives were used: surfactant (plasticizers) hardening accelerator. As surfactants, additives of CBS and potassium citrate were used, as hardening accelerators - sodium sulfate and carbonate: Na2SO4 and Na2CO3.

Table 1. Composition of the mixed cements

Code of structure	Mixed cement composition,% by weight			
	Pc	Domain granulated slag	Ashes ablation	Microsilicon
MC 1	20	50	30	-
MC 2	20	40	30	10
MC 3	10	50	30	10
MC 4	10	60	30	-
MC 5	10	80	-	10

Introduction of plasticizing additives in amount of 0.6 - 1% by weight into binder composition allowed to reduce their water consumption on average by 12-20% with increase of 30-40% strength (Table 2). In addition, the rheological characteristics of the cement slurry are improved.

Table 2. Properties of the mixed cements

Structure	Amount of additive,% by weight		V/V	Terms of curing, min..		Compression strength after WMP (t=95°C,τ = 1,5+6,5+1,5 ч), MPa	Compression strength, MPa, after		
	$C_6H_5K_3O_7 \cdot H_2O$	Na_2SO_4		beginning	end		1 day	7 day	28 day
Ps M400	-	-	0,28	30	-	32	15	30	48
Ps	1	2	0,25	3	10	45	38	46	55
MC 1	-	-	0,32	60	-	24	4	26	31
MC 1	1	2	0,28	13	42	32	2	14	46
MC 2	-	-	0,34	55	-	25	3	20	33
MC 2	1	2	0,25	10	18	50	5	32	60
MC 3	1	2	0,27	35	-	5	1	5	10
MC 4	1	2	0,30	55	-	4	1	2	6
MC 5	1	2	0,25	18	50	15	2	25	42

Microsilica is a very fine additive increasing the V/V ratio, however, the presence thereof in an amount of 10% improves the plasticizing effect of the potassium citrate additive plasticizer and significantly reduces the V/V ratio by increasing the strength of the compositions (MC composition 2). The pozzolanization reaction between silica and calcium hydroxide of cement results in the formation of further low-base calcium hydrosilicates of type CSH (I), as confirmed by the results of physical and chemical analysis. It should be noted that the pozzolane effect of microsilica in mixed cements has a significant effect between 3 and 28 days of hardening. In general, the addition of a complex addition of potassium citrate and sodium sulfate reduces the setting time of mixed cements, which is due to the increase in the hydration rate of the aluminate phase of the cement.

On optimal compositions of mixed cements - MC 1 and MC 2 and on pure CP, studies were carried out to obtain foamed concrete without autoclave hardening according to classical process according to SN 277-80. Multi-factor experiments were carried out to select the composition of foamed concrete, ensuring quality macrostructure of concrete. When determining optimal compositions and manufacturing modes, the influence of the following technological factors was investigated, which have a decisive influence on the quality and construction and technical properties of foam concrete: optimal ratio of silica component and binder in the mixture, fluidity of the solution mixture and its corresponding water-solid ratio, amount of additive - foaming agent. As a result of operation with these factors, quality structure and good construction and technical properties of foam concrete were obtained.

The results of the studies prove the possibility of producing foam concrete on mixed cement (Table 3).

Table 3. Properties of foam concrete based on mixed cements

Structure	Additive, % on weight		Type of FA	Quantity of FA,% by weight	V/H	Average density, kg/m^3	Compression strength, MPa, after	
	$C_6H_5K_3O_7\cdot H_2O$	Na_2SO_4					at a bend	at compression
MC 2	1	2	on the basis of a protein	0,3	0,30	780	-	3,5
				0,5	0,30	580		2,5
				1,0	0,30	400		1,5
MC 2	1	2	Centripor SK 120, 10%	0,3	0,30	1090	1,35	6,5
				0,5	0,30	940	1,1	3,5
				1,0	0,30	580	1,0	3,5
MC 2	1	2	Centripor SK 120, 15%	0,5	0,30	670	1,15	4,2

The use of ash as an aggregate was investigated in the production of PK-based foam concrete. The weight ratio of ash to sand varied from 0.25 to 1. Coarse microsilica was added as an AMD additive to the CP composition in an amount of 0 to 20% by weight of the cement.

Analysis of the results of the studies (Figure 1.2) showed that an increase in ash content in foam concrete resulted in an increase in the strength of the samples. The highest strength was obtained with the same mass content of sand and ash. Application of microsilica increases strength of foam concrete samples (Fig. 2) - strength increase reached up to 60% at content of microsilica 15-20% by weight. At the same time, for equal-plastic concrete mixtures, cement consumption is reduced to 30%, while the same amount of microsilicon in the foam concrete mixture at constant cement consumption significantly increases its plasticity.

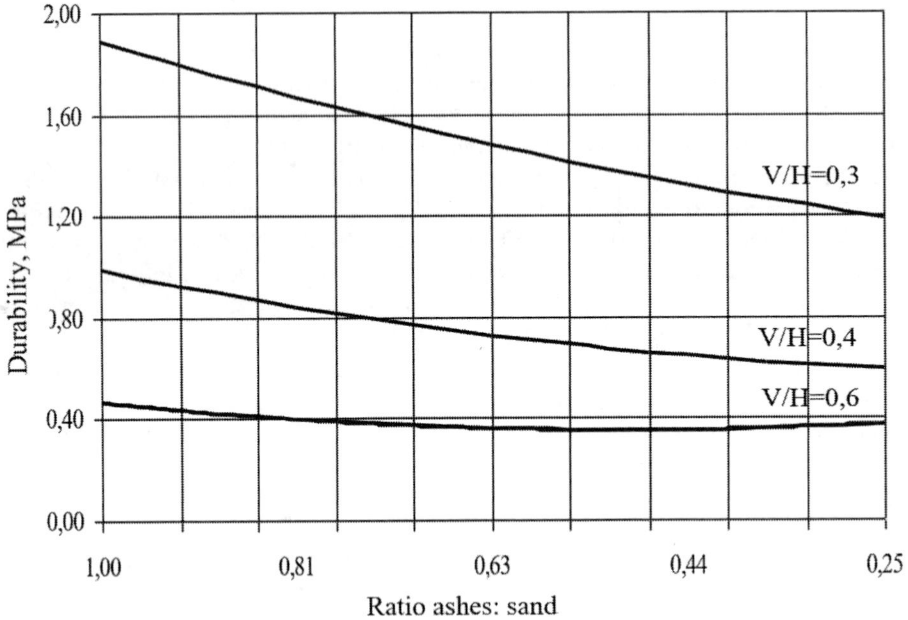

Figure 1. Impact of ash: sand mass ratio on foam concrete strength

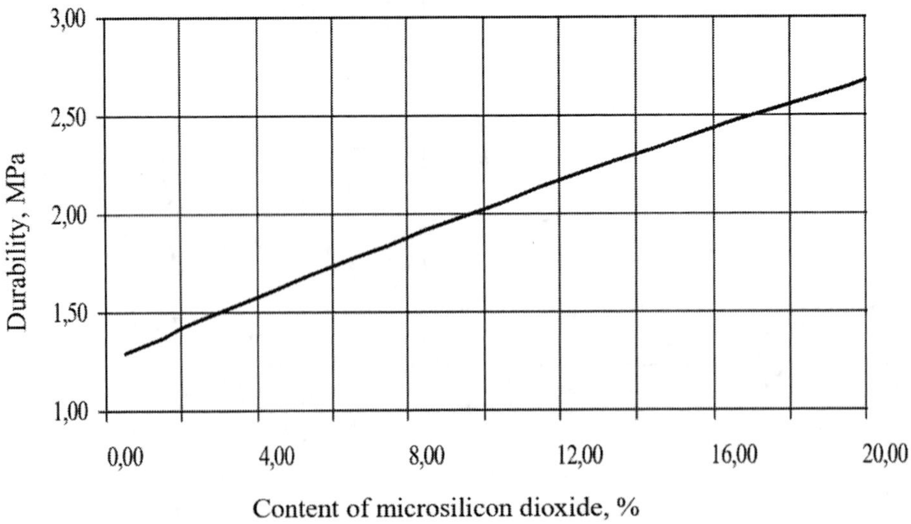

Figure 2. Dependence of foam concrete strength on microsilica content

The results of the studies prove the possibility of using ash, blast furnace granulated slag and microsilica for the production of foamed concrete of non-autoclave hardening. Use of industrial wastes and by-products allows to increase strength of foam concrete, provides possibility to use it as heat-insulating, heat-resistant and fire-retardant material at temperature up to 1200 ^{0}C. In addition, the availability and low cost of raw materials will reduce the cost of production, and with large volumes of foam concrete production - to improve the environmental situation in Buryatia and the Irkutsk region.

The economic efficiency of foam concrete production is calculated in relation to the conditions of the industrial and raw materials base of the Republic of Buryatia and the Irkutsk region with the link of development to the current technologies at the enterprises of the construction industry of the region.

References

[1] Dashitsyrenov D D, Zayakhanov M E and Urhanov L A 2007 Effective foam concrete based on effusion rocks *Construction materials* **4** pp 33
[2] Bozhenov P I 1978 *Technology of autoclave materials* p 368
[3] Chitov Y D, Gusenkov S A and Krasnov M V 2005 Theoretical bases of creation of cellular concrete from dust-like wastes of concrete scrap crushing *International scientific and practical conference. Porobeton* pp 85-89
[4] Laukaitis A A 2014 Study of influence of additives of ground wastes of cellular concrete on its properties *Construction materials* **3** pp 33
[5] Shesterkin M N 2002 Linkless binder and structural-heat-insulating foam concrete of non-Autoclave **1** p 17
[6] Bertov V M and Sobkalov P F 2005 Use of fly ash in foam concrete production *Construction Materials* **5 p** 12
[7] Holin S I 2004 Application of microsilica on concrete production *Popular concrete science* **2** pp 65-69
[8] Luzhnova E V, Tugarin A O and Korchinov A S 2004 Wastes of forest chemistry in the technology of obtaining cellular concrete *Works of the Bratsk State Technical University*. - **2** p 370
[9] Tugarina A O 2006 Way of preparation of the aerated gazozolobeton (patent) No. 2284979, 10.06.2006.
[10] Panova V F 2003 Development of decorative slag cement. NGASU *Modern construction materials and resource-saving technologies* p 168
[11] Avvakumov E G 2001 Soft Mechanochemical Synthesis a Basis for New Chemical Synthesis a Basis for New Chemical Technologies. Hardbound: Kluwer Academic Publishers p 216
[12] Zhigulina A 2012 Building Envelopes: An Objective Measure of Comfort in Residential Buildings *Urban Planning* **1** pp 80-81
[13] Korenkova S 2000 Filled Foam Concretes in Building Envelope Construction *Building Materials* **8** pp 12-14
[14] Tarasova A G, Alalykin A A and Vesnin E S 2017 New heat-insulating materials based on inorganic binder **4**
[15] Shalaginya N A Patent for utility model RU 86613, 10.09.2009. Denunciations No. 2009116591/22 of 30.04.2009
[16] Selezneva O I, Radaev S S and Baranova A A The impact of thermal treatment on the granulometric and chemical composition of diatomite of the Kamyshlov field when obtaininga liquid-glass binder *Investment. Construction. Real estate* **9(3)** pp 550–555
[17] Moskvitin V A., Emelyanova N A. and Mashovich A Y 2019 Experimental studies of air permeability indicators of composite "Poroplast CF" *Investment. Construction, Real estate* 9 (2) pp 342–353
[18] Peshkov V V and Gamayunova E V 2018 Specific features of investment support for the repair and reconstruction of buildings and facilities *Proceedings of Universities. Investment. Construction. Real estate* **8** pp 56-72 DOI: 10.21285/2227-2917-2018-2-56-72

IOP Conf. Series: Materials Science and Engineering **880** (2020) 012021 doi:10.1088/1757-899X/880/1/012021

Foam concrete production with addition of microsilica

S A Sherbin[1] and P S Gorbach[1]

[1] Federal state-funded educational institution of the higher education "Angarsk state technical university", 665835 Angarsk, Russia

E-mail: pgs@angtu.ru

Abstract. The article deals with the influence of microsilica on technological properties of cement matrix and finished foam concrete. The possibility of replacing a part of the binder with microsilica was evaluated.

One of the current trends in the development of the construction market is the wide use of cellular concrete - artificial stone with evenly distributed pores. The pore structure is created mainly with air-absorbing additive (aerated concrete), or with foam (foamed concrete).

Rational use of articles made of foamed concrete is due to relatively high strength and operational properties at low density, which allows to significantly reduce the weight of the structure while ensuring the required structural and heat insulation indicators, which, in turn, creates prerequisites for effective solution of resource and energy saving problems in construction [1,2].

The disadvantage of foamed concrete is low initial strength, in the result of which articles acquire undesirable chips and cracks during production and installation. Analysis of studies and theoretical data [3,4] allowed to determine two main ways of strengthening the material with preservation of tee-technical properties and low density - strengthening of the framework (inter-pore partitions) [5,6] and improvement of the pore structure.

Directed structuring in cellular concretes is achieved due to reduction of the amount of water and introduction of various active additives (mineral, fibrous) [7]. It is obvious that it is most advantageous to use additives with a high specific surface area that do not require complete grinding, for example microsilica.

Microsilica (MS) is a waste from the production of silicon-containing alloys, an ultradisperse material consisting of particles of spherical shape with a size of 0,01 – 0,1 μm. The main component of the material is amorphous modification silicon dioxide containing up to 95% pure amorphous silica, which is able to react actively with lime released by portland cement during its hydration to form water-insoluble substances.

The use of MS can help to reduce the cost of construction materials, as well as improve the environmental situation in the Irkutsk region [8].

In order to determine the amount of microsilica to which cement can be replaced in the foam concrete composition without deterioration of the main physical and mechanical properties of the latter, a series of experiments was carried out to determine the mobility of the cement slurry, the strength of the cement matrix, as well as the density, strength and thermal conductivity of the foam concrete, cast using the MS of the Bratsk Aluminium Plant.

Composition of cement mortar included cement of grade M500D0, water and MS, replacing from 5 to 15% of cement weight. Ratio of water and solid components (cement and MS) - Water/Solid, ranged from 0,4 to 0,6. According to [8], the density of the future material depends on the diameter of a spreading of the initial mixture according to the Suttard viscometer. Thus, in order to obtain a density of 600 kg/m^3, the diameter must be 24 cm. Solutions mobility measurement results are shown in Figure 1.

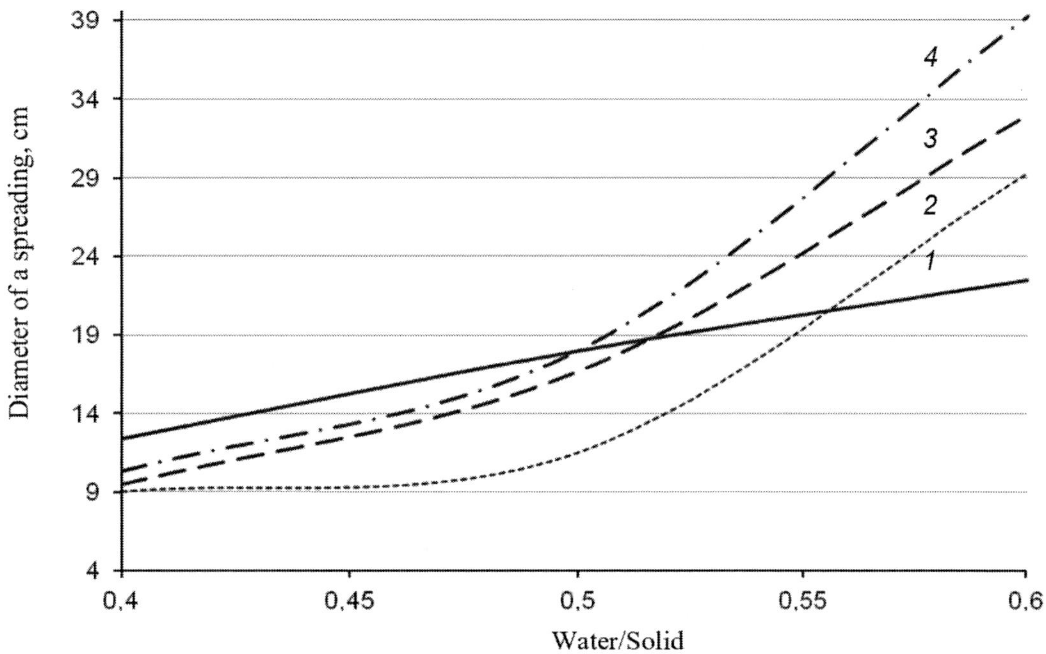

Figure 1. The dependence of diameter of a spreading of the initial mixture on the water-solid ratio: 1 - is a control sample; 2 - with replacement of 5% mass of cement MS; 3 - with replacement of 10% by weight of cement MS; 4 - with replacement of 15% mass of cement MS

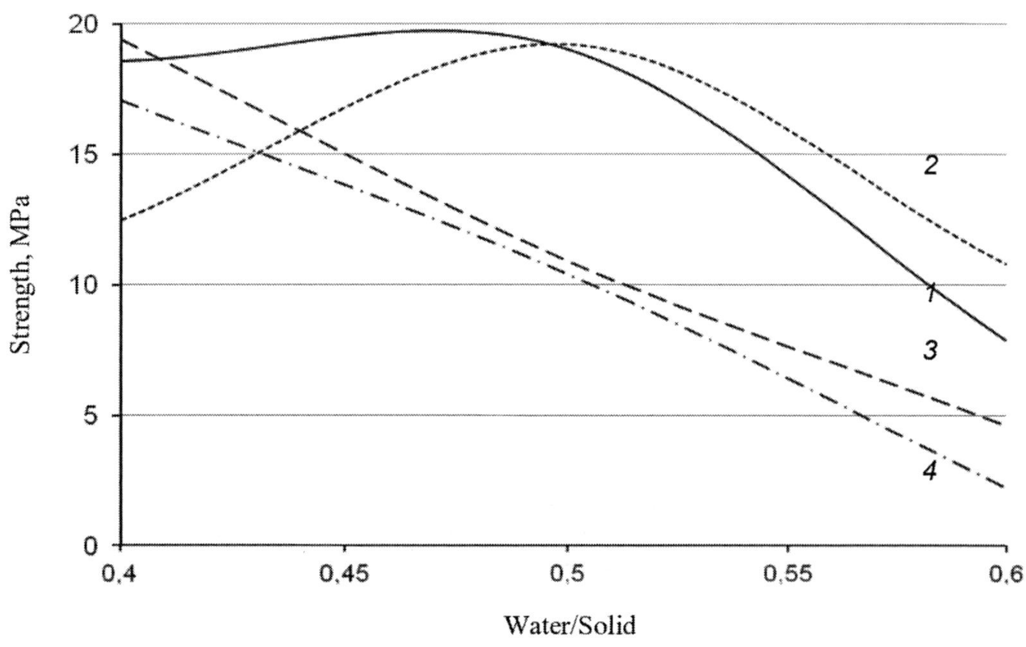

Figure 2. The dependence of the strength of the cement matrix on the water-solid ratio: 1 - is a control sample; 2 - with replacement of 5% mass of cement MS; 3 - with replacement of 10% by weight of cement MS; 4 - with replacement of 15% mass of cement MS

It can be seen that at Water/Solid from 0,4 to 0,5, the addition of MS results in an increase in the water demand of the mixture due to the high specific surface area of the additive. The diameter of a spreading was less than 18 cm, suggesting that the required density of the material could not be obtained at such component ratios. With an increase in Water/Solid, there is an increase in plasticity and, accordingly, mobility of weight. The desired diameter of a spreading (24 cm) at Water/Solid ranging from 0,53 to 0,58, depending on the amount of MS used. But such a large amount of water in the mixture will not produce a material with acceptable density and strength. Therefore, to achieve the desired parameters of foam concrete, it is necessary to use plasticizing additives. Based on the results of the mobility measurement, the rational ratio of components of the foam concrete mixture was determined.

Further from samples of solution beams of 40×40×160 mm in size prepared, and after 28 days their strength at compression (figure 2) was defined.

It can be seen from the graphs that the replacement of more than 5% by weight of the cement with microsilicon at leads to a significant reduction of the matrix strength (lines 3, 4). The strength of the control sample (without the addition of MS) and the sample containing MS in 5% of the weight of the cement had the highest value at Water/Solid of 0,5.

At the next stage foam concrete with design density of 600 kg/m^3 was prepared according to classical technology - cement mortar containing MS in quantity from 5% to 30% of cement weight was mixed with foam.

The effect of the amount of MS added on the density of the material was found: that with the addition of microsilica over 15% by weight, there is a significant decrease in density. In addition, visual analysis of the corresponding samples revealed a disturbance in the uniformity of their structure, which will inevitably have a negative impact on the construction and technical properties of

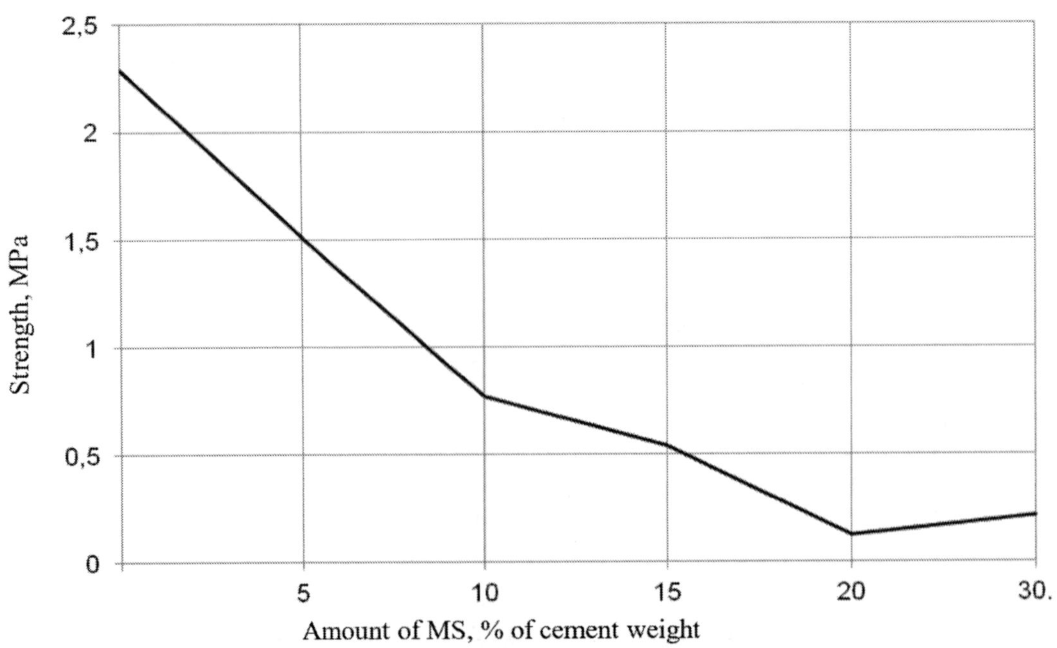

Figure 3. The dependence of the compression strength of the foam concrete samples on the amount of microsilica

The results of determining the compression strength of the foam concrete samples are shown in

Figure 3. It is evident that the addition of MS to the foam concrete composition without the use of plasticizing additives leads to a reduction of this critical parameter. According to [10], at density of 600 kg/m^3 the strength of foam-concrete should not be less than 1,5 MPa. Experiments have found that this requirement can be met by replacing up to 5% by weight of the cement with microsilica.

Analysis of thermal properties of foam concrete shows (Figure 4) that thermal conductivity of samples prepared using MS is on average 2 times less than that of the control sample and varies in the range from 0,12 to 0,08 W/(m•K), corresponding to [11-15].

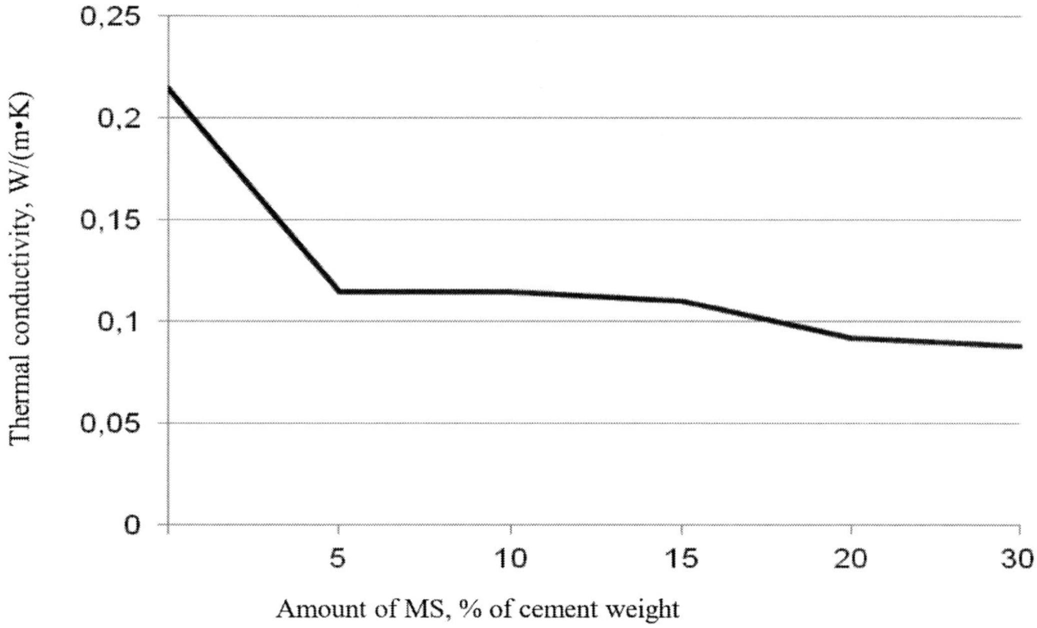

Figure 4. The dependence of the thermal conductivity of the foam concrete samples on the amount of microsilica.

A number of conclusions can be drawn from the studies:

- It is possible to replace up to 5% of cement with microsilica without deterioration of physical and mechanical properties of the obtained foam concrete;
- Introduction of MS into the composition of foamed concrete leads to reduction of its thermal conductivity, which can be used in manufacturing of heat-insulating materials;
- to improve the strength characteristics of foamed concrete made using microsilica, the amount of water in the mixture (water-solid ratio) should be reduced. It will be appreciated that plasticizers are required.

Accordingly, it is necessary to continue research in this direction using modern plasticizers, which may increase the strength of the product.

References

[1] Dashitsyrenov D D, Zayakhanov M E and Urhanov L A 2007 Effective foam concrete based on effusion rocks. *Construction materials* **4** pp 33

[2] Gorbach P S, Novikova L E and Tikhonov E G 2013Effective foam concrete with specified Properties *Problems of energy saving in industrial and social-communal complexes: collection of works of the XIV International Scientific and Practical Conference* pp 53-56

[3] Lot B A and Mitina H A 2000 Orming of optimal porous gas concrete string of non-autoclave Hardening *Eq. higher education institutions. Chemistry and Chemical Technology* **43 (3)** pp 118-119

[4] Laukaitis A A 2004 Study of influence of additives of ground wastes of cellular concrete on its

properties. *Construction materials* **3** p 33
[5] Krivitsky M J, Levin N I and Makarychev V V 1972 Cellular concrete (techno-logic, properties and structures) p 136
[6] Bertov V M and Sobkalov P F 2005 Use of fly ash in foam concrete production *Construction Materials* **5** pp 12
[7] Strokova V V 2007 Non-autoclave cellular concrete based on modified binder *Science and youth at the beginning of the new century* pp 65-68
[8] Luzhnova E V, Tugarin A O and Korchinov A S 2004 Wastes of forest chemistry in the technology of obtaining cellular concrete *Works of the Bratsk State Technical University*. – **2** p 370
[9] Urhanova L A, Gorbach P S, Savenkov A I and Scherbin S A 2008 Use of secondary raw materials for foam concrete production *Construction materials* **1** pp 30-34
[10] Panova V F 2003 Development of decorative slag cement. NGASU *Modern construction materials and resource-saving technologies* p 168
[11] Avvakumov E G 2001 Soft Mechanochemical Synthesis a Basis for New Chemical Synthesis a Basis for New Chemical Technologies. Hardbound: Kluwer Academic Publishers p 216 p
[12] Zhigulina A 2012 Building Envelopes: An Objective Measure of Comfort in Residential Buildings *Urban Planning* **1** pp 80-81
[13] Kazimirov I A and Peshkov V V 2019 Determination of price behaviour in the secondary residential real estate market using a multidimensional regression model *Proceedings of Universities. Investment. Construction. Real estate* **9(3)** pp 476–487 DOI: 10.21285/2227-2917-2019-3-476-487
[14] Korenkova S 2000 Filled Foam Concretes in Building Envelope Construction *Building Materials*, № 8 pp 12-14
[15] Petrov A V and Efimova A K 2019 Nguyen Thanh Tung. Optimisation of technology used for restoring gas-ash concrete outdoor wall panels using modified structural and heat-insulating *concrete Investment. Construction. Real estate* **9(3)** pp 542–549
[16] Tarasova A G, Alalykin A A and Vesnin E S 2017 New heat-insulating materials based on inorganic binder **4**
[17] Shalaginya N A 2009 And. Patent for utility model RU 86613, 10.09.2009. Denunciations No. 2009116591/22 of 30.04.2009

IOP Conf. Series: Materials Science and Engineering **880** (2020) 012022 doi:10.1088/1757-899X/880/1/012022

Whiskers as perspective input material for production of high temperature resistant ceramic building materials

T I Shishelova[1], V V Fedchishin[2] and M A Hromovskih[3]

[1]Department of Physics, INRTU, Lermontov Str. 83, Irkutsk 664074, Russia
[2]Energy department, INRTU, Lermontov Str. 83, Irkutsk 664074, Russia
[3]Department of Construction Production, INRTU, Lermontov Str. 83, Irkutsk 664074, Russia

E-mail: Hramovskih.misha@yandex.ru

Abstract. Production of electric power by nuclear power plants (NPP) in the world and in Russia accounts for about 17%. Much attention is currently paid to safe NPP operation and to enhancement of their reliability. Present-day nuclear power plants with high-temperature gas-cooled rectors require development and application of new composite materials capable to withstand high temperatures and mechanical stresses. Ceramic composite materials where whiskers of heat-resistant chemical compounds are used as reinforcement fibers look promising for those purposes. The paper investigates and analyses the opportunity of applying whiskers as reinforcement fibers for heat-resistant composite ceramic building materials. Studies revealed that whiskers of silicon carbide and aluminium nitride are resistant towards high temperature heating without notable change in their chemical composition. It was found out that mono crystals of silicon carbide have higher mechanical strength owing to lack of defects in its structure. Linear heat expansion factors of silicon carbide that are needed for thermo-chemical calculations of composite materials on its base were determined.

1. Introduction

About 17% of electric power generated in Russia and in the world fall on nuclear power plants. Design of nuclear reactors is continuously enhanced with focus on their safety, reliability, and efficiency. High-temperature gas-cooled nuclear reactors are thought to be more advanced units [1]. Temperature of their fuel elements may be as high as 1600°C [2, 3]. Correspondingly, temperature of the heat transfer medium and of pipes for its circulation, temperature of the gas turbine elements ensuring circulation of heat medium in the loop, and that of a heat exchanger may be higher than 1000°C. The structures of such reactors need high-temperature composite building materials of high chemical resistance and mechanical strength [4]. In this connection the ceramic materials of different structure and chemical composition arouse more and more interest. Whiskers have high mechanical strength owing to peculiarities of their structure, they are chemically inert towards many aggressive media, and have high temperature resistance [5-7]. Whiskers are used in industry as reinforcing fibers for a large range of composite materials. Development of ceramic temperature-resistant composite materials with the use of whiskers would allow production of composites capable to withstand high mechanical loads but maintaining high temperature and chemical resistance [8-11]. Creation of such materials requires studies on changes of whiskers properties under high temperatures.

IOP Conf. Series: Materials Science and Engineering **880** (2020) 012022 doi:10.1088/1757-899X/880/1/012022

2. Methods and materials

Mechanical properties and temperature resistance of silicon carbide (SiC) and of aluminium nitride (AlN) whiskers at different temperatures were studied using the IR spectroscopy. IR spectra were recorded by an IR spectrometer Specord. Factors of heat linear expansion of silicon carbide whiskers were determined using a dilatometer.

3. Results and discussion

Silicon carbide (SiC) and of aluminium nitride (AlN) whiskers demonstrated the highest temperature resistance. The main mechanical properties of whiskers of those compounds are given in Table 1.

Table 1. Mechanical properties of silicon carbide (SiC) and of aluminium nitride (AlN) whiskers

Fiber material	Melting temperature, °C	Density, 10^{12} kg/m^3	Elasticity modulus, 10^9 N/m^2	Ultimate resistance , 10^9 N/m^2	Specific strength, 10^6 N·m/kg	Specific elasticity modulus, 10^6 N·m/kg
SiC	2665	3.32	580	27	8.4	180
AlN	2400	3.30	380	15	4.55	115

Considered chemical compounds of whiskers have high melting temperature corresponding to a class of extremely-high temperatures (2000-3000°C). The 1.8 times difference in the values of ultimate resistance of mono crystals is due to difference of elasticity modulus of those compounds. This fact, in its turn, evidences the differences in the structure of a crystalline lattice of compounds at equal densities, and availability of such defects as brittle fractures in the aluminium nitride structure [12,13]. The data obtained prove more perfect (in terms of strength) structure of silicon carbide mono crystals.

Silicon carbide and aluminium nitride whiskers were also subjected to IR investigation at different temperatures of heating.

IR absorption spectrum of SiC crystals is given in Fig. 1. The main absorption band of silicon carbide is observed at a frequency of (840-920)·10^2 m^{-1}. Moreover, silicon carbide has two low-intensity bands at 670·10^2 m^{-1} and 1090·10^2 m^{-1}, the latter characterizing the variation in Si-C bonds. The band of 920·10^2 m^{-1} rapidly loses its intensity and shifts towards a short-wave area when heated up to 200°C. At the same time a fuzzy band emerges in the range of (1100-1300)·10^2 m^{-1}. At 800°C the band of 920·10^2 m^{-1} grows in intensity, whereas the band of 840·10^2 m^{-1} extends towards a long-wave area. During heating for one hour at a temperature of 1200°C new peaks emerge in the area of (1200-1260)·10^2 m^{-1}.

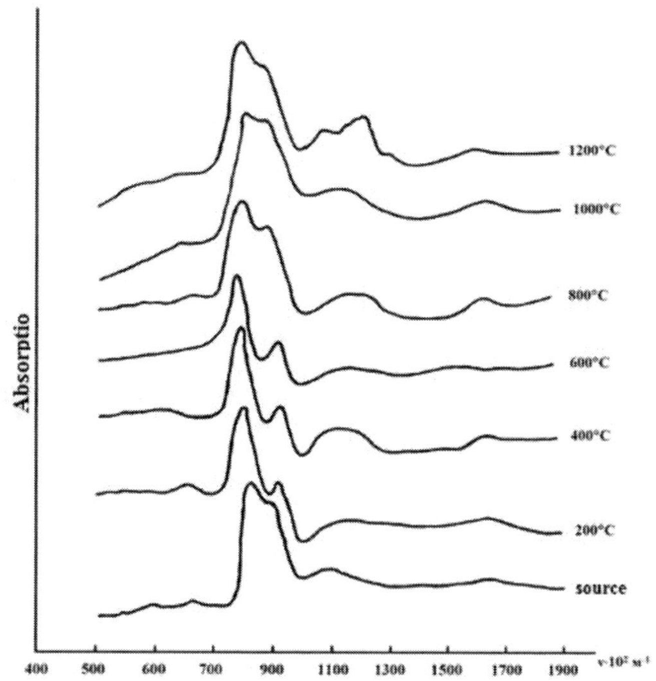

Fig. 1. IR spectra of SiC whiskers at different temperatures.

IR spectra of SiC whiskers are given in Fig. 2. They constitute a system of intense fuzzy bands in the area of KBr and NaCl. Peaks are obvious at $625\cdot10^2$ m^{-1} and $815\cdot10^2$ m^{-1}. With the temperature growth during heating, these bands practically do not change. During further heating the weak bands in the areas of $1160\cdot10^2$ m^{-1} and $1330\cdot10^2$ m^{-1} gradually disappear.

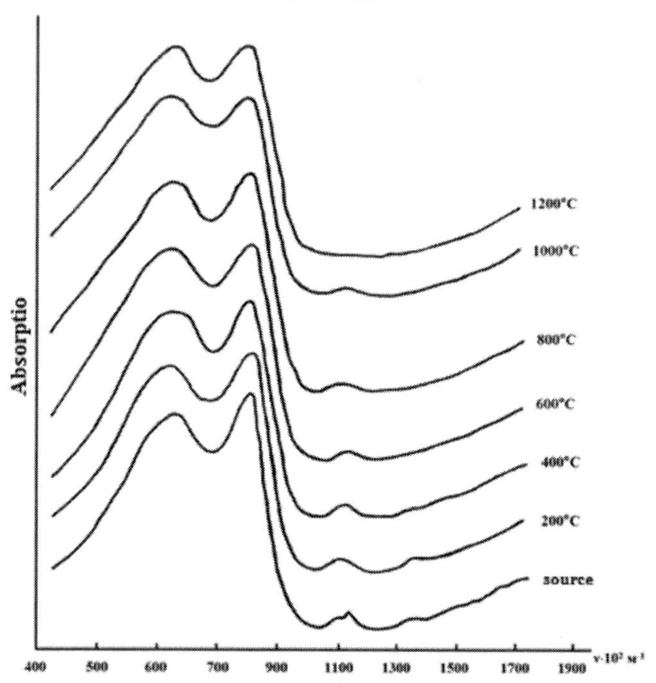

Fig. 2. IR spectra of AlN whiskers at different temperatures.

Under relatively small changes in the spectrum absorption bands at different temperatures of heating the whiskers of investigated chemical compounds, the silicon carbide demonstrates better mechanical strength and has more perfect structure of mono crystals.

When developing a composite material, the thermal expansion factors of materials constituting the composite shall be taken into account along with temperature-dependent changes in its composition [14-18]; for thermo-mechanical calculations of temperature-resistant composite materials the thermal expansion factors of its components shall be considered in a wide range of temperatures. SiC is known to have a wide variety of crystalline structure types. Authors determined the thermal linear expansion factors for two main SiC modifications, namely for cubical (β-SiC) and for hexagonal (α-SiC) ones. Results of studies are given in Table 2.

Table 2. Thermal linear expansion factors for different modifications SiC crystalline structure

Material	Temperature, °C	Thermal expansion coefficient $\alpha \cdot 10^{-6}$ degree^{-1}
Mono crystal α-SiC	25-1000	5.12
	25-1500	5.48
	25-2000	5.77
Mono crystal β-SiC	200	3.8
	400	4.3
	600	4.8
	800	5.2
	1000	5.8
	1400-1800	5.5

Optimum composition of bonding material is selected based on the equality of numerical values of linear heat expansion factors for whiskers applied as reinforcement fibers of composites.

4. Conclusion

Thermal studies of silicon carbide (SiC) and of aluminium nitride (AlN) whiskers confirm the suitability of whiskers for creation of high temperature resistant ceramic building materials and coatings. IR studies of mono crystals have shown that there is no notable change in the chemical composition of the substance during heating. Considerable differences in the mechanical strength of mono crystals have been observed. Silicon carbide whiskers have lesser number of structural defects than aluminium nitride whiskers. Numerical values of linear heat expansion factors have been determined for silicon carbide whiskers that have higher mechanical strength and melting temperature that are needed for thermo-mechanical calculations of composite materials.

References

[1] Patent RU 2481656 C2, A nuclear fuel rod holder in the fuel assembly of gas-cooled fast-neutron reactor with high-temperature gas cooling Lorenzo Deni, Ravene Alen, Garry Roren and Melkh Gui; applicant and patent holder Commissariat a l'Energie Atomic E OZ Energy Alternative – № 2010121168/07; applied on 24.10.08; published on 10.05.13.

[2] Denisov V P and Dragunov Yu G 2002 WWER reactors units for nuclear power plants. 480

[3] Frost B 1986 Fuel elements of nuclear reactors p 221-2

[4] Chernysheva T A, Kurganova Yu A, Kobeleva L.I, Bolotova L K, Kalashnikov I E and Katin I V 2007 Composite materials with an aluminium alloys matrix reinforced with particles, for slipping friction couples *Special-purpose coatings and materials* **3** pp 38-49

[5] Kablov E N 2012 Strategic trends in the development of materials and technologies for their processing for the period till 2030 *Aviation materials and technologies* pp 7-17

[6] Andrievsky R F and Ragulya A V 2005 Nano structural materials p 192

[7] Keller E 1964 *Industr. And Engng. Chem* **56** p 9
[8] Levett A P 1966 *Mater. Res. and Standarts* **6** p 64
[9] McCreight L R, Rauch H W и Sutton W H 1965 *Ceramic and Graphite Fibers and Whiskers* 244
[10] Cratchhley D 1965 *Met. Rev* **10** p 79
[11] Parratt N J 1964 *Powed. Met* **7** p 152
[12] Mikhailov A M 2009 Theory of strength of materials: Textbook for students of higher educational institutions p 448
[13] Samsonov G V 1978 *Nitrides* p 356
[14] Kostikov V I and Varenkov A N 2003 *Extremely high-temperature composite materials* p 560
[15] Kostikov V I and Varenkov A N 2001 *Composite materials on the base of aluminium alloys, reinforced with hydrocarbon fibers* p 528
[16] Vodennikov S A, Skachkov V A, Vodennikova O S and Ivanov V I 2012 Thermal and physical characteristics of multi-component composite materials. *New materials and technologies in metallurgy and machine building* **1**
[17] Vodennikova O S 2012 Study and forecasting the thermal and physical characteristics of multi-component composite materials *East-European journal of advanced technologies* **1/5 (55)** pp 11-3
[18] Skachkov V A, Baglyuk G A, Vodennikova O S and Ivanov V I 2012 On forecasting the heat conductivity and linear thermal expansion factors for multi-component composites. *Tribology issues* **2(64)** pp 108-11

IOP Conf. Series: Materials Science and Engineering **880** (2020) 012023 doi:10.1088/1757-899X/880/1/012023

Use of mica-based composite materials for radiation protection of buildings

T I Shishelova[1], M A Hramovskih[2] and N F Chuvashev[3]

[1]Department of Physics, INRTU, Lermontov Str. 83, Irkutsk 664074, Russia
[2]Department of Construction Production, INRTU, Lermontov Str. 83, Irkutsk 664074, Russia
[3]Department of Physics, INRTU, Lermontov Str. 83, Irkutsk 664074, Russia

E-mail: Hramovskih.misha@yandex.ru

Abstract. Radioactive emission is widely applied in different fields of sciences, in various technologies, and in medicine. For its safe use the protective screens shall be used that limit the impact of radiation on a human being. The paper deals with study and analysis of opportunities for applying different materials for protecting buildings and premises from radiation. Studies have shown that phlogopite micalex has high radiation resistance. Calculations of masses of conventional panels made of different materials ensuring tow-fold and ten-fold attenuation of X-ray and gamma radiation are presented. Studies revealed that micalex is the most promising and economically feasible material that can be used for those purposes. A new area for micalex application, namely, for radiation protection of buildings and structures of different purpose has been proposed.

1. Introduction

"Green" houses are more and more widely recognized by population. The main concept of such buildings consists in reduction of their impact on the environment, human beings, and human health. Sustainable construction is closely related to the concept of people protection from natural or man-induced impact of infrastructure facilities located in the neighborhood of dwelling housing. Mills producing radioactive ores in Irkutsk obalst, in other regions of Russia, and abroad are hazardous for human health. According to data of the World Nuclear Association as of 2015, Russia possesses 9% of the world uranium resources and produces about 3-3.5 thousand ton of uranium annually. The main uranium deposits in Russia are located in the Trans-Baikal area and in the West Siberia. Uranium ore is open-pit mined. Primary uranium treatment is an environmentally hazardous process. Radiation pollution in the areas of uranium ore production and treatment is thousands times higher than the permissible radioactive background. One of the uranium enrichment mills of Russia (out of four) is located in Angarsk, Irkutsk oblast. Uranium enrichment process generates radioactive wastes, a part of them being buried right on the territory of the mill [2-5]. Emissions from chemical processes, use of phosphate fertilizers in agriculture, operating or abandoned nuclear test bases, areas for disposal of nuclear wastes and other chemical substances, and natural emissions of radionuclide in the tectonic faults have negative impact on human health as well [6,7]. In this connection, there is a real demand for structures protecting humans from negative impact of radiation.

2. Methods and materials

Samples of phlogopite micalex (mica - 60%, low-melting glass 203 - 35%, volcanic ash - 5%) were used for studies. Electro physical and mechanical characteristics of micalex were determined following standard techniques set forth in GOST; to determine a linear coefficient of micalex gamma-radiation attenuation the radiometry methods were used.

3. Results and discussion

Screens are the most efficient and universal method of protection from radiation. Materials used for manufacture of screens for different types of radiation differ in composition, structure and properties. Gamma rays have the highest penetrability and, hence, are most hazardous for human health. For protection from gamma radiation the highly dense materials composed of chemical elements with high nuclear weight are used, they are lead, steel, concrete, etc. Thickness of screens made of those materials is rather high and, hence, their manufacture and erection especially for dwelling houses is technologically difficult and economically inefficient [8]. Barite X-ray protection plaster is an alternative for internal protection of premises. It is predominantly used in medical facilities and research laboratories for protection of premises that are close to rooms with X-ray equipment. A drawback of this method of protection is complexity of its application: the layer applied shall not exceed 10 mm, and it needs reinforcement and polishing after hardening. Moreover, hazardous baryte dust ($BaSO_4$) shall be coated with common plaster, painted, covered with wallpaper or plastic panels. In view of the above said, it is necessary to note that ability of mica and mica-based composite materials [9,10] to protect from radiation is underestimated. In the course of the patent search conducted, we came across a patent describing a room for temporary stay of people (a mica room) for relaxation after emotional stress, for protection from hazardous noise and radiation of any kind [11]. The premise was arranged on the base of a maritime container whose walls, floor, ceiling, door and a transparent part of the window were made of mica. Protection of people exposed to higher radiation implies enhancement of radioactive protection of places for durative stay of people, i.e., special protection is required for work places located close to a radioaction source, as well as protection of homes. For siding the premises, it is advisable to use the so-called mica fits. The cost of such fits is high as production of large sheets of mica is labor intensive, and after cutting the shapes of desired size a large amount of scrap is generated which damages environment in the areas of mica production and processing. Micalex is a universal composite material with a number of useful characteristics. It is produced by hot pressing of a mixture of powdered mica and finely dispersed soft glass, and this process practically does not generate scrap [12]. Micalex has high mechanical and dielectric strength, low water absorbency, high heat and chemical resistance, which improves fire safety and sanitary-hygienic micro flora of the building after it has been sided with micalex. Average density of micalex is2800-3000 kg/m^3, i.e., it is similar to that of heavyweight concrete whose fillers are steel, carbonate of magnesia, baryte, and iron ore.

Previous studies [13] on identifying the optimum mix of micalex for radiation protection have shown that samples with the following composition: mica – 60%; low-melting glass 203 – 35%; volcanic ash – 5% have the best radiation resistance characteristics. It should also be noted that a linear attenuation coefficient of phlogopite micalex is higher than that of muscovite one. This difference may be due to different structure of phlogopite and muscovite and due to higher heat resistance of phlogopite mica. Phlogopite dehydroxylation process occurs at higher temperatures, which allows maintenance of the structure and density of the input material for a longer time period.

When X-rays and gamma radiation go through the substance, their intensity attenuates exponentially following the law described by equation (1):

$$I=I_0\cdot e^{-\mu\cdot d} \quad (1)$$

where I_0 – intensity of radiation falling on the surface of a substance;

I - intensity of radiation leaving the substance;

μ - linear coefficient of radiation attenuation in the substance that depends on the substance nature;

d – thickness of absorbing layer of a material.

The linear attenuation coefficient in this equation is the main characteristic of material resistance towards X-ray and gamma radiation that depends on the type of protection material and on energy of penetrated X-ray and gamma radiation.

Linear coefficients of gamma-radiation attenuation for micalex of optimum composition were calculated for radiation energy of 10 and 15 MeV. Table 1 [17] gives values of the most popular radiation protection materials, such as lead, aluminium, steel, concrete, water, and micalex for comparison.

Table 1. Linear attenuation coefficients of different materials vs gamma radiation energy

	Material	Radiation energy, MeV	
		10	15
μ, cm^{-1}	Lead	0.5520	0.6280
	Aluminium	0.0619	0.0584
	Steel	0.2330	0.2410
	Concrete (common)	0.0369	-
	Water	0.0220	0.0193
	Micalex	0.2200	0.1870

Analysis of data in Table 1 shows that gamma radiation linear attenuation coefficient for micalex is 5.96 times higher than that of concrete. The difference is dramatic. Linear attenuation coefficients for micalex are comparable to those of steel, and are inferior to those of lead by a factor of 2.51-3.36 only, whereas lead density is 3.77 times higher than that of micalex.

Physical mass of protection screens is an important parameter to be taken into account during buildings engineering: it shall be minimum subject to maintenance of all the protective properties. To determine the efficiency of micalex use in terms of this indicator, the thickness of protection screens for radiation energy of 10 and 15 MeV was calculated. Thickness of material layers most frequently used in engineering calculations is the one that ensures two-fold of ten-fold attenuation of gamma and X-ray radiation. Mass of materials was calculated for a conventional panel with dimensions of (1 м×1 м×d м), where d is thickness of the material layer needed for two-fold attenuation of the X-ray and gamma radiation.

Table 2. Thickness of two-fold attenuation of X-ray and gamma radiation at different radiation energies; volume of conventional panels and their mass for different materials

Material	Density ρ, kg/m^3	Layer thickness d (cm) at radiation energy of 10 MeV	Volume V (1m × 1m × d m), m^3	Mass m, kg	Layer thickness d (cm) at radiation energy of 15 MeV	Volume V (1m × 1m × d m), m^3	Mass m, kg
Lead	11300	1.256	0.01256	**141.928**	1.104	0.01104	**124.752**
Aluminium	2700	11.198	0.11198	**302.346**	11.869	0.11869	**320.463**
Steel	7890	2.975	0.02975	**234.728**	2.876	0.02876	**226.916**
Water	1000	31.507	0.31507	**315.070**	35.914	0.35914	**359.140**

Concrete (common)	2300	18.800	0.18800	**432.400**	-	-	-
Micalex	3000	3.151	0.03151	**94.530**	3.707	0.03707	**111.210**

It is obvious that for calculated masses presented in Table 2, a conventional micalex panel of equal size and equal radiation attenuation efficiency is lighter than a lead panel by a factor of 1.5, and lighter than a steel panel by a factor of 2.48. A ratio between a concrete conventional panel and a conventional micalex panel is 4.57, which evidences the economic inefficiency of using conventional concrete for protection from radiation. For the case of heavyweight concrete, the mass ratio diminishes, but the cost of concrete would be much higher than that of micalex. Therefore, utilization of micalex for protection from radiation turns out to be most efficient. Results for 15 MeV radiation energy are similar. With the radiation energy growth the efficiency of micalex use becomes lower than that of lead. In the 15-25 MeV range the efficiency of using both materials will be equivalent. At higher energy values, the use of lead becomes more efficient. It should be noted that radiation energy of 20 MeV and higher is not practically used in industries.

Masses of a conventional panel were also calculated for tenfold attenuation of X-ray and gamma radiation. Mass of materials was calculated for a conventional panel with dimensions of (1 м×1 м×d м), where d is thickness of the material layer needed for tenfold attenuation of X-ray and gamma radiation.

Data presented in Table 3 show that ratios between lead and steel mass to micalex mass remained the same, which confirms conclusions given in Table 2.

Thus, authors propose new field of application for mica- and glass-based composite materials, that is, for radiation protection of buildings and structures of different purpose.

Table 3. Thickness of panels for tenfold attenuation of X-ray and gamma radiation at different radiation energies; volume of conventional panels and their masses for different materials

Material	Density ρ, kg/m^3	Layer thickness d_1 (cm) at radiation energy of 10 MeV	Volume V (1m × 1m × d m), m^3	Mass m, kg	Layer thickness d_1 (cm) at radiation energy of 15 MeV	Volume V (1m × 1m × d m), m^3	Mass m, kg
Lead	11300	4.1713	0.041713	**471.357**	3.6665	0.036665	**414.315**
Aluminium	2700	37.1985	0.371985	**1004.360**	39.4278	0.394278	**1064.551**
Steel	7890	9.8823	0.098823	**779.713**	9.5543	0.095543	**753.834**
Water	1000	104.6630	1.046630	**1046.630**	119.3049	1.193049	**1193.049**
Concrete (common)	2300	62.4007	0.624007	**1435.216**	-	-	-
Micalex	3000	10.4663	0.104663	**313.989**	12.3133	0.123133	**369.399**

4. Conclusion

The study presented is devoted to the problem of radiation protection of humans in the places with high radiation background. The subject of analysis was radiation pollution of the Irkutsk oblast. The paper demonstrated the need for engineering and constructing the buildings resistant towards radiation. Linear attenuation coefficients for X-ray and gamma radiation at different values of radiation energy were calculated. Thickness of layers for two-fold and ten-fold attenuation of radiation was calculated for materials most demanded and most frequently used for radiation protection. Masses of conventional panels of equal area for thickness ensuring twofold and tenfold attenuation of radiation were calculated

IOP Conf. Series: Materials Science and Engineering **880** (2020) 012023 doi:10.1088/1757-899X/880/1/012023

and compared. Studies performed allowed conclusions on economic feasibility of using micalex for radiation protection.

5. References

[1] Muratov O E, Tikhonov M N and Petrov E L 2006 Radiation, ecological and resource aspects of uranium-thorium fuel cycle *Ecol. Expertise* **6** pp 99-109

[2] Dovgusha V V and Tikhonov M N 1997 Ensuring the ecological safety when handling radioactive wastes at nuclear and radiation hazardous facilities of Russian Federation *Ecology of Industry* **3** pp 30-46

[3] Tikhonov M N, Petrov E L and Muratov O E 2005 Systems glimpse at nuclear energy and radiation through public outlook *Regional ecology* **1** pp 80-89

[4] Shatalov V V, Brykin S N and Serebryakov I S 2004 Account and control of radioactive substances and radioactive wastes at mills with nuclear fuel cycle *Nuclear strategy–XXI pp* 19

[5] Isaev A N 2005 World experience of storing the waste nuclear fuel *Nuclear technologies abroad* **1** pp17-21

[6] Tikhonov M N and Rylov M I 2007 Complex assessment of nuclear and radiation heritage of Russia *Problems of environment and natural resources* **3** pp 77-110

[7] Tikhonov M N 2007 Radiation geography of Russia *Ecol. Expertise* **3** pp 62-71

[8] Gusev N G, Mashkovich V P and Suvorov A P 1980 *Protection from ionizing radiation* 1 623

[9] Alekseenko V A and Ivanov A B 1995 Geological and hydrological aspects of the problem of liquid radioactive wastes burial *Materials of III International Symposium on the issues of radioactive wastes burial* pp 16-19

[10] Sobotovich E V 2003 *Radioactive wastes of Ukraine: status, problems, solutions.* 400

[11] Patent 2483703 RF, MPK A61G 10/00, A premise for temporary residence (a mica room) Kulikov I N; Applicant and patent holder Kulikov I N – № 2011151707/14; applied on 16.12.11; published on 10.06.13.

[12] Shishelova T I and Shulga V V 2016 Physical and chemical grounds for mica-based composite materials production *Fundamental Research* **3** pp 538-41

[13] Shishelova T I and Zhitov V G 2017 Use of mica-based composite materials for burial of radioactive wastes *XXI century. Technosphere safety* **2** pp 86-92

[14] Bereznyak E P, Saenko L A and Shevyakova E P 2010 Radiation resistance of dark-colored minerals, members of granites *Vestnik HNU* **915** pp 44-47

[15] Neklyudov I M 2006 Handling the waste nuclear fuel and radioactive wastes *Nuclear Energy* p 253

[16] Neklyudov I M, Shevyakova E P and Bereznyak E P 2006 *Physics of radiation damages and radiation materials sciences* p 60

[17] Dubrovsky V B 1997 Radiation resistance of building materials *Stroyizdat* p 256

IOP Conf. Series: Materials Science and Engineering **880** (2020) 012024 doi:10.1088/1757-899X/880/1/012024

The elastic modulus for ash and slag mixtures reinforced with inorganic binders

N Slobodchikova[1] and A Rudykh[2]

[1]Irkutsk National Research Technical University, Department of Automobile Roads, 664073 Irkutsk, Lermontov st., 83, Russian Federation
[2]Irkutsk National Research Technical University, Department of Mechaniks and materials, 664073 Irkutsk, Lermontov st., 83, Russian Federation

E-mail: NSlobodchikova@rambler.ru

Abstract. The article analyses applications of waste produced by thermal power plants and used by the road construction industry. One of the promising areas for the use of ash and slag materials is their reinforcement with inorganic binders and using for the construction of pavement. The Russian regulatory documents governing the use of ash and slag materials have several drawbacks, including the lack of recommendations for determining the elastic modulus for bases and coatings made from ash and slag mixtures reinforced with an inorganic binder in laboratory conditions. The lack of these recommendations does not allow the selection of the mixture with specified design values of the elasticity modulus. It makes it difficult to ensure required values of the modulus of elasticity in the construction of road pavement. The authors developed a method for determining the elasticity modulus in laboratory conditions in order to produce materials with specified values of the elasticity modulus and improve quality of design and construction of road pavement from ash and slag mixtures reinforced with inorganic binders.

1. Introduction

The road industry needs cheap and effective building materials which can improve quality of pavement, increase its turnaround time and reduce the cost of road maintenance. The use of waste produced by thermal power plants for producing road-building materials can help reduce their cost and preserve natural resources [10, 16, 23]. Over 80 million tons of ash and slag materials have been accumulated in Irkutsk region, and their volume is increasing.

One of the promising areas for the use of ash and slag materials in the road industry is their reinforcement with inorganic binders for the construction of base layers of pavements [1-33].

The reinforcement of ash and slag materials with inorganic binders is regulated by the following documents:

- GOST 23558-94 "Mixtures of crushed stone-gravel-sand and soils reinforced with inorganic binders for road and airfield construction. Specifications (as Amended by N 1, 2) ";
- ODM 218.2.031-2013 "Guidelines for the use of fly ash and ash and slag mixtures from coal combustion at thermal power plants in road construction".

These documents have a number of drawbacks, the most important of which is the lack of recommendations for determining the elasticity modulus for ash and slag mixtures reinforced with inorganic binders in laboratory conditions. The lack of these recommendations does not allow the selection of the composition of mixtures with specified design values of the elasticity modulus. It makes it difficult to ensure required values of the modulus of elasticity in the construction of road pavement.

The article aims to develop a method for determining the elasticity modulus for ash and slag materials reinforced with inorganic binders in laboratory conditions.

2. Materials and methods

To determine the elasticity modulus for ash and slag mixtures reinforced with inorganic binders, the following method has been proposed.

1. Production of laboratory samples of reinforced soil with specified values of maximum density and optimal humidity.

Samples are made using a hydraulic press.

To produce samples of reinforced soil, it is recommended to use mold cylinders, the requirements for which can be found in GOST 12801-98 "Materials based on organic binders for road and airfield construction. Test methods" (Table. 1).

Table 1. Dimensions of mold cylinders in accordance with GOST 12801

The largest size of soil grains, mm	Sizes, mm				Area, cm2
	Diameter d	**Height H**	**Top liner height h1**	**Lower liner height h2**	
5	50,5	130	80	50	20
10, 15, 20	71,4	160	100	60	40
40	101	180	110	70	80

The mass of soil required for producing one sample is 600 - 2200 g. Depending on the size of the molds, sample cylinders have a height of 50.5; 71.4; 101.0 mm. If the sample cylinders have a height different from these dimensions, the mass of the mixture has to be adjusted.

$$g = g_0 \frac{h}{h_0}, \quad (1)$$

where g_0 - mass of the test sample, g;
h – required height of the sample, mm;
h_0 – height of the sample, mm.

The required amount of the mixture is determined by formula

$$P_{CM} = \rho_{в} \cdot V \cdot n = \rho_{CK}(1 + 0{,}01W_o)V_n, \quad (2)$$

where P_{CM} – mass of the mixture, g (kg);
$\rho_{в}$, ρ_{CK} - density of the wet mixture and the skeleton of the mixture, g / cm ^{3};
V – volume of the sample, cm^3;
n – the number of samples required to determine physical and mechanical properties of the mixture;
W_o – optimal humidity of the mixture, %.

Water content is determined in accordance with optimal humidity values determined according to PNST 324-2019 "General automobile roads. Soils. Determination of optimal humidity and maximum density by the Proctor method" and GOST 22733-2016 "Soils. Laboratory method for determining maximum density (as amended)."

1. Holding samples of reinforced soil under normal hardening conditions until they reach a design age

Samples are tested at the design age, depending on the type of binder:

- 28 days - Portland cement, slag Portland cement;
- 90 days – slowly hardening binders (fly ash, lime-ash, lime-slag, lime, cement-ash, etc.).

The samples should be stored and hardened at 20 ± 2 ° C and a relative humidity of 95 ± 5%. They are placed in

- a chamber with normal hardening conditions;
- a bath with a hydraulic shutter;
- a desiccator, etc.

Samples are placed on liners so that the distance between the samples, as well as the samples and the walls of the chamber is at least 5 mm. The contact area of the sample with pads on which it is installed should be no more than 30% of the area of the reference face of the sample. The samples should not be moistened. They should be stored under wet sand, sawdust or other moistened hygroscopic materials.

Water saturation of samples

Compressive strength is determined on samples subjected to complete or capillary water saturation in accordance with PNST 322-2019 “Automobile roads of general use. Soils stabilized and fortified with inorganic binders. Specifications“ and GOST 23558-94 “Mixtures of crushed stone-gravel-sand and soils reinforced with inorganic binders for road and airfield construction. Specifications (as Amended by N 1, 2) “.

Compression of samples

The samples are compressed using a press under the static load. When the load reaches a value of 50% of the breaking load, the sample is unloaded and reloaded (cycling).

Data processing

The elasticity modulus is calculated by formula

$$E = tg\alpha = \frac{\sigma}{\varepsilon}, \quad (3)$$

where E – elasticity modulus, Pa;

σ – stress caused by the acting force, Pa

$$\sigma = \frac{F}{A}, \quad (4)$$

where F – acting force, N;

A – force application area, m^2;

ε - elastic relative deformation of the material caused by stress

$$\varepsilon = \frac{l_1 - l_0}{l_0} \cdot 100\%, \quad (5)$$

where l_1 –size after deformation;

l_0 – initial size.

The elasticity modulus is determined by testing a series of laboratory samples. The number of samples should be 6.

To conduct an experiment according to the proposed method, an ash and slag mixture produced by Ust-Ilim TPP was used. Ash-cement binder based on Portland cement and an ash-slag mixture of Novo-Irkutsk TPP were used as binders.

The results of determination of the elasticity modulus are presented in Table 2 and Figure 1.

Table 2. The results of determination of the elasticity modulus for ash-slag mixture reinforced with inorganic binders

Binder	Binder content, %	Age, days	Elasticity modulus E, MPa	Compression strength, MPa	Values of the elasticity modulus E according to ODM 218.2.031-2013
Ash-cement	8	28	1250	3,7	no
Ash-cement	8	90	1500	6,6	no
Hydrated Air Lime	7	90	1200	4,39	350-250

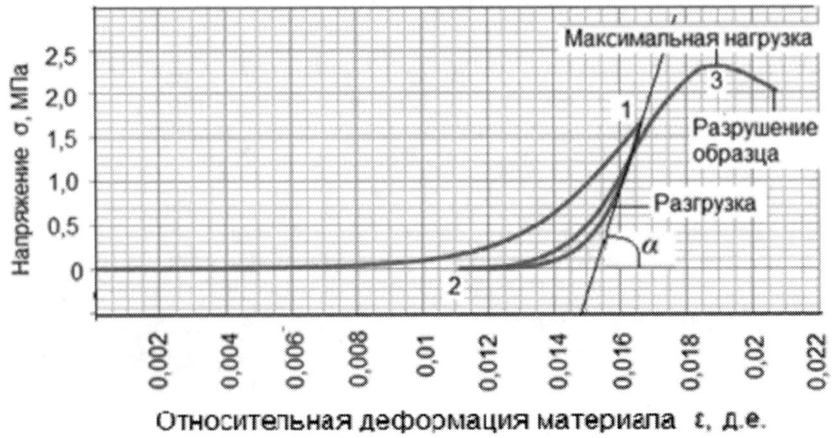

Stress, MPa relative deformation of the material
Maximum load Destruction of the sample Unloading
Fig. 1. Determination of the elasticity modulus for ash and slag mixture reinforced with ash-cement binder in the amount of 8%:
0-1 - loading;
1-2 - unloading;
2-3 - loading

The elasticity moduli of layers of monolithic materials can decrease as a result of cracking and water-frost effects and increase as a result of long-term physical and chemical processes. For materials and soils reinforced with inorganic binders located in the structural layers of pavement, cracking and spatial heterogeneity is characteristic. Therefore, the calculated value of the elasticity modulus is

$$E_{\text{р}} = \text{К}_{\text{к}} \cdot E, \tag{6}$$

where Кк is the coefficient taking into account the change in the modulus of elasticity of the material in the structural layer

The estimated value of the coefficient Kk is 0.4.

3. Conclusion

Determination of the elastic modulus for ash and slag mixtures reinforced with inorganic binders is very important since it allows the construction of pavement with specified elasticity modulus values.

The following conclusions can be drawn:

1. The current regulatory and technical documents governing the design and construction of pavement have several drawbacks, including the lack of recommendations for determining the elasticity modulus for ash and slag mixtures reinforced with inorganic binders in laboratory conditions.

2. The method proposed by the authors makes it possible to produce materials with specified elastic modulus values and improve quality of road pavement made from soils reinforced with inorganic binders.

References:

[1] *Design Procedures for Soil Modification or Stabilization* 2008 Production Division Office of Geotechnical Engineering 120 South Shortridge Road Indianapolis p 13
[2] *Field Manual No. 5-410* Headquarters Department of the Army Washington 1997 p 344
[3] Makusa G P 2012 State of the art review soil stabilization methods and materials in engineering practice *Department of Civil, Environmental and Natural resources engineering Division of Mining and Geotechnical Engineering Luleå University of Technology* p 35
[4] Lime-treated soil construction manual lime stabilization & lime modification *Nacional Lime Association Lime The Versatile Chimikal* 2004 p 326.

[5] *Road Construction Lime for Soil Stabilization and Soil Modification* A Proven Solution Quicklime and Hydrated Lime http://www.carmeusena.com/
[6] Kestler M A 2009 Stabilization Selection Guide for Aggregate and Native-Surfaced Low Volume *Roads Department of transportation* p 93
[7] Untersuchungen zur hygromehanischen Stabilitat von kristallinem Calciumsulfat-Halbhzdrat 2001 p 465
[8] User Guidelines for Waste and Byproduct Materials in Pavement Construction Publication Number: FHWA-RD-97-148.
[9] Highways and bridges. Construction of structural layers of pavement from soils reinforced with cementitious materials 2007 *Overview information* **3**
[10] Vatin N I 2011 The use of ashes and ash and slag waste in the construction industry Civil Engineering Journal **4 (22)** pp 16–21
[11] Vdovin E A 2013 Ways to increase the effectiveness of soil strengthening for the construction of pavement *Bulletin of SibADI* **1 (29)** pp 52–58
[12] Volzhensky A V 1979 *Mineral binders (technology and properties)* p 476
[13] Volkova E V 2015 Consideration of the unstabilized state of soils when assessing spatial stability of slopes of embankments of roads *Proceedings of universities. Investment. Construction. Real estate* **2 (13)**
[14] Goncharova L V 1973 *Fundamentals of artificial soil improvement* p 375
[15] Guryachkov I L 1965 Strengthening loamy soil with cement and addition of fly ash *Materials for the V Meeting on the consolidation and compaction of soils* pp 563–567
[16] Dvorkin L I 2007 *Building materials from industrial wastes* p 368
[17] Zubova O V 2015 Use of ash mixtures reinforced with cementitious materials in forest road construction
[18] Morozov S A 1961 Improving the effectiveness of soil reinforcement with lime in order to create sustainable foundations for *paving Proceedings of meetings on the theoretical foundations of technical soil reclamation* pp 205–215
[19] Nikolaev A A 1980 At the level of modern requirements *Highways* **7** pp 1–4
[20] Osin B V 1954 *Quicklime as a new astringent*
[21] Podolsky V P 2014 About the possibility of expanding the resource base of road construction due to stabilization and strengthening of soils *Scientific Herald of the Voronezh State University of Architecture and Civil Engineering. Construction and architecture* 1 pp 102–111
[22] The use of cements, lime and other binders in road construction *VesBeton* article 208.
[23] Putilin E I 2003 The use of fly ash and ash and slag mixtures in the construction of roads Overview of domestic and foreign experience in the use of waste from the combustion of solid fuels at thermal power plants
[24] Putsenko K N 2019 Review of regulatory documents in the field of design, construction and operation of engineering structures on highways [Electronic resource]. - URL https://elibrary.ru/item.asp?id=26284001 (accessed: 11.24.2019)
[25] Rogalsky B I 1956 The use of ground quicklime in construction *State publishing house of literature on construction and architecture* p 148

IOP Conf. Series: Materials Science and Engineering **880** (2020) 012025 doi:10.1088/1757-899X/880/1/012025

Thermodynamic calculations on system «Portland cement – water – carbon nanomodifier»

N N Smirnyagina[1], B O Tsyrenov[2] and L A Urkhanova[2]

[1]Institute of Physical Materials Science SB RAS, 670047, Ulan-Ude, Russia
[2]East Siberia State University of Technology and Management, 670013, Ulan-Ude Russia

E-mail: bulatzsk@gmail.com

Abstract. The article presents the results of thermodynamic modelling in a cement-water-carbon nanomodifier system. A carbon nanomodifier contains fullerenes C_{60} and C_{70}. Thermodynamic calculations make it possible to evaluate the contribution of a carbon nanomodifier to the hydration process of Portland cement. Using thermodynamic calculations, changes in the phase composition and general properties of the system (total enthalpy, entropy) with the introduction of a carbon nanomodifier were determined.

1. Introduction

Research is relevant for improving the properties of cement composites using carbon-type modifying additives [1-4]. Modifying additives containing carbon and allotropic forms of carbon (fullerenes, nanotubes, fulleroid particles) can improve the properties of the cement composite at small amounts: 0.001 mass% - 0.1 mass%. Modified composites have improved strength characteristics, increased frost resistance properties. The use of carbon nanomodifiers in the future will reduce cement consumption in the production of concrete various classes. The effect of using carbon nanomodifier (CNM) depends on the type and composition of the additive, since allotropic forms of carbon have different physicochemical and adsorption properties. One of the explanations for the effect of using carbon nanomodifiers is the effect of nanomodifiers on kinetics of cement stone hydration, changes in phase composition, and changes in the heat release of cement dough during hydration. The change in thermophysical properties can be determined experimentally (for example, by the calorimetric method) and theoretically, by the method of thermodynamic modeling.

Thermodynamic modeling (TM) is a universal research method. When creating new composite materials and developing new effective, resource-efficient and energy-efficient methods of synthesis, TM is used. The main task of thermodynamic modeling is to determine phase and chemical composition of the system and values of thermodynamic properties of the system. Thermodynamic data allow us to assess the possibility of processes in terms of thermodynamics.

There are many programs for thermodynamic modelling [5,7]. For example, the interactive GEMS-Selektor package for thermodynamic modeling of water (geo) chemical systems by Gibbs Energy Minimization [5-6]. The interactive GEMS-Selektor package can be used in various fields of science depending on the database, particularly in construction chemistry.

In this paper, we used the multi-purpose software package TERRA [8]. Thermodynamic modeling is an important step in creation of composite materials, their production processes, and hydration of the source components. Terra software for calculating chemical reactions uses equilibrium conditions:

Gibbs energy minimization and entropy maximum condition. In the first case, the optimization problem presented by the conditions $\Delta G^{o}_{reaction} \rightarrow \min$; $\sum_{ij}^{N} C_{ij}X_j \leq b_i$, where b_i a certain amount of interacting chemical element B_i ; X_j - stoichiometric coefficient of the compound, C_{ij} C_{ij} - number of Bi atoms in compound j, N - number of chemical elements involved in the reaction; n - number of reaction products. In the second case, the calculation is based on the maximum entropy conditions $S \rightarrow S_{\max}$, constancy of the total internal energy, fulfillment law of conservation mass of all chemical elements [8, 9].

In the problem of thermodynamic modeling, two conditions of equilibrium with the environment were assigned. These conditions can be either numerical values of the thermodynamic characteristics of the equilibrium, or functional relations between the parameters of this state. To describe a system as a material object, it is necessary to know the content of the chemical elements that form it. Internal and interfacial interactions are described by model thermodynamic relations, which are closed using the properties of only individual substances-components of equilibrium.

2. Experimental part

In this paper we used the interfaces of the TERRA software package:

1. *TERRA*. The maximum number of chemical elements that can make up the system under study is 50; the number of condensed phases considered in a single calculation is limited to 200, and the number of components of the gas phase formed in equilibrium (the number of individual substances) can reach 800. When performing calculations for heterogeneous systems, it is possible to use a model of single-component immiscible phases and models of condensed solutions.
2. *INFO*. The INFO helper program is designed to serve the TERRA database. The INFO database contains the thermodynamic data of substances for calculating chemical and thermodynamic equilibrium of arbitrary multicomponent systems.

For thermodynamic calculations hydration of Portland cement in the database of the TERRA software complex (INFO program) were added:

- fullerenes C_{60} and C_{70}, Table 1 [10],
- fullerene C_{60} hydrides [11],
- thermodynamic properties of substances involved in the hydration process, Table 2 [12-14]

Table 1. Thermodynamic properties of fullerenes

	$\Delta_f H^0$ kJ/mol	S^0 J/K/mol	a_0 J/K/mol	a_1 J/K^2 /mol	a_2 J/ K^3/mol	a_3 J/K^4/mol
Fullerene						
C_{60}	2346	427.1	-292.4	3.43	-0.025	$6.36 \cdot 10^{-7}$
C_{70}	2555	452.7	-295.03	3.78	-0.0025	$5.39 \cdot 10^{-7}$
$C_{60}H_2$	2280	395.8	-424.44	4.23	-0.0034	$1.05 \cdot 10^{-6}$
$C_{60}H_{18}$	1561	437.9	-507.17	4.74	-0.0036	$1.06 \cdot 10^{-6}$
$C_{60}H_{60}$	1869	745	-640.67	5.54	-0.032	$6.85 \cdot 10^{-7}$

In Table 1 a_0, a_1, a_2, a_3 are the empirical coefficients of the heat capacity function: a_0, a_1, a_2, a_3 are the empirical coefficients of the heat capacity function: $C_P^{\circ} = a_0 + a_1T + a_2T^2 + a_3T^3$; $\Delta_f H^0$ - enthalpy of formation at 298 K; S^0 - entropy at 298 K [10].

Table 2. Thermodynamic properties of substances added to the TERRA program database

	$\Delta_f H^0$ kJ/mol	S^0 J/K/mol	a_0 J/K/mol	a_1 J/mol/K^2	a_2 J K/mol	a_3 J/K$^{0.5}$/mol	V^0 cm^3/mol
AFt-phases							
(Al-)ettringite	-17535	1900	1939	0.789	-	-	707
$C_6As_3H_{30}$	-16950	1792.4	1452	2.156	-	-	708
$C_6As_3H_{13}$	-11530.3	1960.4	970.7	1.483	-	-	411
$C_6As_3H_9$	-10643.7	646.6	764.3	1.638	-	-	361
Tricarboaluminate	16792	1858	2042	0.559	$-7.78 \cdot 10^6$	-	650
Fe-ettringite	-16600	1937	1922	0.855	$2.02 \cdot 10^6$	-	717
Hydrogarnet							
C_3AH_6	-5537.3	422	290	0.644	$-3.25 \cdot 10^6$	-	150
$C_3AS_{0.41}H_{5.18}$	-5699	399	310	0.566	$-4.37 \cdot 10^6$	-	146
$C_3AS_{0.84}H_{4.32}$	-5847	375	331	0.484	$-5.55 \cdot 10^6$	-	142
C_3FH_6	-4518	870	330	1.237	$-4.74 \cdot 10^6$	-	155
Al-Fe siliceous hydrogarnet (solid solution)							
$C_3FS_{0.84}H_{4.32}$	-4823	840	371	0.478	$-7.03 \cdot 10^6$	-	149
$C_3A_{0.5}F_{0.5}S_{0.84}H_{4.32}$	-5335	619	367	0.471	$-8.10 \cdot 10^6$	-	146
$C_3FS_{1.34}H_{3.32}$	-4994	820	395	0.383	$-8.39 \cdot 10^6$	-	145
AFm-phases							
C_4AH_{19}	-10017.9	1120	1163	1.047	-	-1600	369
C_4AH_{13}	-8262.4	831.5	208.3	3.13	-	-	274
C_4AH_{11}	-7656.6	772.7	0.0119	3.56	$1.34 \cdot 10^{-7}$	-	257
$C_2AH_{7.5}$	-5277.5	450	323	0.728	-	-	180
CAH_{10}	-5288.2	610	151	1.113	-	3200	193
$C_4Ac_{0.5}H_{12}$	-8270	713	664	1.014	$-1.30 \cdot 10^6$	-800	285
$C_4Ac_{0.5}H_{10.5}$	-7813.3	668.3	0.0095	2.836	$1.07 \cdot 10^{-7}$	-	261
$C_4Ac_{0.5}H_9$	-7349.7	622.5	0.0088	2.635	$9.94 \cdot 10^{-8}$	-	249
C_4AcH_{11}	-8250	657	618	0.928	$-2.59 \cdot 10^6$	-	262
C_4AcH_9	-7618.6	640.6	192.4	2.042	-	-	234
C_4AsH_{16}	-9930.5	975.0	636	1.606	-	-	351
C_4AsH_{14}	-9321.8	960.9	1028.5	-	-	-	332
C_4AsH_{12}	-8758.6	791.6	175	2.594	-	-	310
$C_4AsH_{10.5}$	-8311.9	721	172	2.402	-	-	282
C_4AsH_9	-7845.5	703.6	169	2.211	-	-	275
C_2ASH_8	-6360	546	438	0.749	$-1.13 \cdot 10^6$	-800	216
C_2ASH_7	-6066.8	487.6	0.0063	1.887	$7.12 \cdot 10^{-8}$	-	215
$C_2ASH_{5.5}$	-5603.4	454.8	0.0057	1.685	$6.36 \cdot 10^{-8}$	-	213
$C_4As_{0.5}ClH_{12}$	-8472	820	557	1.414	$-1.02 \cdot 10^6$	751	289
$C_4ACl_2H_{10}$	-7604	731	498	0.895	$-2.04 \cdot 10^6$	1503	272
$C_4A(NO_3)_2H_{10}$	-7719.3	821	580	1.02	$-2.77 \cdot 10^6$	872	296
$C_4A(NO_2)_2H_{10}$	-7493.1	799	565	0.99	$-2.24 \cdot 10^6$	703	275
C_4FH_{13}	-7435	630	694	1.113	$2.02 \cdot 10^6$	1600	286
$C_4Fc_{0.5}H_{10}$	-6581	1270	308	1.201	$-9.08 \cdot 10^5$	3200	273
C_4FcH_{12}	-7485	1230	612	1.157	$-5.73 \cdot 10^5$	-	292
C_4FsH_{12}	-7662	1430	577	1.234	$2.02 \cdot 10^6$	-	321

$C_4FCl_2H_{10}$	6528	1286	481	0.961	$-1.61 \cdot 10^4$	1503	278
Sulfates							
Cs(anhydrite)	-1434.60	106.7	70.2	-0.099	-	-	46
CsH_2	-2023.36	193.8	91.4	-0.318	-	-	75
Syngenite	-3172	326	201	0.308	$-1.78 \cdot 10^6$	-	128
(Hydr)oxides							
$Al(OH)_3$(gibbsite)	-1288.7	70.1	36.2	0.191	-	-	32
FeOOH(mic)	-551.1	60	1.25	-0.233	$-3.14 \cdot 10^5$	-	21
CH(portlandite)	985	83	187	-0.022	-	-1600	33
SiO_2(quartz)	-909	41	47	0.034	$-1.13 \cdot 10^6$	-	29
Clinkers							
C_3S	-2931	169	209	0.036	$-4.25 \cdot 10^6$	-	73
C_2S	-2308	128	152	0.037	$-3.03 \cdot 10^6$	-	52
C_3A	-3561	205	261	0.019	$-5.06 \cdot 10^6$	-	89
$C_{12}A_7$	-19414	1045	1263	0.274	$-2.31 \cdot 10^7$	-	518
CA	-2327	114	151	0.042	$-3.33 \cdot 10^6$	-	54
CA_2	-4004	178	277	0.023	$-7.45 \cdot 10^6$	-	89
C_4AF	-5080	326	374	0.073	-	-	130
K_2SO_4	-1438	176	120	0.1	$-1.78 \cdot 10^6$	-	66
Na_2SO_4	-1387	150	58	0.023	-	-	53

In Table 2 a_0, a_1, a_2, a_3 are the empirical coefficients of the heat capacity function: $C_P^{\circ} = a_0 + a_1 T + a_2 T^{-2} + a_3 T^{-0.5}$; «-» = 0. Cement shorthand notation is used: A = Al_2O_3; C = CaO; F = Fe_2O_3; H = H_2O; M = MgO; S = SiO_2; c = CO_2; s = SO_3 [12].

In our research, we used Portland cement (PC) "Timlyui Cement Plant" (Republic of Buryatia, Russian Federation).[16] The initial chemical composition of Portland cement is shown in Table 3. Thermodynamic calculations were carried out in the range of 273-573 K, the amount of CNM was 0.001 wt%, 0.01 wt%, 0.1 wt%, and the amount of water was 30%.

Table 3. Initial composition of Portland cement (PC).

Oxides	Content wt %
SiO_2	21.5
Al_2O_3	4.92
CaO	65.4
Fe_2O_3	4.27
MgO	1.06
K_2O	0.2
SO_3	2.57

3. Results and discussion

General properties (enthalpy, entropy, heat capacity) and phase composition changes with temperature were determined. An analysis of the properties dependence in certain temperature ranges were founded sharp changes. This can attributed to phase and chemical transformations associated with the formation of new or intermediate compounds. The increase in enthalpy and entropy with the addition of CNM in the cement-water system can explain by the influence of CNM on the heat release of the cement paste.

Thermodynamic calculations describes exothermic reactions and thermal effects in the cement - water - CNM system and connected the cement hydration in the presence of CNM.

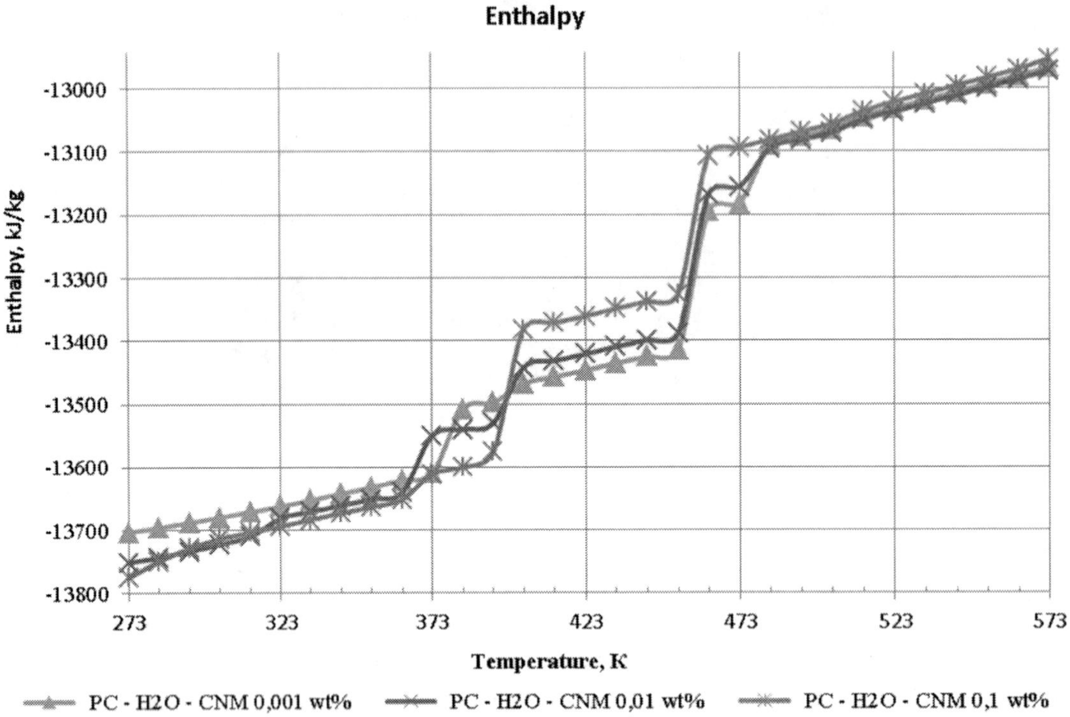

Figure 1. Enthalpy in system PC - water - CNM

The degree of cement hydration can be determined in various ways by measuring: the amount of $Ca(OH)_2$ in cement paste; heat during hydration; specific gravity of cement paste; amount of chemically bound water; the amount of unhydrated cement (using X-ray diffraction analysis), as well as indirectly by the strength of the cement stone [17]. The calculations performed in the program and experimental data [15, 16] on the study of the properties of modified composites using carbon nanomodifier, allow us to evaluate the efficiency of using fullerenes C_{60} and C_{70} as a modifying additive. The amount of $Ca(OH)_2$ phase varies depending on the amount of CNM in the cement-water system as shown by thermodynamic calculations. The change in the yield of calcium hydroxide shows that the carbon nanomodifier acts as an additive-retarder of cement dough setting. Setting is the transition of cement dough from liquid state to solid state. The effect of such additives is estimated by the amount of $Ca(OH)_2$ in the liquid phase of hydrated cement. Increasing the pH of liquid phase cement dough is the effect of such additives [17]. Experimental study of rheological characteristics was presented in [16].

Table 4. Compressive strength of cement stone

Composition	Compressive strength, after	
	7 days of hardening	28 days of hardening
Control	40	61
CNM 0,01%	38	67
CNM 0,001%	44	82

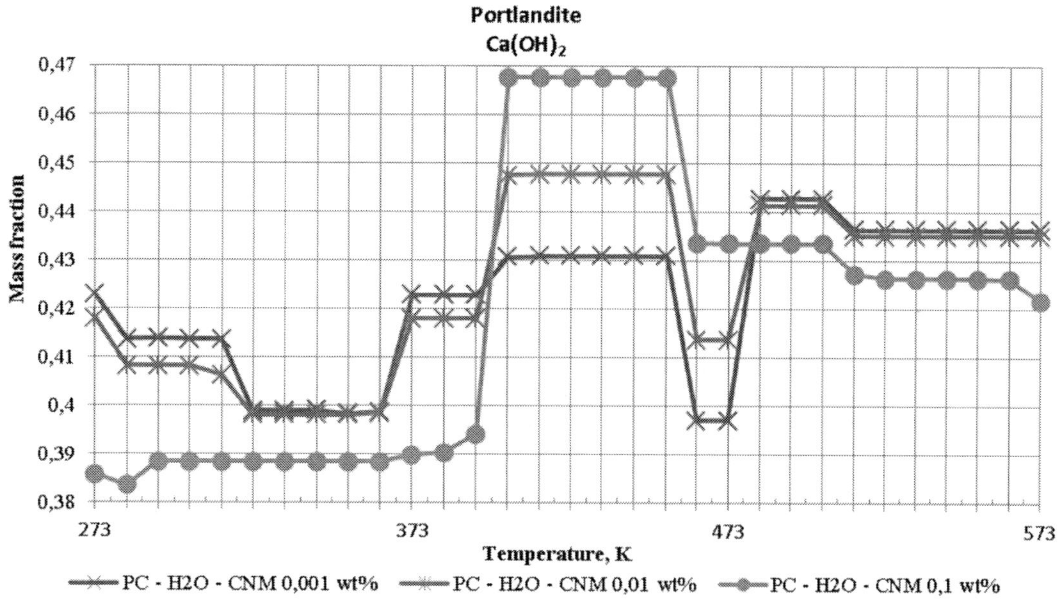

Figure 2. Content of portlandite in system PC - water – CNM

Carbon nanomodifier effect on hydration and kinetics of cement hydration, depending on the amount of carbon nanomodifier. The effect of cement stone modification is shown by thermodynamic calculations and experimental strength data of modified cement stone.

The modification of cement stone helps to deepen the hardening processes, starting from the first hours from the beginning of mixing. The introduction of CNM leads to an increase in the duration of the induction period with subsequent acceleration of the increase in strength. After the initial hardening period (7 days), the strength of the modified cement stone is formed at an increasing rate and by 28 days of hardening exceeds the strength of the control sample by 30% [16].

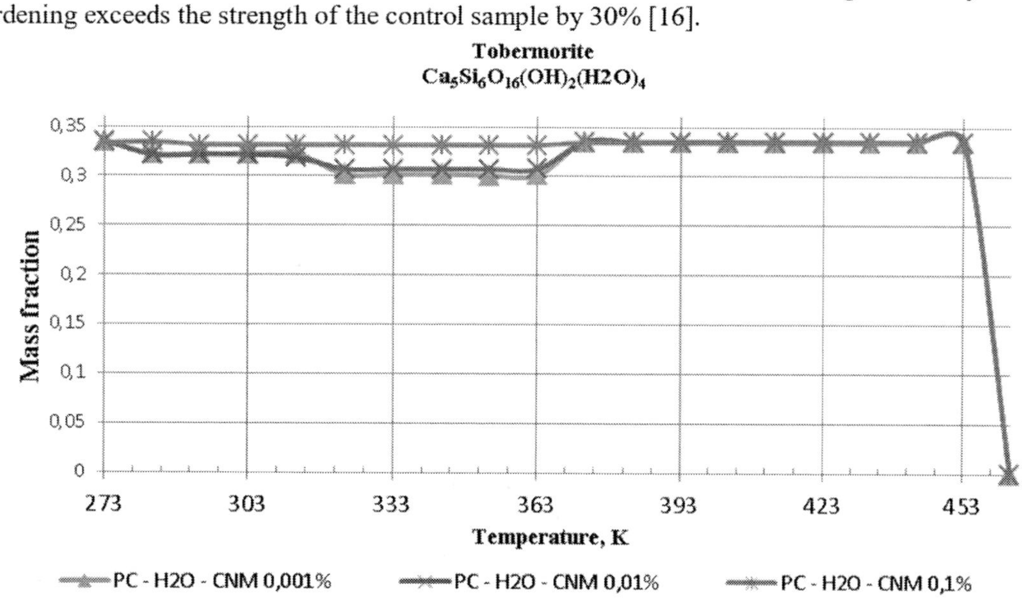

Figure 3. Content of tobermorite in system PC - water – CNM

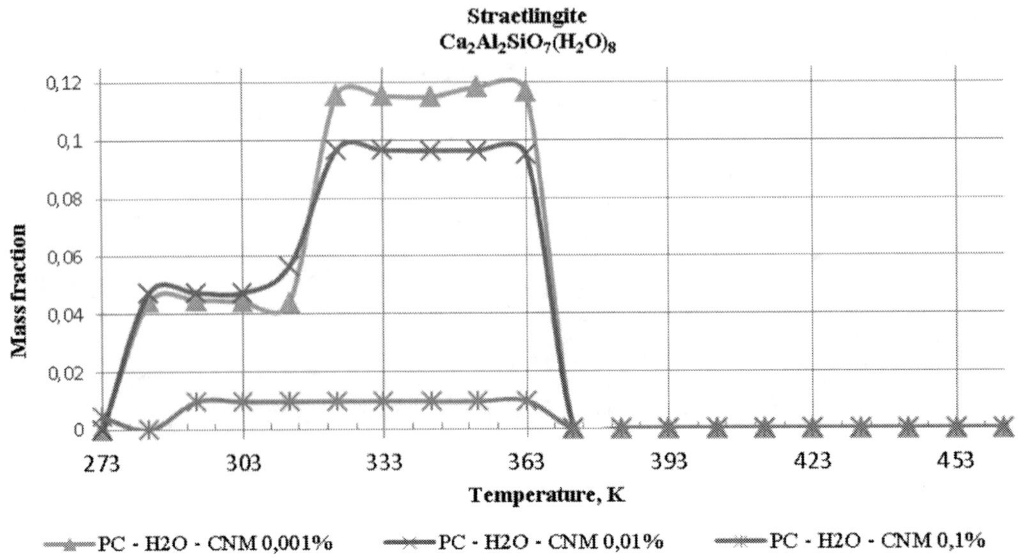

Figure 4. Content of straetlingite in system PC - water – CNM

4. Conclusion

Thermodynamic calculations performed in the TERRA program made it possible to evaluate the effect of CNMs on the hydration of Portland cement. The calculation results are consistent with experimental data and confirm the assumption that CNM has a positive effect on the strength of cement stone by various effects. Thermodynamic calculations made it possible to evaluate effect of CNM on process of hydration Portland cement and the effect of CNM on exothermic reactions in system Portland cement-water - CNM. Thermodynamic modelling can be successfully using to study modification of cement composites with carbon nanomodifiers or other various additives that improve the properties of the cement composite.

5. References

[1] De Ibarra Y S, Gaitero J J and Campillo I 2006 Atomic force microscopy and nanoindentation of cement pastes with nanotube dispersions. *Phys. Status Solidi A* **203** pp. 1076-1081.

[2] Li G Y, Wang P M, and Zhao X 2005 Mechanical behavior and microstructure of cement composites incorporating surface-treated multi-walled carbon nanotubes. *Carbon* **43** pp. 1239-1245.

[3] Falikman V R and Weiner A Ya 2015 New highly effective nano-additives for photocatalytic concretes: synthesis and research. *Nanotechnologies in construction*, **7 (1)** pp. 1828.

[4] Artamonova O V and Sergutkina O R 2013 Construction Nanomaterials: Trends of Development and Prospects. *Scientific Bulletin of Voronezh State Architectural and Construction University* **6** pp. 13-23.

[5] Kulik D A, Wagner T, Dmytrieva S V , Kosakowski G, Hingerl F F, Chudnenko K V, and Berner U R 2013 GEM-Selektor geochemical modeling package: revised algorithm and GEMS3K numerical kernel for coupled simulation codes. *Computational Geosciences* **17 (1)** 1-24.

[6] Wagner T, Kulik D A, Hingerl F F, and Dmytrieva S V 2012 GEM-Selektor geochemical modeling package: TSolMod library and data interface for multicomponent phase models. *The Canadian Mineralogist* **50 (5)**, 1173-1195.

[7] Belov G V, Iorish V S, and Yungman V S 2000 Simulation of equilibrium states of thermodynamic systems using IVTANTERMO for Windows. *High Temperature*, **38 (2)** 191-196.

[8] Trusov B G 2012 Code System for simulation of phase and chemical equilibriums at higher temperatures *Engineering Journal: Science and Innovation* **1** 240-9
[9] Trusov, B G 2002 Program system TERRA for simulation phase and thermal chemical equilibrium. *Proc. XIV Intern. Symp. on Chemical Thermodynamics* 483-484
[10] Diky V V, Kabo G J, 2000 Thermodynamic properties of C_{60} and C_{70} fullerenes *Uspekhi Khimii* **69 (2)** 107–117
[11] Karpushenkava L S, and Kabo G Y 2008 The thermodynamic properties of fullerene hydrides C60H2. *Russian Journal of Physical Chemistry A* **82 (7)** 1170-1174.
[12] Lothenbach B, Kulik D A, Matschei T, Balonis M, Baquerizo L, Dilnesa B, Miron G D and Myers R J 2019 Cemdata18: A chemical thermodynamic database for hydrated Portland cements and alkali-activated materials. *Cement and Concrete Research* **115** 472-506.
[13] Dilnesa B Z, Lothenbach B, Renaudin G, Wichser A, Kulik D 2014 Synthesis and 1characterization of hydrogarnet Ca3(AlxFe1-x)2(SiO4)y(OH)4(3-y). *Cement and Concrete Research* **59** 96-111
[14] Winnefeld F and Lothenbach B 2016 Phase equilibria in the system $Ca_4Al_6O_{12}SO_4$–Ca_2SiO_4–$CaSO_4$–H_2O referring to the hydration of calcium sulfoaluminate cements *RILEM Technical Letters* **1** 10-6
[15] Semenov A P, Smirnyagina N N, Urkhanova L A, Kanakin S V, Lkhasaranov S A, Semenova I A, Dasheev D E, Tsyrenov B O and Khaltarov Z M 2017 Reception carbon nanomodifiers in arc discharge plasma and their application for modifying of building materials *IOP Conference Series: Materials Science and Engineering* **168 (1) 012059**
[16] Smirnyagina N N, Semenov A P, Tsyrenov B O, Dasheev D E, and Khaltarov Z M 2016 Plasma-chemical synthesis of carbon nanotubes and fullerenes to create frost-resistant composite building materials. *International Congress on Energy Fluxes and Radiation Effects* 356-356.
[17] Pukharenko Yu V, Ryzhov D I and Staroverov V D 2017 Peculiar properties of structural formation of Cement Composites in the Presence of Fueleroid Type Carbon Nanoparticles *Proc. of Moscow State University of Civil Engineering* (Moscow: Vestnik MGSU) **12 7 (106)** 718–23
[18] Kozlova V K, Karpova Y V, and Volf, A V 2006 Evaluation effectiveness of additives that slow down the setting of cement dough. Polzunovskij vestnik **2** 230
[19] Melekhov E S and Shlepnev O K 2018 Development and reconstruction of district water supply systems *Proceedings of Universities. Investment. Construction. Real estate* **8(4)** pp 114–127 DOI: 10.21285/2227-2917-2018-4-114-127
[20] Matveeva M 2018 Synchronization of the activity of participants during the implementation of housing construction projects *Proceedings of Universities. Investment. Construction. Real estate* **8** pp 31-41 DOI: 10.21285/2227-2917-2018-2-31-41.
[21] Kazimirov I A and Peshkov V V 2019 Determination of price behaviour in the secondary residential real estate market using a multidimensional regression model *Proceedings of Universities. Investment. Construction. Real estate* **9(3)** pp 476–487 DOI: 10.21285/2227-2917-2019-3-476-487

IOP Conf. Series: Materials Science and Engineering **880** (2020) 012026 doi:10.1088/1757-899X/880/1/012026

Discrete continuous mathematical models in the dynamics of horizontal vibrations of buildings

V I Sobolev[1] and T Y Druzhinina[1]

[1]Department of mechanics and resistance of materials Irkutsk National Research Technical University, Irkutsk, Russia

E-mail: vladsobol@yandex.ru

Abstract. In the design practice of design, dynamic models are used, based on the discretization of structures whose inertial parameters are always distributed. The use of discretized models leads to the formation of large-dimensional problems, and the estimation of discretization errors is extremely complicated by the presence of constructive irregularities and heterogeneities of the boundary conditions of the calculated objects. In view of the foregoing, the proposed work is devoted to the formation of mathematical models that allow combining elements with concentrated inertial parameters (discrete elements) with deformable elements whose inertial parameters are distributed (continuous elements). The need for the formation of such models arises, for example, when calculating buildings for horizontal dynamic effects and also when solving the problems of dynamic analysis of building structures with the presence of vibroactive technological equipment.

Currently, the tasks of assessing the dynamic properties and residual seismic resistance of low-rise buildings are relevant. In the Irkutsk Region alone, there are more than one thousand two hundred (4-5) story buildings of the 335 series, the assessment of the residual seismic resistance of which must be determined. When solving such problems, one inevitably has to simulate the dynamics of buildings. The formation of dynamic models is accompanied by questions of the choice of modeling methods and a description of the nature of the distribution of the inertial parameters of the building.

With horizontal seismic effects, it is advisable to consider the mass of the overlap as concentrated, since taking into account the deformability of the overlap in the plane in this case does not exceed the necessary refinements in the formation of an adequate model, and the complexity of forming the model is great and absolutely not justified. Similarly, the situation is formed when deciding on the need to take into account the deformability of bearing vertical structures, if they are represented by wall elements-diaphragms.

Thus, when describing the dynamic model of cross-wall buildings, it is advisable to use models in which the dynamic parameters of vertical structures are distributed vertically, and the nodes of intersection of vertical and horizontal elements have concentrated inertial elements due to the presence of overlappings.

The description of the elastic properties of vertical elements requires solving the question of the need to take into account various types of deformations that determine with some accuracy the displacement of the nodes of the dynamic model. It is known that the defining types of deformations for bar bearing elements are bending deformations, the account of which for Euler beams [6] gives an accuracy of more than 95% when ignoring shear deformations.

For low-rise wall structures (4-5 floors), the ratio of the sizes of the sections to the longitudinal ones does not fit into the conditions that determine the category of Euler beams, since the *l/n* ratio in this case is close to unity. Here *l* is the height of the vertical bearing element, *h* is the section height of this element. The design diagram of such an element can be depicted as a short beam of length *l*.

Let us determine the displacement of the node *B* when exposed to a unit horizontal force *P* applied to this node, taking into account various options for the deformation of the element.

The movement carried out due to bending:

$$\Delta_u = \int_0^l \frac{\bar{M}\bar{M}}{EJ} dy = \int_0^l \frac{y \cdot y \cdot 12}{Ebh^3} = \frac{4l^3}{Ebh^3}$$

Let be $l = hn$, then $\Delta_u = \frac{4n^3}{Eb}$.

Shifting movement:

$$\Delta_c = \int_0^l \frac{1 \cdot 1}{GF} dy = \int_0^l \frac{1 \cdot 1}{Gbh} dy = \frac{3l}{Ebh} = \frac{3n}{Eb}, \quad \text{here } E = G/3$$

Thus, the deformability of the supporting structures of rectangular sections is largely determined by bending deformations.

To solve the problems of dynamics (direct and inverse), it is inevitable to determine the frequency characteristics of the structure. Consider the definition of the frequency parameters of the structure, taking into account the bending nature of the deformation of the bearing elements.

First, we consider the possibility of determining the natural frequencies of the system with inertial parameters distributed along the length of the supporting structures. To enable the further inclusion of inertial elements with lumped parameters, which can be represented in the form of overlappings, it is advisable to use the harmonic element method [7-9]. Obviously, the direct purpose of the harmonic element method is to design structures for forced harmonic influences. Determination of frequency characteristics - using the method of dynamic compliance and the method of harmonic elements (GAE), it is possible to carry out by taking into account the conditions of resonance manifestations arising from harmonic influences. This approach was implemented for low-rise structures that took into account shear deformations.

To take into account the manifestations of bending deformations in vertical structures, we consider a feature of the formation of harmonic elements using the dynamic displacement method [10].

The development of the dynamic displacement method allows the use of both distributed and concentrated parameters in harmonically loaded structural elements. In this case, it is possible to simulate harmonic oscillations with heterogeneity of boundary conditions and an irregular distribution of physical parameters. The determination of the dynamic properties of mathematical models is possible based on the formation of matrices of an ensemble of such elements.

Some similarity in the construction of solutions allows us to speak about the manifestation of the properties of finite elements, but in contrast to the classical versions specialized for solving stationary dynamics problems and containing as parameters a constant frequency of monoharmonic oscillations and parameters of distributed and concentrated masses.

Given these features, such dynamic elements are called harmonic elements (GaE) [7-9, 11]. The nodal matching of the harmonic solutions of the equations of the dynamic state of the elements obtained in the form of amplitudes of harmonic reactions in the imposed bonds allows for the analytical representation of vibrational forms in the form of functions of amplitudes on the coordinates of the axis of rigidity of the bent element. The latter circumstance greatly simplifies the analysis and generalization of the solution results and allows for non-traditional methods of vibration isolation based on the formation of vibrational forms with predetermined properties [1, 2].

Let us consider forced steady-state harmonic oscillations with a frequency ω of a beam of length a with a linear mass ρ uniformly distributed along the length and bending stiffness *EJ* when it is also subjected to a constant longitudinal force *N*.

Oscillations of the beam are carried out under harmonic influences with unit amplitudes applied at some points (nodes) of the beam. In the absence of interstitial dynamic action, the equation of dynamic equilibrium of the elementary section of the beam in the span between the nodes is described by the

Euler – Bernoulli equation [10], where V is the displacement of the axis of rigidity of the beam along the y axis.

$$EJV_{xxx} - NV_{xx} + \rho V_{tt} = 0. \quad (1)$$

For a given frequency of external action ω, the vibrational forms of the beam are uniquely determined by the amplitude vector of nodal harmonic displacements

$$Y = \left(Y(0), Y_x\big|_{x=0}, Y(l), Y_{xx}\big|_{x=a}\right)^T, \quad (2)$$

in which the components are arranged in the order of numbering of the corresponding bonds. The vector Ysinω defines the boundary conditions for equation (1). Let us imagine a solution of the equation of beam oscillation in the expansion in forced vibrational forms from single harmonic displacements (with unit amplitudes) of the bonds that ensure the fastening of the nodes.

A single harmonic movement of each bond with number i in its direction causes reactions in all superimposed bonds. The amplitudes of the magnitudes of the bonds of these reactions ordered by bond numbers form a certain vector Ri of the amplitudes of the harmonic reactions. Alternate, in the order of numbering, harmonic movements of bonds form a matrix of amplitudes of dynamic reactions

$$R = \{R_1, R_2, R_3\}$$

In the general case, the number of bonds can be equal to two to four (Fig. 1) and depends on the boundary conditions (methods of fixing the boundary nodes).

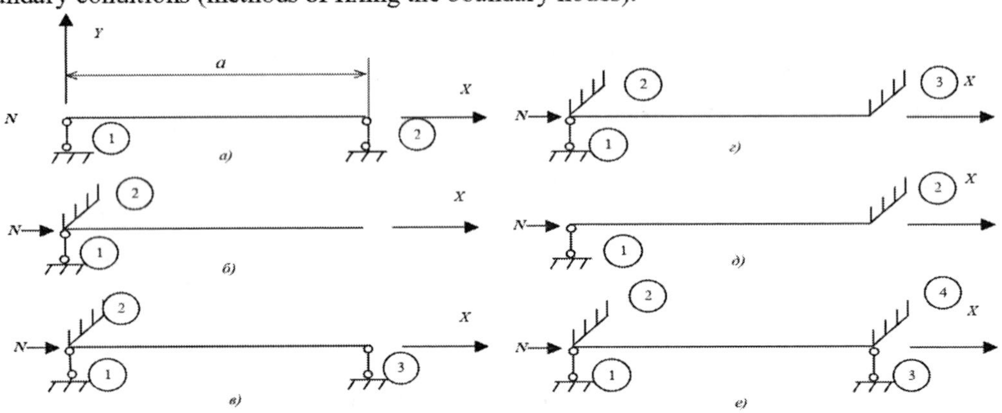

Fig. 1. Design schemes of beams with various options for fastenings;
1, 2, 3, 4 - link numbers

The rules for signs of movements and turns in the directions of bonds, as well as signs of the magnitudes of reactions are shown in Fig. 2

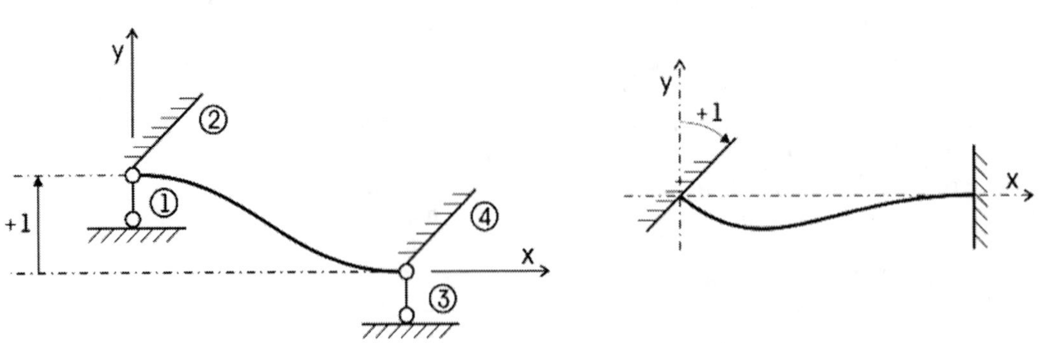

Fig. 2 The rule of signs of displacements and reactions in nodal connections

IOP Conf. Series: Materials Science and Engineering **880** (2020) 012026 doi:10.1088/1757-899X/880/1/012026

To determine the coefficients of the matrix R and the function $Y(x)$ of the amplitudes of the forced vibrations of the beam, we use the method of separation of variables [12, 13]. To do this, imagine the solution in the form:

$$V(x) = Y(x)sin(\omega t)$$

Substitute the desired solution in equation (2). Reducing the resulting equation by $sin(\omega t)$, we obtain the ordinary differential equation.

$$Y^{IV}(x) - \frac{NY^2(x)}{EJ} - Y(x)\frac{\omega^2\rho}{EJ} = 0 \quad (3)$$

The characteristic equation of which is of the form:

$$\mu^4 - \frac{N}{EJ}\mu^2 - \frac{\omega^2\rho}{EJ} = 0.$$

Performing a change of variables in the characteristic equation $\mu^2 = q$, form a quadratic equation:

$$q^2 - \frac{N}{EJ}q - \frac{\omega^2\rho}{EJ} = 0.$$

Having determined the roots of this equation, we obtain:

$$q_{1,2} = \frac{N}{2EJ} \pm \sqrt{\frac{N^2}{4(EJ)^2} + \frac{\omega^2\rho}{EJ}}.$$

Passing to the original variable, we have solutions of this equation in the following forms:

$$\mu_1 = q,\ \mu_2 = -q,\ \mu_3 = is,\ \mu_4 = -is,$$

where

$$q = \sqrt{\frac{N}{2EJ} + \sqrt{\frac{N^2}{4(EJ)^2} + \frac{\omega^2\rho}{EJ}}},$$

$$s = \sqrt{\frac{N}{2EJ} - \sqrt{\frac{N^2}{4(EJ)^2} + \frac{\omega^2\rho}{EJ}}}.$$

Common decision $Y(x)$ equation (3) can be represented as a linear combination

$$Y(x) = H(x)\vec{C} \quad (4)$$

four linearly independent partial solutions forming a basis vector function

$$H(x) = (e^{qx}, e^{-qx}, sin(sx), cos(sx)),$$

where $\vec{C}$ is a vector column of coefficients of a linear combination of the right side of the expression (4) [14].

The boundary conditions for solving equation (3) can be specified as the values of the components of the vector (2) of the amplitudes of displacements along the directions of the bonds.

Consider oscillations of a non-deformable in the plane of the floor disk pivotally supported by a certain number of bent vertical elements having a distributed mass ρ (Fig. 3).

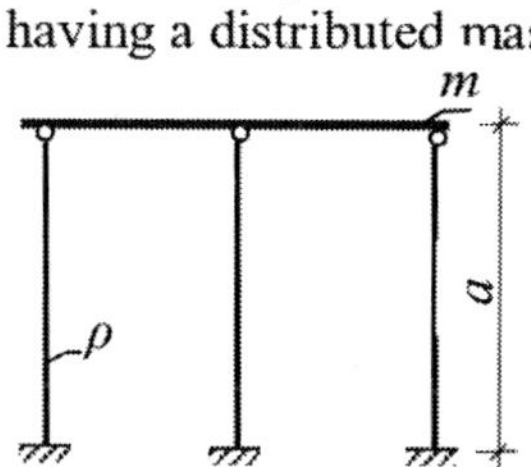

Fig. 3. The design scheme of horizontal vibrations of an undeformable disk on massive bending elements
m - the mass of the overlap. ρ - distributed mass of racks

Such a constructive scheme allows us to analyze the simplest version of the influence of the distributed masses of the supporting vertical structures on the dynamic properties of a system containing a concentrated mass m and distributed masses of bent elements.

For a qiven circuit $Y_{xx}\big|_{x=a} \equiv 0$. When determining *Y(x)*, the coefficients of linear combinations of the fundamental solution are determined in the form of a matrix *C* by alternately setting the unit amplitudes of displacements along the directions of the bonds.

The vectors of amplitudes of unit displacements ordered by bond numbers form the matrix *L*. The matrices C and *L* have the following form:

$$C = \begin{bmatrix} C_{11} & C_{12} & C_{13} \\ C_{21} & C_{22} & C_{23} \\ C_{31} & C_{32} & C_{33} \\ C_{41} & C_{42} & C_{43} \end{bmatrix}; \quad L = \begin{bmatrix} 1 & 0 & 0 \\ 0 & 1 & 0 \\ 0 & 0 & 1 \\ 0 & 0 & 0 \end{bmatrix}.$$

Elements of the matrix *C* are determined from the solution of systems of equations:

$$AC = L, \tag{5}$$

where *A*– matrix formed from the basis vector of the function *H(x)* as follows:

$$A = \begin{bmatrix} H(0) \\ H_x(x)\big|_{x=0} \\ H(l) \\ H_{xx}(x)\big|_{x=a} \end{bmatrix}.$$

The matrix of amplitudes of reactions in bonds with alternating unit harmonic movements of bonds in the order of their numbering (matrix of harmonic reactions) is defined as:

$$R = -EJC^T H, \tag{6}$$

where $H = \left(H^T_{xxx}(x)\big|_{x=0}, H^T_{xx}(x)\big|_{x=0}, H^T_{xxx}(x)\big|_{x=a}\right).$

The vector column *Y* of the amplitudes of the nodal displacements corresponding to the vector *Y* of the amplitudes of the harmonic actions along the directions of the bonds with frequency ω is determined from the solution of the system of equations:

$$RY = F, \quad \text{or} \quad -EJ(A^{-1}L)^T H = F.$$

$$Y(x) = Y^T C^T H^T(x). \tag{7}$$

Taking into account the boundary conditions allows us to obtain the matrix *A*, which in this case has the form:

$$A = \begin{bmatrix} 1 & 1 & 0 & 1 \\ q & -q & s & 0 \\ \exp(a \cdot q) & \exp(-a \cdot q) & \sin(a \cdot s) & \cos(a \cdot s) \\ \exp(a \cdot q) \cdot q^2 & \exp(-a \cdot q) \cdot q^2 & -\sin(a \cdot s) \cdot s^2 & -\cos(a \cdot s) \cdot s^2 \end{bmatrix}$$

Having determined matrix C from equation (5) and substituting it into expression (6), we determine the amplitudes of the dynamic reactions arranged in the form of the matrix R. The magnitude of the amplitude of the dynamic reaction in connection with number 3 has the form:

$$r_{33} = EJ \cdot \frac{(s^2 \cdot \cos(a \cdot s) + q^2 \cdot \cosh(a \cdot q)) \cdot (q^3 \cdot s + q \cdot s^3)}{[(s^2 + q^2) \cdot (q \cdot \sin(a \cdot s) \cdot \cosh(a \cdot q) - s \cdot cos(a \cdot s) \cdot \sinh(a \cdot q))]}$$

The addition of a discrete mass during harmonic action with a frequency ω forms the amplitude of the dynamic reaction in the form of:

$$r = r_{33} - m \cdot \omega^2.$$

References

[1] Berman A 1984 System identification of structural dynamic models – theoretical and practical Bounds pp 123-129
[2] Argyris J H, Boni B and Hinderlang V 1982 Finite element analysis of two- and three

dimensional elastoplastic frames – the natural approach *Comp. Meth. Appl. Mech.* **35 (2)** pp 221-248
[3] Sobolev V I 2000 *SPbGU* pp 2–17
[4] Gallager R 1984 p 428
[5] Argyris J H, Boni B and Hinderlang V 1982 Finite element analysis of two - and three dimensional elastoplastic frames - the natural approach *Comp. Meth. Appl. Mech.* **35 (2)** pp 221-248
[6] Kiselev V A 1969 (Stroyizdat) 1969
[7] Sobolev V I 2002 (IrGTU) p 202
[8] Sobolev V I 2003 Vestnik IrGTU **1 (13)** pp 124–129
[9] Sobolev V I and Chernigovskaya T N 2010 Vyp **1** pp 43-51
[10] Koloushek V 1965 p 632 s
[11] Sobolev V I 2003 Vestnik IrGTU **2 (14)** pp 89–92
[12] Moskvitin V A., Emelyanova N A. and Mashovich A Y 2019 Experimental studies of air permeability indicators of composite "Poroplast CF" *Investment. Construction, Real estate* 9 (2) pp 342–353
[13] Farlow S 1985 *Partial differential equations for scientists and engineers* p 384
[14] Krylov A N 1932 p 480
[15] Dobruskina M A, Petrov A V and Bat-Erdene Z 2019 Improving the technology of retaining walls in the Irkutsk region using gabion baskets *Proceedings of Universities. Investment. Construction. Real estate* **9 (2)** pp 312–323
[16] Selezneva O I, Radaev S S and Baranova A A The impact of thermal treatment on the granulometric and chemical composition of diatomite of the Kamyshlov field when obtaininga liquid-glass binder *Investment. Construction. Real estate* **9(3)** pp 550–555
[17] Hirer E, Nerset S and Wanner G 1990 *Solution of ordinary differential equations* p 512

IOP Conf. Series: Materials Science and Engineering **880** (2020) 012027 doi:10.1088/1757-899X/880/1/012027

Heat insulators based on silica clay raw materials

G I Storozhenko[1] and L K Kazanceva[2]

[1]Novosibirsk State University of Architecture and Construction, Leningradskaya str. 113, 630008 Novosibirsk, Russian Federation
[2]Institute of Geology and Mineralogy, SB RAS, Akademika Koptyuga Ave. 3/1, 630090 Novosibirsk, Russian Federation

E-mail baskey_ltd@mail.ru

Abstract. The paper presents the results of laboratory studies and pilot tests of the production of foamed glass based on the silica clay from Shipovo field (Republic of Kazakhstan). It is shown that the granulated foamed glass can be commercially produced from widespread siliceous raw material, using home equipment.

Development and production of durable, environmentally-friendly and effective heat insulators for Ural, Siberia and Far East regions is still a topical challenge. Experience has proven that as time goes on mineral products on synthetic binders, polystyrene and other artificial organic materials lose their thermal-insulating properties [1]. For this reason science and industry seek for new materials which routinely turn out to be undeservingly dust-down and forgotten. Among these materials there are foamed glass, foamed silicates and foamed-crystal materials, which differ in types of raw material and manufacturing techniques [2]. This light ($\rho_{pour.}$ = 140÷650 kg/m^3), resisting ($R_{compr.}$ = 0.5÷5 MPa), durable and inflammable material with low thermal conductivity (0.045÷0.1 W/m×K) can be produced in the form of either granules or blocks and plates [3]. Today the technology of foamed glass production, developed in Russia by Academician I. Kitajgorodsky, is implemented in the equipment of the German companies «Foamglas» and «Poraver».

Despite the high quality of the foamed glass from glass scrap or specially prepared glass, this production process has not been commonly implemented yet. Main restrictions here are high energy consumptions for glass melting, limitedness of such material source as glass scrap, which is characterized by a wide viriety of chemical compound, and also a high price of German equipment.

In Russia there are not only theoretical but also commercialized projects for the production of foamed materials, involving silica-containing rocks not pre-melted into glass [2].

Main types of silica rocks (tripolith, diatomite, silica clay, etc.) do not contain an inherent source of pore-forming gases and feature quite a high melting point (above 1,100°C). Nevertheless there is a serious motivation to use them as a raw material for the foamed glass production. Silica rocks belong to the opal-crystobalite varieties with a high content of chemically active amorphous silica, which define the main direction of the production process of porous construction materials. The process is based on the interaction between the amorphous silica and NaOH solution resulting in the formation of hydrated polymer sodium silicates, i.e. liquid glass ($Na_2O{\cdot}mSiO_2{\cdot}nH_2O$). Application of sodium hydroxide in foamed materials production from silica rocks solves two problems: melting point decrease and formation of a bloating gas source.

Though the works for foamed substances production from natural raw materials have been performed within more than 10 years, there is still no special operating factory in Russia except for small experimental plants. It results from insufficient academic elaboration of the problems of glass formation and gas elimination in such compositions and also from the conversion complexity of some processes to the industrial level.

The present paper shows the results of the lab-scale technological optimization of the compositions and thermal modes for the production of granulated foamed glass (GFG), followed by pilot tests of the production of this glass from the silica clay from Shipovo field (Republic of Kazakhstan, South Ural).

Complex analysis of the silica clay was performed in the Institute of Geology and Mineralogy, SB RAS, by the X-ray-fluorescence (VRA-20R), X-ray diffractometry (Thermo Scientific ARL-X'tra), thermogravimetry (Mettler TC-10A) and electronic microscopy (Philips XL30-FEG) methods [4]. Chemical compound of the silica clay is presented in Table 1.

Table 1. Chemical compound of the silica clay from Shipovo field

Oxide percentage % per dry substance, %										
SiO_2	Al_2O_3	TiO_2	Fe_2O_3	MnO	MgO	CaO	Na_2O	K_2O	P_2O_5	*n.n.n.*
81.07	7.63	0.43	3.87	0.036	1.08	0.69	0.23	1.66	0.08	3.59

The main component of the silica clay is chemogenic opal; crystal phases are presented by quartz and crystobalite. There are also minor quantities of plagioclase and clay minerals. The rock is a porous conglomerate (porosity rate reaches 85%), which promotes formation of hydrated sodium polysilicates, not only on a particle surface but also inside them.

Results of the lab-scale technological analysis. Optimal compositions and methods of granules preparation were developed under laboratory conditions in order to manufacture the GFG and block-structured foamed glass from silica-clay raw material, as the sodium hydroxide concentration in the compositions (per dry components of the furnish) varied from 17 to 22 mass per cent. With the maximum concentration of $NaOH$ the obtained GFG had the poured density of 80 kg/m^3, block-structured foamed glass density was of 120 kg/m^3. With the same sodium hydroxide concentration in the furnish (17 mass per cent), regarding the peculiarities of granulate preparation technique, we produced the GFG of size -12 +10 mm with the poured density within the range from 135 to 200 kg/m^3.

Results of pilot tests. Laboratory results were tested on the pilot equipment for simulation and tuning of the technological modes of foamed silicates production from silica rocks, company «Baskey Ceramics».

The representative technological samples of the silica clay from Shipovo field had the humidity of 32 – 35%, the size -40+0 mm. It was simultaneously dried up to 6 – 8% humidity and crushed to the size -140+0 μm in the crushing-drying machine USP-S-04.55M with the capacity of 2 tones per hour. The average size range of the crushed silica clay from three tanks of the machine aspiration system approached to the grain range of the silica clay powder used for GFG production in the laboratory conditions.

The produced powder was granulated in the industrial periodical turbo-bladed mixer-granulator TL-100 made by «Dzerzhinsktekhnomash». In order to provide the processes of silicate-forming and granulation the technological parameters such as granulation time, humidity and size range of the granulate were practiced. The granulate of poly-fraction composition with the humidity of 22% was produced in the turbo-bladed granulator.

The granulated silica clay was dried in the industrial heating drum. Before the calcination in an electrical furnace the dry granulate was mixed with a kaolin powder (5 – 10 mass %) in order to prevent granules adhesion during the bloating. The modes of GFG drying and calcination during the pilot tests correlated to the laboratory recipe.

Physical and mechanical properties of the representative sample of the produced foamed glass are presented in Table 2.

Table 2. Physical and mechanical properties of the representative sample of the produced foamed glass

Parameter	Unit	Test result value/size	
Poured-bulk density of fraction 2.5 – 5 mm	kg/m^3	220	D250
Ultimate compressive resistance in the cylinder	MPa	1.54	П100
Resistance to silicate disintegration	%	1.4	–
Frost resistance 15 cycles	%	1.8	–
Humidity absorption, vol.	%	4.2	–
Loss of mass during boiling	%	0.55	
Content of water-soluble sulphides and sulphites in SO_3 equivalent	%	0.22	–

Fig.1 shows the porous structure of the GFG based on the silica clay from Shipov field. It features uniformity and satisfactory vitrification. Smaller pores form in the vitrified walls of pores (Fig. 2). This vindicates the uniform formation of hydrated polymer sodium silicates over the whole volume of the granulated material.

Figure 1. General view of a porous macrostructure of the GFG. Scale rule corresponds to 5 mm.

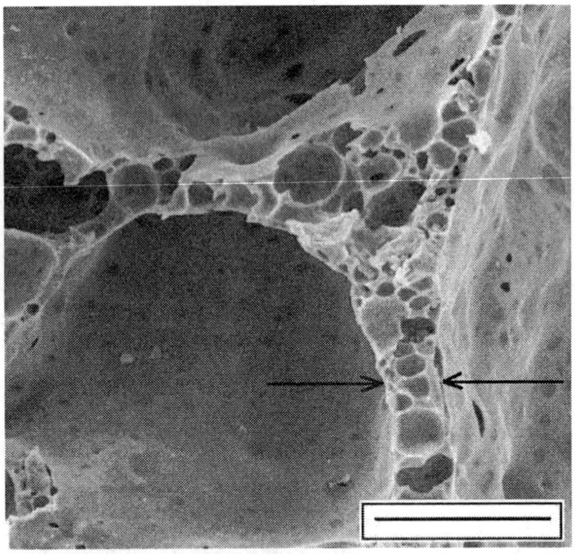

Figure 2. A porous microstructure of the GFG. Scale rule corresponds to 50 μm.

The results of pilot tests show that it is possible to organize the industrial-scale production of granulated foamed glass which would satisfy all reference standards, using widely spread silica rocks and home equipment.

References

[1] Ivanov K S 2011 Insulators for ground thermal stabilization *Cryosphere of the Earth* **4** pp 120–122

[2] Ketov A A 2012 Production of constructional materials from hydrated polysilicates *Constru tional materials* 11 pp 22-24

[3] Goryajnov K E and Goryajnova S K 1982 *Technology of heat-insulating materials and products* (Moscow: Strojizdat) p 296

[4] Kazantceva L K, Zheleznov D V, Seretkin Yu V, Raschenko S V 2012 Formation of a pore-forming gas source at natural alumosilicate wetting with NaOH solution *Glass and Ceramics* 10 pp 37-42

[5] Komarov A K, Ivanov I A and Lundenbazar B 2019 Theory and practice of the use of gabions

for forming protective structures *Proceedings of Universities. Investment. Construction. Real estate* **9(1)** pp 78–89 DOI: 10.21285/2227-2917-2019-1-78-89

[6] Kazimirov I A and Peshkov V V 2019 Determination of price behaviour in the secondary resi dential real estate market using a multidimensional regression model *Proceedings of Universities. Investment. Construction. Real estate* **9(3)** pp 476–487 DOI: 10.21285/2227-2917-2019-3-476-487

[7] Matveeva M 2018 Synchronization of the activity of participants during the implementation of housing construction projects *Proceedings of Universities. Investment. Construction. Real estate* **8** pp 31-41 DOI: 10.21285/2227-2917-2018-2-31-41.

[8] Larina O P and Khalabuda E I 2019 Accident prevention during construction and installation works at height *Proceedings of Universities. Investment. Construction. Real estate* **9(4)** pp 734–741 https://doi.org/10.21285/2227-2917-2019-4-734-741

[9] Peshkov V V and Gamayunova E V 2018 Specific features of investment support for the repair and reconstruction of buildings and facilities *Proceedings of Universities. Investment. Construction. Real estate* **8** pp 56-72 DOI: 10.21285/2227-2917-2018-2-56-72

[10] Yaskova N Yu and Sarchenko V I 2018Development of strategic approaches to the creation of aqua-territorial industrial complexes in the Arctic zone *Proceedings of Universities. Investment. Construction. Real estate* **8** pp 84-93 DOI: 10.21285/2227-2917-2018-2-84-93

[11] Yaskova N Yu 2018 Construction in the context of new views on the goals of economic devel opment. *Proceedings of Universities. Investment. Construction. Real estate* **8** pp 133-142 DOI: 10.21285/2227-2917- 2018-1-133-142

[12] Tyuryukhanov K Yu and Pugin K G 2019 Impact of the surface of particles of moulding sand on the structural formation of asphalt *concrete Proceedings of Universities. Investment. Construction. Real estate* **9(3)** pp 566–577

[13] Dobruskina M A, Petrov A V and Bat-Erdene Z 2019 Improving the technology of retaining walls in the Irkutsk region using gabion baskets *Proceedings of Universities. Investment. Construction. Real estate* **9 (2)** pp 312–323

[14] Moskvitin V A., Emelyanova N A. and Mashovich A Y 2019 Experimental studies of air per meability indicators of composite "Poroplast CF" *Investment. Construction, Real estate* **9 (2)** pp 342–353

IOP Conf. Series: Materials Science and Engineering **880** (2020) 012028 doi:10.1088/1757-899X/880/1/012028

Research of composite binders with nanomodifiers for dry mixes

L Urkhanova[1], S Lkhasaranov[1] and E Badmaeva[1]

[1]Department "Construction Materials, Roads and Woodworking", East Siberia State University of Technology and Management, Klyuchevskaya street, 40 V, building 1, Ulan-Ude, 670013, Russian Federation

E-mail: urkhanova@mail.ru

Abstract. The article discusses the application of various mineral materials for the modification of Portland cement in order to obtain dry mixes. The main component in the composition of dry mixes is Portland cement, affecting its physical, mechanical and technological properties. To modify Portland cement, which is used to obtain dry mixes, various micro- and nano-additives were used: volcanic slag, microsilica, nanosilica. Mechanical parameters of modified Portland cement at various hardening periods are determined. The introduction of micro- and nano-additives positively affects the hydration process of Portland cement, which is confirmed by the study of the microstructure of the modified cement stone using scanning electron microscopy. It has been established that when using various micro- and nano-additives, the amount of low-basic calcium hydrosilicates changes, affecting the increase in the mechanical properties of cement stone. The use of volcanic slag and nanosilica allows to improve the technological and quality indicators of dry mixes. Studies of determining the angle of repose of the composition of composite binders based on Portland cement and volcanic slag, as well as the introduction of nanosilica in Portland cement show a significant change in this indicator indicating an improvement in the flowability of the material.

[1] Introduction

The main research in the design of dry building mixes is aimed at improving the processability and reducing the cost of their production through the use of composite materials of optimal structure based on cement, mixed composite binders, fine aggregate of optimal particle size distribution, finely divided fillers and chemical additives [1, 2].

The use of nanoscale modifiers for the modification of dry building mixtures not only actively affects the structure formation processes during cement hydration, but also improves the technological properties of the mixtures. The introduction of nanomodifying additives leads to a significant change in physical and mechanical properties [3-5].

[2] Materials

In studies to obtain dry mixes on cement and composite binders were used: Portland cement (OPC) CEM I 32.5 N; volcanic slag (VS) of the Khurai-Tsakir deposit (Baikal region), microsilica (MS), nanosilica (NS).

Microsilica is a by-product of the manufacture of silicon metal and ferro-silicon alloys. The process involves the reduction of high purity quartz (SiO2) in electric arc furnaces at temperatures in excess of

2000°C. SF is a very fine powder consisting mainly of spherical particles or microspheres of mean diameter about 0.15 microns.

Nanosilica was obtained by evaporation and condensation for installation at the Institute of Theoretical and Applied Mechanics and the Institute of Nuclear Physics of the Siberian Branch of Russian Academy of Sciences (Novosibirsk city). Nanosilica contains more than 99% of amorphous silica, the content of inclusions, in mass. %: Al - 0.01, Fe - 0.01, Ti - 0.03.

[3] Results and Discussion

The introduction of nanomodifying additives leads to a significant change in physical and mechanical properties. For comparative analysis, experiments were carried out to determine the mechanical parameters of cement stone modified with micro- and nanosilica, volcanic slag (fig. 1).

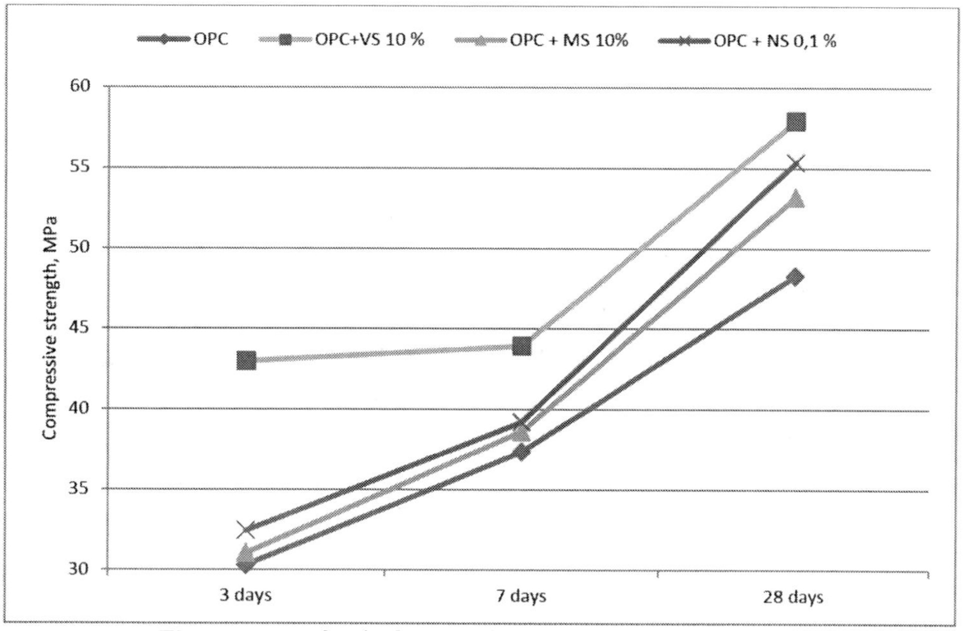

Figure 1. Mechanical properties of modified cement stone

Theoretical studies and experimental results established that the optimal dosage of silica fume lies within 10%. Since the complete binding of calcium hydroxide lowers the pH of the pore fluid and thereby adversely affects the stability of the calcium silicate hydrogel and the corrosion resistance of the reinforcement. The kinetics of the set strength of the modified cement stone shows that when using volcanic slag, maximum values were obtained associated with pozzolanic activity and structure-forming action of slag. The indicators of micro- and nanosilica are somewhat lower, which is possibly associated with an irrational dosing interval. The mechanism of action of modifying additives is associated with a change in the structure of the composite.

Changes in the mechanical properties of modified cementitious compositions are associated with a change in its structure. Microphotographs of the chips of modified composites in comparison with the control non-additive composition showed a significant difference in the formed structure of the composites (fig. 2).

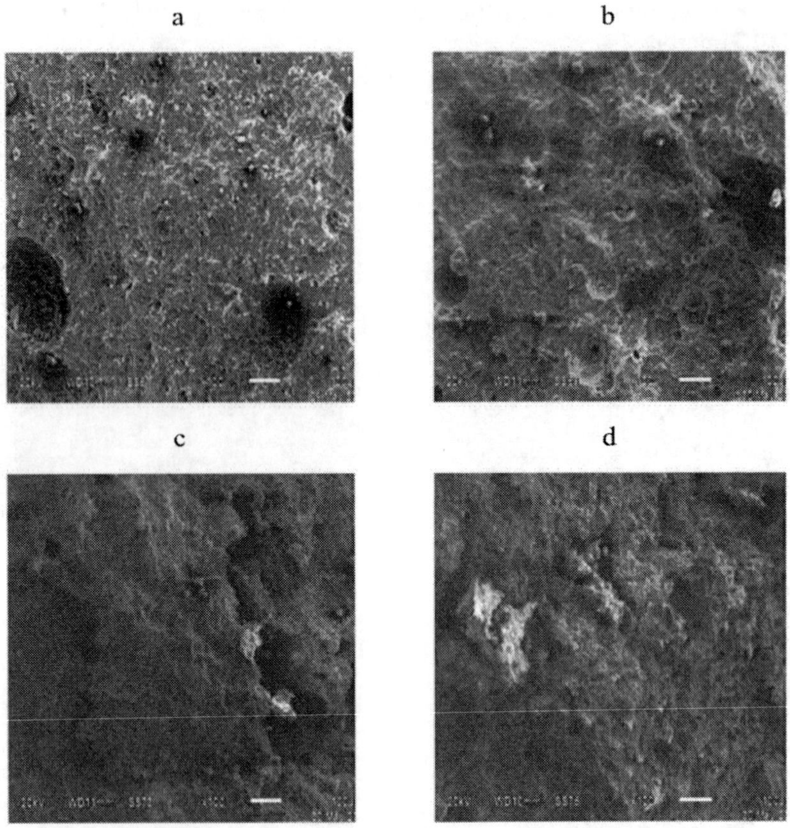

Figure 2. Microstructure (x500) of a modified cement stone at the age of 28 days: a – OPC, b – OPC+MS 10%, c – OPC+NS 0,1%, d – OPC+VS 10%

In the initial cement stone, due to the loose packing of cement particles, there are pores; during hardening, this space is filled with lime crystals of Ca(OH)2. The microstructure of the samples using VS, MS and NS is denser in comparison with the control composition. The introduction of mineral additives leads to a decrease in capillary porosity, and at the same time leads to an increase in the number of tiny gel pores that are part of the calcium silicate hydrogel.

The presence of silica with a defective crystal lattice in the high-level gas; leads to the creation of a high-density structure of cement stone due to the binding of portlandite and the additional formation of calcium hydrosilicates. The content of capillary pores decreases, with an increase in gel porosity. MS has a positive effect on the structure of the transition zone; the thickness is comparable with the size of cement grains. Thick matrix reinforcement is noted, which leads to an increase in the density and strength of the composite. It is worth noting that the structure with the introduction of NS showed a steady tendency to create a strong single conglomerate, there are volumetric germination of calcium hydrosilicates throughout the volume.

The use of nanoscale modifiers for the modification of dry mixes not only actively affects the structure formation processes during cement hydration, but also improves the technological properties of the mixtures.

One of the main technological properties of dry mortar is flowability. Providing a good indicator of the flowability of dry mixes will allow uninterrupted production process to be conducted under production conditions and intensified. To assess the manufacturability of powdered materials, which include Portland cement, as well as to predict the behavior of bulk materials in industrial production, their technological properties are determined: flowability, angle of repose. Studies of determining the angle of repose of the composition of composite binders based on Portland cement and volcanic slag (fig. 3), as well as the introduction of nanosilica in Portland cement (fig. 4) show a significant change in this indicator indicating an improvement in the flowability of the material.

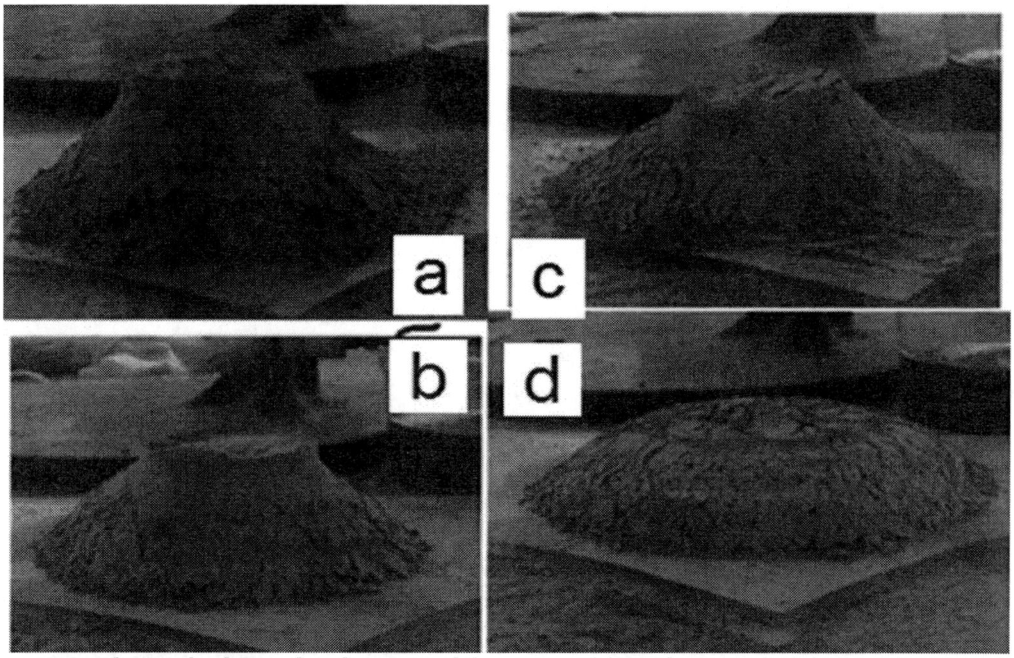

Figure 3. Change in the angle of repose of the composition of composite binders with volcanic slag: a - control (volcanic slag content - 0%), b - 5%, c - 10%, d - 20%

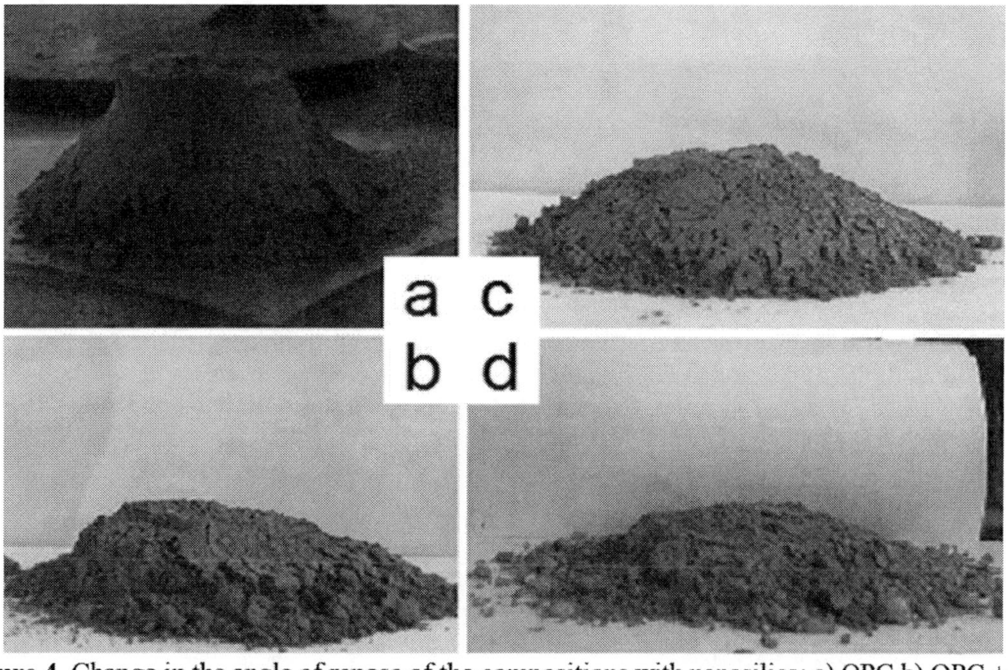

Figure 4. Change in the angle of repose of the compositions with nanosilica: a) OPC b) OPC + NS (0.5%), c) OPC + NS (1%), d) OPC + NS (3%)

Replacing part of Portland cement with volcanic slag leads to improved technological performance. The angle of repose with increasing volcanic slag content decreases by 20-30%, which indicates a better flowability of the material. The flowability of binders with nanosilica improves with an increase in nanosilica from 0.5 to 3%. Since the flowability of the material depends on the size and shape of the particles and is determined by the amount of internal friction between them, the introduction of nanopowder apparently reduces the amount of internal friction between the cement particles, thereby improving the flowability.

An important indicator of the quality of dry mortar is water retention capacity. Water retention capacity with an increase in slag content up to 20% lies in the range of 97-98%, which meets the requirements of standards (fig. 5). An improvement in the water-holding ability of compositions with volcanic slag and nanosilica is associated with a high specific surface area of the particles. Around the particles water shells form and the amount of "free" unbound water decreases. Particles of finely dispersed volcanic slag and nanosilica interact through a solvate shell formed on the surface, consisting of water adsorbed on the surface of the solid phase.

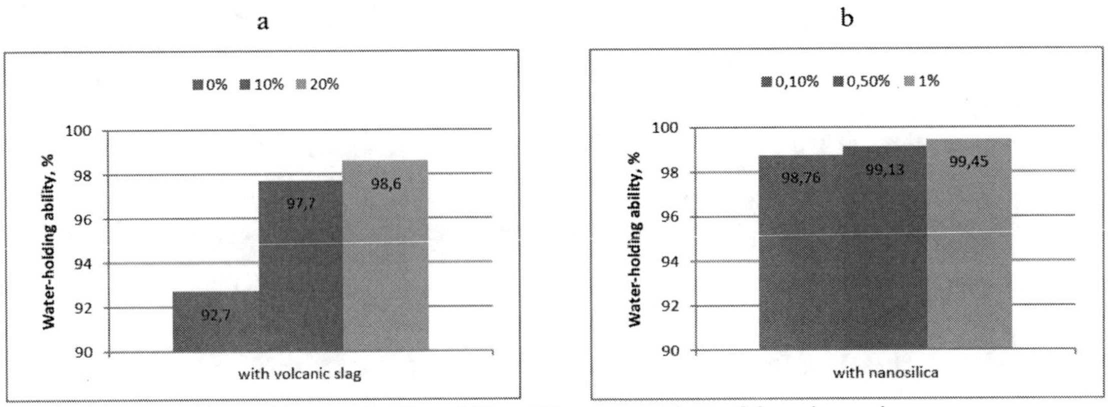

Figure 5. Water-holding ability of the compositions of dry mixes using volcanic slag (a) and nanosilica (b)

[4] Conclusion

Thus, the conducted studies show that the composition of composite binders with volcanic slag is characterized by improved physical and mechanical and operational characteristics, the introduction of nanosilica in the composition of binders has a beneficial effect on the technological and mechanical properties of the composites. The study of the microstructure of modified cement composites showed a difference and favorable structure-forming effect of volcanic slag and nanosilica. The introduction of active mineral micro- and nano-additives leads to the creation of a dense structure and an increase in the number of neoplasms. The use of composite binders using volcanic slag and nanosilica allows to improve the technological and quality indicators of dry mixes.

References

[1] Bazhenov Yu M, Korovyakov V F, Denisov G A 2003 Technology of dry building mixtures (Moscow: ASV) p 96
[2] Kozlov V V 2000 Dry mixes (Moscow: ASV) p 96
[3] Zagorodniuk L, Lesovik V, Shamshurov A, Belikov D 2014 Bulletin of BSTU named after V.G. Shukhov **5** pp 25-31
[4] Lesovik V, Zagorodniuk L, Belikov D, Schekina A, Kuprina A 2014 Construction materials **7** pp 82-85
[5] Lesovik V, Zagorodniuk L, Belikov D 2014 Bulletin of RAASN **18** pp 112-119
[6] Tyuryukhanov K Yu and Pugin K G 2019 Impact of the surface of particles of moulding

sand on the structural formation of asphalt *concrete Proceedings of Universities. Investment. Construction. Real estate* **9(3)** pp 566–577

[7] Dobruskina M A, Petrov A V and Bat-Erdene Z 2019 Improving the technology of retaining walls in the Irkutsk region using gabion baskets *Proceedings of Universities. Investment. Construction. Real estate* **9 (2)** pp 312–323

[8] Moskvitin V A., Emelyanova N A. and Mashovich A Y 2019 Experimental studies of air permeability indicators of composite "Poroplast CF" *Investment. Construction, Real estate* **9 (2)** pp 342–353

[9] Matveeva M V and Kalyuzhnova N Y 2019 Industrial Construction: Investment Support in Entrepreneurship *Investments. Construction. Real Estate: New Technologies and Targeted Development Priorities IOP Conference Series: Materials Science and Engineering* **667** 012062

[10] Matveeva M V, Doroshenko T G and Sukhanova I A 2019 On the issue of cost and timing of contract tenders for industrial and civil construction *Investments. Construction. Real Estate: New Technologies and Targeted Development Priorities IOP Conference Series: Materials Science and Engineering* **667** 012063

[11] Komarov A K, Ivanov I A and Lundenbazar B 2019 Theory and practice of the use of gabions for
forming protective structures *Proceedings of Universities. Investment. Construction. Real estate* **9(1)** pp 78–89 DOI: 10.21285/2227-2917-2019-1-78-89

[12] Kazimirov I A and Peshkov V V 2019 Determination of price behaviour in the secondary residential real estate market using a multidimensional regression model *Proceedings of Universities. Investment. Construction. Real estate* **9(3)** pp 476–487 DOI: 10.21285/2227-2917-2019-3-476-487

IOP Conf. Series: Materials Science and Engineering **880** (2020) 012029 doi:10.1088/1757-899X/880/1/012029

Review and analysis of winter concreting methods, used on construction sites in Russia

E G Zhuravlyov[1]

[1]Irkutsk National Research Technical University, 83 Lermontov st., Irkutsk, 664074, Russia

E-mail: expertiza@istu.edu

Abstract. The features of the application of winter concreting methods on construction sites in Russia are considered: thermos methods (thermos, hot thermos), progressive methods (electrode heating, induction heating, infrared heating, the heating of the heating wires, heated in the mold thermoactive or thermoactive a flexible covering, heating in enclosures), concreting with antifreeze additives. The analysis of the application of winter concreting methods on construction sites in Eastern Siberia is carried out.

In modern construction, the construction of monolithic structures and buildings is carried out year-round. Production of concrete works in winter conditions is always associated with additional material, labor and energy costs. The cost of construction of monolithic structures increases by 1.3-2.0 times compared to summer conditions [3, 4].

When concrete freezes at an early age, irreversible processes occur in its forming structure, causing a significant decrease in the final strength and frost resistance. The final strength is not reduced if the concrete has gained a certain strength, which is called critical, before freezing [5].

Strength of concrete of monolithic and precast-monolithic structures at the moment of freezing (critical strength) [7]:

- for structures operated inside buildings, foundations for equipment that are not subjected to dynamic influences, underground structures, for concrete strength class:
  - up to B10-50% of the design;
  - up to B25-40% of the design;
  - B30 and above-30% of the design;
- for structures subjected to variable freezing and thawing in a water – saturated state at the end of aging-80%;
- for long-span structures:
  - when flying up to 6 m-70%;
  - during the span of over 6 m – 80%;
- in prestressed structures-80%;
- for concrete with antifreeze additives for classes:
  - up to B15-30% of the design;
  - up to B25-25% of the design;
  - B30 and above-20% of the design.

With normal (summer) methods of concreting to obtain the critical toughness at negative temperatures of external air is impossible.

According to SR 70.13330.2012 "Bearing and enclosing structures" [7] during the production of concrete works at the expected average daily outdoor temperature below +5°C (degrees Celsius) and the minimum daily temperature below 0°C (degrees Celsius), special methods of winter concreting should be used to ensure favorable heat and humidity conditions of concrete hardening.

In the arsenal of Russian builders there are the following methods of winter concreting [8, 9, 10]:

1 group. Thermos methods (heat is introduced before laying the concrete mixture in the formwork):

- thermos;
- hot thermos.

2 group. Progressive methods (heat is introduced after laying the concrete mixture in the formwork):

- electrode heating;
- induction heating;
- infrared heating;
- the heating of the heating wires;
- heated in the mold thermoactive or thermoactive a flexible covering (TAFC);
- heating in enclosures.

3 group. Concreting with antifreeze additives.

Thermos method was founded in 1910 by Russian professor Kireenko I. A., is the most simple and economical method of winter concreting.

The essence of the method is as follows. Concrete mix with an initial temperature of 10-20°C (degrees Celsius) is transported to the object and placed in insulated formwork, open surfaces are also insulated. During cooling to 0°C (degrees Celsius) concrete gains a certain strength, which should be no less critical.

The method is based on the use of heat introduced in the manufacture, as well as exothermic heat release of cement.

The thermos method is used for concreting massive structures (e.g. foundations) with a surface modulus up to $M_p = F : V < 8$ m^{-1}, at low negative temperatures (up to -10-15°C) and the required strength is not higher than critical.

where V is the volume of concrete of the constructed structure, m^3;

F – surface area of the structure, m^2.

In the thermos method, it is advisable to use additives of hardening accelerators.

The method of hot thermos or preliminary heating of concrete mixture was proposed in 1962 by the Russian professor A. S. Arbenyev.

The essence of the method is that delivered to the construction site concrete mixture is heated by electric current to a predetermined temperature, placed in insulated formwork and maintained in compliance with the rules of the thermos method.

Electric heating of the concrete mixture is carried out in buckets equipped with plate electrodes.

The duration of forced heating of the concrete mixture is 5-20 minutes, the maximum temperature is 60°C.

Preservation of mobility of the warmed-up concrete mix is reached by introduction of the plasticizing or slowing down setting additives.

Electrode heating was proposed in 1931 by swedish engineers A. Brand and H. Bolin.

Electrode heating-based on the use of heat released in the concrete during the passage of an electric current through it. This is achieved by including freshly laid concrete as an active resistance in the alternating current circuit of industrial frequency using steel electrodes of various types and layouts.

The type of electrodes (rod, plate, strip, floating, string) for heating a particular structure is selected depending on its size and configuration, reinforcement parameters, the location of embedded parts.

Electrode heating of designs is carried out by means of heating oil transformers with voltage on a secondary winding of 55, 65, 75, 85, 95 V.

Heating concrete heating wire.

To warm up the concrete, a cheap single-core insulated wire with a steel core with a diameter of 1.2-2.0 mm is used.

The wire is attached to the reinforcement cage before laying the concrete, after heat treatment remains in the body of the structure.

Heating of the heating wire is used for pouring high-density reinforcement structures, joints of precast concrete.

The step of fastening of a wire in a structures is accepted by calculation and makes 50-150 mm. The wire is wound turns on a reinforcing skeleton or stacked along reinforcing bars.

Heat treatment of concrete is carried out at a voltage of up to 95 V from the heating transformers.

Infrared heating of concrete is developed by the russian professor Danilov N. N., is based on use of energy of infrared radiation submitted on open surfaces of concrete or a surface of not warmed steel timbering.

In the production of concrete works in winter conditions infrared heating is recommended:

- for warming frozen soil and concrete foundations, reinforcement and formwork, removal of snow and ice;
- for pre-warming zone of the joints of precast concrete structures;
- for creation of thermal protection of the surfaces inaccessible for warming;
- for heat treatment of joints of precast concrete and monolithic structures (with the possibility of creating closed volumes, excluding convective heat exchange of heated surfaces with the environment).

Infrared installation consists of a spherical or trapezoidal body-reflector, in the inner cavity of which are placed radiators (TEHs, ceramic radiators, infrared lamps) with mounting elements and current supply.

To avoid intensive evaporation of moisture, the open surfaces of freshly laid concrete are closed with a plastic film and sprinkled with a layer of sand.

Induction heating developed in 1946 by russian professor Netushil A.V., based on the use of heat generated in the reinforcement or steel formwork, located in the electromagnetic field of the coil-inductor, which flows through an alternating electric current. To do this, an insulated inductor wire is laid in successive turns on the outer surface of the formwork.

An alternating electric current of 55-95 V, passing through the inductor, creates an alternating electromagnetic field, which causes eddy currents in the metal located in this field, which leads to its heating. The heat from the reinforcement and metal formwork is transferred to the concrete due to thermal conductivity.

Induction heating is most effective when concreting structures of linear type (columns, beams, crossbars, etc.) densely saturated with reinforcement with, as well as when using metal formwork.

Heating of concrete in thermoactive formwork.

The essence of the method is to transfer heat to concrete from electric heaters installed in insulated inventory formwork. Heat in the concrete is distributed due to thermal conductivity.

The advantages of the method are the exclusion of any additional operations for the preparation and conduct of heat treatment: all includes a standard cycle of formwork. Also the costs of consumables are excluded (electrodes, heating wire).

Large-Board thermoactive formwork consists of a steel formwork shield frame structure, heaters (TEHs, heating cable), screen (aluminum foil), insulation (foam, mineral wool), protective casing.

Small-Board formwork can have the following design solution: plywood formwork frame structure, fixed on the deck removable heating element of conductive sheet or filler plastic (eg, polypropylene with graphite filler) and plastic sheet with a heater fused to the rear of the nichrome wire.

Russian manufacturers (PC "Concrete Heating", PC "Thermal systems") offer small-panel formwork with film heaters and thermal insulation installed between layers of moisture-resistant plywood.

For heating of horizontal surfaces of constructions: overlappings, concrete preparations – thermoactive flexible coverings (TFC) are used, which consist of:

- flexible heaters (as a rule, the heating wire nichrome residential);
- protective moisture-insulating shell (polymer film);
- elastic insulation.

Heating of concrete in enclosures.

Enclosures are temporary structures within which a positive temperature is maintained and either the entire cycle of concrete work is performed, or only concrete aging.

The essence of the method is to create conditions close to summer in the local area around the concreted structure.

The temperature in the enclosures at the bottom of the concreted structure should be at least 5°C. The average height temperature is usually taken in the range of 15-25°C.

By design, dimensions and methods of laying concrete mixture in them, the following types of *enclosures* are used:

- *small enclosures* (installed after laying the concrete mix): caps frame structure, canvas tents, shelters made of polymer film;
- *volumetric* (frame structure or air-supporting shells), inside which the means of mechanized laying of the mixture are placed and the entry of vehicles is provided;
- *mobile* (frame structure with tarpaulin coating), moved along the rails along the concreted extended structures (usually zero cycle): inside the entire cycle of concrete works.

To maintain the required temperature in enclosures, it is recommended to use electric heaters of various designs, heat guns on liquid or gas fuel.

Concretes with antifreeze additives.

Application in winter conditions of concretes with antifreeze additives allows to do without heating of the laid concrete and warming of the basis. Additives are introduced into the concrete mixture during its preparation in the form of aqueous solutions.

In concretes with antifreeze additives at negative temperatures, the liquid phase is preserved, which provides the possibility of hydration processes (interaction of cement with water) and, as a consequence, hardening and strength gain by concrete in the cold.

The scope of application of a particular antifreeze additive is limited by the freezing point of its solutions.

Antifreeze additives by the mechanism of action can be divided into two groups:

Group 1 includes additives that reduce the freezing point of the liquid phase of concrete and practically do not affect the rate of structure formation. This group includes SC, SN and U (sodium chloride NaCl, sodium nitrite $NaNO_2$, urea $(NH_2)_2CO$). Concrete with these additives slowly gains strength at an early age.

Group 2 includes additives that greatly accelerate the setting and hardening, and their solutions have a sufficiently low eutectic temperature. These additives include P, CH, CN, NCN (potash K_2CO_3, calcium chloride Ca_2, calcium nitrate CA $(NO_3)_2$, nitrite calcium nitrate CA $(NO_3)_2$+ CA $(NO_2)_2$).

In connection with the above, it is recommended to use complex additives that allow to obtain concrete mixtures optimized for freezing point, hardening rate and degree of aggressiveness to the reinforcement.

Concrete with antifreeze additives is allowed to be used if the temperature of the concrete with the maximum permissible dosages of additives does not fall below during the holding until the critical strength is acquired:

- -15 °C – for SN;
- -20 °C – for CH+SC, CN+U, NCN, NCN+U;
- -25 °C – for P, CH+SN, SNCH, SNCH+U.

For the design temperature, the average value of the outdoor temperature is taken according to the forecast for the first 20 days from the moment of laying concrete.

The amount of antifreeze additive is determined by the calculated temperature of hardening to a set of critical strength: the lower the outside temperature, the higher the consumption of the additive.

In accordance with GOST 24211-2008 "Additives for concrete and mortars" [2] antifreeze additives are divided into two groups:

- *for "cold concrete"* - provide hardening of concretes and mortars at negative temperatures;
- *for "warm concrete"* - provide protection of concrete mix from freezing for the period from its production to laying and supply of external heat.

As a rule, these are the same additives with different dosages in concrete.

Practice of winter concreting in the siberian region.

The analysis of application of winter methods of concreting in eastern Siberia is given in table 1.

Table 1. Features of application in eastern Siberia

№ s.c.	Method for winter concreting	Features of application in eastern Siberia
1	Thermos's	It is used in the construction of massive structures (ribbon and columnar foundations, grillages, foundation slabs) in the spring and autumn periods
2	Hot thermos	It is not used due to the significant required electrical power and loss of mobility of the concrete mixture during heating and laying
3	Electrode heating	Prevailed in the application of up to 2000 years. In modern construction it is not often used due to the consumption of steel for electrodes and aluminum wire for switching electrodes
4	Induction heating	Not applicable
5	Infrared heating	It is used for heat treatment of precast concrete joints, warming of soil base and formwork cavity
6	Heating by heating wires	Currently the most common progressive method
7	Heated in the mold thermoactive thermoactive flexible coatings (TAFC)	Not applicable due to limited (constructive solutions) offers from domestic manufacturers and the high cost of imported thermoactive formwork and TAFC
8	Heating in enclosures	Enclosures are applied in the form of caps of a skeleton design, tent coverings
9	Antifreeze additives	Antifreeze additives are used mainly for " warm concrete – - provide protection of the concrete mixture from freezing for the period from its manufacture to installation and supply of external heat.

HSV 1.2 wire, which is in demand in construction, is specially designed for winter pouring of concrete with gradual heating of surfaces. The product is a wire in PVC insulation with a thin (1.2 mm in diameter) core of steel alloy. The electrical resistance of the steel wires of 0.15 Ohm/m, working current 15A. The cost of wire 1000-1500 RUB/1,000 m. Switching (cold ends, highway) heating wire is an insulated aluminum wire, the heating is carried out from the step-down transformer.

IOP Conf. Series: Materials Science and Engineering **880** (2020) 012029 doi:10.1088/1757-899X/880/1/012029

Russian manufacturers of chemical additives in concrete: companies “Fort”, “Formula concrete”, “Sonstrumax”, production company “Gidrostroykomplekt”, manufacturer of construction chemicals TM "Sika", PC "POLYPLAST", PC “RAMEKO”, PC “Skytrade”, PC “Chemical holding” and others produce complex antifreeze additives, the composition of which is not disclosed:

- Krioplast P25;
- Krioplast PC;
- Krioplast Premium;
- Krioplast Extra;
- Штайнберг FROST15;
- Штайнберг FROST25;
- Term's;
- Frost 30;
- PUFAS.

Sodium nitrite and sodium formate *HCO_2Na* are supplied from China to Russia.

Here are the main characteristics of one of the complex antifreeze additives *“Krioplast P25”:*
Manufacturer PC "POLYPLAST".

The additive is a mixture of antifreeze and plasticizing component. Available in dry and liquid commodity forms

In case of application by the customer of the additive “Krioplast P25” in dry commercial form it is recommended to observe the following dosages (table 2):

Table 2. Average design temperature of concrete hardening

Average design temperature of concrete hardening, ^{0}C	Dosage range of dry product,%, by weight of cement
on “warm” concrete»:	
-5	0,7
-10	1,0 – 1,5
-15	2,0 – 2,5
-20	2,5 – 3,0
-25	3,5 – 4,5
on the “cold” concrete:	
-5	1,0
-10	1,5 – 2,0
-15	2,5 – 3,5
-20	3,5 – 4,0
-25	5,0 – 6,0

The cost of the solution concentration of 35-40% additives “Krioplast P25” is 40-50 rubles. / kg, dry additives- 80-90 rubles. / kg. manufacturers of ready-mixed concrete prefer to purchase an additive in the form of solutions, which simplifies the technology of feeding additives in the preparation of concrete mix.

In the near future, the specifics of the application of winter concreting methods at construction sites in eastern Siberia, apparently, will remain. Although, it is possible and the appearance of thermoactive formwork and thermoactive heating coatings (TAFC).

References

[1] Afanasyev A A, Matveyev Ye P, Minakov Yu A 1997 Technological efficiency of accelerated concrete hardening methods in monolithic housing construction *Industrial and civil construction* **8** p 36-37

[2] GOST 24211-2008. Additives for concrete and mortars, p 16
[3] Golovnev S G 2010 Modern construction technologies p 262
[4] Golovnev S G, Krasny Yu M, Krasny D Yu 2012 Production of concrete works in winter Conditions *Quality assurance and efficiency* p 336
[5] Krasnovskiy B M 2004 Engineering and physical bases of winter testing methods concretings p 470
[6] Mozgalev K M Intensification of technological processes for winter concreting of monolithic buildings *South Ural state University* p 130- 133
[7] SR 70.13330.2012. Bearing and enclosing structures pp 71-79
[8] R-NP SRO SSK-02-2015. Recommendations for the production of concrete works in the winter Period P 84
[9] Guide to the production of concrete works in winter conditions, areas of the Far East, Siberia and the Far North 1982 p 313
[10] Guide to heating concrete in monolithic structures. Research Institute of concrete and reinforced Concrete 2005 p 276
[11] Telichenko V I, Terentyev O M, Lapidus 2006 Construction technology buildings and structures p 446
[12] Tyuryukhanov K Yu and Pugin K G 2019 Impact of the surface of particles of moulding sand on the structural formation of asphalt *concrete Proceedings of Universities. Investment. Construction. Real estate* **9(3)** pp 566–577
[13] Dobruskina M A, Petrov A V and Bat-Erdene Z 2019 Improving the technology of retaining walls in the Irkutsk region using gabion baskets *Proceedings of Universities. Investment. Construction. Real estate* **9** **(2)** pp 312–323
[14] Moskvitin V A., Emelyanova N A. and Mashovich A Y 2019 Experimental studies of air permeability indicators of composite "Poroplast CF" *Investment. Construction, Real estate* **9** **(2)** pp 342–353

Perfection of technology for manufacture of heating units on the base of mica and glass

V G Zhytov[1] and T I Shishelova[2]

[1]Institute of architecture, construction and design of INRTU, 83 Lermontov street, Irkutsk, 664074, Russia
[2]Department of Physics, INRTU, Lermontova Str. 83, Irkutsk 664074, Russia

E-mail: Tamara.shishelova@gmail.com

Abstract. Implementation of energy saving programs, reduction of energy consumption for heating and development of efficient heating systems is a *challenge*. The use of resistive mica- and glass-based heaters (SKENs) is thought to be a perspective trend in this area. It is a heater of new generation that has no analogues abroad. *Objective* of the paper is enhancement of technologies for manufacture of resistive heaters on the base of mica and glass for improving heater's performances. Consideration is given to development of new compositions of charge material and to treatment of its surfaces. *Results of study*. Technology for manufacture of SKENs with high mechanical strength and low water absorption that are widely used in different areas of building industry has been developed on the base of experimental studies and scientific generalizations.

1. Introduction

Use of heat and power for residential buildings in Russia is currently rather inefficient. The policy of "cheap" energy carriers pursued in Soviet period allowed construction of buildings with low heat insulation level, and absence of heat meters and poor control over heat energy consumption favored its wasteful use. All these factors necessitated development of new technologies for low-temperature heating.

2. Methods and materials

Mixtures of muscovite and phlogopite mica of different percentage were investigated using IR-spectroscopy and thermography methods. Resistive heaters on the base of mica and glass were subjected to mechanical and physico-chemical studies in conformity with GOSTs.

3. Results and discussion

Research workers of INRTU developed a method for manufacture of mica- and glass-based heating elements named SKEN. It is a heater of new generation that has no analogues abroad. A new method for heater manufacture includes the following procedures: a nichrome heating element is placed into a mixture of powdered mica and soft glass to be followed by briquettes production by applying cold pressing method. Briquettes produced are heated to the temperature of 600-700°C. A heated briquette is then hot-pressed and further baked [1-4].

SKEN has good operating performances. It is intended for heating the industrial and residential facilities, cattle-breeding and paltry farms, storehouses, auxiliary premises, car washing premises, as well as for heating the premises with high humidity. It is also recommended for container-type houses [5, 6].

Present-day heating units shall meet a number of requirements, such as higher insulation resistance, maintenance of mechanical strength at higher temperatures, good thermal conductance, and low water absorption [7].

For enhancing the mechanical strength and for reducing the water absorption the charge for the electric insulation layer of a heater is recommended to have the following percentage of components: muscovite mica - 25-35%, phlogopite mica - 25-35%, and low-melting alumina-borosilicate glass - 35-45%. Addition of phlogopite mica into the electric insulation mixture activates inter-phase interrelations of components during heating and hot pressing. As a result, a dense monolith structure is formed [8-10].

During heating the mica is dehydroxylated. IR-spectroscopy and thermography studies revealed that dehydroxylation process in the mixture with approximately equal percentage of mica and phlogopite is intensified. This factor creates conditions for more intensive subsequent interaction of mica and glass. In the course of interaction in such a system, alumomagnesian spinels and olivine are formed on the mica crystals boundaries along with feldspar, sillimante and leucite, which additionally strengthen the material [11-14].

Moreover, presence of magnesium oxide and fluorine within phlogopite contributes to formation of low-melting eutectics with glass components, which favors more intensive glass formation. A part of hydroxylic ions of some phlogopites are substituted by fluorine that is not removed during heating, as chemically bound fluorine is less volatile compared to hydroxylic ions. Fluorine during heating can form fluorides interacting with glass components, particularly with quartz. Lower percentage of high-melting (heat resistant) oxide Al_2O_3 in phlogopite as compared to muscovite also contributes to formation of eutectics in the mica and low-melting-glass system.

Flumomagnesium spinel $MgAlO_4$ makes a major contribution into the structure strengthening. During muscovite dehydroxylation, Al_2O_3 is formed, and MgO is formed during dehydroxylation of phlogopite; their interaction produces spinel [15,16]. Higher mechanical strength and lower water absorption are attained at approximately equal content of the above mentioned mica in the composition of electric insulation material (25-30%). At lower content of phlogopite the output of spinels is negligible, and in case of its absence the spinels are not formed at all as there is magnesium neither in muscovite nor in glass. At low muscovite content the output of spinels drastically falls due to deficiency of Al_2O_3 in the system, which is partially replenished at the expense of glass.

Thus, in the process of resistive heater manufacture, more intensive inter-phase interaction is ensured in the mica-glass system that creates monolith dense material with high mechanical strength and low water absorption. Breaking strength at static bending increases by 40-50%, water absorption decreases fivefold.

Addition of a modifying agent (5% furnace clinker) into the composition of electric insulation charge of a heater allows 50% reduction of water absorption, specific volume resistance in this case increases tenfold [17-19].

Higher water resistance subject to maintenance of insulation resistance during long operation of heaters in humid media is attained owing to surface treatment after baking with 3-5% solutions of organic stillage residues from methylchlorosilane production, that is to be followed by heating at 110-150°C.

During methylchlorosilane synthesis the stillage residue contains considerable amount of hydroxylans, particularly, methyldichlorosilane, trichlorosilane, and methyltrichlorosilane. The proposed oligomethyl-chlor-silane, when interacting with the surface of mica-ceramic material, forms a dense hydrophobial electric insulation layer on the heater surface by substituting the mica hydroxile ions for chlorine ions. Chemical reaction is given in equation (1).

$$KAl_3Si_3O_{10}(OH)_2+(CH_3)_nSiCl_m \rightarrow KAl_3Si_3O_{10}Cl_m(CH_3)_n+2OH \quad (1)$$

It is the presence of hydroxylans and multi-functional organochlorsilanes that causes efficient substitution of hydroxyl groups on the mica surface, creates steric difficulties for moisture penetration to its surface due to formation of branched structures of products of multi-functional organochlorsilanes interaction. This fact predetermines higher efficiency of treatment with stillage residues as compared to the use of common hydrophobisators.

Use of phosphate glass and other compositions in heaters demonstrates good results.

For enhancing the compression resistance and thermal resistance of a composite, the 0.04-0.07 mm glass should be used, rather than previously used 0,10-0,16 mm glass. Baking temperature in this case can be lowered down to 600-700°C, and, consequently, energy intensity of the technological process can be lowered accordingly.

Temperature required for the composite baking is conditioned by physical and chemical interaction of mica particles with softened glass. The smaller the particles, the more intensive and active the interaction at lower temperatures.

Softened glass most actively interacts with mica at glass size of 0.10-0.16 mm at temperatures of 700-800°C. If glass size reduces down to 0.04-0.07 mm, the interaction starts at temperatures close to the temperature of glass softening, i.e., at about 500°C. In this case, for formation of mechanically resistant material, the heating up to 650-700°C is sufficient. Moreover, owing to increased surface of bonding material the interaction between glass and mica is intensified, it is more complete and material obtained has better thermo-chemical performances.

Composites with better dielectric properties can be produced using certain temperature mode of charge treatment for phlogopite heater [20-22].

Micalex formation process is closely related to mica dehydroxylation process. In the course of mica and glass briquette heating the process of mica decomposition in the softened glass medium occurs. Mica decomposition products interact with glass forming new constituents. These processes condition formation of cementing layer on the surface of mica crystals thus ensuring high dielectric and mechanical properties of composite. It was found out that removal of inter-layer water in phlogopite was terminated at a temperature of 800°C. At a temperature of about 850°C the dehydroxylation process started that precedes the micalex formation process. Preliminary thermal treatment of phlogopite at 850-950°C creates conditions for its more complete interaction with softened glass, which ensures high electric insulation properties of the composite. 50-80 minute thermal treatment is due to low rate of dehydroxylation. Table 1 presents properties of composite produced from phlogopite-based charge

Table 1. Dielectric properties of composite made of fillers prepared in different ways

Filler	**Dielectric properties**				
	Tangent of dielectric losses angle tgδ	**Elasticity module E**	**Specific volume resistance ρ_v, Ohm·m**	**Specific surface resistance ρ_s, Ohm**	**Electric resistance E_{str}, MW/m**
Phlogopite after thermal treatment at 850-950°C (for 50-60 min.)	0.0005	6.3-7.5	$1\cdot10^{11}$	$1\cdot10^{11}$	22-31
Thermally non-treated phlogopite	0.005	8	$5.2\cdot10^{11}$	$1.4\cdot10^{11}$	20

It is obvious that thermal treatment of phlogopite as filler produces composite with enhanced dielectric properties. Such a thermal treatment is most advisable for off-grade phlogopites.

Use of fluoric phlogopites demonstrated the advisability of using natural fluorine-containing phlogopites in composites with higher heat resistance. For raising the heater's efficiency (in case of production of a panel-type heater) and reliability it was proposed to use phlogopite with hydroxyllic groups partially substituted for fluorine, its content being 3.0-6.5%. The proposed technical solution allows manufacture of a panel-type electric heaters designed for the temperature of 850-1000°C.

4. Conclusion

Studies on selection of the optimum composition of charge for electric insulation layer of a heater were conducted. Addition of phlogopite mica into the electric insulation mixture intensifies inter-phase interrelations of components during heating. At approximately equal percentage of muscovite and phlogopite mica the interaction of mica with glass is more intensive, which enhances the mechanical strength of a heater. Addition of a modifying agent into the charge considerably reduces water absorption, and increases specific volume resistance. Demonstrated is the reaction of chemical interaction of oligomethyl-chlorine-silane with the surface of mica-ceramic materials, which reduces water absorption. Optimum glass size has been selected the use of which results in higher breaking strength towards compression and higher heat resistance. It was found out that for enhancing the electric insulation properties of mica-ceramic material, plogopite mica should be subjected to preliminary thermal treatment. Advisability of using natural fluorine-containing phlogopites for enhancing the composite heat resistance has been demonstrated.

5. References

[1] A.s. 1139367 USSR. A method for manufacture of a resistive heater / Baiborodin B A, Borzov V G, Starikov V S, Shishelova T I and Leonov S B / # 3025799/24-07; Applied for on 04.11.80.

[2] A.s. 1555922 USSR / A method for manufacture of a panel-type heater / Gavrilov V G, Bobrova G I, Belinsky M E and Shishelova T I / # 4358237/24-07; Applied for on 16.11.87; approved on 25.05.88.

[3] Bobrova G I and Shishelova T I 1992 Heat-resistant mica containing materials for electrothermy *Irkutsk State Univ. Publishing House.* P 114

[4] Kogan V E and Shahparonova T S 2015 Methods for production and properties of composite materials in the glass-mica system *International Research Journal.* **1-1** 45-7

[5] Shelekhov I Yu and Shishelova T I 2007 Development of heating units and investigation of its efficiency in the life support systems *Vestnik IrGTU.* **1** 104-9

[6] Shelekhov I Yu, Shishelova T I and Dukhovny L I 2012 Peculiarities of using heating equipment in buildings with variable heat regime *Fundamental Research.* **3-2** 437-40

[7] Shishelova T I 1993 Physical and chemical principles of mica composites production. P 212

[8] Fedorova S V 2015 Study of physical processes and chemical reactions of mica-glass composite of new chemical composition *Theoretical and Applied Science* **4** 108-12

[9] Liopo V A, Ovchinnikov E V, Sitkevich F A, Sitkevich A L and Struk V A 2015 Structural transformations of laminated silicates under thermal impact (on the example of mica) *Advanced Technologies and Machine-Building Systems.* **1** 90-9

[10] Gershenkop A Sh, Chernyshev A V and Hohulya M S 2013 Chemical and physical treatment of mica before disintegration *Mine surveying and Subsoil Use.* **3** 58-61

[11] Shishelova T I 2018 Physical-chemical quantum-mechanical aspects of mica dehydration and dehydroxylation: monograph *Irkutsk National Research Technical University*. P 178

[12] Avdeychik S V 2007 Polymer-silicate engineering materials: physical-chemistry, technology, application *Technology*. P 431

[13] Lakhwant Singh, Sukhnandan Kaur, Mohan Singh and Navjeet Kaur 2011 Electrical and dielectric characteristics of annealed muscovite rubt mica *Materials Physics and Mechanics.* **11** 60-7

[14] Prost R and Laperche V 1990 Far-infrared study of potassium in micas *Clays Clay Miner.* **4** P 351

[15] Kuznetsova G A, Kalikhman V M, Liopo V A, Metsyk M S and Shzetsov B N 1979 Radiographic study of structural conversions of muscovite and phlogopite at heating up to 1500 °C *Radiographic Investigation of Mineral Raw Material*. 111-6

[16] Fripiat I I, Rouxhet P and Jacobs H 1965 Proton delocalization in micas *Amer. Miner.* **11/12** 1937-58

IOP Conf. Series: Materials Science and Engineering **880** (2020) 012030 doi:10.1088/1757-899X/880/1/012030

[17] Fedchishin V V 2014 Mobility of charge carriers in polymer and mica electrets *Izvestiya VUZov. Physics.* **57** 26-30
[18] Abdulova S R and Olakh N M 2016 Production of electric insulation materials on the base of micaceous laminate *Relations between Science and Society: Problems and Prospects.* 22-3
[19] Shishelova T I and Shulga V V 2016 Physical and chemical grounds for mica-based composite materials production *Fundamental Research.* **3** P 538-41
[20] Sreenivasan H and Kinunen P 2017 Thermally treated phlogopite as a rich in magnesium precursor for alkali activation *Minerals Engineering.* **113** 47-54
[21] Guseinov A A 2015 Influence of temperature dependence of electric conductance in mica as reflection of thermal dehydroxylation processes *Renewable Energy: Problems and Prospects. Challenges of Developing the Renewable Energy Sources.* 289-95
[22] Shishelova T I, Chilikanova L V and Sozinova T V 1999 Electric strength as a quality criterion of mica ceramic electric heating *Intl. Science-tech.*

IOP Conf. Series: Materials Science and Engineering **880** (2020) 012031 doi:10.1088/1757-899X/880/1/012031

Conceptual framework for implementing energy-efficient technologies into construction

E Y Gorbachevskaya[1], O A Bezrukikh[1], T A Maletina[2] and D A Safronov[2]

[1] Irkutsk National Research Technical University, 83 Lermontov str., Irkutsk, 664074, Russia
[2] IIEL ISU, Ulan-Batorskaya, 6, 664082 Irkutsk, Russia

E-mail: olga.bezrukikh11@gmail.com

Abstract. The energy problem is becoming global every year. The reason for this is the growth of the world's population and the intensive development of technologies, which causes a constantly growing level of energy consumption. The construction industry is the largest sector of final energy consumption, accounting for a higher share of consumption than in industries such as industry and transport. The relevance and versatility of the problem of improving the efficiency of energy-efficient technologies in construction determines the need to improve the methodological foundations of the mechanism for managing production and economic processes for the introduction of energy-efficient technologies. Improvement and implementation of current methodological foundations and principles will form a systematic approach to the issue of mass use of energy efficient technologies in construction.

The reduction of natural energy reserves leads to the problem of energy saving. The solution to this problem is directly correlated with an increase in the use of energy-efficient technologies in construction, since the construction industry is the largest sector of final energy consumption, which accounts for a higher share of consumption than in industries such as industry and transport. Based on this, one of the main tasks of today is to find the most effective measures to maintain a modern and comfortable living of citizens, ensuring the rational use of energy resources and increasing the energy efficiency of buildings.

Currently, most of the utility costs are spent on heating buildings and structures and account for more than 60% of the total cost structure. However, the use of energy efficiency technologies will provide an opportunity to reduce the costs of the federal and municipal budgets to 100-150 billion rubles [1]. Despite the possibility of achieving sufficiently high indicators of economic and environmental efficiency, energy efficiency technologies in Russia are still not fully developed. From the Russian Federation development of energy-efficiency technologies are supported through development of regulatory documents, codes of practice etc. However, these measures are not sufficient to pilot projects of energy-efficient houses have become a real energy-efficient buildings, which is a collection of architectural planning, design, engineering solutions, and is aimed at reducing the buildings consumption of energy resources without losing their reliability and comfort.

In the process of developing energy-efficient technologies, the term "energy efficiency" has changed and expanded the boundaries of its influence on the construction industry. Due to the fact that energy efficiency technologies include the use of not only thermal energy, but also the use of other types of energy resources and energy, it is necessary to give a more precise definition of this concept. As the definition of energy-efficient buildings shows, it is a fairly broad concept that includes a large number of methods aimed at minimizing energy costs and saving resources [2].

Issues of energy efficiency and energy saving are now relevant not only in the regions of Russia, but also in all countries of the world. Construction companies that are focused on the future and aimed at creating competitive advantages are moving to a higher quality level of construction by using energy-efficient technologies. Buildings of the present and future must demonstrate a rational approach to the use of energy resources, including respect for the environment, while maintaining the comfort of the consumer's stay in it.

Due to rapid urbanization and industrialization, the demand for energy is growing very strongly. New construction and renovation methods can minimize energy consumption by using more efficient construction methods. However, it is necessary to develop a regulatory framework to facilitate energy decision-making, along with indicators of the life cycle of the deployed technology. The prospects for energy-efficient buildings are to improve the quality of energy services provided, such as lighting, and to ensure minimum economic and environmental costs. Adopting an environmentally sustainable energy efficiency ideology will provide tremendous energy and cost savings in the use and consumption of resources.

Despite the high importance of using energy-efficient technologies, there is no methodological support for the life cycle of energy-efficient buildings. Currently, most of the regulatory documents in the construction industry are Advisory in nature [3,4]. They are not able to clearly regulate the use of energy-efficient technologies in the construction of low-rise buildings. Based on this, we will define the main organizational aspects and regulatory documents for the design, construction and operation of energy-efficient low-rise buildings (table 1).

Table 1. Organizational aspects and regulatory documents for the design, construction and operation of energy-efficient buildings

Stage of life cycle	Organizational aspect	Normative document
Designing	Selection of materials, structures, and equipment that can reduce energy costs Conduct monitoring of compliance with regulatory requirements for energy savings in the project Ensuring the availability of devices in the project accounting and regulation of main energy carriers Availability of the "Energy efficiency" section in the project documentation	Letter from the state committee for construction of the Russian Federation № NK-5607/6 from 10.09.2003. Resolution Government of the Russian Federation № 87 from 16.02.2008 Resolution Government of the Russian Federation № 235 from 13.04.2010 Order of the ministry of economic development of the Russian Federation № 229 from 4.06.2010 [5]
	Filling in the energy passport of the building at each stage of the project Determining the energy efficiency class of buildings	Order of the ministry of regional development № 262 from may 28, 2010 Resolution Government of the Russian Federation № 318 from 25.04.2011

Construction	Certification of the construction production process to obtain the projected energy efficiency indicators Conducting internal supervision of construction process and control of energy efficiency indicators Delivery of the object in accordance with the projected indicators energy efficiency	Law № 190-FL of 29.12.2004 Law № 261-FL from 23.11.2009 [6] Resolution Government of the Russian Federation
Operation	Filling in the energy passport (repeatedly) Ensuring constant monitoring of specific heat energy consumption indicators for heating and ventilation of the building Conducting periodic and unscheduled surveys buildings and their engineering systems Conducting periodic energy audits building and indications of metering devices of energy resources	Order of the ministry of regional development № 262 from 28.05.2010 Order of the ministry of energy of Russia № 182 from 19.04.2010 Resolution of the government of the Russian Federation № 19 from 25.01.2011

As the table shows, in order to increase the energy efficiency of buildings, the life cycle processes must be organized in such a way that a high level of energy efficiency is achieved at each stage of the life cycle. Based on this, it is appropriate to develop and implement a methodology for organizing the life cycle of energy-efficient buildings, together with the use of a systematic approach and international standards ISO 14001, ISO 9001 and ISO 50001 [7].

Experience in implementing building construction projects using energy-efficient technologies shows that at the stage of planning and designing a new facility, the potential for energy saving is higher, and investment costs are significantly lower than when implementing measures to improve the energy efficiency of the facility during operation [8] (Picture 1).

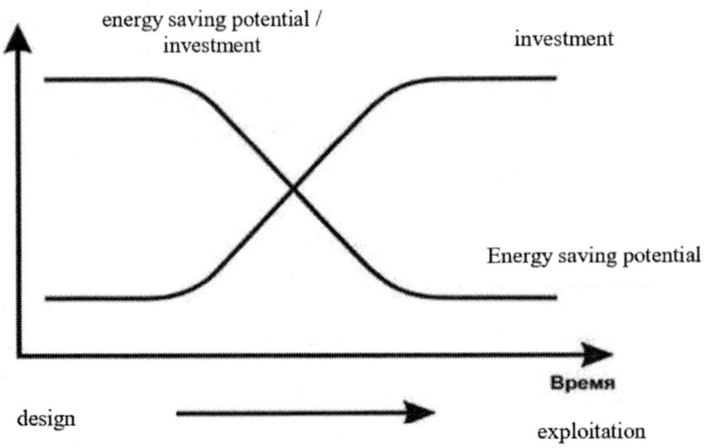

Picture 1. The ratio of the level of energy saving potential and investment volumes at the design and operation stages

Based on the data in table 1, we will design a conceptual diagram of the building life cycle processes using energy-efficient technologies (Picture 2).

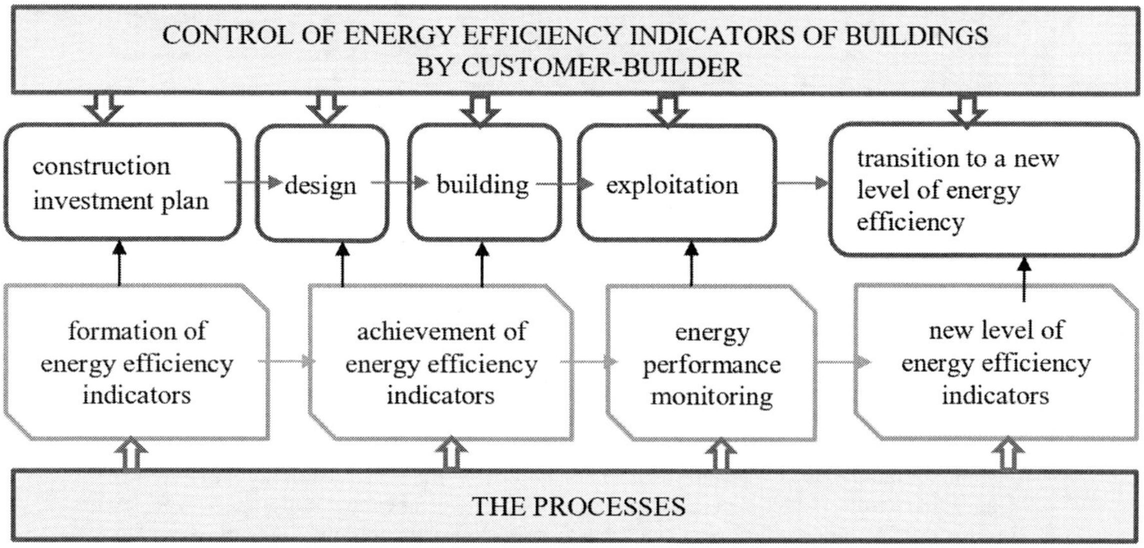

Picture 2. Conceptual diagram of the life cycle processes of low-rise buildings using energy-efficient technologies

Thus, the process of organizing the life cycle of an energy-efficient building will not only achieve a high level of implementation of energy-efficient technologies, but also allow:

•determine the logical relationship between the stages of design, construction, operation, reconstruction, and decommissioning;

•determine the energy efficiency of each stage of the life cycle;

•develop estimates of energy efficiency of buildings depending on the stage of the life cycle;

•determine the timing of current and major repairs of buildings, taking into account the loss of energy resources due to wear and tear of materials and structures.

Based on the life cycle of energy-efficient buildings, it is necessary to form current principles for the use of energy-efficient technologies in low-rise construction. The principles take into account a systematic approach to construction, the specifics and requirements of energy efficiency (Picture 3).

1. Functional and system principle of energy efficiency of buildings
The system-forming factor is the specific result (target function) of the system functioning [9]. It is the result-the achievement of the required level of energy efficiency by buildings is a system-forming factor in the construction industry and requires a reorientation of many organizational, technological and managerial decisions.

2. Probabilistic and statistical principle of energy efficiency
Application of probabilistic models in which the variables considered (energy consumption, thermal protection, etc.) are random variables. At the same time, the value of the target function is expressed by statistical distributions that are in stochastic dependence on all statistical distributions of the system parameter values [10, 11].

3. Simulation-modeling principle of energy efficiency of buildings
The principle is to study complex systems using mathematical modeling methods. It is relevant for energy-efficient buildings, as the systems of design, construction and operation of energy-efficient buildings, of course, belong to the class of the most complex systems, both in their structure and functioning [12]. The application of the simulation-modeling principle is associated with the complexity of construction systems, the organization of their functioning in conditions of energy efficiency requirements, when the number of parameters that most significantly reflect the functioning of the system and the achievement of the desired result increases [13,14]. Functional modeling of the life cycle of energy-efficient buildings is a complex task, the solution of which requires the use of special techniques and tools.

4. Engineering and economic principle of energy efficiency of buildings
According to this principle, it is necessary to evaluate the energy efficiency of buildings not only from a technical, but also from an organizational and economic point of view, since the lack of comprehensive indicators and criteria for evaluating economic efficiency is the most significant methodological disadvantage of the methods used to assess energy efficiency in construction. Economic methods for evaluating efficiency should take into account the economic interests of individual participants in the investment and construction process and the country's construction industry as a whole.

Picture 3. Principles of using energy-efficient technologies in construction

Thus, the implementation of these principles should become a methodological basis for the construction of energy-efficient buildings and contribute to improving energy efficiency not only within the normalized indicators, but also the entire construction industry as a whole.

References

[1] Aloyan R M, Petrukhin A B, Oparina L A 2015 Improvement of organizational and technological solutions for resource and energy saving in construction using synthetic geomaterials *News of higher educational institutions. Technology of the textile industry* № **6** p 9-15

[2] Unified information system of housing construction *[Electronic resource]* URL: https://наш.дом.рф (accessed 19.10.2019)

[3] Order of the ministry of construction of the Russian Federation of 17.11.2017 № 1550 PR About approval of requirements for energy efficiency of buildings, structures, structures (Registered in the Ministry of justice of the Russian Federation 23.03.2018 № 50492)

[4] Timofeev M M, Viniukov O O and Bondareva O B 2017 The Interaction of biogenic and

technical-technological factors at sustainable agro-biocenoses formation *Balanced Nature Management* **1** p 43-49
[5] Reference document on best available techniques for energy efficiency *[Electronic resource]* Access mode: // http://portalenergo.ru/files/articles/portal-energo_ru_dokument_es_po_e_ef.pdf (Date of issue 03.09.2019)
[6] Polezhaeva T O and Flaxman A S 2018 Energy consumption and energy saving in the operation of public buildings *Actual problems of management in the fuel and energy sector-2018: VECTOR 2035* p 238-244
[7] Kazmina A I and Koroy E I 2017 Methods of architectural and construction solutions and thermal efficiency of buildings *Modern concepts of science development collection of articles of the international scientific and practical conference* p 180-188
[8] Urmeeva A R and Markina M V 2018 Development in construction *Collection of scientific articles of the 8th International scientific and practical conference. Institutions and mechanisms of innovative development: world experience and Russian practice-2018* p 327-329
[9] Doleva Z N and Makhov A Yu 2017 Reduce time in construction *A young scientist* № **5** p 37-40
[10] Kazmina A I and Koroy E I 2017 Energy potential of the climate of the construction area and conditions for self-regulation of the thermal regime of the territory *New science: history of formation, current state, prospects of development* p 169-174
[11] Korol O A 2016 Organizational and technological mechanism for implementing energy-saving measures in the construction of monolithic housing construction objects p163
[12] Astafieva O E 2017 Features of the organization of an effective management system for resource saving and energy efficiency in various sectors of the economy *Vestnik universiteta* № **2** p 197-201
[13] Giria M A and Sheina S G 2018 Problems of energy saving in the housing stock of Russian cities *Engineering bulletin of the don* № **3**
[14] Sorrell S and Dimitropoulos J 2008 The rebound effect: Microeconomic definitions, limitations and extensions *Ecological Economics* **65 (3)** p 636–649

Improving the energy efficiency of a residential building

V M Kartavskaya[1] and S A Khoroshikh[1]

[1] Federal State Budgetary Educational Institution of Higher Education Irkutsk National Research Technical University, Institute of Energy, 644074, 83Lermontova str., Irkutsk, Russia

E-mail: vinkar@mail.ru

Abstract. The possibilities of introducing energy-saving technologies to increase the energy efficiency of an apartment building with a central heating system are considered. The efficiency of using heat energy will increase in accordance with the value of reducing heat loss to the environment. When implementing the project of the proposed energy-saving measures with effective overhaul, the energy efficiency indicator in terms of the specific consumption of thermal energy by a residential building can reach a value close to the base, and the efficiency class to normal D and claim to be increased C.

The energy efficiency of buildings is determined by a combination of many factors, the main of which in Russia are harsh climatic conditions in a large territory of the country [1], the relatively low cost of fuel and energy resources and the heat-shielding properties of external building envelopes [2]. Over 60% of the total volume of thermal energy consumed during the operation of residential buildings is accounted for by heating and ventilation [2, 3].

In the countries of the European Union, the largest consumers of energy (more than 40%), and their number is growing, are residential buildings and the service sector, despite the fact that most of these countries are located in temperate and warm climates. In this regard, the need to limit the consumption of fuel and energy resources and increase the energy efficiency of buildings is relevant.

Directives 2002/91 / EC and 2010/31 / EC (updated version) on the energy characteristics of buildings contain requirements for assessing and standardizing energy efficiency (certification - setting energy efficiency classes) of designed, new and operated public, occupied by public authorities and residential buildings [4].

Moreover, they are constantly and consistently improving, increasing demands not only on thermal insulation, but also on engineering systems for heating, ventilation, hot water supply and air conditioning. When designing new buildings, it is required to use local and centralized systems that operate in whole or in part on renewable heat resources and heat pumps. For reconstructed buildings and for overhaul, it is recommended to use the same sources, but with a feasibility study. Preferably, the design of new buildings with almost zero energy consumption.

After 2018, all new public buildings, and after 2020 all new residential buildings must meet the requirements for buildings with almost zero energy consumption. Until 2015, all existing buildings with a useful area of 500 m^2 and higher were certified according to the criteria of energy efficiency, and after that, houses with an area of up to 250 m^2.

Each of the countries belonging to the European Union is developing Energy Saving Programs, which sometimes outperform average indicators not only in energy efficiency, but also in climate protection. Each of the countries that are members of the European Union is developing Energy Saving Programs, which sometimes outperform average indicators not only in energy efficiency, but also in climate protection.

The main goals: reducing energy consumption, carbon dioxide emissions and increasing the share of renewable energy in the overall energy balance by 20% compared to 1990. An example of the EU is followed by Canada, the USA, Japan, Georgia, Armenia and Kyrgyzstan.

In Russia, for newly created buildings (including apartment buildings), structures, structures, the specific characteristic of the consumption of thermal energy for heating and ventilation decreases according to [5]: from January 1, 2023 - by 40%, from January 1, 2028 - by 50 %.

Also, from January 1, 2023, for designed, reconstructed, overhaul buildings, with the exception of apartment buildings, it is recommended (if there is technical feasibility and a feasibility study) to install renewable and alternative sources of energy and secondary energy resources in engineering systems.

Thus, buildings become energy efficient, they provide the necessary level of comfort with standard or lower costs of thermal energy and the use of other types of energy - Figure 1 [6].

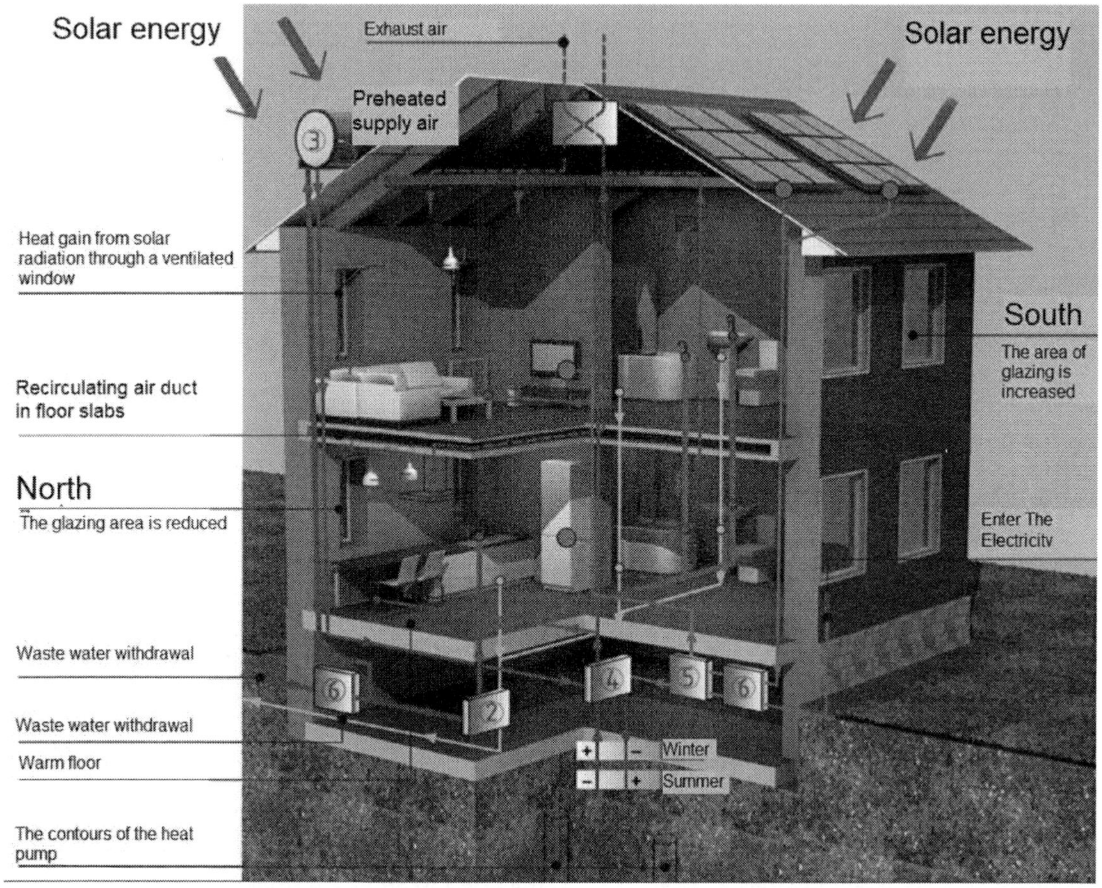

Figure 1. Schematic diagram of an energy-efficient building

The generally accepted energy characteristics of buildings of various categories are thermal protection and the specific annual consumption of thermal energy for heating, ventilation and hot water supply (in the European Union for a year, in Russia for the heating period).

The energy efficiency class of multi-apartment buildings (houses) in Russia is determined by the deviation of the actual total annual specific consumption of energy resources (heat energy for heating, ventilation and hot water supply and electricity supply in terms of electricity consumption for common house needs) per 1 m^2 of the total area of apartments and useful the area of non-residential premises from the baseline for the heating period [3].

The basic levels depend on climatic conditions [1, 3] and the number of storeys of operated buildings (2, 4, 6-12 and higher) and for the city of Irkutsk are 330, 305, 293-275, respectively, including thermal energy for heating and ventilation 216 , 180, 144-126 kWh / m^2 per year.

When establishing the base level of the specific annual consumption of energy resources, the following design conditions were adopted: the temperature of the indoor air in the apartments is 20 ° C, the population of 20 m^2 of the total area of the room per resident, which corresponds to the normative air exchange of 30 m^3 / h per resident, and the specific domestic heat input 17 W / m^2 total area.

Actual values of the specific annual consumption of energy resources are reduced to design conditions, proportionally decreasing or increasing them by linear interpolation from the established design conditions. Depending on the deviation of the actual specific annual energy consumption from the base level, energy efficiency classes are established from A ++ (highest, deviation of –60%) to G (very low, deviation of more than +50%).

Consider the possibility of improving the energy efficiency of a residential multi-apartment five-story brick building located in the city of Irkutsk.

Its main characteristics: external dimensions: length - 30 m, width - 10 m, height - 15 m; the thickness of the walls of the facade (external fencing without thermal insulation) - 0.65 m; total area of apartments - 1395 m^2; windows with wooden bindings; heating appliances - cast iron radiators.

There is a heating unit without automatic control of the temperature of the coolant, depending on the temperature of the outdoor air, but equipped with a common house meter for thermal energy. The heat consumption of a building, as a rule, exceeds that established by the contract with the energy sales company. So in 2019, the heat overrun (heat loss to the environment) amounted to about 9.3 thousand kWh.

Since there are no individual apartment-based heat energy meters in the building, automatic temperature control of the coolant depending on the outdoor temperature, energy-efficient (LED) lighting of common areas, the energy efficiency requirements, therefore, the energy efficiency class of this multi-apartment building cannot be established. Moreover, the energy efficiency class of a building determines the level of energy efficiency of a building's energy consumption, which characterizes its energy efficiency during operation.

To reduce the loss of thermal energy into the environment and ensure comfortable living conditions for residents of an apartment building, it is proposed to prepare a project and carry out the following priority energy-saving measures when preparing and conducting major repairs of the building with mandatory confirmation of energy efficiency requirements [5].

Insulation of the building facade by restoration of damaged layers of the building plaster with painting or using modern thermal insulation materials. So when installing a ventilated (hinged) type of facade, a multilayer structure [7], consisting of a metal frame, a heat insulation layer 0.05 m thick (basalt mineral wool), air gap and cladding (siding), will reduce heat loss by 13.33% per annum thermal energy consumption.

Installation of modern heating devices (high-quality and aesthetic aluminum with anodic oxidation of surfaces or bimetal radiators [8] at the choice of apartment owners). Both types of radiators are designed for maximum pressure in the central heating system up to 1.5 MPa, corrosion-resistant to the quality of the coolant, it is possible to connect a temperature sensor and an automation system. The reduction in heat loss will be 5% of the annual heat energy consumption.

Installation of plastic PVC profile windows [9, 10, 11], providing noise insulation, tightness (heat saving) and a comfortable microclimate. The reduction of heat loss when replacing all the windows in the building will be 12.24%, and the common house windows 1.1% of the annual heat energy consumption.

Installation of an electromagnetic common-house heat energy meter with automatic systems for taking readings and regulating the temperature of the coolant depending on the outdoor temperature [12,

13]. Installation and maintenance of a common house metering device is carried out at the expense of the supplier of resources [12]. The reduction in heat loss will be 10% of the annual heat energy consumption. In total, when implementing the proposed energy-saving measures, thermal energy savings will amount to 29.43% of its annual consumption, and when replacing all the windows in the building, 40.57%.

Specific brands of heat-insulating materials, types of windows, heating devices, heat metering systems and other installed equipment for the implementation of a project to improve the energy efficiency of a building are discussed and accepted by all owners of the apartment building.

In Russia, in accordance with the Federal Law [14,15,16], a Program of state support for energy efficiency projects for apartment buildings during capital repairs has been adopted and is working. A reimbursement of 70-80% of the cost of implementing energy-saving measures with an expected efficiency of at least 10% is envisaged.

For this, the meeting of owners should determine and unanimously adopt the list and cost of the work, select the contracting organizations for their implementation and submit an application to the Government.

When implementing the project of the proposed energy-saving technologies during energy-efficient capital repairs, the energy efficiency indicator of the building in terms of specific heat consumption can reach a value close to the base (see above), and the efficiency class [3] to normal D and claim to be higher C.

Confirmation of the energy efficiency requirements of an apartment building [5] is provided by managing organizations of homeowners, housing and other specialized cooperatives that manage apartment buildings using the instrumental calculation method (energy examination with the development of an energy passport of the building).

References

[1] Code of rules SP 131.13330.2018 "SNiP 23-01-99 *. Construction climatology. " URL: http://sniprf.ru/sp131-13330-2018 (01/30/2020).
[2] Code of rules SP 50.13330.2012 "SNiP 23-02-2003. Thermal protection of buildings. " URL: http://docs.cntd.ru/document/1200095525 (01/30/2020).
[3] Order of the Ministry of Construction and Housing and Communal Services of the Russian Federation of June 6, 2016 N 399 / pr "On approval of the Rules for determining the energy efficiency class of apartment buildings". URL: https://demo.garant.ru/#/document/71462244/paragraph/1:07 (01/30/2020).
[4] Nikolaev E.N. Energy performance of the building: foreign experience. URL: https://www.c-o-k.ru/archive-cok?num=12&year=2011 (01/30/2020).
[5] Order of the Ministry of Construction and Housing and Communal Services of the Russian Federation of November 17, 2017 N1550 / pr "On approval of energy efficiency requirements for buildings, structures, structures." URL: https://demo.garant.ru/#/document/71868640/paragraph/1:07 (01/30/2020).
[6] Oparina L A 2010 The definition of "energy-efficient building" *Housing* **8** pp 2-4
[7] WorlDes Construction and repair. URL: http://worldes.net/stroitelstvo-i-remont/ (01/30/2020).
[8] Heating radiators which are better. URL: https://otoplenie-expert.com/radiatory-otopleniya/radiatory-otopleniya-kakie-luchshe-dlya-kvartiry.html (01/30/2020).
[9] First by the windows. URL: https://1pooknam.ru/ustanovka/kakie-byvayut-okna.html (01/30/2020).
[10] Plastic windows. Informational portal. URL: https://okna-biz.ru/praktika/zamena-obychnogo-okna-na-plastikovoe/ (01/30/2020).
[11] Resistance to heat transfer of windows: we carry out the calculation ourselves. URL: http://jsnip.ru/normy/soprotivlenie-teploperedache-okon.html (01/30/2020).
[12] The use of ODPU in apartment buildings. URL: http://propertyhelp.ru/kvartira/kommunalnye-uslugi/odpu-rasshifrovka.html (01/30/2020).

IOP Conf. Series: Materials Science and Engineering **880** (2020) 012032 doi:10.1088/1757-899X/880/1/012032

[13] PromService-MSC. URL: http://promservis-msk.ru/prod/blochnye-moduli-reshenija/blochnyj-modul-regulirovanija-bmr/ (01/30/2020).
[14] Federal Law "On the Fund for Assistance to the Reform of Housing and Communal Services" of July 21, 2007 N 185-ФЗ (as amended). URL: https://normativ.kontur.ru/document?moduleId=1&documentId=353285 (01/30/2020).
[15] Sedash T.N. Energy Saving Experience. URL: http://www.vce34.ru/experience/ (01/30/2020).
[16] Energy efficiency in Russia and abroad. URL: https://maxelektro.ru/news/energoeffektivnost_v_rossii_i_za_rubezhom_/ (01/30/2020).

IOP Conf. Series: Materials Science and Engineering **880** (2020) 012033 doi:10.1088/1757-899X/880/1/012033

Assessing the energy efficiency of air-to-water heat pumps

S I Kuzmin[1] and A A Zelenina [1]

[1] Federal State educational institution of higher education budget "Angarsk State Technical University, Russia, Angarsk, 665835

E-mail: pgs@angtu.ru

Abstract. This paper considers the model of assessing the energy efficiency of energy generation by air-to-water heat pumps in the changing temperatures of heat-containing and heat-perceived environments. The results of the assessment of the heat transformation rate for different climatic areas are presented.

1. Introduction

The presence of huge reserves of low potential thermal energy in the environment contributes to the development of methods of its use in heat-consuming systems of buildings (heating, hot water) [1-3]. The most promising and already quite common technique of implementing this "philosophy" is to use devices that extract heat from environments with the conversion of capacity to the required level [4-6] – heat pumps.

The extraction of heat from the environment usually occurs in conditions of changing energy potential. Even in relatively stable environments - soil, reservoirs - the temperature does not remain constant, and in the most accessible resource - the outdoor air this change occurs continuously and with large (up to several tens of degrees) amplitudes. As the analysis of the thermodynamic cycle of the heat pump works, the efficiency of its operation depends significantly on the stability of the thermal potential of the heat-containing environment [7]. It is therefore useful to assess the extent to which energy is transformed in a changing capacity.

IOP Conf. Series: Materials Science and Engineering **880** (2020) 012033 doi:10.1088/1757-899X/880/1/012033

2. Materials and methods

This paper considers a model for assessing the energy efficiency of heat energy extraction by air-to-water heat pumps in changing heat-containing and heat-perceived environments. The methodology is based on the use of statistical models of outdoor air parameters in different climatic areas of the country.

3. Results

The most affordable and environmentally friendly way to generate thermal energy is the air-to-water heat pump system, in which heat is extracted from the outside air, transferred to the heat-carrier heating system and returned to the atmosphere in the form of transmission heat flows through the external fencing and exhaust ventilation systems of buildings. The energy efficiency of heat pumps is estimated by the heat transformation factor– *COP* [8]:

$$COP = \frac{W_{in}}{A}, \tag{1}$$

where is W_{in} - heat energy transferred by the pump to the heat-consuming system, W*hour;

A - energy spent on the heat pump, W*hour;

The main operational feature of the system in question is the dependence of the heat performance of the pump on the temperature of the outside air. And since the heating system is operated precisely in the conditions of variable temperatures, for the objective performance of the heat pump the transformation factor will be imagined as a weighted average in the temperature of the outside air and the duration of its standing in the period in question:

$$COP_T = \frac{W_{in}}{A} = \frac{\sum_{tн}^{tк} \frac{Q_{in.t}}{Q_{o.t}} * \tau_t}{T_{tн.н-tн.к}} = \frac{\sum_{tн}^{tк} COP_t * \tau_t}{T_{tн.н-tн.к}}, \tag{2}$$

where is COP_T - the transformation factor weighted average in the temperature of the outside air and the duration of its standing in the period in question;

$Q_{in.t}$ - thermal power generated by the pump at the current temperature of the outside air, W;

$Q_{o.t}$ - the electrical power of the heat pump at the current temperature of the outside air, W;

τ_t - the duration of the period of standing temperature of the outdoor air, hour;

$T_{тн-тк}$ - the duration of the period for which is determined is hour.

COP_t - the current value of the transformation factor that corresponds to the conditions of heat extraction and transmission.

$t_{н.н}$ и $t_{н.к}$ - outdoor air temperatures corresponding to the beginning and end of the determination period COP_T, °C.

Current Transformation Rate COP_t it is difficult to determine with sufficient accuracy from the analysis of the heat pump's thermodynamic cycle, as the real cycle depends on many factors that cannot be taken into account [8].

Therefore, data of natural tests of heat pumps of type conducted at the manufacturer are used to solve the problem. [9] and presented in figure 1. It is fair to assume that the change in the characteristics of pumps of other brands has the same tendency as those shown in Picture 1. According to the

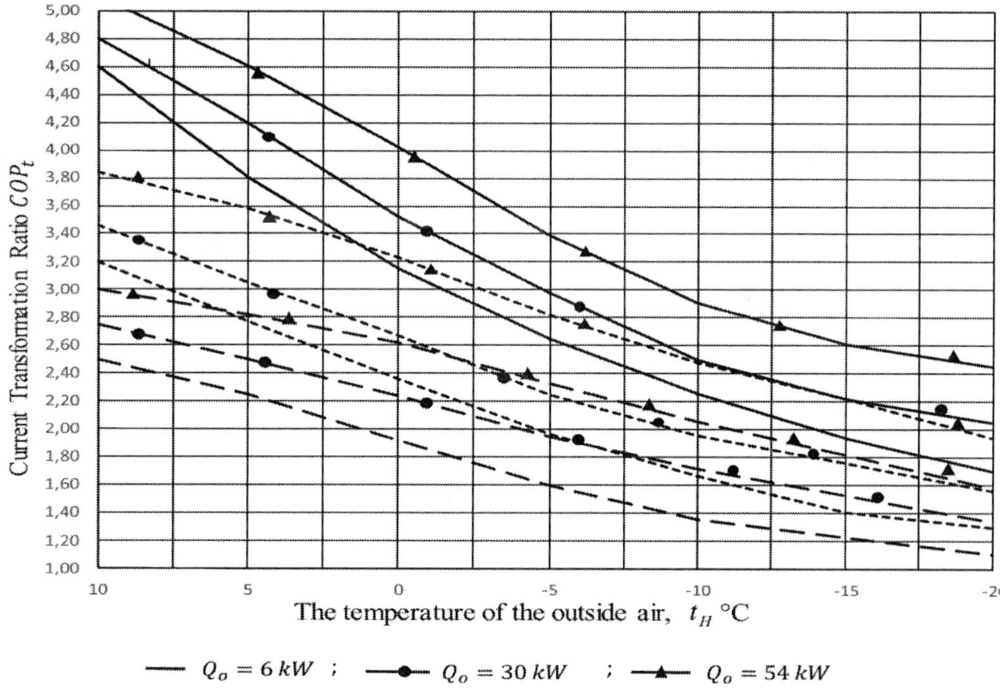

Picture 1. Dependence of the current transformation factor on temperature conditions

Analysis of the available results, the current efficiency of the heat pump depends not only on the temperature of the heat-containing environment t_H, but also from the temperature of the heat-perviating contour t_w and the estimated power of the device Q_o.

The presented graphic dependence with an accuracy of up to 2% is apraxiimized by the following model:

$$COP_t = \exp(2{,}8746 + 0{,}00157Q_o - 0{,}064t_W - 0{,}0417t_H + 0{,}00007Q_o * t_W - \\ - 0{,}000153Q_o * t_H - 0{,}00022t_W * t_H + 0{,}0000135Q_o^2 + 0{,}00041t_W^2 + 0{,}000067t_H^2) \quad (3)$$

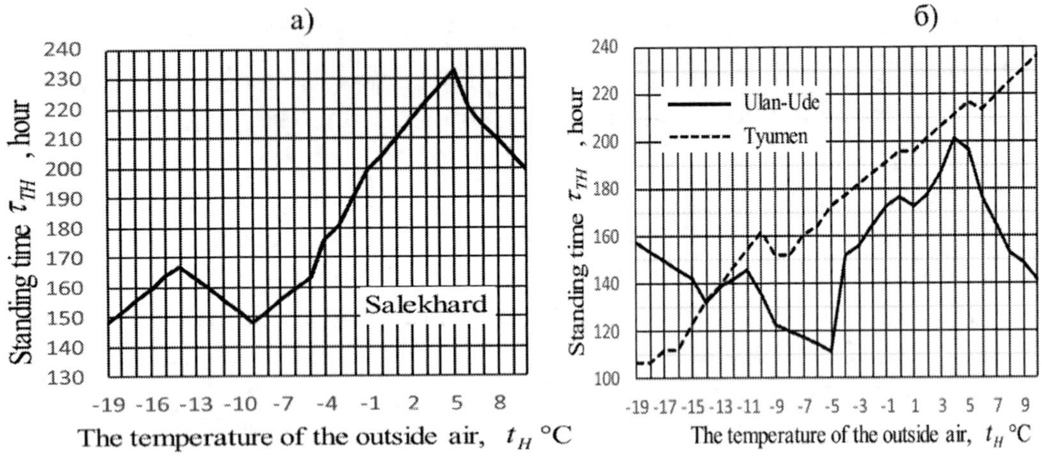

Picture 2. The duration of the temperature of the outdoor air for the "cold" a) and "moderate-cold" b) climate.

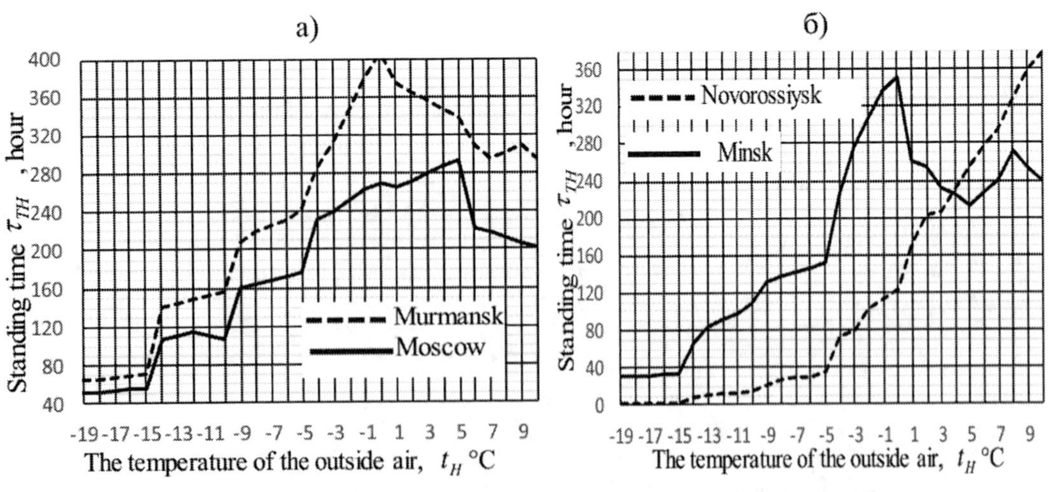

Picture 3. The duration of the external air temperature for the "moderate" (a) and "moderate-warm" b) climate.

The duration of periods of standing of outdoor air temperatures τ_t is cited in [10] for the characteristic points of climatic zones. Picture 2-3 provides the distribution of temperature standing periods for geographic points of "cold," "moderately cold," "moderate" and "moderately warm" climatic zones in the working ranges of heating and heat pump temperature systems (from q 10 degrees Celsius to -20 degrees Celsius).

As can be seen from the graphs, the nature of the distribution of temperature periods is almost incalding to any generalization even for the same zone. Therefore, the definition of the weighted average energy conversion rate by heat pumps was conducted by (2) taking into account (3) for each of these points.

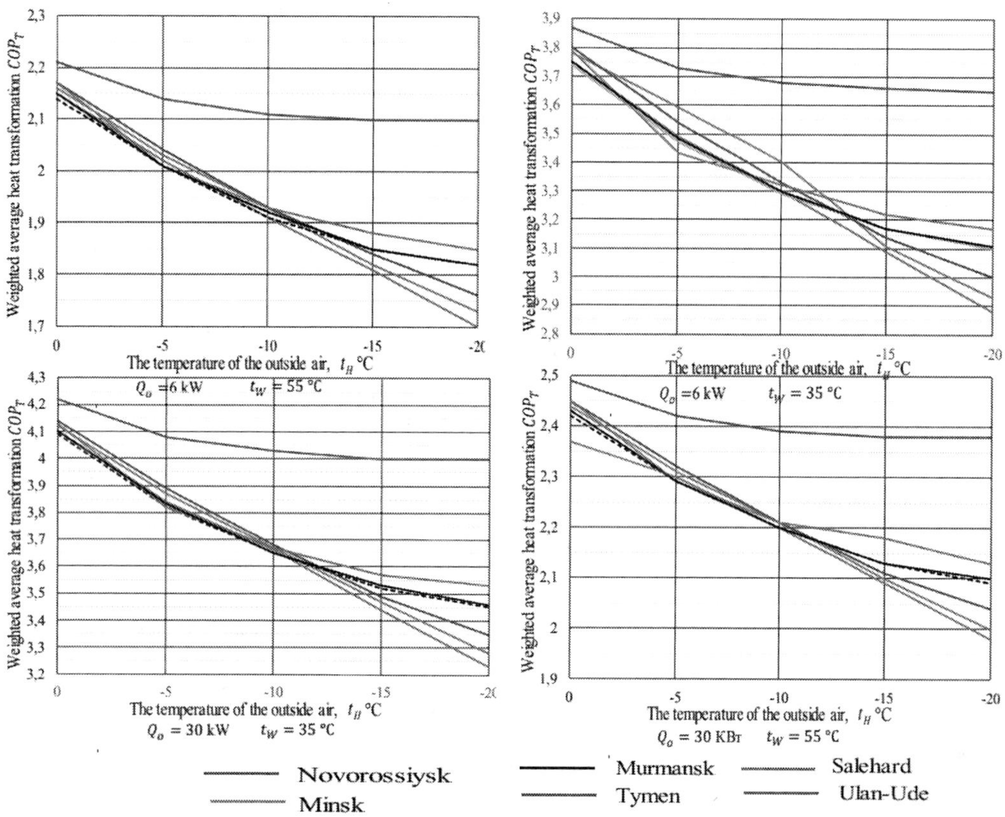

Picture 4. Change in the weighted average heat transformation rate during different heat pump periods

Settlement results COP_T made for intervals of external temperatures starting with $t_{н.н} = 10°C$ Until the appropriate pump temperature is running - t_{HK} , Picture 4.

IOP Conf. Series: Materials Science and Engineering **880** (2020) 012033 doi:10.1088/1757-899X/880/1/012033

4. Debate

These calculations show a steady trend of reducing the energy efficiency of heat pumps when working in conditions of reduced outdoor temperature potential. Despite the different periods of the duration of standing of lowered and negative temperatures, the value COP_T roughly the same in any interval of the possible pump period of operation for almost all climatic zones, except for the point in the "moderate-warm" but humid zone (Novorossiysk). In general, the efficiency of pumping heat is reduced from a range of outdoor air temperatures (10 degrees Celsius - 5 degrees Celsius) to (10 degrees Celsius - -20 degrees Celsius) by (18-14) %., which is significantly less than the variation of the transformation factor itself COP_t (50-60%)%.

The absolute value is mainly determined not by the temperature range of the heat-containing environment, but by the level of thermal energy of the locomotive-taking environment - t_w and the power of the device Q_o. At the same time for powerful pumps $Q_o \geq 30$ Kw ratio COP_T retains a value of about 3, even at low outdoor temperatures, which allows us to talk about their effectiveness for relatively cold regions. Of course, for less powerful systems (up to 10 kW) the efficiency of heat conversion under these conditions is much lower - at the level (2,5-1.7), but it should be borne in mind that in such areas the use of low-power pumps is not relevant because of the high heat need of systems heating even in individual homes of small heated volume.

Interestingly, in the presented model COP_T, its values intersect at about one point corresponding to the period. This indicates a certain semblance of climatic conditions and an uncritical difference in the energy efficiency of heat pumps when used in most regions.

Naturally, the most stable value COP_T corresponds to a warm climate with a relatively short heating season and with small temperature amplitudes (Novorossiysk, $T_{+10-(-20)} = 3392\,h$). Energy efficiency COP_T remains within 95% of the maximum value during any operational period, and the weighted average transformation rate is between 82% and 92% compared to the maximum (with $t_H = +10°C$). At the same time, it should be borne in mind that the distribution of periods of standing of outdoor air temperatures can have a significant impact on the efficiency of the pump. For example, the nature of the change COP_T for a point related to the "moderately warm" climate, but with a longer heating season (Minsk, $T_{+10-(-20)} = 5229\,h$) has the same tendency and about the same values as in colder regions.

5. Conclusion

This analysis shows that it is not correct to assess the effectiveness of heat pumps by the passport value of the heat transformation factor. A more reliable estimate can only be obtained using the weighted average of its value focusing on the possible duration of the pump's operation period and the distribution of air temperatures during this time.

Another conclusion from the results of the calculations is the presence of a similarity of change for different climatic zones. This feature can be taken into account when selecting a heat pump and assigning its operating mode.

References

[1] JV 50.13330.20112 Building Heat Protection.
[2] JV 60.13330.2012 Heating, ventilation and air conditioning.
[3] Vakhitova R I 2015 Sarychev D A To question about saving energy in Buildings Materials *Scientists and runs the State oil Institute* 2 p 65-68
[4] Hu Y. 2016 Analysis on thermal performance of composite wall of energy saving building *Chemical Engineering Transaction* **51** p 1183-1188
[5] Luneva S K, Chistovich A S and Emirov I H 2013 To the issue of the use of heat pumps *Journal of Technical and Technological Service Problems* **2** p 15-18
[6] Petin Y M 2004 New generation of heat pumps for heating and efficiency in Russian Energy Outiooc p 175
[7] Kochev A, Sokolov M, Kocheva E and Moskaeva A 2018 Energy-saving measures in buildings considered as monuments of history and architecture *MATEC Web of conferences «International conferences on Research in Mechanical Engineering Sciences,* RiMES2017», 2018 C. 04021.
[8] Vasilyev G P 2006 Heat-cooling supply of building and structures using low-potential thermal energy of the earth p 176
[9] Graphic performance of DHP-A-type heat pumps. Test materials at the manufacturer
[10] GOST 16350-80 Climate of the USSR. The zoning and statistical parameters of climatic factors for technical purposes *conferences on Research in Mechanical Engineering Sciences* p 04021

Justification of the estimated difference in the potential of the water autonomous heating system

S I Kuzmin[1] and A V Zateeva[1]

[1] Federal State educational institution of higher education budget "Angarsk State Technical University, Russia, Angarsk, 665835

E-mail: pgs@angtu.ru

Abstract. This paper provides a method for justifying the cost-effective ratio of heat supply and the surface of the heat output of heating system appliances. The results of a comparative assessment of changes in operating costs from the parameters of the heating system are presented.

1. Introduction

The cost of the heating system and the cost of its operation are determined by the main characteristics of the system - estimated power, the surface of heat recovery of heating devices, hydraulic resistance of the system [1,2]. The latter two characteristics initially depend on the temperature difference taken during the design of the system between the hot and the cooled heat carrier $\Delta t_{co} = t_г - t_o$. In systems connected to a centralized heating supply, this difference is determined by regulatory values. But in independent heating facilities, the parameters of the heating system can be assigned from the conditions of optimal operating costs and individual elements or equipment (heating appliances, circulating pumps).

2. Materials and methods

This paper considers a possible way to justify the estimated temperature differences of the coolant in the autonomous heating system (if you have your own heat generator) from the condition of minimizing the cost of heating devices, circulating pumps and energy to carry out the circulation of the coolant. In the autonomous heating system of a given thermal power, the change in the estimated temperature difference of the coolant is accompanied by effects affecting one-time and operating costs: -change in the heat transfer ratio of heating appliances and, as a result, to change the required surface of heat, and hence The cost of these items The need to provide an estimated consumption of the coolant, which entails the use of different characteristics of circulating pumps, with the corresponding cost of circulation. At the same time, the consequences of changing the parameters of the coolant have a different direction - if the reduction of the estimated temperature difference increases the heat output ratio of heating devices, the estimated consumption increases.

It is therefore useful to study this issue in terms of minimizing the cost of these components of the heating system. The method of selecting the optimal estimated temperature differences of the heat carrier of the heating system is based on the use of analytical expressions and statistical models of the characteristics of the selected elements, determining the cumulative effect of the optimization.

3. Results

Effect of changes in estimated temperature difference Δt_{co} we will evaluate by the size of the unit relative costs $\overline{C}_{np}$:

$$\overline{C}_{np} = \frac{C_t}{C_o}, \quad (1)$$

where is C_t и C_o - specific costs for performance-changing elements and system operation, respectively, at the current Δt_{cot} and basic $\Delta t_{co.o}$ temperature differences, ruble/year.

Expression for unit costs, with high-quality regulation of heat consumption, we will present in the following form [3]:

$$C = K_\tau * c_A * A_{HП} + c_э * N_{ц.н} * \tau_{co}, \quad (2)$$

where is K_τ – ratio of the estimated period of self-sufficiency of capital investments in the heating system, year^{-1}; c_A - the specific cost of the heating system elements, which is given to the unit of the surface of the heating devices of the system, rub./m^2; $A_{HП}$ – the calculated surface of the heating system heating devices, m^2; $c_э$ - the cost of electricity rubles/

(W*h); $N_{ц.н}$ - the power of the circulation pump, W; τ_{co} - duration of the heating period, h.

One-time costs c_A we will divide the heating system into two components: - the cost of the circulation pump, - $c_{цн}$; - cost of heating appliances - $c_{H.A}$.

Then the specific cost of the heating system will be expressed by the equation:

$$c_A = \frac{c_{ц.н} + c_{H.A}}{A_{НП}} = \frac{c_{ц.н}}{A_{НП}} + \frac{c_{H.A}}{A_{НП}} = \frac{c_{ц.н}}{A_{НП}} + \frac{c_{H.A} * A_{НП}}{A_{НП}} = \frac{c_{ц.н}}{A_{НП}} + c_{H.П}, \quad (3)$$

where is $c_{H.П}$ - the cost of a unit of the heating device area, rub./m^2.

Using known ratios between the heat consumption of the system and the parameters of the coolant [2] let's say the equation (3) through the basic characteristics of the heating system:

$$A_{НП} = \frac{Q_o}{\Delta t_m * K_{НП}}; \qquad N_{цн} = \Delta P_{co} * G_{co} * \rho_w^{-1} * \eta^{-1}; \qquad \Delta P_{co} = S_{co} * G_{co}^2;$$

$$\Delta t_m = \bar{t}_{НП} - t_в; \qquad G_{co} = \frac{Q_{co}}{\Delta t_{co} * c_w}; \quad \Delta t_{co} = t_г - t_o; ,$$

where is Q_o - estimated thermal power of the heating system, W; ΔP_{co} - hydraulic resistance of the heating system, Pa; η - efficiency pump; Δt_m - The temperature pressure of the system's heating devices,°C; Δt_{co} - average temperature difference of the coolant in the heating system, °C; $t_{НП}$ - average surface heating temperature,°C; $t_г$, t_o and $t_в$ - estimated temperatures, respectively, direct and return heat and internal air,°C; G_{co} - heat consumption in the heating system, kg/sec; ρ_w - heat density, kg/m^3; S_{co} - hydraulic characteristic of the heating system, (Pa*sek^2)/kg^2; c_w - heat capacity of water, J/ (kg*°K); $K_{НП}$ - heat transfer ratio of heating surface, W/ (m^2 °K).

Hydraulic characteristics of the heating system s_{co} can be determined from the estimated resistance values $\Delta P_{co.o}$ and expense $G_{co.o}$, determined by the estimated varying level $\Delta t_{co.o}$:

$$S_{co} = \frac{\Delta P_{co.o}}{G^2_{co.o}} = \frac{\Delta P_{co.o} * c_w^2 * \Delta t^2_{co.o}}{Q^2_{co}} = S_{co.t} = \frac{\Delta P_{co.t}}{G^2_{co.t}} = \frac{\Delta P_{co.t} * c_w^2 * \Delta t^2_{co.t}}{Q^2_{co}} . \quad (4)$$

And accordingly hydraulic resistance of the system at the current temperature differences $\Delta t_{co.t}$:

$$\Delta P_{co.t} = \frac{S_{co.o} * c_w^2 * \Delta t^2_{co.t}}{Q^2_{co}} = \Delta P_{co} \frac{\Delta t^2_{co.o}}{\Delta t^3_{co.t}} . \quad (5)$$

Then, accordingly, the expression (1) will take the form of:

$$\overline{C}_{co} = [K_\tau * (\frac{c_{ц.н.t}}{Q_o * c_{Н.П.}} + \frac{1}{K_{НП.t} * (t_г - 0{,}5 * \Delta t_{co.t} - t_в)}) + (\frac{c_э * \Delta P_{co,o} * \Delta t^2_{co.o} * \rho_w^{-1}}{c_{Н.П.} * \Delta t^3_{co.t} * c_w * \eta_t}) * \tau_{co}]:$$

$$: [K_\tau * (\frac{c_{ц.н.o}}{Q_o * c_{Н.П.}} + \frac{1}{K_{НП.o} * (t_{г.o} - 0{,}5 * \Delta t_{co.o} - t_в)}) + (\frac{c_э * \Delta P_{co,o} * \rho_w^{-1}}{c_{Н.П.} * \Delta t_{co.o} * c_w * \eta_t}) * \tau_{co}] \quad (6)$$

where is t and o - indices corresponding to the values of the parameters at the current Δt_{cot} and base $\Delta t_{co.o}$ temperature differences.

For water heating systems, the heat transfer factor $K_{НП}$ the surface of the heat output depends on the type of heating appliances and temperature pressure Δt_m

For aluminium radiators, two-tube systems and $\Delta t_m \geq 5°C$ this dependency can be expressed by the following ratio:

$$K_{НП} = 8{,}39 + 0{,}052 * (t_г - 0{,}5 * \Delta t_{co} - t_в) . \quad (7)$$

In (7) the temperature pressure of the heater and the temperature of the cooled water is presented through the desired temperature difference Δt_{co}:

$$\Delta t_m = (t_г - 0{,}5 * \Delta t_{co} - t_в); \quad t_o = t_г - \Delta t_{co} .$$

From the analysis of the existing prices for aluminum radiators of the most well-known suppliers, the cost of a unit of the heating device area (at the beginning of 2020), you can take an equal to 1640 rubles/m^2 [4].

And accordingly hydraulic resistance of the system at the current temperature differences $\Delta t_{co.t}$:

$$\Delta P_{co.t} = \frac{S_{co,o} * c_w^2 * \Delta t^2_{co.t}}{Q^2_{co}} . \quad (8)$$

The cost of circulating pumps $c_{ц.н}$ depends on the prices of the manufacturers and the speculative component of the sellers and can therefore be considered as a statistical value. Taking into account that the systems currently use pumps of well-known brands (Grundfos, Wilo) the average cost of this equipment will be presented in conditional units (i.e.) [5]:

$$c_{ц.н} = 69{,}4 * \exp[0{,}167 * (H_N - 2)] , \quad (9)$$

where is H_N - nominal (passport) pump pressure, m of water. Art.

Keep in mind that pump selection is made according to the performance of pumps (pressure H_p and performance Q_p) at maximum efficiency. And as a rule, the ratio between H_p and H_N c leaves 0,75. Then the expression (8) can be converted:

$$c_{ц.н} = 69{,}4 * \exp\lfloor 0{,}222 * (H_p - 1{,}5) \rfloor. \quad (10)$$

Performance H_p usually corresponds to the hydraulic resistance of the heating system $\Delta P_{co.o}$. Figure 1 shows graphs of changes in the specific given costs of the building's water heating system at different characteristics at a basic estimated temperature difference $\Delta t_{co.o} = 40°C$ and (i.e.=62 rub.).

A common trend for any initial level of heating system parameters is the greater impact of estimated cost on more powerful systems. Maximum reduction $\overline{C}_{co}$ is almost 20% in systems with an estimated thermal capacity of 20 kW, against 15% in systems with a capacity of 10 kW.

a)

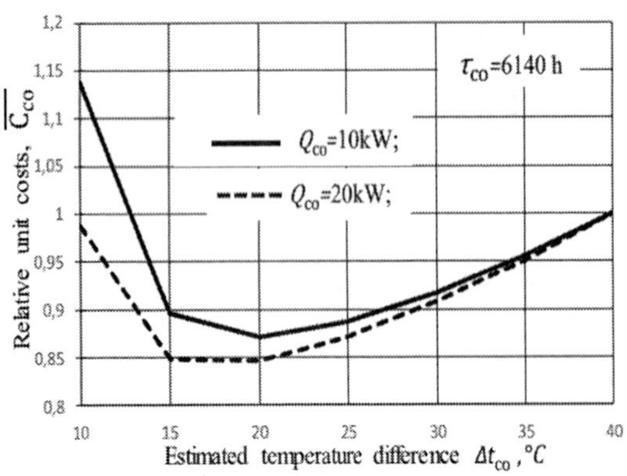

b)

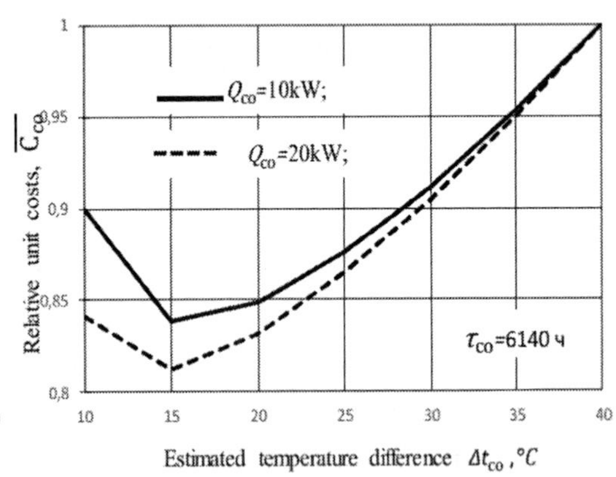

a)

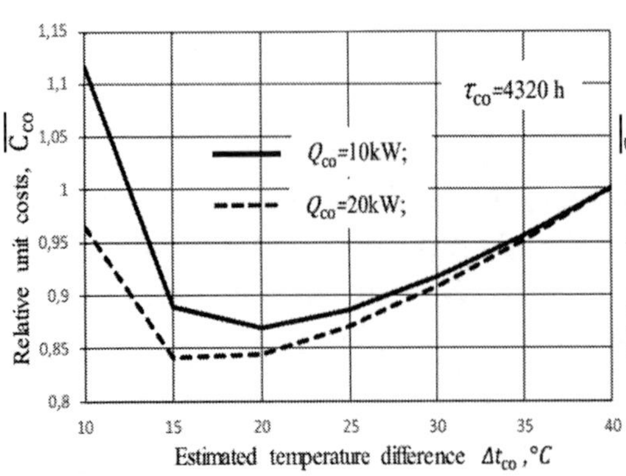

b)

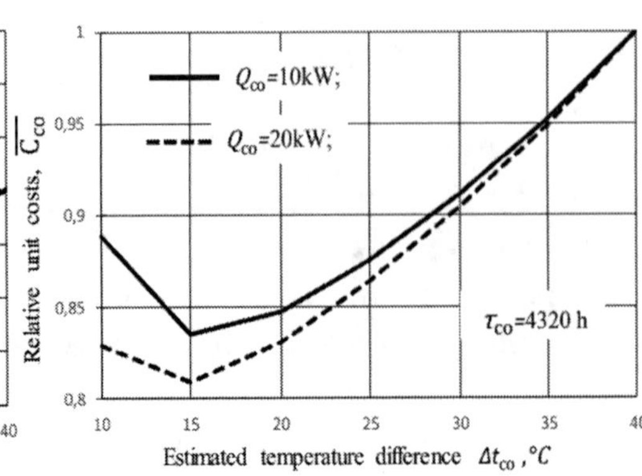

Figure 1. Changing the specific costs from the estimated temperature difference of the heating system for K_{τ} =0,1 year^{-1};
$c_{ц.н}$ = 2 rub./kW*h a) $\Delta P_{co.o} = 10$ kPa; б) $\Delta P_{co.o} = 5$ kPa

The second common factor is the significantly superior impact of the cost of the circulation pump over the cost of moving the coolant and the change in the total cost of heating devices. With the constant hydraulic performance of the system S_{co} correspond to their temperature changes depending on the parameters of the system.

Reducing the estimated temperature differences below 15 degrees Celsius leads to a sharp increase in the resistance of the system and, accordingly, an increase in the cost of the pump, which is practically not compensated by the reduction in the required area of heat return of the devices, even when the relatively long lifespans of systems. Nevertheless, under the conditions discussed, there are quite pronounced minimums of unit costs. The magnitude and conditions of the minimum $\overline{C}_{co}$ correspond to their temperature changes depending on the parameters of the system.

a)

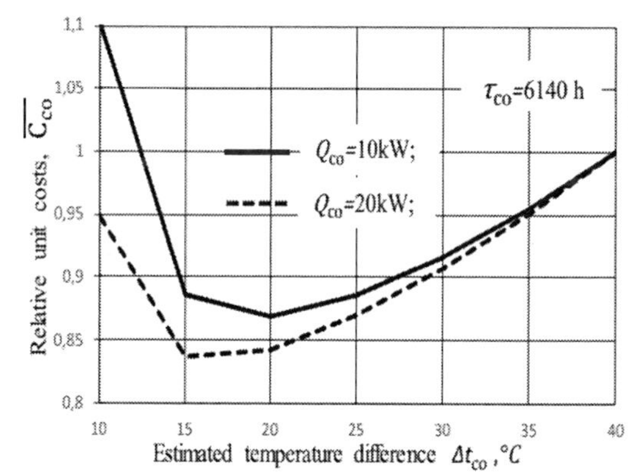

b)

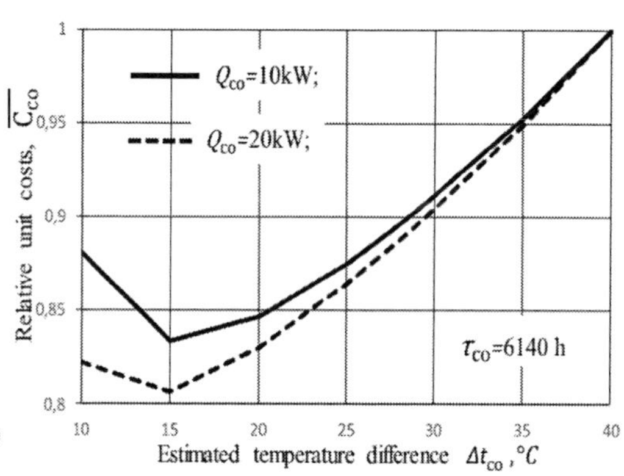

a)

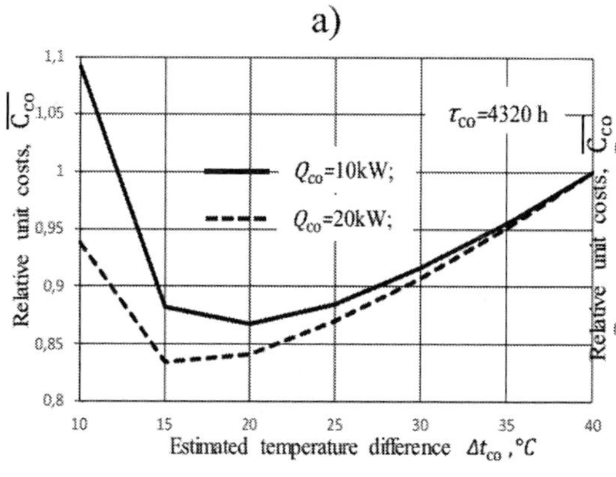

b)

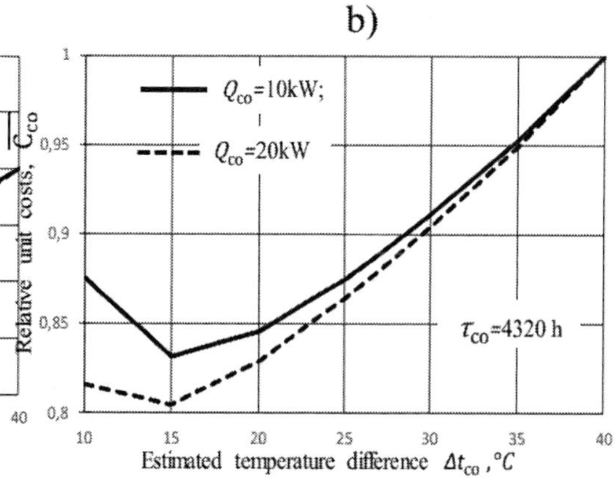

Figure 2. Changing the specific costs from the estimated temperature difference of the heating system for K_{τ}=0,2 year^{-1}; $c_{ц.н}$=2 rub./kW*h
a) $\Delta P_{co.o}=10$ kPa; б) $\Delta P_{co.o}=5$ kPa

So with relatively cheap electricity (2 rub./kW*h) the minimum cost comes at a time when $\Delta t_{co.o}=15°C$ and $\Delta t_{co.o}=20°C$ for systems with high initial hydraulic resistance and a capacity of 20 kW and 10 kW respectively (see Figure 1 a). But with less, this minimum is the same for systems with different power (see Figure 1 b). And this trend is manifested regardless of the life of self-sufficiency of the systems (see Figure 2). In addition, under these conditions, the values of the lows $\overline{C}_{co}$ almost do not depend on the length of the heating period.

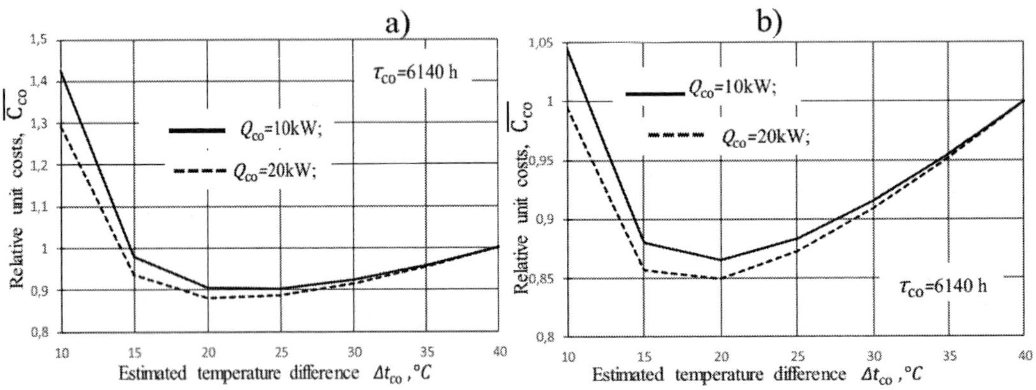

Figure 3. Changing the specific costs from the estimated temperature difference of the heating system for K_{τ}=0,1 year^{-1};
$c_{ц.н}$=10 rub./kW*h a) $\Delta P_{co.o}=10$ kPa; б) $\Delta P_{co.o}=5$ kPa

As the cost of electricity increases, the minimum cost shifts towards an increase in the estimated temperature differences, but the impact of this parameter on improving the efficiency of the system becomes less (Figure 3).

4. Conclusion

IOP Conf. Series: Materials Science and Engineering **880** (2020) 012034 doi:10.1088/1757-899X/880/1/012034

Analysis of the results obtained from the proposed target function model (2-9) shows the presence of a certain optimal value of the estimated difference in the potential of the coolant. The spread of the economic indicator of the heating system reaches more than 35%, which confirms the conclusion that it is necessary to make a reasonable choice of estimated temperature difference for the design of heating. Of course, individual indicators in the model can be specified based on the existing policy of equipment prices, but this can be taken into account by the introduction of adjustment ratios or the introduction of the current cost of elements of the system.

References

[1] JV 60.13330.20112 Heating, ventilation, air conditioning.
[2] *Designer's handbook. Internal sanitary and technical systems* 1993 p. 896.
[3] Kuzmin S I 2019 Justification of the level of parameters of the autonomous heating system p 91
[4] Plumbing-ONLALINE/htths: santehnika-onlane/catalog/.
[5] Price list Grundfos//https://www/c-o-k.r/catalog/grundfos/22772/91/
[6] Chupin V R and Dushin A S 2019 Parameter optimisation for new and reconstructed systems of water supply and distribution for ensuring consumer's uninterrupted water supply *Proceedings of Universities. Investment. Construction. Real estate* **9(4)** pp 790–803
[7] Kulkov V N, Solopanov E Yu and Kamalov R T 2019 Using immobilised sludge for wastewater treatment and its air regeneration *Proceedings of Universities. Investment. Construction. Real estate* **9(3)** pp 522–529
[8] Pupyrev E I and Chupin V R 2019 Features of regional development of water disposal systems in the central ecological zone of the Baikal natural territory *Proceedings of Universities. Investment. Construction. Real estate* **9** pp 354–363
[9] Chupin V R and Dushin A S 2019 Assessment of the reliability of water supply to consumers: water supply reliability indicators Proceedings of Universities. Investment. Construction. Real estate **9(3)** pp 578–593
[10] Lavygina O L and Grebneva O A 2019 Environmental technologies in the housing and communal service system of the Baikal natural territory *Proceedings of Universities. Investment. Construction. Real estate* **9(4)** pp 726–733

IOP Conf. Series: Materials Science and Engineering **880** (2020) 012035 doi:10.1088/1757-899X/880/1/012035

Research of the possibilities of SmartEnergyCoating applying for thermal insulation of engines in harsh climate

I Y Pospelova[1] and D A Kornlov[1]

[1]Irkutsk National Research Technical University, 83, Lermontov Str., Irkutsk, 664074, Russia

E-mail: pospelova@istu.edu

Abstract. The investigated SmartEnergyCoating is able to generate electricity independently, provide a stable temperature regime for some types of technological coatings, elements of machines and control systems. The article presents the results of investigation to determine the time of the efficiency of the composition elements for the possibilities of generating electricity by the temperature difference between the technological coating and an environment.

1. Introduction

The developed material SmartEnergyCoating is an innovative form of energy production at the surface of machine elements and control systems along with effective insulating properties. This coating has proven effective in the laboratory [1-3]. It can be used to convert low-grade heat from various sources into high-quality energy, which can be used to transfer signals and other purposes, as well as to use and store energy in several forms. The authors propose to modify the material with a layer containing the phase transition material and thermoelectric elements to improve its thermotechnical properties.

In some technical processes, there is an excess of waste heat. This energy is lost in an environment or gives heat or cooling to the surrounding elements, which can convert for future use. The developed material does not require high labour requirement and consists of a simple structure. The material can use in extreme temperatures and harsh climate, especially in harsh climatic conditions, where temperature fluctuations during the day can reach more than 20-25 degrees. Also the material can be used in emergency conditions when other sources of energy cannot work, when structures, pipelines or installations is used in non-stationary temperature conditions or processes with excessive heating and loss to the environment.

Earlier, the advantages of the developed material were noted [4,5,8]. It is a best way to get energy in a small space. The absence of moving parts makes the system very reliable. There is no need to change the refrigerant (for example, freon, because it is away), no sensitivity to vibrations, what is important for transport. Thermoelectric devices do not contain toxic refrigerants. This is an environmental friendliness [6,7,9]. There is the possibility of smooth and precise temperature regulation [10-17]. The material can have an arbitrary orientation in space and a gravitational field and can apply to any surface. The elements for thermal insulation are securely fixed in their layer, can be defined in compliance with working conditions and provide a thermal mode. The material has a slight inertia and transforms from cooling to heating mode easily.

In previous studies, dependences of the geometry, shape, and composition of the material that could affect the temperature profiles and behaviour were determined. Also it has been proven that thermoelectric effects are associated with temperature lines and a temperature gradient. The distribution

IOP Conf. Series: Materials Science and Engineering **880** (2020) 012035 doi:10.1088/1757-899X/880/1/012035

gradient and coating behavior can model for specific operating conditions. The possibility of theoretical control of thermoelectric effects with thermal properties of materials was been proven also. Next research is aimed at the identifying of the dependences of the heat-insulating materials included in the composition on the thermal characteristics and properties of the material layers. For the study, we took the case of using Smart Energy Coating for generating excess heat from a motor or warm surface.

In some technical processes, there is an excess of waste heat. This energy is lost in an environment or gives heat or cooling to the surrounding elements, which can convert for future use. The developed material does not require high labour requirement and consists of a simple structure. The material can use in extreme temperatures, especially in difficult climatic conditions, where temperature fluctuations during the day can reach more than 20-25 degrees. Also the material can use in emergency conditions when other sources of energy cannot work, when structures, pipelines or installations is used in non-stationary temperature conditions or processes with excessive heating and loss to the environment.

2. Theoretical description of the process

In the developed coating the thermal insulation materials and thermal mass play a certain part. The model contains the heat source, what accumulates heat mass with a specified heat capacity and environment with variable dynamic parameters.

For the generality of the processes description for the model, the heat transfer is composed of heat conduction and convection. The radiation component is been excluded, considering that in the process the radiation losses are minimal.

The study was conducted the Simulink model using. For simplicity, it accepted the model consist of a heat source with a total thermal power P. The temperature inside the coating is Ti, the ambient temperature is Tamb. We are interested in how the temperature Ti changes when the power P changes. Since the ambient temperature is negative, the heat flux distributes from a heated source to a cold one. (Fig. 1).

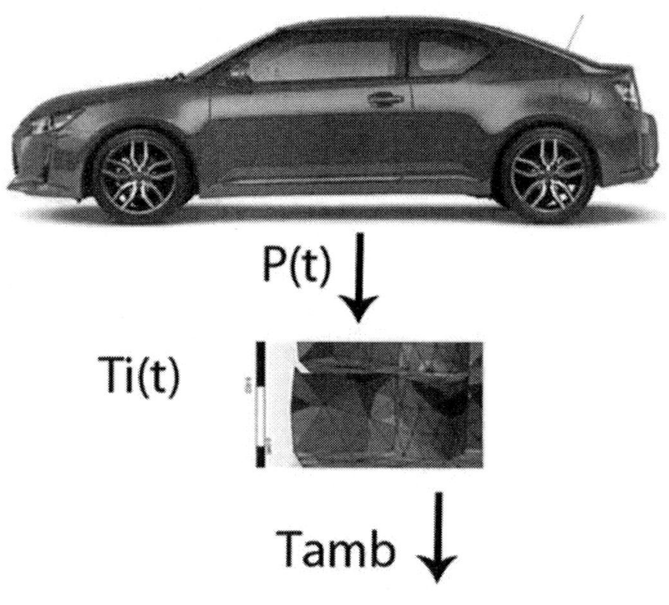

Figure 1. Model of the coating energy flows along the input-output during engine operation in the energy storage mode.

Before modelling, it should to consider some properties. Firstly, if the heat source is turned on, then the temperature will rise and stabilize. In this case, a thermal equilibrium has been reached between the

external heat and accumulated by the heat mass. Then it will be dissipate into the environment. If the heat source is turned off then temperature and energy decreases. Finally, the temperature of the coating brings closer the ambient temperature, but it has time to accumulate a determine energy already. In this case the essential parameters of the model are:
Ambient temperature - the lower it is, the more heat leaves by the coating and the more heater power need to insulation. The bed thermal insulation - the more heat has been releasing into the environment.

Heat capacity and mass inside the coating. The larger this parameter, the more energy is required and the heating time to a predetermined temperature is longer. Also the coating saves the heat during long time when the source of heating is turned off.

In heat engineering, there are models that simulate heating and cooling processes with varying order of accuracy. Further, we will consider the simplest of them. For this, it is necessary to introduce the concept of the heat amount. It is the energy necessary to change the thermodynamic parameters of a body (for example, temperature).

The amount of heat Qi that comes from a heat source or heater with a power P during time τ is the time integral:

$$Q_i(\tau) = \int_0^{\tau} P(t)dt \tag{1}$$

To determine a waste heat to the environment, it is necessary to use the concept of heat flux Qamb(t)–- the amount of heat passing through the surface per time unit. The unit heat flux is proportionally to the temperature difference:

$$Q_{amb} = -k\,(T_i - T_{amb}) \tag{2}$$

Over a particular period of time:

$$Q_{amb}(\tau) = -k \int_0^{\tau}(T_i(\tau) - T_{amb})d\tau \tag{3}$$

The general heat balance equation is:

$$Q = Q_i + Q_{amb} \tag{4}$$

By differentiating both parts by time, we can write down a differential equation of relating the dynamics of the temperature $Ti\ (\tau)$ to the heater power:

$$cm\frac{dT_i}{dt} = k\big(T_{amb} - T_i(\tau)\big) + P(\tau) \tag{5}$$

Designate the coefficient cm = a and dividing the variables for integration, we can write:

$$Q(\tau) = a\int_0^{\tau}(k\big(T_{amb} - T_i(\tau)\big) + P(\tau))d\tau \tag{6}$$

The last expression is a model of the heat transfer process, when heat is received from the engine and the cold from the environment. To control the properties of the coating, we cannot directly affect the ambient temperature parameter T_{amb}, so we can monitor the parameter P.
Let's to see how this system was simulated using Simulink.

3. Modelling

The part of the model diagram in which the connections between the blocks are present and their settings are shown in Figure 2.

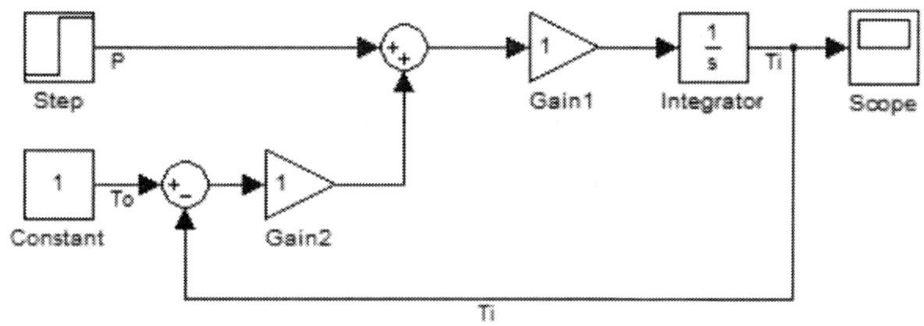

Figure 2. Preliminary model diagram.

The temperature inside the coating drops from a given temperature to a certain temperature. In this case, the specific heat of the coating for calculations is C = 3.2 kJ / kg°C, the mass of the material is 1 kg, the heat transfer coefficient is k = 0.34 W / m^2°C, taking into account the heat of the phase transition.

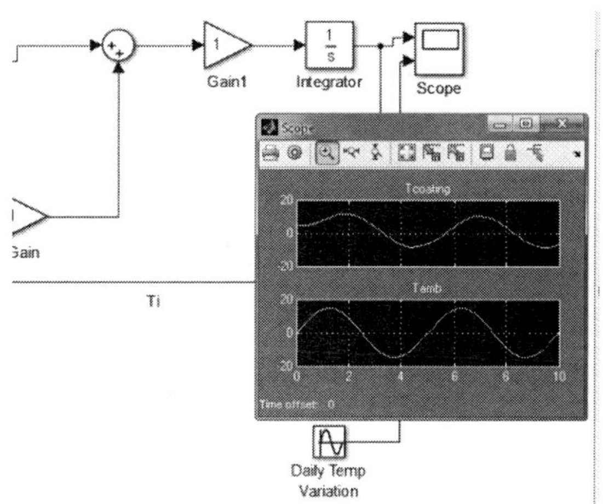

Figure 3. The final model diagram and the result for 2 days.

The upper graph shows the expected temperature of the coating, the lower graph shows the daily temperature difference. The initial temperature of the coating has been equal +5 degrees, because the motor has not cooled down yet.

From the results of the graphs, it can see that we can wait for the Smart Energy Coating to cool for some time without reducing the power generation. But it is necessary to provide for the engine starting. After the heater starts the temperature parameters reach the required values in short time.

Those, we found that the design parameters of the coating composition to eliminate the need of the heater activation during 4 hours after temperature reduction. This time uses to the thermodynamic equilibrium between heat supplied by the heater and outward. With these composition parameters SmartEnergyCoating, we simulated the time of thermal energy receipt and its probable transition into electrical energy. The model allows making the dependence of the composition cooling time without heating at certain temperature parameters and determining conclusions about the parameters of coating composition.

By thermal conductivity change and different heat capacity of SmartEnergyCoating we can maintain a certain temperature level of the coating depending on the temperature (material properties), to improve the properties.

4. Implementation

After simulation results, a prototype SmartEnergyCoating with certain properties was created. SmartEnergyCoating features were improved. The phase transition structure has been composing in such

IOP Conf. Series: Materials Science and Engineering **880** (2020) 012035 doi:10.1088/1757-899X/880/1/012035

a way, that its melting point was equal to the work temperature (ambient). The phase material was resolved on which provided fora synthetic paraffin substance and adaptable physical properties.

The generated electricity can use for pipelines with automation elements, heat-insulating composite materials and electricity generating. The paper presents the effect of thermal insulation and independent energy production by the coating. Most of the mass of the phase-transition material is located on the outer surface. This allows ensuring the damping of temperature deformations and independent power supply to remote elements.

One of the goals of this study was to modelling the coating installation under the hoods of cars in freezing conditions to prevent prolonged warming up during "cold start" of the car. Installation of the coating allowed reducing the time between engine starts, since heat accumulation takes place. In addition, a small amount of current can use to add power to the starting torque. Flexible conductive compounds constantly supported the work of the phase transition material and allowed to work with a sufficient level of quality. In this case, linear elongation of the base material did not occur, and the coating worked as a heat-insulating layer with electricity generation.

References

[1] Pospelova I, Pospelova M, Bondarenko A and Kornilov D 2018 Results of thermal modeling of Smart Energy Coating with phase-transition material for independent electricity generation. *J. Phys.: Conf. Ser.* **1015** 032108

[2] Pospelova I, Kornilov D, Bondarenko A, Pospelova M 2018 Thermal Insulation of Smart Energy Coating for Energy Production *Proceed.of the Int.Conf. AVENT* 18.2018.64

[3] Pospelova I Y, Pospelova M Y and Kornilov D A 2019 Smart energy coating for independent power generation in pavement and machine elements. *IOP Conf. Series: Materials Science and Engineering* 632 012018

[4] Pospelova I 2017 Determination of Optimal Heat-Storage Thickness of Layer for "Smart Wall" by Methods of Nonlinear Heat Conduction Equations for Phase-transition IOP Conf. Ser.: Mater. *Sci. Eng.* 262 012031

[5] Pospelova I and Filatov V 2020 Investigation of feasibilities of heat pipe using in recuperation systems with heat excess of paper industry enterprises *IOP Conf. Ser.: Earth Environ. Sci.* **408** 012010

[6] Peskina S 2018 Immersion electronics cooling, J. *Components and technologies* 2 80

[7] Osinin I, Shatohin A 2016Technical aspects and economic effect of immersion cooling system *Russian Supercomputing Days* (Moscow) 1008

[8] Pospelova I, Daniletz T, Pospelova M 2014 Problems of engineering systems in special conditions *Sc.j. Proc. of Univ. Inv. Constr. Real Est.* **9** (Irkutsk, ISTU) pp 90-97

[9] Kalnin Y, Makagonov V, Sitnikov A 2015 Thermoelectric properties of thin-film composites *J. Solid state physics* **57**

[10] Changcheng M, Jing H, Jingjing L, Jie Z and Zuoqi H 2018 Modeling of the Temperature Profiles and Thermoelectric Effects in Phase Change Memory Cells *Appl. Sci.* **8** 1238 MDPI

[11] Hidekazu T, Hidefumi T, Teruo K 2016 Local Peltier-effect-induced reversible metal-insulator transition in VO2 nanowires *AIP Advances* **6** 065118

[12] Marun I, Berg J, Dejene F, Wees J 2016 Direct electronic measurement of Peltier cooling and heating in grapheme *Nature Communications* **7** 11525

[13] Cho P, S. et al. 2013 Thermoelectric imaging of structural disorder in epitaxial graphene *Nat. Mater.* **12** , pp 913–918.

[14] Heremans JP , Dresselhaus, MS , Bell, LE & Morelli, DT 2013 When thermoelectrics reached the nanoscale *Nat. Nanotechn.* **8** pp 471–473

[15] Khvesyuk V, Ostanko D, Scriabin A, Tsygankov P, Chelmodyev R, Chirkov A 2016 Limiting efficiency of thermoelectric heat conversion in high-temperature power plants *Science &*

Education of Bauman MSTU **3** pp 81-105
[16] Kulbachinskiy V, Kytin V, Blank V, Buga S, Popov M 2011 Thermoelectric properties of nanocomposites of bismuth telluride with fullerenes *Physics and Technology of Semiconductors* **45(9)** pp1241-1245
[17] Dubrovin G, Gerasimova E, Shupenev A 2015 Calculation of a thin-film thermoelectric module produced by pulsed laser deposition *Engineering Journal Science and Innovation* **12**

IOP Conf. Series: Materials Science and Engineering **880** (2020) 012036 doi:10.1088/1757-899X/880/1/012036

Energy efficient methods of protecting buildings from cold air flows

I Yu Shelekhov[1], N L Dorofeeva[1], E I Smirnov[2], M I Shelekhov[1] and I A Dorofeev[1]

[1] Irkutsk National Research Technical University, 83 Lermontov str., Irkutsk, 664074, Russia
[2] LLC "Thermostat +", Irkutsk, Russia (664011, pereulok 8 Marta, 4-25)

E-mail: promteplo@yandex.ru

Abstract. In this work, we study new innovative methods for creating an additive temperature environment, study the influence of a new type of infrared devices such as air and air-thermal curtains on the living conditions and infiltration parameters of buildings and structures from penetrating cold air flows. For calculating these structures, the engineering and technical methods of V.M. Elterman was used, based on the methods of calculating heated air curtains of Soviet scientists V.V. Baturin, I.A. Shepelev, S.E. Butakov. Additionally, the method of calculating air curtains by M.E. Diskin was used. We studied the distribution of temperature fields in the gate area of the building, in which road transport is being repaired. The experiments showed that the best indoor microclimate is achieved by using dual-flow systems consisting of a heated air curtain and an air curtain. The study of thermoelectric systems showed that it is possible to recuperate about 500 watts of electrical energy in this case. Studies have shown that there are modern, innovative developments, use of which improves the microclimate in the rooms without spending additional energy resources.

1. Introduction

Creating comfortable conditions in rooms where, depending on the specifics of operation, the infiltration of cold air flows occurs is a complex engineering task. Usually, this problem is solved by installing heated air curtains, the energy costs of which, in some cases, are similar to the costs of heating the entire room. Energy efficiency of the entire building as well as the microclimate of the room depends on the right choice of measures to ensure the protection of open doorways from the infiltration of cold air streams [1-4].

Usage of heated air curtains creates a separation of two air environments with different temperatures. Experimental data show that the temperature gradient directly at the output grille can reach a 160°C. This allows to create thermoelectric systems to recuperate some amount of electricity, and thereby offset some energy costs used to operate the air-heat curtain [5]. Data on the temperature gradient were obtained in the study of heated air curtains used to heat trams and trolleybuses in the city of Irkutsk [6].

2. Materials and methods

Heated air curtains (HAC) are intended for the separation of air zones with different temperatures, while the operational efficiency is primarily assessed by the absence of influence of the zones on each other [7].

The foundations of the calculation methods of HAC were developed by Soviet scientists V.V. Baturin, I.A. Shepelev, S.E. Butakov, who formed the basis of the engineering technique of V.M. Elterman [8]. He considered the work of HAC as an additional resistance, reducing the amount of air passing through the opening, using the momentum theorem of P.N. Kamenev [9]. Subsequently, theoretical calculations were confirmed by experiments, but in underestimated rates [10].

An alternative method of calculating HAC was proposed by M.E. Diskin [11], arguing that if the curtain is rationally executed, only the air stream of the curtain passes through the opening. Even if part of the jet passes outside the opening, the temperature of the air passing through this opening will not be lower than the average temperature in that section of the jet that first contacts the opening enclosure [12].

Of particular interest, in our opinion, are the designs of dual-flow HAC, which have two separated air flows [13]. The first high-speed air flow without heating, carries out the primary barrier function. The second air stream is carried out by a heated air stream. In this case, two flows move parallel to each other, creating ideal conditions for the use of thermoelectric generators [14]. This design was made for a truck repair company. At the confluence of the two air streams, thermoelectric modules were installed, made according to the patent for the invention No. 2611562 [15]. Experiments was carried out using OWEN TPM138 device with a set of thermocouples for temperature and OWEN "ИМС-Ф1.Щ1" for voltage, current and power measurement.

3. Results

In the first experiment, at an external temperature of -12°C, the floor temperature was measured at a distance of 5 meters from the gate, with the air and heated air curtains turned off. It can be seen from the graph shown in Figure. 1 that when the gate is opened, the temperature at the entrance decreases to negative values. The microclimate in the room is not satisfactory, which makes work impossible. It is necessary to heat the room to positive temperatures.

The second experiment was carried out with the heated air curtain turned on, the outlet air temperature was 60°C, and the air flow velocity was 8 m/s. The graph in Figure 2 shows that the favorable temperature at which the microclimate parameters pass into a satisfactory condition is already at a distance of 3-4 meters from the gate. Despite the fact that the total temperature in the room dropped by 3°C, it is possible to carry out repair work on vehicles without preliminary measures.

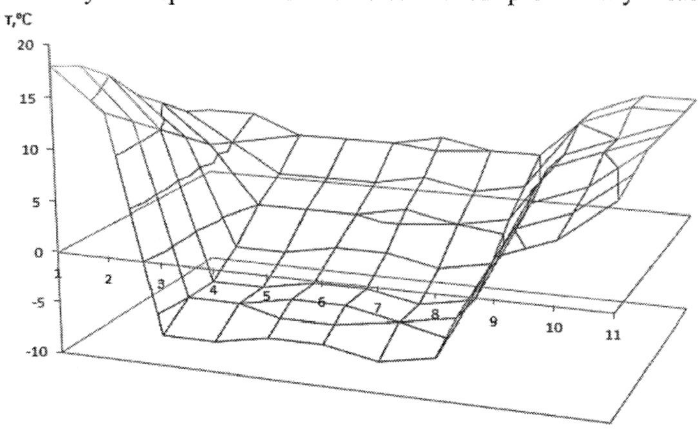

Figure 1. Temperature distribution at the gate with air curtains turned off.

In the third experiment, an air curtain with air intake from the upper part of the building was additionally included; the total height of the building is 7 meters. The velocity of the unheated airflow at the exit of the curtain was 10 m/s.

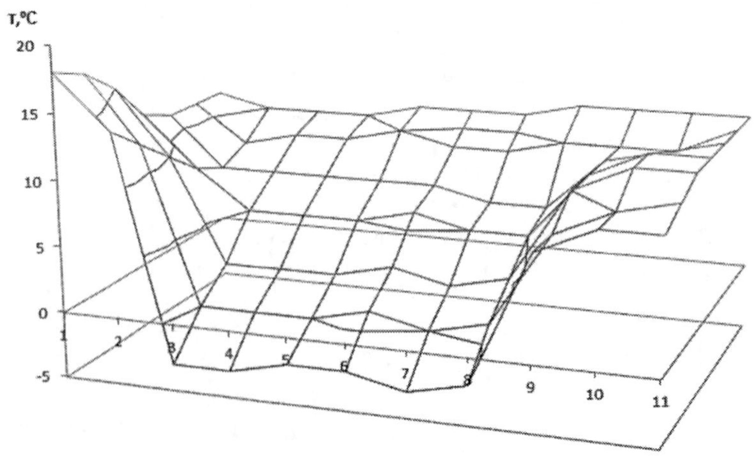

Figure 2. Temperature distribution at the gate with a working heated air curtain.

The graph shown in Figure 3 shows that there are no negative temperatures in the gate area and temperature in the main room have not changed. The value of the generated voltage from one module was 18 V at a load of 1 ohm, which corresponds to the generated power of 18 watts. Air curtain is powered by 180W engine. In this design, it is possible to install up to 25 thermoelectric modules, which are made in the form of additional screens. In conventional designs in heated air curtains these screens serve to distribute air flow to improve efficiency. The installation of thermoelectric generators does not impair the operation of the system of protecting openings from the infiltration of cold flows, but it can generate electrical energy up to 500 W at an external temperature of -12°C, which blocks the need for electricity for the operation of the air curtain.

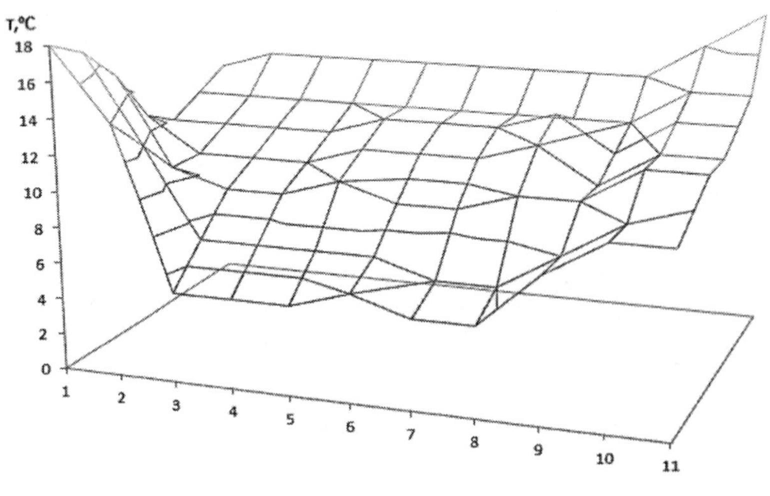

Figure 3. Temperature distribution at the gate with a working air and heated air curtains.

4. Conclusion

Studies have shown that the use of new innovative solutions in the design of systems for protecting openings from infiltrating cold flows helps to reliably protect rooms from temperature extremes, provide favorable climatic conditions throughout the room, and reduce the recovery time of microclimate parameters when changing external meteorological conditions. Research in the field of creating thermoelectric systems showed that using the latest developments in this area, it becomes possible to implement energy-saving measures without attracting additional energy resources.

The study was carried out with the financial support of the Russian Foundation for Basic Research (RFBR) and the Government of the Irkutsk Region as part of a scientific project No. 20-48-380002.

References

[1] Proposals of "VTI" OJSC on improving the reliability and cost-effectiveness of the operation of heat supply facilities for housing and communal services in cities and towns 2010 *Outreach material* (Moscow: All-Russian Thermotechnical Institute) p 10

[2] Terentyev D 2015 Improving the energy efficiency of buildings, structures and structures. Tasks of the Russian Ministry of Construction *«Energosberezhenie» Magazine* **3** (LLC IIP "AVOK-PRESS") pp 18-21

[3] Gasho E, Puzakov V and Stepanova M 2012 Features of the implementation of energy conservation policies in the regions *Analytical collection* (Analytical center under the Government of the Russian Federation) p 43

[4] Tulikov A 2016 Improving state regulation in the field of increasing the energy efficiency of buildings *«Energosberezhenie» Magazine* **2** (LLC IIP "AVOK-PRESS") pp 4-11

[5] Shelekhov I, Shishelova T and Smirnov E 2018 Application of new technical solutions in the design of thermoelectric systems *Bulletin of the Mordovian University* **28** pp 48-61

[6] Shelekhov I and Kobak M 2013 Features of the use of urban electric transport in winter time *University News. Investments. Construction. Real Estate* **2 (5)** (Irkutsk: Publishing House of ISTU) pp 126-130

[7] Elterman V 1966 *Air curtains* (Mechanical Engineering) pp 164

[8] Kruglova E, Kruglov G and Ryspaeva A 2009 Regulation of air-thermal curtain parameters based on monitoring of disturbing influences *Materials of the XLVIII international scientific and technical conference "Achievements of science - agricultural production"* **4** p 120

[9] Kruglova E, Kruglov G and Rostislavov O 2009 Adjustable distributor for air-curtain *Materials of the XLVIII international scientific and technical conference "Achievements of science - agro-industrial production"* **4** p 118

[10] Kruglova E 2006 Development of resource-saving air-thermal curtains to maintain normalized microclimate parameters in the industrial premises of the agricultural sector. *Abstract of the dissertation of PhD Kruglova E 05.20.01* (Chelyabinsk)

[11] Diskin M 2004 Once again on the calculation of air curtains *ABOK 2004 No 4*

[12] Diskin M 2003 On the issue of calculating air curtains *ABOK 2003 No 7*

[13] Kruglov G and Kruglova E 1998 Reducing the resource intensity of the air-thermal curtain due to the improvement of the structure of the jet *Safety of vital activity: Collection of scientific papers* (Chelyabinsk: CSTU) pp 24-28

[14] Shelekhov I, Smirnov E, Inozemtsev V and Fedorova E 2015 Analysis of the possibility of using thermoelectric generators in ventilation systems *«Scientific Review» Magazine* **8** pp 67-75

[15] Shelekhov I, Smirnov E, Kashko K, Shelekhova I Patent for invention No. 2611562 of 02/28/2017 Spatially oriented thermoelectric module and method for its manufacture

IOP Conf. Series: Materials Science and Engineering **880** (2020) 012037 doi:10.1088/1757-899X/880/1/012037

Research of changes in the efficiency of renewable energy depending on external influences

I Yu Shelekhov[1] and E I Smirnov[2]

[1]Department of Urban Construction and Economy, Irkutsk National Research Technical University, 83, Lermontov Str., Irkutsk 664074, Russia
[2]JSC "Termostat+"

E-mail: promteplo@yandex.ru

Abstract. The article presents the results of a study of renewable energy sources (RES) based on solar panels, the operation of which took place in the Irkutsk region. The results of changing the efficiency of the solar panels depending on the use of light energy concentrators and the temperature of the panel itself are presented. The authors set the task of increasing the efficiency of the solar panels through the use of technical and constructive solutions.The experiments were carried out in the laboratory of IRNITU "Modern Heating Equipment". For testing, we used a bench based on the TPM138 measuring regulator, an OVEN device of the IMS-F1.SH1 brand with a set of thermocouples. , The subject of the study was two solar panels, on one of which a Fresnel lens was installed.The article formulates the purpose and objectives of the study, conducted research on the parameters of solar panels depending on the influence of the light energy concentrator and temperature changes on the panel itself, depending on external influences.As a result of the work, it was found that the efficiency of solar panels in the Irkutsk region in the summer period of time decreases by an average of 14.5%. And in the winter time by 7.5%. It is shown that to increase the efficiency of solar panels it is necessary to apply technical and constructive solutions that will reduce the payback period of solar engineering systems by 10-18%.

1. Introduction

Unlike most developed countries, for our country the issue of developing and introducing new energy sources is far from relevant. In our country, extensive research is traditionally carried out in the use and production of oil, natural gas, coal, electricity. Despite the fact that oil and gas reserves are not unlimited, there is a potential reserve for our country, which allows us not to actively search for alternative energy sources. On the other hand, the real consumer is faced with a constant increase in the cost of traditional sources of heat and electricity and their unstable operation, which forces him to look for more promising and reliable options. An analysis of existing life support systems in the Irkutsk region showed that, in parallel with traditional energy sources, solar energy is preferred as a backup energy source. Basically, in these systems, solar hot water systems and solar power supply systems worked well. Despite our climatic conditions, a certain difficulty in operation, high cost and the likelihood that the profit from the introduction of such systems will not be able to cover costs, the devices are widely used [1-2].

Solar backup systems are very popular in our region, despite the fact that we have the cheapest electricity. Vast territories, climatic conditions, branched and outdated networks do not always guarantee a stable supply of energy to consumers. Therefore, studies in the field of the efficiency of solar cells, depending on various factors, are relevant, and these works should be carried out in the aspect of practical application.

2. Literature review

For a number of reasons, including because of the huge reserves of energy raw materials, the issue of using renewable energy sources in the energy complex of Russia was practically not considered. In recent years, special attention has been paid to the use of renewable energy sources [3]. This is mainly justified by the need to fight for new opportunities to improve the quality of life of people living outside the city, the desire to improve the energy efficiency of the equipment used. Renewable energy sources are seen as a significant addition to traditional ones, with particular attention being paid to solar energy.

Solar energy has been used in engineering systems since the inception of mankind, when the concept of "engineering system" has not yet been invented. The International Energy Agency predicts that by 2050, solar energy will account for 27% of global energy consumption, and it will become the largest source of energy. In the coming years, the total capacity of photovoltaic stations in the world may double [4]. The leader in commissioning new solar power capacities is China. According to the closest forecast, by 2020 the total capacity of solar power plants will reach 110 GW. Analysts already predict a 10 percent decline in the cost of solar energy per year. Many world scientific schools are engaged in increasing the efficiency of solar installations. The main scientific direction is associated with improving the characteristics of semiconductor materials used in the manufacture of solar cells [5]. In addition, there are many works where the efficiency of solar cells is increased through the use of new design solutions and additional equipment [6].

One of the most promising areas of research in this area is the use of solar energy concentrators using Fresnel lenses. The Fresnel lens is one of the first devices whose operation is based on light diffraction. Despite a number of significant drawbacks that limit the range of application of these lenses in industry, they find their application in devices for the concentration of light energy.

Scientists from NASA conducted a series of experiments with these lenses to create miniature solar systems called SolarVolt [7].

A new installation using Fresnel lenses increased the luminous flux by twenty times. At the same time, final electricity production increased by 15 percent. In solar panels, such lenses were used with a thickness of only 0.2 millimeters.

When using lenses, the concentration of sunlight occurs on a smaller area, which allows to reduce the overall dimensions of installations without losing the amount of generated electricity. Due to new technologies, the size of solar panels is significantly reduced, and during the reconstruction of existing stations, their productivity will increase with the same occupied area [8].

Homemade solar panels using lenses were installed in California back in 1998 to provide electricity to a private home. Over the past 10 years, scientists from different countries have made great strides in the technology of manufacturing thin lenses for solar electric panels [9-12].

The most famous large company that tried to introduce lens solar cells was SolFocus (California, USA). In 2006, they were able to develop compact solar cell lenses. To create them, much less silicon was used than on conventional panels, since they used lenses and mirrors to concentrate sunlight.

In addition to the Americans, Russian and foreign scientists worked in the direction of creating solar concentrators for generating electric energy [13-14]. A large contribution in this direction was made by the Russian scientific school from Rostov-on-Don [15].

Using the technology of concentration of solar energy, Russian scientists managed to increase the efficiency of solar panels to 40%, and in field trials the efficiency of the installation reached 28.5%. In addition, the researchers noted that the use of solar energy concentrators is limited by the amount of solar energy entering the conditional area of the photodetector, with a significant concentration of energy, the panel overheats, and as a result, its efficiency decreases. Research in this area is carried out by many Russian scientific schools, mainly their research is geographically related to the location of the scientific organizations in which they work. But, based on the results of them research, it is necessary to carry out further work in the field of technical and structural solutions that will favorably affect the efficiency of solar energy systems [16-18].

3. Materials and methods

The experiments were conducted on the territory of the Irkutsk National Research Technical University in the laboratory "Modern Heating Equipment".

To measure the parameters of solar engineering systems, a bench was used based on the TPM138 counter-regulator, a universal eight-channel counter-regulator designed to build automatic control systems and regulation of production processes. Signals from thermocouples were recorded by this device using standard software supplied with the devices (OwenProcessManager).

The obtained graphs of changes in the parameters of solar engineering systems were imported into the MS Excel program. Monitoring of voltage, current, power was carried out by an OVEN company instrument of the IMS-F1.SH1. brand.

The OVEN device of the IMS-F1.SH1brand is designed to measure electrophysical parameters such as voltage, current, frequency, apparent, active and reactive power and power factor.

For the experiments, a solar panel and a panel with Fresnel lenses from Delta were used.

4. The results of the study

The dependence of the efficiency of a solar panel on the temperature on its surface is a known fact. In the city of Irkutsk, the average temperature range extends from -20 to + 30°C. The measured surface temperature of the solar panel showed that in winter on sunny days, the surface temperature reaches 40 °C, and in summer up to 70°C. In order to measure the real values of the dependence of the efficiency of the solar panel on temperature, with a stable light flux, we heated a pre-cooled panel from a temperature of -20 ° C to a temperature of 110 ° C. The change in voltage as a function of temperature is presented in Table 1 and in Fig. 1.

Table 1. dependence of the output voltage of the solar panel on temperature

Temperature, °C	-15	0	15	30	45	60	75	90	105
Voltage, V	2	1,9	1,85	1,75	1,65	1,5	1,2	0,7	0,4

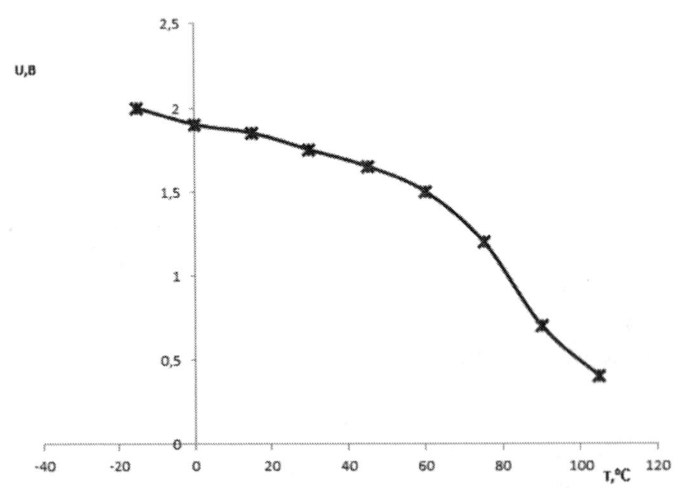

Figure 1. The graph of the dependence of the output voltage of the solar panel on temperature.

On a sunny summer day at 10 a.m. two solar panels were installed, one of which had a Fresnel lens. The temperature change on the panel depending on the operating time is shown in Fig. 2.

The graphs show that, being under constant solar radiation, the panel heats up the entire period of time, the heating process does not reach a stationary state. With a Fresnel lens, the heating process is faster, over the same period this panel heats up to 55 ° C, and without a Fresnel lens to 42 ° C.

When measuring temperature, was also measured the generated voltage on the solar panels at a current load of 1A. The experimental data are presented in graphical form in Fig. 3, table 2 presents the numerical values of the generated voltage and changes in efficiency (K,%) depending on the time of day.

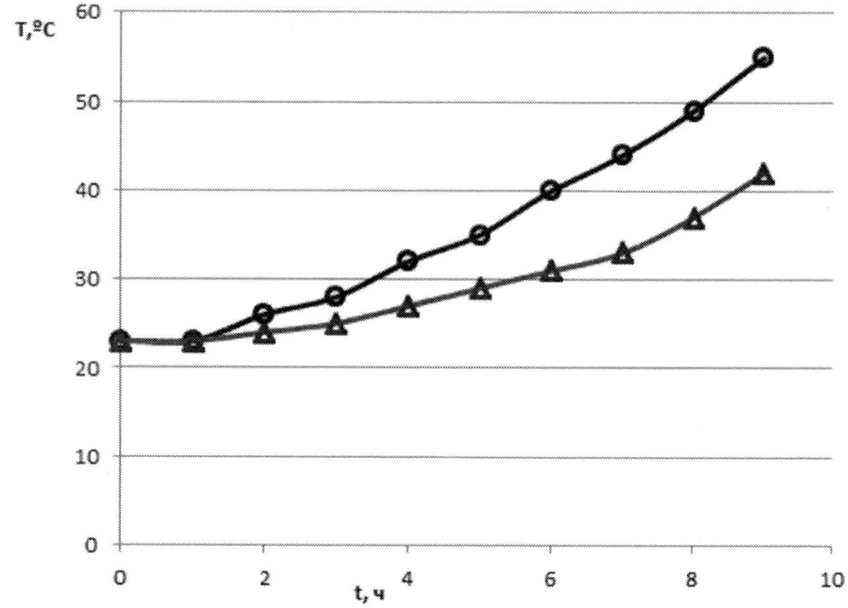

Figure 2. A graph of temperature versus operating time of the solar panel. Θ - with a Fresnel lens. Δ - without a Fresnel lens.

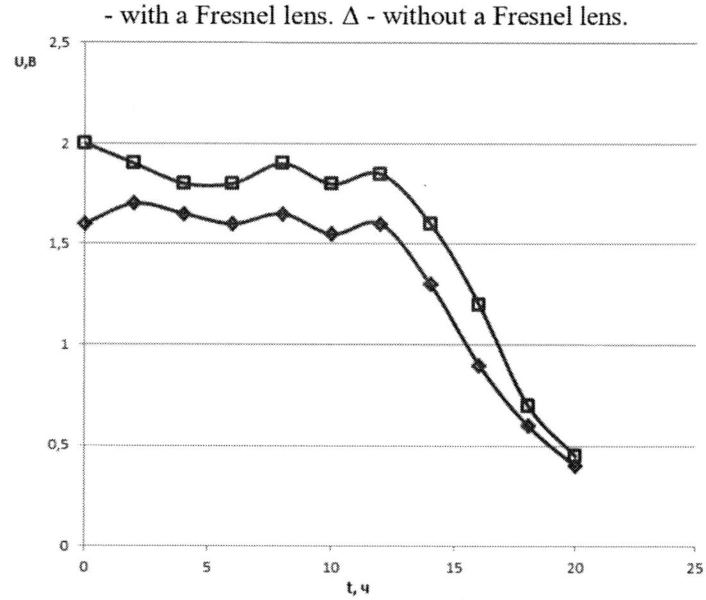

Figure 3. Graph of the dependence of the voltage change on the solar panels at a current load of 1A. □ - with a Fresnel lens. ◊ - without a Fresnel lens.

Table 2. Dependence of the voltage change on the solar panels at a current load

U1,B	1,6	1,7	1,65	1,6	1,65	1,55	1,6	1,3	0,9	0,6	0,4
U2,B	2	1,9	1,8	1,8	1,9	1,8	1,85	1,6	1,2	0,7	0,45
K, %	20	10,5	8,3	11,1	13,2	13,9	13,5	18,8	25,0	14,3	11,1

IOP Conf. Series: Materials Science and Engineering **880** (2020) 012037 doi:10.1088/1757-899X/880/1/012037

5. Discussion and conclusions

Analyzing the results, we can conclude that additional measures to remove excess thermal energy from the solar panel will help increase efficiency by about 20%. Analysis of the voltage dependence graphs on solar panels with Fresnel lenses installed shows that at the initial moment the efficiency of a solar panel with a Fresnel lens is much higher than the efficiency of a panel without a Fresnel lens. The installation process was carried out at 10 o'clock, by 12 o'clock in the afternoon, despite the increase in the intensity of solar radiation, the efficiency of the panel with a Fresnel lens decreased, and the efficiency of the panel without a Fresnel lens increased. This effect is easily explained by the fact that with the installation of a Fresnel lens, the mass of the device increased, and, accordingly, its heat capacity increased. The heating process continued over time, in addition, as the solar radiation increased, the temperature of the solar cells increased, while the heat transfer region calculated for a certain amount of solar radiation remained. Table 2 and Figure 3 show that in the main period of time, when most of the electric energy is generated, the average value of the efficiency of the solar panel is 10.8%. But at the same time, it is clear that if additional cooling of the panel is provided, then its effectiveness will also increase. The results showed that seasonal changes in the weather in sunny weather affect the energy parameters of solar panels, and the temperature of the panels in the summertime heats up to a temperature of 70 ° C and higher.

The data obtained showed that the efficiency of the solar panel in the hot season decreases by an average of 14.5%. In winter, solar panels heat up to a temperature of 40°C, and the efficiency of solar panels decreases by an average of 7.5%.

By organizing activities for the concentration of solar energy and providing stabilization systems for temperature parameters on solar panels, it is possible to increase the efficiency of these devices. In accordance with this, the payback period is reduced.

When using Fresnel lenses, efficiency increases by an average of 10% with minimal investment. It is also seen that, by changing the design of the solar panel, increasing only the heat transfer area on the back side, it is possible to increase the efficiency by at least 8% when implementing the measures described in this article, the payback cost of solar engineering systems can be reduced by 10-18%.

References

[1] Grigorash OV 2012 *Renewable energy sources* p 272

[2] Sokolov M M 2015 *Use of renewable and non-traditional energy sources textbook* (NNGASU) p 116

[3] Fortov V E, Popel O S 2010 Renewable energy sources for energy supply to consumers in Russia *Energy Bulletin* **1 (8)** p 9-29

[4] Imbert M, Papió A, Flaviis F, Jofre L and Romeu J 2015 Design and Performance Evaluation of a Dielectric Flat Lens Antenna for Millimeter-Wave pplications *IEEE Antennas And Wireless Propagation Letters* **14** pp 342-345

[5] McManus T, Mittra R, Pelletti C 2012 A Comparative Study of Flat and Profiled Lense *IEEE International Symposiumon Antennasand Propagation* pp 1-2

[6] Petosa A, Ittipiboon S Thirakoune 2003 Shadow Blockage Improvement Using a Perforated Dielectric Fresnel Lens *IEEE Antennas and Propagation Society International Symposium* pp 514-517

[7] Yakimov A I, Dvurechenskii A V, Nikiforov A I, and Proskuryakov Yu Yu 2001 J. Appl. Phys **89** pp 5676-5681

[8] Röpcke I. 2011 est labs and certification bodies: towards one standard *Sun & Wind Energy* **1** pp 54–63

[9] The NASA Surface Meteorology and Solar Energy Data Set 2009 http: //eosweb.larc.nasa.gov/sse

[10] Klychev Sh I, Bakhramov S A, Klychev Z Sh and Kharchenko V V 2012 Efficiency of the

concentration of solar radiation in Fresnel lenses. *Proceedings of the international scientific and technical conference Energy supply and energy saving in agriculture* **4** pp 118-122
[11] Gremenok V F 2006 Thin film solar cells based on Cu (In, Ga) Se2 *Proceedings of the VI International Youth Environmental Forum* pp 24–28
[12] Strebkov D S, Bobovnikov N Yu and Filippchenkova N S 2017 Hybrid sunny modules with a hub *Actual directions of scientific research of the XXI century: theory and practice* **5 (31) pp** 280-287
[13] Mayorov V A 2012 Research of linear solar battery modules with frenel lenses viesh *Bulletin* **1 (6)** pp 13-16
[14] Khimich A P 2011 Analysis of economic efficiency of solar power plants with photoelectric converters when using hubs. *Construction and industrial safety* **40** pp 184-188
[15] Manuilov B D, Manuilov M B, Strelchenko S A and Chernykh V B 2016 Comparative characteristics of plane zoned lenses and frenell lenses *26th International Crimean Conference "Microwave Technology and Telecommunication Technologies" (KryMiKo'2016) Conference materials* pp 1093-1099
[16] Taran A A and Voronin S M 2011 Efficiency of photoelectric converters in concentrated solar radiation *Bulletin of Agricultural Science of the Don* 3 pp 35-40
[17] Baineke A A, Zheksembieva N S 2017 Indicators of efficiency of photoelectric converters with concentrated solar radiation *Modern scientific bulletin* **2 (269)** pp 094-099
[18] Arbuzov Yu D, Evdokimov V M, Mayorov V A and Saginov L D 2017 Research of photoelectric transformation of bodies used in an experimental solar installation. Bulletin VIESH **4 (29)** pp 90-96

Research was conducted with financial support from the Russian Foundation for Basic Research and the Government of the Irkutsk Region in the framework of the scientific project No. 20-48-380002.

IOP Conf. Series: Materials Science and Engineering **880** (2020) 012038 doi:10.1088/1757-899X/880/1/012038

Technology of hybrid telemetry network of the Internet of things "KUMIR-Net"

R Belousov[1], A Gapeenko[1], A Nikolaev[2], E Fiskin[2] and M Fiskina[2]

[1]LLC "Scientific and Technical Center" KUMIR ", 83a Lermontov street, office 201, Irkutsk, 664074, Russia
[2]Irkutsk National Research Technical University, 83 Lermontov street, Irkutsk, 664074, Russia

E-mail: belousov@ntckumir.ru

Abstract. This article is devoted to the development and implementation of information-measuring and analytical systems for enterprises in the energy sector and housing and communal services, specialized in accounting and monitoring the production, transmission and consumption of energy resources.

It is predicted that the global market for smart energy accounting systems as an important component of Smart Grid technologies will continue to show steady growth in the medium term. At the same time, the high growth rates of this market in the EU, China and the United States will be the key factors contributing to the global growth of demand for smart energy accounting systems.

The situation in the Russian Federation is developing with a certain delay in relation to global trends in the field of intelligent information and measurement systems. So, after a long discussion in the expert community, the law of 27.12.2018 N 522-FZ "on amendments to certain legislative acts of the Russian Federation in connection with the development of electric energy (power) metering systems in the Russian Federation" was adopted, which prescribes that from 2022, all electricity meters for commercial accounting in retail markets and utilities should provide free of charge a minimum set of functions for intelligent electric energy (power) metering systems. Requirements for an intelligent electricity metering system are formulated in the draft resolution of the government of the Russian Federation "Rules for providing access to a minimum set of functions of intelligent electric energy (power) metering systems". Directly to the system for collecting and transmitting data from metering devices are requirements for the functionality, procedure and protocols for the exchange of information, access, storage and data security. The development of such systems, taking into account the number of metering devices in Russia of about 100 million units, will probably be based on Internet of Things (IoT) technologies and will contribute to the development of these technologies both in Russia and on the world market. Overview of existing Internet of things technologies.

6LoWPAN. Standard for IPv6 communication over low-power wireless personal networks of the IEEE 802.15.4 standard. The purpose of the IEEE 802.15.4 standard is to offer the lower layers of the network base for networks such as wireless personal networks focused on low cost, low speed ubiquitous communication between devices. The emphasis is on the very low cost of communicating with nearby devices, with very little basic structure, in order to operate at a low energy level. The IEEE 802.15.4 standard does not define other higher layers and the compatibility of intermediate layers — these are other solutions and standards that 6LoWPAN belongs to. 6lowpan's goal is to provide Internet of things devices with a cheap interface for connecting to the Internet. Application developers will determine the exchange protocols.

Advantages of the technology: an open standard for which cheap transceivers are produced by many electronics manufacturers, the frequency range of 800-900MHz is harmonized by regulators in many countries and one or another part of the range can be used in different countries of the world without obtaining permits for the use of radio frequencies. The competition of chip manufacturers, hardware and software solutions allows us to expect a reduction in the cost of Internet connection for Internet of things devices in the future.

Disadvantages: the end device requires support for the entire TCP/IP stack, which requires an advanced microcontroller / processor in the device, and also provides increased memory requirements. For cheap sensors, this can greatly increase the cost of devices. Another drawback is the low penetration of IPv6 networks both in the world and in Russia. According to the monitoring of IPv6 connections by Google, the penetration of IPv6 networks in the world in 2019 is 25%, in Russia-4.3%, in China-1.3%, in the CIS countries <0.3%. Therefore, traffic tunneling may be required to connect 6LoWPAN networks to the outside world, which is an additional cost.

LoRaWAN. Patented high-budget radio channel technology from the American company Semtech. It allows you to achieve a long communication range and high connection density (up to 5000 devices per BS). LoRaWAN is one of the wireless technologies that provide an environment for collecting data from various equipment: sensors, counters, and sensors, i.e. Internet of things devices. In this case, the specification does not define data representation models — this is left to the manufacturer of the final solution. Advantages of the technology: Long range of radio signal transmission compared to other wireless technologies used for telemetry (up to 10-15 km). High network scalability over large territories. No need to obtain a frequency resolution and pay for the radio frequency spectrum. Disadvantages: With a variety of manufacturers of base stations and modules for terminal equipment, everything is based on Semtech microcircuits, which presents some concerns when introducing any restrictions on their supply. The lack of competition in the production of chips can lead to an increase in their cost in comparison with similar technologies. The technology provides high characteristics of the radio channel only at low data transfer rates (tens of bits per second). At speeds of 9600 bps or more, the performance of the radio channel is comparable to **6LoWPAN**. Nevertheless, in electricity meters operating under the **SPODES** protocol **(DLMS / COSEM)**, it is necessary to transmit data packets that are rather large by the standards of LoRa, and long delays can lead to breaks in sessions.

NB-IoT. Internet of things technology aimed at mobile operators. Cellular communication standard for telemetry devices with low data exchange volumes. Developed by the 3GPP consortium as part of work on new-generation cellular network standards. Designed to connect a wide range of Autonomous devices to digital communication networks. For example, medical sensors, resource consumption meters, smart home devices, and so on. Deployed over existing LTE or GSM (EC-GSM-IoT) networks. Advantages of the technology: flexible management of power consumption of devices, huge network capacity (up to 100,000 devices per 1 BS), low cost of devices, increased sensitivity of receivers compared to LTE. The coverage area during deployment will be equivalent to the entire service area of the mobile operator. In a mass application, the communication quality will obviously be better than that of **6LoWPAN** or **LoRaWAN** due to the licensed frequency range.

Disadvantages of the technology: higher power consumption and higher cost of modules than **6LoWPAN** and **LoRaWAN** due to more complex circuitry, the subscription fee is "for the device" and not for the "base station".

Summarizing the above, it can be argued that the currently known IoT technologies are focused on high-density radio communication systems with low energy consumption so that the final device can operate on batteries up to 10 years. The KUMIR-Resource information-measuring system has been developed by the team of the KUMIR Research and Development Center since 2014 and is the product of a deep modernization of the KUMIR-TeploCom Information-Measuring System, developed in 2008. However, more than 10 years of experience in the field of automation and dispatching utilities and energy, as the most popular category of customers, shows that the use of the radio channel in absolutely all cases is an economically unjustified strategy. Often, metering devices can be grouped into a wired cluster and already connect it wirelessly. The use of the 868 MHz radio channel in the city is often limited by the terrain and in many cases this justifies the use of mobile communications as a better and more reliable communication channel. When automating industrial facilities, it is often necessary to derive the readings of dozens of devices from one or more technological rooms within the same building - a hybrid (wired + wireless) connection is also advisable here. In Yakutia, for dispatch projects, we had to use VSAT technologies several times, because the quality of mobile Internet did not allow us to achieve the desired quality of communication with the object. Thus, to reduce the cost of connecting and servicing IoT devices and applications, it is necessary to develop a hybrid network architecture, on the one hand transparent for IoT applications, and on the other hand, efficient and reliable in any communication environment.

Also, existing IoT technologies do not offer specifications for typical applications (sensors, sensors, switches, servos, etc.). This forces equipment manufacturers to develop their own protocols even for typical applications, which results in the incompatibility of products from different manufacturers among themselves.

The product being created is the technology of the KUMIR-Net Internet of Things hybrid telemetry network. The main purpose of the product is a tool for creating large platforms of the Internet of things with different physical levels of data transfer. The prerequisites for the creation of the technology were the fact that most developers of technologies for the Internet of things are guided by the same principles of the functioning of the network, the same requirements as for conventional Internet devices, but with a focus on low power consumption. This approach does not fully meet the current state of affairs and inhibits the development of IoT technology. The main problem of modern IoT technologies is that due to the presence of the IoT interface, the device increases in price several times, for example, electricity meters with a wired RS-485 interface, which measure not only electricity consumption (kW * h), but also quality parameters (voltage, current, frequency, power factor, etc.) cost in Russia 2500-3500 rubles. depending on the manufacturer, and with IoT interface (GPRS, NBIoT, Ethernet) - 5500-6500 rubles. The situation is even sadder in the segment of cheap devices (temperature sensors, pressure sensors, motion sensors, etc.) - from the price without IoT - up to 500 rubles, with IoT - from 2500 rubles.

The thing is that they are trying to turn an IoT device into an Internet site with all the associated costs. The creation of our technology is aimed at solving this problem. In the sense of creation, it should provide interaction between terminal devices and IoT applications in a single infocommunication space based on the KUMIR-Net protocol, while application developers can abstract from data transfer technologies and focus on describing application logic, and equipment developers - on the device's hardware. The introduction of technology will generally help to reduce the cost of creating and managing large networks of process equipment, telemetry and telealarm systems. The technology also includes the possibility of building hybrid networks: on the basis of generally accepted Internet technologies, local field networks of bus topology with random access, as well as on the basis of wireless LPWAN and PAN networks.

Methods and methods for solving problems to obtain the expected characteristics

1. Creating a simple protocol for exchanging data with IioT technological devices with low requirements for end devices, due to which their low cost will be ensured. Creating transport layers for the protocol on top of the CAN field bus and on top of the radio channel of the IEEE 802.15.4 standard - as the most common and cheapest way of radio channel data transmission.

2. Creation of the implementation of a fail-safe cluster with the ability to scale horizontally with a capacity of at least 5 million end devices, which ensures reliable functioning of the IoT ecosystem.

3. Creating simple API libraries for integrating end devices into the network to reduce the labor costs of developers of Internet of things equipment, metering devices, etc.

4. Creation of simple API libraries for integration into the network of applications for processing data from end devices and metering devices.

The goal of STC Kumir LLC's efforts to develop a protocol for the Internet of things, a field network protocol, an 868 MHz radio protocol and integrate Kumir-Resource IMS into the developed technology is to create a distributed intelligent information-measuring system (IMS), intended primarily for energy distribution organizations of Russia and neighboring countries. But this does not exclude the possibility of using the KUMIR-Resource platform as a combination of hardware and software technologies in various IoT applications in the future.

The leaders in the smart metering market are currently leading international companies such as Siemens, Itron, Landis + Gyr, Honeywell, and American ElectSolve. In Western Europe, German Cuculus, Dutch Energywox, Italian Ferranti are active. Due to the similarity of working conditions and goals, the Indian provider Fluentgrid, offering its products (IMS MDMS) and services in Central Asia and EMEA (markets of the Middle East, Europe and Africa), can become a definite beacon in the field of market promotion efforts. In Russia, the companies offering AIIS KUE for OREM and the retail market are: NPO Mir, AIIS KUE of the Energomera concern, LLC Matritsa (AIIS KUE Smart Ims for the retail market), CJSC Telecommunication and Telemechanics Systems (integrated system AIIS KUE / ASDTU), EnergoKapitalServis LLC (AIIS KUE AlfaCenter), etc. As a top-level software, energy sales organizations use widely distributed products such as AlfaCenter, Pyramid, Energosfera, as well as solutions of their own development. The requirements put forward by the "Rules" for intelligent accounting systems, in terms of data collection, transmission and storage systems, are most closely aligned with the modernized information-measuring system IMS "Kumir-Resource".

IMS "Kumir-Resource" can be used as a vertical industrial IoT platform for energy retail organizations in Russia and neighboring countries - in particular, those consumers who are looking for simpler and cheaper solutions that differ from the offers of large industrial IoT vendors (such as GE, Itron or Siemens).

Wireless communication networks and the upgraded Kumir-Resource platform can become the basis for large-scale IoT solutions, including smart metering and smart networks, street lighting, electric transport and security, etc. This is possible because most of the IoT applications are related to different parts of the technological stack of IMS "Kumir-Resource" (radio components, sensors, network hubs, databases, servers).

Competitive advantages of the created product, comparison of technical and economic characteristics with domestic and foreign analogues:

1. Wireless solutions. 868 MHz radio channel transceiver as part of the KUMIR-Resource system

The decision to develop an inexpensive domestic radio channel transceiver in the public range of 868 MHz was made primarily based on the conditions for leveling non-trade restrictions and risks. The concept of the transceiver was developed taking into account the analysis of the existing wireless systems on the market (Sigfox and its Russian counterparts, LoRaWAN and NB-IoT) and the possibility of further transferring the development of STC Kumir LLC to the domestic element base.

The design took into account both the strengths and limitations of the radio channel transceivers and systems on the market. So:

Sigfox has a unidirectional transmission channel, which does not allow it to be used in systems with accurate time synchronization and organization of balance accounting of resources.

LoRaWAN is the closest to the radio channel solutions of Kumir-Resource LLC. The system also provides bidirectional signal transmission, but has several disadvantages:

- the time synchronization procedure of base stations does not allow making them inexpensive, which makes the system as a whole more expensive;
- in systems on the market with LoRaWAN radio modules (868 network, etc.), the signal from the primary devices is transmitted to the concentrators exclusively via the radio channel. Such a solution in the conditions of a complex urban radio environment makes the installation of monitoring systems for meter readings and IoT unpredictable and costly;
- licensing the production of microchips of the LoRa standard in the USA and the supply of microchips from a single manufacturer (Semtech) does not allow us to talk about import substitution

of systems based on this standard and carries the corresponding limitations and risks;

The method of increasing the noise immunity and stability of communication by expanding the dynamic range is a strong point of the LoRa standard. But in tests, NTC" KUMIR" LLC achieves similar results using noise-resistant coding based on reed-Solomon codes, without expanding the signal spectrum. At the same time, the effective channel speed is 37.5 kbit / s versus 30 kbit/s for LoRa.

Mobile operators with **NBoT** solutions, with appropriate modernization of the software of mobile base stations, can significantly expand their sales due to the scope of smart measurements and IoT. But here they encounter several significant obstacles, the main of which is the need to equip primary devices with SIM cards and pay for the mobile traffic of each of the end devices, which significantly increases their price and, ultimately, the cost of the system as a whole.

All of the listed competitive systems are also characterized by a lack of specification of the “Internet of things” itself as an infocommunication object platform, which gives rise to incompatibility between devices of the same purpose produced by different manufacturers.

2. Multiresource execution of IMS "KUMIR-Resource"

IMS "Kumir-Resource" was originally designed as a system for remote monitoring of measurement data of electric and heat energy, gas, hot water and cold water. At present, the Kumir-Resource system supports more than 80 metering devices with the potential for introducing almost any meter with open protocols. This opens up opportunities for the potential Customer to choose metering devices for future use.

Those. the energy sales organization, deploying IMS "KUMIR-Resource" in the housing estate to monitor the readings of electricity metering devices, gets the opportunity to include monitoring of gas, hot water, cold water and heat consumption in its data transmission system. At the same time, it can transfer data to specialized resource providers for an additional fee, earning additional money. Such a solution is a serious competitive advantage of IMS Kumir-Resource among Russian companies developing the IMS KUE direction and not having the ability to include monitoring of gas, heat and water meter readings in their systems.

In general, multi-resource execution of information-measuring systems is a development trend in this area abroad.

3. Hybrid solution for circuit implementation of IMS "KUMIR-Resource"

The possibility of using wired and radio channel solutions in the IMS “Kumir-Resource” allows you to flexibly vary design decisions.

In some cases, based on the configuration, relative position of buildings and radio coverage, the most appropriate option is to use the radio channel as the main method of transmitting data from primary metering devices. These are mainly open spaces - rural settlements, SNT, garage cooperatives, etc.

In conditions of dense urban development, radio noise and reinforced concrete walls create significant difficulties in the operation of 868 MHz radio channel systems, such as LoRaWAN (868 network) and LPWA (Sigfox and Russian counterparts - Strizh and Vaviot). In this case, the most appropriate and least costly hybrid solution is to transfer the data of the home metering devices to the lifting concentrators using a radio channel (PAN system) and further transmitting the information to a GSM modem or the Internet using a wired bus.

Circuitry IMS "KUMIR-Resource" allows you to apply and combine radio and wire solutions, depending on the situation. A specific placement option is developed when analyzing the radio environment, geometry and material of buildings, based on the choice of the most economical option.

IMS "Kumir-Resource" has been in commercial operation in its current form since 2012. Time has shown the correctness of the basic decisions on the architecture of the system, the choice of the base operating system and database management system, the "cloud" solution of the interface.

The proposed integration of IMS Kumir-Resource into the technology being created will not affect the foundations of existing software, but will add new communication protocols and support for hybrid technology, improving fault tolerance and service availability.

4. Integration of lower and upper level software and promising services for energy retail organizations

At present, over 80 metering devices are integrated in the IMS "Kumir-Resource". In this case, the integration of meters with open protocols in the system, as a rule, is resolved within 2 weeks.

NTC Kumir LLC flexibly solves the problems of end-to-end data transmission to existing systems of commercial electricity metering (AIIS KUE) and billing. For this, the customer is offered an open API for developing interaction drivers with existing systems through the Kumir-Resource data hub or can independently develop interaction drivers with customer's top-level applications that are currently in operation or promising.

NTC LLC Kumir is a developer and manufacturer of software and hardware for remote monitoring systems of metering devices. Because of this, it sells both software and devices of its own production on the market.

Conclusions

Thus, the modernization of the architecture of the IMS "Kumir-Resource" will make it possible to scale the system to the requirements of electricity suppliers for large systems. This will make it possible to deploy IMS "Kumir-Resource" as a basic system of intellectual resource accounting when implementing the Law of December 27, 2018 No. 522-FZ.

References

[1] Overview of wireless communication protocols for the Internet of things *https://iot.ru/promyshlennost/obzor-protokolov-besprovodnoy-svyazi-dlya-interneta-veshchey*
[2] Moskalenko T A, Kirichek R V and Kucheryavy A E 2017 Review of Internet of things protocols *Information Technologies and Telecommunications* **5** ISSN 2307-1303. http://www.itt.sut.ru
[3] Aleshin C V, Balaev M O and Vlasyuk V V Software and hardware solutions for the Internet of things technology *https://cyberleninka.ru/article/n/programmno-apparatnye-resheniya-dlya-tehnologii-internet-veschey*
[4] Novitsky N 2015 *Pipeline energy systems: methodological and applied problems of mathematical modeling (*Novosibirsk) p 476
[5] Belousov R A, Pinkin A A, Fiskin E M and Fiskina M M 2015 The possibilities of "Kumir-Resource" system for SMART GRID and apartment level energy accounting *EURO-ECO Das Internationale Symp. «Ökologische, Technologische und Rechtliche Aspekte der Lebensversorgung»* pp 15-16
[6] Amelin A A, Belousov R A, Fiskin E M 2016 IMS "KUMIR-Resource" as the basis for Smart Grid *Improving the efficiency of energy use in Siberia* **2** pp 461-466 ISBN 975-5-8038-1120-0 (T2) 978-5-8038-0837.
[7] Belousov R, Fiskin E and Fiskina M 2018 Integrated Systems for Commercial and Technological Control of Energy Resources Consumption in Urban Areas and Businesses in Russia *E3S Web of Conf. Scopus* **39**
[8] Nikolaev A, Belousov R, Fiskin E and Fiskina M 2019 Development of equipment for monitoring the lake Baikal water area *E3S Web of Conferences Scopus Volume* **102**
[9] Chupin V R and Dushin A S 2019 Parameter optimisation for new and reconstructed systems of water supply and distribution for ensuring consumer's uninterrupted water supply *Proceedings of Universities. Investment. Construction. Real estate* **9(4)** pp 790–803
[10] Kulkov V N, Solopanov E Yu and Kamalov R T 2019 Using immobilised sludge for wastewater treatment and its air regeneration *Proceedings of Universities. Investment. Construction. Real estate* **9(3)** pp 522–529
[11] Pupyrev E I and Chupin V R 2019 Features of regional development of water disposal systems in the central ecological zone of the Baikal natural territory *Proceedings of Universities. Investment. Construction. Real estate* **9** pp 354–363

IOP Conf. Series: Materials Science and Engineering **880** (2020) 012039 doi:10.1088/1757-899X/880/1/012039

Comprehensive optimization of the route, composition and parameters of transporting pipelines for various technological purposes

V R Chupin[1] and D I Maizel[2]

[1]Irkutsk National Research Technical University, 83 Lermontov street, Irkutsk, 664074, Russia
[2]Verkhnechonskneftegaz, Irkutsk, Bolshoi Liteiny, 3, Russia

E-mail: chupinvr@ex.istu.edu

Abstract. Pipeline transport is the most common in our country and abroad. All of Russia is entangled in a single system of oil and gas transport, which already permeate many European and Asian countries. At the same time, pipelines pass through swamps, pass through mountain ranges, cross seas and lakes. Group and district water pipelines cover thousands of populated areas, heat supply, water supply and sanitation systems cover cities and urban agglomerations. When designing such capital-intensive linear objects, one of the most important tasks is to trace them. The paper proposes to solve this problem in conjunction with the optimization of parameters of pipelines, pump and compressor substations, throttling devices and mating structures. The technique is based on a combination of discrete optimization methods and reflects the real processes of substantiating the structure and parameters of structures. The calculations of real pipeline systems showed high computational and economic efficiency of the proposed approach and its software implementation.

1. Introduction

Often, in the practice of designing, reconstructing, and developing pipeline systems for various technological purposes (transporting liquids, gases, multiphase media), the problem arises of optimizing the route and parameters of individual pressure and pressureless pipelines connecting two nodes, one of which is an intake node, the other is dumped. To solve this problem, a new approach is proposed, based on an algorithm for constructing shortest circuits and a dynamic programming methodology [1-10]. The essence of the proposed approach is to build conditionally optimal solutions between two given (initial - source and final - consumer) peaks in the graph of possible pipeline paths and its parameters (diameters and wall thicknesses of pipes, slopes, pressure boosting, pumping and compressor stations, differential and throttle structures). After reaching the final peak, from all conditionally optimal solutions, the best one is selected and the optimal pipeline route and the optimal parameters of pipelines and conveying structures are restored by reverse movement.

2. Methods

If we fix the topology of the pipeline network, then the problem arises of choosing the optimal number of discrete values: the diameters of the pipelines, the pressure of the pumping stations and throttle devices. Essentially, the problem arises of finding the optimal piezometric surface of a moving medium (water, gas, oil product).

The search for such an optimal piezometric can be represented as a multi-step control process, characterized by the following values.

The control variables are the pressure losses hi, the pressure of the pumping stations Hi (-), throttle devices H i (+):

The phase variables are nodal piezometric heads (free surface marks) at the beginning and at the end of the calculation section i (there will be Z and C, respectively):

$$Pi(н) = \{Pz(н)\}; \qquad z \in Z_i,$$

$$Pi(k) = \{Pc(k)\}; \qquad c \in C_i,$$

The sequence of which P (n) = {P1 (-), ..., Pn (-)}, P (k) = {P1 (k), ..., Pn (k)} is the set of possible states of the systems under consideration.

It should be noted that for section i, the initial variables are associated with the following final relations:

$$Pc(н) = Pz(н) + hy - Hu; \qquad Pc(к) = Pz(н) + hy + Hv. \tag{1}$$

In formulas (1), hy represents the pressure loss values over the network sections, which for different transported media they have different hydraulic ratios.

Due to the discreteness of the control variables, the transition from Pz (n) to Pc (k) cannot always be carried out, i.e. there is not always such a combination hy,, Ni,; Hv, so that condition (1) is satisfied. In this case, it is advisable to take as phase variables not their point values, but some subsets:

$$P_i^{(н)} = \bigcup_{c=1}^{c} P_c^{*(н)} \text{, so that } \bigcap_{c=1}^{c} P_c^{*(н)} = 0; \tag{2}$$

$$P_i^{(k)} = \bigcup_{z=1}^{z} P_z^{*(k)} \text{, so that } \bigcap_{z=1}^{z} P_z^{*(k)} = 0.$$

Algorithmically, this is done by dividing P i (n) ≤ Pc (n) ≤ and Pi (k) ≤ Pc (k) ≤ into z and from intervals. If Pz (n) Pz * (n), then it is considered that a transition has been made from the initial (input) to the final (output) phase variables.

Under the condition of discrete values of the controlled variables hi, H and (-), H y (+), the objective function in the form of life cycle costs [11,12] has the following form:

$$F(P,h,H) = \sum_{i=1}^{n} F_i\left(h_i,\ H_i^{(+)}, H_i^{(-1)}, P_i^{(н)}, P_i^{(k)}\right), \tag{3}$$

and is represented as an additive function of many variables.

We consider in more detail one step i of the control process, first in hi and then in Hi.

It is necessary to choose a control hy, that would translate the initial state Pz (n) to the final Pc (k) and minimize the objective function Ψi (Pc (k)), which consists of the costs obtained in the previous optimization step and the costs associated with transition from Pz (n) to Pc (k) according to (1):

$$\Psi_i^*\left(P_c^{(к)}\right) = \Psi_i\left(P_z^{(н)}\right) + F_i^*\left(h_y, P_c^{(к)}, P_z^{(н)}\right).$$

Since there are ci final and zi initial states, the optimal transition to any final Pc (k) from all possible initial states (Pz (n), z = 1, ..., zi) by analyzing all admissible controls (hy, y = 1, ..., Yi) will be called conditionally optimal solutions.

In view of the foregoing, the functional equation for calculating conditionally optimal solutions for hy at step i is written as follows:

$$\Psi_i\left(P_c^{(\kappa)}\right)=\min_{\substack{h_y, y\in Y_i \\ z\in Z_i}}\left\{\Psi_i^*\left(P_z^{(n)}\right)+F_i\left(h_y, P_c^{(n)}, P_z^{(k)}\right)\right\} \tag{4}$$

According to this expression, the conditionally optimal solution corresponding to the phase variable Z at the beginning of the section is obtained by minimizing (or finding the best) transition to Pc from any admissible states Pz (n) (z runs through the values z = 1, ..., zi) using controls hy (y runs through y = 1, ..., Yi). Moreover, for each analyzed Pz (н) and hy the numerical value of the cost function F (hy, Pc (н), Pz (k)) is determined according to (2) in the pipeline construction. Only those controls for which condition (2) holds are taken into account:

$$P_c^{*(\kappa)}=P_z^{*(n)}+h_y,\quad z=1,\dots,Z_i,\ y=1,\dots,Y_i \,.$$

It is determined in which of the intervals of the initial phase variables each of the obtained Py (s) falls.

The best decision to remember:

$$\Psi_i\left(P_c^{(\kappa)}\right)=\min\left\{\Psi_i\left(P_c^{(\kappa)}\right), \Psi_y^{(c)}\right\}.$$

As a result, functional equations can be written separately for pump and throttle structures:

$$\Psi_i^{(HC)}\left(P_c^{(\kappa)}\right)=\min_{H_u, u\in U}\left\{\Psi_i\left(P_y^{(\kappa)}\right)+f_i^{(HC)}\left(H_u\right)\right\}; \tag{5}$$

$$\Psi_i^{(\partial)}\left(P_c^{(\kappa)}\right)=\min_{\substack{H_v, v\in V_i \\ \mu=c-1,\dots,1}}\left\{\Psi_i\left(P_\mu^{(K)}\right)+f_i^{(\partial)}\left(H_v\right)\right\}. \tag{6}$$

The best decisions are remembered:

$$\Psi_i\left(P_c^{(\kappa)}\right)=\min\left\{\Psi_i\left(P_c^{(\kappa)}\right), \Psi_u^{(c)}, \Psi_v^{(c)}\right\}$$

In view of (5), (6), the optimal transition from any Pz (n) to the initial state Pc (k) by controlling with respect to hu, Ni, Hv is carried out:

$$\Psi_i^*\left(P_c^{(\kappa)}\right)=\min\left\{\Psi_i\left(P_c^{(\kappa)}\right), \Psi_i^{(HC)}\left(P_c^{(\kappa)}\right), \Psi_i^{(\partial)}\left(P_c^{(\kappa)}\right)\right\}$$

As a result of a multi-step process of minimizing the cost function, the following parameters are determined: the diameters of new pipelines and the thickness of their walls; areas where the construction or modernization of pumping stations and throttle devices is necessary.

If, for example, a sewer collector is designed that has one node for the flow of wastewater and one discharge node, then the multi-step optimization process can be organized either from the subscriber node, since the ranges of phase variables are usually determined for it, or from the discharge and sink unit on CBS, for which the values of Pi (k) are fixed. Figure 1 shows the process of building up conditionally optimal solutions according to (4), (5), (6) and the optimal territory is shown in pink. Figure 2 shows the optimal profile obtained as a result of the computational process according to the specified formulas.

Based on the foregoing methodology for optimizing the parameters of the pipeline and pipeline structures, we propose a methodology for the comprehensive optimization of the route and parameters of the pipeline.

At the same time, the concept of a redundant design scheme is introduced - a graph of possible options for passing pipeline routes and possible locations for pipeline structures. Formed by the designer.

The first step of the method is to break down all nodes of the redundant graph into two subsets: Jg (v) - nodes that form the subtree and Jx (v) - nodes of the subgraph outside the tree, v = 1, ..., m. Next, an iterative calculation process is organized, consisting of six stages.

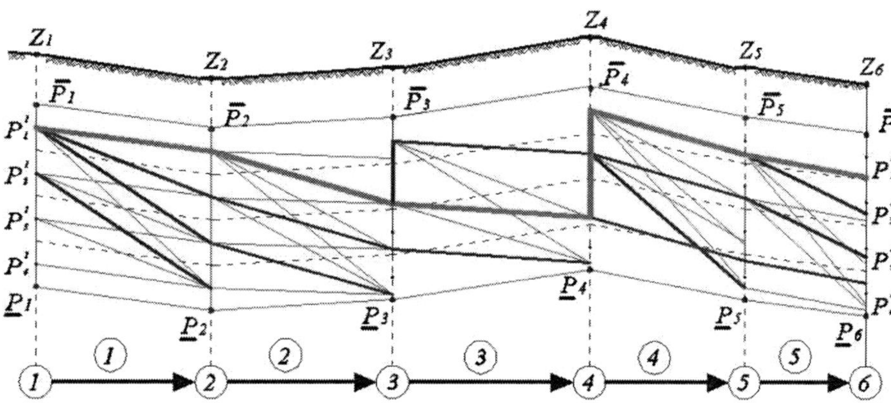

Figure 1. Building conditionally optimal solutions

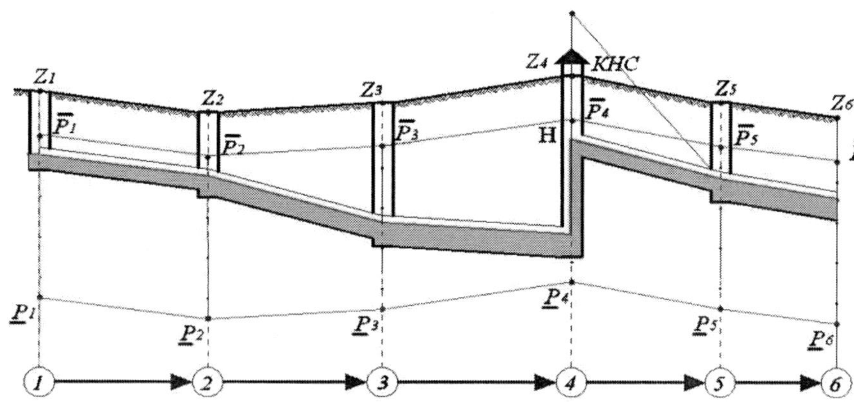

Figure 2. Optimal collector profile

1. In the array Jx (v) the numbers of all nodes of the circuit are entered, and in Jg (v), initially filled with zeros, only the node j corresponding to the source is entered. For all nodes, two arrays of phase variables are formed. The first of them contains C pairs of elements Pjc and Ψjc (Pjc), and the second consists of z groups of elements, each of them includes the node number j, from which the transition to the node j + 1 takes place; the number of the corresponding interval z, with which we arrive in the interval with; hy control number and Ni and Hv control number. For the subset j є Jg (v), the values Ψjc = 0, and for j є Jx (v), Ψjc = ∞. At this stage, v = 0.

2. In the subset Jx (v), nodes j + 1 are defined that have connections with the node j є Jg (v), and conditionally optimal solutions are built up for the branches corresponding to them.

3. For each interval C, the conditionally optimal solutions obtained in stage 2 are compared with the existing solution at j + 1 є Jx (v):

$$\Psi jc = min\ \{\Psi j(Pc(H),\ \Psi jc)\}$$

The best of them are remembered, and the corresponding values are recorded in the information arrays. Then, for each node, the maximum gain is determined:

$$\Psi jc = \frac{\min}{c}\ \{\Psi jc\},\ j \in Jx(v)$$

4. Among the optimums for subgraph nodes outside the tree, the best one is selected:

$$\Psi = \min_{j \in J_x^{(V)}} \left\{ \Psi_j \right\}$$

and the corresponding node goes into the subset Jq (v).

5. We take v = v + 1 and calculations 2-4 are repeated until v becomes equal to m – 1, which will correspond to the transition of all nodes from the subset Jx to Jg.

6. For the drain node, among the conditionally optimal solutions, the best one is selected and, by returning through the information arrays, the optimal values of the system parameters and the path leading from the source node to the drain node are restored.

As an illustration of the operation of the described algorithm, we consider the redundant network (Figure 3, a) with the dimension: m = 9; n = 12. The source in it is node 1, and the sink is 9. Then j є Jx (0) will correspond to node 1, Ψ1 = 0, Ψj = ∞, j = 2, ..., 9. Nodes 2 and 4 are adjacent to 1 , and therefore, for conditional plots 1-2 and 1-4, conditionally optimal solutions are built up (Figure 3, b). For the final phase variables of the analyzed sections, the maximum payoffs Ψ4 and Ψ2 are determined, which, in turn, are compared with each other: Ψ = min {Ψ4, Ψ2} and node 4 goes to the subset Jg (1).

The next step is to identify the nodes adjacent to the 4th node. They will be 7 and 5. Further, conditionally optimal solutions are already being built up for sections 4-7 and 4-5, maximum wins are determined and a comparison is made:

Ψ = min {Ψ2, Ψ5, Ψ7} = Ψ2.

Node 2 goes into the subset Jg (2) (Figure 3, c). For node 2, 5 and 3 will be adjacent, therefore, the building up of conditionally optimal solutions is carried out in sections 2-3 and 2-5. Since for node 5, Ψ5c is different from ∞, then for each phase variable c: c = 1, ..., c, the conditionally optimal solutions obtained by branches 4-5 and 2-5 are compared. The best values and the corresponding paths from 4 or 2 to the source are stored in information arrays. Then, according to nodes 3, 5, 7, the maximum wins are determined, which are compared among themselves:

Ψ = min {Ψ3, Ψ5, Ψ7} = Ψ

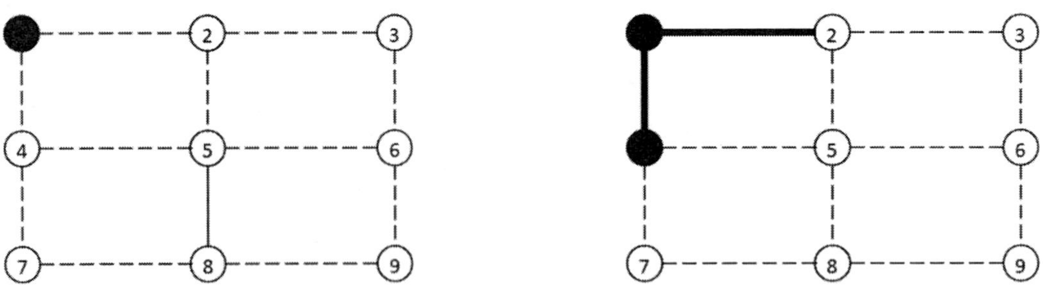

a) в)

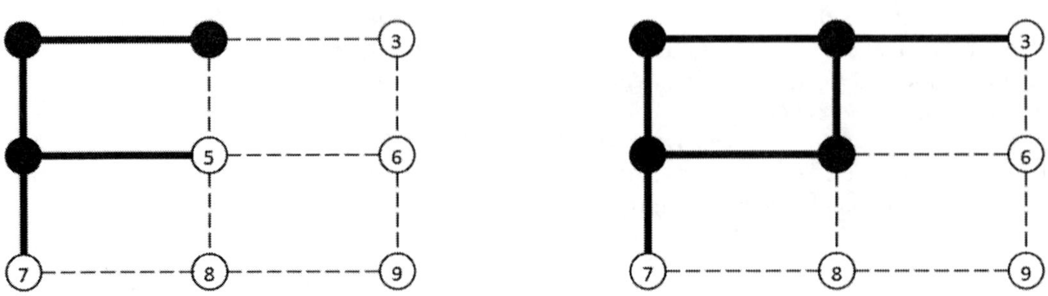

c) d)

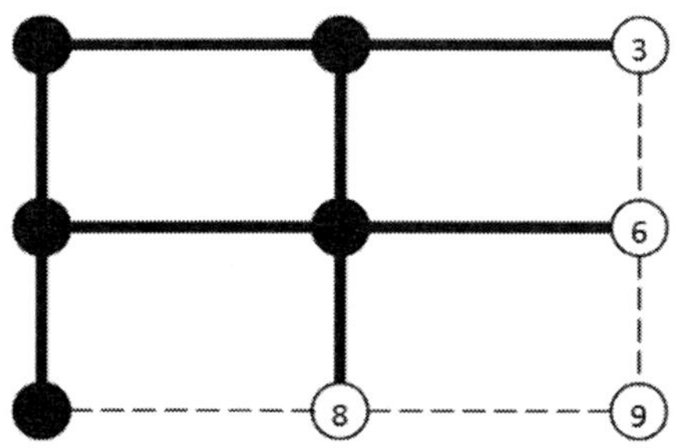

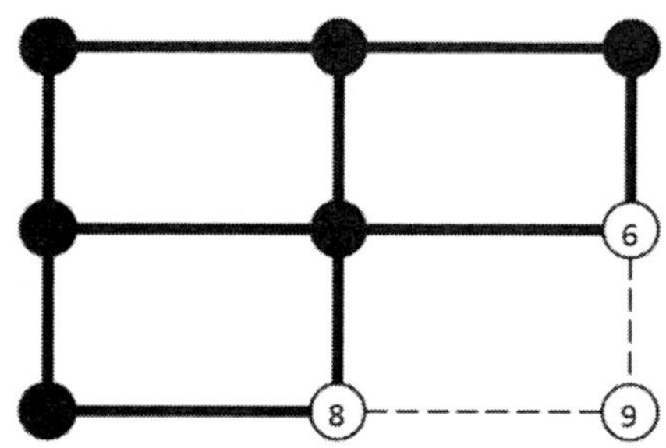

e) f)

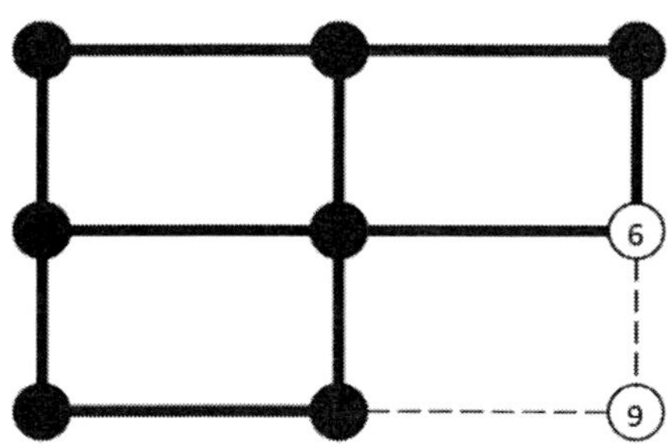

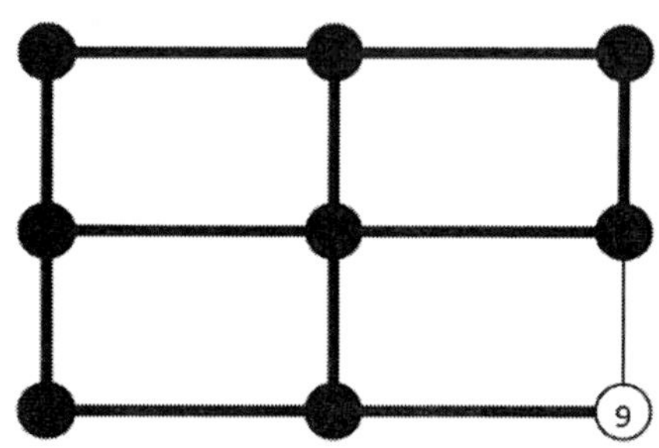

g) h)

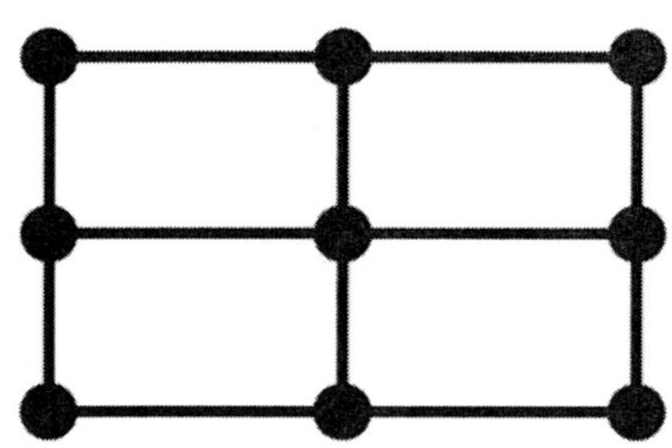

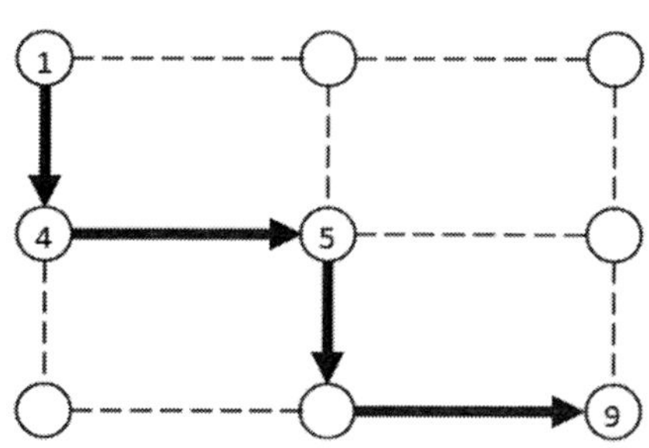

i) k)

Figure 3 - Algorithm for optimizing the route and parameters of the main collector

Node 5 goes into a subset of Jg (3) (Figure 3, d), etc. Figure 3, k shows the optimal path obtained as a result of the return stroke, by restoring the optimal values of the parameters and the path leading from 1 to node 9.

3. Results and discussion

We show the effectiveness of this methodology on the example of justifying the route of a promising water supply scheme for the city of Irkutsk and the Irkutsk region from lake. Baikal. As an alternative option for laying the reservoir, a variant of its arrangement along the bottom of the Irkutsk reservoir arose. The field of hydrological research, analysis of the formation of the banks and the movement of the pumps identified five possible routes for the passage of the collector with a deepening in the bottom of the reservoir. Based on these options, a redundant water supply scheme has been compiled, presented in Figure 4.

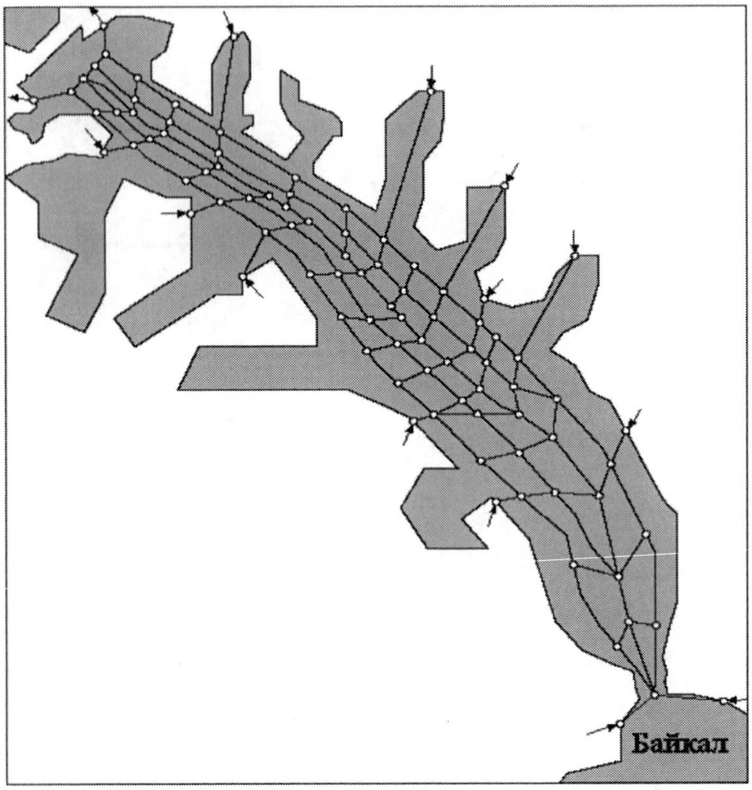

Figure 4. Excessive scheme for optimizing the pipeline along the bottom of the Irkutsk reservoir

The length of the reservoir along the fairway is 57 km. to deepen the pipeline into the bottom of the reservoir to a depth of 1.5 m. For some sections, it is necessary to provide measures to prevent leaching of soils, the formation of pumps, measures to protect against sludge, etc. As a result, the specific costs of laying in different sections are different.

Taking into account the constructed redundant scheme and cost indicators for each possible section of the collector route, calculations were carried out according to the method of complex optimization. As a result, the optimal route option passed almost along the fairway of the Irkutsk reservoir and is shown in Figure 5. Polyethylene pipes with a total length of 62 km and a pipeline diameter of 1.2 m. The cost was 4.2 billion rubles. The residual pressure in the pipeline near the platinum of the hydroelectric station was 0.28 MPa.

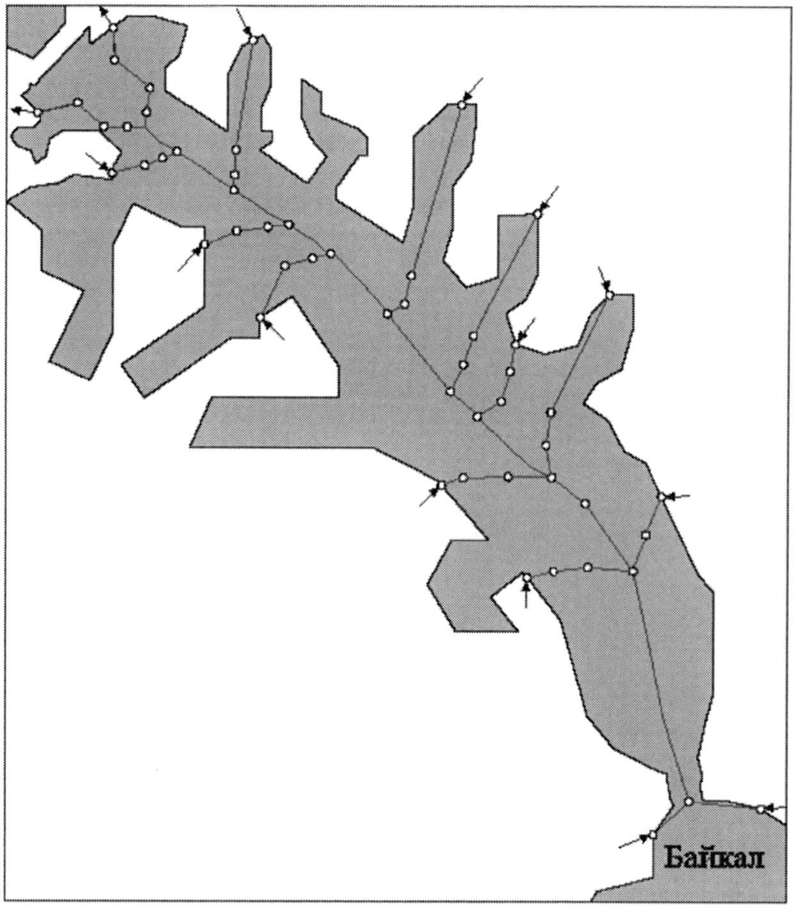

Figure 5. The optimal route of the water supply pipeline from about. Baikal to the city of Irkutsk but to the bottom of the Irkutsk reservoir

4. Conclusions

The developed methodology for the comprehensive optimization of the route and parameters of the pipeline intended for transporting liquids and gases was implemented in the Trace – VK software package [13], which is an effective tool for substantiating the structure and parameters of designed, reconstructed, and developing pipeline systems for various technological purposes.

References

[1] Ford L R 1963 Streams in networks p 216
[2] Wenzel E S 2000 Problems and exercises in probability theory Textbook for universities 3rd ed Higher school p 366
[3] Merenkov A P 1985 Theory of hydraulic circuits p 278
[4] Chupin R V 2017 Pressure-flow and Free-flow Discharge Modes in Closed-Loop Sewage Systems IOP Conf. Series: Materials Science and Engineering **262** 012079 doi: 10.1088 / 1757-899X / 262/1/012079
[5] Chupin R V 2017 Trajectory of Sewerage System Development IOP Conf. Series: Materials Science and Engineering **262** 012080 doi: 10.1088 / 1757-899X / 262/1/012080.
[6] Chupin R V 2019 Optimization of development schemes for group water supply systems under uncertainty of prospective water consumption E3S Web of Conferences: *Mathematical*

IOP Conf. Series: Materials Science and Engineering **880** (2020) 012039 doi:10.1088/1757-899X/880/1/012039

Models and Methods of the Analysis and Optimal Synthesis of the Developing Pipeline and Hydraulic Systems **102**

[7] Chupin R V 2019 Optimal reconstruction of water supply network under conditions of reduced water consumption E3S Web of Conferences: *Mathematical Models and Methods of the Analysis and Optimal Synthesis of the Developing Pipeline and Hydraulic Systems* **102**

[8] Chupin V R 2019 Optimization of developing district water supply systems taking into account variability of perspective water consumption IOP Conf. Series: Materials Science and Engineering **667** 012018 doi: 10.1088 / 1757-899X / 667/1/012018

[9] Chupin V R 2019 Optimization of the sewerage systems scheme of cities and populated IOP Conf. Series: Materials Science and Engineering **667** 012017 doi: 10.1088 / 1757-899X / 667/1/012017

[10] Surin A A 1927 The choice of water supply scheme p 126

[11] Chupin R V 2011 Development of theory and practice of modeling and optimization of water supply and sanitation systems Monograph Publishing House of ISTU p 323

[12] Chupin R V 2015 Optimization of developing drainage systems Monograph Publishing house of ISTU p 418

[13] Certificate of state registration of the computer program No.2016615463 TRACE-VK dated May 25 2016

IOP Conf. Series: Materials Science and Engineering **880** (2020) 012040 doi:10.1088/1757-899X/880/1/012040

Comprehensive optimization of water supply and sanitation systems

R V Chupin[1], I V Maizel[2] and I G Maizel[3]

[1] LLC Vortex, 664039, Irkutsk, Lermontov str., 297 b
[2] Federal State Budgetary Educational Institution of higher Professional Education INRTU, IACD, chair of urban construction and farm, 83 Lermontov st., Irkutsk, Russia, 664074
[3] Federal State Budgetary Educational Institution of higher Professional Education INRTU, IAET, chair technologies and equipment for machine-building industries, 83 Lermontov st., Irkutsk, Russia, 664074

E-mail: ch-r-v@mail.ru

Annotation. Because of the uneven distribution of water resources across the territory of our country, many localities and even cities do not have their own sources of water supply and reservoirs where it would be possible to discharge treated wastewater. To solve this problem, group and district water supply and sanitation systems are being designed, built and developed. The length of such systems is hundreds or even thousands of kilometers, and they include many pumping stations, regulating tanks, storage ponds, local and centralized treatment facilities. Significant financial investments are required annually for the construction and operation of such structures [1-6]. Therefore, the choice of routes, structure of structures, ways of transporting water and waste water, justification of the location of water intakes, treatment facilities are relevant and require special attention and technical and economic analysis of design options. The paper proposes a comprehensive approach to optimizing the structure of structures and methods for supplying water to consumers and transporting waste water to treatment facilities on the basis of the redundant design methods.

1. Introduction

The following types of transport can be attributed to the methods of water supply and drainage: pipeline, road, rail, water and even air.

Road transport is the leading one in the Russian Federation. It accounts for 68% of the total cargo transported. The second place is taken by pipeline transport, slightly ahead of railway transport.

In housing and communal services (HCS), pipeline transport is dominant, with the help of which heat, water, and gas are supplied to residents, and sewage and storm water are diverted. The length of this type of transport in HCS is: water pipelines- 523 ths. km; sewerage-163 ths. km; heat networks-183 ths. km.

Road transport in housing and communal services is used to provide water and remove waste water from individual consumers in the private sector who are not connected to centralized water supply, sanitation, heat and gas supply systems.

IOP Conf. Series: Materials Science and Engineering **880** (2020) 012040 doi:10.1088/1757-899X/880/1/012040

Railway transport is used to deliver bottled and mineral water. Water transport is used very rarely, mainly in marine areas, as well as for the collection and delivery to treatment facilities of shale water, household waste water from protected and protected areas.

When justifying promising schemes for the development of water supply and sanitation systems in populated and sparsely populated areas, territories with a special status of ecological zones and national parks, the question arises which type of transport to use in order to minimize one-time and operational costs, and not to cause environmental damage to nature.

For localities that do not have their own drinking water supply sources and are located far from each other, water delivery options may be as follows:

1) group water supply from a surface or underground source of drinking water to the reservoirs of localities;

2) delivery of water by road from a surface or underground source of drinking water to each subscriber of localities;

3) delivery of water by road from a surface or underground source of drinking water to each subscriber of localities;

4) if the water sources do not meet the GOST "Drinking water", then it is necessary to install water treatment facilities. In this case, there may be the following options:

- treatment facilities are set up at each source of water, which is then delivered to localities by pipeline or road transport;
- technical water is delivered by road or pipeline transport to localities where it is purified to the required standards for drinking water supply purposes. At the same time part of the process water without treatment can be used for agricultural and industrial needs;

5) if there are underground sources of drinking or technical water in localities, alternative options may be their purification and delivery by road or pipeline transport to each subscriber within the locality, as well as to other localities.

Pic. 1a shows nine localities that require water to be supplied from an open water source-a river. Pic. 1b shows possible options for a centralized water supply system by pipeline (solid lines) and road transport (dashed lines) from a surface water source. Pic. 1c shows the same thing, but with a possible device for an underground source of drinking water located at some distance from populated areas (a rectangle with dots). In pic. 1d the same, but for the case of the need for water treatment facilities.

Pic. 1e shows a redundant diagram in the form of a graph with a single vertex, from which fictitious branches go to possible water sources. These fictitious branches model the economic performance of possible water sources.

In pic. 1f is the same, but with the possible arrangement of water intakes and treatment facilities in each locality. This option includes all possible options 1-6 and their combinations.

Of all the proposed alternatives, it is necessary to find the optimal solution for cost, technical and environmental factors. Pic. 1g shows a possible option, according to which water will be delivered by pipeline to settlements 1, 2, 4. Further from the village settlements of 2 to 3 and 5 of the road transport. In village 8, the underground source of technical water, after its purification, will be used in the village itself, and delivered by road to villages 7 and 9. In the village 6, water from an underground source of technical water will be used only in the village itself after its purification.

When transporting water by road, special requirements are imposed on the containers in which liquids are transported:

- the water must not be heated or cooled to standard values;
- water should not lose its original properties and quality.

2. Methods

To find the optimal variant, we propose an approach based on purposeful search of redundant scheme tree variants using the contour minimization method (coordinate with respect to contour variables).

The essence of this approach is as follows: a redundant scheme is formed, which is obtained by superimposing pre-planned options and can represent a graph of possible routes and locations of various structures.

For example, the graph of the diagram shown in pic. 1b. If there are several possible sources of water, all of them are closed to a common node by fictitious branches that model the cost functions of possible sources (pic. 1e).

We believe that the solution to the problem will be some branched scheme in the form of a tree of a redundant graph. At the same time, the reliability of water supply will be provided by parallel laying of two water pipelines. Taking into account the assumptions made, it is necessary to delete some edges in the redundant graph so that the graph becomes a connected tree of the minimum values of the cost function.

To find such a solution, the following computational process is organized. On the redundant graph, the initial approximation tree is selected, which is taken as the tree of shortest distances (minimum estimated costs). Then, to generate new trees, the tree branches are replaced with chords (edges that are not part of the tree).

To illustrate the computational process, pic. 2.1 shows a redundant diagram in the form of a graph that had 9 vertices, 12 edges, and four fundamental independent contours. For example, a tree is selected as the initial approximation (see pic. 2.2). From an algorithmic point of view it is easy to write this tree as a single array:

IOP Conf. Series: Materials Science and Engineering **880** (2020) 012040 doi:10.1088/1757-899X/880/1/012040

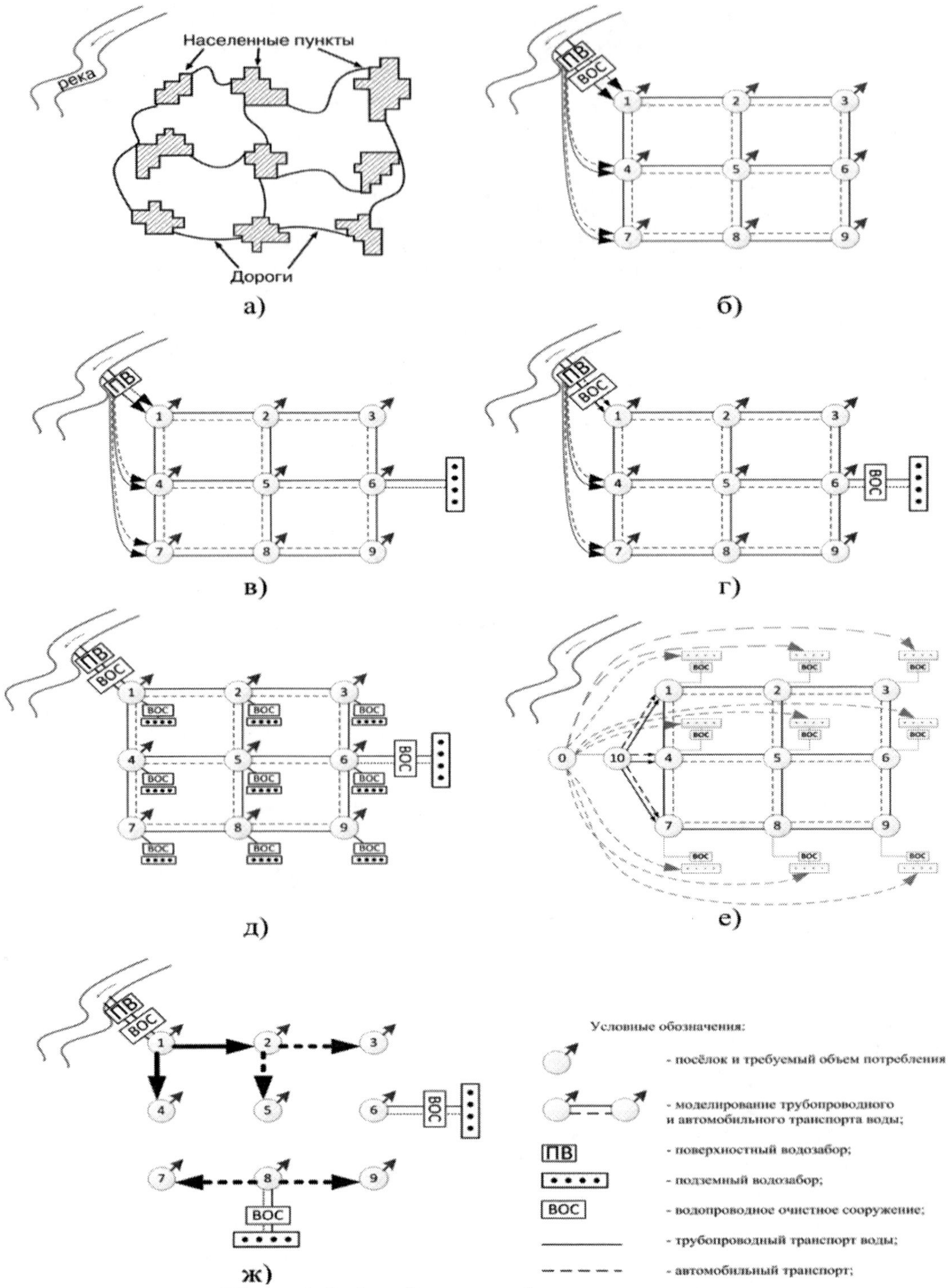

Figure 1. Options for transporting water to consumers

IWAY

0	1	2	1	4	3	4	7	6
1	2	3	4	5	6	7	8	9

Where the second line is the number of the array element, or the number of the final vertex of the arc.

The elements of the array represent the outgoing vertices of the tree, that is, we get to vertex 2 from 1, vertex 3 from 2, 4 from 1, 5 from 2, and so on. As already noted, the choice of the tree uniquely determines the fundamental cycles.

Chorda $x_2 - x_5$ forms a cycle $x_2 - x_5$; $x_1 - x_2$; $x_4 - x_5$; $x_1 - x_4$. *I*

Chorda $x_5 - x_6$ forms a cycle $x_5 - x_6$; $x_4 - x_5$; $x_1 - x_4$; $x_3 - x_6$; $x_2 - x_3$; $x_1 - x_2$ *II*

Chorda $x_5 - x_8$ forms a cycle $x_5 - x_8$; $x_4 - x_5$; $x_7 - x_8$; $x_4 - x_7$ *III*

Chorda $x_8 - x_9$ forms a cycle $x_8 - x_9$; $x_7 - x_8$; $x_4 - x_7$; $x_1 - x_4$; $x_6 - x_9$; $x_3 - x_6$; $x_2 - x_3$; $x_1 - x_2$ *IV*

In this case, the direction of the contour traversal will coincide with the direction of the chord.

For subsequent generation of new trees, the first contour is selected. The contour will consist of two chains: $x_2 - x_5$; $x_1 - x_2$ and $x_4 - x_5$; $x_1 - x_4$, i.e. the contour will always have the vertex of "exit" 1 and the vertex of "entry" 5.

Therefore, if the chord $x_2 - x_5$ becomes a branch of a tree, that is, a branch $x_4 - x_5$ become a chord (Pic. 2,3). In this case, to form a tree, it is enough to replace 4 with 2 in the *IWAY* array at position 5 (the original vertex of the chord):

0	1	2	1	**2**	3	4	7	6

Then the $x_1 - x_4$ branch is tried as a chord and the array is corrected again *IWAY*:

0	1	2	**5**	2	3	4	7	6

After viewing all branches of the first contour, the second contour is selected, and all operations for generating new trees are repeated (pic. 2.4, pic. 2.5) Then the third contour is selected and trees are generated (pic. 2.6, pic. 2.7). Then the fourth circuit (see pic.2.8-2.11). Then the first contour is formed again, and all operations are repeated until the trees that have already been built appear. In this case, the best option is remembered. If the best solution is not found after viewing all the contours, the computing process ends.

The drainage options for the localities shown in pic. 1a are as follows:

1) A centralized system of waste water diversion from all settlements to centralized IOWs is organized, followed by discharge into a flowing reservoir. Wastewater at IOW can be delivered by pipeline and road transport.
2) In each locality, a centralized system of wastewater disposal and treatment is organized. Drains on the IOW of each village can be delivered from subscribers by pipeline and road transport.
3) The Treated effluents in each settlement are discharged into a flowing reservoir (if there is one), or into specially equipped storage ponds, filtration and evaporation, or delivered to a common storage pond for all settlements or a specially organized discharge into a flowing reservoir (river).
4) There may also be combinations of the following options for wastewater disposal and treatment.

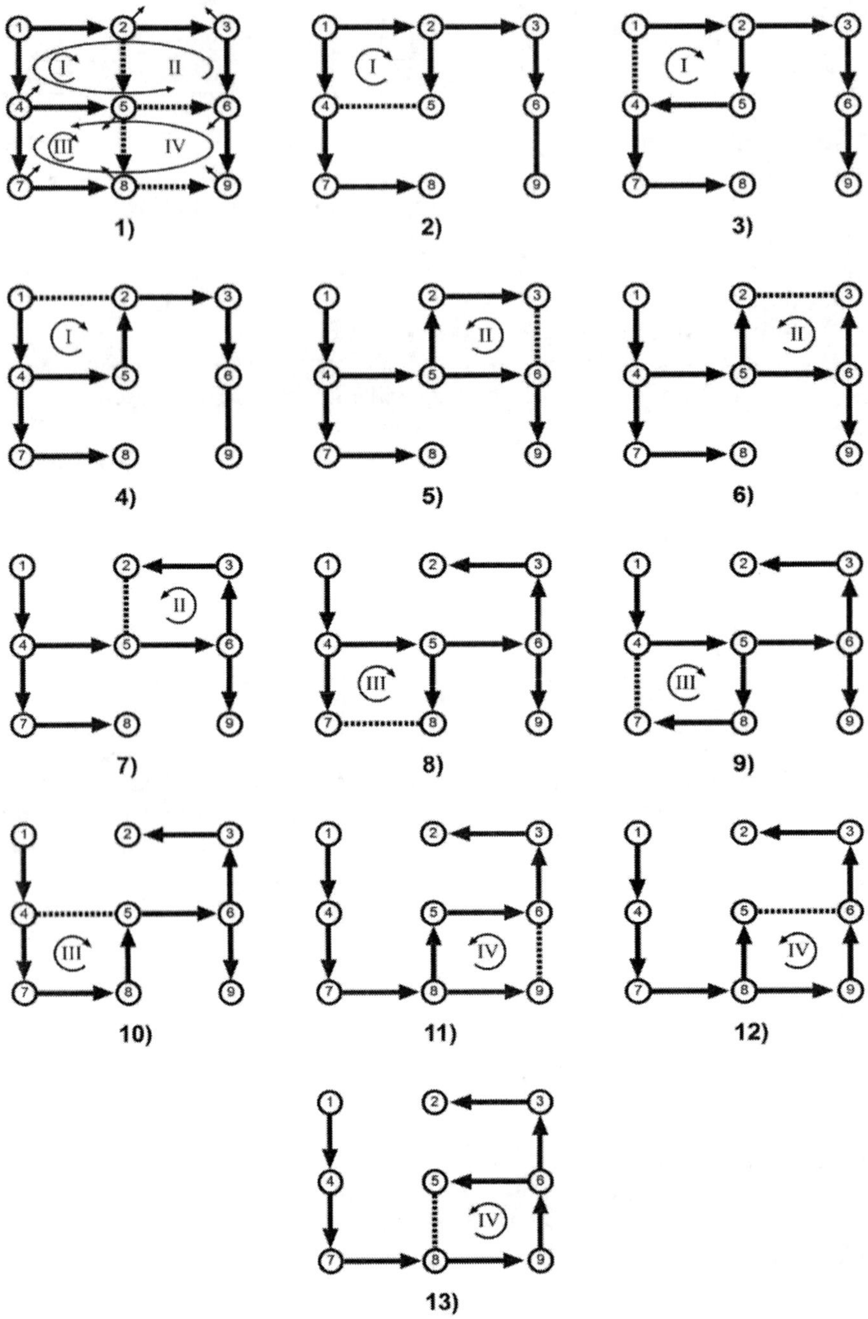

Figure 2. Illustration of a method for purposefully iterating over variants of redundant schema trees

3. Results and discussions

Pic. 3a shows options for a centralized water disposal system (in the form of a redundant scheme). Wastewater is collected from each locality and transported to IOW by road or pipeline. After cleaning, they are dumped into the reservoir.

Pic. 3b shows options for the possible installation of IOW in each locality with the delivery of wastewater from subscribers to these treatment facilities by pipeline and road transport. In this case, the effluents are discharged into evaporation and filtration ponds after treatment.

Pic. 3c shows the final redundant scheme, which has a common flow discharge node from each possible node of the wastewater treatment device.

Pic. 3d shows the best possible option for transporting and treating waste water. According to this option, within localities, the effluents from the subscribers cesspits will be delivered by road to receiving tanks, from which they will be diverted by pipeline to centralized IOWs.

Considering the options for centralized and decentralized water supply and sanitation systems, it is not difficult to understand that they are interconnected and decisions on water disposal systems depend on the accepted water supply options. There are common corridors for laying pipelines, common roads for transporting water and wastewater, and other mutually beneficial situations. Therefore, there is a problem of complex optimization of schemes and systems of water supply and sanitation. To solve this problem, it is proposed to combine two redundant schemes pic. 1e and pic. 3c. As a result, a redundant scheme is formed, as shown in pic. 4a.

It is not difficult to notice that the resulting redundant scheme is nothing more than a transport network [11-13], consisting of a node S of the input of flows and a node t of the output of flows. Branches in the transport network simulate the actual process of flow movement, its transformation (consumption, discharge, cleaning, etc.), and the cost value of structures (the cost per unit of flow). Also, each arc and edge has bandwidth restrictions.

For example, the arc S-1 (see pic. 4a) simulates a surface water intake from a river, a water treatment plant, and has a flow unit cost-the cost of building these structures, assigned to 1 m^3 of water taken from the river and treated. It has a capacity limit equal to the maximum possible volume of water taken from the river. Arc S-3 simulates the device of a local underground water source in village 3 with the device of treatment facilities, has the cost per unit of flow, i.e. the cost of all structures is calculated for 1 m^3 of water, the flow limit is the maximum possible intake of water from an underground source. Arcs 1-11, 2-12, ... - simulates the use of water by the subscriber and its transformation into waste water. Arc 20-t -simulates wastewater treatment on centralized IOW and discharge of treated wastewater into a reservoir. Arcs 11-t, 12-t-model wastewater treatment in populated areas and discharge it into storage ponds, have specific costs and flow restrictions [10-15].

The most effective method of finding the optimal solution based on the transport scheme is the method described in the works [7-10]. In the work on the basis of the developed software package Track-VK [16] optimization calculations are performed and the results shown in pic. 4b are obtained. it is clear from the calculations that the solutions obtained differ and their total costs are lower than the sum of the costs for the solutions in pic.1zh and pic. 3g.

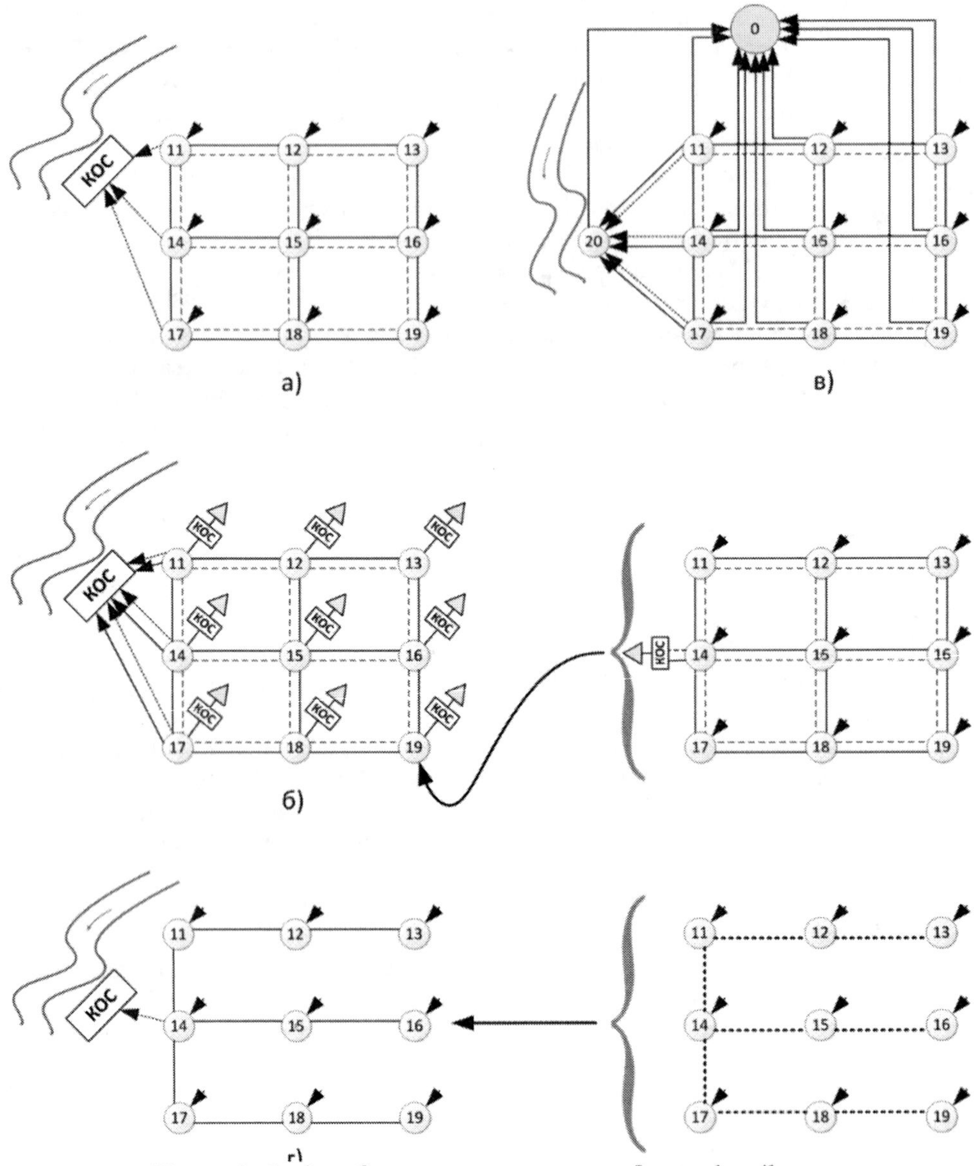

Figure 3. Options for wastewater transport from subscribers

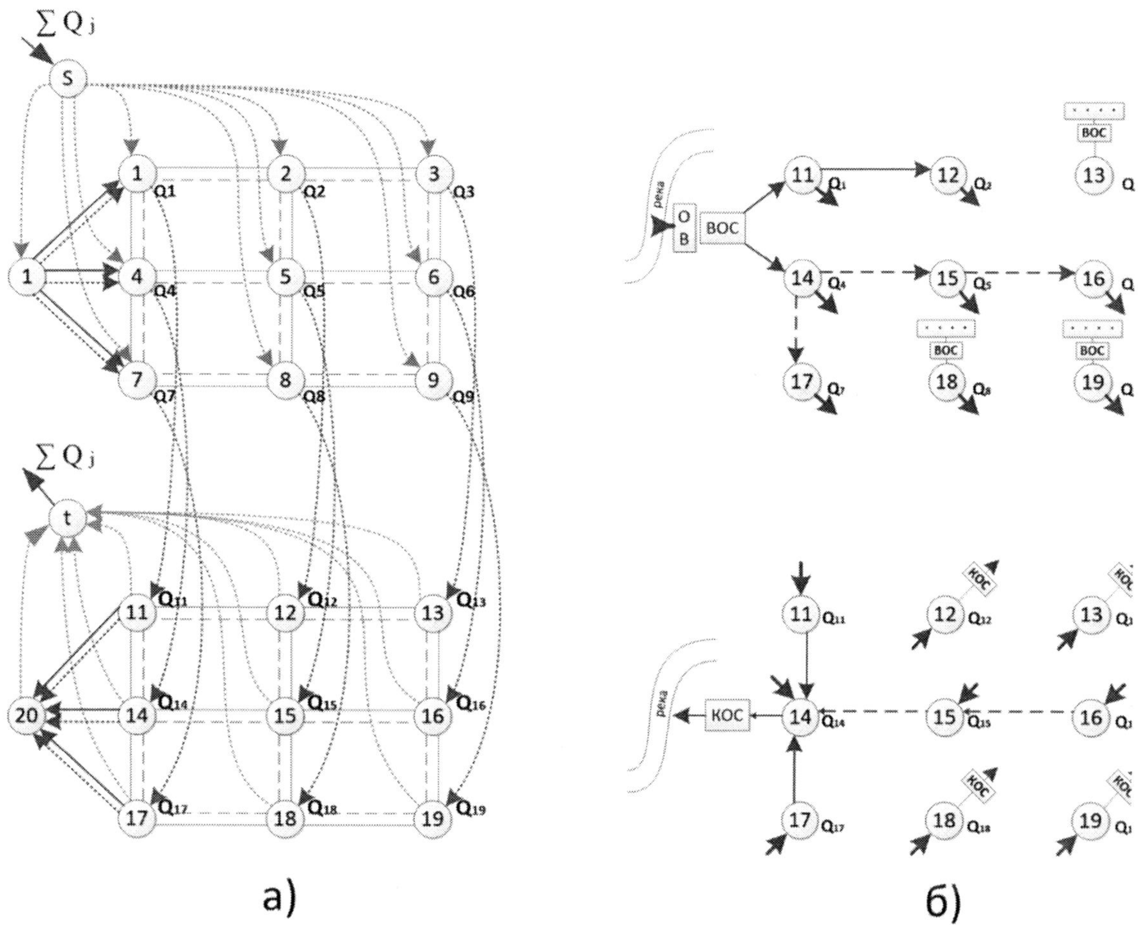

Figure 4. The diagram of the district water supply and sanitation system is shown

Conclusions

The proposed approach to the complex optimization of the structure of structures and methods for supplying water to consumers and transporting waste water to treatment facilities is based on the method of redundant design schemes. This approach and its implementation in the form of a software package Track-VK allow you to quickly justify the newly designed, reconstructed and developing water supply and sanitation systems in urban areas. On its basis, it is not difficult to form regional programs for the integrated development of engineering infrastructure in cities and localities.

IOP Conf. Series: Materials Science and Engineering **880** (2020) 012040 doi:10.1088/1757-899X/880/1/012040

References

[1] Cemilev A A 1971 Group systems of rural water supply p 192
[2] Pupyrev E I Primin O G 2013 Methodological support for the development of schemes for centralized water supply and sanitation systems. *Water treatment. Water preparation. Water supply* **5(65)** p 4-8
[3] Primin O G, Bortkevich V S, Mirkis V I, Kantor L I and Vinarsky S L 2014 About the developmentof water supply schemes for Russian cities. Water supply and sanitary equipment **7** p 26-32
[4] Karambirov N A 1978 Agricultural water supply p 438
[5] Gogina E S Gurinovich A D 2016 Application of the LCC methodology for evaluating the effectiveness of wastewater treatment plant projects *Water supply and sanitary engineering* **9** p 36-41
[6] Bazhenov V I Pupyrev E I Sambursky G A and Berezin S E 2018 Development of a method for calculating the life cycle cost of equipment for systems and structures for water supply and sanitation **2** p 10-19
[7] Chupin R V Pham N M 2019 Optimization of the structure and parameters of developing systems of group water supply Water supply and sanitary equipment **1** p 30 – 36
[8] Chupin R V 2018 Optimization of promising schemes for the development of Sewerage systems in conditions of limited financing Water supply and sanitary technique **2** p 67-74
[9] Chupin R V Melekhov E S 2011 Development of the theory and practice of modeling and optimization of water supply and drainage systems: Monograph Irkutsk INRTU publishing House p 323
[10] Chupin R V 2015 Optimization of developing water disposal systems Irkutsk Irkutsk state technical University Press p 418
[11] Taha H 1985 Introduction to operations research p 476
[12] Ford L R and Fulkerson D R 1963Flows in networks p 216
[13] T Hu 1974 Integer programming and threads in networks p 520
[14] Chupin V R 2019 Optimization of developing district water supply systems taking into account variability of perspective water *consumption IOP Conf. Series: Materials Science and Engineering*. **667** 012018 doi:10.1088/1757-899X/667/1/012018
[15] Chupin V R 2019 Optimization of the sewerage systems scheme of cities and populated areas IOP Conf. Series: Materials Science and Engineering. **667** 012017 doi:10.1088/1757-899X/667/1/012017
[16] Melekhov E S Chupin V R Chupin R V 2016 Certificate of state registration of the computer program № 2016615463 TRACE-VR dated May 25

Schematic and structural optimization of group water supply systems

V R Chupin[1] **and A S Dushin** [2]

[1]Irkutsk National Research Technical University, 83 Lermontov street, Irkutsk, 664074, Russia.
[2]Community with limited liability "Network company "IRKUT", 664020, Irkutsk, aircraft Manufacturers str., 28 A

E-mail: chupinvr@ex.istu.edu

Abstract. Group water supply systems (GWSS) are designed for centralized water supply of settlements located in waterless areas and remote from drinking water sources. GWSS are an object of increased attention, have a great social and environmental significance, require significant capital investments and financial resources for operation, ensuring the required reliability and manageability. Therefore, the issues of optimization of design solutions for the structure and parameters of structures are relevant and require improvement and development of methodological and software solutions to these problems. The paper proposes a method of redundant design schemes and a method of schematic and structural optimization based on a sequential and iterative solution of two problems: for a fixed value of flows in the graph of the redundant scheme, network parameters are optimized (pipeline diameters, heads, and the composition of pumping stations); with fixed network parameters, the problem of the distribution of water flows is solved in the GWSS.

1. Introduction

Group water supply systems (GWSS) are designed for centralized water supply of settlements located in waterless areas and remote from drinking water sources. GWSS are an object of increased attention, have a great social and environmental significance, require significant capital investments and financial resources for operation, ensuring the required reliability and manageability. Therefore, the issues of optimization of design solutions for the structure and parameters of structures are relevant and require improvement and development of methodological and software solutions to these problems. The analysis of existing GWSS indicates that many systems today do not meet the requirements of efficiency, reliability and manageability. The main reasons are insufficiently justified decisions at the design level on the choice of structure of structures, pipeline routes, and their parameters. Increasing reliability by backing up and ringing networks requires additional investment. It is proved that taking into account reliability in the form of parallel laying of pipelines, installation of emergency water storage tanks on the network and in localities leads to an optimal solution of the network structure in the form of a tree for new networks, and in the presence of existing pipelines-in the form of a dominant graph with ringed sections of the network [1-16]. The creation of an extensive network with the strengthening of the graph tree trunk makes it possible to increase the speed of water movement in pipes, reduce the formation of stagnant zones, preserve water quality, and reduce the immediate threat of bacterial and chemical contamination. In addition to improving water quality, capital costs for the construction of reserve sites are reduced, the probability of accidents is reduced, and so on.

At the same time, we need to develop a method that allows us to optimize the structure of this tree and the ring sections of the network.

2. Methods

Many approaches and methods have been developed to find the optimal network structure in the form of a tree [12-18]. However, all of them are reduced to a purposeful and limited search of variants of trees of the redundant design scheme and do not view the ring graphs of the network. Traditionally, in project practice, this task is solved as follows [17-25]:

- several variants of routes are planned;
- for each option, the parameters of the pipeline, pumping stations and control structures are determined;
- each option is evaluated for reliability, manageability and economic costs of its implementation. The best one is selected. Almost no optimization of design solutions is performed.

At the same time, the task of optimizing the structure and parameters of the GWSS is to find such a pipeline route, places where water intakes, reservoirs, and pumping stations are installed that would correspond to the minimum of the total reduced costs.

From a mathematical point of view, this problem has a multi-extreme character, is difficult to formalize and implement [13]. In most cases, the available approaches and methods for solving this problem are reduced to a step-by-step and iterative procedure for optimizing the route and parameters of transport structures.

This paper proposes to solve this problem in a single iterative process based on the methodology of coordinate minimization. One coordinate is the vector of distribution of expenses in the network, the other is the vector of losses of pressure in pipelines and pump stations, which determines the composition of transportation structures. This is done algorithmically as follows:

- assigned all sorts of connections between the possible sources of consumers (in the General case this can be a complete graph).
- on this graph, the initial flow distribution is assigned (for example, using the proportional flow division method);
- for fixed flow distribution, the optimal diameters and parameters of transport structures are determined and the cost of the option is determined.
- with known diameters and parameters of water pumping stations, the problem of flow distribution in the network is solved (for example, by the method of nodal pressures [1]);
- next, the diameters and parameters of the pumping stations are selected again for the newly obtained water flow values;
- such procedures are performed for new networks until the water flow rate and, accordingly, the pipeline diameters are close to zero in some parts of the network. The number of such sections will be: n-m+1, where n is the number of sections of the redundant scheme, m is the number of nodes.
- as a result, we get a variant of the network in the form of trees. This tree (or network route) will be the solution to the problem. If there are already built sections and structures in the network, then the computing process will end on some solution that will correspond to the graph with the ring sections of the network.

For simplicity and clarity of the proposed approach, we accept the speed of water movement in pipelines equal to 1 m / s. Although this parameter also requires optimization and depends on the cost of electricity in the design area (for example, for the Irkutsk region, the optimal value of water speed in pipelines is 2.7 m / s). With this in mind, the diameter (d) of the pipelines is determined by the following formula:

$$d = (1/30)\sqrt{Q} \tag{1}$$

where Q is the water flow rate on the pipeline section of the network.

The costs shown $З_i$, taking into account (1) calculated using the following formula:

$$З_i = \sum_{i=1}^{n} K_{уд}(D_i) * L_i * E + \frac{\gamma * 365 * 24 * m}{102 * \eta} * Q_i * H_i$$

(2)

where γ - specific weight (for water =1), number of hours per year 365*24, Hi-head of pumping stations, mV.

Q_j - consumption in the node l / s, m-unit electricity costs = 1.067 rubles/KWh (for Irkutsk), η - pump efficiency 0,7

E= the coefficient of comparative efficiency of capital investments is the inverse of the payback period, which is recommended to be determined on the basis of forecast tariffs for utilities; $K_{уд}(D_i)$ - specific capital investments, RUB/m, L_i - the length of channel, m.

On the basis of (2) level of each iteration, the optimal scheme structure and its parameters are evaluated

The computational process ends if there is no decrease in the value of the reduced costs during the transition to subsequent iterations.

Thus, the option in which the value of the reduced costs will have the smallest size and will be optimal.

3. Results and discussions

Let's illustrate this approach using the example of a redundant network shown in figure 1, containing 12 nodes and 17 network sections. The source information for network of nodes is shown in table 1, and the source information for network of sections is shown in table 2

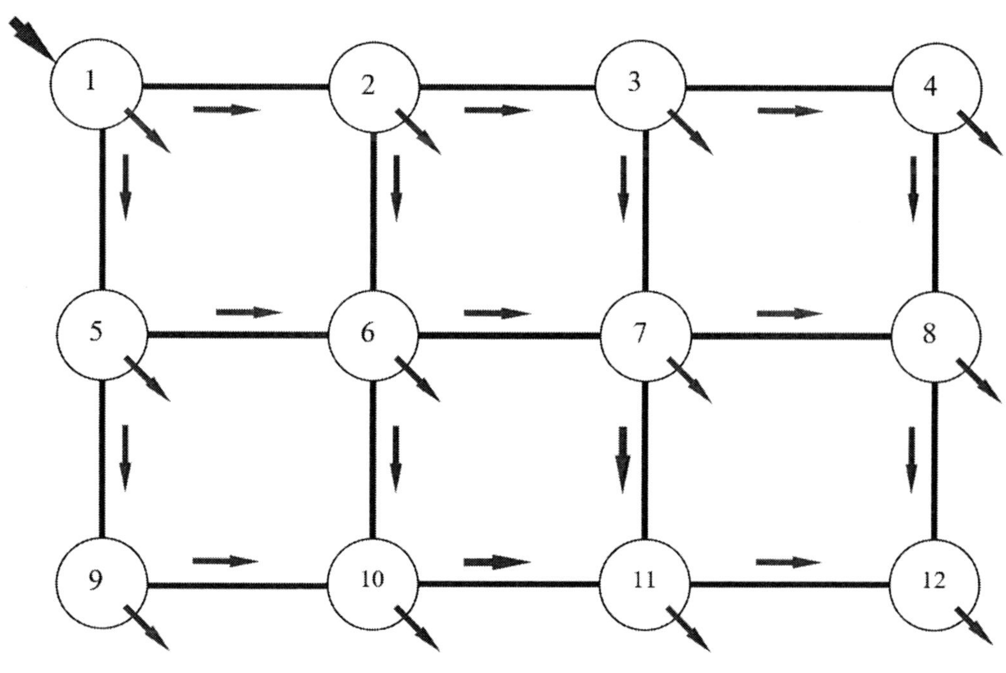

Figure 1. Redundant (ring) scheme of the group water supply system

Table 1. Source information for nodes

node number	mark in node, m	selection in node Q, l / s	node number	mark in node, m	selection in node Q, l / s
1	400	30	7	400	30

2	400	30	8	400	30
3	400	30	9	400	30
4	400	30	10	400	30
5	400	30	11	400	30
6	400	30	12	400	30
in total		**180**			**180**

Table 2. Source information for sites

node number		the length of the DL, m	tube material	node number		the length of the DL, m	tube material
the beginning of the segment	the end of the segment			the beginning of the segment	the end of the segment		
1	2	1000	new	7	8	1000	new
2	3	1000	new	5	9	1000	new
3	4	1000	new	6	10	1000	new
1	5	1000	new	7	11	1000	new
2	6	1000	new	8	12	1000	new
3	7	1000	new	9	10	1000	new
4	8	1000	new	10	11	1000	new
5	6	1000	new	11	12	1000	new
6	7	1000	new				

The results of iterative calculations to determine the optimal parameters of the group water supply system are summarized in tables 3.1, 3.2, and 3.3.

Table 3.1. The result of the iterative calculation of diameterм

node number		the number of the iteration															
the beginning of the segment	the end of the segment	0	1	2	3	4	5	6	7	8	9	10	11	12	13	14	15
1	2	500	472	472	477	484	491	499	507	515	523	531	539	545	547	548	548
2	3	400	370	370	378	386	395	404	411	415	416	414	413	411	409	409	408
3	4	250	260	260	268	278	289	300	308	314	317	316	316	316	316	316	316
1	5	400	379	379	373	364	355	344	332	319	306	291	276	265	259	258	258
2	6	250	229	229	228	227	227	228	233	243	259	277	295	307	314	316	316
3	7	200	191	191	193	196	198	200	201	201	199	196	192	188	185	183	183
4	8	200	185	185	197	210	224	237	248	256	259	258	258	258	258	258	258
5	6	250	219	219	210	199	186	171	155	137	116	91	63	35	14	3	**0**
6	7	200	176	176	159	138	112	83	53	28	12	3	0	0	0	0	**0**
7	8	150	128	128	114	95	73	49	27	12	3	0	0	0	0	0	**0**

5	9	250	250	250	248	245	241	236	230	223	216	207	197	188	183	183	183
6	10	200	191	191	194	197	200	203	205	209	217	228	240	250	256	258	258
7	11	150	132	132	128	122	115	106	96	87	79	71	60	46	30	16	**0**
8	12	150	131	131	135	141	149	159	171	180	183	183	183	183	183	183	183
9	10	200	170	170	167	163	157	149	139	128	115	98	75	46	16	5	**0**
10	11	200	179	179	179	179	177	173	168	164	164	168	172	177	180	182	183
11	12	150	127	127	123	116	105	89	65	32	14	6	0	0	0	0	**0**

Table 3.2. Result of iterative calculations on expenses, l / s

node number		the number of the iteration															
the beginning of the segment	the end of the segment	0	1	2	3	4	5	6	7	8	9	10	11	12	13	14	15
1	2	197	201	205	210	217	224	231	238	246	254	261	267	270	270	270	270
2	3	120	123	128	134	141	147	152	155	156	155	153	152	151	150	150	150
3	4	53	61	65	70	75	81	86	89	90	90	90	90	90	90	90	90
1	5	133	129	125	120	113	106	99	92	84	76	69	63	60	60	60	60
2	6	47	47	47	46	46	47	49	53	60	69	78	85	89	90	90	90
3	7	38	33	34	34	35	36	37	36	36	34	33	32	31	30	30	30
4	8	23	31	35	40	45	51	56	59	60	60	60	60	60	60	60	60
5	6	47	43	40	36	31	26	22	17	12	7	4	1	0	0	0	**0**
6	7	38	28	23	17	11	6	3	1	0	0	0	0	0	0	0	**0**
7	8	23	15	12	8	5	2	1	0	0	0	0	0	0	0	0	**0**
5	9	56	56	55	54	52	50	47	45	42	39	35	32	30	30	30	30
6	10	26	33	34	35	36	37	38	39	42	47	52	56	59	60	60	60
7	11	23	16	15	13	12	10	8	7	6	5	3	2	1	0	0	**0**
8	12	15	15	16	18	20	23	26	29	30	30	30	30	30	30	30	30
9	10	26	26	25	24	22	20	17	15	12	9	5	2	0	0	0	**0**
10	11	23	29	29	29	28	27	25	24	24	25	27	28	29	30	30	30
11	12	15	15	14	12	10	7	4	1	0	0	0	0	0	0	0	**0**

Table 3.3. Result of iterative calculations based on reduced costs

the number of the iteration	0	1	2	3	4	5	6	7	8	9	10	11	12	13	14	15
Amount of reduced costs, RUB.	58 115,08	57 817,68	55 690,35	55 554,14	55 290,46	54 697,32	53 588,77	52 042,02	50 361,29	49 261,08	48 377,92	47 007,39	45 646,41	44 346,30	43 611,20	43 257,75

The diagram of the group water supply system corresponding to the most optimal structure is shown in figure 2.

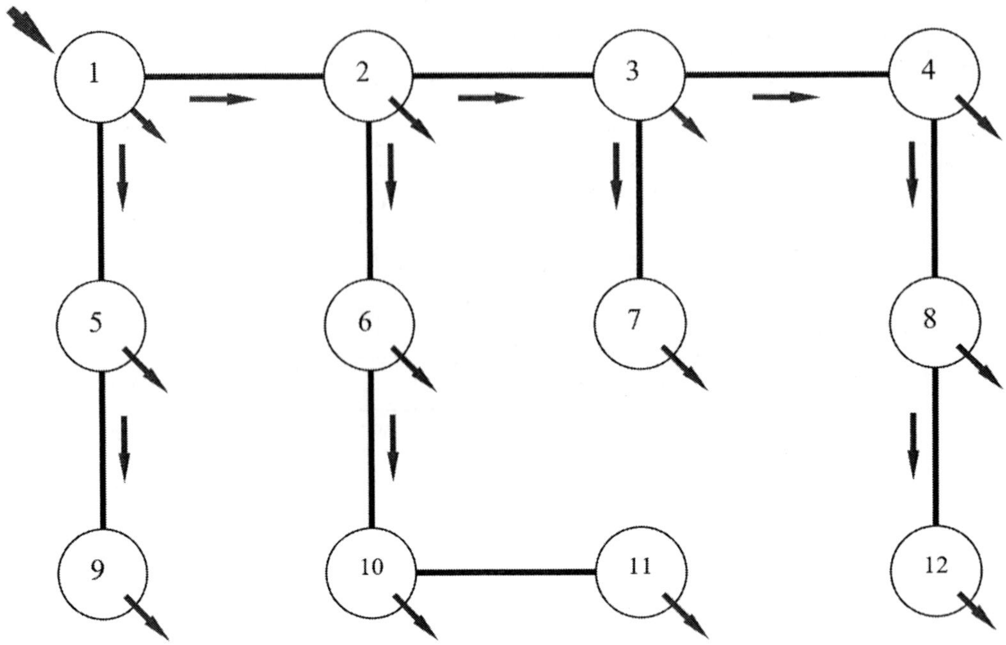

Figure 2. Optimal (branched) scheme of the group water supply system

Figure 2, as a result of the above example, shows that the redundant (ring) circuit has become a branched network as a result of optimization. Part of branches (5-6, 6-7, 7-8, 7-11, 9-10, 11-12) it will be rejected, and the remaining cost values will correspond to the optimal distribution of flows in the network.

Conclusions

A new method of optimization of GWSS based on the reduced costs of their construction and operation is proposed. This method is an effective tool for justifying the structure of the network and structures of water supply systems and can be useful for decision-makers to optimize the structure and development of a promising water supply scheme for urban area

References

[1] Merenkov A P Khasilev V Ya 1981 Theory of hydraulic circuits p 320
[2] Chupin R V Melekhov E S 2011 Development of the theory and practice of modeling and optimization of water supply and sanitation systems p 23
[3] Methodology for assessing the capacity of existing water supply networks and developing measures to intensify the Methodology for assessing the capacity of existing water supply networks and developing measures to intensify their work using computers 1972
[4] Mints D M 1951 On technical and economic calculation of water supply networks **1** pp 48 52
[5] Moshnin L F 1940 Choice of diameters of water supply lines **2/3** pp 48-55
[6] Moshnin L F 1995 Improving the efficiency of water supply and distribution systems *Water supply and sanitary equipment* **11** pp 23-26
[7] Nechaev A P Moshnin L F 1998 Optimization of water supply and distribution systems Water supply and sanitary equipment. **8** pp 12-14
[8] Sirotin V P 1951 Economic calculation of water pipes *Construction industry* **5** pp 8-12
[9] *SNiP 2.04.02-84. Water supply. Outdoor networks and structures* 1985 Gosstroy of the USSR (Moscow: Stroizdat) p 136
[10] *SNiP 2.05.06-85. Main pipelines* 1988 Gosstroy of the USSR (Moscow: stroizdat) p 56

[11] Moshnin L F 1984 Modern methods for calculating water supply and distribution systems *Water Supply and sanitary engineering* **10** pp 7-9
[12] Sumarokov S V, Khramov A V 1978 Questions of optimal synthesis of water supply systems taking into account reliability Questions *of reliability of water supply systems* pp 36-44
[13] Sumarokov S V 1927 *Mathematical modeling of water supply systems* p 167
[14] Surin A A 1927 *Choice of water supply scheme (*State science.- land reclamation Institute) p 126
[15] Chupin V R 1991 *Optimization of developing water supply and distribution systems* Abstract. the dessert. on the job. academic step. d-RA tekhn. Sciences (Moscow: VNIIVODGEO) p 41
[16] Chupin V R, Maizel D I and Chupin R V 2008 Modeling and optimization of utility pipeline systems *Vestnik Irgtu* **1** pp 169-180
[17] Shevelev F A 1973 Tables for hydraulic calculation of steel, cast iron, asbestos-cement, plastic and glass water pipes (Moscow: Stroizdat) p 114
[18] Hasilev V 1966 Questions of mathematical modeling in optimization of hydraulic systems using computer *Methods of mathematical modeling in power engineering* pp 343-348.
[19] Alekseev M I 1990 *Urban engineering networks (*Stroizdat)
[20] Abramov N N 1974 *Water Supply* (Moscow: Stroizdat) p 480
[21] Abramov N N 1973 On problems of reliability of water supply systems *Tr.MISI im.* p 138
[22] Dushkin S S, Kraev I O 1993 *Operation of water supply and sanitation networks* (ISIO) p 160
[23] Evdokimov A G, Dubrovsky V V and Tevyashev A D *1979 Flow Distribution in engineering networks* (Moscow: Stroizdat) p 200 PP
[24] Evdokimov A G and Tevyashev A D 1980 *Operational control of flow distribution in engineering networks* (Kharkiv: Vyshcha SHKOLA, Ed. at Kharkiv University) p 144
[25] Evdokimov A G, Tevyashev A D and Dubrovsky V V 1990 Modeling and optimization of flow distribution in engineering networks (Stroyizdat) p 368

IOP Conf. Series: Materials Science and Engineering **880** (2020) 012042 doi:10.1088/1757-899X/880/1/012042

Development of closed systems of oil- field water treatment as a basis of ecological safety of oil deposits settlements maintenance

O G Dubrovskaya [1], **D I Dubrovskiy**[1], **E V Zhmakov**[1] **and A I Matushenko**[1]

[1] Federal State Autonomous Educational Institution of Higher Education "Siberian Federal University" pr. Svobodny, 82, Krasnoyarsk, 660041, Russian Federation

E-mail: dubrovskayaolga@mail.ru

Abstract. Abstract. The methods used for oil-containing sewage treatment and the efficiency of an existing water treatment system are dealt with in the paper. The importance of work is stipulated by the necessity of improvement of technical sewage conditioning systems of oil complex enterprises. To optimize the operation of the waste water treatment closed system, a block of sorption structures is suggested to be included in the filtering stage. This method allows to achieve the required quality both for using this water as return oil – field water and technical one. The use of Unisorb-Bio sorbent is substantiated to implement the treatment process of heavily polluted oil-containing sewage water. In the course of an experiment the optimal temperature of maximum sorption treatment is revealed. Sorption filtration rational modes are selected and scientifically based. The data obtained can be used in the design of closed systems structures of oil-containing sewage treatment.

1. Introduction

The relevance of research aimed at the problem of development of a closed technological scheme of sewage treatment in oil and gas fields is stipulated by the necessity to reduce environmental loads and to realize resource-saving principles on the territories with complicated climatic conditions.

The basic approach in the construction of oil wells is a concept of "zero discharge". For the phased implementation of the principles of this concept, identification of waste liquids generated during construction and operation of oil fields is of key importance. The main technological waste of oil drilling includes drill dirt, used drilling fluids and sewage formed in the rinsing wells.

Mineralogical composition of the drill dirt is determined by the lithological composition of rocks and can significantly change with the deepening of the well. Chemical composition of the drill dirt depends both on its mineral composition and the properties of the rinsing fluid. In the process of drilling, in addition to this fluid, other technological fluids are used, e.g., buffer, perforating fluids. After use they are fully or partially transferred to the category of waste. Used drilling rinsing fluid (UDRF) is mostly formed.

The second category of sewage is the drilling sludging wastewater generated during the drilling site and equipment washing, the equipment cooling system.

The third category is the waste fluid for the well inflow and muffling, and the fluids (oil - field water, oil, gas) obtained during the test.

Environmental hazard of all sewage generated only at the stage of construction of oil and gas wells is primarily determined by the high toxic effects on organisms of any classification.

IOP Conf. Series: Materials Science and Engineering **880** (2020) 012042 doi:10.1088/1757-899X/880/1/012042

As for the concept of "zero discharge", development of waste treatment technology and its reuse as technical water for maintaining oil –field water pressure is the only possible principle of its implementation.

The required amount of return oil – field water is quite large. An estimated water consumption for the production of one ton of oil is on average 1.6 - 2.5 m3 with peripheral water flooding and 10 -15 m3 with the site flooding.

Water injected into the formation, requires special preparation, and since 40-50 % of it is sewage of different origin, it is necessary to look for the treatment equipment with high removal efficiency of such pollutants as oil products, suspended solids, iron, calcium, magnesium, potassium (carbonate, bicarbonates, sulfates), microorganisms.

Particles of algae, silt, iron compounds contained in the inlet water, hardness salts clog the pore channels of the formation, reducing the capacity of forcing wells. Microorganisms available in the injected water can form undesirable compounds. For example, sulphate restorable bacteria in the process of life produce hydrogen sulfide in an amount up to 100 mg/L. later this corrosive gas destructs pipelines, devices and equipment. In addition, these bacteria provoke the release of calcium carbonate and iron sulfide, which in turn form the mineral deposits in pipes, structures, and equipment.

According to Standard 39-225-88 "Water for waterflooding of oil formations. Quality requirements'', the water intended for injection is to contain not more than 2 mg/l of suspended solids with a particle size of 1-5 mcm, 0.3 mg/l of iron, 0.5 mg/l of oil with the size of the emulsified particles of not more than 5 microns, 0.05 mg/l of dissolved oxygen.

The existing technological schemes of sewage treatment, with the condition of reuse, provide for the application of the following processes:

1) Coagulation; 2) Decarbonization; 3) Iron removal; 4) Decontamination.

Coagulation generally refers to the preliminary stage of water treatment. This step is necessary to improve the efficiency of subsequent treatment steps. Without this phase the use of some treatment methods presents significant difficulties. For example, the presence of organic substances in water leads to changes in technological ammonites, contributes to their aging and sharp (4-8 times) reduction in service life [1 – 3].

Water subjected to pre-treatment, almost does not contain coarse impurities, but it still contains a significant amount of emulsified organic compounds, a variety of fine impurities, as well as various bio-organisms, some of which may have pathogenic viruses and bacteria. To remove these contaminants various treatment methods are used: oxidation, ion exchange, sodium-cation exchange, sodium-chlorine-ionization, hydrogen-sodium-cation exchange, ammonium-sodium-cation exchange, anation, baromembrane, electromagnetic, etc.

The most common methods of the organic compounds physical-chemical treatment are oxidizing methods, that are used in case of impossibility or economic inexpediency of impurities extraction from sewage, and in case the runoff compounds are biochemically insoluble or toxic. Such methods of treatment are often referred to as destructive, because as a result of water treatment from impurities, organic compounds are broken down and the decomposition products are removed from the water or harmless for aquatic bodies compounds are formed. These include biological and chemical oxidation of contaminants, chemical recovery by hydrogen, electrochemical destruction, etc. [1, 3, 4 – 12].

Rreagents oxidation is most widely used nowadays as a method of sewage treatment from oil products. Chlorine and its compounds, ozone, oxygen, hydrogen peroxide, copper sulfate, potassium dichromate, etc. are used as oxidizers [13].

When traditional methods of mechanical, biological and physical-chemical treatment of water are insufficient or cannot be used, because of the deficit of production areas, the complexity of delivery and the use of reagents or for other reasons, electrochemical methods for wastewater treatment are applied.

Electrochemical methods can be used to adjust physical-chemical properties of the treated effluent, they have a high bactericidal effect, allow to oxidize organic compounds to the required standards, simplify the technological scheme of treatment. In many instances electrical and chemical methods eliminate secondary pollution.

Electrical and chemical destruction is based on the oxidation and recovery of inorganic and organic compounds. Moreover, these reactions include a wide variety of processes starting from the simplest ionic recharge to complex transformations that underlie organic synthesis.

The mechanisms of electrical and chemical oxidation (or recovery) depend on the electrode material, the nature of substances subjected to electrolysis, temperature and electrolyte composition.

Currently, based on anodic oxidation and cathodic recovery, technology of sewage treatment has been developed from phenols, thiocyanates, nitro compounds, formaldehyde, methanol, dyes, simazine, cinar chloride, derivatives of anthraquinone, ethylene glycol, 2.4-D-acids, peroxide organic compounds, sulfur –containing and other organic pollutants present in industrial effluents [2, 4, 14 – 16].

The application of membrane techniques for treatment of oily waste is increasingly used especially abroad [4, 17 – 21].

One of the most effective methods of advanced sewage treatment from dissolved organic contaminants in petrochemical industry is sorption [22, 23]. As sorbents different porous materials are used: ash, coke waste peat, silicagels, alumogels, active clay, materials on the base of intercalated thermoexpanded graphite and its modifications, sorbents based on carbonized rice peelings, etc. Activated carbons of different brands are also effective adsorbents.

The main advantage of the sorption method is the repeated use of the sorbent in the «sorption-desorption" cycle, which significantly increases the economic efficiency of sorption treatment. Besides solving the problem of sewage treatment to maximum permissible concentration it is possible to make extraction of a valuable component from the surface of the sorbent.

The use of coal-sorption technology for additional treatment of sewage waters will enable the use of treated water in recycled water production, improve ecological situation and help preserve the mineral resources, which are currently subject to high depletion [22, 24].

The use of microorganisms immobilized biocenosis on the surface of sorption particles [1, 6, 8, 25] is a new development in the field of sorption conditioning of oily sewage for various purposes.

2. Description of research

Open and closed types of diagrams are used for sewage treatment in oil and gas fields. Diagram of water treatment of an open type enables to treat technical and storm waste water in the same flow regardless of composition, pressure and water gas saturation, and together upload them into injection wells. It is usually recommended to be used for sewage with a high content of hydrogen sulfide and carbon dioxide. However, the efficiency of drain treatment according to this diagram does not exceed 60%, and the economic costs of its layout and operation are relatively high. In addition, as a result of contact with oxygen the corrosiveness of water increases.

A schematic diagram of the treatment unit of the oil – field water of the closed type is shown in Figure 1.

Separated from the oil in the sedimentation chamber of preliminary discharge (SCPD) the water along line 1 goes into the reservoir – chamber 2 and then partially dehydrated oil (till 5 %), having passed the oil complex preparation unit (OCPU) goes to thermally insulated sedimentation chambers 3. The process of water separation in them is accelerated due to the OCPU heating and introduction of surface-active substances. Separated hot water is supplied to pump intake 4 and then fed back to the SCPD of the OCPU, which allows to reduce the consumption of demulsifier and the emulsion heating temperature. From the reservoir – chamber 2 the oil – field water is drawn by pump 5 and is fed to the sewerage pump station (SPS).

The application of a closed system of treatment allows intensifying the process of water preparation with the use of sedimentation and filtration under pressure, significantly reducing the aggressiveness of the sewage water by eliminating its contact with oxygen, to use the residual pressure existing in the system of oil preparation. The disadvantage of the closed system is the necessity of building a block for parallel treatment of surface stormwater runoff.

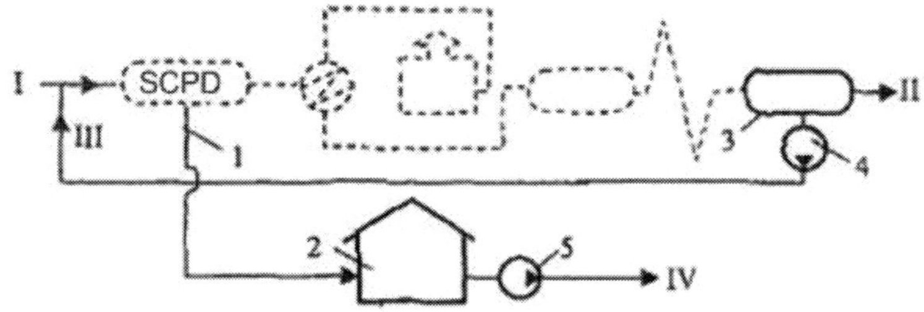

Figure 1. A schematic diagram of the treatment unit of the oil – field water of the closed type: 1- the water discharge line from sedimentation chamber; 2 - reservoir-sedimentation chamber; 3 — thermally insulated sedimentation chamber; 4,5 - pumps; I - cold raw oil; II - dehydrated oil; III - hot water with surfactants; IV – water for SPS

To optimize the operation of the closed system treatment of sewage water, a block of sorption structures is proposed to be included in the filtering stage. This method allows to achieve the required quality both for return oil –field and technical water.

A Unisorb-BIO, as a sorbent is suggested to use which is a complex with an addition of oil-oxidizing cultures of microorganisms fixed in the pores of the installation. One of the main advantages of the resulting sorbent is that it not just accumulates oil as other sorbents, but exposes it to degradation under the influence of immobilized microflora (Table 1). This sorbent belongs to the class of bio-sorbents. Granules have microporous, mesoporous and layered flaky macroporous structure with the surface coated with a carbon water –repellent film. The investigated sorbent is non-flammable, non-explosive, with a long operating life (up to 3 years), is bioregenerated, utilized according to the IV class of hazard.

Table 1. The Unisorb – BIO sorbent characteristics

Bulk density, kg/m^3	18…25
Flakes (granulated material), mm	3…10
Oil sorption capacity, kg of oil / specimen kg	30…60
Flotation capacity for a month, %	90…100
Reduction of interfacial tension (water - oil), erg/ cm^2	2,5…3,5
The share of processed oil by microorganisms after 7 - 14 days	
a) biodestructable activity under aerobic conditions, %	
Above more than 10 °C	50…70
At 0 ... +10 °C	25…40
b) biodestructive activity under anaerobic conditions, %	
More than 10 °C	20…35
at 0 ... +10 °C	10…15
Decrease in the activity after three years of storage, %	30…40

In the course of research there were 7 lines of experiment with model and field water with different concentration of oil (10 – 20mg/l). The water was filtered through the Unisorb-Bio sorbent at different temperatures. The concentration of surface and emulsified oil infiltration was monitored (Table. 2) in the experiment.

Table 2. Results of the experiment

№ of lines	Original concentration of oil products in model water	Conditions of filtration	Final concentration (surface/ emulsified)	Treatment effect, %
1	10 mg/l		0.12 mg/l /1.3 mg/l	99,98/99,87

2	15 mg/l	Normal conditions (20 ±2 ^{0}C)	0.08mg/l /1.320 mg/l	99,98/99,8
3	20 mg/l		0.08 mg/l /2.3 mg/l	99,94/99,76
4	10 mg/l	Heating up to 40-60 ^{0}C	0.6 mg/l /2.4 mg/l	99,94/99,76
5	15 mg/l		0.9 mg/l /4.2 mg/l	99,86/99,37
6	20 mg/l		1.280 mg/l /3.840 mg/l	99,74/99,23
7	10 mg/l	Cooling to +0.1 +4 ^{0}C	2.8 mg/l /8.755mg/l	72 / 12,4

The experiment has shown that when the temperature of sewage is 30 – 60^{0}C, the effect of sorption treatment is maximum, and when the sewage temperature is reduced, the indicators are considerably worsened (Figure 2).

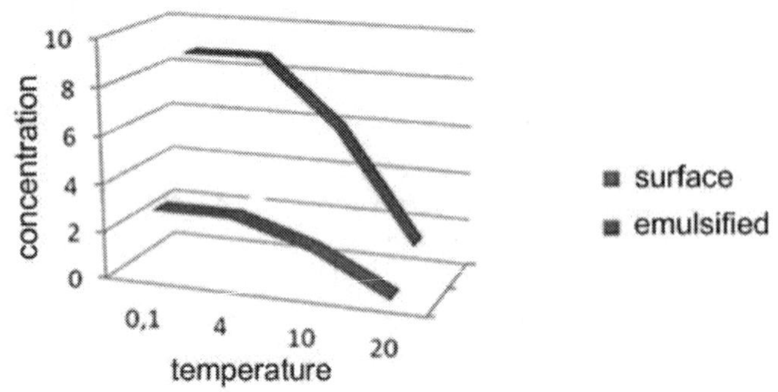

Figure 2. Effectiveness of oil-containing sewage treatment under different temperatures of original water

3. Conclusion

The relevance of research aimed at the issue of technological scheme composition and equipment selection for oil-containing sludging sewage is stipulated by the necessity of reuse of treated water as formation water for technical purposes of enterprises and reduction of environmental loads.

3.1 It has been experimentally shown that the high effect of sewage treatment demonstrates the feasibility of introducing a host of sorption filters for the treatment and conditioning of industrial water in closed circulating cycles.

3.2 The main characteristics and trends of the sorption processes with the use of Unisorb-Bio sorbent have been identified, which is a complex with the addition of oil-oxidizing cultures of microorganisms fixed in the pores of the device.

3.3 It is revealed that the used sorbent not just accumulates oil products as other sorbents, but exposes them to destruction under the influence of immobilized microflora, which enhances its competitive advantages from the point of view of sustainability of a full cycle of use.

This work was funded by RFBR and the Government of Krasnoyarsk Territory in the framework of research project 18-41-242008.

References

[1] Pinkin A V and Kasatkin A V 2012 New solutions for the collection of oil and fat production *Ecology of production* 2012 **4** pp 70-72

[2] Koganovsky A M, Klimenko N A, Levchenko T M, Marutovsky R M and Roda I G 1983 Purification and use of wastewater in industrial water supply *Chemistry*

[3] Alekseev M I, Ivanov V G, Kurganov A M, Medvedev G P, Mishukov B G, Feofanov Y A,

Tsvetkova L I, Chernikov N A 2007 Engineering manual of water treatment *New Journal* p 360

[4] Chernenko T V, Imatullina G K 2011 Methods of wastewater treatment of oil products *Scientific works collection of the Fourth International Teleconference "Basic Science and Practice" (Problems and perspectives of modern science)* **3**

[5] Zhao D, You F 2015 Mechanisms of chemical cleaning of ion exchange membranes: a case study of plant-scale electrodialysis for oily wastewater treatment *Journal of Membrane Science* **496** pp 310-317

[6] Guo H, Yu S, Li L, Zhao D and You F 2015 Mechanisms of chemical cleaning of ion exchange membranes: a case study of plant-scale electrodialysis for oily wastewater treatment *Journal of Membrane Science* **496** pp. 310-317

[7] Strepetov I V and Moskvicheva E V 2006 The use of sorbents based on polymer waste materials for waste water treatment of oil contamination *Internet-Herald VolgGASU* **1**

[8] Hua F L, Tsang Y F, Wang Y J, Chan S Y, Chua H and Sin S N 2007 Performance study of ceramic microfiltration membrane for oily wastewater treatment *Chemical Engineering Journal.* **128** pp 169-175.

[9] Kovaleva N G, Kovalev V G 1987 *Biochemical waste water purification of chemical industry* (Chemistry) p 160

[10] Lü T, Zhao H, Qi D, Chen Y 2015 Synthesis of a novel amphiphilic and cationic chitosan-based flocculant for efficient treatment of oily wastewater *Advances in Polymer Technology* **34 (3)** pp 21502

[11] Azimov Yu I, Savdur S N and Fesina E L 2014 Ensuring environmental safety based on the modelling of biological process of oily sewage *Mediterranean Journal of Social Sciences* **5 (24)** pp 372-376

[12] Stepanov S V 2014 Features of the calculation of structures of biological wastewater treatment refineries *Water supply and sanitary engineering* **3** pp 49-56

[13] Salu O A, Adams M, Robertson P K J, McCullagh C and Wong L S 2011 Remediation of oily wastewater from an interceptor tank using a novel photocatalytic drum reactor *Desalination and Water Treatment* **26 (1-3)** pp 87-91

[14] Dubrovskaya O G, Evstigneev B B and Kulagin V A 2011 Conditioning of Waste Water of Energy Systems *Journal of Siberian Federal University. Engineering & Technologies* **4** pp 665–675

[15] Sobgayda N A, Ol'shanskaya L N and Makarova Y A 2012 Filters from the waste effluent treatment Egelssky *Institute of Technology. Industrial ecology* **3** pp 40 - 41

[16] Smirnov A N, Stepanchikova I G, Baryshenko A V, Kuznetsov Y N and Kotljarova N B 2008 New technology for industrial wastewater treatment from oil products *Ecology and Industry of Russia* **4** pp 4-5

[17] Maksimov E A, Vasiliev V I 2015 Innovative methods and apparatus for cleaning oily waste water treatment *Water treatment. Water supply* **10 (94)** pp 28-39

[18] Padaki M, Surya Murali R, Abdullah M S, Misdan N, Moslehyani A, Kassim M A, Ismail A F and Hilal N 2015 Membrane technology enhancement in oil-water separation. A review *Desalination* **357** pp 197-207

[19] Li X, Ang W L, Liu Y and Chung T S 2015 Engineering design of outer-selective tribore hollow fiber membranes for forward osmosis and oil-water separation *AIChE Journal* 61 (12) pp 4491-4501

[20] Zhang Y, Gan F, Li J, Li S, Gao Y and Li M 2010 Oily wastewater treatment by porous ceramics mediums with surface lipophilic modification *Journal Wuhan University of Technology, Materials Science Edition* **25 (5)** pp 807-810

[21] Peng H, Tremblay A Y 2008 Membrane regeneration and filtration modeling in treating oily wastewaters *Journal of Membrane Science* **324 (1-2)** pp 59-66

[22] Li Y S, Yan L, Xiang C B and Hong L J 2006 Treatment of oily wastewater by organic-

inorganic composite tubular ultrafiltration (UF) membranes *Desalination* **196 (1-3)** pp 76-83
[23] Dvadnenko M V, Privalov N M, Lyavina E B, Protsay A A and Dinchenko Y V 2009 The use of sorption technology for oily wastewater *Basic Research* **5** pp 45-46
[24] Nikolaeva L A and Golubchikov M A 2016 Cleaning of manufacturing waste water from oil modified sorption materials based on carbonaceous sludge // *Water supply and sanitary engineering* **7** pp 51-58
[25] Trusov V V 2014 Coal sorption technology oily waste and recycled water *Water Treatment* **12** pp 32-41
[26] Klimov E S and Buzaeva M V 2011 *Natural sorbents and chelators in wastewater treatment* Monograph (Ulyanovsk, Ulyanovsk State Technical University) p 201

IOP Conf. Series: Materials Science and Engineering **880** (2020) 012043 doi:10.1088/1757-899X/880/1/012043

Study of the possibility for reducing of environmental damage when the transition to alternative sources of heat supply systems in housing and communal services

O Grebneva[1,2] **and O Lavygina**[2]

[1]Melentiev Energy Systems Institute of SB RAS, Department of pipeline systems of Energy, 664033, 130, Lermontov street, Irkutsk, Russia
[2]Ikutsk National Research Technical University», Institute of Architecture and Construction, Department of Urban development and Economy, 664074, 83, Lermontov street, Irkutsk, Russia

E-mail: olgakot81@mail.ru

Abstract. Modern socio-economic conditions dictate increasingly stringent requirements for housing and communal infrastructure. When choosing a source of energy in heat supply systems, variants having maximum environmental efficiency are preferable. For an objective assessment of increasing of environmental safety degree, it is advisable to use a complex indicator that reflects the level of reduced environmental damage to the environment, in particular atmospheric air. In this work, the calculation of the prevented environmental damage for two types of fuel is made: natural gas and wood fuel (pellets). The calculation results showed when using natural gas instead of wood fuel for heat source that environmental and economic damage will be 64% less.

1. Introduction

Environmental issues are becoming increasingly relevant in the system of housing and communal services, as they are an integral part of heat supply systems. The heat supply system includes heat sources (thermal power plants, boiler houses, etc.), which is one of the main sources of negative impact on environmental components. Therefore, among the main criteria when choosing equipment for heat sources of heat supply systems today, not only technical, economic, but also environmental ones are distinguished. It should also be noted that the reconstruction of heat supply systems (in particular, heat sources) is no exception. When choosing a source of heat energy in heating systems of settlements, it is advisable to compare not only the number of emissions of harmful substances, but also the level of damage to the environment.

In the Baikal Natural Territory, technical projects in the field of reconstruction of heat supply systems are subjected to environmental review, during which the degree of environmental efficiency of the accepted technical decisions is identified [1]. However, to compare variants, it is often not enough to compare quantitative indicators for pollutants entering the environment. The most effective should be recognized environmental and economic criteria, in particular the assessment of environmental damage. Environmental damage takes into account not only damage to atmospheric air as a component of the environment, but also as an indicator reflecting the recreational potential of the territory, a decrease in bio-productivity, a possible increase of sickness rate, etc.

2. Statement of the problem

The main priority types of negative impact of equipment of heat supply system are emissions of pollutants into the atmosphere [2]. Assessment of environmental and economic damage is a complex indicator that allows taking into account not only the amount of emitted pollutants, but also the environmental situation, as well as the environmental significance of the state of atmospheric air in the territory under consideration.

The determination of the amount of environmental damage is carried out in accordance with the "Methodology for the determination of prevented environmental damage" [3]. The prevented environmental damage is determined individually on each territory of a separate constituent entity of the Russian Federation in accordance with the amount of negative impact and the value of the indicator of specific environmental damage caused by a unit of the reduced mass of air pollution.

When conducting an integrated assessment of the prevented environmental damage, the following formula is used:

$$D = D_s \times M_1 \times C_{env} \times C_{env}^{aa} \times J_d,$$

где D_s – specific damage, rubles / standard fuel ton (for the East Siberian region D_s = 46.9, rubles / standard fuel ton);

M_1 – reduced mass of emitted pollutants during operation of the heat supply system, tons / year;

C_{env} – coefficient of relative environmental and economic hazard of pollutant interring into the air is taken according to Appendix 2 of the calculation method [3];

C_{env}^{aa} – coefficient of the ecological situation and the environmental significance of atmospheric air state, for the Irkutsk region C_{env}^{aa} =1.4;

J_d – index-deflator by industry, for 2020 the deflator (Electricity, gas and steam supply) will be 104.1 [4].

3. Numerical research

In the framework of this study, for the schema of heat supply system a comparison of the indicators of environmental and economic damage when using natural gas and wood fuel (pellets) as a source of energy was made. Schema of heat supply system is shown in Fig. 1.

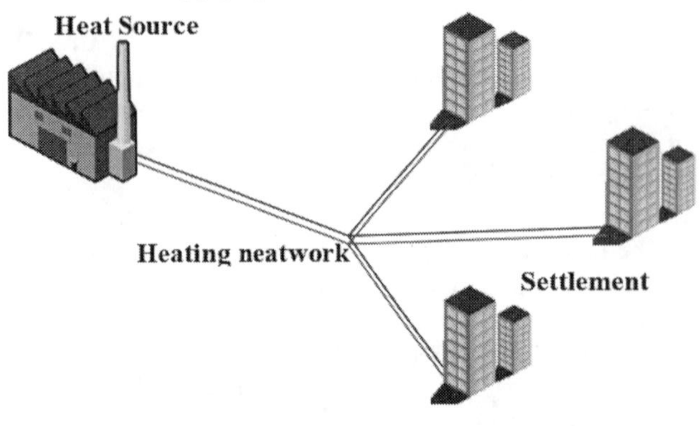

Figure 1. The schema of heat supply system

The article considers the possibility of using boiler using both natural gas and wood fuel (pellets) at a heat source. The characteristics of gas and wood fuels are given in table 1.

Table 1. The characteristics of gas and wood (pellets) fuel

Parameter	Gas Fuel	Wood Fuel (Pellets)
Ash, %	0.00	1.5
Sulfur content (total), %	0.00	0.03
Net calorific value, MJ / kg	32.500	17.322

At the initial stage, the gross emissions of pollutants into the air were calculated for two variants: when using gas and when using wood fuel in the form of pellets as heat source fuel. The calculation was carried out using the "Methodology for determining of pollutants emissions into the atmosphere when burning fuel in boilers with a capacity of less than 30 tons of steam per hour or less than 20 Gcal per hour" and the "Methodological manual for calculating, standardizing and controlling of pollutants emissions into the atmosphere" [5,6]. The calculation is based on the fuel consumption in accordance with their lower calorific value (Q_{gas}= 32500 kJ / kg; Q_{pellet} = 17322 kJ / kg). At the same time, the calculation of the amount of natural gas burned was 731.375 thousand cubic meters, and 1100 tons of wood fuel (pellets). Table 2 shows a comparison of gross emissions from the use of gas and wood (pellets) fuels. Table 3 shows the calculation results of environmental damage for different types of fuels. The name of the pollutants is given in accordance with the Hygienic Standards of the Russian Federation [7].

Table 2. Results of gross emissions calculation for gas and wood (pellets) fuel

Code	Emission name	Gross emission (t / year) with operation of boiler using natural gas as fuel	Gross emission (t / year) with operation of boiler using wood (pellets) fuel
0301	Nitrogen (IV) oxide	0.72214486	0.973138459
0304	Nitrogen (II) oxide	0.11734854	0.158135
0330	Sulphur dioxide	-	0.051377
0337	Carbon oxide	2.37696875	0.867846
0703	Benz / a / pyrene	0.00000125	0.000201881
2902	Suspended matter	-	0.266893

Table 3. Results of environmental damage for gas and wood (pellets) fuel

Code	Emission name	Environmental damage with operation of boiler using natural gas as fuel (thousand rubles)	Environmental damage with operation of boiler using wood (pellets) fuel (thousand rubles)
0301	Nitrogen (IV) oxide	81.44415	109.7514
0304	Nitrogen (II) oxide	13.23467	17.83461
0330	Sulphur dioxide	0	7.023448
0337	Carbon oxide	6.498828	2.372762
0703	Benz / a / pyrene	0.1068	17.24873

IOP Conf. Series: Materials Science and Engineering **880** (2020) 012043 doi:10.1088/1757-899X/880/1/012043

2902	Suspended matter	0	4.925525
Total		101.2844	159.1565

Thus, despite the «environmental friendliness» of wood fuel [8], a lower value of environmental damage corresponds to the variants with boiler using natural gas.

The calculation results showed that when using natural gas as boiler fuel, the amount of environmental damage will amount to 101.28 thousand rubles, and when using wood fuel (pellets) – 159.16 thousand rubles (in prices for 2020). Therefore, a more «environmental friendliness» variant is heat supply system with a natural gas-based heat source.

A graphical representation of calculation results is shown in Fig. 2.

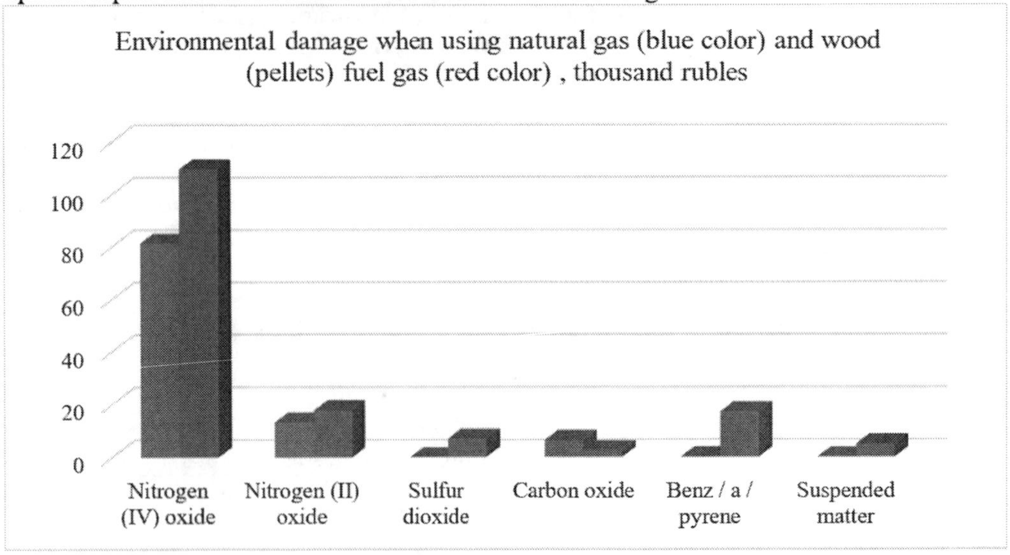

Figure 2. Bar chart of environmental damage calculation results

As can be seen from fig. 1. the main contribution to the environmental and economic damage (in the case of using natural gas as a fuel) to the atmosphere is carbon monoxide, and when using wood fuel (pellets) – nitrogen dioxide. It should also be noted that when using natural gas, such pollutants as sulfur dioxide and suspended solids are absent in the emissions.

4. The practical relevance of numerical research and conclusions

Environmental and economic damage is a comprehensive indicator that assesses the damage to atmospheric air not only as an object of the national economy, but also takes into account possible consequences for public health. The results of environmental and economic damage can be used when choosing alternative variants for the reconstruction of heat supply systems in housing and communal services [8].

Tightening of requirements in the field of environmental safety of the Russian Federation [9,10,11] leads to the need to search for new renewable energy sources [12], and accept of increasingly energy-efficient and environmentally friendly solutions. While the sphere of housing and communal services are no exception.

In this paper, we assessed the environmental damage when using alternative variants of heat energy source of in heating systems of settlements. The calculation method proved that environmental and economic damage when using natural gas instead of wood fuel (pellets) will decrease by 64%. It was revealed that during the operation of equipment using natural gas, there are no such pollutants as sulfur dioxide and suspended solids in the emissions.

Acknowledgments

The studies were carried out in the framework of project III.17.4.3 of the Program of basic research of the SB RAS (AAAA-A17-117030310437-4)

References

[1] Khan V.V., Dekanova N.P., Khan P.V. 2017 *IOP Conference Series: Materials Science and Engineering* **262(1)** 012081

[2] Lavygina OL, Grebneva OA. 2019 *University News. Investments. Construction* **9(4)** p 726 [In Russian]

[3] Methodology for determining prevented environmental damage (Approved Chairman of the State Committee for Environmental Protection of the Russian Federation on November 30, 1999, Moscow) [In Russian]

[4] Letter of the Ministry of Economic Development of the Russian Federation of October 1, 2019 N 33198-PB / D03i "On the Application of Forecast Indicators for the Socio-Economic Development of the Russian Federation for the Pricing of Products Delivered by State Defense Order" [In Russian]

[5] Method of Determining Atmospheric Emissions of Pollutants Produced by Burning Fuel in Boilers Having Capacity of 30 t/h of Steam or Less Than 20 GCal/h 1999 (Moscow: State Committee for Ecology of Russia) [In Russian]

[6] Method for calculating, standardizing and controlling emissions of pollutants into the atmosphere (supplemented and processed) 2012 (St. Petersburg: Research Institute Atmosphere) [In Russian]

[7] **GN 2.1.6.3492-17** 2017 [In Russian]

[8] Olga Lavygina, Oksana Grebneva, and Irina Maizel 2019 *IOP Conf. Series: Materials Science and Engineering* 667 012056

[9] Federal Law of July 21, 2014 N 219-FZ (as amended since January 1, 2020) [In Russian]

[10] Federal Law of January 10, 2002 N 7-FZ (as amended from January 1, 2020) [In Russian]

[11] Federal Law of May 4, 1999 N 96-FZ (edition as of November 1, 2019) [In Russian]

[12] Shelekhov I. Yu., Shishelova T.I., Smirnov E.I. Bulletin of the Mordovian University = Bulletin of the Mordovian University 2018 **28(1)** p 48 [In Russian]

[13] Pupyrev E I and Chupin V R 2019 Features of regional development of water disposal systems in the central ecological zone of the Baikal natural territory *Proceedings of Universities. Investment. Construction. Real estate* **9** pp 354–363

[14] Chupin V R and Dushin A S 2019 Assessment of the reliability of water supply to consumers: water supply reliability indicators Proceedings of Universities. Investment. Construction. Real estate **9(3)** pp 578–593

IOP Conf. Series: Materials Science and Engineering **880** (2020) 012044 doi:10.1088/1757-899X/880/1/012044

Actual problems of the operation of apartment buildings

V V Khan[1], N P Dekanova[2] and P V Khan[3]

[1] Irkutsk National Research Technical University, Lermontov str. 83, Irkutsk, Russia
[2] Department of information systems and information security, Irkutsk State Transport University, Chernyshevskogo str., bld. 15, Irkutsk, Russia
[3] Melentiev Energy System Institute, Lermontov str. 130, Irkutsk, Russia

E-mail: khan@istu.edu

Abstract. The issues related to the occurrence of accidents during the operation of multi-apartment residential buildings, and the contradictions between property relations, rights of use and liability of business entities in the current legal environment are considered. The analysis of typical accidents leading to significant losses is conducted. The significance of the main factors influencing the occurrence of accident situations, the amount of damage from accidents and the degree of responsibility attributed to various categories of business entities in terms of compensation for material damage are assessed. Options for solving problems are proposed, including technical solutions in terms of diagnosing and localizing accident situations, and measures to increase the efficiency of the organizational and legal basis for the operation of multi-apartment residential buildings.

1. Introduction

Low efficiency and a high level of costs with low quality of management during the operation of multi-apartment residential buildings are steadily a headache for management bodies and executive bodies of all levels, from the government to management companies and homeowners associations. The low reliability of housing and communal services, primarily heating systems, hot and cold water supply and sanitation, lead to significant material damage, as well as to a threat to the health, and sometimes the lives of residents. This issue is paid the closest attention to government bodies and the public every year, many decisions are made, but the severity of the problem does not decrease. Moreover, due to the increasing complexity of engineering equipment, the frequency of accidents per one building increases, and with the increase in the number of stores of buildings, the size of material damage increases.

According to the authors, the main reason for this situation is the totality of contradictions between the relations of ownership, rights and responsibilities of business entities in multi-apartment residential buildings, combined with insufficient equipment of buildings with diagnostic and safety systems. It is now becoming apparent that the existing regulatory framework for the technical regulation of the work of construction companies and management company in the field of housing and communal services requires serious analysis and changes. The analysis of the main conflicts leading to accidents is carried out. The consequences of accidents, the degree of responsibility of business entities and the degree of damage to each of the parties are assessed.

2. The problem of the diversity of ownership types

There is only one object - an apartment building, and there are many owners. The apartment building is a single complex of building structures and engineering systems. At the same time, there are various types of premises in the house, in relation to which the ownership right is different, including individual ownership of individual premises and collective ownership of common property. As a real estate object, the premises exist as part of another real estate object - buildings, forming two categories depending on the purpose:

- main premises, which are non-residential premises in a non-residential building or residential (apartments) and non-residential premises in an apartment building;
- auxiliary premises, which are auxiliary premises in a non-residential building or common premises in an apartment building.

Each of these categories of premises has its own legal regime. Thus, the assignment of premises to the number of objects of civil rights, in fact, leads to the formation of several objects with different legal regimes: 1) the building as a whole, including an apartment building; 2) the main premises, which may belong to various owners on the basis of individual ownership; 3) common premises and land, which are a common property owned by all owners of premises on the basis of the right of common shared ownership. According to paragraph 2 of Art. 290 of the Civil Code of the Russian Federation, a specific feature of the concept of an “individualistic approach” to property existing in Russia was formed. In fact, individual property on its parts and common shared property, the elements of which are inseparable, are established on a single real estate object (building and the land occupied by it). In an apartment building, you can distinguish the main types of premises:

- residential premises;
- non-residential common premises in an apartment building, building infrastructure (stairwells, elevators, halls, passages, technical rooms, basements, ground floor);
- non-residential premises used for commercial or other purposes.

At the same time, various forms of ownership can be distinguished. Residential and non-residential premises may be individually owned by any subjects of the right of ownership, regardless of its form, and auxiliary property - only in the common shared ownership of these entities. The common property of the owners of premises in an apartment building are those parts of the house that have auxiliary, serving value and are not individual property objects. An approximate list of objects of common property is established in Part 1 of Art. 36 of the Code [1].

A diversity of forms of management and the right to use causes a number of conflict situations, leading to significant losses, and sometimes posing a threat to the safety and health of inhabitants. Two or more houses connected to the heat distribution node and, respectively, to one node of the commercial metering of thermal energy and heat carrier flow. As a rule, the owners of each house form a separate legal entity. Moreover, each of the houses can be served by different organizations. Existing normative acts [2] (Rules for the commercial accounting of thermal energy, heat carrier flow); [3] (Decree of the Government of the Russian Federation of May 6, 2011 N 354 "On the provision of communal services to owners and users of premises in apartment buildings and residential buildings") do not clearly regulate the methods for distributing energy costs between consumers, the costs of maintaining property servicing different houses , as well as losses during the transport of thermal energy between houses. In addition, resource-supplying organizations often require payment of consumed resources in full from residents of a house in which metering devices are installed, and calculation according to standard loads, from houses connected to this heat point.

3. The problem of the rights and responsibilities of business entities

The owners assume the ownership of the premises includes the right to carry out work in them without proper experience and knowledge, or to hire contractors for low prices, as well as to use cheap equipment, which leads to serious accidents. Such situations arise due to the fact that with the current practice, any people can perform work in the engineering systems inside the apartments, including those without training and knowledge of safety rules. The consequence of the poor quality of work and materials is accidents leading to pipe leaks in apartments caused by the pipe joint sealing damage and/or the destruction of fittings, hoses, pipelines, radiators. In [4] it is noted: “According to statistics, the damage caused to property by flooding is three times higher than the loss from apartment thefts.

According to the Moscow Property Department, the vast majority of accidents (89%) occur in the flooding". The analysis of arbitration cases in the field of housing and communal services leads to the same conclusion. When flooding the premises, almost everything suffers: ceilings, walls, floor coverings, doors, windows, furniture and household appliances. In the event of a short circuit due to dampness, even a complete burnout of the apartment is possible, sometimes with fatalities, as a rule these are elderly people and small children. The situation of flooding apartments with hot water from the heating system to such an extent that residents receive burns incompatible with life is also fatal.

A common cause of accidents is a defect made during the construction phase, which manifests itself during operation, including after the warranty period has expired, as well as the use of low-quality appliances and equipment in pursuit of savings. Figure 1 shows typical examples of the breakage of low-quality equipment. The warranty period is too short, as a rule - no more than 5 years. At the end of the warranty period the responsibility for the defect falls on the property owners, or on the management company, if the accident occurred in their area of their responsibility.

Figure 1. Breakage of low-quality faucet and flexible hoses.

Another common cause of emergencies is negligent attitude and / or low qualification of performers of repair and other works. However, often these "masters" remain unknown. Stimulated by existing legislation [5], a mentality has developed in Russia, the main requirement of which is the purchase of goods and services at the lowest prices, often below cost. Such prices are possible only when using low-quality materials, violation of technology and low-skilled labor. One of the fundamental principles formulated by E. Deming, a classic of quality management more than 60 years ago [6], states: "End the practice of awarding business on the basis of a price tag. Instead, minimize total cost. Move towards a single supplier for any one item, on a long-term relationship of loyalty and trust". Ignoring the fundamental principles of management in modern Russia leads to significant damage [7, 8].

4. The problem of restricting the right of access to areas of operational responsibility for the management company personal

Often, accidents arise in the area of responsibility of the management company, while its' personal does not have the right of free access to this zone, since the zone is located in a room that is owned or used by another individual or legal entity. This situation occurs when communications pass through warehouses, shops, pharmacies, cafes, fitness centers, and other institutions located in the basement or ground floors of apartment buildings. The management company in this case is not able to carry out normal maintenance in the area of its responsibility, but in case of an accident is found guilty [9-11].

Table 1 presents the most common types of accident situations, the main causes of their occurrence and the weight fraction of each of the causes in the occurrence of the accident [12].

Table 1 The types of accident situations and the causes of their occurrence.

Types of accident situations	Causes of accident situations, their weights, %				
	Manufacturing defects	Construction defects	Residents actions	Physical wear	Improper maintanance
Damage to fittings inside the apartment	50	20	20	10	10
Pipe joint sealing damage inside the apartment	-	80	20	-	-
Flexible hoses damage	50	20	30	-	-
Damage of radiator	50	10	10	20	10
Indoor pipes damage	30	30	20	20	-
Pipes, fitting, hoses or joint seal damage in the responsibility area of management company	30	20	10	20	20

The weight of each of the factors under consideration (the causes of the accident) was adopted by experts on the basis of an analysis of judicial practice and a sample survey of management companies.

Table 2 presents the average size of damage from different types of accident situations per year and per an average apartment building. The calculations take into account that the apartments are located on different floors and during an accident, as a rule, two to three lower apartments are poured. Accidents in the ground floor do less damage. In case of accidents in the distribution pipes passing between floors and inside the premises belonging to different owners, the damage may turn out to be significant due to the longer duration of the accident (the management company does not have the right of free access to such zones).

Figure 2 shows the weighted average distribution of liability for accidents of the parties - business entities, as well as assessing their material damage from accidents.

Table 2 The average size of damage from accident situations.

Types of accident situations	Frequency, 1/year	Damage per one accident, thousand rubles	Average annual damage, thousand rubles
Damage to fittings inside the apartment	0.2	300	60
Pipe joint sealing damage inside the apartment	0.4	50	25
Flexible hoses damage	0.3	300	90
Damage of radiator	0.05	500	25
Indoor pipes damage	0.05	500	25
Total	-	-	225

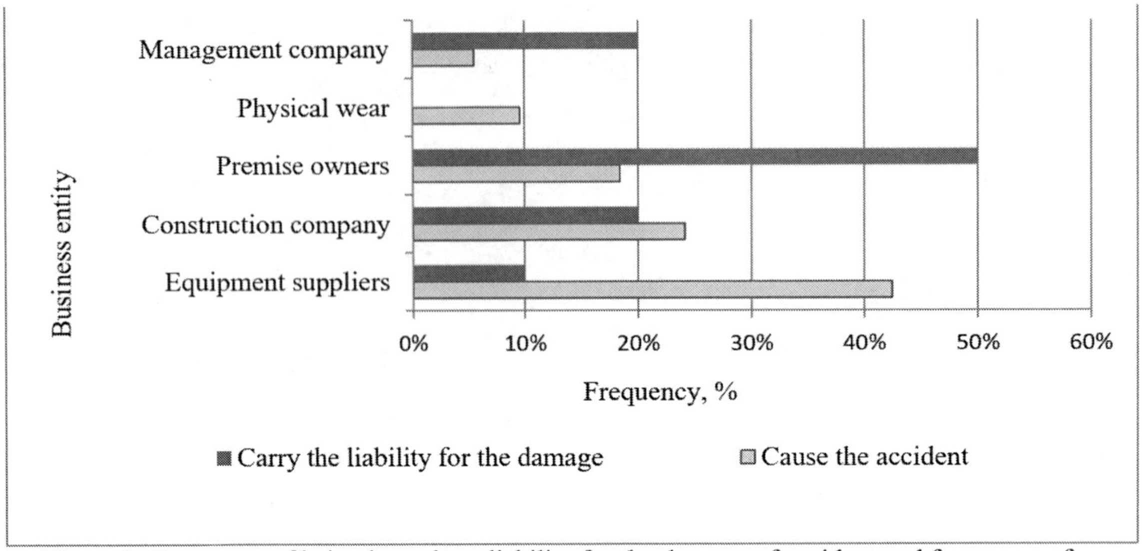

Figure 2. Frequency of being brought to liability for the damage of accident and frequency of causing the accident

As a rule, suppliers of low-quality equipment cannot be brought to full liability for the damage caused. Initially, the charge is brought against the owners of the premises in which the accident occurred. The owners of the premises, in turn, are trying to bring charges against the management company. When considering such cases, the court may conclude that the accident occurred due to the poor quality of the equipment installed or poor quality of construction work. But a fairly common phenomenon is the termination of the activities of construction organizations working poorly, or it turns out that the warranty has expired. In such cases, there is no one to recover damage and material damage remains with the owners.

5. Conclusions and suggestions

The contradictions between the relations of ownership, rights and responsibilities of business entities in multi-apartment residential buildings, combined with the insufficient equipment of buildings with diagnostic and safety systems, are the main reason for the low efficiency of housing and communal services in terms of operating multi-apartment residential buildings. Despite the fact that a significant part of the problems is caused by the low quality of equipment and materials, as well as the low quality of construction works, the main damage is suffered by the owners and service organizations.

The extent of damage can be significantly reduced due to operational diagnostics and localization of accidents [13]. It seems important to use, along with technical means, modern methods of analysis [7, 9, 14]. It is necessary to increase the degree of responsibility of suppliers of low-quality equipment, construction organizations and organizations that carry out repairs in buildings [15]. One of the means to increase the responsibility of construction organizations can be a significant increase in the insurance premium to a self-regulatory organization (SRO) and the practice of compensation for damage due to accidents in cases where the guilt of the construction organization is proved, even if this construction organization ceased to exist. If the accident arose as a result of the fact that the owners of the premises allowed random persons who are not members of the SRO to carry out work on the engineering equipment, or themselves carried out the work, then the responsibility should lie entirely with them.

Acknowledgements

This work was carried out as part of research on the topic No. 30/19 of Irkutsk National Research Technical University, with partial financial support from Melentyev Energy Systems Institute of SB

RAS in the framework of the scientific project III.17.1.3 of the program of basic research of the SB RAS, reg. No.AAAA-A17-117030310443-5.

References

[1] The Housing Code of the Russian Federation dated December 29, 2004 N 188-FZ (as amended on February 6, 2020) http://docs.cntd.ru/document/901919946.

[2] Decree of the Government of the Russian Federation of November 18, 2013 N 1034 "On the commercial accounting of thermal energy, heat carrier [O kommercheskom uchete teplovoj jenergii, teplonositelja]" (as amended). https://base.garant.ru/70511954/

[3] Decree of the Government of the Russian Federation of May 6, 2011 N 354 "On the provision of utilities to owners and users of premises in apartment buildings and residential buildings [O predostavlenii kommunal'nyh uslug sobstvennikam i pol'zovateljam pomeshhenij v mnogokvartirnyh domah i zhilyh domov]" (as amended) https://base.garant.ru/12186043/.

[4] Otstavnov A A Protection of apartment buildings from leaks [Zashhita mnogokvartirnyh domov ot protechek] https://www.c-o-k.ru/articles/zaschita-mnogokvartirnyh-domov-ot-protechek.

[5] Federal Law "On Procurement of Goods, Work, Services by Certain Types of Legal Entities [O zakupkah tovarov, rabot, uslug otdel'nymi vidami juridicheskih lic]" dated July 18, 2011 N 223-FZ (latest revision) http://www.consultant.ru/document/cons_doc_LAW_116964/.

[6] Kruglov M, Shishkov G 2007 *Quality Management As It Is [Menedzhment kachestva kak on est']*(Moscow: EKSMO) c. 544.

[7] Khan V V, Dekanova N P , Romanova T A and Sharaeva S A 2017 Complex analyses of efficiency of energy effectiveness measures for the objects of social sphere in the Eastern Siberia on the basis of system approach. *Izvestiya vuzov. Investitsii. Stroitel'stvo. Nedvizhimost* (Proceedings of Universities. Investments. Construction. Real estate vol 7) **1** pp. 84–93. (In Russian)

[8] Lavygina O, Grebneva O and Maizel I 2019 Environmental aspects for the reconstruction of housing and communal services in the village Listvyanka of Irkutsk region. *IOP Conf. Series: Materials Science and Engineering* 667 012056 IOP Publishing doi:10.1088/1757-899X/667/1/012056

[9] Khan V V, Dekanova N P and Khan P V 2017 Heat Supply of Villages and Towns of Eastern Siberia. Problems and Ways of Development *International Conference on Construction, Architecture and Technosphere Safety* (ICCATS. IOP Conference Series: Materials Science and Engineering vol 262) **1**. ID: 012081. DOI: 10.1088/1757 899X/262/1/012081.

[10] Shcherbakov V, Akulshin A, Chizhik K and Tolstoy M 2018 Design of interacting wells for optimization of investments and operating costs while constructing water-diverting structures *6-th International Scientific Conference on Integration, Partnership and Innovation in Construction Science and Education,* IPICSE Moscow; Russian Federation; 14-16 November Код 143421. DOI: 10.1051/matecconf/201825103037 *MATEC Web of Conferences* **251** *03037*

[11] Chupin V R, Pham N M and Chupin R V 2019 Optimization of developing district water supply systems taking into account variability of perspective water consumption *IOP Conference Series: Materials Science and Engineering* **667** 012018 doi: 10.1088/1757 899X/667/1/012018/

[12] Population Survey. Heat and Heating Problem 2003 FOM Database https://bd.fom.ru/report/cat/cata_st/dd030427

[13] Shonina N A 2019 Flood protection systems. *Сантехника* **3** https://www.abok.ru/for_spec/articles.php?nid=7246.

[14] Skibo D, Sudnikovich V and Tolstoy M 2018 Technical innovations as a means of increasing the efficiency of technological equipment in modern production conditions *International Scientific Conference "Investment, Construction, Real Estate: New Technologies and Special-*

IOP Conf. Series: Materials Science and Engineering **880** (2020) 012044 doi:10.1088/1757-899X/880/1/012044

Purpose Development Priorities" ICRE Irkutsk; Russian Federation; 26-27 april Код 142654 **212** 010142018

[15] Federal Law of December 27, 2002 N 184-FZ (as amended on November 28, 2018) "On Technical Regulation [O tehnicheskom regulirovanii]" http://docs.cntd.ru/document/901836556

Assessment of the energy performance of buildings

V V Khan[1], N P Dekanova[2] and P V Khan[3]

[1]Irkutsk National Research Technical University, Lermontov str. 83, Irkutsk, Russia
[2] Department of information systems and information security, Irkutsk State Transport University, Chernyshevskogo str., bld. 15, Irkutsk, Russia
[3]Melentiev Energy System Institute, Lermontov str. 130, Irkutsk, Russia

E-mail: khan@istu.edu

Abstract. The problems of the heat engineering condition of buildings during operation diagnos are considered. A significant problem is the difficulty of obtaining measurement and other information for complete decision making. As a result of this, it is often necessary to make technical decisions regarding the long-term operation of facilities in conditions of incompleteness and fuzzy information about their functioning and their condition The use the methods of the theory of fuzzy sets. is proposed when making technical decisions to increase the energy efficiency of objects.

Currently, the energy efficiency survey of a large number of buildings and structures is being carried out as part of the energy audit of organizations in accordance with Federal Law No. 261 "On Energy Saving and Energy Efficiency Improvement and Amending Certain Legislative Acts of the Russian Federation", as well as Programs to promote housing and communal services reform after major repairs or reconstruction of residential buildings.

An important component of the energy inspection of buildings is the diagnosis of their energy performance, which consists of the condition of the building envelope, the condition of the internal heating pipes, ventilation and hot water supply, the availability and quality of work of regulation systems [1, 2]. Operational diagnostics of the energy performance of buildings and structures is also necessary for the organization of their high-quality operation, which is especially important for areas with a long heating period and high tariffs for electricity and thermal energy [3-8].

In the ideal case, for a full-fledged diagnosis, it is necessary to have, along with technical documentation, the ability to take air temperature measurements in all rooms of the building while simultaneously recording the outside temperature; thermograms of all internal and external surfaces obtained simultaneously; heat carrier temperatures in all sections of the internal heating pipes. It is important to obtain the results of such measurements at different outdoor temperatures. Obviously, taking measurements in such a volume requires significant costs and is advisable in exceptional cases [8-11]. In most cases, when examining buildings and structures, a very limited scope of measurements is available. At the same time, it is important to maximize the use of additional information obtained from surveys of operating personnel and users of the premises. Information of this kind usually falls into the category of fuzzy information and is presented in the form of intuitive or expert judgments. For example, "hot", "warm", "cool", "cold" - if this relates to indoor air temperature. Regarding the quality of regulation of indoor microclimate parameters, one can obtain estimates of this kind: "good", "normal", "satisfactory", "unsatisfactory". In terms of assessing the condition of building envelopes, one can also obtain judgments of the type: "normal", "satisfactory", "unsatisfactory". Some of the information is fuzzy in nature. For example, heat carrier flow in open heat supply networks depends on many factors, while constantly changing over time. The imbalance of heating networks often leads to the fact that in peripheral objects the available pressure drop between the supply and return pipelines is insufficient to ensure normal circulation of the heat carrier. In such cases, users often resort to draining the coolant into the sewer [4,6]. This, in turn, affects the distribution of flows in the heating networks and the hydraulic modes of their operation. Obviously, user behavior can only be described on the basis of fuzzy set theory approaches. Usually, qualitative indicators of the thermal conductivity characteristics of building envelopes obtained as a result of surveys are taken into account on an intuitive level, or based on expert judgment [1-3]. At the same time, modern methods of the theory of fuzzy sets are widely and effectively used for decision-making in a wide variety of fields. [2-5].

Given the above factors, it is advisable to use an integrated approach based on the combined use of measurement results, analysis of balance relations of objects of different hierarchical levels and methods of the theory of fuzzy sets [1, 6] to diagnose the state of engineering systems of buildings. At the same time, it seems important to develop a formalized decision-making algorithm taking into account fuzzy information. The aim of the present work is to develop an algorithm for diagnosing the heat engineering state of a residential building based on the approaches of the theory of fuzzy sets [12-15].

Usually, when examining heat consumption objects, their design characteristics and contractual loads are known, the results of flow measurements and the parameters of the heat carrier at the entrance to the building are available, the results of measurements of the temperature of the heat carrier along the risers are known. However, the available information is insufficient to determine the actual thermal and hydraulic modes of the heating system, to assess the energy performance of the building envelope and the actual loss of thermal energy. Additional sources of information may be the results of thermography of the building, as well as conclusions obtained on the basis of a rule base based on fuzzy information [12,13].

Based on the results of a survey of experts, qualitative scales are formed for three characteristics: the indoor air temperature, the deviation of the indoor climate parameters from the normative and the deviation of the state of the building envelope from the design. The indoor air temperature is evaluated according to the following quality scale: "very cold", "cold", "cool", "normal", "warm", "hot". Membership functions are formed for each indicator according to the experts judgments. The membership functions are linear for the extreme indicators "very cold" and "hot":

- "very cold"

$$\mu^l(x) = \alpha(x)\cdot\frac{x^l_{max}-x}{x^l_{max}-x^l_{min}},\ \text{где}\ \ \alpha(x) = \begin{Bmatrix}1, & x^l_{min} \le x \le x^l_{max}\\ 0, & \text{иначе}\end{Bmatrix}, \tag{1}$$

- "hot"

$$\mu^p(x) = \alpha(x)\cdot\frac{x-x^p_{min}}{x^p_{max}-x^p_{min}},\ \text{где}\ \ \alpha(x) = \begin{Bmatrix}1, & x^p_{min} \le x \le x^p_{max}\\ 0, & \text{иначе}\end{Bmatrix}. \tag{2}$$

For intermediate indicators "cold", "cool", "normal", "warm" the membership functions are non-linear and are determined by the following relation:

$$\mu^k(x) = e^{-\left[\frac{2.1\cdot(x-\alpha(x))}{x^k_{max}-\alpha(x)}\right]^2},\ \text{где}\ \ \alpha(x)=\frac{x^k_{min}+x^k_{max}}{2},\ \ k=1,\dots,4. \tag{3}$$

In relations (1)-(3) x - current value of the characteristic; $x_{\min}$ and $x_{\max}$ - limit values of relevant indicators.

Graphs of membership functions of indicators of internal air temperature are given in the figure. 1.

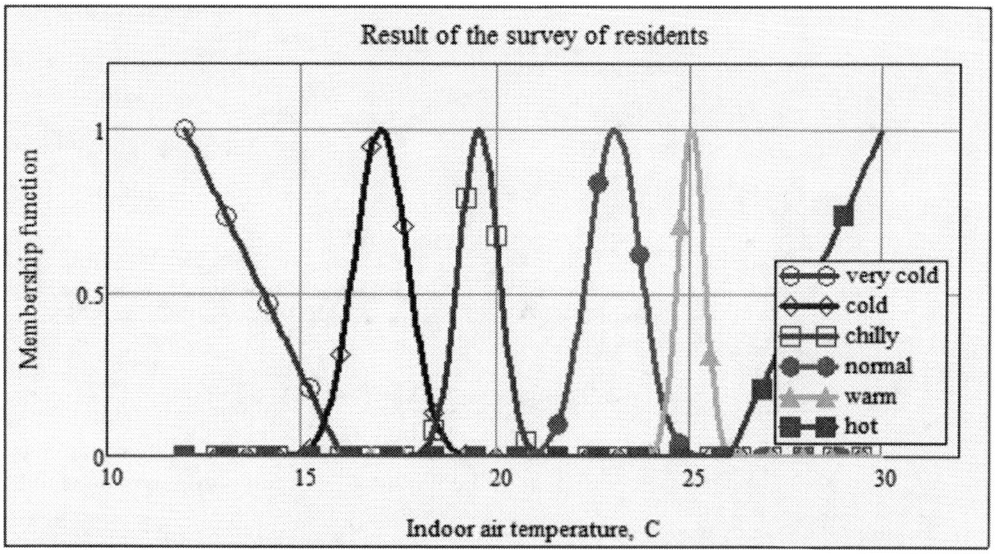

Figure 1. The membership functions for indoor air temperature

The control of the indoor climate parameters according to the results of a survey of residents is evaluated on a scale of "good", "normal", "satisfactory", "unsatisfactory". When forming membership functions for the indicators “good” and “unsatisfactory”, formulas (1) and (2) are used, respectively, and for the indicators “normal” and “satisfactory”, the formula (3) is used. Membership functions corresponding to such judgments are shown in the figure. 2..

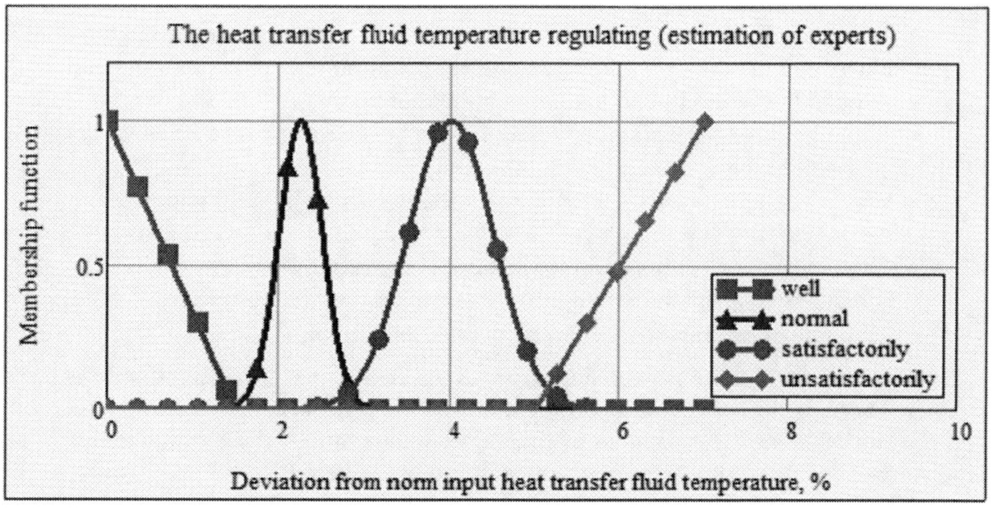

Figure 2. The membership functions for the deviation of the indoor climate parameters from the normative

The condition of the building envelope is estimated by the deviation of the value of the overall heat transfer coefficient of the building from the normative indicator in three gradations: “normal”, “satisfactory”, “unsatisfactory”. Membership functions formed in accordance with relations (1) - (3) are shown in Figure 3.

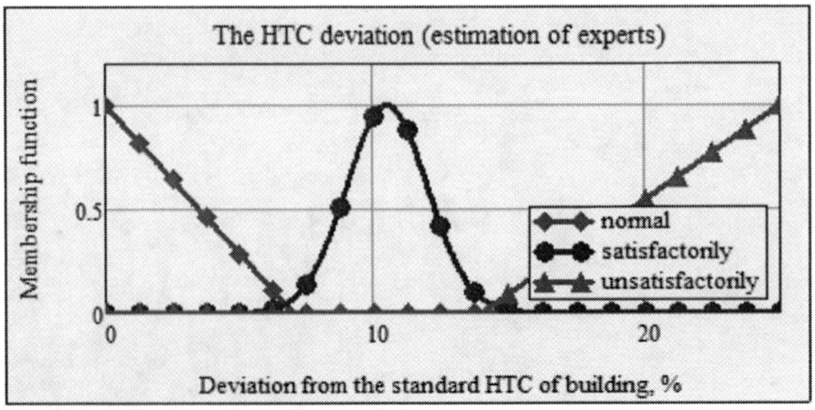

Figure 3. Membership functions of the deviation of the value of the overall heat transfer coefficient of the building from the normative indicator.

To diagnose the energy performance of buildings, we define the following rule base:

1. Poor quality of service by the operating organization if
   - the indoor temperature is rated as very cold, cold or hot;
   - regulation of the water temperature in the heat network is rated as very good or normal;
   - building heat transfer coefficient is rated as satisfactory or unsatisfactory.
2. Unsatisfactory quality of heat carrier supply by heating network company if^
   - the indoor temperature is rated as very cold, cold or hot;
   - regulation of the water temperature in the heat network is rated as satisfactory or unsatisfactory;
   - building heat transfer coefficient is rated as good, normal or satisfactory.
3. Good quality of service by the operating organization if
   - the indoor temperature is rated as normal or warm;
   - regulation of the water temperature in the heat network is rated as satisfactory or unsatisfactory;
   - building heat transfer coefficient is rated as normal or satisfactory.
4. All services work fine
   - the indoor temperature is rated as normal or warm;
   - regulation of the water temperature in the heat network is rated as very good or normal;
   - building heat transfer coefficient is rated as normal.

The algorithm for obtaining a conclusion about the energy performance of buildings and the quality of service, based on the results of the observation and the rule base, is as follows.

At the current time, the indicators of the indoor temperature , the characteristics of regulating the temperature of the network water and the actual heat transfer coefficient are estimated. We call such a set of parameters "a situation". Further, for each situation, the following steps of the algorithm are performed.

1. For each rule i, the value of the membership function of each characteristic j is determined in accordance with the union operation based on the s-norm (algebraic sum):

$$\mu^i(x_j) = \bigvee_{\forall k \in K_j^i} \mu_k(x_j), \quad i = 1, \dots, 4; \quad j = 1, \dots, 3.$$

For each rule i, the value of the generalized membership function is determined in accordance with the intersection operation on the basis of the t-norm (average):

$$\mu^i(x_1, x_2, x_3) = \mu^i(x_1) \bigwedge \mu^i(x_2) \bigwedge \mu^i(x_3), \quad i = 1, \dots, 4.$$

2. To determine the resulting value of the membership function and select the output corresponding to the situation in question, the accumulation operation is used:

$$D = \max_{i=1,\dots,4} \mu^i(x_1, x_2, x_3).$$

We illustrate the operation of the algorithm in the following example. Suppose that during monitoring of heat consumption objects 10 different situations are recorded (table 1, figure. 4).

Table 1. Initial observational data

Situation number	Air temperature, °C	Network water temperature deviation, %	Building heat transfer coefficient deviation, %
1	12	1	12
2	16	3	20
3	27	2	10
4	12	4	5
5	16	6	8
6	27	5	11
7	23	4	3
8	25	7	10
9	23	1	2
10	25	2	5

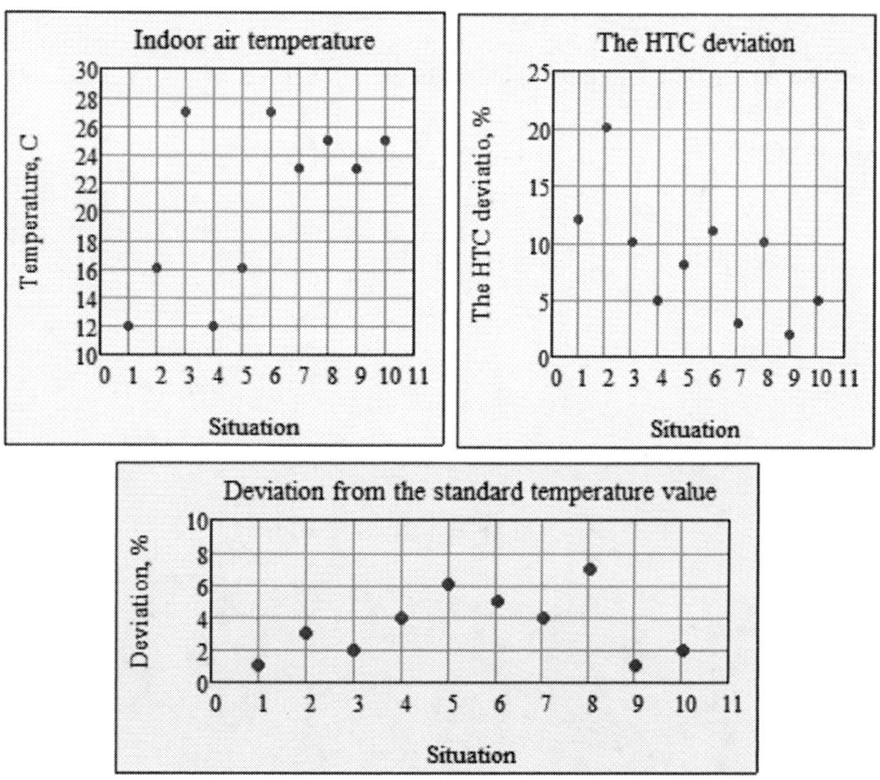

Figure 4. Observation data.

Based on the membership functions, qualitative estimates of the observed situations were obtained (table 2).

Table 2. Qualitative assessment of the observed situations.

Situation number	Air temperature	Network water temperature deviation	Building heat transfer coefficient deviation
1	very cold	very good	satisfactory
2	cold	satisfactory	unsatisfactory
3	hot	normal	satisfactory
4	very cold	satisfactory	normal
5	cold	unsatisfactory	satisfactory
6	hot	satisfactory	satisfactory
7	normal	satisfactory	normal
8	warm	unsatisfactory	satisfactory
9	normal	very good	normal
10	warm	normal	normal

A graphical representation of the values of the membership function for each rule is given in Figure 5. Based on the graph, for each situation, not only the value of the membership function of the dominant rule, but also the ratio of the values of the membership functions of each rule to each other can be evaluated.

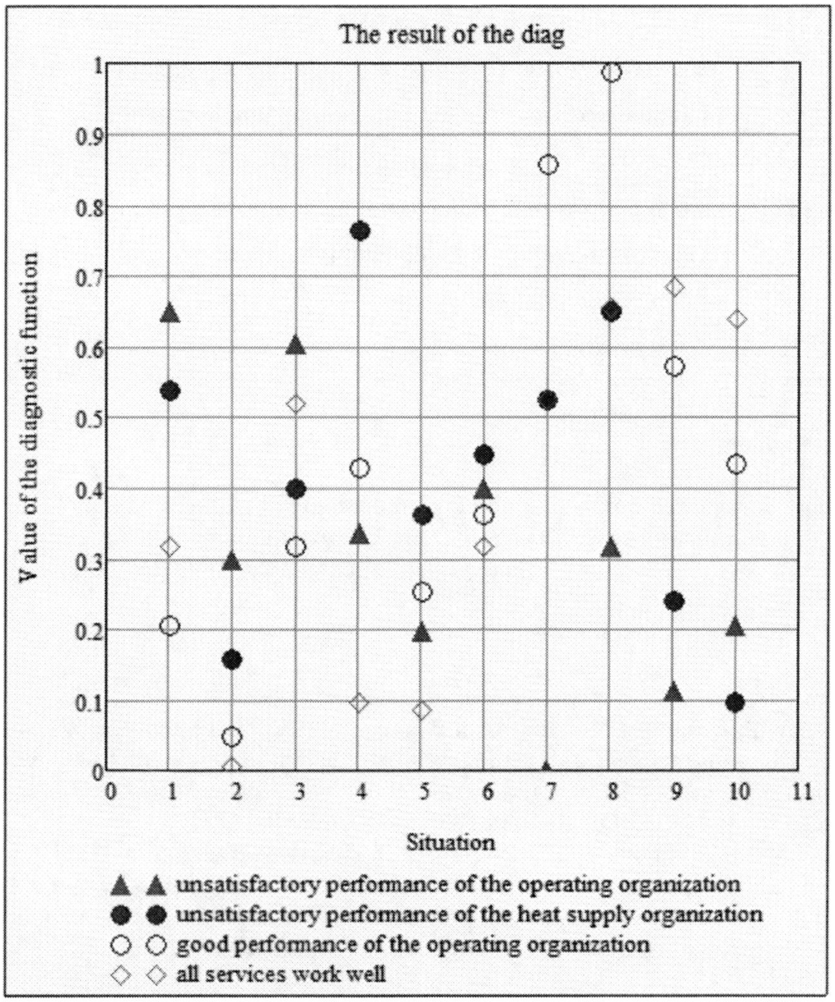

Figure 5. Values of membership functions of each of the four rules for the situations under consideration.

The result of the algorithm execution is the conclusion based on the rule base on the quality of the heating network company and the quality of servicing heat consumption objects from the operating organization and the values of the resulting membership function for the situations under consideration (table 3).

The results obtained make it possible to take into account the oral information obtained by interviewing operating personnel and/or residents when making decisions. Further development of the proposed approach is associated with the development of a model that takes into account the full range of information obtained during measurements and during surveys.

Table 3. Qualitative and quantitative assessment of the situations under consideration, obtained on the basis of membership functions and formulated rules.

Situation	Diagnos	Value
1	Unsatisfactory performance of the operating organization	0.649
2	Unsatisfactory performance of the operating organization	0.297

3	Unsatisfactory performance of the operating organization	0.603
4	Unsatisfactory performance of the heat supply organization	0.762
5	Unsatisfactory performance of the heat supply organization	0.363
6	Unsatisfactory performance of the heat supply organization	0.446
7	Good performance of the operating organization	0.857
8	Good performance of the operating organization	0.986
9	All services work well	0.683
10	All services work well	0.637

Acknowledgements

This work was carried out as part of work on the topic of research No. 30/19 of IRTU with partial financial support from Melentyev Energy Systems Institute of SB RAS in the framework of the scientific project III.17.1.3 of the program of basic research of the SB RAS, reg. No.AAAA-A17-117030310443-5

We thank all contributors for their cooperation.

References

[1] Khan V V, Dekanova N P and Khan P V 2017 Comparative analysis of heat supply options for small and middle-sized settlements of Eastern Siberia by using uncertain and fuzzy information *Journal of Physics: IOP Conference Series* Materials Sc. and Engineering. **262 (1)** ID: 012081. DOI: 10.1088/1757 899X/262/1/012081.

[2] Khan V V, Dekanova N P , Romanova T A and Sharaeva S A 2017 Complex analyses of efficiency of energy effectiveness measures for the objects of social sphere in the Eastern Siberia on the basis of system approach. *Izvestiya vuzov. Investitsii. Stroitel'stvo. Nedvizhimost Proceedings of Universities. Investment. Construction. Real estate* **7(1)** pp. 84–93. (In Russian)

[3] Dekanova N P Khan V V 2011 Approaches of the theory of fuzzy sets in the problems of diagnostics of heat networks and heat consumption objects *Modern technologies. System analysis. Modeling* **4(32)** pp. 96-102

[4] Chupin RV and Pham N M 2019 Optimization of the structure and parameters of developing systems of group water supply *Water supply and sanitary equipment* p.30-36

[5] Shcherbakov V, Akulshin A, Chizhik K and Tolstoy M 2018 Design of interacting wells for optimization of investments and operating costs while constructing water-diverting structures *6-th International Scientific Conference on Integration, Partnership and Innovation in Construction Science and Education,* IPICSE Moscow; Russian Federation; 14-16 November Код 143421. DOI: 10.1051/matecconf/201825103037 *MATEC Web of Conferences* **251** *03037*

[6] Chupin V R, Pham N M and Chupin R V 2019 Optimization of the sewerage systems scheme of cities and populated areas *IOP Conference Series: Materials Science and Engineering* **667** 012017. doi: 10.1088/1757 899X/667/1/012017/

[7] Lavygina O L and Grebneva O A. 2019 Environmental technologies in the housing and communal service system of the Baikal natural territory *Izvestiya vuzov. Investitsii. Stroitel'stvo. Nedvizhimost Proceedings of Universities. Investment. Construction. Real estate* **9(4)** pp 726–33. https://doi.org/10.21285/2227-2917-2019-4-726-733

IOP Conf. Series: Materials Science and Engineering **880** (2020) 012045 doi:10.1088/1757-899X/880/1/012045

[8] Lavygina O, Grebneva O and Maizel I 2019 Environmental aspects for the reconstruction of housing and communal services in the village Listvyanka of Irkutsk region. *IOP Conf. Series: Materials Science and Engineering* IOP Publishing doi:10.1088/1757-899X/**667**/1/012056

[9] Chupin V R, Pham N M and Chupin R V 2019 Optimization of developing district water supply systems taking into account variability of perspective water consumption *IOP Conf. Series: Materials Science and Engineering* **667** 012018 doi: 10.1088/1757 899X/**667**/1/012018/

[10] Skibo D, Sudnikovich V and Tolstoy M 2018 Technical innovations as a means of increasing the efficiency of technological equipment in modern production conditions *Int. Scientific Conf. Investment, Construction, Real Estate: New Technologies and Special-Purpose Development* Priorities" ICRE Irkutsk; Russian Federation; 26-27 april Код 142654 **212** 010142018

[11] Tunik A A, Tolstoy M Y and Kalashnikov M P 2020 The complex mobile independent power station for the recreational areas *IOP Conf. Series: Earth and Env. Science* **408(1)** 012012

[12] Kobersi I S, Finaev V I 2013 Control of the Heating System with Fuzzy Logic. World Appl. Sc. J. **23** (**11**): 1441-47.

[13] Piegat A 2001 Fuzzy Modelling and Control Physica-Verlag Heidlberg p 798

[14] Rutkovska Д Pilinski M and Rutkovski L 2008 Neural networks, genetic algorithms and fuzzy systems (Moscow: TeleCom) p 462

[15] Taha H 1985 Introduction to operation research. In 2 books (Moscow: Mir) book I p 476, book II p 496

[16] Kulkov V N, Solopanov E Yu and Kamalov R T 2019 Using immobilised sludge for wastewater treatment and its air regeneration *Proceedings of Universities. Investment. Construction. Real estate* **9(3)** pp 522–529

[17] Pupyrev E I and Chupin V R 2019 Features of regional development of water disposal systems in the central ecological zone of the Baikal natural territory *Proceedings of Universities. Investment. Construction. Real estate* **9** pp 354–363

[18] Chupin V R and Dushin A S 2019 Assessment of the reliability of water supply to consumers: water supply reliability indicators Proceedings of Universities. Investment. Construction. Real estate **9(3)** pp 578–593

IOP Conf. Series: Materials Science and Engineering **880** (2020) 012046 doi:10.1088/1757-899X/880/1/012046

Construction of waste incineration thermal power plants for a comprehensive solution to the problems of solid waste disposal in the Irkutsk region

A N Kudryashov[1], E V Samarkina[1] and A N Petrova[1]

[1]Irkutsk Research and Development Technical University, Lermontov St., 83, Irkutsk, 664074, Russian Federation

E-mail: ekatsamar@yandex.ru

Abstract. Proposed is a method of processing and recycling solid household wastes (SHW) by combustion in boiler furnaces together with irring - Borodin coal. The technology of preliminary sorting with the list of necessary equipment and works is presented. Thermal calculations of the boiler were performed BBP-420-140-6 on combined combustion of SHW and coal in different ratios. Technological solution is proposed, which allows to dispose of sorted solid household wastes during combustion in furnaces together with coal in different ratios.

1. Introduction

In the age of industrialization and mass consumption of raw materials, one of the major problems is the issue of generation and accumulation of garbage, as well as the search for methods of its disposal. The situation in the Russian Federation and the Irkutsk region in the field of education, use, decontamination, storage and disposal of waste leads to negative pollution of the environment, unsustainable use of natural resources, significant economic damage and poses a real threat to the health of the population of the Russian Federation.

According to statistics, the Irkutsk region is ranked ninth in the Russian Federation and third in the Siberian Federal District by the number of SHW formed. The volume of waste is 2.5% of the same indicator for the Russian Federation. In terms of the volume of processing and disposal of garbage the region occupies one of the last revenge. If in the country this indicator is 60%, in the Irkutsk region so far 27.7%. Some of the waste can be processed to produce secondary products. However, solid household waste (SHW) is also generated, with no applications that can be an affordable and cost-effective renewable energy resource. Unlike natural fossil resources, whose location is independent of man, solid household wastes are formed where electrical and thermal energy are most needed, i.e. in large cities. Also important is the fact that SHW is independent of different climatic and geographical factors, unlike other secondary and renewable sources of energy.

Given the characteristics of our region and the world experience in the field of processing and recycling of SHW, one of the directions today is the joint burning of SHW and fossil fuels in powerful energy boilers. It should be noted that in the construction and operation of thermal electric stations (TES), SHW not only generates and usefully uses electric and thermal energy, but also solves an important social task - waste landfills are eliminated, polluting the environment and removing huge areas from the sphere of useful use.

2. Materials and methods

The existing SHW disposal system in the Russian Federation is based on the disposal of the vast majority of waste (about 98%) at landfills and unorganized landfills. This situation has a negative impact on the state of the natural environment and the sanitary and epidemiological situation both in the country as a whole and in the vicinity of major cities in particular. Existing polygons built decades ago are almost completely filled. Therefore, many large cities, megacities and agglomeration systems of settlements face a "garbage collapse" with a projected social explosion. The Irkutsk region is no exception. Figure 1 shows a map of landfills in the Irkutsk district.

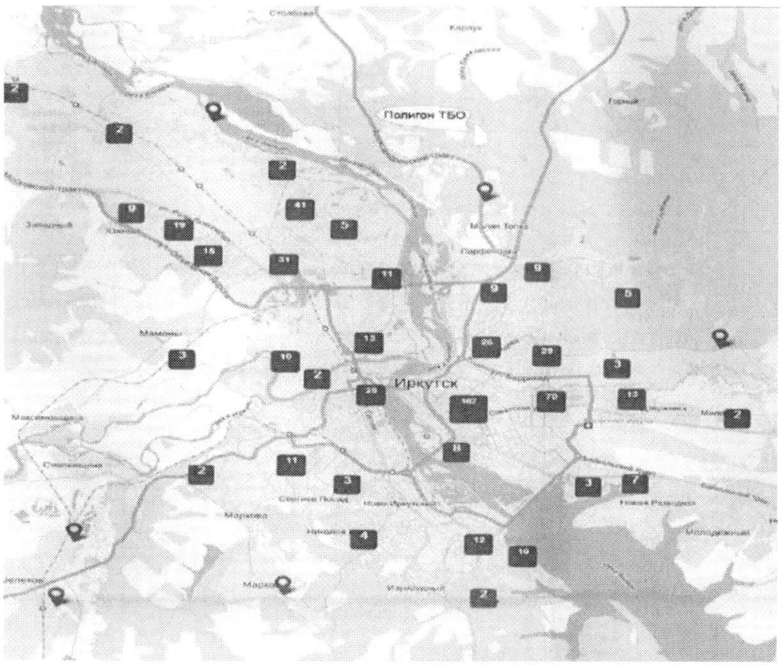

Figure 1. Map of landfills in the Irkutsk region

The paper offers a comprehensive solution to the problem of solid waste disposal, which consists of several stages: preliminary sorting of solid waste with the selection of secondary raw materials; drying, grinding of sorted solid waste; burning of solid waste together with coal in the furnace of an energy boiler. At the same time, the technology of direct combustion of solid waste, the implementation of thermal methods for burning 1000 kg of solid waste will allow you to get thermal energy equivalent to burning 250 kg of fuel oil. However, the real savings will be greater, since the fact of preserving primary raw materials and the costs of its production are not taken into account.

The proposed technological scheme for generating heat and electricity in combination with a thermal power plant (reconstruction of an existing thermal power plant) is shown in figure 2.

IOP Conf. Series: Materials Science and Engineering **880** (2020) 012046 doi:10.1088/1757-899X/880/1/012046

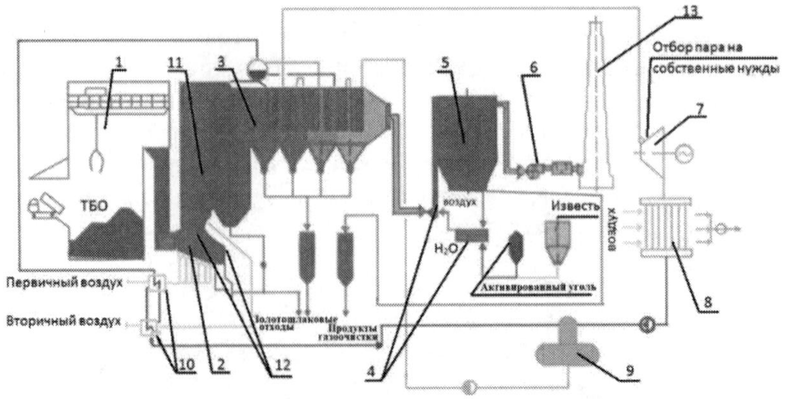

1-receiving hopper; 2-firebox; 3-heat recovery boiler; 4-reactor with mixer; 5-bag filter; 6-smoke pump; 7-turbogenerator; 8-air condenser; 9-deaerator; 10-air heaters; 11-reagent input; 12-gas burners; 13-chimney

Figure 2. TES burning solid waste with a capacity of 150-380 thousand tons.

Pre-sorting involves separating solid waste into fractions at waste processing plants manually or using automated conveyors. This includes the process of reducing the size of garbage components by grinding and sifting them, as well as extracting larger or smaller metal items, such as cans. Selecting them as the most valuable secondary raw materials precedes further disposal of solid waste. To implement pre-sorting of garbage, the following equipment: sorting cabin area of 140 m2, equipped with 20 work stations for sortie-toos 16 kinds of products [10]; frameless hangar, with an area of 427 m2 (28,5x15 m), where the domestic premises on the basis of two sea containers with an area of 62 m2, which houses changing rooms with showers and toilets, an office, recreation room workers; landscaping an area of 165 m2 (137 road plates 6x2 m and the cage from the scattering of debris); construction of a frameless sorting building that does not require "serious" foundations on "standard" soils.

The estimated estimated cost, according to the calculations received, is about 30 million rubles. Of these, 10 million 404400 rubles – the total price of the equipment (crane departures 9.7 m EPSILONM100L97, specialized and/m KAMAZ-6520 with Multift bunker under cullet V=20 m3, feed conveyor and sorting conveyor, compactor and more), the rest of the cost of machinery and vehicles.

The paper calculates the main technical and economic indicators of a thermal power plant with a capacity of 380 thousand tons per year using solid waste operating independently (option 1) and in conjunction with an energy thermal power plant (option 2). The results are presented in table 1.

Table 1. Main economic indicators of a thermal power plant with a capacity of 360 thousand tons per year using solid waste operating independently (option 1) and in conjunction with an energy thermal power plant (option 2)

№	Name of the indicator	Option 1	Option 2
1	Cost of equipment, mln. RUB.	2600	2230
2	Cost of construction and installation works, mln. RUB.	1000	860
3	Cost of project work, mln. RUB.	415	360
4	Cost of commissioning, mln. RUB.	215	175
5	Total capital expenditures, mln. RUB.	4230	3625
6	Specific capital expenditures, million rubles per 1 Megawatt	170	98
7	The cost of disposal of solid household waste, RUB/tons	1200	1020
8	Cost of electricity generation, RUB / kWh at:		
	$Q_н^р$ = 1300 kcal / kg	1,44	0,86
	$Q_н^р$ = 2000 kcal / kg	0,89	0,48

The considered option of disposal of solid waste after sorting when burning in the furnaces of power boilers of TES, will reduce fuel costs and, consequently, the cost of energy produced [11].

3. Results

It should be noted that for reliable combustion of solid waste together with coal in boilers (in particular BBP-420-140-6), it is necessary to provide for their pre-drying and grinding separately from the fuel preparation path of the TES. BBP boiler layout-420-140-6 it is made according to the P-shaped scheme and designed for operation in seismic areas [12].

In order to choose the optimal ratio of Irsha-Borodinskiy brown coal and SHW, several variants of the composition of the solid fuel and SHW mixture were calculated. The main calculated parameters are shown in table 2.

Table 2. Final table of completed calculations

Name of the value	Irsha-Borodinskiy brown coal	10% SHW and 90% of coal	20% SHW and 80% of coal	30% SYM and 70% of coal
the efficiency of the boiler, %	92,97	92,853	92,687	92,55
Total fuel consump-tion, kg / s	20,17	20,88	21,66	22,49
Estimated fuel con-sumption, kg / s	20,11	20,82	21,6	22,43
Exhaust gas temperature, ^{0}C	147	148	150	151
Temperature of gases at the outlet of the fur-nace, ^{0}C	1063	1063	1063	1062
The quantity of exhaust gases, kg / h	660612,3	668720,9	678711,4	688070,9

The optimal ratio is proposed: 10% of solid waste and 90% of Irsha-Borodinskiy coal.

To increase the efficiency of a thermal power plant using solid waste operating independently, it is necessary to provide for an increase in steam parameters (above 420 ^{0}C). This will be achieved by using corrosion-resistant steels and protective coatings; using intermediate superheated steam: the steam at the exit of the high-pressure stage of the turbine is sent back to the boiler, where it is heated to the initial temperature, then enters the pressure stage; using external superheating on heat exchange surfaces heated by the combustion products of organic fuel.

The combination of a Thermal power station using Solid household waste with traditional thermal power plants and CCGTs will allow:

- increase the efficiency of power generation, in relation to waste processing, from about 20 % to 30-35 % and thereby increase the installed capacity;
- to minimize the cost of construction and installation works;
- to save fossil fuels.

4 Discussions

Thermal utilization of sorted SHW with the release of heat and electricity to the urban consumer is a more preferable option than generating the same amount of energy at a conventional thermal power plant plus processing, and even more so, disposal, of an equivalent amount of waste [22]. That is why it will be more efficient to use sorted SHW as fuel together with coal. In addition, the analysis of the waste composition showed that 35 tons of 100 tons of solid waste will be recycled. During sorting, hazardous waste is also removed, which will be sent to recycling companies.

An important issue that requires a separate serious study is the issue of the formation of pollutants into the environment, as well as the development of technologies for their disposal. Table 3 shows the regulatory emissions of pollutants from outgoing gases of power plants that burn solid municipal waste and organic fuel.

Table 3. Some regulatory emissions of pollutants from outgoing gases of power plants that burn solid municipal waste and organic fuel

	Energy technology installations that burn solid waste (EU 94/67/EEC)		Power plants for burning solid fuels (GOST 50831-95)	
	Daily average	Average half-hours	α=1,4	in recalculation on α=2
Solid particle	10	30	150-250	100-165
Carbon monoxide, CO	50	100	300-400	200-260
Nitrogen oxides, NO2	200	400	300-640	200-420
Sulfur oxides, SO2	50	200	1200-1400	800-920
Hydrogen chloride, HCL	10	60	not normalized	not normalized
Hydrogen fluoride, HF	1	4	not normalized	not normalized

Given that at the moment, most thermal power plants that use coal and solid waste as fuel have a negative impact on the environment, which is unacceptable in many cases today, and even more so in the future. Therefore, promising energy technologies and equipment should be developed in such a way that their impact is minimal and meets modern and planned sanitary standards. Meeting these requirements by economically justified means is very difficult. To overcome them, a balanced concept of creating large environmentally friendly power plants that run on low-grade coal is necessary.

In this paper, the calculation of harmful emissions into the environment is performed, the results of which are presented in table 4.

Table 4. Comparative calculation of emissions of pollutants into the atmosphere

	Irsha-Borodinskiy brown coal	10% SHW and 90% of coal
Ash particles	480,98 tons / year	520,54 tons / year
Sulfur oxides	1415,93 tons / year	1465,78 tons / year
Nitrogen oxide	1281,82 tons / year	1274,38 tons / year
Benzo(a)pyrene	0,56 tons / year	0,546 tons / year

From the calculations carried out, it can be seen that when solid waste is burned together with coal, emissions of Benz(a)pyrene and nitrogen oxide into the atmosphere are reduced through the use of reagent technologies. However, this increases the number of emissions of ash particles and sulfur oxides.

5 Conclusions

In this work, we selected one of the most effective and appropriate methods for processing and disposing of SHW in our region – burning pre-sorted SHW together with Irsha-Borodinskiy coal. The paper presents a pre-sorting technology with a list of necessary equipment and works. Thermal calculations of the BBP boiler unit were made-420-140-6 on Irsha-Borodinskiy coal and on the joint combustion of solid waste and coal in different ratios. The optimal ratio is proposed: 10% of solid waste and 90% of Irsha-Borodinskiy coal. A comparative calculation was performed for various fuel combustion technologies.

Thus, the study showed that the development and implementation of this technological project requires significant capital investment at all stages of the life cycle of this type of production, and this, in turn, should be accompanied by a thorough technical and economic justification of innovations with the determination of the economic effect of the project, as well as competent environmental proposals.

References

[1] *Order Of the government of the Russian Federation "On the Energy strategy of Russia for the period up to 2030"* dated 13.11.2009 N 1715-p.

[2] *Concept of solid waste management in the Russian Federation. Resolution of the state Committee of the Russian Federation on construction and housing and communal services* No. 17 of December 22, 1999.

[3] *Federal law* "On energy saving and energy efficiency improvement and on amendments to certain legislative acts of the Russian Federation" dated 23.11.2009 N 261-FZ.

[4] *Catalog-directory of enterprises operating in the field of waste management of production and consumption in the Irkutsk region 2011* **3** p 43

[5] Tugov A N 2014 Prospects for energy utilization of solid waste *Energosovet* **4 (35)** pp 31-35

[6] Touš M., Pavlas M, Putna O and Stehlík P 2015 Combined heat and power generation planning in a waste-to-energy plant on a short-term basis *Energy* **90** pp 137-147

[7] Teixeira S, Monteiro E, Silva V and Rouboa A 2014 Prospective application of municipal solid wastes for energy production in Portugal *Energy Pol.* 71 pp 159-168

[8] Tan H, Zhao Y, Ling Y, Wang Y and Wang X 2017 Emission characteristics and variation of volatile odorous compounds in the initial decomposition stage of municipal solid waste *Waste Manag* **68** pp 677-687

[9] Ilinykh G V 2013 Evalization of Thermotechnical Properties of Solid Waste from Their Morphological Composition *Urbanistika* **3** pp 125-137

[10] Edjabou M E, Martín-Fernández J A, Scheutz C and Astrup T F 2017 Statistical analysis of solid waste composition data: arithmetic mean, standard deviation and correlation coefficients. *Waste Manag.* **69** pp 13-23

[11] Zelikov E N 2008 Ash Properties, Deposit Formation and Corrosion of Superheater Tubes of Mimicipal Solid Waste Firing in Swirl Fluidized Bed Furnaces *Circulating Fluidized Bed Technology IX in conjunction with the 4-th Int. VGB Workshop "Operating Experience with Fluidized Bed Firing Systems* pp 887-892

[12] Urbancl D, Zlak J, Anicic B, Trop P and Goricanec D 2016 The evaluation of heat production using municipal biomass co-incineration within a thermal power plant *Ener-gy* **108** pp 140-147

[13] Shvartz A L 2001 *Cooperation of VTI with boiler-building plants in the development of domestic energy* Heavy engineering **6** pp 5-8

[14] Ugnachev V I 2001 Control of operation of gas-cleaning equipment on installations for burning solid household waste *Teploenergetika* **12** pp 52-56

[15] Tugov A N Designs of boilers for burning solid household waste and the specifics of their thermal calculation *Electric Power stations* **9** pp 21-25

[16] Tugov A N 2008 Methods for reducing emissions of nitrogen oxides at thermal power plants that burn solid household waste *Electric stations* **11** pp 13-17

[17] Kurakov V M 2009 Technology of a full cycle of waste management *Housing and communal service* 2 (1) pp 55-58

IOP Conf. Series: Materials Science and Engineering **880** (2020) 012047 doi:10.1088/1757-899X/880/1/012047

Determining the cost-effectiveness of thermal energy production by air-to-water heat pumps

S I Kuzmin[1] and N A Ivshina[1]

[1] Federal State educational institution of higher education budget "Angarsk State Technical University, Russia, Angarsk, 665835

E-mail: pgs@angtu.ru

Abstract. This paper provides a methodology for identifying the cost-effectiveness of thermal energy generation by air-to-water heat pumps. The results of a comparative estimate of the cost of energy received and the profitability of the method under consideration are presented.

1. Introduction

In solving the problem of energy conservation, one of the most promising methods to improve the energy efficiency of buildings is the use of thermal energy of the environment - atmosphere, sources of natural water, soil, having a comparatively low temperature potential. The thermal energy contained in these sources is extracted by heat pumps with increased temperature capacity for use in the building's heat-consuming systems - heating, hot water, ventilation.

The use of heat pumps as "miners" of energy from "gift" sources is an attractive idea in itself, but its implementation is associated with the high cost of heat extraction equipment, which calls into question the economic feasibility of this way of heating. On the other hand, the constant increase in energy prices with irritating consumers inexplicable speculative component of tariffs gives some positive emotional effect from the "independent" energy supply. Therefore, many consumers are willing to take one-off economic costs for the sake of gaining independence and relative freedom in providing thermal energy "their object" from a "no man's" source: air, soil, reservoir.

At the same time, it is necessary to have a fairly objective idea of the price, which will cost the use of an alternative method of energy generation. Moreover, traditional heating and heat generation systems have spent technology (whole industry), use fossil fuels, as yet inexhaustible reserves, established infrastructure that ensures stable energy supply of guaranteed capacity and volume.

The transition from traditional sources of heating to alternative (renewable) should be accompanied by an objective understanding of the ratios of acquisitions and losses from such replacement, especially since the economic component in long-term operation object may prove to be a priority when deciding on the maintenance of engineering equipment and building systems.

2. Materials and methods

This paper considers the provisions of the methodology and the results of the evaluation of the effectiveness of heat pumps in the conditions of the changing temperature of the heat-containing environment - outdoor air. The methodology is based on the use of analytical expressions and statistical models of system parameters and related elements that determine a comprehensive assessment of the effectiveness of the heating method.

3. Results

The most affordable way to generate thermal energy is the air-to-water heat pump system, in which heat is extracted from the outside air and transferred to the heat-carrier heating system. The main operating feature of such a system is the dependence of the heat performance of the pump on the temperature of the outside air and the required temperature of the coolant. It therefore seems necessary to take this factor into account when choosing equipment and analysing the system. The main criterion for the effectiveness of the "heat source" is the cost of the thermal energy generated (RUB/MW/h):

$$C_{пр.Q} = (\frac{C_K}{T_Э} + C_Э)/W_{TH}, \quad (1)$$

where is C_K and $C_Э$ - costs respectively for the heat pump and its installation (rub.) and operational, (rub./year);

$T_Э$ - the period of operation of the system, the year;

W_{TH} - the amount of thermal energy generated by the (pumped) system,

MW/year.

The cost of the heat pump system will be presented by dependence on its calculated (passport) thermal performance:

$$C_K = K_{монт} * c_q * Q_{TH.0}, \quad (2)$$

where is c_q - Value (unit value) related to the unit of the estimated thermal power of the pump, ruble/kW;

$K_{монт}$ - factor that takes into account the cost of installing the system.

Accepted the proportional cost of the heat pump.

Operating costs are mainly related to the consumption of electricity for pumping the working environment by the pump compressor $C_{э.эл}$ and ongoing repairs (maintenance) $C_{э.TP}$:

$$C_Э = C_{эл} + C_{TP} . \quad (3)$$

$$C_{э.эл} = N_{TH.0} \sum (N_{tt.i} * \tau_{tt.i}) * c_{эл}, \quad (4)$$

where is $N_{TH.0}$ - installation capacity of the heat pump compressor, kW;

N_{tt} и τ_{tt} -accordingly, the relative power consumption of the heat pump under current working conditions and the duration of the period of standing of these conditions, an hour;

$c_{эл}$ - cost of electricity, ruble/kW*h.

The annual cost of ongoing repairs and maintenance is usually tied to the cost of the equipmen [5] and is (3-8)% of C_K. Then addiction (3) will take the form:

$$C_Э = N_{TH.0} \sum (N_{tt.i} * \tau_{tt.i}) * c_{эл} + k_{TP} * K_{монт} * c_q * Q_{TH.0}, \quad (5)$$

where is k_{TP} - the rate of deductions for the current repair, year^{-1}.

As the analysis of the prices of equipment presented on the market shows, the specific cost of heat pumps is inversely proportional to their estimated thermal performance. Figure 1 is dependent c_q from maximum (passport) thermal power for well-known heat pump brands (according to the Price-list of manufacturers and the retail chain at the end of 2019).

As can be seen from the chart, the price range fluctuates significantly, especially for devices with a capacity of up to 15 kW - the difference in price can reach 600%! Therefore, in order to implement different comparison options, it would be appropriate to introduce a distribution c_q (ruble/kW) 2 parametric model (6): for maximum price (argument) and minimum (argument) $\overline{P} = -1$).

$$c_q = \exp(4{,}461 + 0{,}998\overline{P} - 0{,}0571Q_{TH.0} - 0{,}0093Q_{TH.0} * \overline{P} + 0{,}00051Q_{TH.0}^2) * 10^3.$$
(6)

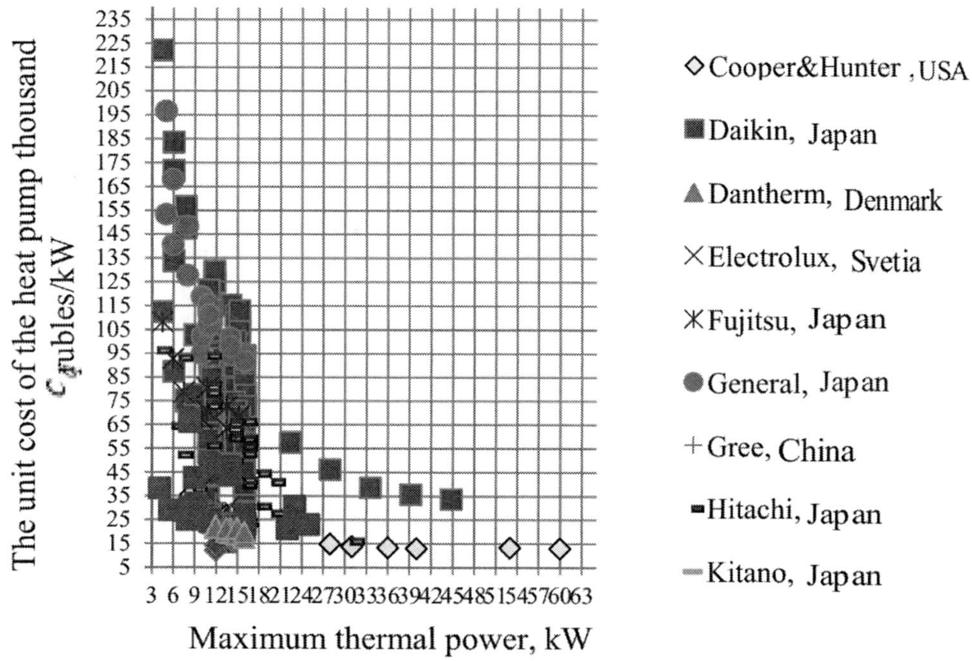

Picture 1. Value dependency C_q, thd. rub./kW. from the estimated thermal power $Q_{TH.0}$, kW

Equation (4) adequately characterizes price distribution to 5% accuracy (statistical convergence criteria: Fisher's calculation criterion for adequacy variance is 0.0022 at table value $F_{ad}^{reg.} = 3{,}84$ 0.95 per cent).

The cost factor for the installation of the heating system $K_{монт}$ proportional to the power of the heat pump, being for air-water systems ranging from 1.8 to 1.3 and slightly decreases with growth $Q_{TH.0}$. This dependency can be imagined as an equation (6):

$$K_{монт} = 1{,}826 - 0{,}0087 * Q_{TH.0}. \quad (6)$$

The amount of heat energy pumped by the heat pump (MW)hour depends on its power $Q_{TH.0}$, duration τ_{TH}, temperature of the heated environment t_w and outdoor air $t_н$:

$$W_{TH} = \sum (Q_{TH.0} * k_{tt.i} * \Delta\tau_t)_i, \quad (7)$$

where is k_{tt} - temperature factor taking into account the change in the heat performance of the pump from the temperature t_w и $t_н$;

$\Delta\tau_t$ - period during which the temperature of the outside air corresponds to a certain value, an hour.

The temperature of the heated environment is determined by the type of heating system. For a water system with compact heating devices (radiators or convectors) it is accepted at the level of 55 degrees Celsius, for heating systems of the type "warm floor" - 35 degrees Celsius [6].

The heat conversion rate by the pump and the electricity consumed depend on the design features and the type of refrigerant and is the working characteristic of the pump. For pumps with inverter compressors, this characteristic is graphically presented in Figure.2 [7].

These characteristics are universal and express a change in the parameter relative to the estimated power of the heat pump. To solve the problem, it is convenient to present them in the form of dependencies:

$$k_{tt} = \frac{Q_{tt}}{Q_{TH.0}} = \exp(0{,}0333 * t_н - 0{,}0062 * t_w - 0{,}1289) \quad (8)$$

- for $t_w = 55°C$: $\overline{N}_{TH.tt} = \frac{N_{tt}}{N_{TH.0}} = 0{,}899 + 0{,}0125 * t_н$

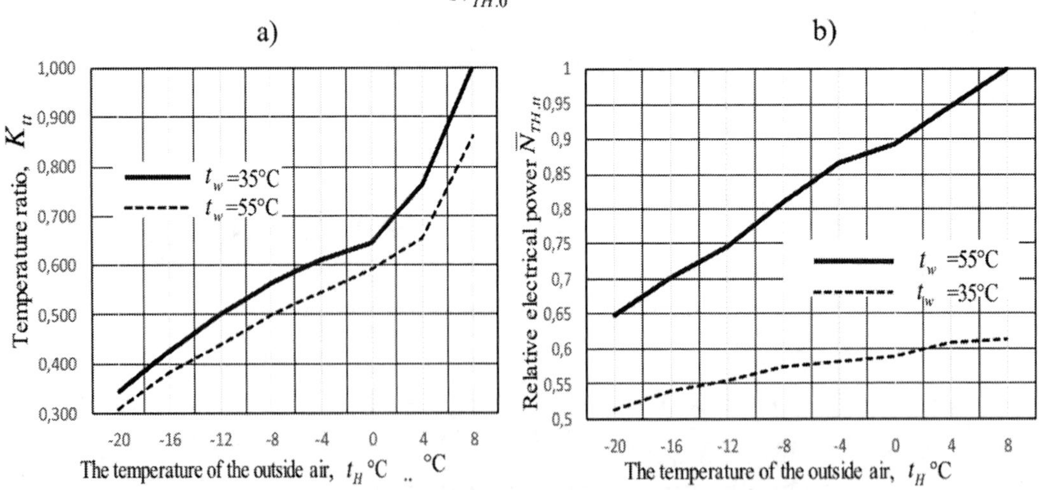

Picture 2. Changing temperature k_{tt} (a) and the electricity consumed $\overline{N}_{TH.tt}$ b) from working temperatures

- for $t_w = 35°C$: $\overline{N}_{TH.tt} = \frac{N_{tt}}{N_{TH.0}} = 0{,}585 + 0{,}0036 * t_н \quad (9)$

where is Q_{tt} - current heat pump power at appropriate temperature values, W;

N_{tt} - current electricity pump power consumption at appropriate

IOP Conf. Series: Materials Science and Engineering **880** (2020) 012047 doi:10.1088/1757-899X/880/1/012047

temperature values, kW;

The distribution of periods of standing of outdoor air temperatures is given in [8] for the characteristic points of climatic zones. Figure 3 shows graphs of the distribution of temperature standing periods for geographic points "moderate" (Moscow) and "moderately cold" (Ulan-Ude) climate in the working ranges of heat pump temperature (from 8 degrees Celsius to -25 degrees Celsius).

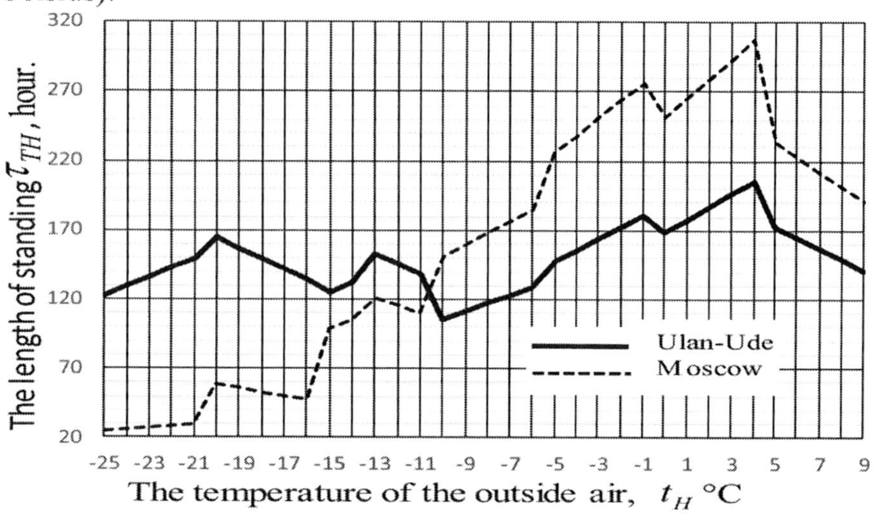

Picture 3. The duration of the temperature of the outdoor air

Installation power of heat pump compressors $N_{TH.0}$ proportional to nominal thermal power (Figure 4) and this dependence can be represented by formula (10).

$$N_{TH.0} = 0{,}0277 + 0{,}257 * Q_{TP.0} - 0{,}00015 * Q_{TH.0}^{2} \cdot \quad (10)$$

Electricity tariff $c_{эл.\tau}$ not constant and increases in time. To take this factor into account, we will use the expression [9]:

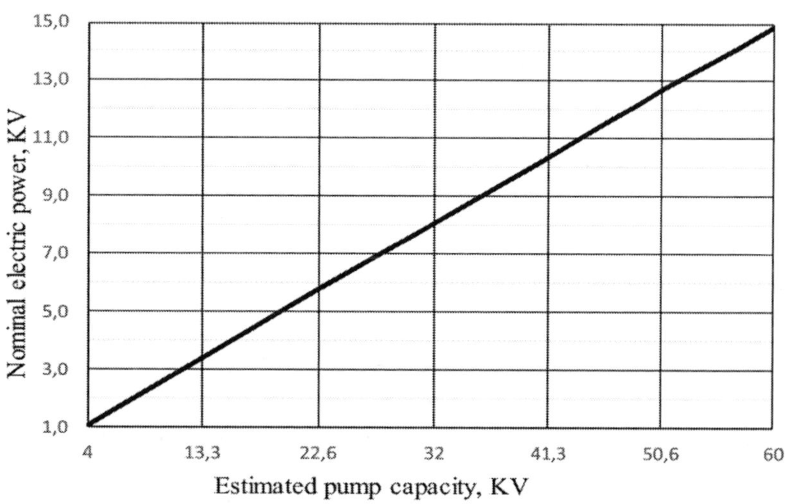

Picture 4. Compliance with the installation electrical and thermal capacity of the pumps.

$$c_{эл.\tau} = c_{эл.0} * (1+\bar{c}_W)^{(\tau_R - 1)} \quad , \tag{11}$$

where is $c_{эл.\tau}$ the current cost of electricity, rub/kW*h;

$c_{эл.0}$ -the cost of electricity at the time of the project, rub/kW*h;

$\bar{c}_W$ - tariff increase in shares from the previous year;

τ_R - period from the implementation of the project to the calculation of the calculations, the year. In this case, it is accepted equal $T_э$

Thus, equations (2-13) form a system that fully describes the specifics of the production of thermal energy by the pump in the conditions of changing economic and technical parameters.

4. Debate

The steady trend of rising energy tariffs contributes to the search for technical solutions that provide cheaper energy with ways of heating independent from network companies. On rice. 5 the results of the analysis of the impact of individual characteristics of the system on the cost of thermal energy produced by heat pumps (with a minimum price level) calculated on the proposed model (2-13) (at the level of electricity tariff of 1.1 rubles/kWh for Ulan-Ude and 5.47 rubles/kWh for Moscow and 2 percent annual increase in the cost of electricity). To estimate this indicator, the graphs show a line reflecting the tariff on thermal energy in the central heating systems as of 2019 for the relevant item.

The cost of obtaining air-water pumps depends significantly on its estimated capacity and the duration of operation of the equipment before it is replaced or overhauled. The high proportion of one-time costs in the cost of a system with a relatively small capacity (up to 15 kW) means that the cost of

heat produced does not become less than the district tariff, even during the long period of operation of the system in temperate-cold climate. And only due to the high level of heat prices in the temperate climate, the cost of production of low-potential heat ($t_w = 35°C$) the pump begins to compete with the traditional method at the (7-8-year)-year period of operation and further becoming a cost-effective and therefore more attractive way for the consumer. The use of more powerful heat pumps (30-60 kW) looks quite optimistic.

Here, the share of capital expenditures in these drastically reduced, which affects the cost of the final product - $C_{np.Q}$. The cost of thermal energy generation becomes less costly after the (6-7) years of operation of the equipment, even at a relatively low tariff (moderately cold area) and almost

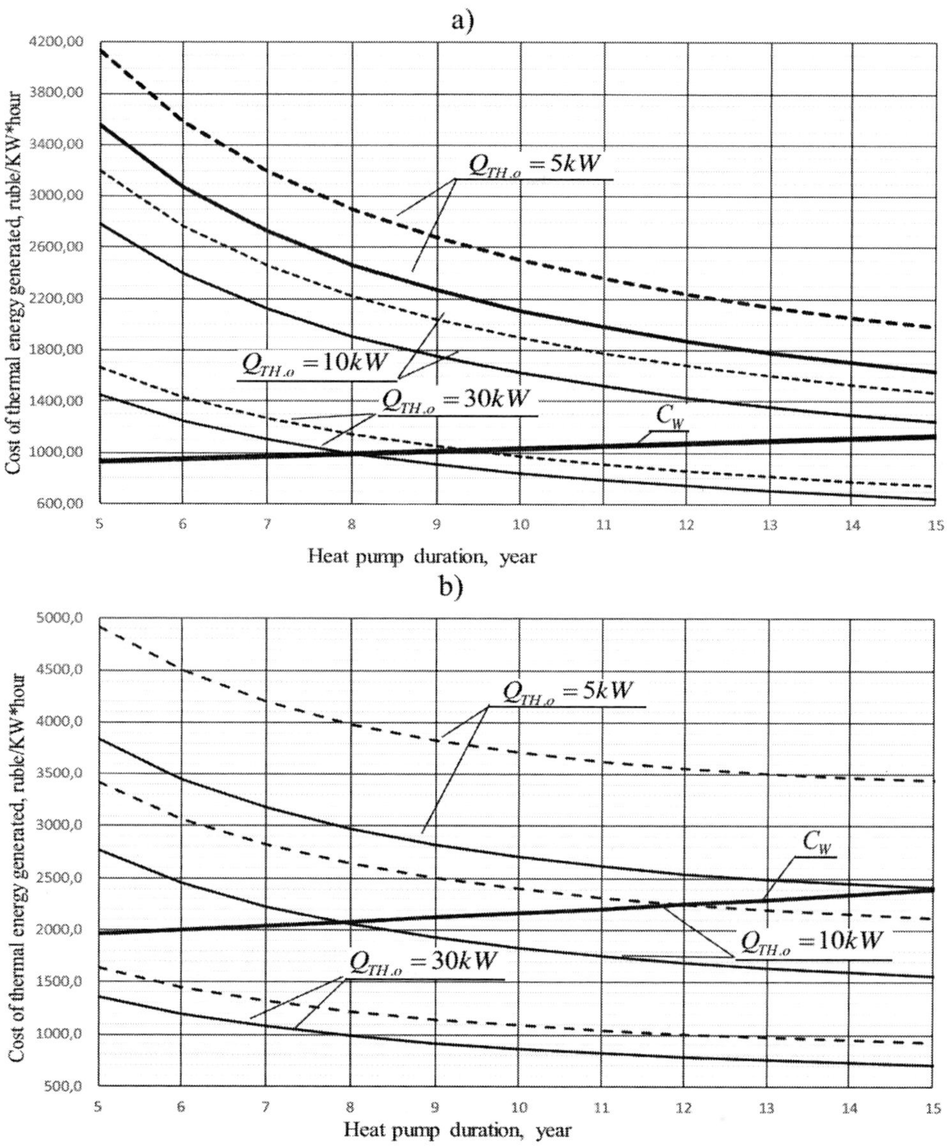

Picture 5. Change in the cost of thermal energy generated by heat pumps for the conditions of Ulan-Ude a) and Moscow b).

immediately (3-4) years) for items with high prices for Energy. At the same time, it should be kept in mind that the life of such equipment (similar to household refrigerators) can be 20-25 years and, accordingly, you can eventually get thermal energy at a price almost 2 times lower than the current tariff. Therefore, it makes sense to focus on the use of one more powerful heat pump for more objects than to supply each object with a low-power machine.

Of course, it should be borne in mind that these conclusions follow from the analysis of graphs (Figure 5) oriented to the minimum level of prices (P= -1 in the equation (6)). And as a rule, cheap equipment is less durable than expensive. Therefore, the cost and timing of achieving competitive-capable costs will be different. But using the proposed dependency can easily identify these metrics for a specific facility and operating conditions.

5. Conclusion

Analyzing the resulting expressions (2-13) to determine the cost of thermal energy production, it can be noted that the proposed methodology takes into account as fully as possible factors that contribute to obtaining objective baseline information for evaluation efficiency of air-to-water heat pumps for heat heating facilities. At the same time, the proposed approach applies to any thermal consumption capacity facility located in any climate zone. The proposed method, although it has an "estimated" purpose, allows you to predict the possible effects of the use of heat pumps and choose a satisfying way to provide the object with thermal energy.

References

[1] JV 50.13330.20112 Building Heat Protection.

[2] Groshev A E and MiroshnichenkoT A 2018 Analysis of energy-saving standards and energy-efficient buildings exhibition centres *Collection of scientific works "development of modern science: theoretical and applied aspects* p13-17

[3] Vakhitova R I and Sarychev D A 2015 To question about saving energy in buildings *Materials of the scientific session of scientists and runs the State oil Institute* **2** p65-68

[4] Hu Y 2016 Analysis on thermal performance of composite wall of energy saving building *Chemical Engineering Transaction* **51** p 1183-1188

[5] Romanova A A, Rymkevich P P and Groshev A S 2014

Methodology Calculation the projected payback term of energy-saving measure on insulation of buildings *Technical and economic problems of the*
[6] *service* **9** p 68-74
[7] JV 60.13330.2012 Heating, ventilation and air conditioning.
[8] Graphic performance of DHP-A-type heat pumps. Test materials at the manufacturer.
[9] GOST 16350-80 Climate of the USSR. The zoning and statistical parameters of climatic factors for technical purposes.
[10] Kuzmin S I 2013 Efficiency of heat-saving activities for a large-panel apartment building *Collection of scientific works by AGTA* p101-108
[11] Kochev A, Sokolov M, Kocheva E and Moskaeva A 2018 Energy-saving measures in buildings considered as monuments of history and architecture *MATEC Web of conferences «International conferences on Research in Mechanical Engineering Sciences* p 04021
[12] Chupin V R and Dushin A S 2019 Parameter optimisation for new and reconstructed systems of water supply and distribution for ensuring consumer's uninterrupted water supply *Proceedings of Universities. Investment. Construction. Real estate* **9(4)** pp 790–803
[13] Kulkov V N, Solopanov E Yu and Kamalov R T 2019 Using immobilised sludge for wastewater treatment and its air regeneration *Proceedings of Universities. Investment. Construction. Real estate* **9(3)** pp 522–529
[14] Pupyrev E I and Chupin V R 2019 Features of regional development of water disposal systems in the central ecological zone of the Baikal natural territory *Proceedings of Universities. Investment. Construction. Real estate* **9** pp 354–363

IOP Conf. Series: Materials Science and Engineering **880** (2020) 012048 doi:10.1088/1757-899X/880/1/012048

Study of changes in emissions into the atmosphere with the reconstruction of heat supply systems

O Lavygina[1] and O Grebneva[2]

[1]Ikutsk National Research Technical University», Institute of Architecture and Construction, Department of Urban development and Economy, 664074, 83, Lermontov street, Irkutsk, Russia
[2] Melentiev Energy Systems Institute of SB RAS, Department of pipeline systems of Energy, 664033, 130, Lermontov street, Irkutsk, Russia

E-mail: olgakot81@mail.ru

Abstract. In today's world, environmental issues are becoming increasingly important and relevant. The housing and communal services system is often one of the main sources of anthropogenic impact. So, when receiving thermal energy by burning of organic or mineral fuel, various types of pollutants enter the atmospheric air. In this paper, a comparison of gross emissions into the air during operation of a heat source using fuel oil and wood fuel (pellets) is made. An analysis of the results was carried out. It showed that the most "environmentally friendly" are heat sources, the main equipment of which includes boiler units burning wood fuel (pellets). With the operation of this type of boiler, the level of anthropogenic load in the form of pollutants emissions is significantly reduced, in particular, there is a multiple decrease in emissions of pollutants of hazard class III.

1. Introduction

Currently, on the territory of the Russian Federation, the infrastructure of the housing and communal services system has a high degree of moral and physical deterioration. This fact indicates the need for reconstruction of many objects of the heat supply system. In modern socio-economic conditions, the key factor determining the need for the implementation of innovative projects is not only technical and economic indicators, but also environmental ones. So, when choosing the main equipment of sources (for example, boiler units) in heat supply systems, such criteria as the cost of equipment are considered; efficiency and flow rate of resources to ensure their work; the unit cost of the generated target product (thermal energy), taking into account the fuel and amortization components; the range of regulation of equipment power without a significant reduction in efficiency [1]. In addition to technical and economic indicators, environmental aspects play a significant role. The main types of negative impact of heat supply system facilities are emissions of pollutants into the atmosphere [2]. As is known, the main factor determining the quantitative and qualitative composition of emissions is the type and properties of the burned fuel [3,4].

In connection with the latest changes in environmental legislation, the state of the environment and the factors of negative impact are receiving increasing attention [5-9]. All of the above necessitates the research presented in this article.

2. The relevance of research

Modern requirements in the sphere of environmental safety determine the conduct of research in the field of improving the quality of environmental protection in reconstruction of heat supply systems. In addition, given the natural value of the Irkutsk region, due to its location near the Lake Baikal, technical solutions that have an environmental advantage are priority tasks in modern conditions.

3. Content statement of the problem and initial data

A study of impact of replacing the main equipment of the heat source when the reconstruction of the heat supply system for one of the settlements of the Irkutsk region on the change in harmful emissions into the environment was carried out.

The reconstruction of the heat source of the heat supply system consisted in the replacement of the hot water boiler KVA 3.0 G / L [10], which runs on liquid fuel (fuel oil), with the Pyrolysis Master boiler, industrial series PELLET PRO [11], which runs on wood fuel (pellets). The characteristics of the working fuel of boiler units are given in table 1.

Table 1. The characteristics of the working fuel

Indicators	Fuel oil	Pellets
Ash, %	0.03	1.5
Sulfur content (total), %	1.22	0.03
Nnet calorific value, MJ / kg	38.77	17.,316

To estimate the calculation variants, the main environmental criterion was used. It is gross emissions of harmful emissions into the air. The calculation of gross emissions was carried out for two variants of fuels for heat source: 1) liquid fuel (fuel oil); 2) wood fuel (pellets).

Calculation of gross emissions was carried out using the «Methodology for determining of pollutants emissions into the atmosphere when burning fuel in boilers with a capacity of less than 30 tons of steam per hour or less than 20 Gcal per hour» [12,13]. When calculating gross emissions fuel combustion indicators are taken into account. These indicators are given in table 2.

Table 2. Indicators of fuel combustion

Indicators	Fuel oil	Pellets
Heat loss from mechanical incomplete combustion q_4, %	0.08	0.5
Heat loss due to chemical incomplete combustion of fuel q3 (общ), %	0.2	0.5
Flue gas recirculation rate r, %	0	0
The sulfur content of the fuel on the working mass Sr, %	1.22	0.03

4. Numerical research

To estimate the degree of negative impact on the atmospheric air, a comparison of the gross emissions of pollutants was made. The calculation results for the two variants of fuels used on the heat source are shown in table 3.

Table 3. Results of gross emissions calculation for combustion of fuel oil and wood fuel (pellets)

Code	Emission name	Gross emission (t / year) with operation of a fuel oil boiler	Gross emission (t / year) with operation of a pellet boiler
0301	Nitrogen (IV) oxide (Nitrogen dioxide)	5.422760	0.973138459
0304	Nitrogen (II) oxide (Nitrogen oxide)	0.881199	0.158135
0330	Sulfur dioxide (Sulphurous anhydride)	35.114772	0.051377
0328	Carbon (Soot)	1.393727	-
0337	Carbon oxide	7.395466	0.867846
0703	Benz / a / pyrene (3, 4-Benzpyrene)	0.00000143689	0.000201881
2902	Suspended matter	-	0.266893
2904	Fuel oil ash of thermal power plants (in terms of vanadium)	0.092996	-

An analysis of the data showed that a significant reduction in gross emissions is noted when using a pellet boiler. When using a fuel oil boiler, more than 35 tons of sulfur dioxide enter the atmospheric air annually. When using wood fuel (pellets), nitrogen dioxide is the main to air pollution.

A feature of environmental performance assessment is the need to take into account the degree of hazard of the emitted substances. Thus, it is more expedient to compare the amount of substances emitted, based on the degree of their danger to human health, which was done at the next stage of numerical studies. So, extremely dangerous substances (having a carcinogenic risk) carry the maximum danger, i.e. substances of hazard class I, and substances of hazard class IV are of low hazard [14].

Thus, taking into account the hazard classes of all substances (given in the Hygienic Standards [14]), it should be noted that when using fuel oil and wood fuel (pellets), substances belonging to hazard classes III and IV are released into the atmosphere. However, when using wood fuel (pellets), their amount is several times less than when using fuel oil. So, for example, when burning fuel oil, more than 42 tons of hazard class III substances will be released into the atmosphere annually, and when burning wood fuel (pellets) – 1.4 tons annually. Figures 1 and 2 show a graphical representation of the distribution of emissions of hazard classes III and IV for different types of fuels.

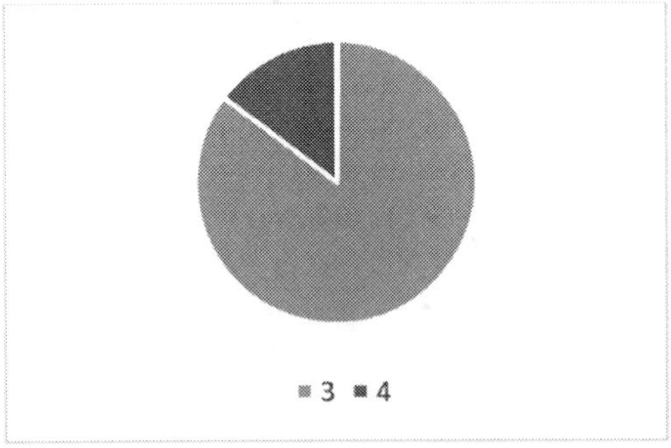

Figure 1. Diagram of total gross emissions with operation of a fuel oil boiler (t/year)

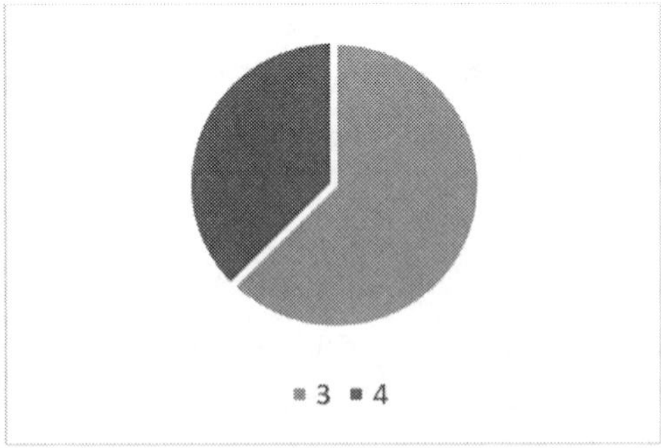

Figure 2. Diagram of total gross emissions with operation of a pellets boiler (t/year)

The total indicator of gross pollutant emissions is:

- when using fuel oil as fuel it is more than 50 tons per year with a predominance of emissions of hazard class III;
- when using wood fuel (pellets) it is 2.3 tons per year, while the share of pollutants of hazard class III is reduced by almost 30 times.

Thus, the analysis of the obtained during the numerical studies results of the distribution of gross emissions of harmful substances showed that when using of boilers for various fuels for heat source wood fuel (pellets) is the most «environmentally friendly».

5. The practical relevance of experimental research

In this work, the efficiency of using boiler on wood fuel (pellets) is shown using a example of a real heat supply system. The obtained results proved that during the operation of pellet boilers, the level of anthropogenic load in the form of pollutant emissions decreases significantly.

The decrease in gross emissions indicates the environmental and economic efficiency of the reconstruction of the investigated energy object. So, with a decrease in gross emissions into the atmosphere, firstly, environmental damage for the environment is directly reduced and, secondly, there is a reduction in payments of enterprises for negative impact. Thus, the transition to the use of wood-burning (pellets) boilers at heat sources will reduce the negative impact on the environment, and enterprises operating housing and communal services will reduce the financial costs of the environmental component.

6. Conclusions

The calculation method proved the feasibility of reconstructing of heat supply system with the transfer of the heat source from fuel oil to wood fuel (pellets). The environmental and economic efficiency of the use of pellet fuel is due to a decrease in gross emissions of pollutants into the atmosphere. The results showed that during the operation of a pellet boiler, there is reduction in emissions of pollutants of hazard class III by 30 times compared with a fuel oil boiler.

Acknowledgments

The studies were carried out in the framework of project III.17.4.3 of the Program of basic research of the SB RAS (AAAA-A17-117030310437-4)

References

[1] Khan V.V. 2016 *University News. Investments. Construction* **4 (19)** p 146 [In Russian]
[2] H. Zhou, Q. Yang, Sh. Zhu, Y. Song, D. Zhang 2019 *Resources, Conservation and Recycling* **144** p 74
[3] Hu Wang, Xumin Zhao, Laihui Tong, Mingfa. 2018 Fuel **227** p 457
[4] H. Chen, J. He, X. Zhong. 2018. Journal of the Energy Institute, corrected proof [In press]
[5] Federal Law of July 21, 2014 N 219-FZ (as amended since January 1, 2020) [In Russian]
[6] Federal Law of January 10, 2002 N 7-FZ (as amended from January 1, 2020) [In Russian]
[7] Federal Law of May 4, 1999 N 96-FZ (edition as of November 1, 2019) [In Russian]
[8] Lavygina OL, Grebneva OA. 2019 *University News. Investments. Construction* **9(4)** p 726 [In Russian]
[9] Olga Lavygina, Oksana Grebneva, and Irina Maizel 2019 *IOP Conf. Series: Materials Science and Engineering* 667 012056
[10] KVA-3.0 boiler characteristic (KVA-3.0-95 boiler) http://www.energokomplekt.net/kotel-kva-3.pdf [In Russian]
[11] Boiler catalog of Pyrolysis Master https://pirolizmaster.ru/kotly/ [In Russian]
[12] Method of Determining Atmospheric Emissions of Pollutants Produced by Burning Fuel in Boilers Having Capacity of 30 t/h of Steam or Less Than 20 GCal/h 1999 (Moscow: State Committee for Ecology of Russia)
[13] Method for calculating, standardizing and controlling emissions of pollutants into the atmosphere (supplemented and processed) 2012 (St. Petersburg: Research Institute Atmosphere) **GN 2.1.6.3492-17** 2017

IOP Conf. Series: Materials Science and Engineering **880** (2020) 012049 doi:10.1088/1757-899X/880/1/012049

Improving the quality of the air in modern energy-efficient buildings

S S Makarov[1] and E E Baimachev[2]

[1]Institute of High Technologies and department Engineering Systems, Irkutsk National Research Technical University, Lermontova 83, Irkutsk 664074, Russia
[2]Department of Railways, Bridges and Tunnels, Irkutsk State Transport University, Chernyshevsky 15, Irkutsk 664074, Russia

E-mail: slavaclimat@istu.edu

Abstract. When operating modern buildings with natural ventilation, it is difficult to maintain the required microclimate balance in the buildings, so mechanical ventilation must be used. This article presents the results of modeling the operation of a supply and exhaust unit with a plate heat exchanger. Based on the analysis, data on energy consumption depending on the method of preheating the supply air are obtained. It is concluded that the least energy-consuming option is to use an air-to-air heat pump.

The operation of a modern energy-efficient building is impossible without properly designed and operated ventilation systems. The increase in the cost of energy (electric, thermal) gives an impetus to the improvement of technological and structural systems of buildings. Installation of airtight windows with increased heat transfer resistance reduces energy consumption during operation by increasing the heat transfer resistance of building envelopes. The rest of the building systems (ventilation, energy, hot water), which affect their energy characteristics, remain at the same level. At this stage, the development of building technologies in relation to engineering systems and building envelopes is a drawback. Approaches to designing an individual building element in isolation from others does not allow optimizing the building structure as a single energy system. First of all, the problems of the building associated with the development of structural, technological and space-planning solutions are solved. By increasing the thermal resistance of the building envelope, a reduction in the level of operating costs of thermal energy for heating is achieved. And despite significant progress in this area, engineering equipment is lagging behind in its development [1, 2].

The use of window structures with increased tightness, while maintaining natural ventilation in the building, leads to negative consequences in the state of the indoor microclimate, which does not meet the norms of humidity and temperature of the indoor air during the heating season, air quality deteriorates. Up to 60% of heat losses account for air exchange in buildings under construction according to existing standards [3, 4]. In the applied design and technical solutions using natural ventilation, this problem cannot be solved. It is necessary to switch to supply and exhaust ventilation with artificial motivation and the use of heat energy recovery.

There is a need to solve the following problems:

IOP Conf. Series: Materials Science and Engineering **880** (2020) 012049 doi:10.1088/1757-899X/880/1/012049

- Sealed window structures, well-insulated walls almost completely seal the rooms, preventing the air exchange level from being maintained within the normal range. The constant opening of windows for ventilation devalues the original meaning of producing windows with higher thermal resistance. In this case, the use of supply ventilation valves is not a way out. The task of choosing an effective ventilation scheme must be solved by a different approach. The ventilation scheme of the building in the city, by opening the windows for ventilation, leads to a deterioration in air quality in residential premises (especially in apartments on lower floors) and to an increase in the noise level above the established limits.
- Requirements for increasing the heat transfer resistance of building envelopes have led to the fact that the maximum component of the heat loss of buildings during the heating period is air exchange (more than 50% of the total level). Saving thermal energy for heating buildings is associated with the return of heat leaving the air from the premises [4].
- Solving the problem of air intake using various technical means, for example, local heat recovery systems while maintaining the natural ventilation system, does not lead to improved comfort. The task of maintaining the normative value of air exchange is not properly solved due to the dependence of the level of air exchange on many multidirectional reasons (level of wind load, opening windows, degree of compaction of windows, condition of exhaust shafts, apartment height, outdoor temperature) [5].

While maintaining the natural air exchange scheme, the problem of rational use of energy, including solar energy entering the room, is not solved as a share in the building's overall energy supply system. Avoiding acceptable microclimate parameters during prolonged and systematic exposure to a person can cause a general and local discomfort, poor performance and poor health while strengthening thermoregulation mechanisms, although the state of health does not worsen [6, 7].

Currently, there is no systematization of the experience of using forced ventilation systems with the recovery of thermal energy of air removed from the premises of multi-storey residential buildings, as well as the features of using regenerative heat exchangers in residential buildings.

The transition to systems with mechanical ventilation of rooms using plate heat exchangers has one characteristic drawback: freezing of the heat exchanger from the exhaust air side with an increase in moisture content in the air. This is due to an increase in relative humidity to saturation when the temperature drops below the dew point, when crystallization of moisture begins. In parallel with the formation of condensate, two effects are observed (an increase in heat transfer due to the heat of the phase transition; a decrease in the heat transfer coefficient due to the liquid layer formed on the plates and a decrease in the cross section of the air channels, which leads to an increase in static pressure losses).

The crystallization temperature at which the water vapor contained in the exhaust air is frozen depends on a number of factors (thermophysical parameters of the exhaust air, the ratio of air flows on the supply and exhaust parts of the heat exchanger, design features of the heat exchanger, heat transfer efficiency).

Currently, there are several approaches to solving the problem of preventing frost recuperators. This is defrosting the entire heat exchanger, partial defrosting (stratification), installing an additional air heater on the supply air side and installing a bypass. All of these methods have both individual and general shortcomings leading to both reduced energy efficiency and increased investment and operating costs.

Using a heat pump allows you to use low-potential energy of the exhaust air, transmitting it to the supply air, Figure 1 [6].

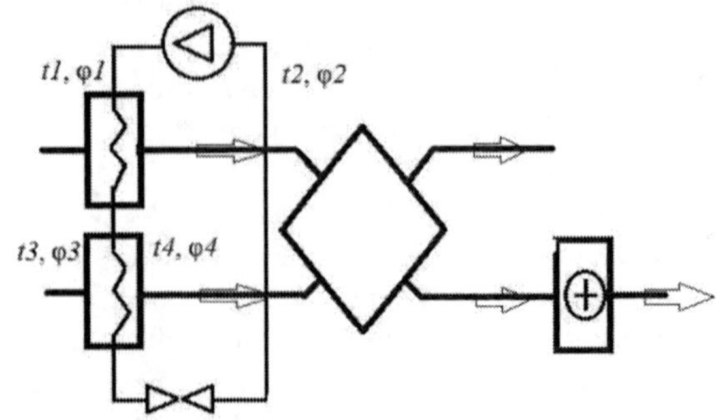

Figure 1. Forced-air and exhaust installation with a heat pump.

To consider the optimal use of a supply unit with a plate heat exchanger, three options were simulated and calculated (a supply and exhaust unit with a bypass, preheating using an additional heater, preheating using a heat pump):

1. The calculation for the existing supply and exhaust installation of production VTS (1500 m3/h) with a plate recuperator (VTS VS21PCR), with accessories for air filters of cleaning class G4, a supply fan with a capacity of 0.23 kW, an exhaust fan of 0.24 kW and a generated pressure in the ventilation network equal to 100 Pa. The plant efficiency is 57%, when the outside air is heated from minus 14 ^{0}C (minus 14 ^{0}C is the calculated freezing temperature of this heat exchanger) to plus 4 ^{0}C (φ = 13%) and cooling the exhaust air to plus 2 ^{0}C (φ = 100%). At minus 36 ^{0}C (outdoor temperature), the supply air temperature after the recuperator drops to minus 3,3 ^{0}C (φ = 4%), the temperature of the exhaust air decreases to minus 6,9 ^{0}C (φ = 100%), the freezing condition is met recuperator. In this installation, to avoid freezing, a bypass channel with a variable width is used. The result is a decrease in the supply air after the recuperator to minus 19.80 ° C, exhaust after the recuperator, plus 3.8 ^{0}C. To warm the air up to plus 20 ^{0}C, it is necessary to spend 20.91 kW, the total energy consumption is 21.38 kW.
2. Using preheating of the supply air (additional air heater to the recuperator), under the previously specified conditions (minus 36 ^{0}C), it is necessary to expend 11.56 kW of power. Outside air is heated to plus 4 ^{0}C (φ = 3%), the exhaust air is cooled to plus 2 ^{0}C (φ = 100%), the heat exchanger efficiency is 56%. To warm the air up to plus 20 ^{0}C, it is necessary to spend 8.4 kW, the total energy consumption is 20.44 kW.
3. When using a heat pump, the outdoor air is heated to a temperature of minus 14 ^{0}C (the condition of freezing of this heat exchanger) to the heat exchanger, after it has a temperature of minus 4.7 ^{0}C (φ = 3%), the exhaust air cools down to minus 5.5 ^{0}C (φ = 100%). The necessary costs for heating the supply air are 12.97 kW. The additional costs that go to the heat pump compressor when using Freon R410A (tc = 6 ^{0}C) are 0.126 kW. The total energy consumption (necessarily taken into account the fan group) is 13.58 kW [7.8].

The data obtained are summarized in table 1 and presented graphically in the Figure 2, where t1, φ1 - temperature and relative humidity of the supply air to the condenser; t2, φ2 - temperature and relative humidity of the supply air after the condenser; t3, φ3 - temperature and relative humidity of the utilized air to the evaporator; t4, φ4 - temperature and relative humidity of the utilized air after the evaporator.

Table 1. Experimental temperature and humidity data.

№	t_1, ^{0}C	$φ_1$, %	t_2, ^{0}C	$φ_2$, %	t_3, ^{0}C	$φ_3$, %	t_4, ^{0}C	$φ_4$, %

1	-24	76	-18	73	21	3	18	3
2	-10	60	-5	49	23	5	20	6
3	-4	80	1	59	23	10	20	11

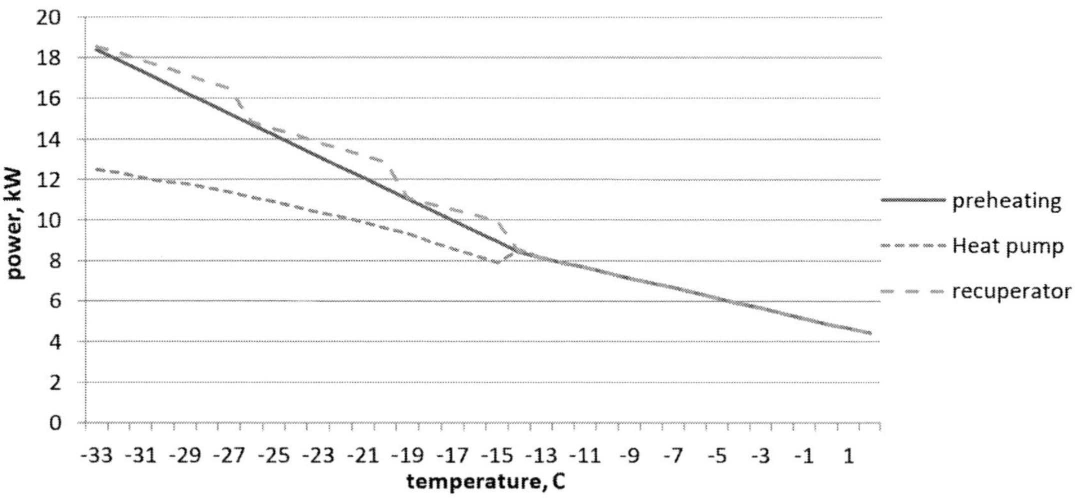

Figure 2. Dependence of power on the outside air temperature when preheating the air with an additional heater and heat pump.

The use of a heat pump in forced ventilation systems of buildings allows solving the problem of expanding the temperature range of the heat exchanger, which will allow rational use of energy conversion. Laying into the project a new approach to ventilation systems will not lead to a significant increase in price at the time of construction.

Reference

[1] Danilevskij L N 2014 Compulsory ventilation systems with recovery of thermal energy of the removed air for residential buildings (Minsk)

[2] Alo Mikola 2015 Problems with Using the Exhaust Air Heat Pump for Renovation of Ventilation Systems in old Apartment Buildings *Danish Journal of Engineering and Applied Sciences* (Tallinn: Tallinn University of Technology) p 44-55

[3] Yue Qi, Junjie Liu, Xilei Dai, Lei Zhao, Dayi Lai and Shen Wei 2019 Investigation of Ventilation Behaviors in Mechanically Ventilated Residential Buildings in China, *E3S Web conference 111 (06048)*

[4] Eberhard Paul 2001 Lüftungsgeräte mit Wärmerückgewinnung für Passivhäuser *5.Passivhaustagung* p 103 – 112

[5] Danilevskij L N 2011 Design principles and engineering equipment of energy-efficient residential buildings (Minsk) p 375

[6] Besant RW and Allan B. Johnson 1995 Reducing energy costs using run-around systems *ASHRAE Journal* **37** ISSN 0001–2491

[7] Poreckij V V, Berezovich I S and Stomahina G I 2003 Heating, ventilation and air conditioning (Moscow)

[8] Baimachev E E, Makarov S S, Sharova O V and Vygonets A V 2013Installation for ventilation

and air conditioning systems (options) Patent - 131458 RF IPC F24F
[9] Bajmachev E E and Makarov S S 2014 Modeling the thermodynamic cycle of a heat pump to expand the temperature range of an air heat exchanger (Irkutsk: Bulletin of Irkutsk State Technical University) **6**
[10] Stepanov V S 2009 Methods for assessing the thermodynamic effectiveness of microclimate maintenance systems *University News Construction* **10**

IOP Conf. Series: Materials Science and Engineering **880** (2020) 012050 doi:10.1088/1757-899X/880/1/012050

Justification of the parameters of developing water supply and sanitation systems based on their electronic models

E S Melekhov[1] and R V Chupin[2]

[1]Irkutsk National Research Technical University, 664074, Irkutsk, Lermontov st., 83, Russia

[2] VORTEX Limited Liability Company, 664039, Irkutsk, Lermontov st., 297, Russia

E-mail: melechov@istu.irk.ru

Abstract. Most of the water supply and sanitation systems of modern cities and populated areas of the Russian Federation are characterized by significant physical depreciation, which is estimated at 60%., Significant moral depreciation, since some structures and pipelines are almost a hundred years old or more. At the same time, the replacement of old equipment is carried out annually at a rate of 2%. Such unsatisfactory rates of renovation are due to the lack of adequate funding and their irrational allocation to the construction of new ones, reconstruction and development of existing pipelines and structures. Obviously, to solve this problem, new and integrated approaches, new technologies for the design, construction and operation of water supply and sanitation systems are needed. One of such approaches is proposed in the work, based on digital technologies using electronic models of water supply and sanitation systems, on the basis of which methods for modeling operating modes and optimization methods for design solutions for the development of networks and structures have been developed. The results of applying the proposed technology on the example of water supply and sanitation in the city of Nizhneudinsk, Irkutsk region.

1. Introduction

Most of the water supply and sanitation systems in Russia currently do not fully comply with modern technological, environmental and sanitary-hygienic requirements, which are set out in the relevant state standards, technological regulations and standards. The reason for this was the lingering nature of the transition of communal systems to market relations.

The rapid scientific and technological progress brought new technologies and materials, so necessary for the construction and reconstruction of wastewater systems. The emergence of geographic information systems, the creation of a global communication and information transfer system, the development of intelligent control systems - all this contributes to the reassessment of the existing technology for designing water supply and sanitation systems and the development of a new methodology for the formation and optimization of their prospects for reconstruction and development schemes.

Despite the large number of works on this issue, modern information technologies are not used enough in the practice of modeling and designing water supply and sanitation systems.

This work proposes a new approach to optimizing the reconstruction and development of water supply and sanitation systems, based on digital technologies for modeling the systems themselves, their operating modes, mathematical methods for optimizing design decisions, taking into account increased reliability, seismic resistance, operational controllability, risk assessment methods from decisions made, methods for overcoming the uncertainties in the behavior of systems in future periods of its operation. The proposed approach and the above methods [12] are implemented in the environment of the TRACE-VK software package [13], on the basis of which electronic models of water supply and sanitation systems in the cities of the Irkutsk region: Irkutsk, Angarsk, Shelekhov and Shelekhovsky district, Baikalsk, Ust-Ilimsk, have already been developed. Nizhneudinsk and others, developed a program for the integrated development of engineering infrastructure and water supply and sanitation schemes. In this work, using the example of the city of Nizhneudinsk in the Irkutsk region, the possibilities of the proposed approach are demonstrated and recommendations are given for intensifying the work, reconstruction and development of water supply and sanitation systems.

2. Methods

2.1. *The method of forming an electronic model of water supply and sanitation.* The method is based on a digital representation of the pipeline network, water intakes, water treatment plants, pumping stations, control tanks, sewage treatment plants, pressure and flow regulators. This method is based on the principles of graph theory [1,2], vector-matrix notation and representations [3]. The vertices of the graph are used to simulate manholes, wells for dividing and distributing flows, wells for water withdrawals and wastewater inflows, water intake points from surface and underground water sources, and sewage discharge points. The edges of the graph represent the pipeline sections of the network, the pumps and the characteristics of their impeller, water intake facilities, treatment facilities, pressure and flow regulators. Each vertex has absolute and relative coordinates, geodesic marks, for the tops of subscribers - schedules of wastewater intake and water withdrawals. For visualization of mappings, a specialized graphic editor has been developed for inputting and presenting an electronic model in the form of a geographic information system with N number of layers, such as houses, streets, rivers, ponds, terrain, networks, individual buildings and structures, etc. [4].

2.2. *Methods of distributing the flow of water and wastewater in the network.* Based on the theory of hydraulic circuits [4], methods for solving linear and nonlinear systems of equations [3], special-purpose algorithms [5]. At the same time, on the basis of the electronic model, the following matrices are formed: the adjacency of sections (edges of the graph) and network nodes (vertices of the graph); adjacencies of sections and network contours, cyclic schemes [6]. Based on these matrices, the conditions of material and energy balance are formed separately for the water supply and wastewater system in the form of the first and second laws of Kirchhoff [4]. The linear and nonlinear equations are solved by the method of I. Newton [3]. For water supply systems in automatic mode, in the presence of several water sources, regulating tanks, an expanded graph is constructed by introducing a dummy node with atmospheric pressure, from which dummy arcs are formed that are closed with the water source nodes. Effective pressures equal to the geodetic marks of the water sources and the regulating capacities are attributed to these arcs. On the basis of the network graph expanded in this way, matrices are formed, and an expanded system of nonlinear equations is constructed and solved. The result will be a load distribution between the water sources and the regulating tanks. For the calculation of drainage systems, a new technique is proposed, based on the construction of cyclic circuits with the allocation of passive, active and dummy branches. Passive sections of the arc model the reservoir and wells of the network. All wells are closed with dummy branches on a common node with atmospheric pressure. Fictitious sections simulate the flow of wastewater into the sewage system and the discharge of wastewater into the WWTP collection tank. Each fictitious branch is assigned acting pressure equal to the geodetic marks of the top of the wells. On the basis of the graph expanded in this way, matrices are constructed, a system of nonlinear equations is formed and solved. As a result, if all the fictitious branches are directed to the nodes - wells, and the costs on the passive branches are less than the costs for their full cross section, then the network will establish a gravity flow mode. If on some sections of the network the flow rate will be greater than or equal to the flow rate for their full cross-section, then the pressure-pressure-free mode will be established in the network. If some fictitious branches will be directed from the wells to a common node with atmospheric pressure, then the mode of effluent exit to the earth's surface will be established in an amount equal to the flow rate of this fictitious branch.

.

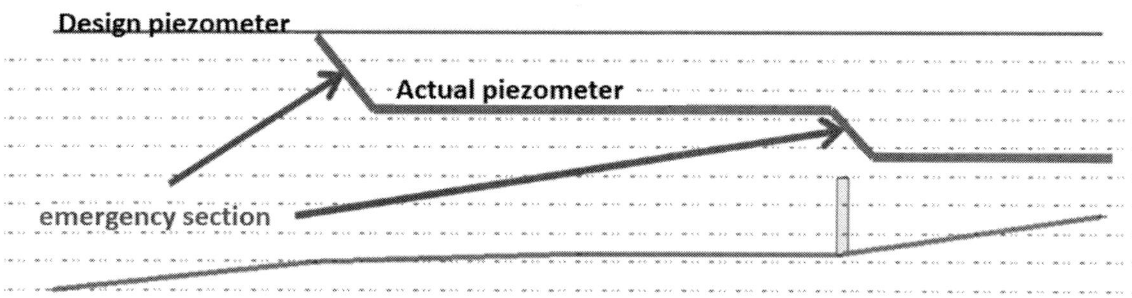

Figure 1. Calibration of the hydraulic model of the water supply network

2.3. *Methods for detecting blockages and sunken valves*. Such a task usually arises at the network calibration level, i.e. checking the mathematical model of the network for its adequacy to real parameters. This problem is solved as follows, hydraulic calculation of the water supply network is performed. Free pressures in the nodes of the circuit are determined and compared with the actual measured values. For example, in Figure 1, the calculated piezometer is shown in red, and the actual one is shown in blue. If the calculation model is adequate to the real network, then the red and blue piezometers coincide. In this example, in the second section, the actual piezometer "breaks"; in this section, greater than the calculated pressure losses occurred. Consequently, the section has blockages, or the actual diameter is less than the calculated one, or has valves with sunken working bodies. The second section in terms of pressure losses coincides with the calculated one, therefore there are no blockages. The third section also has a "scrapping" and, accordingly, blockages, etc. The described procedure automatically issues such suspicious network sections.

2.4. *Methods for optimizing design decisions*. The problem is solved on a previously constructed redundant scheme. For example, the designers, taking into account possible development, surveys, terrain, etc., outlined two options for tracing the drainage system, which are presented in Figure 2a, b. These two options can be replaced with a single graph, shown in Figure 2c, for which it is already possible to count 8 options for tracing the water supply and sanitation system. As part of urban development, you can build a graph that will contain many possible options.

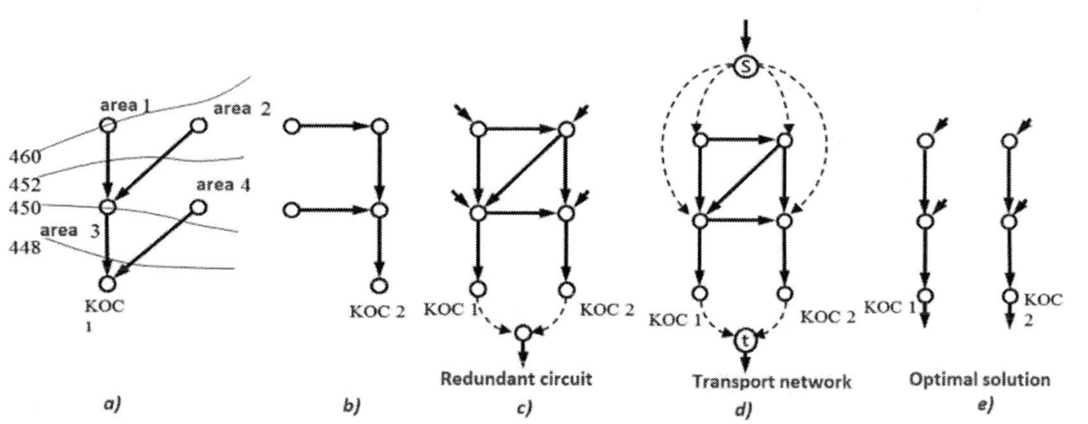

Figure 2. Illustration for the optimization of water supply and sanitation systems on graphs and networks

At the same time, possible ways of their reconstruction (parallel laying, relocation, laying of a new collector, arrangement of pumping stations, etc.) can be indicated on various sections of the network, and the location of new water intake facilities, treatment facilities with various water treatment technologies can be outlined in separate nodes and wastewater. The task will be to reject inefficient (according to a given cost criterion) sections and nodes of the excess scheme of the water supply and sanitation system. In the TRACE-VK software package, several optimization methods have been proposed and implemented. The first is a modified method for finding the maximum flow of minimum cost [6-8]. To implement this method, a transport network is constructed on the basis of the redundant circuit [8-10], which consists of a stream entry node and a stream exit node and a redundant network scheme. For drainage systems, the transport scheme is formed as follows. The flow inlet node is closed by arcs with all subscribers from which waste water enters the drainage system. These arcs are assigned the maximum throughput equal to the estimated performance of the subscribers. All nodes - possible sewage treatment plants are closed by arcs to a common outlet outlet. Each arc is assigned the maximum throughput equal to the maximum productivity of the treatment plant and the cost of a unit flow equal to the cost of cleaning one cubic meter of waste fluid. For network sections simulating existing network sections, two-sided flow restrictions are imposed, lower ones based on non-silting flow rates, and upper ones based on the inadmissibility of the work of gravity collectors in pressure mode. For sections of the network simulating the laying of new pipelines and structures, the cost of a flow unit is assigned. For water supply systems, the inlet of flows is connected to nodes that are possible sources of water supply (surface and underground), and the outlet of flows is closed by arcs with all subscribers. The flows of these arcs are subject to restrictions equal to the estimated loads of subscribers. Taking into account the constructed transport network, using the modified search algorithm for the maximum flow of minimum cost, the structure of structures, the route of pipelines, ways of laying new and reconstruction of existing sections of the network are determined. The type of transport is determined - pipeline, or automobile.

In detailed design, when local conditions, soils, intersections with other communications are already defined, the task of specifying the parameters of the designed water supply and sanitation systems arises. To solve this problem, a method has been proposed that implements a multi-step process of targeted sorting of possible slopes and diameters of pipelines (from among the standard series), pressure of pumping stations and parameters of mating structures. The multi-step process is carried out according to the dynamic programming scheme and starts from the subscriber units to which water is supplied and from which the wastewater enters the wastewater system to the nodes - water intakes and wastewater intake units at the WWTP. From the set of variants of piezometric surfaces of pipelines constructed in this way, the best of them is selected by the criterion of reduced costs. Figure 3a shows the multi-step process of forming the piezometric surface of five pipelines, where the solid bold line indicates the best option for the given costs. This option corresponds to the optimal profile shown in Figure 3b.

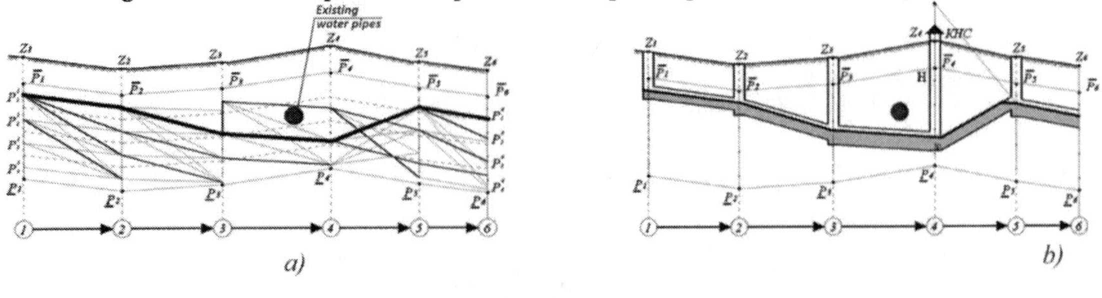

Figure 3. Optimization of sewage network parameters

Optimization Criteria. The generally accepted criterion is the present discounted costs and life cycle costs. Studies have shown that if the problem of the long-term development of water supply and sanitation systems is being addressed, then the life cycle cost criterion provides a solution with minimal operating costs, and for short-term projects, the presented costs give a solution with minimal one-time costs (capital investments).

2.5. *Methods of dealing with the uncertainty of promising water consumption*
Currently, in almost all cities of the Russian Federation there is a significant decrease in specific water consumption and sanitation. For example, for the city of Irkutsk over the past 17 years, it has decreased from 300 to 191 l / person per day. It is possible that by the end of the implementation of the approved water supply and sanitation scheme (by 2035), it will either decrease and reach the limit of 100 l / person per day, or it will return to its previous values.

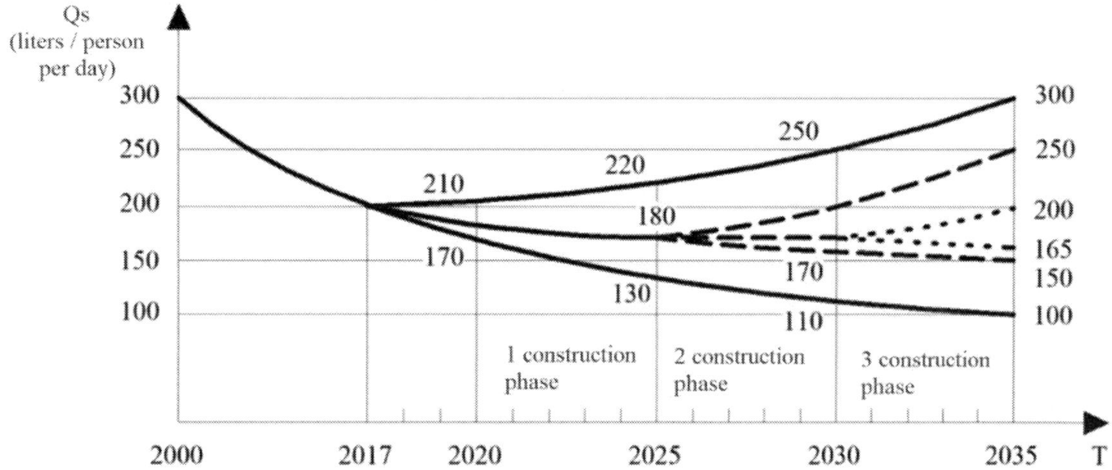

Figure 4. Possible intervals for the change in specific water consumption for the period of the implementation of the water disposal scheme

Figure 4 shows the possible intervals and trajectories of changes in the values of specific water consumption at the stages of development of the wastewater system. For the first phase of construction, the possible range of specific loads will be (130-220 l / person per day), for the second (110-250 l / person per day), for the third (100-300 l / person per day) . The TRACE-VK software package implements an approach based on the use of the theory of fuzzy sets and decision theory [11-12]. Moreover, for each construction phase, the intervals of possible loads are divided into n values. As a result, taking into account the membership function (preference), n variants of development of the water supply and sanitation system are formed. For each option, justification (optimization) of the parameters of the water supply and sanitation system is carried out immediately for all (in this case, three) construction phases. Moreover, the option for which the estimated investment in the construction of the first stage exceeds the possible funds generated in the investment programs is excluded from consideration. For each of the investigated options for the development of a drainage system, financial risks are assessed. For this, a risk matrix is built. In this matrix in the first row and first column presents the values of the estimated estimated costs. On the diagonal are located "zero" values of costs - risks, which means that the accepted value of costs coincides with those that will be after the implementation of the project (100% match option). The values to the right of the diagonal indicate the risk values from the fact that the actual value of the wastewater discharge after the project will be greater than their values assigned in the project. Consequently, additional reconstruction and costs — financial risks — will be required. To the left of the diagonal in the matrix will be the risk values associated with overstating the parameters and, consequently, with excessive investment. The last column of the "risk matrix" gives the maximum risks for each option for making an estimated expense. The last element of this column corresponds to the minimum value of the maximum risks.

3. Results and discussion
We illustrate the work of the proposed complex of models and methods on the example of water supply and sanitation in Nizhneudinsk, Irkutsk region. The city of Nizhneudinsk, with a population of 34 thousand people, is located along the Trans-Siberian Railway on the right and left banks of the Uda

River. In the summer of 2109, due to flooding, the city was partially flooded. At the same time, sand and silt got into the city's pipeline water supply system, which affected the water quality and hydraulic parameters of the system.

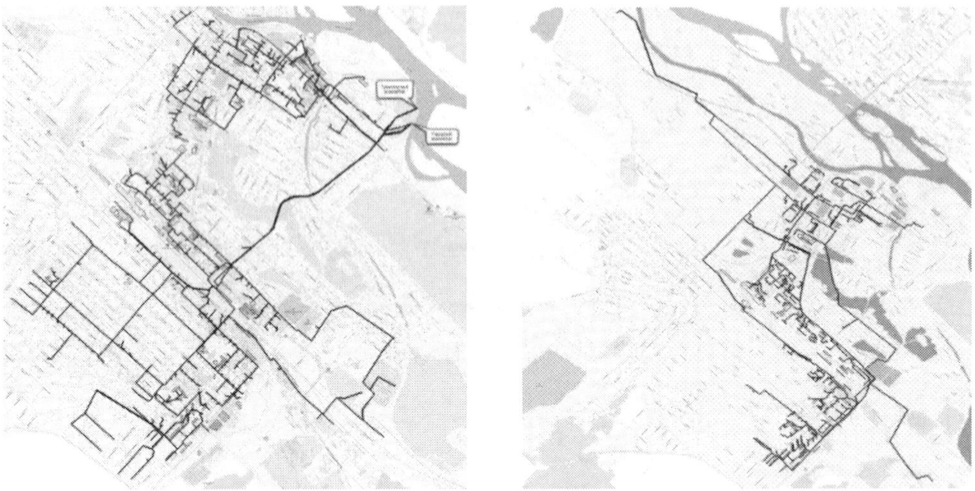

Figure 5. Hydraulic model of the water supply and sanitation system in Nizhneudinsk

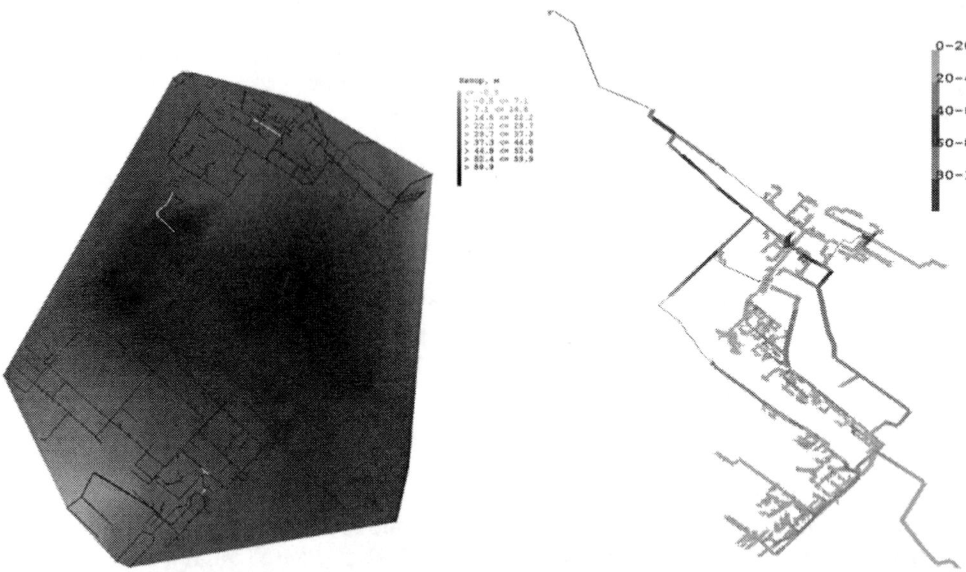

Figure 6. Pressure map for water supply and filling map for water disposal

Figure 7. Modeling of promising city development

Figure 5 shows the hydraulic models of the city's water supply and sanitation systems. Figure 6 shows calibrated models in the form of calculated and actual pressure maps for water supply and filling maps in gravity collectors. During the calibration, it was shown that the sand that got into the pipelines shifted to the entrances to the buildings and the dead ends of the network. Some sewer collectors were completely clogged with sludge from the flood. Figure 7 shows illustrations for optimizing the connection of a new residential community to existing water supply and sanitation networks.

4. Conclusions

A set of models and methods is proposed for assessing the health, optimization and reconstruction of developing water supply and sanitation systems. These models and methods are implemented in the TRACE-VK software complex, the use of which in the practice of operation and design of water supply and sanitation systems has shown its economic efficiency and efficiency in substantiating design decisions.

References

[1] Hu T 1974 *Integer programming and flows in* networks p 520
[2] Ford L R 1963 *Streams in networks* p 216
[3] Wenzel E S 2000 *Problems and exercises in probability theory* Textbook for universities. 3rd ed Erased (Higher. School) p 366
[4] Merenkov A P 1985Theory of hydraulic circuits p 278
[5] Chupin R V 2017 Pressure-flow and Free-flow Discharge Modes in Closed-Loop Sewage Systems *IOP Conf. Series: Materials Science and Engineering* **262** 012079 doi: 10.1088 / 1757-899X / 262/1/012079
[6] Chupin R V, Mayzel I V Chupin V R 2017 Trajectory of Sewerage System Development Optimization *IOP Conf. Series: Materials Science and Engineering* **262** 012080 doi: 10.1088 / 1757-899X / 262/1/012080
[7] Chupin R V, Chupin R V, Pham N M and Chupin V R 2019 Optimization of development schemes for group water supply systems under uncertainty of prospective water consumption *E3S Web of Conferences: Mathematical Models and Methods of the Analysis and Optimal Synthesis of the Developing Pipeline and Hydraulic Systems* **102**
[8] Chupin R V, Pham N M and Chupin V R 2019 Optimal reconstruction of water supply network under conditions of *reduced E3S Web of Conferences: Mathematical Models and Methods of the Analysis and Optimal Synthesis of the Developing Pipeline and Hydraulic Systems* **102**
[9] Chupin, V R, Chupin R V and Pham N M 2019 Optimization of developing district water supply systems taking into account variability of perspective water consumption *IOP Conf. Series: Materials Science and Engineering* **667** 012018 doi: 10.1088 / 1757-899X / 667/1/012018
[10] Chupin, V R, Chupin R V, Pham N M 2019 Optimization of the sewerage systems scheme of cities and populated areas *IOP Conf. Series: Materials Science and Engineering* **667** 012017 doi: 10.1088 / 1757-899X / 667/1/012017
[11] Chupin R V 2011 *Development of theory and practice of modeling and optimization of water supply and sanitation systems* p 323
[12] Chupin R V 2015 *Optimization of developing drainage systems* Monograph p 418
[13] Certificates of state registration of computer programs:
[14] Melekhov E S, Chupin V R, Chupin R V Certificate of state registration of the computer program No.2016615463 TRACE-VK dated May 25 2016

IOP Conf. Series: Materials Science and Engineering **880** (2020) 012051 doi:10.1088/1757-899X/880/1/012051

Variable geometry impeller centrifugal pump

A Pankin[1] and L Makotrina[1]

[1]Irkutsk National Research Technical University, 83 Lermontov Str., Irkutsk, 664074, Russian Federation

E-mail: alexeypankin@gmail.com

Abstract. The article compares the energy efficiency of the pump installation in a changing flow rate for options without flow control and with regulation by reducing the shaft speed and trimming the impeller. A comparison of these methods. A description is given of the operation of the pumping unit with a self-trimming impeller and an algorithm for controlling it.

Pumping stations and pumping units installed in them are designed to supply water to piping networks for various purposes and, as a result, are the most critical elements. When designing water supply systems, engineers are faced with the technical task of linking the joint work of pumping stations, a network of piping lines and structures on them. The goal of the created system is to provide consumers with water with a given minimum guaranteed pressure and the required flow rate.

The water flow in the pipeline network is variable and varies during the day depending on the need for water, which is represented by various modes of operation of the network. These modes are dictated by the uneven hour and nature of water consumption by various network subscribers. To reduce the irregularity of water flow at the supply section from the pumping station to the network, pressure-regulating structures are installed - water towers, accumulating tanks. Therefore, for systems with variable flow rate to ensure the supply of all network subscribers, it is necessary to provide a pressure control system at the dictating point.

The distribution of water flow by the hours of the day is set by the unevenness coefficient determined by [1] and depends on the number of consumers. In a particular case, the non-uniformity is determined by the probability of action of devices that are consumers of the system in question, and depends on their number in the hydraulic system.

The coordination of the joint work of the pumping station and the water supply network is mathematically described by the system of equations:

$$\begin{cases} H_{pump} = f(Q) \\ H_{pipe} = H_g + SQ^2 \end{cases} \qquad (1)$$

Where H_{pump} – head developed by a pump or group of pumps;
H_{pipe} – pressure losses in the pipeline network;
Q – water flow in sections of the pipeline network;
S – resistance of individual sections of the pipeline network to the dictating point.

In the mode of minimum hourly consumption, unregulated operation of the pumping station leads to an increase in excess pressure in the water supply system, since the operating point on the joint pump-feed network schedule changes its position relative to the calculated one. This operating mode is characterized by excessive pressure, i.e. increase in free pressure relative to the minimum guaranteed (Fig. 1).

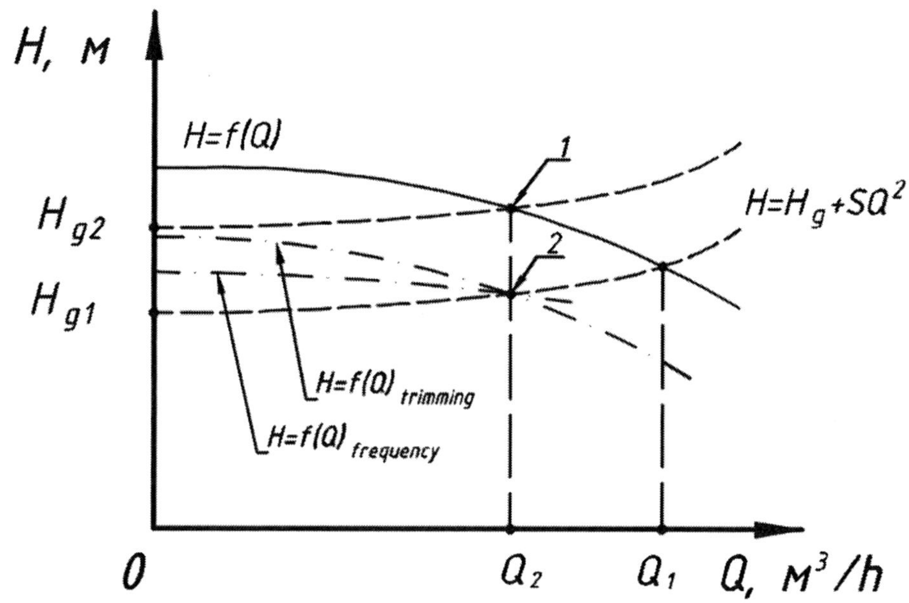

Figure 1. Characteristic operation of the pump-water supply system
1 - position of the operating point with uncontrolled pump operation
2 - position of the operating point with adjustable pump operation

Figure 1 shows that the unregulated operation of a centrifugal pump or a group of them in variable flow systems leads to an increase in guaranteed pressure for consumers, which negatively affects the energy efficiency of this state of the system, since the amount of electrical energy consumed per unit of pumped liquid can be reduced. To do this, use methods of regulating the operation of pumping units.

Under the regulation of the operation of pumping units in the technical literature [2, 3] we mean a change in the characteristics of the external elements of the water supply system that are directly related to their work in order to maintain the required pressure and at the same time reduce the power consumption. These include the throttling method for artificially changing the resistance of the pipeline network and reducing the flow of incoming water, the method of changing the speed of the pump drive shaft and the method of trimming the impeller.

The change in the frequency of rotation of the impeller is achieved through the use of a frequency converter in the wiring diagram of an asynchronous motor, which changes the frequency of the alternating current supplying the motor windings.

Since the characteristic of the pumping equipment is the passport data obtained as a result of a bench test, this complicates the research task by the fact that the Q-H, Q-N, Q-η dependency curves cannot be described as mathematical dependencies. Therefore, the selection of pumping equipment and recalculation of characteristics is carried out by the graphoanalytical method.

The parameters curves of the characteristics of a centrifugal pump with frequency regulation of the drive is described by the equation:

$$n_1/n_2 = Q_1/Q_2 = (H_1/H_2)^{0,5} = (N_1/N_2)^{0,33} \quad (2)$$

IOP Conf. Series: Materials Science and Engineering **880** (2020) 012051 doi:10.1088/1757-899X/880/1/012051

The parameters curves of the characteristics of the centrifugal pump when trimming the impeller is described by the equation:

$$H_2/H_1 = Q_2/Q_1 = N_2/N_1\ (D_2/D_1)^2 \qquad (3)$$

Where n_1; n_2 – drive shaft rated and reduced speed,
Q_1; Q_2 – nominal and reduced pump flow,
H_1; H_2 – pump head rated and reduced,
N_1; N_2 – power consumption of the pump rated and reduced,
D_1; D_2 – outer diameter impeller nominal and trimmed respectively.

The difference between the frequency regulation and trimming of the impeller from throttling is that the pump characteristics are adjusted to such a state of the pump-water supply system when the pressure levels of the dictating consumer are reduced to the minimum guaranteed (Fig. 1)

This difference provides greater energy efficiency of these control methods than throttling. For example, table 1 shows the calculated values of the specific energy consumption for each option for regulating the operation of the Grundfos NB 40-200 / 206 pumping unit with the joint operation of a steel conduit with a nominal diameter of 175 mm and a length of 1 km based on data published by the manufacturer.

Table 1. Comparison of specific energy consumption per unit of pumped liquid for different regulation methods

Q, м^3/h	H, м	N, кW	Efficiency, %	N/Q, кW/м^3/h
without regulation or throttling				
55	46	10,7	68,2	0,19
50	47	10,5	67,5	0,21
45	50	10,2	67	0,23
40	52,5	9,48	65,3	0,24
35	53	8,85	62,5	0,25
30	55	8,2	58,7	0,27
25	56	7,6	54	0,30
20	56,5	6,9	48	0,35
15	56,9	6,2	40	0,41
frequency regulation				
55	46	11,25	61	0,20
50	45,15	10	61	0,20
45	44,3	8,95	61	0,20
40	43,6	8	59,3	0,20
35	43	7,14	57,3	0,20
30	42,4	6,36	54,5	0,21
25	41,98	5,67	50,5	0,23
20	41,56	5,05	45,2	0,25
15	41,22	4,48	38,1	0,30
impeller trimming				
55	46	10,7	68,2	0,19
50	45,15	9,67	67,5	0,19
45	44,3	8,60	67	0,19

40	43,6	7,60	66,3	0,19
35	43	6,74	64,5	0,19
30	42,4	6,06	60,7	0,20
25	41,98	5,42	56	0,22
20	41,56	4,80	50	0,24
15	41,22	3,97	45	0,26

Analyzing dependencies (2) and (3), we can conclude that the cutting of the impeller has an advantage over frequency regulation due to the possibility of reducing the supply by half more without reducing the minimum guaranteed head while maintaining a higher efficiency value. According to the materials of the catalog of the manufacturer of pumping equipment [4], the addition of a frequency converter to the motor circuit of the motor windings reduces the efficiency of the pumping unit from 3 to 10%.

The disadvantage of trimming the impeller is the irreversibility of the changes, and if it is necessary to increase the pump capacity, it will be necessary to stop the operation of the pump unit and replace the impeller.

Therefore, to increase the energy efficiency of regulating the operation of the pumping unit, it is necessary to formulate and solve the technical problem of ensuring the possibility of changing the outer diameter of the impeller during its operation.

The problem can be solved as follows, to change the design of the impeller blades and to make them consisting of two parts, movable and stationary, to move the moving part of the blade from the edge to the center of rotation (Fig. 2).

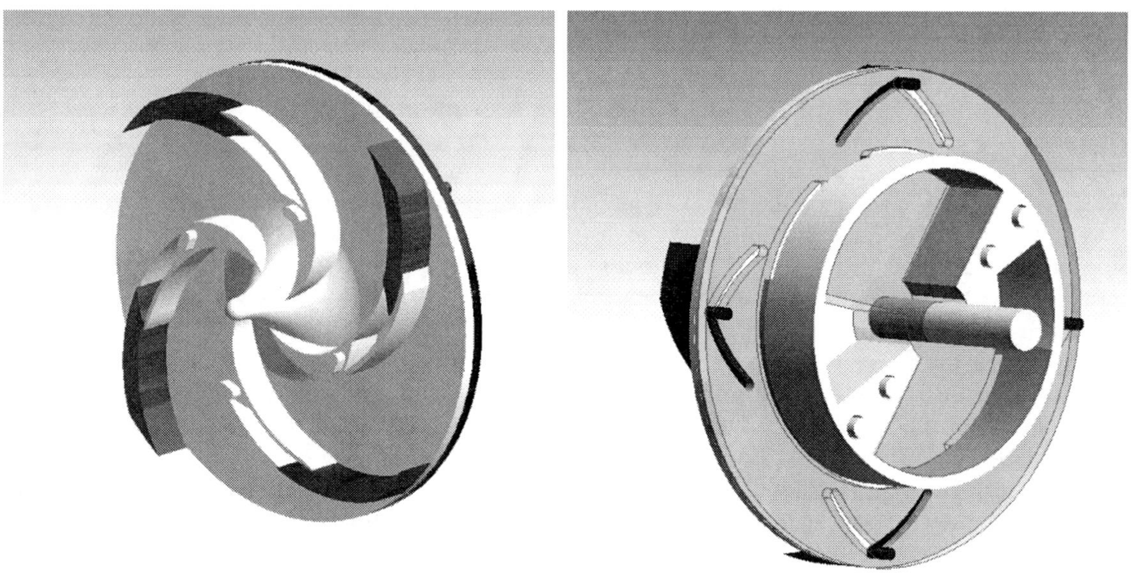

Figure 2. Sketch of self-trimming impeller
a) front view b) back view

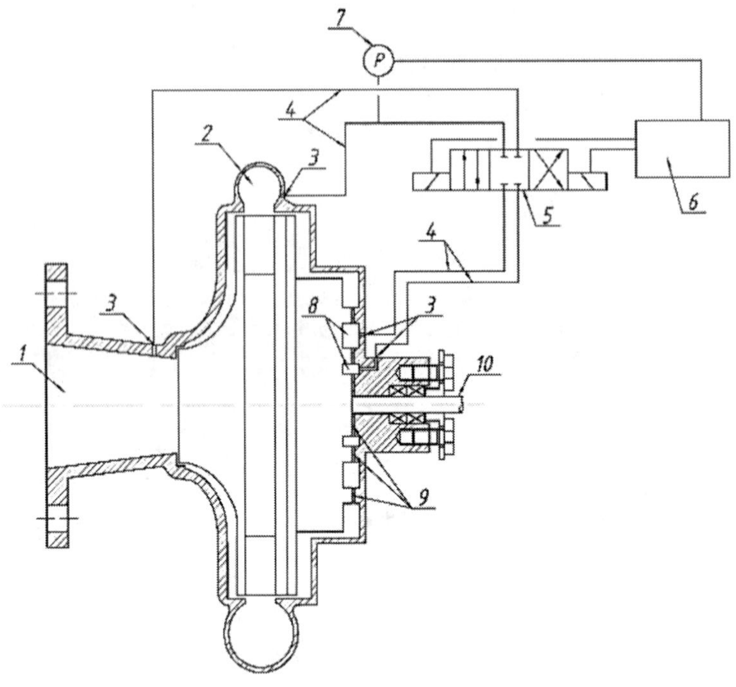

Figure 3. The proposed pump operation scheme

The centrifugal pump (Fig. 3) works as follows: when excessive pressure occurs between the suction zone 1 and discharge 2, which corresponds to a non-productive mode, the pressure sensor 7, using the control unit 6 and the electromagnetic valve 5, changes the pressure ratio of the pumped liquid in the hydraulic lines 4 supplying the pumped liquid into cavities 8 separated by mechanical seals 9 between the pump casing and the impeller. When the pressure changes in the cavities of the impeller casing, an angular displacement of the actuating element occurs, leading to the displacement of the pins of the moving blades to a position corresponding to the required outer diameter. The outer diameter of the impeller is thus variable. Upon reaching the required pressure in the network supplied by the pump, which corresponds to optimal operation, the control unit through the valve distributes the pressure in the hydraulic actuators by a signal from the pressure sensor, fixing the position of the blades corresponding to the required impeller diameter.

If the proposed changes in the design of the flowing part of the impeller will not lead to a deterioration in efficiency within 10%, then in systems with variable flow, the pump according to the described design without a frequency drive will be more energy efficient than a similar one with a frequency drive. The use of pumping units with the described design can allow abandoning the operation of regulatory structures in the water supply system, as well as reducing water losses in pipelines caused by excessive hydrostatic pressure, which in turn will reduce operating costs and reduce the accident rate of water supply networks and unproductive losses of water during its transportation.

References

[1] SR 31.13330.2012 Water supply. Pipelines and portable water treatment plants.
[2] Lobachev *Pumps and pumping stations 1990* (M.: Stroyizdat) pp 108-110
[3] Karelin and Minaev 1986 *Pumps and pumping stations* (M.: Stroyizdat) pp 54-55
[4] *Grundfos product center* URL: https://product-selection.grundfos.com (date of the application: 02.02.2020)

[5] Chupin V R and Dushin A S 2019 Parameter optimisation for new and reconstructed systems of water supply and distribution for ensuring consumer's uninterrupted water supply *Proceedings of Universities. Investment. Construction. Real estate* **9(4)** pp 790–803
[6] Kulkov V N, Solopanov E Yu and Kamalov R T 2019 Using immobilised sludge for wastewater treatment and its air regeneration *Proceedings of Universities. Investment. Construction. Real estate* **9(3)** pp 522–529
[7] Pupyrev E I and Chupin V R 2019 Features of regional development of water disposal systems in the central ecological zone of the Baikal natural territory *Proceedings of Universities. Investment. Construction. Real estate* **9** pp 354–363
[8] Chupin V R and Dushin A S 2019 Assessment of the reliability of water supply to consumers: water supply reliability indicators Proceedings of Universities. Investment. Construction. Real estate **9(3)** pp 578–593
[9] Lavygina O L and Grebneva O A 2019 Environmental technologies in the housing and communal service system of the Baikal natural territory *Proceedings of Universities. Investment. Construction. Real estate* **9(4)** pp 726–733

IOP Conf. Series: Materials Science and Engineering **880** (2020) 012052 doi:10.1088/1757-899X/880/1/012052

A promising water supply scheme for Khomutovo village (Irkutsk district, Irkutsk region)

A A Shcheglov [1] and V G Sudnikovich [1]

[1] Irkutsk National Research Technical University, 83 Lermontov street, Irkutsk, 664074, Russia

E-mail: shegloff8@gmail.com

Abstract: The article proposes a water supply scheme for Khomutovo village. The authors describe regulation options for domestic and drinking water supply from the existing water intake facility in Dzerzhinsk under uniform operation of pumps using a regulating capacity and a frequency converter without intermediate tanks.

1. Introduction

Currently, the urgent problem is the lack of centralized water supply in settlements of Irkutsk region. A rather intensive increase in the number of residents of suburban villages created a problem of water supply. No exception is Khomutovo village, as well as settlements that are part of Khomutovo municipal district. In 2031, the number of residents in Khomutovo will be 17,900 people.

Sources of centralized water supply are groundwater intakes. Available local sources located in the territories of departmental enterprises supply water to their own facilities. They do not supply water to the population and public institutions. For this reason, they are not dealt with in the present paper.

Surface sources of water supply (the Kuda river - Khomutovo, Kuda, Gorny; the Ushakovka river - Plishkino) are not analyzed, since they have a low flow rate, especially in winter. In addition, they are susceptible to organic and bacterial contamination from upstream settlements and horticultural partnerships [1].

2. The water supply scheme developed for Khomutovo

The water supply system is a system of engineering structures designed to produce, improve water quality and supply water of required quality and under required pressure to the consumers. The sequential arrangement of natural water intakes, its purification and disinfection, pumping, transportation to consumers, and storage are determined by the water supply scheme.

There are several water supply schemes developed for Khomutovo by employees of the technical department of the village administration and specialists from third-party organizations.

The main problem is the choice of a water supply source, because, according to hydrogeological studies, the quality of underground water is unsatisfactory due to the increased content of salts and impurities. River pollution is caused by the unorganized flow of melt and storm water from the entire catchment area. It washes away the soil, organic and inorganic fertilizers when passing through settlements, especially through territories of buildings, gardens, etc., located in water protection zones. The use of underground sources is not possible, because the rivers are hydraulically connected to groundwater horizons.

Currently, water supply to the villages of Khomutovo municipal district is carried out from municipal and private wells. That is why it is difficult to control water quality. Groundwater deposits near the boundaries of the settlement have been understudied. According to hydrogeological zoning, the territory is located in Irkutsk artesian basin. Groundwater was discovered in the Quaternary and Jurassic sediments [2].

In total, territories with a centralized water supply system make up about 10%. The rest of the settlement occupied by individual residential buildings lacks a centralized water supply system. Water supply is carried out from individual and local water sources (wells). There are several municipal (for the needs of boiler rooms) and individual (water supply of private houses) decentralized water supply systems. According to the data provided, water quality does not meet the standards established for quality of drinking water. It is used for technical needs [3].

The management company LLC Ushakovskoye developed the following water supply scheme for Khomutovo: water intake facilities take water from an underground source; pressure conduits transport water to a clean water tank (CWT); water is stored in a pure water tank; water is supplied to Khomutovo and other settlements through pressure lines; in Khomutovo, there is a distribution network (Fig. 1).

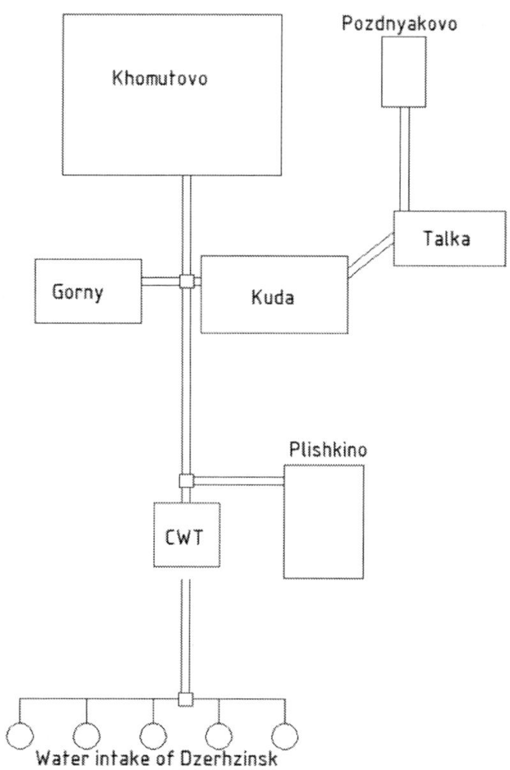

Figure 1. The water supply scheme developed for Khomutovo village

The water supply scheme has a control tank at the entry of the network. This scheme is characterized by the high-altitude location of the water storage tank (on a hill near Plishkino). Water is supplied from the tank to consumers due to the difference in elevations.

Water intake facilities in Dzerzhinsk and additional wells can be used as a source of water supply. The existing licensed underground water intake at 33 Naberezhnaya Street which was put into operation for the needs of Dzerzhinsk and Sovremennik has a productivity of 1000 m³ / day. It consists of two artesian wells with a productivity of 40.0 m³/h. The water intake meets the sanitary standards. Groundwater reserves have been approved; all the permits have been issued:

- the sanitary and epidemiological inspection report according to which water corresponds to the sanitary-epidemiological standards of the design of the sanitary protection zone of the ZSO of the first t belt, No. 38. ITS.06.T.000641.07.15 of 07.23.2015.
- the sanitary and epidemiological inspection report according to which water can be used for drinking and household purposes, No. 38. ITC.06.000.M.000604.12.15 of 01.12.2015;
- the report approving category B underground water reserves in the amount of 1000 m³ / day "Conducting prospecting and assessment work on Dzerzhinsky site aimed to organize drinking and technological water supply in Dzerzhinsk of Irkutsk region". The report has been examined and reserves have been approved;
- the license for the use of subsurface resources IRK No. 03303 VE of July 25, 2016 for supplying drinking water to industrial facilities in Dzerzhinsk;
- quality of groundwater is assessed by quarterly monitoring within the production control program approved by the Federal Service for Surveillance on Consumer Rights Protection and Human Wellbeing.

According to [4], taking into account the development of Khomutovo and improvement of residential buildings equipped with internal water supply and sewerage systems, daily water consumption is

$$Q = (q_r \times N_r)/1000 = (190 \times 17900)/1000 = 3401\ m^3/day \quad (1)$$

where q_r – water consumption per 1 person per day, l / day;

N_r is the number of residents;

According to [5], to calculate the water flow for fire fighting needs, we should use the number of residents equal to 17,900 people, the number of storeys which is no more than three, the estimated number of simultaneous fires equal to 2, the water flow for external fire fighting needs for one fire equal to 15 l/s. According to [6], the flow rate for internal fire extinguishing is 2.5 l/s in one stream. The estimated water consumption for firefighting needs is

$$q_{пож} = n \times (q_i + q_o) = 2 \times (2,5 + 15) = 35\ l/d \quad (2)$$

where n – number of simultaneous fires;

q_i – consumption for internal fire extinguishing, l / d;

q_e – consumption for external fire extinguishing, l / d.

The existing water intake in Dzerzhinsk has one working well and one reserve well with a depth of 75.0 m. The water flow rate of one well is 10-35 l / s depending on the pump load and its supply [2]. According to experimental pumping, the recommended load is 50.0 m.

To supply water to Khomutovo and nearby settlements, five additional working wells having the same characteristics as the existing ones are required. The ECV 10-65-225 NRK* pump with a nominal flow rate of 65.0 m³ / h and a pressure of 225.0 m is installed in each well.

Water from the water intake is transported along two lines of water pipes to a clean water tank located on the hilltop near Plishkino. The pipe material is PE 100 SDR 7.4 polyethylene (GOST 18599-2001). The length of the pipelines (from the water intake in Dzerzhinsk to the tank) is 16,500 m, the elevation of the water intake facilities is 450.00 m, the mark of the location of the tank is 593.00 m, the geometric elevation is 143.00 m. The diameter of the pipelines is 280 mm.

The length of the pipelines from the CWT to the first well in Khomutovo is 15.2 km. The diameter of the water conduits is 355 mm.

3. Options for regulating water supply for the needs of Khomutovo

Let us analyze the cost of electricity for water supply by pumps during their step-by-step operation under the maximum water consumption mode. All 5 ECV pumps 10-65-225 operate simultaneously. To

IOP Conf. Series: Materials Science and Engineering **880** (2020) 012052 doi:10.1088/1757-899X/880/1/012052

regulate the flow, a clean water tank near Plishkino is used. The pumps operate for 21 hours a day, the break in supply is between 1:00 and 4:00 p.m. The pumps lift water to a height of 188.0 m, the flow rate of the pumping station is 326.18 m^3 / h.

The easiest way to change the rotational speed of the rotor is to change the current frequency. Frequency drives with semiconductor converters whose application improves the economic efficiency of pump parameters have been developed [7].

To calculate the power consumption when using a frequency converter for pump motor revolutions, we use the following method [8]:

- the nominal data of the pump are recorded: flow rate Q, m^3 / s; pressure N, wcm; efficiency; frequency ω, rpm; engine power P, kW;
- at the operating pump, power P, kW and capacity Q, m^3 / h within the limit of the flow control limit allowed by the manufacturer are established or calculated. The minimum and maximum powers are 42 and 59 kW (for open and close valves, respectively). According to the pump certificate [9], the minimum and maximum costs are 70-120% of the nominal flow, that is, 46.0 and 78.0 m^3 / h;
- the graph of the dependence of power P on the relative flow rate determined by the following formula is built:

$$Q' = Q / Q_{\max} \tag{3}$$

where Q - current consumption, m^3 / h;;
Q_{max} - maximum consumption, m^3 / h;.

- the required power of the frequency converter is determined by formula:

$$P_{fc} = (1,1 \div 1,2) \times P_{\max} \tag{4}$$

P_{fc}- the dependence of power consumption P_d when controlling the pump flow by throttling the valve on the relative flow rate Q ’ obtained by connecting P_{max} and P_{min} is built

- the dependence of power consumption P_{fr} when using frequency regulation on relative flow rate Q' is built by calculating expression (5) and substituting P_{max} and several values of Q' into it (from 0 to 1 with a step of 0.25) [10].

$$P_{fr} = P_{\max} (Q / Q_{\max})^3 \tag{5}$$

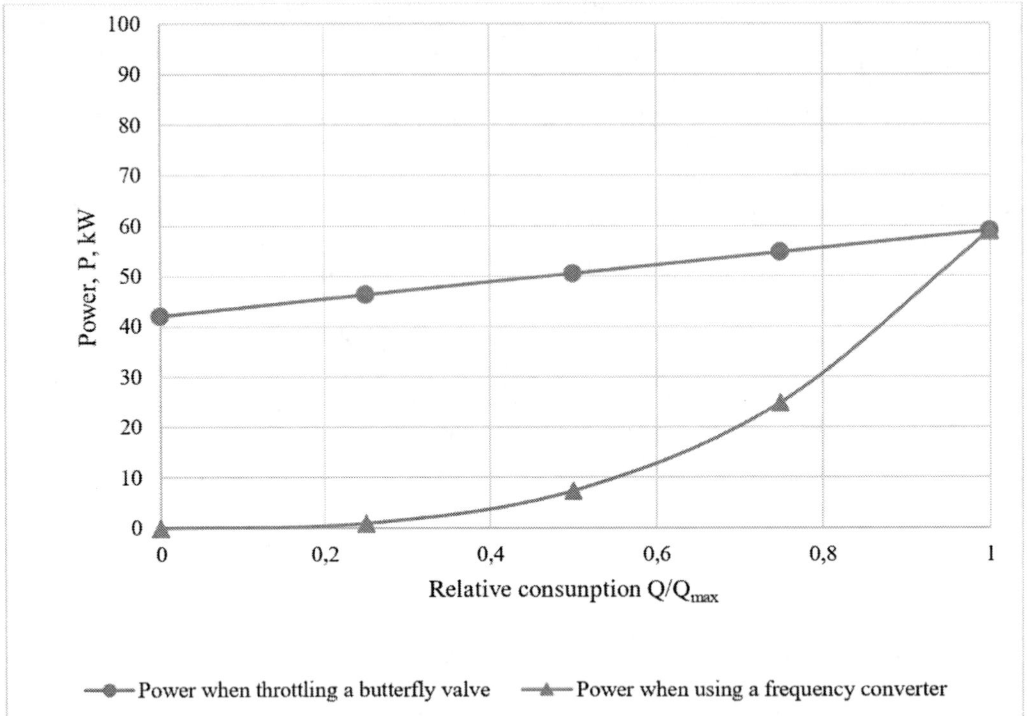

Figure 2. The graph of dependence of power consumption on the pump flow control method depending on the relative consumption

However, the above graph can be misleading, as it is based on the rated power, flow rate and pressure of the pump, and does not take into account the actual graph of water consumption in Khomutovo.

Based on real costs, it is more convenient to interpret the energy consumption of pumping units using a daily graph (Fig. 3).

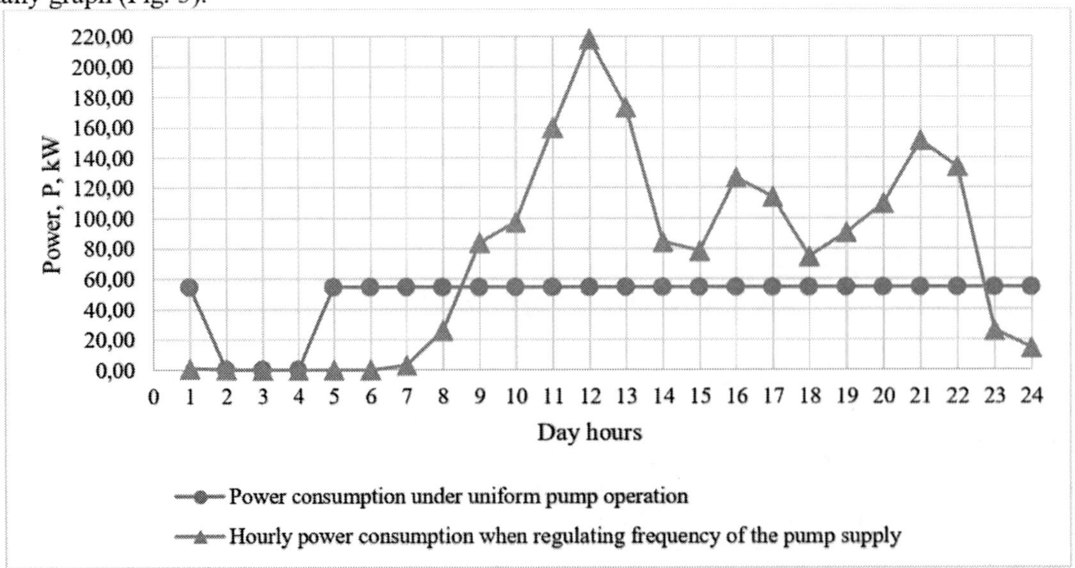

Figure 3. The graph of power consumption on the method of regulating the pump supply by daily hours

Analyzing the graphs, several facts can be emphasized:

IOP Conf. Series: Materials Science and Engineering **880** (2020) 012052 doi:10.1088/1757-899X/880/1/012052

- according to the data obtained, when using a frequency converter, energy savings are due to low energy consumption. However, this requires the pump to be proportionally slow that is impossible according to the manufacturer's passport, since proper cooling cannot be ensured by the flow of pumped water;
- due to the small number of residents, there is increased unevenness in hourly water consumption. Increased water consumption increases the number of revolutions of the unit and the consumption of electricity;
- too large unevenness of water consumption makes pumps regulated by the frequency converter ineffective and increases pressure losses in the water conduits during hours of maximum water consumption.

Conclusion

Currently, the sanitary and epidemiological well-being of sparsely populated areas depends on a well-organized water supply system. The supply of water whose quality meets the requirements of current regulatory documents is of great technical and social importance.

To choose the most optimal water supply system, the network diagram, the type of a pumping unit, the technical assessment is required.

The authors analyzed options for regulating the water supply from the water intake facility in Dzerzhinsk to Khomutovo under uniform operation of pumps using a regulating capacity and a frequency converter without intermediate tanks.

An analysis of the power consumption graphs showed the inappropriateness of using a frequency converter for the developed water supply system in Khomutovo.

References

[1] Protasova M V, Goryachaya M V 2018 *Draft amendments to the General Plan of Khomutov Municipal District of Irkutsk Region* (Irkutsk: Planning Workshop Master Plan") p 405

[2] Safonchikov A P 2015 *Exploration work on Dzerzhinsky site to organize drinking and process water supply in Dzerzhinsk of Irkutsk region and assess groundwater reserves as of 01/01/2015* (Irkutsk: Ministry of Natural Resources of the Russian Federation Federal State-Funded Public Institution Hydrospetsgeologiya, Irkutsk Branch) p 60

[3] Pavlov P P 2019 *Updated water supply and sanitation scheme for Khomutovo, Irkutsk District, Irkutsk Region for the period up to 2031 Volume 1. Water supply scheme* (Irkutsk: Administration of the Khomutov Municipality - Administration of a rural settlement) P 69

[4] CR 31.13330.2012 Water supply. External networks and facilities. Updated version of SNiP 2.04.02-84 * (as amended by N 1, 2, 3, 4 from 07/22/2019). - Introduction. 01.01.2013. - M.: Publishing House of the Ministry of Regional Development of the Russian Federation, 2012. p 128

[5] CR 8.13130.2009. Collection of rules. Fire protection systems. Sources of external fire water supply. Fire Safety Requirements "(approved by Order of the Ministry of Emergencies of Russia dated March 25, 2009 N 178) (as amended on December 9, 2010)

[6] CR 10.13130.2009. Collection of rules. Fire protection systems. Internal fire water supply. Fire Safety Requirements "(approved by Order of the Ministry of Emergencies of the Russian Federation dated March 25, 2009 No. 180) (as amended on December 9, 2010)

[7] Karelin V Y, Minaev A V 1986 *Pumps and pumping stations* (Moscow: Stroyizdat) p 320

[8] Shakaryan Yu G 1997 *Instructions for calculating economic efficiency of the use of a frequency-controlled electric drive* (Moscow: VNIIE, MPEI)

[9] ECV. Well pumps. Engineering solutions for flow control. JSC "Hydromashservice" 2012

[10] Erk A F, Maksimov S V 2010 Methodology for evaluating the effectiveness of the use of frequency controllers as part of hydraulic equipment (Moscow: SZNIIMESKh)

The study of the use of "Dezolak" for disinfection and deodorization of sewage sludge

V I Sherbakov [1], K I Chizhik [2], V V Pomogaeva [1] and O Y Tararykov [3]

[1] Voronezh State Technical University, Moscow Avenue, 14, 394026, Voronezh, Russia
[2] Moscow State University of Civil Engineering, 26 Yaroslavskoye Shosse, 129337, Moscow, Russia
[3] Voronezh NPPI, Lenin Str.et, 73, Voronezh, 394026, Russia

E-mail: irkyt-44@yandex.ru

Abstract. Experimental studies on the decontamination and stabilization of sludge were carried out using the product "Desolak", which is calcium oxide treated with a disinfectant. As a result of the research, it was found that the sewage sludge of Voronezh is safe and can be used as organic fertilizer for growing agricultural plants. The determination of the content of mineral nutrients in the soil and the soil reaction showed that the application of organic fertilizers in the form of sludge in comparison to the variant without fertilizers increases the content of nitrogen nitrate, exchange potassium, and mobile phosphorus in the soil. At the same time, the mass fraction of impurities of toxic elements in the soil decreases when sewage sludge is added as an organic-lime fertilizer.

1. Introduction

More than a hundred million tons of organic sediment is stored at water and utility enterprises in Russia, excluding waste from livestock farms. Major of sewage sludge is delivered to sludge dewatering sites. Currently, most silt sites are overflown and are often located in residential areas of cities.

Organic sediments belong to the fourth or fifth hazard class, therefore, the problem of their disposal is relevant from an environmental, economic and social point of view. According to Dovgan S.A. “The task of utilizing sludge from sewage treatment plants in Voronezh is generally one of the main environmental problems (exhaustion of the capacity of sludge maps, odor effects) [1].

Sediment of urban wastewater is a valuable organic mineral product used for growing agricultural products in various countries of the world. This is due to the fact that the solid phase of sediment contains more than 50% of organic substances, as well as complexes of nitrogen, phosphorus and potassium. Organic and mineral substances increase the yield of cultivated plants when they are brought to the soil [2]. The use of sediment is constrained by the possibility of the presence of pathogenic microorganisms and heavy metals, which is unacceptable when applied to the soil. “The correct use of sediment will increase soil fertility and crop yields, and protect the environment [3]."

The safety of sewage sludge according to sanitary and hygienic indicators is determined by the presence of pathogenic microorganisms and helminths’ eggs. Disinfection of sewage sludge can be carried out by adding lime to it, which is common in Europe and the USA. “The method of disinfecting sewage sludge with quicklime is used at individual treatment facilities in Finland, Germany, Sweden, the USA and other countries” [4].

Safety in terms of sanitary and hygienic indicators of sewage sludge is assessed by the presence and viability of pathogenic microorganisms and helminth parasites' eggs. Liming can significantly reduce the content of pathogenic microorganisms that is a potential hazard. Of the existing reagents, lime is most widely used due to its low cost. When quicklime is added, alkalinity increases, due to the process of extinguishing lime by the water contained in the sediment, the temperature of the sediment increases. According to the studies of Ignatenko A., when a dose of lime is administered up to 30% by dry matter of the sediment, deformation and death of fecal streptococci, Salmonella bacteria occurs. As the pH rises above 11, the content of coliform bacteria decreases from 109 to 103 pcs. per 1 g of dry matter [5, 6].

As a result of studies conducted by various authors [2, 5, 6], it was determined that the treatment of dehydrated sewage sludge with quicklime (CaO) has a longer effect. With a lower water content in the sediment, it is more resistant to the development of acid fermentation processes. When using quicklime, depending on the content of free water (20-30%) and the initial temperature of the sludge, the dosage should be from 100 to 150 kg of CaO per ton of sludge or from 400 to 500 g of CaO for every kilogram of free water to reach a temperature above 50 ° C [7, 8].

Mixing sewage sludge on a twin-screw mixer with quicklime with an activity of 55-58% containing 6-25% CaO by weight of sludge is shown in Fig. 1. Adding lime in an amount of 23.5% allows to increase the temperature of the mixture to 78-82 ° C for 10 minutes, then the temperature drops to 46 ° C, within 40 minutes [5, 9].

The required dose of quicklime is determined from the condition of increasing the temperature of the sediment more than 60 °C and depends on the composition of the sediment and processing methods. The process of extinguishing lime occurs according to the following equation:

$$CaO+H_2O=Ca(OH)_2+65kJ. \tag{1}$$

Consequently, when extinguishing 1 kg of chemically pure lime containing 100% CaO, 1152 kJ of heat is released, with 320 g of water required.

The amount of heat needed to heat the sediment with quicklime can be determined by (2) [2, 5, 7]:

$$Q_r = (M_{sed} \cdot C_{sed} + M_{lim} \cdot C_{lim})\Delta T, \tag{2}$$

where M_{sed}, M_{lim} sediment and lime mass, kg; C_{lim}=0,92 – specific heat of lime, kJ/(kg °C); ΔT – temperature difference required for heating the precipitate and the initial, °C.

The specific heat of the sediment is determined by the formula:

$$C_{sed} = 1{,}8(1 + 0{,}85 \cdot W_{sed}), \tag{3}$$

where 1.8 is the heat capacity of dry sediment with a moisture content of 5-10%, kJ/(kg °C); W_{sed} – sediment moisture, fractions of a unit.

The amount of heat obtained by slaking lime, taking into account its CaO activity, will be:

$$Q_r = 1152 \cdot A \cdot M_{lim}, \tag{4}$$

where A is lime activity, fractions of a unit.

From the equation of material balance, you can preliminarily determine how much the temperature can increase when a certain dose of lime is added to the sediment:

$$\Delta T = \frac{1152 \cdot A \cdot M_{lim}}{M_{sed} \cdot C_{sed} + M_{lim} \cdot C_{lim}}. \tag{5}$$

By setting the required temperature, it is possible to determine the required mass of lime for sludge:

$$M_{lim} = \frac{M_{sed} \cdot C_{sed} \cdot \Delta T}{1152 \cdot A - C_{lim} \cdot \Delta T}. \tag{6}$$

Based on the conditions of complete quenching of active lime in the sewage sludge, it is theoretically possible to determine the sludge moisture:

$$W_c = \frac{1000 \cdot W_{sed} - 0{,}32 \cdot A \cdot M_{lim}}{M_{sed} + M_{lim}}, \tag{7}$$

where W_c – humidity of the precipitate obtained after adding lime to it, a fraction of units.

The above formulas (2-7) can be used for approximate calculations, which is due to the heterogeneity of lime by size and quality of particles, the presence of bound moisture in sediments with various forms of solid particles. As a result of this, the calculated dependences require some adjustment. Since the consumption of lime significantly depends on humidity, it is advisable to carry out preliminary thickening or dewatering of sediments.

The content of heavy metals in the sewage sludge depends on the enterprises discharging water to the treatment plant. For each city such indicators will differ. The MPC (maximum permissible concentrations) of heavy metals in sewage sludge is strictly regulated. Table 1 shows the MPC values for different countries and Russia according to regulatory data.

Table 1. The MPC value of heavy metals in sewage sludge.

TM	MPC TM in sediment (mg / kg dry matter)								
	Germany	Finland	Austria	France	Sweden	Holland	Switzerland	EEC Directives	Russia
Pb	1200	1200	500	300	300	500	900-1500	750	1000
Cd	20	30	10	15	15	10	25-40	20	30
Cr	1200	1000	500	200	1000	500	900-1500	750	1200
Cu	1200	3000	500	1500	3000	500	900-1500	1000	1500
Ni	200	500	200	-	-	100	150-300	300	400
Hg	25	25	10	8	8	10	8-15	16	15
Zn	3000	5000	2000	3000	10000	-	2500-4000	2500	4000
Mn	-	3000	-	500	-	-	-	-	2000
Co	-	100	100	200	-	-	80-150	-	-
Ag	-	-	100	-	-	-	-	-	-
As	-	-	-	-	-	-	-	-	20

Analyzing this table, we can conclude that the Russian standards comply with generally accepted international standards. In terms of arsenic content, they are higher than in the above countries.

2. Experimental part

Experimental studies on the disinfection and stabilization of sludge sediments were carried out using the product "Desolac", which is calcium oxide treated with an ovicidal substance. "Dezolak" products are manufactured by Pridonkhimstroy Lime LLC in accordance with TU 21.23.25-004-00121270 -2016. "Desolak" is intended for disinfection and stabilization of sewage sludge, sewage from livestock and poultry enterprises for the purpose of their further use in agriculture, industrial floriculture, landscaping and landscaping of settlements, for biological reclamation of disturbed lands, solid waste and solid waste landfills, as well as restoration of soil fertility in agriculture and forestry.

The disinfecting reagent is SE-1 (ROSUN). It is produced in the form of a white or gray powder with a moisture content of 0.5%, a bulk density of at least 0.9 g/cm^3, and solubility in water up to 20 g/l. It is a strong oxidizing agent that provides, in addition to exposure to pollutants, a disinfecting effect. The composition of the drug includes the following substances:

- potassium monopersulfate (double salt up to 18-24%),
- sodium chloride (1-5%),
- active oxygen (7-9%).

There is a passport on the reagent, which reflects 17 items required for the corresponding document in the Russian Federation. According to the safety data sheet, the product undergoes biological oxidation and does not accumulate biologically. The disinfecting effect of the drug is also provided by its pronounced oxidizing ability. The mechanism of the bactericidal effect is reached by blocking the transmission of ribonucleic acid and destroying the genetic system. When dissolved in water, active oxygen, a hydroxyl free radical, hydrogen peroxide and hypochlorous acid are formed. All components act simultaneously on organic substances of technogenic and natural origin and inorganic substances with variable valency.

Depending on the percentage of disinfecting reagent, the product is divided into:

- "Desolak - B" calcium oxide with an additive in an amount not exceeding 2% by weight;
- "Desolak - F" calcium oxide with an additive in an amount not exceeding 1.5% by weight.

The main indicators are presented in table. 2.

The difference between these products is the percentage of disinfecting additives, which is associated with the quality of disinfecting sediments:

- "Dezolak-B" is intended for disinfection, stabilization and deodorization of sewage sludge from municipal wastewater treatment plants (a mixture of domestic, industrial and storm sewage), and can also be used at local treatment facilities of various industrial enterprises in combination with other treatment technologies.
- "Desolak-F", in addition to the application for the previously listed sediments, can be used for disinfection, stabilization and deodorization of effluents of livestock and poultry complexes.

Table 2. Key indicators of the quality of the product "Desolac".

Quality indicators	**"Desolak - B"**	**"Desolak-F"**
Active CaO	not less than 80%	not less than 80%
Temperature of the reaction with water	70-85 ^{0}C	70-85 ^{0}C
Reaction Time	3-15 min.	3-15 min.
Disinfecting additives	not more than 2%	not more than 1.5%
Fractional composition of grains	not more than 200 microns	not more than 200 microns
Bulk density	not less than 700 kg/m3	not less than 700 kg/m3

"Desolak" is made of pure carbonate rocks that have undergone thermal heating at a temperature of 900 to 1100 C to produce calcium oxide. Next, grinding to 200 microns occurs. It is mixed with disinfecting additives in an amount of 1.5-2.0%.

For the use of this product there is a Certificate of Conformity ROSS RU. ПЩ01.Н11108№ 2210018 "Passport of compliance of chemical products". For the disinfecting reagent there is an "Expert Opinion No. 6/2-P dated 02/12/2015" and a Quality Certificate ISO 9001: 2008. It is a product of the second hazard class according to GN 2.2.5.1313-03 and GOST 12.1.007.

Studies on the addition of a disinfectant were carried out at the treatment facilities of Voronezh. The sludge was processed in a plant (Fig. 1), consisting of a Desolak feed hopper (point2), a sludge screw mixer with lime (point 3). The precipitate was pre-dehydrated on filter presses. Quicklime from the silo through the auger is fed to the screw mixer, where lime is mixed and dosed with sludge (Fig. 2). The treated sludge is discharged in an automated mode. The experimental setup is shown in Fig. 3.

IOP Conf. Series: Materials Science and Engineering **880** (2020) 012053 doi:10.1088/1757-899X/880/1/012053

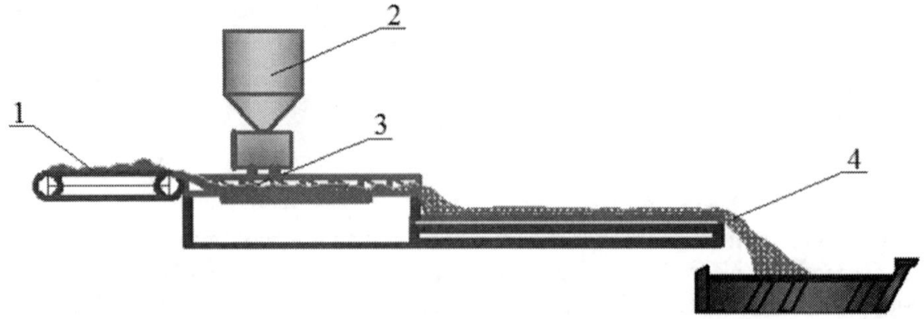

Figure 1. The scheme for the treatment of dehydrated sludge with the Desolak product: 1-dehydrated sludge; 2- lime supply; 3-screw lime sludge mixer; 4- discharge of treated sludge/

To determine the content of toxic elements in the treated sludge, experimental studies were carried out at the Left-bank treatment facilities of Voronezh (LLC "LOS"). The crude sludge from the primary sumps and were mixed with quicklime and dehydrated on a Flottweg decanter (Fig. 2).

Figure 3. Obtaining processed sludge:
a) Flottweg Decanter Centrifuge; b) sludge discharge; c) type of precipitate obtained

3 Evaluation of the results

When conducting experiments on the decontamination of sewage sludge, problems arose in the reaction of lime. Completely dehydrated sludge does not fully react, and the desired decontamination effect is not achieved. Based on the experiments, the required water content in the sediment was set. The experiments were carried out with crude sludge and a mixture of crude sludge and excess activated sludge. To obtain generalized research results in various proportions were mixed: sediment, excess activated sludge and the product "Desolak", the data on the samples are presented in table. 3.

Table 3. Sludge samples from wastewater treatment plants.

Sample	Sample Content		
	crude sediment, %	excess activated sludge, %	Desolac, %
№1	100	-	5

№2	100	-	7
№3	30	70	5
№4	10	90	5
№5	30	70	3

The best option for ading the treated sediment into the soil is sample No. 3. To determine the amount of heavy metals in the sediment and soil, when applying it, an experiment was conducted for two samples with and without "Desolak".

The amount of added sediment was determined by the formula (8), taking into account the content of heavy metals in the soil and sediment:

$$D_{tm} = \frac{(0{,}8 \cdot PDC - F) \cdot 3000}{C_{tm}}, \quad (8)$$

where: D_{tm} - theoretically permissible sediment rate, t / ha dry weight;

PDC- maximum permissible concentration of heavy metal in the soil, mc / kg;

F - actual heavy metal content in soil, mg / kg;

C_{tm} - the content of heavy metal in the sediment, mg / kg dry weight;

3000 - mass of arable soil layer in terms of dry matter, t / ha.

For dehydrated sludge (30% crude sludge, 70% excess activated sludge) of treatment facilities treated with "Desolak" in the amount of 5%, the application dose calculated for 60 kg of active substance per hectare based on nitrogen content was 3.78 t / ha.

For dehydrated sludge (30% crude sludge, 70% excess activated sludge) of treatment facilities, the application dose calculated for 60 kg of active substance per hectare based on nitrogen content was 4.45 t / ha.

The sediment was put under the main tillage.

Studies conducted at treatment facilities have shown that in the sediment after treated with "Desolak" product, the mass fraction of toxic element impurities is not significant. The determination of heavy metals in samples was carried out at the Voronezh State Center for Agrochemical Service. The mass fraction of heavy metal impurities in the dehydrated sludge treated with "Desolak" product and the standard content are shown in Table 4.

Table 4. Physical and chemical indicators of sewage sludge

Name of indicator	**Regulatory requirements for sewage sludge**	**Permissible total impurity content of heavy metals in mg/kg of dry matter**		**Value of indicator**	
		I category of soil	**IIcategory of soil**	**Sludge**	**Sludge with Desolac**
1	**2**	**3**	**4**	**5**	**6**
Moisture content, %	<<82	70	70	36,4	63,6
pH	5,5-8,5	6,0-8,0	6,0-8,0	6,4	12,2
Mass fraction of **organic matter** per dry matter, %	40-60	>30	>30	72,9	66,1
Mass fraction of **nitrogen,%**	1-3	>0,6	>0,6	1,84	2,6
P_2O_5, %	1-4	>0,7	>0,7	1,14	1,4
K_2O, %	0,2-0,7	>0,1	>0,1	0,1	0,13
Hg, mg/kg	15	<2,1	<7,5	0,45	0,41

Cr, mg/kg	<1200	<80	<200	48,3	20,7
Pb, mg/kg	<1000	<130	<250	114,9	97,1
Ni, mg/kg	<400	<132	<750	27,2	20,4
As, mg/kg	<20	<2	<10	0,5	0,6
Specific activity of technogenic radionuclides (ACs/45*ASr/30), dry matter	1,0			0,6	0,8
Mass concentration of pesticide residues in dry matter, mg / kg - HCH (total isomers) - DDT and its metabolites (total amount)	<0,1 <0,1			<<0,007 <<0,005	<<0,005 <<0,005
The presence of viable eggs and larvae of helminths, including nematodes (ascaridate, trichocephalus, strongylates, strongyloids), trematodes, cestodes	not allowed	not allowed	not allowed	not detected	not detected
Tsistys of intestinal pathogenic protozoa	not allowed	not allowed	not allowed	not detected	not detected
The presence of larvae and pupae of synanthropic flies	not allowed	not allowed	not allowed	not detected	not detected

The use of lime-stabilized sludge on agricultural land is constrained by the presence of toxic elements and parasitological parameters in sewage sludge. However, the experiments and the results of the analyzes show that the use of such a sedimnt is safe.

As can be seen from the table. 4 (column 5, 6) in the treated sludge, the mass fraction of heavy metal impurities is much lower than the normative ones. For example, for group I sediment, nickel contains 0.07-0.1 MPC, chromium 0.04 MPC, lead 0.2-0.45 MPC, mercury 0.01-0.06 MPC, arsenic 0.05-0 07 MPC, cadmium 0.09-0.17 MPC. The MPC of the mass fraction of toxic element impurities for group II sediments is twice as large as group I. Sediment of group I refers to fertilizers of group I, and can be used:

- Fertilizers of group I: fertilizers based on sewage sludge used for growing industrial, fodder, grain and green manure crops in a personal subsidiary farm for growing seedlings of vegetable and flower crops;
- Fertilizers of group II: fertilizers based on sewage sludge used for planting forestry crops along roads, in nurseries of forest and ornamental crops, floriculture, for cultivating depleted soils, reclamation of disturbed lands and slopes of roads, and reclamation of solid household waste dumps.

Accordingly, the resulting sediment can be applied to the reclamation fields with a frequency of no more than 2-3 years, in the amount of 5-15 t/ha in terms of dry matter, according to formula (8).

When conducting studies to determine the effectiveness of the use of sewage sludge in conjunction with the Scientific Research Institute of Agriculture of the Central Black Earth Strip named after V.V. Dokuchaev [2, 9, 10] obtained results proving the effectiveness of winter rape cultivation Adriana for increasing the green mass of plants. The concentration of substances necessary to improve soil fertility increases when the sediment is treated with lime [2, 9, 10]. A decrease in the content of heavy metals in the treated sludge was noted, compared with the initial concentration in the sludge of sewage inlet.

One of the main factors influencing the applied treated sediment on soil fertility is the content of minerals in the soil and the reaction of the soil environment. In the soil, when conducting research, there was an increase in the content of metabolic potassium by 3.5 mg / 100 g abs. dry soil or 18.6%. The increase in the content of mobile phosphorus varied: with the addition of dehydrated sludge, it increased by 4.5 mg / 100 g abs. dry soil or 7.6%, and with the introduction of sludge sediment treated with the drug "Desolak" practically remained at the control level. An increase in the content of nitrate nitrogen in the soil with silt sediment by 0.8 mg / kg abs. dry soil or 11.9%, with "Desolac" 1.5 mg/kg abs. dry soil or by 22.4%. The results are shown in table 5.

Table 5. The content of mineral nutrients and the reaction of the soil in the soil layer of 0-20 cm when applying sludge

Option	Indicator			
	$N\text{-}NO_3$*	P_2O_5**	K_2O**	pHсол
The control	6,7	59,5	18,8	7,24
Silt sediment treated with "Desolac"	8,2	57,0	22,3	7,35
Silt sediment	7,5	64,0	22,3	7,27

Note: * mg / kg of soil
** mg / 100 g of soil according to Chirikov

Accounting for the yield of green mass of winter rape and statistical processing of the results showed that the introduction of sludge from sewage treatment plants under the main tillage increases the productivity of winter rape. Excessive yield of green mass of winter rape when applied:

- silt sewage sludge treated with "Desolak" - 7.87 t / ha or 20.1% more than the control option;
- silt sludge - 9.56 t / ha or 24.4% more than the control option.

The use of sludge from wastewater treated with "Dezolak" slightly changes the reaction of the soil medium in the direction of increasing the alkalinity of the soil solution by 0.11 pH units compared to the control variant, which is significantly less than the design and operation accepted in practice [8]. Thus, the use of sludge from sewage treated with "Desolac" will be more effective in soils with acidic reaction.

4 Conclusions

1. Wastewater sludge in Voronezh treated with the "Desolak" preparation is safe and, according to microbiological indicators, refers to clean soils in accordance with regulatory requirements.

2. In accordance with regulatory requirements, sludge from Voronezh sludge sites are safe in terms of the content of toxic elements significantly less than standard ones and can be used as organic fertilizers for use both in grain and industrial crops, and in vegetable growing.

3. As a result of the use of organic fertilizers - silt sediment treated with "Desolak", pathogenic microflora, viable helminth eggs, cysts of pathogenic intestinal protozoa, enterococci and other dangerous biological agents are destroyed.

4. Dehydrated sludge treated with the product "Desolak" can be used in industrial floriculture, forest and decorative nurseries, for biological reclamation of disturbed lands and landfills. The introduction of such a precipitate increases soil fertility, due to an increase in the content of nitrogen, phosphorus and potassium in the soil, leads to an increase in the yield of winter rape by 20.1%. The use of sludge sediment as organic fertilizer does not affect the quality of cultivated plants. All indicators are within the maximum permissible concentration (MPC) or significantly lower.

5. The use of treated sediments of the city of Voronezh as organic-lime fertilizers for soils with pH less than 5.5 in doses calculated taking into account the calcium content in the composition of the

introduced sediment is recommended. When applying sediment as fertilizers, regional and local conditions should be taken into account, including soil properties and hydrological regime, normalized pollution, total and mineral nitrogen, phosphorus, potassium in sediment and soil, crop cultivation characteristics, and crop rotation.

References

[1] Shcherbakov V I, Pomogaeva V V, Chizhik K I and Koroleva E 2019 Biomass resource of domestic sewage sludge. Advances in Intelligent Systems and Computing. T. **983**, pp 361-372

[2] Vasilenko T A and Mohammed A Kh. 2016 Application of sludge for mechanical and biological treatment of domestic and industrial wastewater as fertilizer. *Herald GBTU them. V.G. Shukhov* **6** pp 211-219

[3] Vasilevich E E and Rudenko R N 2016 New approaches to the disposal of sewage sludge pp 53-55

[4] Ignatenko A V 2016 Decontamination of sewage sludge and methods of its control. *Proceedings of BSTU* **4** pp 210-213

[5] Kalimullina D D and Bagautdinov I Z 2016 Features of reagent stabilization of sewage sludge. *Innovative science* **5** ISSN 2410-6070.

[6] Turovsky I S 2018 Sewage sludge. Dehydration and disinfection. *Moscow: DeLi print* p 376

[7] Danilovich D A 2018 Modern methods of disinfection of sewage sludge. *NDT* **6** pp 50-54

[8] Shcherbakov V I and Pomogaeva V V 2018 The use of sewage sludge for growing crops.

[9] *Yakovlevskie readings collection of reports of the XIII International Scientific and Technical Conference dedicated to the memory of academician of the RAS S.V. Yakovleva. National Research Moscow State University of Civil Engineering* pp 155-160 Dovgan S A 2012 Ecological, technological and economic aspects of the use of sewage sludge. *Ecology and industry of Russia* **5** pp 25-30

[10] Matveeva M V, Doroshenko T G and Sukhanova I A 2019 On the issue of cost and timing of contract tenders for industrial and civil construction *Investments. Construction. Real Estate: New Technologies and Targeted Development Priorities IOP Conference Series: Materials Science and Engineering* **667 012063**

[11] Baranova A A and Bobrova A A 2019 Dispersed reinforcement of cellular and fine-grained concrete based on silica fume *Proceedings of Universities. Investment. Construction. Real estate* **9(4)** pp 694–703.

[12] Petrov A V and Efimova A K 2019 Nguyen Thanh Tung. Optimisation of technology used for restoring gas-ash concrete outdoor wall panels using modified structural and heat-insulating *concrete Investment. Construction. Real estate* **9(3)** pp 542–549

[13] Selezneva O I, Radaev S S and Baranova A A The impact of thermal treatment on the granulometric and chemical composition of diatomite of the Kamyshlov field when obtaininga liquid-glass binder *Investment. Construction. Real estate* **9(3)** pp 550–555

Device to perform primary damper tank wastewater treatment at sewage pumping stations. Static calculation

D V Skibo[1] and A G Kunitsyn[2]

[1] Vodokanal Irkutsk Municipal Unitary Enterprise, Stanislavsky st.,2, Irkutsk 664081 Russia
[2] Irkutsk National Research Technical University, Lermontov st.,83, Irkutsk 664074 Russia

E-mail: d.skibo2013@yandex.ru

Abstract. The design of "Device to perform primary wastewater treatment of suspended solids at sewage pumping stations" utility model is considered to be implemented in existing and newly designed sewage pumping stations, as well as in automated damper tanks. Exemplified by real conditions, static loads on the main structural elements are modeled. The construction arrangement is evaluated in the terms of possibility to be used as a base for the development of the model range. The possibility of developing an industrial design is determined based on the results of analysis of the strength of the units.

1. Introduction

Primary treatment of entering the sewage pumping station wastewater is a preparatory stage of further transportation of the water in pressure mode. The content of wastewater mechanical impurities varies greatly in composition and quantity that complicates the operation of pumps and pressure pipelines.
An effective wastewater treatment process is of special importance at the stage of passing through a damper tank of a sewage pumping station [1] during the adjustment or accident period. To perform such tasks, the following technical devices are known: all kinds of gratings, screens and strainers, however, the low level of automation, absence or insufficient level of hardware control were noted during the analysis process , what is more, the operation involves significant labor costs of the personnel [1, 2, 3, 4, 5, 6, 7, 8, 9, 10, 11, 12, 13, 14, 15, 16, 17, 18, 19, 20, 21, 22, 23].

2. Materials and experimental techniques

To improve the efficiency of the mechanical wastewater treatment, the team of authors proposed an utility model, "Device to perform primary wastewater treatment of suspended solids at sewage pumping stations", figure 1 [2]. The initial stage of research, focused on the wastewater treatment issues, determined the practical feasibility of performing this operation using the device as a part of a sewage pumping station and a damper tank of a sewage pumping station [3].

The design of the prototype requires the calculation of support frames cross-sections, slewing rings, drive rods, rotary platforms, as well as modeling sliding bearings and the selection of actuators. The design technique will optimize the operating parameters to determine the model range of industrial products.

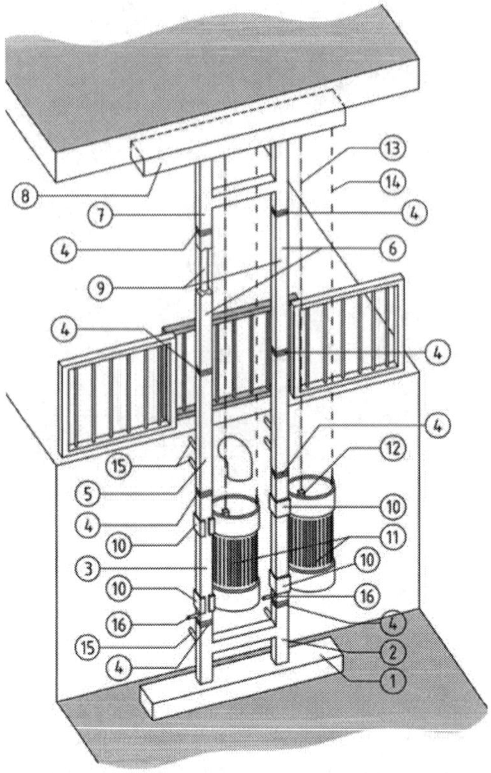

Figure 1. Device to perform primary wastewater treatment of suspended solids at sewage pumping stations [2]: 1is a foundation slap; 2 is a lower support frame; 3 is a lower rotary platform; 4 is a slewing unit; 5 is intermediate slides; 6 is an upper rotary platform; 7 is an upper support frame; 8 is multifunctional operating module; 9 is actuating rod; 10 is a support; 11is a filtration container; 12 is a monitor sensor; 13 is a communication cable; 14 is a load sling; 15 is a metal bracket; 16 is a running limiter [2]

In the design of the utility model, the basic elements to receive static load are slewing units, including the upper and the lower support frames, actuating rods, rotary platforms, slider bearing and supports.

Design solutions in development of a pre-production model should be based on the engineering analysis.

The model of a typical sewage pumping station of modular type with 2.2 m diameter of the receiving tank and 8.5 m height; the inlet gravity pipe diameter is 0.3 m; the daily wastewater flow rate is 1250 m^3 / day per a residential settlement with an estimated population of 5 000 thousand people [3] is the basis to choose the dimensions of the swivel block for a pre-production model, figure 2.

3. Rotary platforms design model

We consider rotary platforms as a structural element that perceives the concentrated load of the weight of the filter container and its contents, figure 2 [2], where G is the total weight of the filtration container (if lattice gap width is 16 mm, in condition of garbage loading equal to 50% of the daily value = 10 + 57 kg ≈ 1 670 *H*.

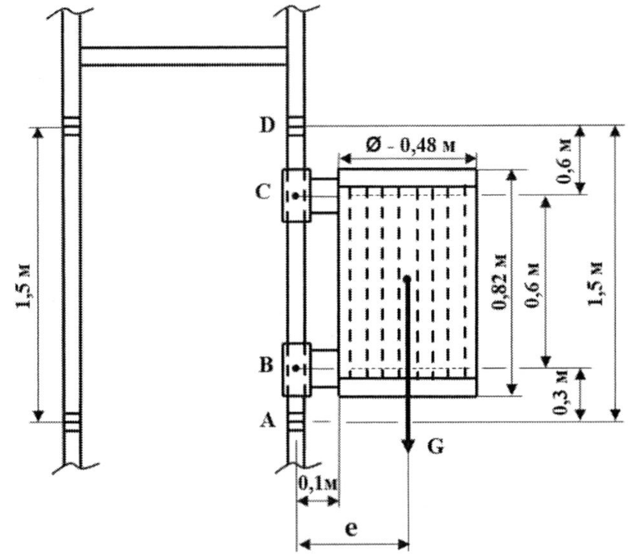

Figure 2. Design model of a swivel block

Section D – A of a rotary platform, figure 3, is presented in the form of a design diagram, and N epure: axial forces, M: bending moments, and Q: cross-axis forces [24].

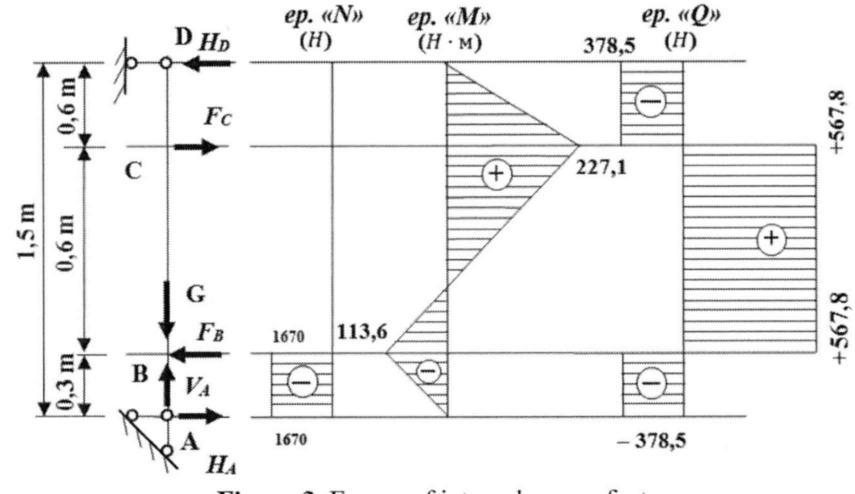

Figure 3. Epures of internal power factors

Longitudinal section is presented as *N epure*:

$$N_{AB} = G = 1\,670\ H\ .$$

Load action line G is at the distance:

$$e = \frac{\varnothing}{2} + 0.10\ \text{м} = \frac{0{,}48}{2} + 0.1 = 0.34\ \text{m}\ .$$

3.1 Rated force determination

The horizontal components of the upper F_C and lower F_B are equal: $F_C = F_B$. It is proved by the equations of the sum of the projections of all forces on the horizontal axis X, figure 4.

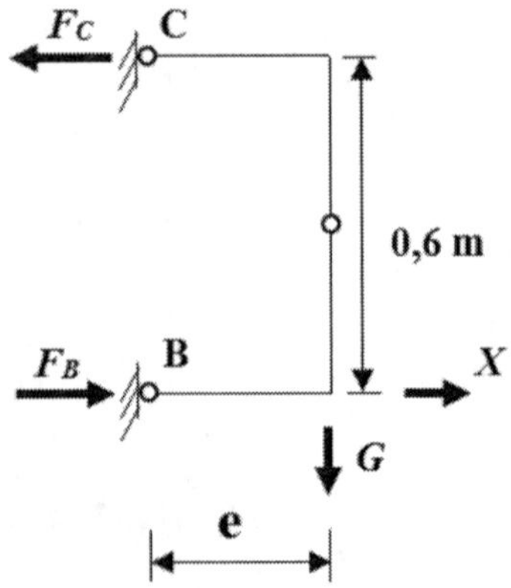

Figure 4. Analytic model

The sum of the moments relative to the point **B,** figure 4, to determine forces $\boldsymbol{F}_C$ and $\boldsymbol{F}_B$ is presented as follows [24]:

$$\sum M_B = 0;$$

$$G \cdot e - F_C \cdot 0{,}6 = 0, \Rightarrow F_C = \frac{G \cdot e}{0.6} = \frac{1670 \cdot 0{,}34}{0.6} = 946.3\ H;\ F_C = F_B = 946.3\ H.$$

Determination of the support reactions [24], figure 3:

а) $\sum M_A = 0;\ -H_D \cdot 1.5 + F_C \cdot 0.9 - F_B \cdot 0.3 = 0; \Rightarrow H_D = 378.5\ H;$

б) $\sum M_B = 0;\ -F_C \cdot 0.6 + F_B \cdot 1.2 - H_A \cdot 1.5 = 0; \Rightarrow H_A = 378.5\ H;$

в) Check: $\sum x = 0;\ -H_D + F_C - F_B + H_A = 0; -378.5 + 946.3 - 946.3 + 378.5 = 0.$

Design the epures of bending moments (*M epure*) and cross-axis forces (*Q epure*) [24].

M epure:

$$M_{T.A} = 0;\ M_{T.C} = H_D \cdot 0.6 = 227.1\ H \cdot \text{m};\ M_{T.B} = -H_A \cdot 0.3 = -113.6\ H \cdot \text{m};\ M_{T.A} = 0.$$

Q epure:

$$Q_{A.B} = -H_A = -\ 378.5\ H;\ Q_{B.C} = -H_A + F_B = +\ 567.8\ H;\ Q_{D.C} = -H_D = -378.5\ H.$$

3.2 Static calculation of the H - shaped part of the frame

Unfavorable position in slewing units is formed by 90° rotation of rotary platforms with filtration containers placed on them, figure 5, where the reaction is transmitted at D point ⊥ plane of the frame.

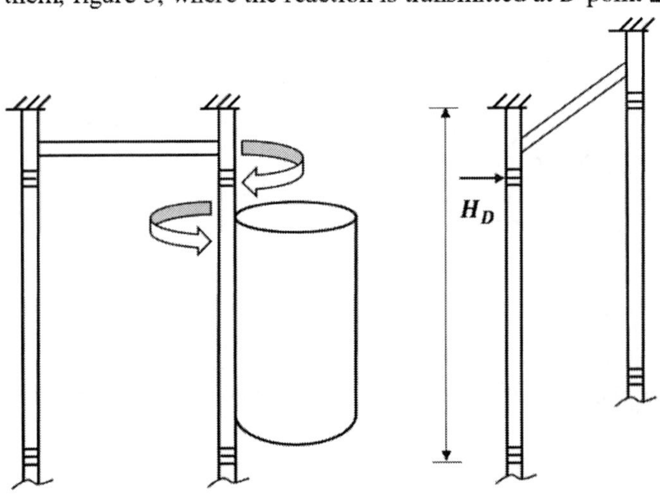

Figure 5. Analytic model

Design the epures of bending moments (*M epure*) and cross-axis forces (*Q epure*) [24], figure 6.

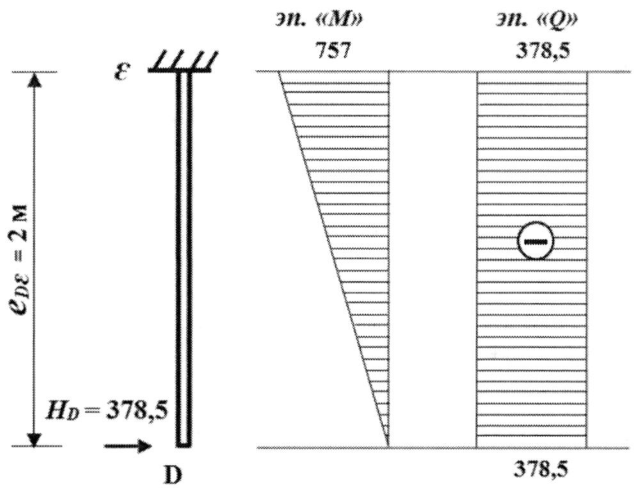

Figure 6. Epures of internal power factors

$M\ epure$: $M_{T.D} = 0$; $M_{T.\mathcal{E}} = H_D \cdot e_{D\mathcal{E}} = 378.5 \cdot 2 = 757\ H \cdot \text{m}$.
$Q\ epure$: $Q_{D.\mathcal{E}} = -H_D = -378.5\ H$.

3.3 Design calculation

The design calculation is performed to determine the safe cross-sectional dimensions of the support-guide frame.

Due to the condition of bending strength [24]:

$$\sigma_{max} = \frac{M_{max}}{W_x} \leq [\sigma],$$

where: σ_{max} is maximum bending stress ($H \cdot \text{m} = \text{Pa}$);
$[\sigma]$ is allowable stress for the material. For steel grades Ст3 $[\sigma]$ = 160 MPa;
M_{max} is maximum moment in the section area
$M_{max} = 757\ H \cdot \text{m}$. (т. $\mathcal{E}$ $M_{T.\mathcal{E}} = 757\ H \cdot \text{m}$);
W_x is section modulus.

$$W_x \geq \frac{M_{max}}{[G]} = \frac{757\ H \cdot \text{m}}{160 \cdot 10^6\ \text{Pa}} = 4.73 \cdot 10^{-6}\text{m}^3 = 4.73\ \text{sm}^3.$$

Accepted by assortment [26] $W_x^{\text{табл}} = 4.81\text{sm}^3$; A = 40 mm; S = 3 mm

$$\text{«tube } \frac{40\ \times 40 \times 3\ \times 1250\ \text{кр ГОСТ } 8639-82.}{\text{В } 10\ \text{ГОСТ } 13663-86}\text{» [25]figure 7.}$$

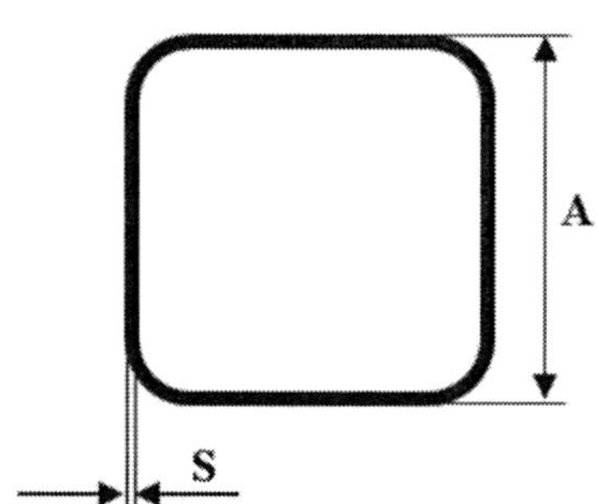

Figure 7. Cross-section

According to the design reasons A = 100 mm; S = 5 mm. Figure 7 (100×100×5×1 250).
The calculation of stiffness is made in order to limit elastic displacements [24].
A = 100 mm; S = 5 mm; $F = 18.57\text{sm}^2$; $J_x = J_x = 276.3\ \text{sm}^4$

$$f = \frac{H_D \cdot e^3}{3EJ_x} = \frac{378.5 \cdot 2^3}{3 \cdot 2 \cdot 10^5 \cdot 276.3 \cdot 10^{-8}} = 1.826 \cdot 10^{-3}\text{m} = 1.826 \text{ mm}.$$

Analyzing stiffness:

$$\frac{f}{e} = \frac{1.826}{2 \cdot 10^3} \approx 0.001.$$

We conclude that the stiffness is sufficiently high,
where: E – Young's modulus [26] (steel grades Ст3 $E = 2 \cdot 10^5$ MPa).

4. Actuating mechanism drive design model

The shaft with length L, figure 8, is an actuating mechanism drive.

Shafts are calculated simultaneously for strength and stiffness.

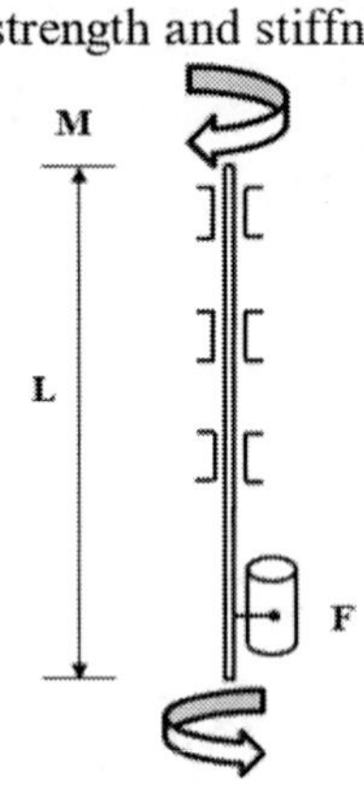

Figure 8. "Swivel" shaft L = 15m. Design diagram

Due to the insignificant resistance created by the frictional forces in the bearing of a slewing unit of the lower support frame and the lower rotary platform, as well as by the forces of internal friction between the moving fluid layers when the filter container is rotated in an aqueous medium, the torque M_{kp}, is taken with a margin of $20\ N \cdot$ m.

4.1 Selection of section due to strength

Torsional strength condition [24]: $\tau_{max} = \frac{M_{kp}}{W_\rho} \le [\tau]$,

where: $[\tau] = 90$ MPa is torsional strength limit (MPa);
W_ρ is polar resistance limit (m^3).

$$W_\rho = \frac{M_{kp}}{[\tau]} = \frac{20 \cdot H \cdot \text{м}}{90 \cdot 10^6} = 2.222 \cdot 10^{-7} \text{sm}^3.$$

a) for round section (solid):

$$W_\rho = \frac{\pi d^3}{16}; \Rightarrow d = \sqrt[3]{\frac{16 \cdot W_\rho}{\pi}} = \sqrt[3]{\frac{16 \cdot 2.222 \cdot 10^{-7}}{3.14}} = 1{,}04 \text{ sm rounded to } 1{,}2 \text{ sm} = 12 \text{ mm}.$$

б) for tubular section:

$$W_\rho = \frac{\pi d^3}{16}(1 - \alpha^4),$$

where: $\alpha = 0{,}8$ is ratio of inner to outer diameter $\frac{d_0}{d}$.

$$D = \sqrt[3]{\frac{16 \cdot W_\rho}{\pi(1 - \alpha^4)}} = 1.11 \text{ sm} = 12\text{mm}\ \varnothing.$$

To check the stiffness, we take structurally the shaft diameter equal to 28 mm.

4.2 Strain determination

IOP Conf. Series: Materials Science and Engineering **880** (2020) 012054 doi:10.1088/1757-899X/880/1/012054

Shaft twist angle [24]:

$$\varphi = \frac{M_{\text{кр}} \cdot L}{G \cdot J_\rho},$$

where: G is shear modulus;

$G = 0.8 \cdot 10^5$ NPa (Ст3);

J_ρ is polar moment of inertia.

a) for round cross-section:

$$J_\rho = \frac{\pi d^4}{32} = \frac{3.14(2.8 \cdot 10^{-2})^4}{32} = 6.03 \cdot 10^{-8}\ \text{m}^4;$$

$$\varphi = \frac{20 \cdot 15}{0.8 \cdot 10^5 \cdot 10^6 \cdot 6.03 \cdot 10^{-8}} = 62.2 \cdot 10^{-3}\ \text{rad}.$$

In degrees:

$$\varphi' = \varphi \frac{180°}{\pi} = 3.56°.$$

б) for ring cross-section:

$$J_\rho = \frac{\pi d^4}{32}(1-\alpha^4) = \frac{3.14(2.8 \cdot 10^{-2})^4}{32}(1-0.8^4) = 3.561 \cdot 10^{-8}\ \text{m}^4;$$

$$\varphi = \frac{20 \cdot 15}{0{,}8 \cdot 10^5 \cdot 10^6 \cdot 3.561 \cdot 10^{-8}} = 105 \cdot 10^{-3}\ \text{рад}.$$

In degrees:

$$\varphi' = \varphi \frac{180°}{\pi} = 6.04°.$$

5. Calculation of a swivel block for a cut

Determination of cut-off area [24], figure 9:

$$F_{\text{ср}} = F_{\text{прям}} - F_{\text{кр}} = a^2 - \frac{\pi d^2}{4} = 9^2 - \frac{3.14 \cdot 3^2}{4} = 73.935\ \text{sm}^2.$$

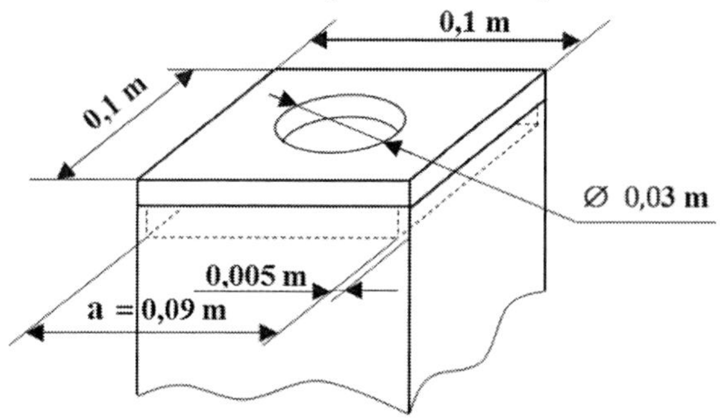

Figure 9. Swivel mechanism of support unit design model

Design power $P = H_D = 378.5\ H$.

Shearing test:

$$\tau_{max} = \frac{P}{F_{\text{ср}}} \leq [\tau],$$

where: $[\tau]$ is allowable shear stress for bronze: $[\tau] = 60$ MPa [24],

$$\tau_{max} = \frac{378.5}{73.935 \cdot 10^{-4}} = 5.14 \cdot 10^{-4} = 5.14 \cdot 10^{-4}\text{MPa}.$$

Strength conditions are observed.

6. Calculation of the suspension support (verification of strength and stiffness)

The support, shown in figure 10 "a", is presented in the form of a design diagram. The support top view, figure 10 "b", is presented in cases of adverse application of load.

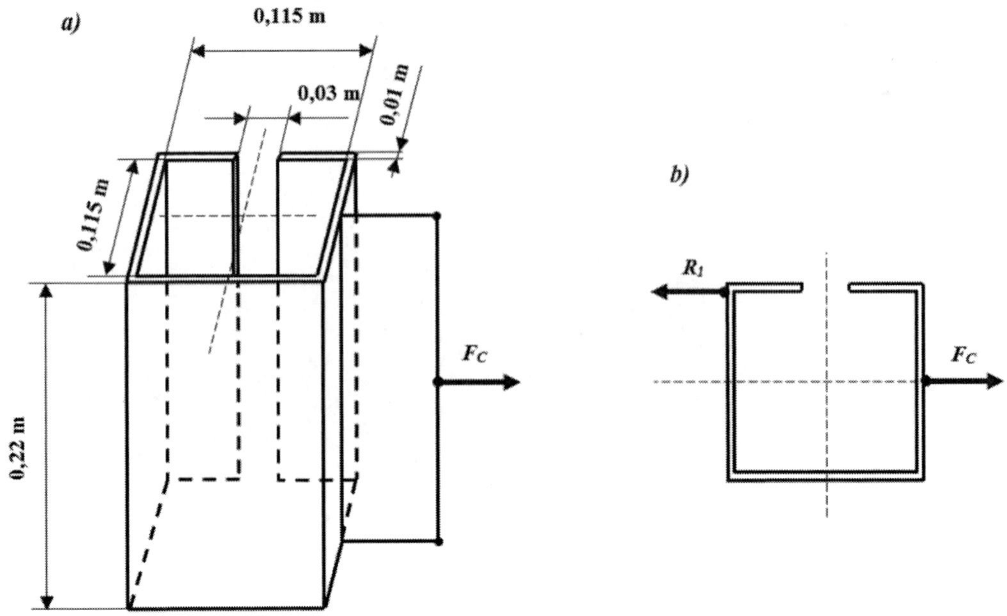

Figure 10. Support design model

It follows from the equitation of projection of forces on the *X* axis [24]:

$$\sum X = 0; \Rightarrow R_1 = F_C = 946.3\ H.$$

We design the bending moment epure by points, figure 10.

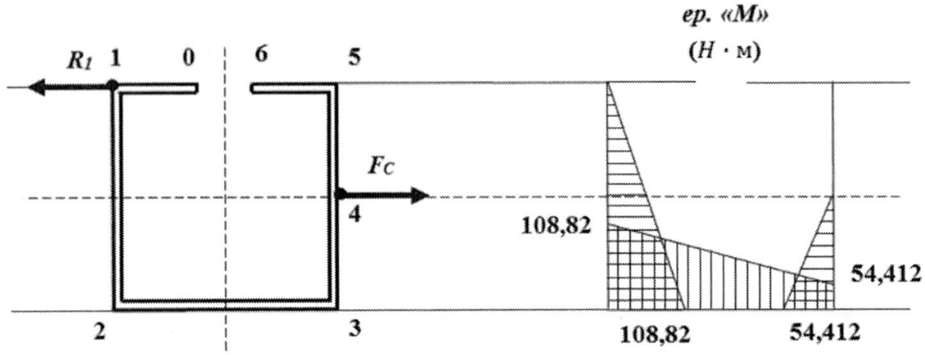

Figure 11. Epure of support bending moments

$M\ epure\colon M_{T.2} = R_1 \cdot 0.115\ \text{м} = 108.82\ H \cdot \text{m}.$

$M_{T.3} = F_C \cdot \frac{0.115}{2} = 54.412\ H \cdot \text{m}.$

Strength test [27]:

$$\sigma_{max} = \frac{M_{max}}{W_x} \leq [\sigma],$$

where: $W_x = \frac{220 \cdot 10^2}{6} = 3\,667\ \text{mm}^3 = 3.667 \cdot 10^{-6}\ \text{m}^3$ – section 2 resistance moment;

$M_{max} = M_{T.2} = 108.82\ H \cdot \text{m}$ – maximum bending moment.

$$\sigma_{max} = \frac{108.82}{3.667 \cdot 10^{-6}} \approx 29.7\ \text{MPa} < [\sigma] = 160\ \text{MPa}.$$

Strength conditions are observed.

6.1 Stiffness calculation

Using the same technique as when calculating the frame for stiffness [27], we write the deformation t. 1:

$$f_1 = \frac{R_{D1} \cdot e_{1-2}}{3E\,J_x} = \frac{946.3 \cdot 115}{3 \cdot 2 \cdot 10^5 \cdot 18\,333} = 9.89 \cdot 10^{-6}\ \text{mm},$$

where: $e_{1-2} = 115$ mm;

$$E = 2 \cdot 10^5\text{MPa} = 2 \cdot 10^{11}\text{Pa} = 2 \cdot 10^5 H/\text{mm}^2;$$

$$J_x = \frac{220 \cdot 10^3}{12} = 18\,333\ \text{mm}^4.$$

The deformation of t. 1 is negligible, respectively, the support stiffness is provided with a large margin.

7. Conclusion

The results obtained in the process of calculating the units of "Device to perform primary wastewater treatment of suspended solids at sewage pumping stations" prototype give reason to consider the design diagram acceptable to be used in a damper tank of a sewage pumping station, pumping stations, as well as in emergency control tanks. The calculation is a theoretical justification for the further development of the device. Analyzing the obtained data, we can conclude that the design has sufficient margin of safety and can be considered as basic one to develop a model range of industrial products.

This article is one more step in the innovative project implementation. In the next work, the authors will consider the optimization of operating parameters, as well as the selection and adaptation of actuators.

References

[1] Skibo D V, Tolstoy M Yu 2019 Automated damping tank of sewage pumping stations Utility Model Patent RU No. 186682 U1, Published: January 29, 2019 Bull. 4

[2] Skibo D V, Sudnikovich V G, Tolstoy M Yu, Kurtin A V 2018 Device to perform primary wastewater treatment of suspended solids at sewage pumping stations Utility Model Patent RU No. 179 790 U1, Published: May 24, 2018 Bull. 15

[3] Skibo D V, Tolstoy M Yu 2018 Mechanical wastewater treatment at sewage pumping stations as a way to reduce anthropogenic environmental impact. *Construction and industrial safety.* 10 (62) pp 117-125

[4] Federal Law No 7 On Environmental Protection dated 01/10/2002 (Electronic materials) http://www.consultant.ru/document/cons_doc_LAW_34823/ (accessed November 1, 2019)

[5] Karmazinov F V, Melnik E A, Probirsky M D, Pankova G A, Mikhailov D M, Ilyin Yu A, Ignatchik V S, Ignatchik S Yu 2013 Technical inspection of pumping stations of St. Petersburg sewage system. *Water supply and sanitary technique* **1** pp 20–28

[6] Salomeev V P 2017 Restoration and reconstruction of pumping stations of water disposal. *Scientific Review* 15 pp 35-38

[7] Salomeev V P, Ryzhkov A D 2017 Recovery and reconstruction of pumping stations in emergency conditions. *Water Magazine* **3 (115)** pp 30-33

[8] Chupin R V 2015 Optimization of developing drainage systems. Monograph (Irkutsk: Publishing House of ISTU) p 418

[9] Chupin V R, Melekhov E S, Chupin R V 2011 Development of the theory and practice of modeling and optimization of water supply and sanitation systems. Monograph (Moscow: ACB) p 325

[10] Skibo D V, Tolstoy M Yu 2018 Study of problems in the operation of the sewage pumping station of the Beryozovy microdistrict of the city of Irkutsk and methods to solve them. *Construction and industrial safety* **12 (64)** pp 123-132

[11] Salomeev V P 2009 Reconstruction of engineering systems and water disposal facilities. Monograph (Moscow: Association of construction universities Publishing house) p 192

[12] Natural and wastewater treatment reference book. Water supply and sewerage (Text) Spellman F R (St. Petersburg: TsOP Profession) p 283

[13] Skibo D V, Tolstoy M Yu, Vasilevich E E, Kurtin AV 2019 Method for wastewater treatment of suspended solids Patent for invention RU No. 2707342 C1 Published: November 26,2019 Bull 33

[14] SP32.13330.2012 Sewerage. External networks and facilities. Updated edition SNiP 2.04.03-85 (including Amendment 1) (Electronic materials) http://docs.cntd.ru/document/1200094155 (accessed December 05, 2019)

[15] MDK 3-02.2001 Rules for the technical operation of systems and facilities for public water supply and sanitation (Electronic materials) http://docs.cntd.ru/document/1200025707 (accessed December 05, 2019)

[16] Voronov Yu V, Shirkova T N 2015 The design of drainage systems of flood territories on the basis of numerical modeling *Water supply and sanitary technique* 4 pp 48-54

[17] Gogina E S, Shmalko V M, Tolstoy M Yu 2018 Technologies for wastewater treatment in the conditions of the Baikal natural territory *Ecology and Industry of Russia* **22 (7)** pp 11-15

[18] Alekseev M I, Ermolin Yu A 2015 On the reliability indicators of water disposal systems *Yakovlev readings*. X Scientific and technical conference dedicated to the memory of the academician of the Russian Academy of Sciences Sergey Vasilievich Yakovlev Collection of reports pp 211-214

[19] Alekseev M I, Baranov L A, Ermolin Yu A 2017 Estimation of the life time of water supply facilities under the influence of a periodically changing flow of failures *Water supply and sanitary technique* **4** pp 50-54

[20] Ermolin Yu A, Alekseev M I 2016 Considering the "aging" of the object while assessing its reliability *Water supply and sanitary technique* **5** pp 68-71

[21] Ignatchik S Yu 2012 Energy conservation and reliability in the reconstruction of sewage pumping stations *Water supply and sanitary technique* **12** pp 37-43

[22] SanPiN 2.1.5.980-00 Hygienic requirements for the protection of surface waters (Electronic materials) http://docs.cntd.ru/document/1200006938 (accessed December 05, 2019)

[23] Spellman F R 2014 Handbook for the treatment of natural and wastewater. Water supply and sewerage (St. Petersburg: TsOP Profession) p 283 ill tab 24 cm

[24] Anuryev V I 2006 Handbook of a designer - machine builder (Moscow: Mechanical Engineering) p 960

[25] GOST 8639-82. Square steel pipes. Assortment (including amendments 1, 2, 3, 4) (Electronic materials) http://docs.cntd.ru/document/gost-8639-82 (accessed on December 07, 2019)

IOP Conf. Series: Materials Science and Engineering **880** (2020) 012055 doi:10.1088/1757-899X/880/1/012055

Development and reconstruction of district water supply systems

O K Slepnev [1], I V Maisel [1] and V R Chupin[1]

[1] Irkutsk National Research Technical University, 83 Lermontov str., Irkutsk, 664074, Russia

E-mail: shlepnev38@mail.ru

Abstract. The aim of the study is to develop a methodology for the comprehensive optimization of promising schemes of water supply systems, ensuring the selection of optimal water withdrawals taking into account water quality, structure and parameters of transporting facilities, as well as operating modes. An algorithm of theoretical, practical and methodological recommendations on optimizing promising schemes of water supply systems is presented. A new optimization technique for group water supply systems is proposed, which allows the complex to solve the following problems: to optimize the location and performance of water intake structures from surface and underground water sources; to optimize installation sites, productivity and composition of water treatment facilities; choose the route of group pipelines and their optimal diameters.

1. Introduction

Our country has huge water resources. However, due to their uneven distribution in many regions, there is an acute shortage of water, and where it is available, its quality does not always meet the requirements of SanPiN 2.1.4.1074-01 and GOST “drinking water”. As a result of man-made impacts, the hydrological situation deteriorates every year, water sources are polluted with untreated sewage, and due to significant fluctuations in the level of surface and underground water, many water intakes operate on the verge of failure or are completely exposed. One of the ways to solve this problem is the construction and development of group water supply systems (WSS). A large number of such systems have already been built in the southern and central regions of our country. Many of them are complex, capital-intensive water management systems designed to transport and supply water over long distances to heterogeneous and dispersed consumers over a wide area. When designing and developing such structures, the most important tasks are:

– justification of the location and performance of water intake and treatment facilities, parameters of pumping stations and control tanks;

– selecting the optimal route and diameter of the pipeline system, taking into account the increased reliability, manageability, and seismic stability of individual structures and WSS as a whole.

– optimal allocation of investments in the reconstruction of existing and construction of new group water supply facilities.

To date, we have accumulated a wealth of experience in the design and construction of WSS, developed various methods for justifying their structure, parameters and modes of operation. However, all of them do not allow us to fully solve these problems in a comprehensive manner and therefore require further development and development of new approaches.

The existing methodology for optimal design is based on a comparison of cost, technical, reliability, and environmental characteristics of the analyzed options for the construction and operation of WSS. The main technological parameters of the designed water supply systems are taken into account through their cost in the optimization criteria, which is used to select the best option for the subsequent construction and operation of the WSS. Reliability, manageability and seismic stability are taken into account in the form of restrictions of a certain type, both on technical parameters and on investments in the construction and operation of WSS.

2. Methods

A common criterion for optimization and comparison of options is the cost of construction and operation of the WSS (by one year), which is determined by the following process [1]:

$$З_{пр} = K \cdot E + Э_з \tag{1}$$

where $З_{пр}$ – given the costs; K – investments in the network; E – the coefficient of efficiency of investment, which in the conditions of market economy is identified with bank interest; $Э_з$ – annual operating costs (thousand rubles/year), defined as follows [2,3,4]:

$$Э_з = C_{ам} + C_{к.р} + C_{т.р} + C_{элк} + C_{фзп} + C_{св} + C_{н.в} + C_{пр.}, \tag{2}$$

where $C_{ам}$ – amortization charge (takes 0.05*K through pipelines, 0.09*K for pumping stations and surface water intake, 0.14*K for wells and underground water intakes); $C_{к.р}$ – the cost of capital repairs (0.046*K), rubles/year; $C_{т.р}$ – cost of current repairs (0.01*K),rubles/year; $C_{элк}$ – cost of electricity, thousand rubles/year:

$$C_{элк} = 108 \cdot z_э \cdot H \cdot x, \tag{3}$$

where $z_{элк}$ – unit cost of electricity (rub per kWh), x - the amount of conveyed water (m^3/s), H – the pressure pumping station (m. the height of a column.).

$C_{фзп}$ – payroll of service personnel per 1 km of water pipe length:

$$C_{фзп} = 0{,}75 \cdot ЗП_{ср} \cdot x^{0{,}3}, \tag{4}$$

where $ЗП_{ср}$ - average projected salary for the company, thousand rubles/month.; x, system performance, in m^3/s.

$C_{св}$ – consists of insurance contributions to the pension Fund (0.22*$C_{фзп}$), social insurance fund (0.029*$C_{фзп}$) and the health insurance fund (0.051*$C_{фзп}$); $C_{н.в}$ – the tax on water (thousand rubles/year) is determined:

$$C_{н.в} = 31536 \cdot C_{н.ст} \cdot x,$$

where $C_{н.ст}$ – tax rate in rubles for 1 m^3 waters, collected from surface and underground sources (in Irkutsk in 2018, this rate is 0.122 rubles per 1 m^3) ; $C_{пр}$ – other expenses, thousand rubles/year, accepted in the amount of 20% of the amount of depreciation charges ($C_{ам}$) and the cost of staff salaries ($C_{фзп}$),

$$C_{пр} = 0{,}2 \cdot (C_{ам} + C_{фзп}), \text{ or } C_{пр} = 0{,}01 \cdot K + 0{,}15 \cdot ЗП_{ср} \cdot x^{0{,}3}$$

Taking into account (2), the operating costs will be as follows:

$$Э_з = 0{,}116 \cdot K + C_{элк} + 1{,}125 \cdot ЗП_{ср} \cdot x^{0{,}3} + 31536 \cdot C_{н,ст} \cdot x. \tag{5}$$

To specify one-time (capital) costs, we will use the information provided in the consolidated construction price standards [5,6]. For outdoor water supply networks with a length of 1 km made of polyethylene pipes, when developing dry soil to a depth of 3 m (which is typical for the Irkutsk region), it is not difficult to obtain the following dependence of investment on the diameter of the pipe wire:

$$K = (27329 \cdot d^2 + 7399{,}5 \cdot d + 2537{,}9) \cdot L, \tag{6}$$

K - capital investment, in thousand rubles., d – diameter, in m.

Taking into account $d=\sqrt{\frac{4x}{\pi v}}$; v – speed in m/s; x – flow rate in m³/s. (7)

$$K = 34796{,}4x\cdot v^{-1} + 8346{,}6x^{0,5}\cdot v^{-0,5} + 2537{,}9\,. \quad (8)$$

Calculation of energy costs requires separate consideration. In the theory of hydraulic circuits [7], the law of energy conservation for an arbitrary hydraulic circuit is formulated as follows: "All the energy brought in, minus the energy for pouring water, is spent on overcoming friction forces":

$$\sum_{j=1}^{J_1} Q_j\cdot H_j - \sum_{j=1}^{J_2} Q_j\cdot P_j = \sum_{i=1}^{n} h_i\cdot x_i\,, \quad (9)$$

where J_1 – a set of nodes of pumping stations, J_2 – a set of nodes of water consumption, H_j – piezometric heads of pumping stations, P_j – piezometric water heads of consumers, in m. h_i – head losses on sections of the water supply network, x_i - the flow of transported water on the network sections, Q_j. – water consumption of the consumer j, m³/s.

Therefore, the annual cost of electricity for the water supply system can be represented as the sum of the cost of electricity for each section of the network:

$$C_{элк} = 108\cdot Z_{элк}\cdot Э_i\,,\quad Э_i = h_i\cdot x_i\,,$$ for x in m³/s.

For simplicity of presentation and calculation of pressure losses along the length of the pipeline we use the formula of F. A. Shevelev for plastic pipes [8]:

$$h_i = 1{,}1\cdot Y_i\cdot L_i\;,\quad Y_i = 0{,}001052\cdot\frac{x_i^{1,774}}{d_i^{4,774}}\,,$$

where Y_i – hydraulic gradient, L_i – network segment length, in m.

$$h_i\cdot x_i = 0{,}0011572\cdot L_i\cdot\frac{x_i^{2,774}}{d_i^{4,774}}\,,$$ or taking into account (7):

$$h_i\cdot x_i = 0{,}000642\cdot L_i\cdot x_i^{0,378}\cdot v_i^{2,387}\,.$$

$$C_{элк} = 108\cdot Z_{элк}\cdot\left(0{,}000642\cdot x_i^{0,387}\cdot v_i^{2,387}\cdot L_i\cdot 1000\right),$$ L *in* km.

The given estimated costs for the section of the network i in thousand rubles.:

$$З_{пр} = 0{,}236\cdot K + C_{элк} + 1{,}125\cdot ЗП_{ср}\cdot x^{0,3}\,. \quad (10)$$

As shown by the research, the optimization criterion in the form of the estimated costs shown is mainly aimed at minimizing one-time capital investments. Although it is known that the operating costs for the life cycle of water disposal systems exceed one-time investment by 10-20 times. Obviously, when justifying the project for the reconstruction and development of the WSS, it is necessary to take into account the costs of the entire life cycle of the system (LCC). For a WSS consisting of n sections, these costs over the lifetime of the system can be represented as the following terms [9-11]:

$$C = \sum_{i=1}^{n}\left(C_i^{TP} + \frac{T}{(1+r)^T}\cdot(C_{ЭКСi}^{TP} + C_{оч})\right) + \sum_{i=1}^{n}\left(C_i^{HC} + \frac{T}{(1+r)^T}\cdot(C_{ЭКСi}^{HC} + C_{ЭЛКi})\right) + \sum_{i=1}^{n}\left(\frac{C_{Di}}{(1+r)^T}\right), \quad (11)$$

where C_I^{TP}, C_{II}^{HC} – costs associated with the design, purchase of equipment, construction and installation of water supply systems; $C_{ЭКС}^{TP}, C_{ЭКС}^{HC}$ – operating costs for pipelines, pumping stations, and other structures. These costs are calculated according to the guidelines [12]. $C_{оч}$ – annual costs for restoring an emergency structure or network section, including the cost of delivering water to consumers during the emergency response; $C_{ЭЛК}$ – annual energy costs for pumping water; C_D – costs for disassembly and disposal of expired equipment; T – life cycle time (the time interval corresponding to the equipment with the longest service life is accepted); r – discount rate (the value of the refinancing rate CB RF, which in 2018 is 7.25%), plays the role of the base level, in comparison with which the economic efficiency of the project option is evaluated. It should be noted that for equipment with a shorter service life, a one-time investment is required C_I^{TP}, C_{II}^{HC} increase by the multiplicity factor in relation to the equipment with the longest service life. With constant operating costs, the life cycle costs can be represented as follows:

$$LCC = K + T \cdot Э_З \,, \tag{12}$$

where T - the service life of the water supply system in years is assumed to be equal to the service life of the longest-lasting element of the water supply system, K – investments equal to one-time costs, plus the cost of restoring individual elements of the system whose life span is less than the estimated (multiple costs).

Taking into account the values of operating costs (4), the life cycle costs (in thousands of rubles) can be represented:

$$LCC = K + T \cdot \left(0{,}116 \cdot K + C_{элк} + 1{,}125 \cdot ЗП_{ср} \cdot x^{0,3}\right), \tag{13}$$

or taking into account (8):

$$LCC = (34796{,}4 \cdot x_i \cdot v_i^{-1} + 8346{,}6 \cdot x_i^{0,5} \cdot v_i^{-0,5} + 2537{,}9) \cdot L_i + T \cdot [0{,}116 \cdot (34796{,}4 \cdot x_i \cdot v_i^{-1}$$
$$+ 8346{,}6 \cdot x_i^{0,5} \cdot v_i^{-0,5} + 2537{,}9) \cdot L_i + C_{элк} + 1{,}125 \cdot ЗП_{ср} \cdot x^{0,3}]$$

As a result, the given costs and life cycle costs are obtained as a function of the flow rate and speed of the transported water, the length of the pipeline and the specific cost of electricity. SP 31.13330.2012 and a number of guidelines recommend that the speed of water movement be taken in the range of 1.5-3 m/s. However, in each case, the optimal speed can be determined by calculation. To do this, you need to take a partial derivative of the reduced costs and life cycle costs by speed and equate it to zero. Then the resulting equation is resolved relative to the speed.

To select the optimal speeds, we will conduct numerical experiments on the reduced costs and life cycle costs, varying the unit values of electricity costs and pipeline lengths. The unit cost of electricity will change from the lowest for the Irkutsk region – 1.06, to the maximum for Chukotka – 8.2 rubles per kWh. the results of numerical experiments are presented in table. 1, which implies that the electricity tariff has a significant impact on the optimal speed values. The length of the networks practically does not affect the speed of water movement. The life cycle cost criterion reduces the optimal speed but 30%.

Table 1. Estimation of optimal water flow rates through pressure pipelines

Cost of electricity, rub per KWh $Z_{элк}$	Optimal water flow rate (m/s) for a flow rate of 1m^3/s					
	For the reduced cost criterion			For life cycle costs (T = 50 years)		
	L =1 km	L = 50 km	L = 100 km	L = 1 km	L = 50 km	L = 100 km
1	3,40	3,35	3,30	2,85	2,80	2,80
2	2,78	2,74	2,72	2,31	2,30	2,30
3	2,40	2,40	2,37	2,08	2,07	2,07
4	2,23	2,20	2,20	1,89	1,89	1,89
5	2,10	2,07	2,04	1,78	1,79	1,79
6	2,00	2,00	1,98	1,70	1,68	1,68
7	1,90	1,88	1,88	1,60	1,58	1,58
8	1,80	1,80	1,78	1,54	1,50	1,50
9	1,78	1,77	1,77	1,49	1,49	1,48

Therefore, optimal speeds can be taken in the range from 1 to 3 m/s, depending on the cost of electricity in the region for which the water supply system is being designed.

Consider the construction region of the Irkutsk region. The length of the water supply network is 1 km made of polyethylene pipes. Let's construct the functions of the reduced costs as a function of water flow, while varying the speed from 0.5 to 3.5 m/s. These functions and their linear approximations are presented in picture 1.

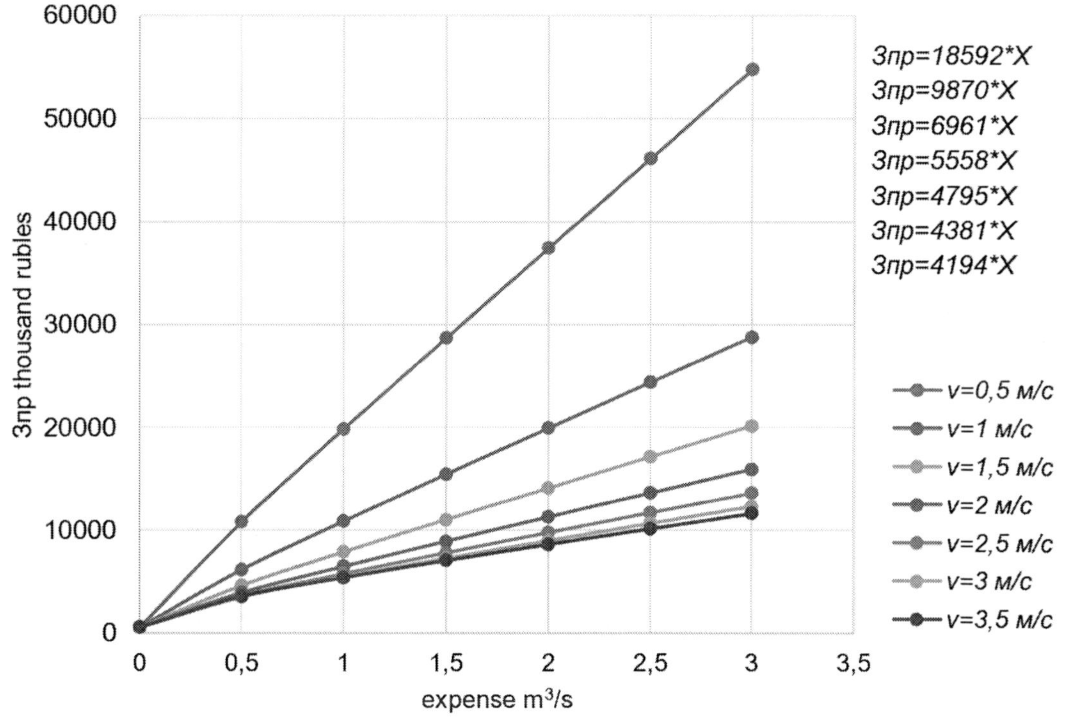

Picture 1. Dependence of the reduced costs on the flow rate and speed water flows through pipelines and their linear approximations

For life cycle costs, these dependencies are also shown in picture 2.

Similar calculations and transformations can be made for underground water intakes with treatment facilities. In recent years, with the advent of new technologies for water treatment, pumping equipment, block configuration of treatment facilities and pumping stations, their pricing policy and efficiency have changed significantly, expensive auxiliary rooms and enclosing structures are no longer required, and maintenance personnel and operating costs have been significantly reduced due to automatization. At the same time, the configuration and cost of treatment facilities depend significantly on the quality of the source water and, therefore, must be determined for each specific case.

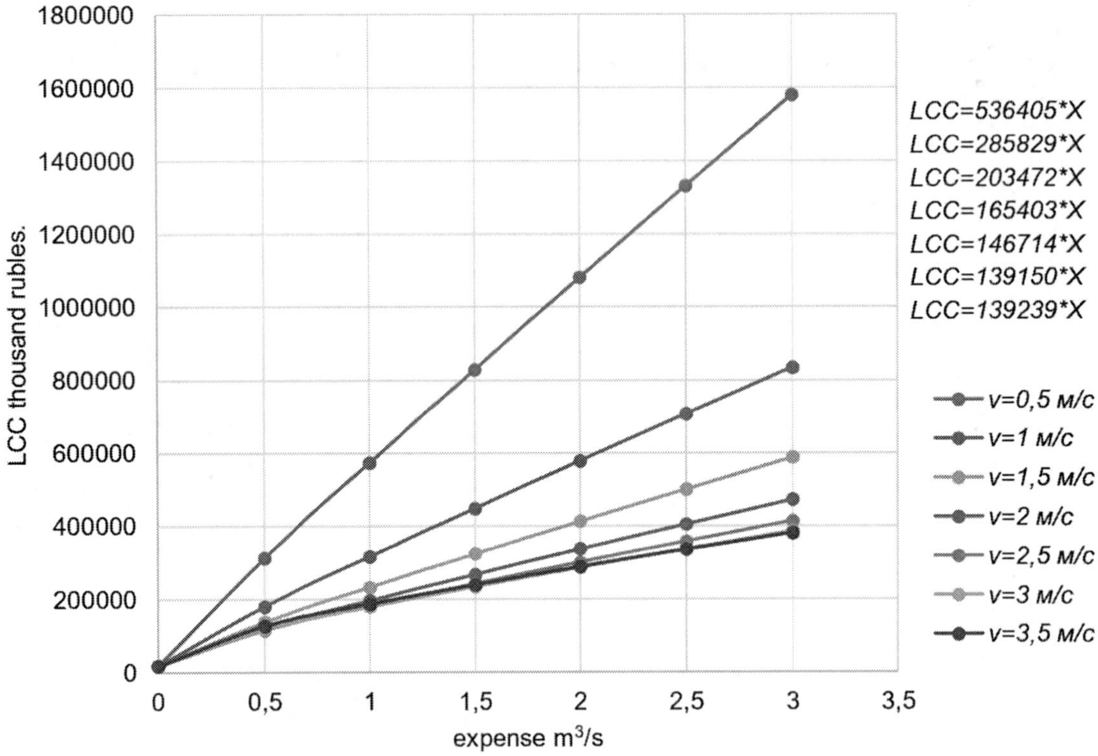

Picture 2. Dependence of life cycle costs on consumption and speed water flows through pipelines and their linear approximation

Determine the unit cost for **underground water intake from an artesian well** with a depth of 100 m with a maximum flow rate of 40 m^3/hour. As a result, we get a function of the following type:

$K = 31500,0 \cdot x$.

At the same time, operating costs:

$$Э_з = 0,2 \cdot K + C_{элк} + 15 \cdot ЗП_{ср} \cdot x^{0,3}.$$

Energy costs will be as follows:

$$C_{элк} = 108 \cdot z_э \cdot 100 \cdot x = 10800 \cdot z_э \cdot x.$$

The estimated costs for network section i, in thousands of rubles:

$$З_{пр} = 51840 \cdot x_i + C_{элк} + 15 \cdot ЗП_{ср} \cdot x_i^{0,3}$$

Life cycle costs:

$$LCC = 162000 \cdot x_i + T \cdot \left(32400 \cdot x_i + C_{элк} + 15 \cdot ЗП_{ср} \cdot x_i^{0,3}\right).$$

Consider a **sewage treatment plant for iron removal and softening** (removal of calcium and magnesium salts) produced by CC "Rosa" - typical water treatment stations based on pressure filters for de-Ironing and softening. The principle of degreasing-softening is based on catalytic oxidation and filtration of impurities in the loading layer.

$K = 178085 \cdot x$

Operating cost:

$$Э_з = 0,2 \cdot K + C_{элк} + 15 \cdot ЗП_{ср} \cdot x^{0,3}.$$

In this case, the energy costs will be as follows:

$$C_{элк} = 108 \cdot z_э \cdot 50 \cdot x = 5400 \cdot z_э \cdot x.$$

The estimated costs for network section i, in thousands of rubles:

$$З_{пр} = 56987 \cdot x_i + C_{элк} + 15 \cdot ЗП_{ср} \cdot x_i^{0,3}.$$

Life cycle costs:

$$LCC = 178085 \cdot x_i + T \cdot \left(35617 \cdot x_i + C_{элк} + 15 \cdot ЗП_{ср} \cdot x_i^{0,3}\right).$$

For **underground water intake with water treatment facilities**:

$K = 209585,0 \cdot x.$

Operating cost:

$$Э_з = 0,2 \cdot K + C_{элк} + 30 \cdot ЗП_{ср} \cdot x^{0,3}.$$

In this case, the energy costs will be as follows:

$$C_{элк} = 108 \cdot z_э \cdot 150 \cdot x = 16200 \cdot z_э \cdot x.$$

The estimated costs for network section i, in thousands of rubles:

$$З_{пр} = 67067 \cdot x_i + C_{элк} + 30 \cdot ЗП_{ср} \cdot x_i^{0,3}$$

Life cycle costs:

$$LCC = 209585 \cdot x_i + T \cdot \left(41917 \cdot x_i + C_{элк} + 30 \cdot ЗП_{ср} \cdot x_i^{0,3}\right).$$

Pumping plant. Capital investment:

$K = 48053 \cdot x.$

Operating cost:

$$Э_з = 0,2 \cdot K + 15 \cdot ЗП_{ср} \cdot x^{0,3}.$$

The estimated costs for network section i, in thousands of rubles.:

$$З_{пр} = 15377 \cdot x_i + 15 \cdot ЗП_{ср} \cdot x_i^{0,3}$$

Life cycle costs:

$$LCC = 48053 \cdot x_i + T \cdot \left(9610 \cdot x_i + 15 \cdot ЗП_{ср} \cdot x_i^{0,3}\right).$$

Such calculations and transformations can be made in relation to underground water intakes with or without treatment facilities and expressed as functions of the form:

$$З_{пр} = C \cdot x \text{ and } LCC = C \cdot x,$$

For pipelines and structures on them, the C values are shown in table 2.

Table 2. Specific flow costs by value 1m^3/s

	For the criterion of reduced costs, thousand rubles.						
v (m/s)	0,5	1,0	1,5	2,0	2,5	3,0	3,5
C, for pipelines, per 1 km	18592,0	9870,0	6961,0	5558,0	4795,0	4381,0	4194,0
C, operating costs, per 1 km	9155,0	4886,0	3491,0	2858,0	2562,0	2463,0	2503,0
C_{well}				21224,0			
$C_{treat.\ fac.}$				62731			
$C_{well} + C_{treat.\ fac.}$				83954			
$C_{н.с.}$				15721			
				For life cycle costs			
V (m/s)	0,5	1,0	1,5	2,0	2,5	3,0	3,5
C, for pipelines, per 1 km	536405,0	285829,0	203472,0	165403,0	146714,0	139150,0	139239,0
C_{well}				903675,0			
$C_{treat.\ fac.}$				2246100			
$C_{well} + C_{treat.\ fac.}$				3149800			
$C_{н.с.}$				545728			

Taking into account the obtained dependencies, we can formulate a mathematical formulation of the problem of optimizing the structure and parameters of group water pipes as a problem of finding the maximum flow of the minimum cost [13,14,15]:

$$\sum_{i=1}^{n} C_i \cdot K_{\text{gen } i} \cdot x_i \rightarrow \min \text{, by, } \underline{\underline{в}}_i \leq x_i \leq \overline{\overline{в}}_i \text{ , } A \cdot x = q'_{cp} \tag{14}$$

when n – number of network sections, C_i - cost per unit of flow. For example, at a speed of 3 m/s for the reduced cost criterion, the specific flow costs will be as follows: for new pipeline sections of the network *4381,0* thousand rubles per 1m^3/s on 1 km length; for existing pipeline sections of the network (operating costs) *2463.0* thousand rubles per 1m^3/s on 1 km length; for the device of wells *21224.0.0* thousand rubles per 1m^3/s; for underground water intakes with water treatment system *1165500.0* thousand rubles per 1m^3/s. $K_{общ,\ i}$ – total coefficient of unevenness on the network section i, obtained on the basis of actual and forecast schedules of water consumption; x_i - the desired flow on a branch of a redundant or transport network; $\underline{\underline{в}}_i$, $\overline{\overline{в}}_i$ - the lower and upper limits on the flux x_i; A – matrix of connections of nodes and branches of the network; q'_{cp} - vector of average water consumption expenditure, m^3/s.

To solve this problem, the following method is proposed in this paper [16, 17, 20]. A redundant scheme is being constructed, which is formed on the basis of the overlay of several pre-developed options for the construction of the WSS. The redundant scheme can represent: all possible places for water intakes (from open and underground sources) with local and centralized water treatment facilities, and without them (if the sources have drinking water quality); all possible routes for passing and supplying water to consumers by pressure pipeline sections.

The transport network is built on the basis of the received redundant scheme. To build it, all water-consuming nodes are closed using fictitious branches to a common node t – flow outlet, and water-source nodes are closed to a common fictitious node s-flow inlet. at the same time, restrictions on their throughput (upper and lower) are assigned for each section of the transport network. For fictitious stream entry branches, the upper limits correspond to the maximum possible water intake from water sources, and for fictitious stream exit branches, the upper limits correspond to the water needs of localities or subscribers. Restrictions on the flows of projected pipeline sections are not assigned. For existing sections of the network, the upper limit on flows is determined based on the optimal values of water flow rates:

$$\overline{\overline{в}} = V_{opt} \cdot \frac{\pi \cdot d^2}{4}. \tag{15}$$

For each section of the network, the cost per unit of flow is determined and assigned. For fictitious branches of input streams, this is the cost of intake and water treatment of one m^3 of water, for real pipeline sections of the network, this is the cost of transporting one m^3 of water per 1 km of pipe-water. For fictitious sections of the flow outlet, costs are not assigned, or are accepted as costs for further transportation of one m^3 of water to specific subscribers. For existing sections of the network, the flow unit costs correspond to operating costs.

Taking into account the transport network built in this way the problem of finding the maximum flow of the minimum cost is solved.

3. Results and discussions

As a result of optimization, the location and performance of water intake and treatment facilities, the optimal route of the pipeline system and its parameters are determined. At the same time, there are options for a fully centralized or decentralized water supply system.

Based on the proposed method, optimization calculations were performed and research was performed on the example of the Cheremkhovsk group water pipeline in the Irkutsk region. As possible sources of water supply, the following are considered: a new surface water intake from the

Belaya river (Belsk village), the water in which requires the installation of a de-Ironing, clarification, discoloration and disinfection station; an existing Cheremkhovsk channel-type water intake from the Angara river with water treatment facilities with a capacity of 85 thousand m^3/day (Svirsk); a new underground water intake in the village of Rysevo, where the water chemical composition mostly meets the requirements of GOST, but the flow rate of wells is insignificant (about 2.52 m^3/h). There is water in each locality at a depth of 80-100 m and it is possible to organize an underground water intake, but with the installation of water softening and de-Ironing stations. Taking into account possible water intakes (in almost every locality), possible pipeline sections that run along existing roads, the redundant scheme presented in picture 3 (number of consumers-33, total network consumption-0.2 m^3/s). Taking into account the redundant scheme, the transport scheme is made (see picture 4) and made optimization calculations. The optimal variant is obtained, which is shown in picture 5.

Picture 3. Redundant scheme of the Cheremkhovsk group water supply system

IOP Conf. Series: Materials Science and Engineering **880** (2020) 012055 doi:10.1088/1757-899X/880/1/012055

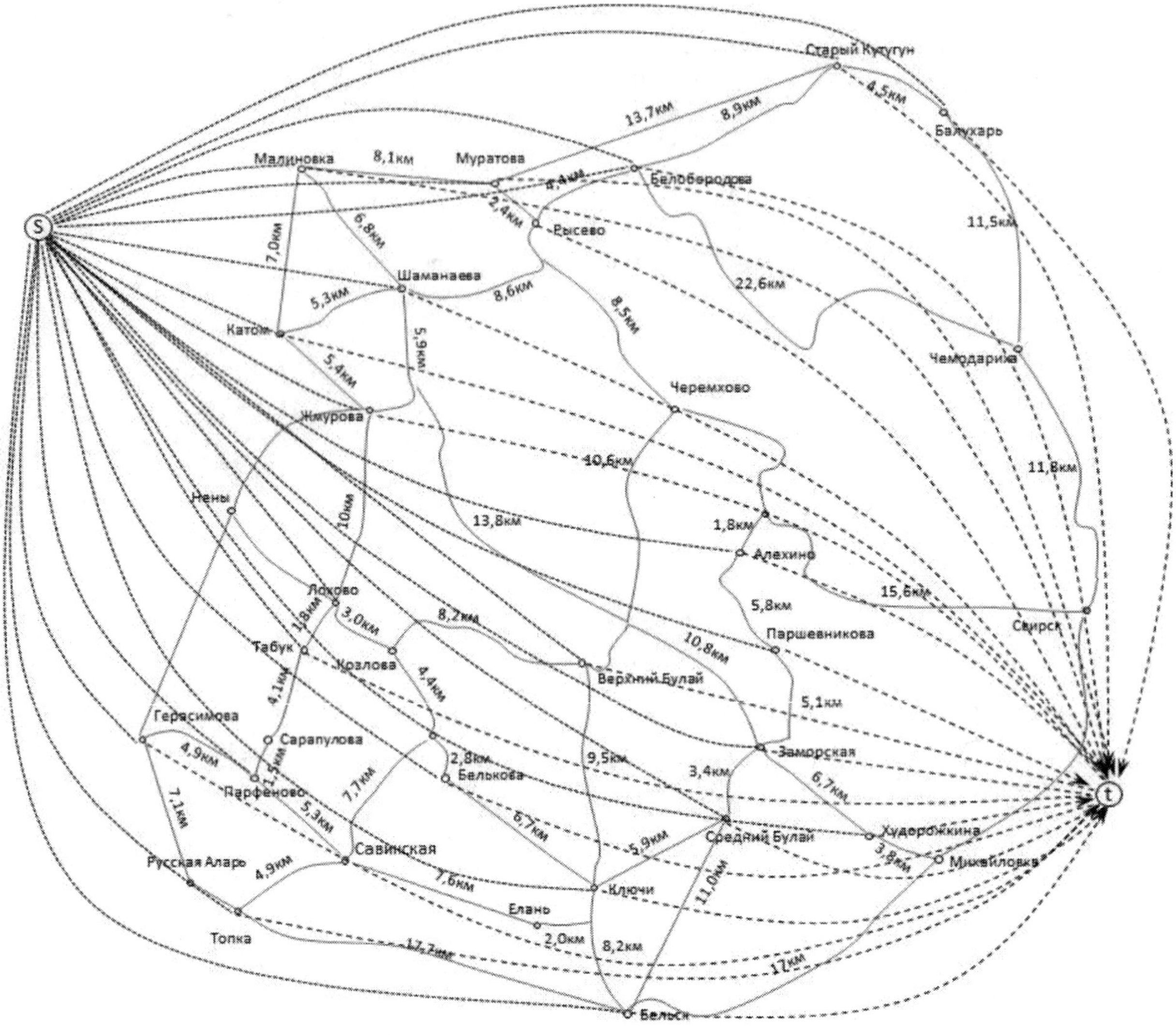

Picture 4. Transport scheme of the Cheremkhovsk group water pipeline

According to picture 5, to the existing water supply system from Cheremkhovsk intake connect two additional localities: D. Camadarie and Hodorogea; he is new in water intake in p. Bielsk connect three settlements: D. Elan, Average Bulay, Keys; from the new water intake in p. Ricevo connects 8 settlements. All other localities must have local water intakes with water treatment facilities. For picture 6 shows the option chosen by the designers based on water intakes from the river. Hangar with water treatment facilities (already built for a capacity of 80 thousand m^3/day). For the given costs, the option shown in picture 5 was 21% more economical than the one proposed by the designers.

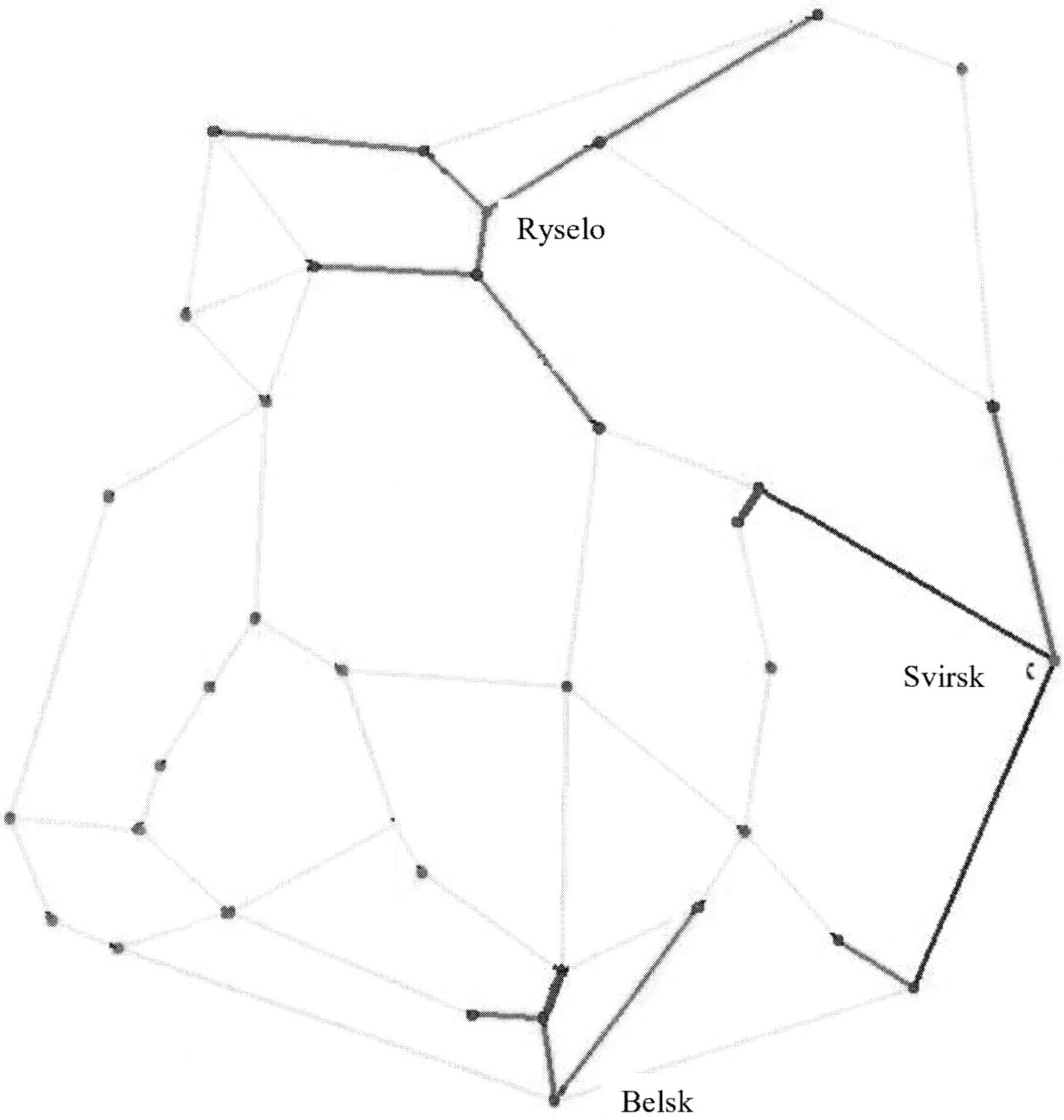

Picture 5. Optimal scheme of the Cheremkhovsk group water supply system

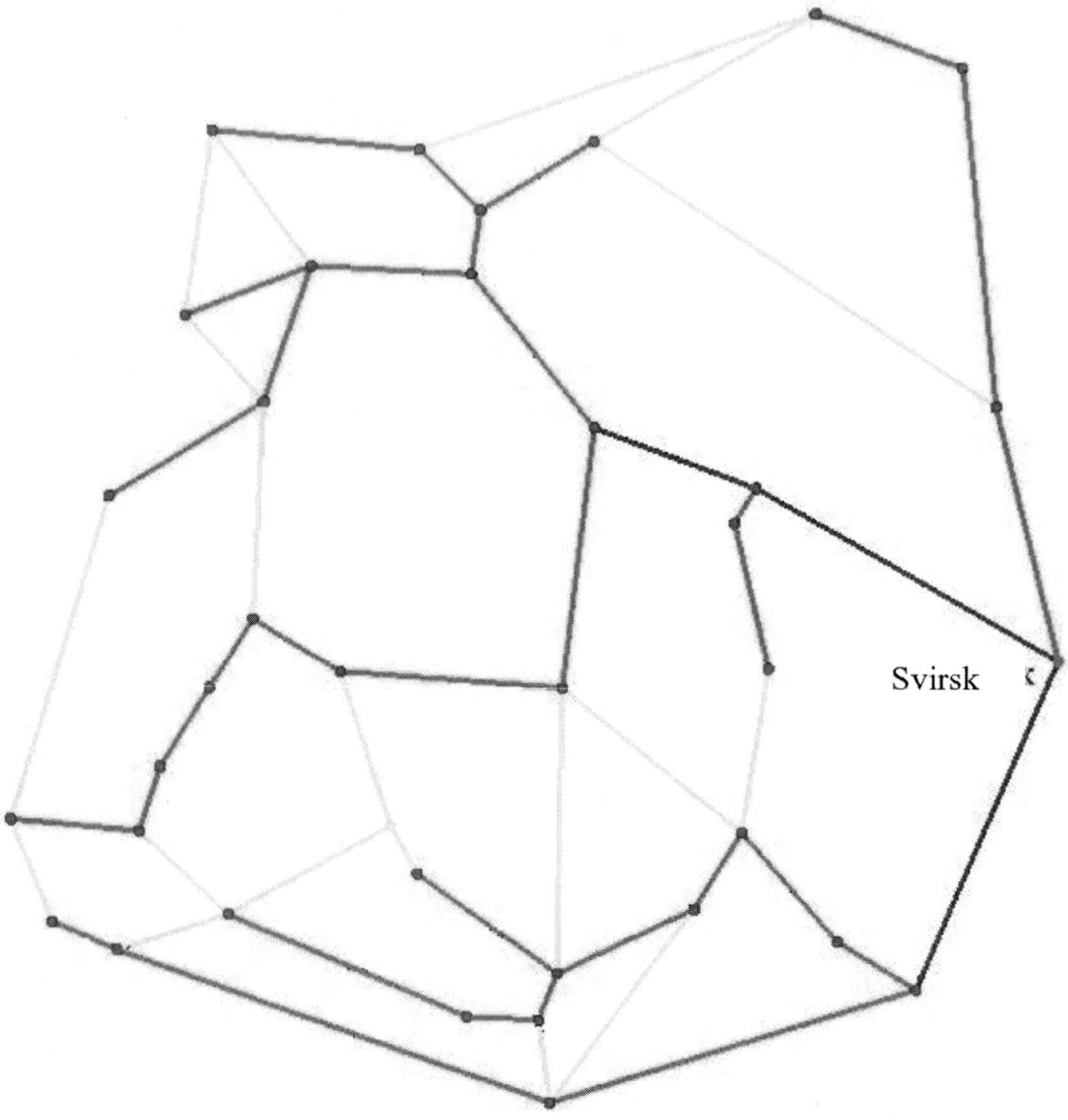

Picture 6. Project version of the Cheremkhovsk group water pipeline

Conclusions

We propose a new method for optimizing group water pipes, which allows us to solve the following problems in a complex:

- optimize the location and performance of water intake structures from surface and underground water sources;
- to optimize the place of installation, performance and composition of water treatment facilities;
- choose the route of group pipelines and their optimal diameters.

The developed methodology and software package [19] are an effective tool for substantiating promising schemes for the development of district water supply systems and can be useful for both designers and decision-makers on the development of urbanized territories.

References
[1] Repin B N 1995 Water supply and sanitation. Outdoor networks and structures p 431
[2] Shifrin S M 1982 Economics of water supply and sanitation p 319
[3] Simonova A A 1990 Economics of engineering equipment systems p 344
[4] Boguslavsky L D and Simonova A A 1984Technical and economic bases of designing water supply and Sewerage systems p 325
[5] Serdyuk E I, Kuznetsov V I and Artemonova L G 1984 p 255
[6] Salomeev V P 2009 Reconstruction of engineering systems and structures of water disposal p 192
[7] Merenkov A P and Khasilev V Ya 1985 Theory of hydraulic circuits p 278
[8] Shevelev F A 1973 Tables for hydraulic calculation of steel, cast iron, asbestos-cement, plastic and glass water pipes
[9] Bazhenov V I, Pupyrev E I, Sambursky G A and Berezin S E 2018 Development of a method for calculating the cost of the life cycle of equipment for systems and structures for water supply and sanitation № **2** pp 10-19
[10] Gogina E S, Gurinovich A D 2016 Application of the LCC methodology for evaluating the effectiveness of wastewater treatment plants *Water Supply and sanitary engineering* № **9** pp 36-41
[11] Preparation of the technical and economic part of projects for off-site water supply and Sewerage systems Soyuzvodokanalproekt 1991 p 80
[12] Chupin R V and Pham N M 2016 Optimization of options for the development of water disposal systems *Izvestiya vuzov. Investment. Construction. Realty* № **3(18)** pp 101-113
[13] Chupin R V and Melekhov E S 2011 Development of the theory and practice of modeling and optimization of water supply and drainage systems *Monograph* p 323
[14] Christofides N 1978 Graph theory. Algorithmic approach p 432
[15] Chupin R V 2015 Optimization of developing water disposal systems *Monograph* pp 418
[16] Chupin R V 2018 Indicative and adaptive management of wastewater system improvement *News of universities: Investments. Construction. Real estate.* 2018 Vol **8** pp 94-108
[17] Chupin R V and Maizel I V 2013 Investments in the development of water supply and sanitation systems *Investment. Construction. Realty* № **1 (4)** pp 126-133
[18] Melekhov E S, Chupin V R and Chupin R V Certificate of state registration of the computer program № 2016615463 TRACE-VR dated May 25, 2016.
[19] Mushik E and Muller P 1990 Methods of making technical decisions p 208

IOP Conf. Series: Materials Science and Engineering **880** (2020) 012056 doi:10.1088/1757-899X/880/1/012056

Using of the complex mobile station for heat and energy supply of residential villages, located on the protected areas

A A Tunik[1] and M Yu Tolstoy[1]

[1]Irkutsk National Research Technical University, Department of engineering communications and life-support systems, 664074, 83, Lermontov st., Irkutsk, Russian Federation

E-mail: alextun@mail.ru

Abstract. The results of studies of a complex mobile station for life support systems are presented. This station has the electric energy and thermal energy production functions, so as wastewater treatment and drinking water preparation. The production of electrical energy occurs through the conversion of solar energy and the use of microbial fuel cells. Thermal energy for the hot water system is obtained using solar collectors. A wastewater treatment plant is proposed based. The required number of complex mobile stations for life support systems, namely for wastewater treatment and hot water supply of the sports and recreation camp «Polytechnic», which is located at 17 km of the Baikal Highway, has been calculated. The calculations took into account the number of residents per season, water consumption and wastewater rate, the geographic location of the camp, the average number of cloudy days in the area and the number of sunny hours. The developed complex mobile station will ensure energy supply and disposal of wastewater from facilities located in recreational and protected areas without environmental consequences and violation of environmental and sanitary requirements. At the same time, the installation can be used on various types of objects.

1. Introduction

The Baikal Natural Territory is the recreational area and also it is the main object in Eastern Siberia for research and application of high-tech, environmental and energy-saving technologies.

Development of the infrastructure of the Baikal Natural Territory is necessary because of the presence of residential villages and the development of its tourism potential, however, there are many problems with energy supply and wastewater disposal in protected areas, associated with increased environmental requirements and restrictions. Using of traditional energy technologies and treatment plants often has environmental consequences, therefore, a number of restrictions existing in protected areas greatly complicate the energy supply and disposal of wastewater from residential areas in these territories. Not to mention the villages located along the Baikal Highway (the route from Irkutsk to Listvyanka), which belongs to the water protection zone due to the presence of the Irkutsk water intake downstream of the Angara river.

The use of renewable energy sources [1-14] can solve many of the above problems, but even they alone cannot solve the problems of mobility and dependence on external factors [15-17]. The complex mobile station, at the Irkutsk National Research Technical University (INRTU), can help to solve such difficulties [18].

2. Complex mobile station

The developed power plant allows to provide hot water supply, electricity supply and wastewater treatment of individual houses, residential villages and tourist camps, even located in water protection zones, recreational and protected areas. The use of renewable energy sources will make it possible to obtain environmentally friendly generation of heat and electricity, as well as provide wastewater treatment, that meets the requirements of protected areas.

The station may consist of a various equipment combination, which depends on the requirements of the end user, on the environmental conditions of the location zone of the customers, on the environmental and sanitary requirements applied to the customers, as well as on available energy sources. It can consist of solar collectors and panels, wind generators, a heat pump, microbial fuel cells, a water treatment and wastewater utilization plants, and even a diesel or gasoline generator.

The station has a modern automation system that provides the effective interaction of all engineering systems used in it, as well as full monitoring and dispatching.

3. Using of the energy station on the protected area

To test the complex mobile station, it was decided to use the territory of the Irkutsk National Research Technical University. INRTU has at its disposal a sports and recreation camp "Polytechnic", which is located at 17 km of the Baikal Highway. The camp has all the conditions for a good rest: summer houses, a club, a gym, a football field, volleyball and basketball courts, a cross-country track. The camp is ready for children, the necessary sanitary standards have been met, the area has been treated with anti-mite drugs. There is a shower near the football field.

Since the "Polytechnic" camp is located in the protected area, and the Central Irkutsk water intake is located downstream of the Angara river, there is no centralized disposal of sewage in the camp, electricity is supplied from the general electricity network running along the Baikal Higway, the camp is supplied with water, and hot water is provided by electric heaters. Given all the above circumstances, it was decided to develop a power station that works exclusively from renewable energy sources, which will reduce energy costs and ensure environmentally friendly wastewater disposal.

The lack of the necessary wind power in the available area makes it impossible to use wind generators. Due to the fact that the camp operates exclusively in the summer, the use of heat pumps is also impractical. As a result, it was decided to use the combination of systems shown in Figure 1 [19-26].

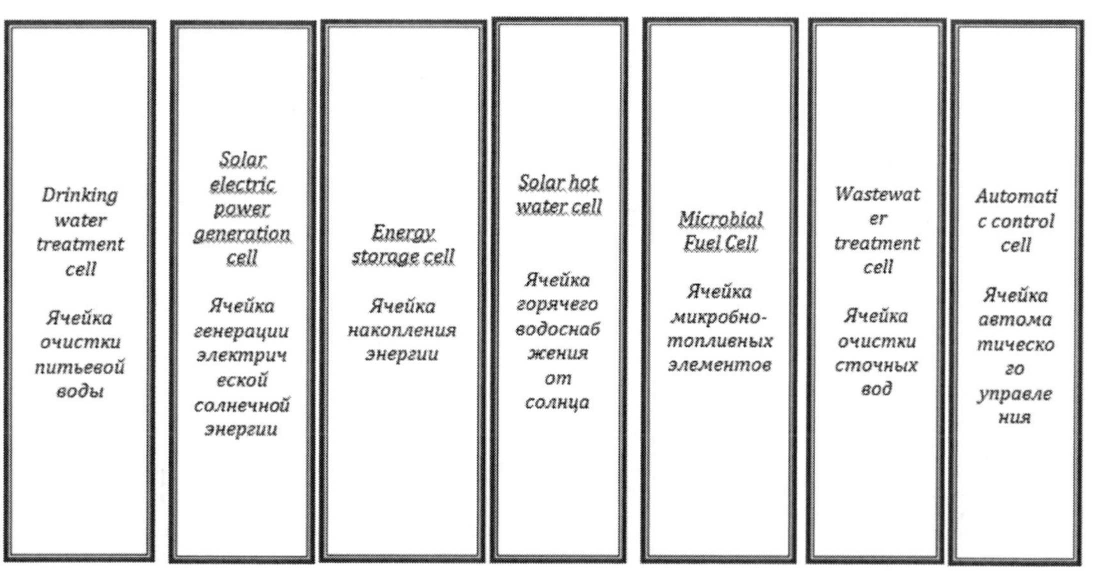

Figure 1. Mobile station of life support systems for recreational areas, consisting of 7 cells.

IOP Conf. Series: Materials Science and Engineering **880** (2020) 012056 doi:10.1088/1757-899X/880/1/012056

The complex mobile station for life support systems can be placed between the showers and canteen. This place is located near the road, that's why it will be easy to operate and drive up the equipment.

About 300 students and another 30 children of employees will stay in the camp for three arrivals. The average number of people per arrival will be 115 people, including employees.

Water consumption per person is taken 60 l per day. (including 25.5 l per day of hot water), in accordance with the regulatory documentation of the Russian Federation. The flow rate of cold and hot water was calculated according to formulas (1) and (2), respectively:

$$Q_1=\frac{q_1\cdot m}{1000}=\frac{34{,}5\cdot 115}{1000}=3{,}97\ m^3/day \quad (1)$$

$$Q_2=\frac{q_2\cdot m}{1000}=\frac{25{,}5\cdot 115}{1000}=2{,}93\ m^3/day \quad (2)$$

q_1, q_2 - water consumption /wastewater level per person for cold water and hot water;
m – number of people.

In winter, the calculation was performed according to formulas (3) and (4), respectively:

$$Q_1=\frac{q_1\cdot m}{1000}=\frac{34{,}5\cdot 30}{1000}=1{,}04\ m^3/day \quad (3)$$

$$Q_2=\frac{q_2\cdot m}{1000}=\frac{25{,}5\cdot 30}{1000}=0{,}77\ m^3/day \quad (4)$$

The calculation of the number of installations is carried out according to the summer period, since during this period the maximum water consumption occurs. One installation has a capacity of 5 kW, which allows to treat 4 m^3 of waste water per day. Then, 2 installations are required for treatment, 6.9 m^3 per day.

For to calculate the required number of solar collectors for 115 people per race, it will be needed to take into account a number of amendments, such as: the number of cloudy days per month and the average monthly length of a sunny day. The season in the camp lasts for 3 months. The calculations used the flat solar collector SUN 1 [27], developed and patented at INRTU. In the future, these installations will also use SUN 3 solar collectors [28].

The calculation results are shown in table 1.

Table 1. The results of calculating the required number of solar collectors

Month	June	July	August	Total
Duration of month, days	30	31	31	92
Monthly average solar energy values falling on an optimally oriented surface, kW·h/m²·day	5,4	4,7	4,31	4,80
Amount of cloudy days	8,1	11,1	9,5	28,7
Average sunny day duration, hours	8,8	7,8	7	7,9
The total value of solar energy per month, kW·h/m²	118,26	93,53	92,67	304,05
The average hourly value of solar energy per month, kW·h/m²	0,61	0,60	0,62	0,61
Hot water consumption per hour, l/hour	122,19	122,19	122,19	122,19
Daily hot water consumption, l/day	2 932,50	2 932,50	2 932,50	2 932,50
Daily heat load on hot water supply, kW/day	126,10	126,10	126,10	126,10

Required solar collectors performance, kW/h	14,33	16,17	18,01	16,03
The average monthly SUN 1 solar collector performance, kW/h	0,75	0,74	0,75	0,74
Required number of SUN 1 solar collectors	19	22	24	22

One station can contain from 2 to 10 solar collectors, depending on the configuration. For "Polytechnic" camp, the optimal number of solar collectors per installation is 6. It will increase the number of equipment for generating electricity in the station, such as microbial fuel cells and solar panels. Thus, for the wastewater treatment and heating water, 4 mobile stations will be needed.

4. General conclusions

As a result of research, a complex mobile energy station was developed. It additionally purifies wastewater to an utilization quality based on patented energy-efficient wastewater treatment technologies. It was accounted the number of facilities for the camp "Polytechnic" of INRTU, which is located on protected area, where there are more stringent requirements for engineering systems mounting. The station is an autonomous system, that provides power for two-three-phase alternating current and heat supply for objects using wind, solar and internal combustion engines. In addition to major renewable energy sources, such as the sun and wind, the use of microbial fuel cells is contemplated.

A distinctive features of the station are to increase the reliability and quality of energy supply, reduce fuel consumption when using a diesel or gasoline generator set by optimizing its operation in the system, increase the efficiency of the system by controlling the operation mode of a complex mobile energy station using an automatic control system.

References

[1] Obama B 2017 Science №55 (6321) *The irreversible momentum of clean energy*, **4**, pp. 126-129

[2] Tene H P, Tetang F A, Edoun M and Kuitche A 2019 Indian Journal of Science and Technology, Vol 12(38), *Numerical Study of the Greenhouse Effect in a Flat-plate Double Glazing Solar Heat Collector,* **9**

[3] Menni Y, Azzi A and Chamkha A J 2018 Journal of Applied and Computational Mechanics, Volume 4, Issue 4, *A Review of Solar Energy Collectors: Models and Applications,* **27**, pp. 375-401

[4] Jacobson M Z and Delucchi M A 2011 En. Pol. № 39, *Providing all global energy with wind, water, and solar power, Part I: Technologies, energy resources, quantities and areas of infrastructure, and materials,* **16**, pp. 1154–1169

[5] Kalaiarasi G, Velraj R and Swami M V 2016 Energy, *Experimental energy and exergy analysis of a flat plate solar air heater with a new design of integrated sensible heat storage*, **111**, pp. 609-619

[6] Hoekman S K, Broch A, Robbins C, Ceniceros E and Natarajan M 2012 Ren. and Sus. En. Rev. № 16 Issue 1, *Review of biodiesel composition, properties, and specifications,* **27**, pp. 143-169

[7] Kalogirou S A 2004 Progr. in En. and Com. Sc. № 30 Issue 3, *Solar thermal collectors and applications*, **65**, pp. 231–295

[8] Mirzaei A, Tangang F and Juneng L 2015 Ren. En. № 80, *Wave energy potential assessment in the central and southern regions of the South China Sea,* **17**, pp. 454-470

[9] Penchahpa M M and Malakooti H 2017 Ren. En. № 111, *Evaluation of planetary boundary layer simulations for wind resource study in east of Iran*, **10**, pp. 1-10

[10] Connolly D, Lund H, Mathiesen B V, Pican E and Leahy M 2012 Ren. En. № 43, *The technical and economic implications of integrating fluctuating renewable energy using energy storage,* **14**, pp. 47-60

[11] Chatzitakisa P and Dawoud B 2017 Ren. En. № 110, *An alternative approach towards*

absorption heat pump working pair screening, **12**, pp. 47-58

[12] Bajpai P and Dash V 2011 Ren. and Sus. En. Rev. № 16 Issue 5, *Hybrid renewable energy systems for power generation in stand-alone applications: A review,* **14**, pp. 2926-2939

[13] Karim M A, Perez E and Amin Z M 2014 Renewable Energy, vol. 67, Mathematical modelling of counter flow v-grove solar air collector, **10**, pp. 192-201

[14] Chaibi M T, Jilar T 2005 Biosystems Engineering,. vol. 90, n°3, *Effects of a Solar Desalination Module integrated in a Greenhouse Roof on Light Transmission and Crop Growth*, **12**, pp. 319–330

[15] Khan V V, Dekanova N P, Romanova T A and Sharaeva S A 2017 Pr. of Universities. Investments.Construction. Real Estate , no. 1, vol.7, *A comprehensive analysis of the effectiveness of energy saving measures for social infrastructure in Eastern Siberia on the basis of system approach,* **12,** pp.101-112

[16] Khan V V, Dekanova N P and Khan P V 2017 International Conference on Construction, Architecture and Technosphere Safety 2017, ICCATS 2017. IOP Conference Series: Materials Science and Engineering. Vol. 262, *Heat Supply of Villages and Towns of Eastern Siberia. Problems and Ways of Development,* **7**

[17] Dekanova N, Khan V and Romanova T 2016 Pr. Universities. Investments. Construction. Real Estate, no. 4 (19), Development *and technical and economic proof of the events to Increase energy effectiveness of the systems of energy consumption of recreational objects of Pribaikalie*, **11**, pp. 146–156

[18] Tunik A A and Tolstoy M Y 2017 IOP Conf. Ser.: Mat. Sc and Eng. № 262, *The Complex Mobile Independent Power Station for the Urban Areas,* **6**

[19] Belokaya N V, Popov V S, Popova E M, Tolstoy V M and Tolstoy M Yu 2018 Construction and industrial safety № 12 (64), *Study of the application of energy-saving installations of life support systems for environmental protection areas*, **7** , pp. 91-97

[20] Konovalova E Yu, Stom D I, Balayan A E, Protasov E S, Tolstoy M Yu and Tyutyunin V V 2014 International Conference on Industrial, Mechanical and Manufacturing Science (ICIMMS 2014) *Using microbial fuel cells for utilization of industrial wastewater*, **4**, pp. 71-74

[21] Konovalova E Yu, Stom D I, Bybin V A, Ponomareva A L, Zhdanova G O and Tolstoy M Yu 2015 Patent RF no 153691 U1, *Microbial fuel cell*, Bul. № 21

[22] Stom D I, Konovalova E Y, Zhdanova G O, Vyatchina O F and Tolstoy M Y 2017 International Multidisciplinary Scientific GeoConference Surveying Geology and Mining Ecology Management, SGEM 17, Energy and Clean Technologies, *ACTIVE SLUDGE AND STRAINS ISOLATED FROM IT AS BIOAGENTS IN BIOFUEL CELLS,* **8**, pp. 19-26

[23] Ponomareva A L, Stom D I and Tolstoy M Yu 2015 In the World of Scientific Discoveries № 8.1 (68), *Intensification treatment of wastewater of oil pollution using the catalysts "Katan" and strains of biodegraders oil*, **7**, pp. 327-333

[24] Kalmynina N P, Kamalov R T and Yakhimbaev A N 2018 ISTU Bulletin of Youth B. 8. № 3, *Use of activated sludge of treatment plants as a bioagent in microbial fuel cells*, **7**, pp. 78-84

[25] Tunik A, Tolstoy M, Stom D, Popova E and Popov V 2019 IOP Conference Series: Materials Science and Engineering (ICRE 2019), Volume 667, *Use of microbial fuel cells and solar collectors in the multifunctional energy container,* **12**

[26] Tolstoy M, Belookaya N, Tolstoy V, Leontyev A and Tunik A 2019 IOP Conference Series: Materials Science and Engineering (ICRE 2019), Volume 667, *The multifunctional power container. Water treatment,* **14**

[27] Tolstoy M Y, Akinina N V and Tunik A A 2012 Patent RF no. 112364 U1 F24J2/ 24 (2006.01), *Solar collector,* Bul. № 1

[28] Tolstoy M Y, Tunik A A and Lapkovskiy A A 2015 Patent RF no.153795 U1, F24J 2/24 (2006.01), *Solar collector,* Bul. № 22.

IOP Conf. Series: Materials Science and Engineering **880** (2020) 012057 doi:10.1088/1757-899X/880/1/012057

Management of the housing and utilities sector in small municipal districts

G F Zharkaya[1], T N Alekseeva[1], D A Tsyrempilov[1] and O A Krapivkina[2]

[1] Department of Investment and Real Estate Management, East Siberian State University of Technology and Management, 26 Smolina street, Ulan-Ude, 670000, Russia
[2] Department of Foreign Languages No 2, Irkutsk National Research Technical University, 83 Lermontov street, Irkutsk, 664074, Russia

E-mail: galina5122@mail.ru

Abstract. The article deals with the issue of management of the housing and utilities sector in small municipalities, trends in the development of housing and communal services at the regional and municipal levels. It analyzes the literature on the research issue, presents problems of the housing and utilities sector, and analyzes the issues of wear of fixed assets in the housing and utilities sector. It assesses fixed asset renewal activities in the industry, identifies problems and determines development trends for the stagnant state of the housing and utilities sector in small municipalities.

In modern conditions, the housing and utilities sector is a combination of social and economic relations providing the population with the engineering infrastructure. The industry is multifunctional. It involves production of various types of energy and resources and provision of housing and utilities services [1, p. 201].

Previously, the housing and utilities sector was regulated by the government represented by municipal bodies. During the period of economic restructuring, the municipal bodies partially transferred engineering networks to private companies. It has caused physical and moral depreciation of equipment in this industry.

According to statistics for the end of 2018, about 50% of fixed assets in the housing and utilities sector are obsolete, which causes up to 25% of losses of the supplied resources (Figure 1)

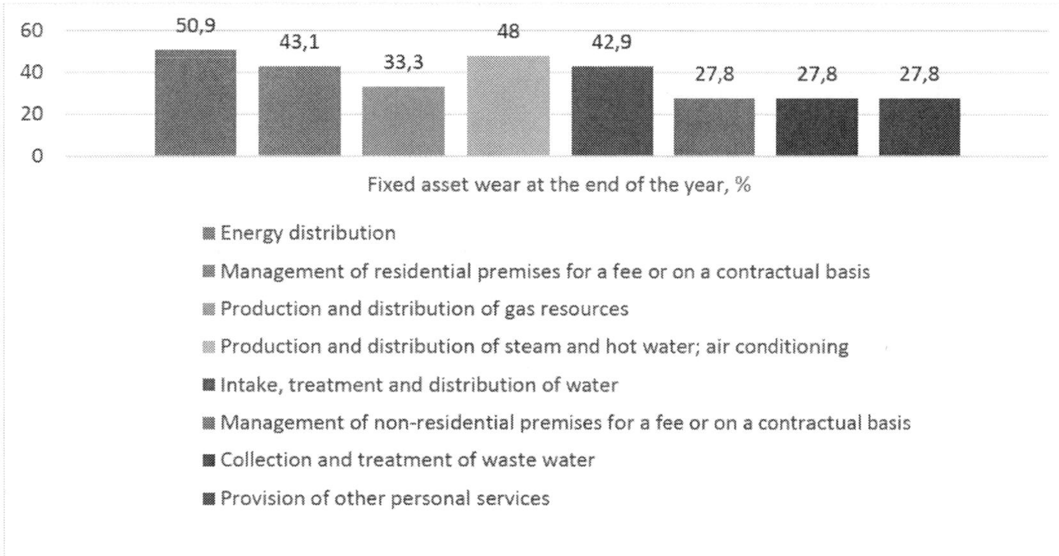

Figure 1. The degree of wear of fixed assets in the housing and utilities sector for the end of 2018. [3, p. 26]

Municipal bodies have not created mechanisms for interacting with the population in the housing and utilities sector. In turn, the literacy of the population is critically low. The solution to this problem is of temporary nature, and does not develop into full-fledged partnerships [1, p. 202].

The main problems of small municipalities reflected in all reports of regional authorities in the field of housing and utilities services are low budget financing and wear of the material component of the housing and utilities sector. Municipal bodies and private investors are trying to solve problems caused by worn out networks and create public-private partnerships. However, these are insignificant tools.

In Russia, the housing and utilities industry is unstable. One might even say that it is in a crisis. Zh. Kolycheva identified the following problems of the housing and utilities sector:

- the conflict between housing and utilities companies and consumers of services;
- high cost of utilities and their low quality;
- large debts of the population.

These problems are caused by social rather than economic factors. Since they are in the center of public attention and more politicized. Housing and communal services are one of the priority programs of the national socio-economic development. But the situation is getting worse. [2, p. 38].

Reformation of the housing and utilities sector was not efficient. Each round of reforms of the housing and communal services discarded the municipal bodies to the initial stage. For example, according to the results of the reforms, the system for providing benefits for certain categories of citizens is inefficient.

The regional authorities should deal with the issues of transforming the housing and utilities sector, form the regional budgets taking into account the individual approach to each municipality, and make decisions at the federal level level in order to create conditions for the long-term systemic development of the housing and utilities industry.

The housing and utilities sector requires new solutions that allow the monetization of costs and improvement of the quality of utilities services. The effective mechanisms of socio-economic development can be created by combining efforts of all participants and users of the housing and utilities sector. One of the tools to attract private investors in the housing sector is improvement of the investment attractiveness by providing benefits.

According to the report of the Ministry of Regional Development of the Russian Federation "On measures aimed to improve quality of housing and utilities services", the coefficient of renewal of fixed assets was 8.2% of available fixed assets, the capital consumption ratio was 0.4%, and the share of completely worn-out fixed assets was 14.1%. (Figure 2)

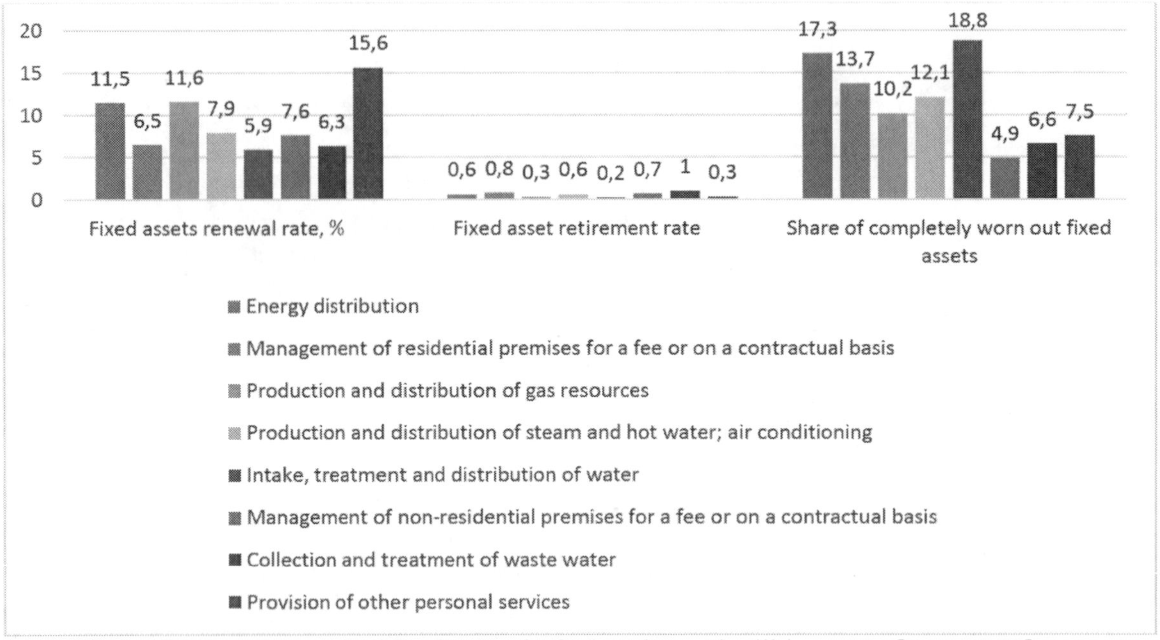

Figure 2. The movement of fixed assets in the housing and utilities sector [3, p. 26-27]

Thus, small municipalities have to eliminate worn out fixed assets in the housing and utilities sector, bringing budgets to the pre-bankrupt level.

As a result of the analysis of activities of the municipal bodies in the housing and utilities sector, the following provisions can be formulated:

- it is necessary to control the housing and utilities sector; at least two interested parties are required (municipal authorities-management companies, municipal authorities-the population);
- it is necessary to create councils of apartment buildings (ABs), neighboring communities, housing cooperatives, local communities, and and hold courses aimed at improving housing literacy of the population;
- it is necessary to ensure financial influences of private investors and budget financing, increase production potential of the industry, eliminate wear of the material base, hire efficient and competent specialists;
- it is necessary to update technical documents on engineering networks, housing cooperatives, buildings, structures; to carry out technical inventories and their digitization;
- it is necessary to implement such technologies as "Smart City", "Smart Village", "Smart Community";
- it is necessary to establish responsibility for effective control of companies providing housing and utilities services.

The system analysis in planning and forecasting activities of municipalities will make it possible to improve the efficiency of housing and utilities sector management and development. The main goal of the ongoing reforms and transformations is sustainable development of the quality of housing services, and improvement of living conditions of the population in their houses and apartments, rational and energy-efficient consumption of resources.

IOP Conf. Series: Materials Science and Engineering **880** (2020) 012057 doi:10.1088/1757-899X/880/1/012057

References

[1] Kozyrev M S 2017 Features of housing and communal services as an object of municipal management *Materials of Ivanovo Readings* **1–2 (11)** pp 200–206

[2] Kolycheva Zh Ya 2018 Interaction of public and business structures in the housing and communal sector of municipalities *Science and Education: Economics and Economics; entrepreneurship; law and management* **2 (93)** pp 37–41

[3] Statistical compilation "Housing in Russia 2019". [Electronic resource: https://www.gks.ru/storage/mediabank/Jil-kom_xoz-vo%202019.pdf]. Accessed on 01.21.2020

[4] Chupin V R and Dushin A S 2019 Parameter optimisation for new and reconstructed systems of water supply and distribution for ensuring consumer's uninterrupted water supply *Proceedings of Universities. Investment. Construction. Real estate* **9(4)** pp 790–803

[5] Kulkov V N, Solopanov E Yu and Kamalov R T 2019 Using immobilised sludge for wastewater treatment and its air regeneration *Proceedings of Universities. Investment. Construction. Real estate* **9(3)** pp 522–529

[6] Pupyrev E I and Chupin V R 2019 Features of regional development of water disposal systems in the central ecological zone of the Baikal natural territory *Proceedings of Universities. Investment. Construction. Real estate* **9** pp 354–363

[7] Chupin V R and Dushin A S 2019 Assessment of the reliability of water supply to consumers: water supply reliability indicators Proceedings of Universities. Investment. Construction. Real estate **9(3)** pp 578–593

[8] Lavygina O L and Grebneva O A 2019 Environmental technologies in the housing and communal service system of the Baikal natural territory *Proceedings of Universities. Investment. Construction. Real estate* **9(4)** pp 726–733

IOP Conf. Series: Materials Science and Engineering **880** (2020) 012058 doi:10.1088/1757-899X/880/1/012058

Techniques for designing a mini house on a garden plot

A G Bolshakov[1]

[1] Irkutsk National Research Technical University, 83 Lermontov str., Irkutsk, 664074, Russia

E-mail: andreybolsh@yandex.ru

Abstract. The problem is that comfortable housing is usually designed in such forms that make it inaccessible to people even with average incomes by russian standards. This leads to the task of reducing the volume and area of the house while maintaining the requirements for completeness of functions and architectural expressiveness. This social order is reinforced by the fact that the housing problem for people with middle and low incomes in Russia is solved in suburbia, on the territory of garden associations, where housing becomes an alternative to urban housing. The article solves the following problems: minimization of floor space, landscape-like architecture in conditions of a significant slope; development of expressiveness techniques for log cabins in combination with a wooden frame; architectonics of parts of a wooden building while ensuring maximum compactness.

Design techniques and the author's experience in using them in the con-struction of a small residential building made of timber with stained-glass windows on the terrace are given. Small house-on a site of 6X6 meters performs all the main residential and additional guest functions. The house was built according to the author's project and carried out by the builder under constant supervision of the author.

The topic of this work is relevant for two reasons. First, modern set-tlement processes include the development of suburbia. As part of this process, numerous garden associations are gradually turning into residential settlements. Round-the-year water supply and electricity networks are being laid, and sewage issues are being resolved autonomously through the construction of septic tanks. People get a residence permit and live in gardening all year round. This is reinforced by the new legislation in force. Secondly, the city life is available for the reason that in the city, citizens with ordinary Russian income can not afford a large comfortable apartment. A mini-house in the suburbs on its site is more comfortable than a single-room apartment in the city. An apartment of 36 m2 cannot be developed to the size of 108 m2 because of neighbor cells in a multi-story building.

Therefore, it is important to study and develop architectural principles that allow you to develop the space of a residential building without going beyond the boundaries of a square of 6X6 meters. And as a result, the house should get a set of necessary functions and even architectural representation. This work shows the possible order of project actions in this direction.

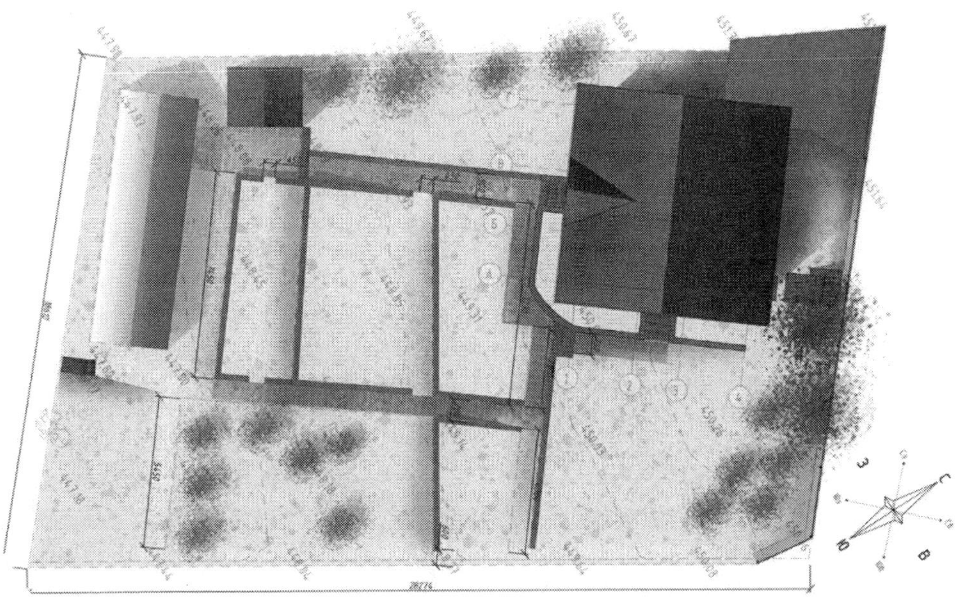

Figure 1. General plan of the plot with a house, a barn and a greenhouse

General topography. The plot of 5.2 acres is located on the South-Eastern slope of the Yershov mountain, 500 m from the bay of the same name. The contour of the site has the form of a parallelogram with sides of 28 m and 18 m. the relief drop is 5 m diagonally, drawn from the northern top of the parallelogram (452 m above sea level) to the southern top (447 m). That is, for each meter of the diagonal, the relief drop is approximately 17 cm.

The principle of terrain organization is terracing. The issues of terrain organization for functional requirements and for reclamation purposes are discussed in [3] and [4]. On the plot formed six terraces. The highest terrace on the northwest section. This is the smallest terrace – 15 m2. there is an entrance to the site from the street side and a concrete septic tank with a depth of 3.5 m is installed under the parking lot. The retaining walls of the parking lot terrace for 1 car are also reinforced concrete.

The principle of terracing is combined with the principle of maintaining a smooth slope along the edges of the site: partially on the eastern and western borders. That is, the section along the slope is divided into three longitudinal panels parallel to the longitudinal axis: Western, Central and Eastern. The Western panel is divided into three terraces, the Eastern panel is divided into 4 terraces. The Central panel has 6 terraces (including the greenhouse). The side panels (West and East) in their topography are inclined terraces with an almost unmeasured smooth slope. The central strip contains six terraces that have retaining walls and horizontal work surfaces. Thus, the plot is divided into 13 cells.

The house and parking lot occupy two adjacent northwest cells. Together with the parking lot in front of the street fence, a green room is formed, with plants that can withstand the start (rhododendron Daursky, chubushnik).

The South-Eastern cell of the site is a natural landscape. There are pine trees and a mixed-grass lawn.

The central terraced strip is a vegetable garden for growing vegeta-bles. On the southern border of the central strip – there is a greenhouse 9X2,8 m.

Western panel – trees and shrubs. At the bottom of the panel, the cell is occupied by a shed with a toilet.

Eastern panel – shrubs and low-growing trees. In the lowest cell of the Eastern panel is a utility area with lumber and a compost box.

Functional program of the house. The dependence of the housing structure on the demographic composition of the family and on the schematic scenario of its life is considered in many works, for

IOP Conf. Series: Materials Science and Engineering **880** (2020) 012058 doi:10.1088/1757-899X/880/1/012058

example [2]. The house is designed to accommodate a single family and guests, for which a guest room is included in the premises. In fact, the older generation-husband and wife – live in the house on a permanent basis. The guest room is intended for adult children, other relatives and friends who come occasionally during the summer.

Nuclear functions for the family: kitchen, dining area, bed, bathroom with shower - located on the first floor. See figure 2. The glazed terrace serves as a vestibule and stairwell. Stairs lead to the second floor. There is also a wardrobe for outerwear and shoes on the terrace. A special purpose of the terrace is a summer tea party.

The second floor is reserved for guests. See figure 3. A staircase from the first floor leads to the gallery. From the gallery, the entrance to the second floor room. The gallery also serves as a viewing platform, since the house is located on a high slope facing the bay.

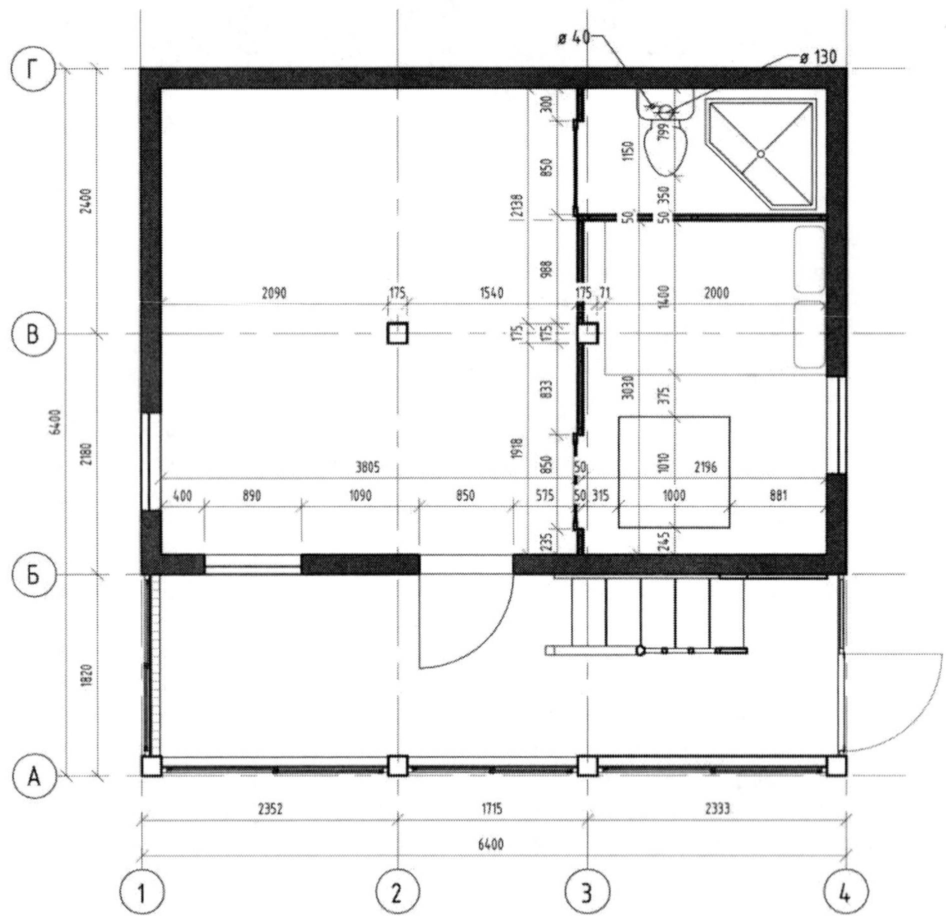

Figure 2. Ground floor plan

The second-floor room has two beds and a work desk in the bright south corner. There are two 90X100 windows in this corner.

At the junction of the roof and side walls (North-East and South-West), the upper-side glazing is built in. See figure 11 and 12.

Spatial and design solutions. Internal space. The house is located on a hillside. The slope faces South-East. The difference in marks between the extreme North-Western point (the corner of the parking lot) and the South-Eastern corner of the house is one and a half meters (451.5 – 450.0). The spatial structure is based on a crate – a two-story frame house made of timber on the underlay in combination with a frame structure.

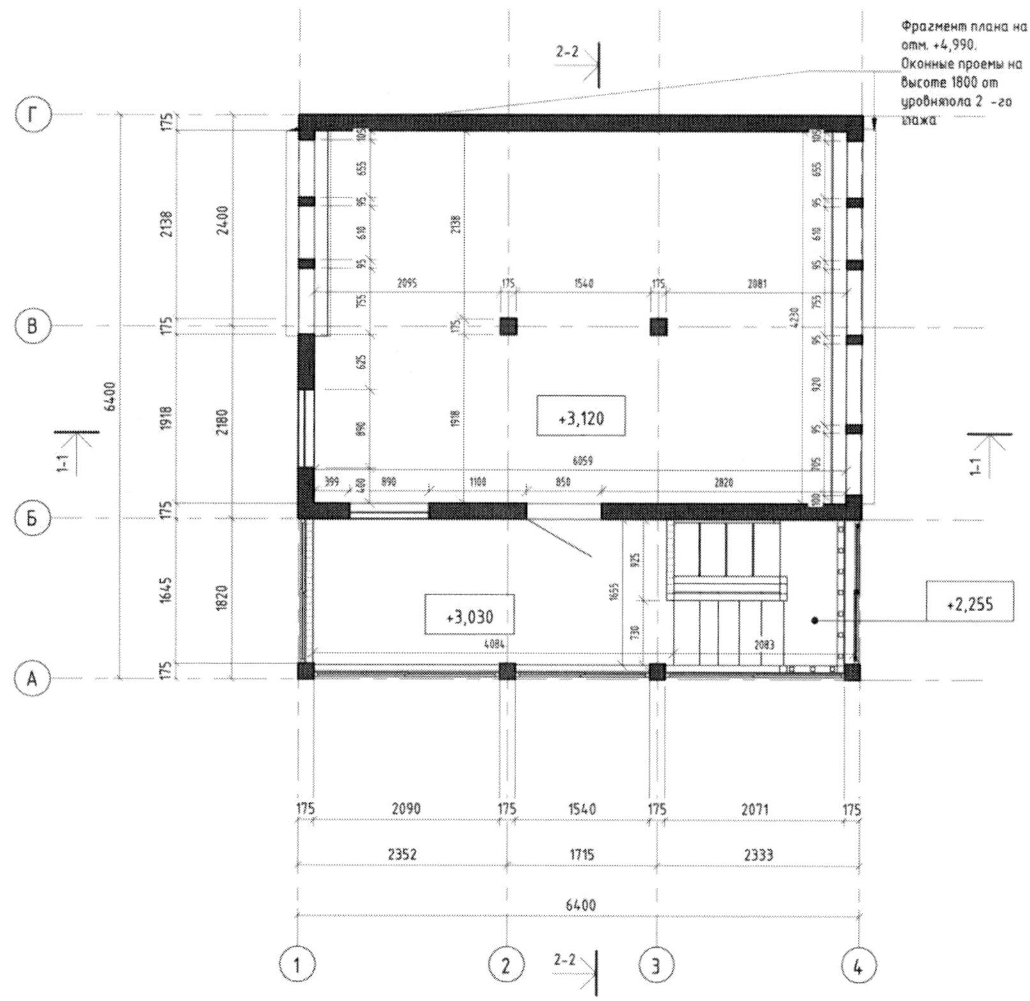

Figure 3. Second floor plan

The underlay is a ribbon concrete Foundation. See figure 4. The depth of the foundation is 2.4 meters at the highest North-Western point. There is a concrete floor on the bottom of the foundation. The foundation square of 6, 4X6, 4 m is divided into three sections so that the walls are separated from each other by 1900, 1800 and 1450 mm in internal dimensions. The basement sections are connected by passageways. The picture in figure 4 was made before the openings were cut. The North-West wall of the foundation is connected to a septic tank for wastewater disposal. Under the north wall, a water pipe

IOP Conf. Series: Materials Science and Engineering **880** (2020) 012058 doi:10.1088/1757-899X/880/1/012058

is connected to the basement. On the middle inner wall, a pylon of 800X700 mm was made for the stove, which later had to be temporarily abandoned.

Figure 4. Ribbon foundation-underlay. View from the northwest corner, from the septic tank wall and parking lot

Many authors, including [7], have considered low-rise rural housing construction in its engineering and structural aspects.

For figure 5 shows a section 1-1. The middle wall is supported by two wooden columns made of timber with a cross section of 18X18 cm and a height of 5.72 m. the columns are located on axes 2 and 3 at their intersections with axes B and C, respectively (see figure 2 and figure 3).

The peculiarity of inter-floor ceilings on beams in this house is that they are solved by short spans. The ratio of the cross-section of the beam (18 cm) to the length of the span does not exceed the ratio of 1:10. This makes the overlap rigid and does not allow for breaks.

The basement ceiling is made with the support of beams on the foundation walls. The length of the beams in this case fit into the specified pro-portion when using a 10x18 bar placed "on the butt".

The overlap of the first floor is solved with deliberately excessive ri-gidity. On the long side of the log (6 m), the main beam-matrix is laid. Traditionally, matrix plays an important constructive and spatial role in the russian home [1]. It is a double beam, each rod of which consists of two 10x18 bars compressed by pins. In addition, matrix is supported by two columns on the axes "2", "3" and "B" and "C" of the first floor plan. See figure 6.

The combination of the spatial system of the frame with the wall structure of the log house is most evident in the portico, which in the course of the evolution of the house turned into a two-story gallery, see figure 7. The interior of the gallery and the view from the gallery to the garden see figure 15.

IOP Conf. Series: Materials Science and Engineering **880** (2020) 012058 doi:10.1088/1757-899X/880/1/012058

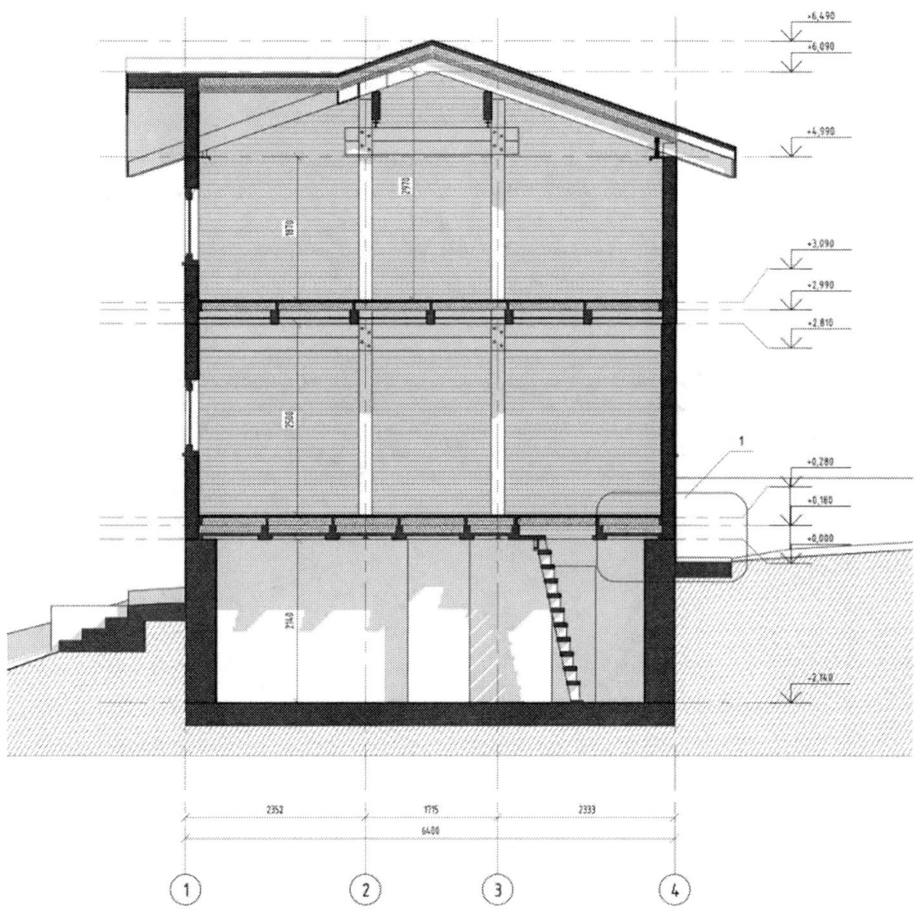

Figure 5. Section 1-1

The facade in figure 7 is the main architectural property, since it faces the bay. There is a glazed terrace facing the bay. Accentuated roof over-hangs are visible. Unfortunately, an unscrupulous neighbor put his house right on the bridge and 90% blocked the view from the terrace to the bay.

Figure 6. Load-bearing system for covering the first floor. Interior

The architectural properties and dimensions of the facade in the or-thogonal drawing are shown in figure 8.

The frame rack-and-beam system is carried out from the inside of the cage to the outside, thus forming a portico and at the same time a two-story glazed gallery.

For figure 2, photos 7 and drawing 8 show that the entrance is made under the stairs from the border of the site to the opening adjacent to the column on the fourth axis. The staircase could be moved to the first axis, but this would completely block the view from the window to the bay. Therefore, the staircase is located between axes 3 and 4, the window is saved next to the first column.

Connection of internal and external spaces. The connection between the internal and external space of the house for low-rise construction on a private plot is of particular importance [6]. The first and main block that implements the transition from the interior space to the garden space is the glazed terrace. The portico on the facade turned into a two-story gallery during the five-year construction. The frame system from the inside of the house came out on the exterior facade. The same block - glazed terrace - implements the connection of the first and second floors. The exit from the warm contour of the cage is carried out on the terrace.

Figure 7. East facade, axis 1-4

The second element that connects the space of the house with the space of the garden is an open terrace patio. The outdoor patio terrace is raised by retaining walls encircling the house, the height of the walls varies from the thickness of the blind to 80 cm. the patio terrace has a concrete surface in which lawns are inscribed. Entrance to the house from both the street and the garden is via a terraced patio. Since the plot has a slope of about 17%, the patio surface is inclined, but the slope is halved to 8%. Due to the noticeable slope, a decision was made on the location of the entrance to the house. The entrance to the house is made on the edge of the foundation, which, first, is located below the other above ground level, and secondly, is closest to the entrance to the site from the street. See figure 8.

For this reason, the connection of the house with the garden is a "fork". From the entrance, a path along the top of the patio leads to a path along the west panel. After descending from the patio via a staircase between axes 2 and 3, the path leads to a path along the east panel.

Both paths, paved with red brick, are closed by a greenhouse in the south of the garden plot and lead to its opposite sides and to a grass path along the southern border of the plot. See figure 1. Thus, the topological scheme of the internal and external space is a ring around the garden with a linear section attached to it along the patio and the node of the entrance to the house [8].

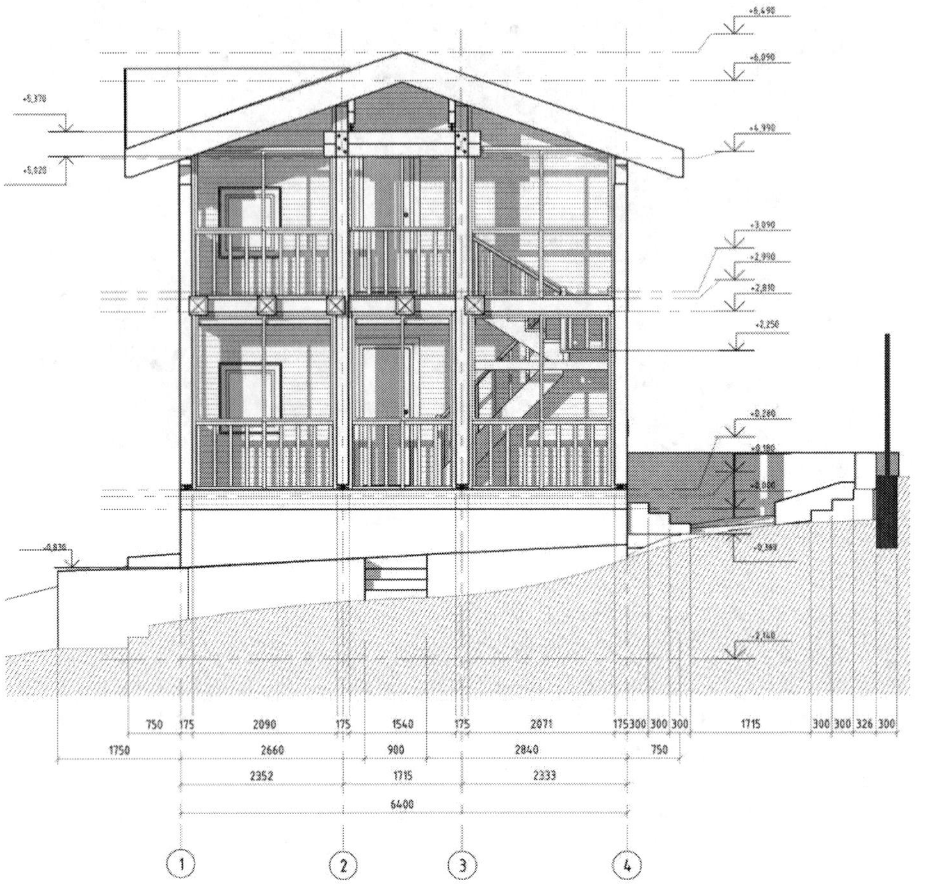

Figure 8. Facade in axes 1-4

Artistic image. The basis of the artistic image is the idea of a brutal roof over the house. Roof slopes are supported by rafters of increased cross-section, roof overhangs have a removal of more than 70 cm. One of the two large sloping ramps is paired with a small gable. The use of multi-gable roofing is traditional and is described in [5]. In the interior of the second floor, the inner surface of the roof (ceiling) gets a complex "CES-zoned" surface.

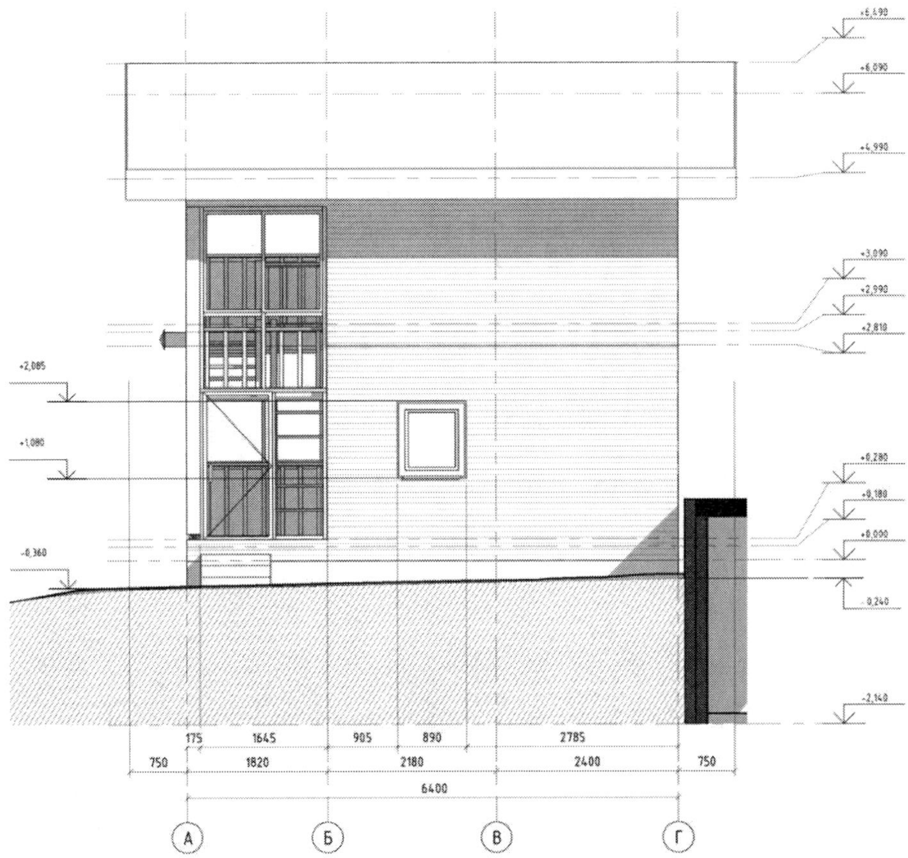

Figure 9. The entrance facade in the axes A – G. The glazed terrace at the side, the axes A – B

The edges of the gable's interface with the main ramp resemble the ribs of a "Gothic" vault. See figure 10.

The structural system laid down on the first floor-columns, the main beam-matrix and ordinary beams is repeated in a modified form on the second floor by the upper part of the same columns, closed by a bridge of double beams, double sub-rafters and rafters ribs. Along the line of conjugation of the roof slopes with the upper crown of the frame, the upper side glazing is arranged. See figure 11.

IOP Conf. Series: Materials Science and Engineering **880** (2020) 012058 doi:10.1088/1757-899X/880/1/012058

Figure 10. Interior of the second floor

The edges of the interface of the small gable with the main surface of the slope. In the foreground on top of the sub-truss beam and the bridge between the columns, on which the beam rests.

Figure 11. Interior of the second floor

The upper side glazing creates a line of visual support of the roof slope on the light strip above the upper crown of the wall. Thanks to this, the heavy roof seems to hover over the wall.

Section 2-2 figure 12 shows two floors and a roof slope. The transition link between the interior and exterior spaces is a two-story glazed terrace. The main staircase on the terrace connects the two floors. Under the stairs is a wardrobe for outerwear. In the depth of the section on the first floor, a sliding partition is visible. The far left entrance leads to a bathroom with a shower. The right entrance leads to the bedroom. See also plan figure 2.

Log house-the whole on figure 13 and pictured figure 14 is paired with a glass body that complements the main volume to cubic proportions. The curved slant line at the base shows the interface between the patio wall and the slope surface.

For figure 16 shows the branching paths from the house to the garden to the west (the top patio) and medium (through natural cell) garden lanes (see general plan figure 1)

IOP Conf. Series: Materials Science and Engineering **880** (2020) 012058 doi:10.1088/1757-899X/880/1/012058

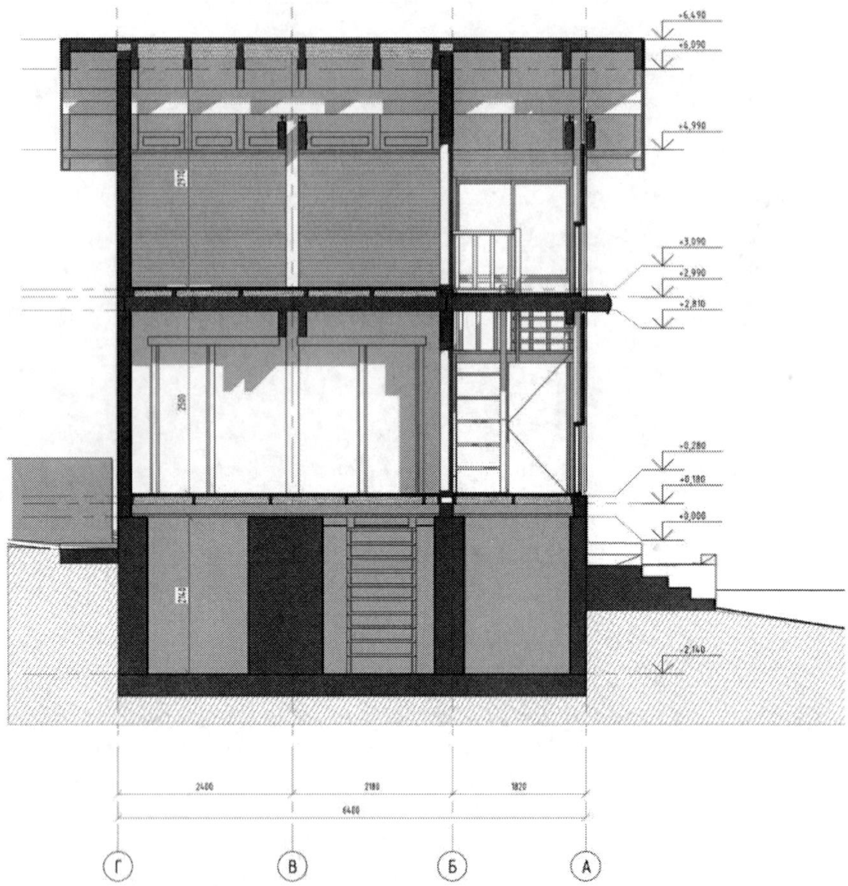

Figure 12. Section 2-2

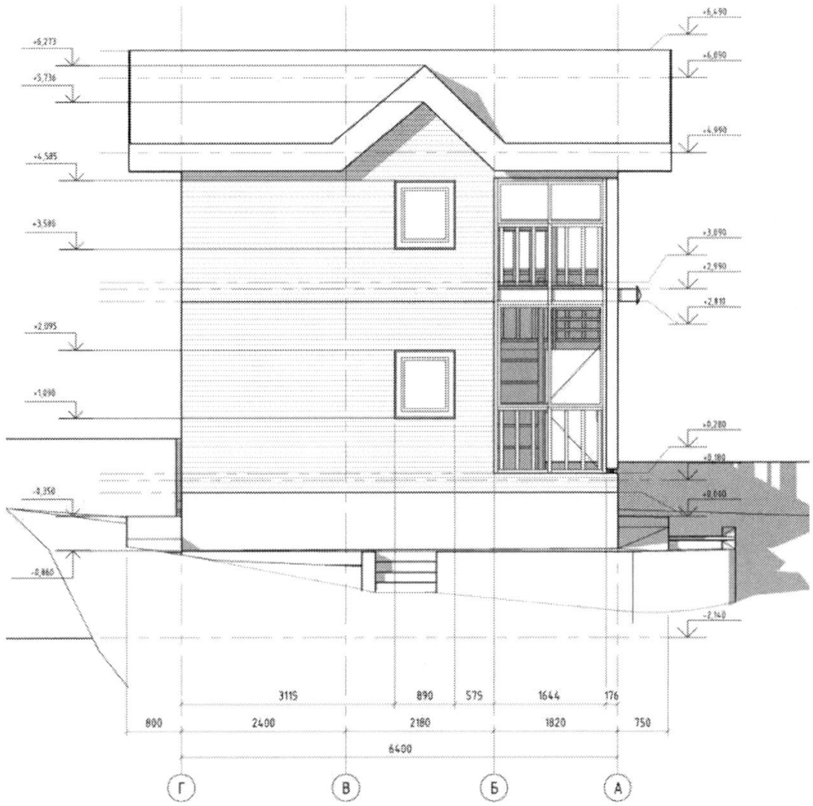

Figure 13. Facade in G-a axes

Figure 14. Photo of the southern facade (in g-a axes)

Figure 15. The first floor of the glazed gallery. Interior. A view of the garden

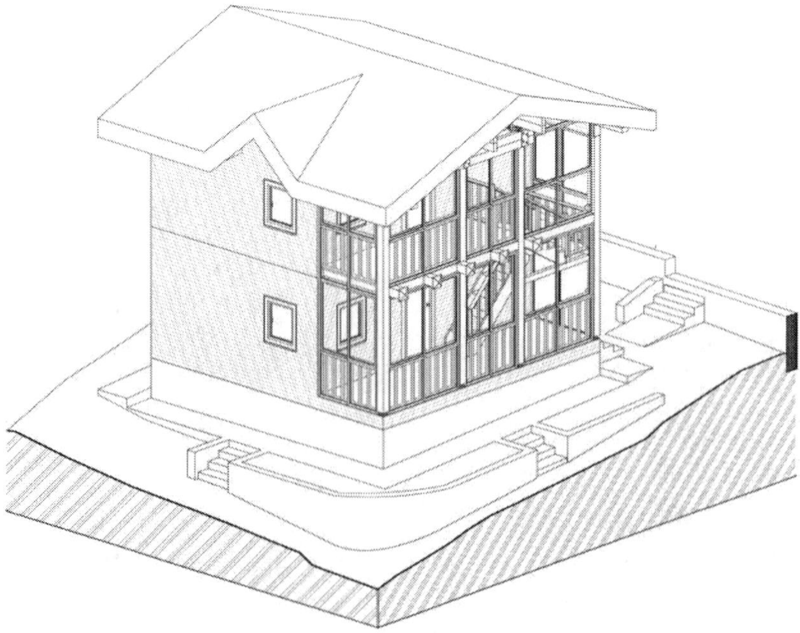

Figure 16. Axonometry of a mini-house on a garden plot with a slope of 17%

1. The task of achieving the fullest possible implementation of residential functions on a site with a minimum size requires compact solutions and a flow of spaces based on the "Studio" principle.
2. The need to separate technical, residential, and guest functions on a small site required vertical zoning and the construction of three floors.
3. Complex node: portico, aka two-storey glazed terrace performs at least four functions: summer recreation with a view of the landscape and tea; winter vestibule and thermal protection of the house; vertical communication between the two floors by stairs; entrance hall and wardrobe for outerwear.
4. The combination of the log wall configuration with the frame sys-tem makes it possible to make the boundaries between functional zones (kitchen, dining, entrance and living areas) permeable. The borders are marked with columns-supports. The frame system allows you to make a portico-terrace. Elements of the frame system on the second floor allow you to make upper-side glazing.
5. The main beams resting on the pillars by holding down the columns with two sides of the beams with the device pin long cords allow you to run a bearing without loosening of the sections and carried by bearing elements, i.e. without frames.
6. The use of a system of consecutive supports of the matrix on the pillars, secondary beams on the matrix allows to achieve an architectonic effect in the composition of a wooden house with a combined (frame and wall) structure.
7. In addition to the above-mentioned architectonics, the artistic image uses the technique of a large roof overhang with the strengthening of the truss beams that are cantilevered behind the frame. The image of the "floating roof" is the final one in the artistic theme of the house.

References

[1] Aschepkov E A 1953 Russian folk architecture in Eastern Siberia p 276
[2] Blagova M V 2015 Architectural formation of commercial housing on a social and functional basis (on the example of Rostov-on-don) p 28
[3] Bolshakov A G 2000 Geoplastics in architecture and landscape planning: a textbook p 171
[4] Bolshakov A G 2015 Restoration of disturbed and degraded territories. History of the author's method of geoplastics *Architecture and Modern Information Technologies* № **2 (31)** p 15
[5] Opolovnikov A V 1983 Russian wooden architecture p 286
[6] Radulova Ya I 2017 Features of integration of internal and external space in the architecture of low-rise housing p 27
[7] Ukrainchenko D A, Zhadanov V I, Likhnenko E V, Pinaykin I P 2018 State and prospects of a wooden low-rise building *Industrial and civil construction* **№ 5** p13-17
[8] Hiller B, Hanson J 1984 The social logic of space p 281
[9] Tkachenko S B 2019 Zaryadye development concept: from hotel to park *Proceedings of Universities. Investment. Construction. Real estate* 9(1) pp 196–213
[10] Khokhrin E V, Smolkov S A 2019 Urban interior. General characteristics of methods and approaches. *Proceedings of Universities. Investment. Construction. Real estate* **9(1)** pp 214–231
[11] Sysoeva E V, Anishchenko I O, Mosolova A S, Remizova A A, Terekhova E S 2019 Rationale for the reconstruction of old cultural buildings on the example of the "Pobedinsky" House of Culture, Skopinsky District, Ryazan Region *Proceedings of Universities. Investment. Construction. Real estate* **9(1)** pp 186–195

Architectural features and problems of preserving the monument of Orthodox construction in Transbaikalia - the Church of the village of Konduy

E Drobotushenko[1], P Novikov[2], S Karasev[3] and Y Lantsova[4]

[1] Faculty of history and philology, Transbaikal State University, 672039, 30 Aleksandro-Zavodskaya str., Chita, Russia
[2] Department of history and philosophy, Irkutsk National Research Technical University, 664074, 83 Lermontova str., Irkutsk, Russia
[3] Department of history and philosophy, Irkutsk National Research Technical University, 664074, 83 Lermontova str., Irkutsk, Russia
[4] Faculty of history and philology, Transbaikal State University, 672039, 30 Aleksandro-Zavodskaya str., Chita, Russia

E-mail: drzz@yandex.ru

Abstract.The article is devoted to the construction of one of the little-known monuments of Orthodox Church architecture on the territory of Eastern Transbaikalia-the Church of the Kazan icon of the mother of God and the Holy Martyrs Kirik and Iulita in the village of Konduy in the Borzya district of the Chita region. The features of using building materials and architecture are described. The question of the origin of building materials is analyzed. It is noted that the Church stands out among Orthodox churches by its external decoration. The question is raised about the need to preserve such an unusual monument of Church architecture.

1. Introduction

On the Trans-Baikal territory (Eastern Transbaikalia), units of Orthodox architecture and construction of the pre-Soviet period have been preserved. Among them, Orthodox churches stand out: the assumption of the blessed virgin Mary of the village of Kalinino in the Nerchinsky district of the TRANS-Baikal territory (1712), the Archangel Michael in Chita (1776), the Holy cross Church of the village of Zakharovo in the Krasnochikoysky district (1892).

You can make an assumption about the reasons for this situation. First, the remoteness from the center of Russia, the peripheral, border, provincial position of the region determined the simplicity of the approach to the construction of the bulk of churches and chapels. In other words, there was no need for serious architectural refinements. We can observe this, for example, on the example of the Chita women's Bogoroditsky monastery of the late XIX – early XX centuries. Its architectural ensemble was incredibly simple and this despite the fact that it was located in the regional center – Chita. Secondly, the significance of the territory, the weakness of its settlement and the dispersion of settlements, caused difficulties with construction. The third component is the legacy of the 20-30s of the XX century-the time of destruction of temples. The region has lost a significant part of cathedrals, churches, chapels and houses of worship. Despite this, some of the churches built in the XIX – early XX centuries have been preserved. There are only a few restored and preserved ones left. This is Holy cross Church in the village of Zakharovo Krasnochikoyskoe area TRANS-Baikal territory, the Archangel Michael (also called «the Church of Decembrists») and Holy resurrection Church (former Catholic Church of Peter and Paul) in the city of Chita.

One of the monuments of Orthodox architecture of the region's pre-Soviet period is the Church of the Kazan icon of the Mother of God and Svyatykh Muchenikov kirika I iulitty Condui in the village of Borzya district Chita region. We often find in the literature that this is a Church in honor of the Nativity of the most Holy Theotokos and the Holy Martyrs Kirik and Iulitta [3; 7; 10; 11; 12; 19 etc.]. In common parlance it is called simply «Konduy Church». Note that knowing it as such, the majority of transbaikalians do not know that this is a Church in the name of the Kazan icon of the mother of God and in the name of the named Holy Martyrs.

This article aims to characterize the architectural features of the named Church and, once again, to focus on the problem of its preservation.

2. Materials and methods

The work is based on two approaches of Russian historical science, formational and civilizational. The first made it possible to divide the history of the Konduy Church into pre-Soviet, Soviet, and post-Soviet periods. The second one made it possible to consider the Church as an expression of the Russian Orthodox civilizational identity with an unusual mixture of elements of Eastern civilization through the use of architectural elements of buildings of other cultures, which is unusual for Orthodoxy.

The paper uses the fundamental principles of historical science. The principle of historicism was given the opportunity to describe the history of the Konduy Church in its historical development. The principle of objectivity allowed the authors to start only from facts, distancing themselves from subjective assessments of various events.

The study used both General scientific and special scientific methods. The first are methods of collecting, analyzing and synthesizing material, as well as methods of induction and deduction, which allowed us to build logical chains from the General to the particular and Vice versa. Among the special methods of historical science can be called comparative-historical, which made it possible to consider the history of the Konduy Church in its development through comparison.

3. Results

In the passport of the object of the Ministry of culture of the Transbaikal region, citing data provided by the State archive of the Transbaikal territory, the Church called of the Nativity of the blessed virgin Mary [4]. According to N.P. Kradin, the Church of the Nativity of the most Holy Theotokos and the Holy Martyrs Kirik and Ulita [11, p. 172]

It should be noted that the Church of the village of Konduy has not become, until today, the subject of serious scientific study with the involvement of a significant amount of sources on its history. At the same time, individual notes appeared at different times. Perhaps the first is an article by the founder of the local history Museum in Chita, A.K. Kuznetsov, which was published in 1925. However, it is mainly devoted to the Konduysky Palace. However, the issues of removal of bricks, stone and «statues»from the territory of the Palace ruins are discussed. Under «statues», perhaps, A.K. Kuznetsov understood the sculptures of stone dragons, some of which formed the Foundation of the Konduy Church [13].

In 2002, an article by N.P. Kradin «architectural monuments of Konduy» was published, which, among other things, refers to the Church being characterized [10]. Small notes on the Church are available in the large four-volume Encyclopedia of Transbaikalia, as well as in special volumes of the small encyclopedia of Transbaikalia («Culture», «Archaeology», Architecture and construction») [1; p. 275; 10; 19, p. 346]. Mention of the fact that the material of the former Konduy town was used for the construction of the temple is also found in a number of notes of the named publications on the Konduy town [8; 9]. The main problem with all these publications is that they are small and simply repeat each other.

IOP Conf. Series: Materials Science and Engineering **880** (2020) 012059 doi:10.1088/1757-899X/880/1/012059

Despite the fact that the Church is quite well-known in the TRANS-Baikal territory, it has not received proper fame outside of it, although in architectural terms it is so unusual that it can claim an independent place in the list of monuments.

The main feature of the Konduy Church consists of two components. The first is that it is built of stone that was left over from the Konduy Palace. It is famous monument of medieval architecture, located in the valley of the Urulyunguy river. In domestic archaeology it is known as «Konduy town» or «Konduy settlement». This is a small copy of the Palace in the capital of the Mongol Empire Karakorum on the Orkhon river. the monument dates back to the XIV century. [9, p. 377] The fact that the Orthodox Church was built from stone from the construction of pagan Mongols is surprising. The second component that makes this monument unusual is that stone statues of dragons from the same Konduy Palace were installed at the base of the Church. Actually, dragons are an element of pagan perception of the world. These are mythical creatures. There is evidence that in addition to stone and dragons from the ruins of the Palace, bricks, stone slabs and sculptures of «men» were used [15]. Also, stone circles were used for the construction of the Church, which, most likely, were the basis for the columns of the Palace. In total, about seventy sculptures were used during construction [15].

This raises a number of obvious questions. Who came up with the idea of using dragons and «little men» to build an Orthodox Church? How did the Russian Orthodox Church perceive this fact? And the most important question is, how was such a Church consecrated?

The history of the construction of the temple is complex. There is evidence that the initiators of the construction were Epov family – residents of Nerchinsk.

They also financed the construction. The Builder was Kirill Sukhanov, Archpriest of the Nerchinsky Cathedral. The beginning of construction, after the blessing of the diocesan authorities, dates back to 1806. The Blessing was given by Bishop Benjamin of Irkutsk and Nerchinsk (Bagryansky). D.E. Tsyrenzhapova writes that the Church was built by the Cossack constable N. Epov with the permission of the Nerchinsky spiritual authorities, however, what is understood by the latter is not quite clear. They write that the Church was built by the efforts of the widow of the Cossack constable Semyon Vasilyevich Epov and her son. According to some references, the Church was built in 1805 [4; 6; 19, p. 346]. The Church is located in the center of the village.

In General, the exterior of the Church building is traditional. The building of the temple is elongated due to an extension on the entrance side, a five-sided apse. According to various characteristics, the altar, refectory and bell tower or apse, the temple room, refectory, bell tower are located along the axis of the Church [4; 11, p. 172; 19, p. 346; 20]. Unfortunately, we do not quite understand what the presence of a refectory in the center of the temple means in the first case.

In the building of the Church, there are no proportions in volume. The Church is squat. The plane of the walls on the outside of the Church is divided by pilasters. They correspond to divisions within the internal space [20].

The temple is two-tiered. As noted by D.E. Tsyrenzhapova, the octagon on the quadrangle is covered by an 8-tray closed vault [19, p. 346].

The Church had two limits. One, the main, in honor of the Kazan icon of the mother of God (or in honor of the Nativity of the most Holy Theotokos), the second, respectively, in honor of the Holy Martyrs Kirik and Iulitta. The second limit was completed and illuminated in the same year, and the main limit, like the Church itself, was only nine years later. The Church was surrounded by a fence, eventually made of stone. To date, no remains of the fence have been preserved.

The Church had a bell tower. Today, it is not possible to observe this visually, but there are mentions of bells that were ordered in Moscow. There were four of them, the largest was weighing up to 500 kg. the Interior of the Church was painted by masters invited from Moscow. A chandelier brought from the same place is mentioned [15].

The most interesting element that adorns the Church is such an element of the entrance extension as columns. There are four of them. The building is decorated with semi-oval Windows along the perimeter.

The most complete description of the architectural elements of the Church is given by N.P. Kradin. He gives the number of decorative elements in the form of dragons and stone circles: 19 elements on the North side, 13 on the East side, 18 on the South side, and 14 on the West side. All named elements are arranged in two rows. One on the plinth, the second - in the middle of the walls, in special niches. There are 34 elements in the lower row and 30 in the upper row, respectively. The lower row of decorative elements is along the entire perimeter of the building, the upper one is along the part [11, p. 172]. In one newspaper publication, the total number of dragon statues embedded in the wall is 69, which is obviously not true [17].

In publications, there are sizes of dragons used in the Foundation of the temple: 76x25x25 sm [19, p. 346]. The location of dragons as a decorative element on the walls of the Church is not the same. So, on the West side there are only two statues, which are embedded in the wall on both sides of the entrance. These dragons are larger than the others. As it is assumed in the passport of the Church on the website of the Ministry of culture of the Trans-Baikal territory, these are dragons from the corners of the balustrade of the Konduy Palace [4].

The thickness of the walls of the Church is within one meter. At the time of construction, and at a much later time, this Church was the only stone and so unusual in its architectural elements. The nearest, perhaps – is the Church of the assumption of the blessed virgin Mary, currently in the village of Kalinino in the Nerchinsky district of the Trans-Baikal territory.

There is an answer to the question of how the diocesan authorities allowed coverage of such a Church. This happened only after the dragon statues were covered with clay [15]. For today, when visually examined, nothing reminds that they were covered up. This is understandable, more than two hundred years have passed. However, the Church in an abandoned state has been standing for a long time, on the walls from the outside you can see the remains of whitewash. There is reason to assume that they are from pre-Soviet times. That is, the Church with visible statues of dragons may have been working and the diocesan authorities turned a blind eye to this.

The significance of the Church for Eastern Transbaikalia can be accessed through visits by its diocesan bishops. In different years in the nineteenth century it was visited by a number of them. These Are The Most Reverend Nil (Isakovich), Eusebius (Orlinsky), Parthenius (Popov), And Miletius (Yakimov) [15].

There are some controversial points about the Church. In one of the publications we meet «... they Say, all of white granite» [20]. At the same time, other data suggest that up to three-quarters of the bricks from the ruins of the Konduy Palace were used in the construction, and partially with bricks, at the expense of the organizer M.S. Epova [15].

Since there are no materials for a serious study of the Church in the public domain, it is difficult to say how much granite was used in the construction of the Church. Probably, the round bases of columns and dragons were granite.

It is known that in the 1920s the priest, for some time, was a priest Luka Epov [6].

Without reference to sources, one of the publications mentions that the Church was closed in 1929 [20]. The bell tower was dismantled in 1930 or «in the 1930s» [19, p. 346]. There is evidence that for some time the Church premises were used as a warehouse, and after that it was abandoned [5]. In the newspaper publication mentioned above, it is noted that the Church has been abandoned since the 60s of the XX century [17].

The building of the Konduy Church is an object of cultural heritage of Federal significance, but this fact does not affect its condition in any way and, quite likely, will not affect it in the near future. It is an object of cultural heritage under the Decree of the President of the Russian Federation No. 176 of February 20, 1995 [18].

In the middle of the first decade of the XXI century, a statue of one of the dragons was removed from the wall of the Church. His fate is not known. There is a destruction of the brickwork of the walls from moisture.

It should be mentioned that stone sculptures of dragons are kept in the State Hermitage Museum in St. Petersburg and in the local history Museum of Nerchinsk in the Trans-Baikal territory.

4. Conclusion

For not even years but decades, there has been talk about the restoration of the monuments of Church architecture. During the visit to Konduya, the authors of these lines heard a story about the money collected for restoration more than once. In some cases, dishonest people, in others, sincere active public figures. However, to date, no work is underway. This does not even mean restoration work, but work on creating a project for restoration or examination of the state of the building to date. At the same time, it should be noted that in 2006 under the guidance of the famous scientist N.P. Kradin, a project for the restoration of the Church in the village of Konduy was created as part of the thesis. However, in the project, the Church is called Christmas [16]. It is obvious that we are talking about one object. The main thing here is that the project was made less than fifteen years ago and it is obvious that today, it can become the basis for a new project.

It is known that a Board of Trustees for the restoration of the Church was established in the Borzinsky district of the TRANS-Baikal territory. By the middle of the second decade of the XXI century, even the amount required for restoration was called-about 40 million rubles. It was noted that after the examination, the cost may increase. However, we have not received any information about the examination as of today [2]. Alexander Lesnyansky on his page in LiveJournal calls the amount of 4 million rubles [5]. It is obvious that the first amount seems more correct.

The Church as an object of cultural heritage was included in the program of the Ministry of culture of the Russian Federation. This was stated in January 2016 by the Minister of culture of the TRANS-Baikal territory V. K. Kolosov [14]. However, it is not clear what, at least, the interim result of the work is to date. In the summer of 2020, the authors of this article visited Konduy again and can state that the situation has not changed. The only thing that distinguishes the picture from that of a dozen years ago is that a small area around the temple is surrounded by a fence that does not allow the cattle of local residents to enter the building.

The authors of these lines wrote about the Church of the Nativity of the blessed virgin Mary and the Holy Martyrs Kirik and Iulitta earlier, but nothing has changed over the years. This publication again focuses on the problem of the importance of restoring the temple and preserving it as an unusual architectural monument, such as, perhaps, cannot be found anymore.

From a scientific and historical point of view, it is obvious that the history of the Konduy Church has not been studied enough. There are many «white spots», contradictions and inconsistencies. And in General, there is not much information about the history of its construction and existence. Archival research is important here.

References

[1] Barinov A O 2006 Church of the name of the Holy Mother and Holy Martyrs of Kirik and Iulitta (Novosibirsk: Nauka Press) *Encyclopedia of Transbaikalia* **IV** p 103

[2] *In Transbaikalia, restore the Temple built by the Cossacks from the remains of the residence of Genghis Khan's nephew* Church of the assumption of the virgin. Demyanovo Manor. G. Klin. URL: http://klin-demianovo.ru/http:/klin-demianovo.ru/novosti/82647/v-zabaykale-vosstanovyat-hram-postroennyiy-kazakami-iz-ostatkov-rezidentsii-plemyannika-chingishana/ (14.01.2020)

[3] *Kazan Temple, Borzinsky district, Konduy village* Donskoy monastery. Moscow Patriarchate. URL: http://russian-church.ru/viewpage.php?cat=Zabajkalskij&page=43 (14.01.2020).

[4] *Konduy Church* Ministry of culture of Zabaikalsky Krai. URL: http://минкультура.забайкальскийкрай.рф/okn/okn_12.html (10.01.2020)

[5] Konduy Temple // LIVEJOURNAL. URL: http://lesnyanskiy.livejournal.com/219730.html (10.01.2020)

[6] Konduy Kazan Temple // TREE. Open Orthodox encyclopedia. URL: https://drevo-info.ru/articles/25316.html (accessed: 14.01.2020)

[7] Konduy Temple, July 24, 2004 // Epov. Our kind. URL: http://www.epov.ru/index.php?section=places (14.01.2020)

[8] Konstantinov A V 2006 Konduy town (Novosibirsk: Nauka Press) *Encyclopedia Of Transbaikalia.* Chita region chapter **III И-Р** p 113

[9] Konstantinov A V 2009 Konduy town (Novosibirsk: Nauka Press) *Small encyclopedia of Transbaikalia. Culture chapter* I p 377

[10] Krandin N 2002 Architectural monuments of Konduy *Archaeology and cultural anthropology of the Far East: collection of reports* (Vladivostok: DVO RAN Press) p. 233-244.

[11] Krandin N P 2011 Konduy Church *Small encyclopedia of Transbaikalia Archaeology* (Novosibirsk: Nauka Press) pp 172-174

[12] Krandin N P Tsyrenzhapova D E 2016 Church of the name of the Holy Mother and Holy Martyrs of Kirik and Iulitta // Small encyclopedia of Transbaikalia. Architecture and construction (Novosibirsk: Nauka Press) pp 442-443

[13] Kuznetsov A K 1925 The ruins of the Konduy town and its surroundings (Vladivostok: Far Eastern Siberian joint-Stock company «Knizhnoe Delo» Press) p 68

[14] Lyashenko V Residents of the village of Konduy can not restore the old Church without the approval of experts // БЕЗФОРМАТА.RU. URL: http://chita.bezformata.ru/listnews/konduj-ne-mogut-otrestavrirovat-starinnuyu/42975088/ (accessed 10.01.2020)

[15] Essays on the history of the village Conduy (since Ancient times till 1914) // APAVI. Our kind. URL: http://www.epov.ru/?section=memoirs&subsection=history-konduy (12.01.2020).

[16] Project of restoration of the Nativity Church in the village of Konduy, Chita region-Diploma: Grebenkin A.S. // Kradin Nikolai Petrovich. URL: http://pnu.edu.ru/ru/library/i-resourses/bibliography/kradin/ (14.01.2020)

[17] Trukhina O October 26, 2010 Legends of Konduy // Gudok newspaper

[18] Decree of the President of the Russian Federation No. 176 of February 20, 1995 «On approval of the List of objects of historical and cultural heritage of Federal (all-Russian) significance» // Garant Information and legal support. URL: http://base.garant.ru/10104140/ (14.01.2020)

[19] Tsyrenzhapova D E 2009 Church in the name of the Holy Mother of God *Small encyclopedia of Transbaikalia*. Culture (Novosibirsk: Nauka Press) Chapter **2** p. 346

[20] Yurinskaya D And the Church does not want to die *Chita and Krasnokamensk diocese*. URL: http://www.chita-eparhia.ru/enapx/new_eparh/ned/kazan/ (14.01.2020)

IOP Conf. Series: Materials Science and Engineering **880** (2020) 012060 doi:10.1088/1757-899X/880/1/012060

The use of standard series plans in designing unique buildings: the case of V. A. Pavlov's residential house in Irkutsk

I Druzhinina[1]

[1]Department of architectural design, Irkutsk National Research Technical University, 83 Lermontov Street, Irkutsk 664074, Russia

E-mail: irk.allegro.id@gmail.com

Abstract. The article analyzes the documents of the Ministry of Construction of Russia, which contain the methods for calculating the index of urban environment quality. It reveals the lack of the housing quality indicator in the sum of the indicators that compose this index. Rethinking of the experience of the standardized construction can be one of the ways to restore the balance of the quality of housing and public spaces. Using the example of V. A. Pavlov's residential buildings in Irkutsk, the article shows a possibility to design high-quality modern dwelling, taking the basis of space-planning decisions for standard series as a starting point.

In the framework of implementation of the national project "Dwelling and Urban Environment", the Ministry of Construction of Russia has worked out the methods for calculating the index of urban environment quality. 36 indicators used for measuring living environment "make it possible to provide a complex assessment of the condition of urban environment" [1]. The methods embrace a wide range of parameters of the urban environment; however, the document lacks such crucial criteria of socially important assessment of residential areas as design decisions and aesthetic attractiveness of residential buildings. To evaluate the quality of housing, the Federal Project "Formation of the Comfortable Urban Environment" considers six criteria: safety, comfort, environmental friendliness, identity and diversity, modernity of the environment and management efficiency of authorities, but also pays no attention to the aspects of space-planning decisions for housing. The lack of these indicators significantly decreases the accuracy of the calculated index of urban environment quality. Taking into account these methods, project designers put a priority on public spaces and urban land improvement, while the quality of housing takes second place. A remote cause of today's increased attention toward the aesthetics of public spaces can also be the fact that for a long time the standardized house-building, which managed to quickly provide housing for citizens during the industrialization period in the USSR [2], produced public spaces with poor appearance. The mass construction of standardized series in Oktyabrsky district and microdistricts of Irkutsk such as Yubileiny, Pervomaisky, Universitetsky, Novo-Lenino and others confirm that. The issue of an individual's comfort in standardized and unitized architecture has been debated for a long time and is still topical nowadays. Russian and foreign practices in adaptation of standardized housing are widely studied. For example, the authors E. A. Repina and D. I. Rybakova pose questions about the role of the individual: "... can we say that standardized architecture is beyond the individual?" [3, 4]. S. A. Makotina's article features social dwelling through the example of small-family hostels in Irkutsk and evaluates the historical and social significance of the mass standardized housing of the 1970-80s. "The commune model of the 1980s could have transformed into a Russian co-housing model" [5]. The evolution of town-planning patterns of Irkutsk's microdistrics and the principles of their development in the 1960-70s were studied by P. Dorofeev in his master's thesis [6]. Considering the above, the residents' social and psychological comfort and the investment attractiveness of the urban area depend not only on the favourable public spaces, but, first of all, on the quality of dwelling. To achieve the greatest possible comfort of the urban environment, it is necessary to balance these two characteristics.

Half a century of experience in designing and building series of standardized housing in Irkutsk has formed a rich database of architectural and planning concepts for residential buildings. This is not only of academic interest as a resource for studying the Soviet architecture, but can also be a starting point for present-day designers. Standard series can be the basis for creation of unique architecture. In this light it is noteworthy to consider several implementations of this task in the projects of a famous Irkutsk architect V. A. Pavlov, such as residential buildings on Gorky, Kievskaya and Lermontov streets in Irkutsk.

One of the first projects worked out by Vladimir Azarievich Pavlov in Irkutsk was an individual project for a five-storey residential building of Irgiredmet and the CPSU District Committee at 1 Gorky Street (Figure 1). The building consists of two brick sections turned 90 degrees in relation to one another with a "connection" between them. A medium-rise residential building included 20 apartments, 10 in each section. Since the order was fulfilled for the officers of the CPSU District Committee, there are two apartments with improved layout on each floor. The area of these apartments exceeds the standard that existed in that period. The storey height is 3 m, against the standard of 2.7 m [7]. The building was designed at Irkutskgrazhdanproject in 1964, built in 1965 and put into operation in 1966. The project was so successful that it was several times repeated for the party "elite" in different versions: at 9 Kievskaya St. (Figure 2) and 2-2a Gorky St. (built in 1972) (Figure 3), in Akademgorodok at 295 and 295a Lermontov St. (built in 1977) (Figure 4) and at 295b Lermontov St. (built in 1982). It should be noted that, despite maintaining common proportions and parameters, every subsequent building had an individual appearance harmoniously integrated in the urban surroundings. For example, architecture of the buildings at 2-2a Gorky St. was inspired by a trip to Japan. The almost continuous glass ribbon of the façades with outstanding elements of the balcony enclosures along the perimeter of the building, as well as bearing beams and pedestals discernible through the piers in combination with the columns contribute Japanese features to the buildings' appearance, while at the same time encompassing the five principles of Le Corbusier.

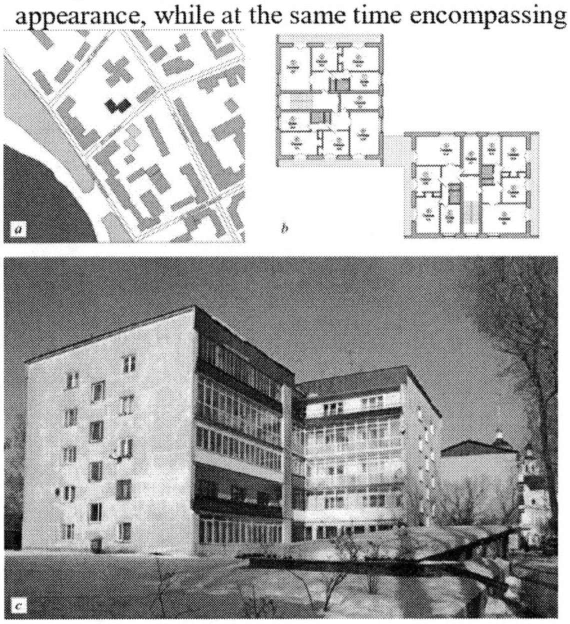

Figure 1. Residential building at 1 Gorky St.: a) layout, b) standard floor plan, c) view from Gorky St.

Figure 2. Residential building at 9 Kievskaya St.: a) layout, b) photo of a facade fragment, c) view from Kievskaya St.

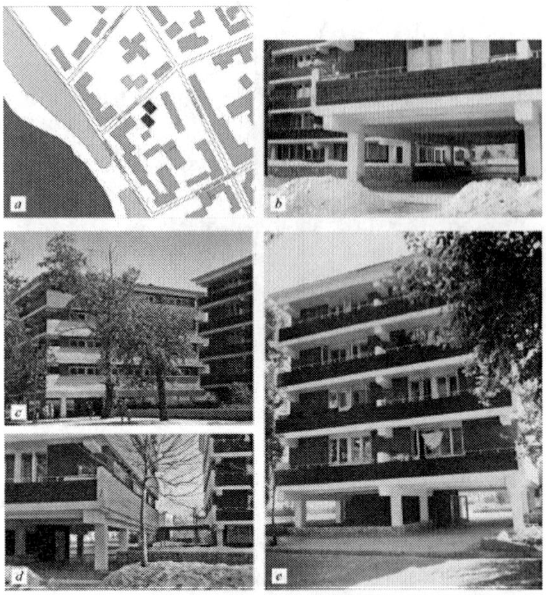

Figure 3. Residential building at 2–2a Gorky St.: a) layout, b), d), e) photo of a facade fragment, c) view from Gorky St.

Figure 4. Residential buildings at 295, 295a, 295b Lermontov St.: a) layout, b), c) photo of a facade fragment, d) view of a block of houses

V.A. Pavlov's plans of the above-mentioned residential buildings have something in common with the plans of standard row sections of series II-03 (Figure 5), which were widely used in construction of 3-5-storey brick residential buildings and series 111–135–60/1.2 (Figure 6). However, they also differ significantly from each other. The arrangement of bathrooms in the spacing adjoining the stair flight on two sides made it possible to unleash a light front along the perimeter of the building. It significantly increased the living comfort in regards to the provision of natural light from three cardinal directions and standard insolation in all dwelling units. This method allowed for unlimited orientation of residential buildings both in the town-planning context and in the volumetric and spatial composition. In its turn, it offered opportunities to use a wide range of structural solutions and finishing materials. The aspect ratio of dwelling units provided a higher variability of furniture arrangement. Besides, the master made provisions in his plan for the presence of utility spaces, which are very important for the inhabitants of Siberia.

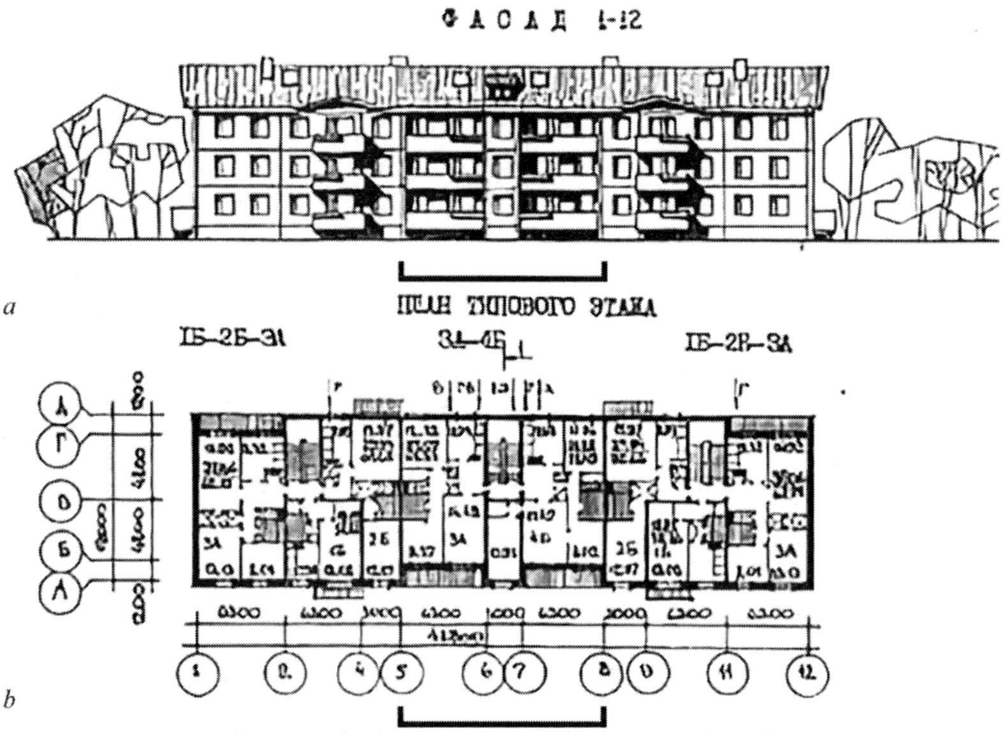

a

b

Figure 5. A diagram of a facade and a standard row section of series II-03

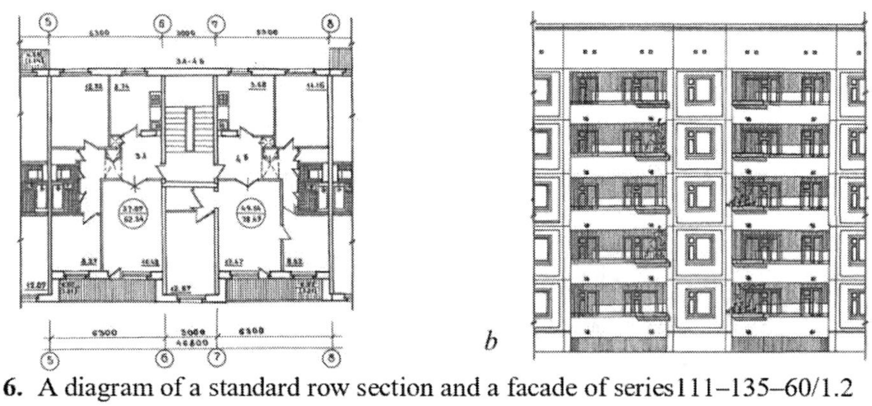

a *b*

Figure 6. A diagram of a standard row section and a facade of series111–135–60/1.2

Regardless of the fact that "standardized architecture is always beyond the individual! It becomes typical (and even one-type), standardized and unitized only because it is formed in accordance with a standardized and unitized prototype of a customer" [4], it is a sort of information system (a database) of space-planning decisions that serves for rethinking and formation of new architecture, which is, on the one hand, no stranger to mathematically precise parameters and rational use of resources, and on the other hand, oriented toward the individual.

In the Soviet period, the prevalence of standardization in architecture and the limitation of opportunities caused progressive-minded architects to search for unique decisions, while using standard elements. In that way, freely navigating in his architectural surroundings and basing on his deep knowledge of related professions (structural arrangement, utility systems etc.) and his innovative architectural and town-planning experience, V. A. Pavlov proposed forward-looking solutions, many of which are still of great current interest.

IOP Conf. Series: Materials Science and Engineering **880** (2020) 012060 doi:10.1088/1757-899X/880/1/012060

References

[1] The Government Of The Russian Federation march 23, 2019 **No** 510-r Method of formation of the quality index of urban environment

[2] Meerovich MG 2017 October 1917 in the history of architecture and fine art of Siberia *Selected theses of the All-Russian Scientific Conf., timed to the centenary of the October Revolution of 1917* Ed Garkusha DD, Orlova EYu, Dukhanov SS Published: Novosibirsk State University of Architecture, Design and Arts (Novosibirsk) pp 88-93

[3] Repina EA, Rybakova DI Human and neighborhood: Problems 2016 *Innovative Project.* **Vol 1 No2** 70-1 DOI: 10.17673/IP.2016.1.02.9

[4] Meerovich MG 2016 Man and neighbourhood: History of Khrushchev reforms in the USSR *Innovative project* **Vol 1 No4** pp 8-12

[5] Makotina SA 2018 Morphology of household hostels of the 70-80th years in Irkutsk *Actual problems of architecture, urban planning and design: Theory, practice, education* Materials of the Int. Scientific Conf. Ed by N.V. Ivanov Published: Volgograd State Technical University (Volgograd) pp 65-9

[6] Dorofeev P 2014 Neighbourhoods structures of the 1960s and 1970s in Irkutsk *project baikal* **No 39–40** pp 230–51

[7] Buh VF, Grigorieva EI 2010 *Vladimir Pavlov. Architecture of Soviet Modernism. Masters* Yekaterinburg TATLIN Publishers p134

[8] Grigorieva E, Tkacheva M 2013 Public spaces and their systems in Irkutsk: Destruction and creation. Dialog between the architect and the culturologist during a walk around the city *project baikal* **No 35** pp 52–6

[9] Kosolapova A R, Antonov A V 2019 20-th session International Baikal Winter University of urban planning: updated methodology of research and design sessions *Proceedings of Universities. Investment. Construction. Real estate* **9(1)** pp 166–173

[10] Bannoud Ghaleb 2019 Features of the historic urban living environment in the city of Aleppo *Proceedings of Universities. Investment. Construction. Real estate* 9(1) pp 158–165

[11] Makotina S A, Daggers T, Karepova A I 2019 Population mobility in a contemporary near-water city (on the example of the cities of Amsterdam and Irkutsk) *Proceedings of Universities. Investment. Construction. Real estate* **9(4)** pp 842–857

IOP Conf. Series: Materials Science and Engineering **880** (2020) 012061 doi:10.1088/1757-899X/880/1/012061

City identity: mechanisms of formation of design of the urbanized territories

E V Hokhrin[1] and S A Smolkov[1]

[1] Institute of architecture, construction and design, Irkutsk national research technical university, 83 Lermontov St., Irkutsk, 664074, Russia

E-mail: hohrin@gmail.com

Abstract. In article theoretical basics of the mechanism of formation of city identity and a condition of realization of ways of broadcast of specific lexicon of shaping and cultural tradition are covered. Due attention is paid to problems of preservation and development of city identity and also the solution of problems of formation of the modern urbanized environment. The methodology of a research of mechanisms of formation of design of unique units of an original image of the historical environment is formed on the basis of results of the analysis of specifics of activity and types of the spatial and functional organization of the downtown as style carrier. At the same time empirical researches of different types of identity are characteristic, owing to specifics of the territory and the multinational population of the country, transformations˘ society in the conditions of which the priority directions a research˘ are national and ethnic identity are characteristic of the period socio-political. Other local types of spatial identity remain almost poorly studied though questions of specifics and development of the cities and the territory˘ are key in a modern urbanism. The solution of the listed tasks allows to be objecting the choice of the generated options that is especially necessary in a designing process and design of the urbanized territories. The aspects of city identity of the historical environment of the cultural and public center of the Siberian settlement considered by us allow to perform the described operations over various types of dimensional sizes, from formation of a range of parameters to their comparative assessment and to receive as a result indicators of compliance of the studied structure of composition to the set standard. Results of a research confirmed the existing theoretical provisions, having proved thereby efficiency of a research.

1. Introduction

Change of a role of the cities in a global context, the involvement of settlements into the competition, continuous transformation of sociocultural space, long-distance migrations cause the need for judgment of a phenomenon of city identity of residents. For increase in level of objectivity of assessment by the government criteria of quality of the urbanized environment and the mechanism of formation of territorial identity of the Siberian city stated in "A technique of assessment of quality of the urban environment of accommodation" [9] are developed. In the same place the concept - "index of quality of the urban environment" on which the rating of the city settlement among the cities of Russia is established is entered.

In recent years the subject city a research˘, for a long time theoretically both methodologically reasonable and empirically approved in the West entered the Russian research space. First of all, it is necessary to decide on concepts, to study essence of a phenomenon of city identity and also factors of its design and designing. Having analyzed materials according to the theory of identity and the approaches to definition of the concept "city identity" which are available in literature, it is necessary to consider the next moments: [6]

– the city identity is a product of a collective mentation activity, is connected with inclusiveness in long-term stable social groups and defines emotional fullness of self-determination of the individual in relation to such group in general, to a group name, to "" and "others" and also group values, social norms, stereotypes, behavioral patterns, etc.

– the city identity is a sociocultural construct which is formed as a result of socialization and adaptation in the concrete territorial settlement and is determined by assimilation and reproduction of cultural symbols, accepted standards, traditions, separate patterns and in general - lifestyle of residents of this settlement.

– the city identity is result of identification of the person with city community as the big social group different from a number of other categories of the population which are marked out on the basis of territorial accessory.

The city identity is a component of social identification of the personality, the sociocultural construct formed as a result of self-identification of the person with concrete city community and determined by assimilation and reproduction of the symbolical capital of the city, the sociocultural norms and lifestyle uniting residents.

Classification of factors which influence process of formation of city identity:

a) stable structural factors: location, climate and history of the city;

b) the quantitative and qualitative factors changed in the long term: size and population of the city, appearance of the city, welfare of inhabitants, cultural traditions of local community;

c) symbolical factors – urban symbolism, political climate, cultural codes of behavior of inhabitants, milestone events, sign persons, fashion for separate goods and services, the nature of communication in community. [7]

2. Materials and Methods

General provisions of the organization of process of formation of city identity set area of a concept about logic of design and the mechanism of formation of the city identity which is the cornerstone of development of technology of strategy of creative search. It is a difficult procedure which not only precedes, but also accompanies all process because the character and the sequence of steps are in many respects predetermined by a ratio between the results and new contradictions received already during work which will become a basis of future developments in the sphere of an ideation and conceptualization of an image of the city and ways of its broadcasting in daily practice. Nevertheless, the main work on designing of logic of formation of city identity needs to be done at the very beginning of process, relying on the principles of modeling of an ideal image and the available ideas of tools of a research, design and designing which will provide its achievement. Shevelyov I.P. described it so: "We look for geometrical structure which at a boundless variety of probabilistic configurations is of the previous existence which is selected by history and creation of set of initial components. It is useful for architect, artist, engineer to know rules of the "correct" partitioning of space. These rules are expressed in the system of orthogonal projections, a proportion and a metrics" [1].

In the organization of a designing process the major place for ensuring optimum operation of the design mechanism it is necessary to consider a designing process as the phenomenon realizing itself at the same time in two plans: plan of the organization of design activity and plan of the organization of a subject and scope of design activity. This very important remark which is essentially influencing an essence and forms of the organization both project work, and subject to the design and actually end result proving and in change of forms of design thinking and in the relation to a subject of city design what will be told [3] below about. Besides, as a rule, allocate three stages of designing of logic of design: production, actually design and design and implementation [2] which are described in "A technique of assessment of quality of the urban environment of accommodation" allowing to define quality of life in the urbanized territories. Let's consider further basic provisions of the document.

IOP Conf. Series: Materials Science and Engineering **880** (2020) 012061 doi:10.1088/1757-899X/880/1/012061

The index of quality — the tool which will help heads of the city administrations, to the research and design organizations to reveal key problems of the urban environment and to begin work on their elimination. The index is approved by the order of the Ministry of Construction, Housing and Utilities of the Russian Federation and applied to all 1112 cities of Russia and consists of two blocks:

- the quality of the environment on 30 indicators is estimated, each of which is responsible for defined from 6 types of city space and displays degree of quality of the environment by one of 5 criteria.

- the efficiency of use of the subsidy allocated within implementation of the priority project "Formation of the Urban Environment" on 11 indicators is estimated.

The main objectives established in "A technique of assessment of quality of the urban environment of accommodation":

1. Creation of the Technique on the basis of Rating for the purpose of providing objective complex assessment of municipal units by the criteria defining the level of development of all spheres of activity of the city.

2. Creation of the effective applied tool:

2.1. for determination of negative and positive aspects in development of municipal units;

2.2. for decision-making on strengthening of competitive advantages and decrease in negative factors in development;

2.3. for leveling of disproportions in territorial structure of municipal units on a number of socio-economic factors.

3. Definition of poles of growth in the territory of the Russian Federation - in terms of the conditions favorable for life of the population, business, investments into the real estate, etc.

4. The description of the existing hierarchical structure of the cities and their typology.

5. The organization of an integrated system of the analysis of socio-economic indexes of the cities which does possible their comparison and also carrying out analytical researches on identification of trends of social and economic development.

As main objectives are formulated in theoretic-applied aspect, at their development semantic and empirical interpretation of initial concepts according to the allocated scientific and practical perspective, the logical analysis of working hypotheses and a possibility of applied application is carried out. [9]

The main objectives established in "A technique of assessment of quality of the urban environment of accommodation":

1. Carrying out the analysis of the existing domestic and foreign techniques and the practical of comparison of subjects of administrative-territorial division of macro regions and countries.

2. Carrying out a comprehensive investigation and determination of criteria and basic indicators of assessment of the cities on means of the selected theoretical and methodological disciplines connected with assessment of qualitative and quantitative characteristics of the studied objects.

3. Creation of the identical methodological device for drawing up rating, being based on the best domestic and foreign practices, using own developments.

4. Development of a technique of complex comparative rating of the cities on the basis of application and calculation of a system of economic indicators.

6. Formation of a system of the economic indicators which are comprehensively characterizing functioning of the cities.

7. Calculations of integrated indicators, indexes and groups of indexes according to the developed methodical device.

8. Allocation of typical groups (typology) of objects on a number of signs, the indication of the general trends and differences in their development, allocation of characteristic individual distinctions, the analysis of the received results.

9. Creation of the mechanism of a monitoring system of economic and social situation of the studied objects.

10. Creation of demonstration analytical material by results of the done work.

11. Granting materials of researches to all interested persons.

Evaluation criteria:

"Safety" is responsible as for determination of the existing level of danger of certain spaces, and for assessment of such parameters of the environment which can potentially promote infliction of harm of life and human health.

"Comfort" helps to estimate convenience of the urban environment and also its visual, acoustic and climatic characteristics.

"Environmental friendliness" defines in what measure use of the city corresponds to the principle of preservation and improvement of the environment and also estimates a condition of separate elements of the urban environment.

"The identity and a variety" are responsible for assessment of an originality and recognition of urban areas and also defines variability spatial the decision˘and a functional variety of the urban environment.

"The present of the environment" measures the parameters describing the city in terms of opportunities which it gives to inhabitants. Indicators help to define whether the city uses outdated model of production of the environment or new model in which Wednesday — one of basic values of development of the city.

"Spaces":

- Housing and adjacent spaces
- Houses and house adjoining territories.
- The planted trees and shrubs and water spaces

"Green zones" make the territories of the city including parks, gardens, squares, forests and also the vacant territories adjacent to city reservoirs.

"Street infrastructure" - city streets of different type, including territories from a facade to a building facade, except for a roadbed. Street infrastructure is estimated separately and regardless of arrangement in this or that space of the city.

"Social and leisure infrastructure and adjacent spaces"

The territories adjoining educational and educational (schools, higher education institutions, day nursery), medical and public health (hospitals, improving the camp, dispensaries) to institutions, to objects of sport and a recreation (recreation facilities, stadiums, sports grounds and complexes), to cultural institutions (movie theaters, the museums, exhibition platforms) and also to subjects to religious appointment.

"Public and business infrastructure and adjacent spaces"

The territories adjoining administrative, business, shopping facilities and also places of public catering, objects of service and services.

"City space"

All territory of the city in administrative borders of the settlement. Introduction of this space is necessary for assessment of the city characteristics of quality of the environment which are not relating directly to one of types of spaces.

3. Results

The area and subject of design and research of city identity are not identical to the concept "objective reality" or the concept "object field of design". Communications, the relations, borders, a framework of consideration and property of a real object which are included in studying process act as area of a research. A design subject in this case are certain parties' structure of properties, qualities and the relations which exists irrespective of learning, but has reflected them, serves as the concrete field of design activity.

Subject to design of city identity in the conditions of the historical environment of the city is a certain phenomenon, some phenomenon which exists irrespective of the subject of design and to which the attention of the designer is drawn. The concept of a subject of design of city identity is even more concrete according to the contents: in an object of research that property or the relation in a designed project which in this case is subject to deep special studying and a research for the purpose of identification of the contradictions and problems forming it is fixed. In the same object various parties, aspects and subjects of design can be allocated.

The purposes and problems of formation of city identity are a reasonable idea of the general end or intermediate results of predesign researches within design process. In our case - justification and development of a methodical system of formation of the comfortable urban environment meeting the requirements and requirements of today. An important and necessary design stage is designing and a specification of a common goal in the system of design tasks. [1]

The design task represents a link, a step, a stage of achievement of the goal. On the other hand, the task is the purpose of transformation of a concrete situation or, otherwise, a situation, i.e. the multiple factor set of specific conditions demanding the transformation for achievement of certain meanings. The task always contains the known information determined by living conditions of a situation and the unknown component calculated on commission of specific actions, efforts for advance to the purpose for permission of the put problem. The task realized taking into account the current conditions synthesizes the substantial, motivational and incentive and operational parties of design activity and is its peculiar "focus". It is very important to distinguish priority from a considerable number of the tasks which are subject to the decision. It is recommended to allocate in general no more than 6 tasks of three main groups.

The first of the main group of tasks - historical and diagnostic - is connected with studying of history of emergence and the current state of a problem, definition or specification of concepts, general scientific, professional and cultural and historical bases of design;

the second - theoretic-modeling - with disclosure of structure, essence of the studied object, factors of its transformation, a form of structure and functions of city identity and ways of its theoretical transformation and creation of model of development;

the third – practical and converting - with development and use of methods, receptions, means of the rational organization of research and design process, their alleged transformations with development of practical recommendations and theoretical generalizations. Other, more private tasks carry most often as subtasks to the main. [1]

Idea, plan and hypothesis as theoretical kernel of formation of city identity.

Statement of tasks and ways of their decision prepares to the designer "break" in the field of formation of city identity, and all subsequent logical stages will present to a step of the movement to tool check of the validity and the embodiment of this complex of the ideas. As a form of such anticipation of results of operation of the mechanism of formation of design of the urbanized territories the design hypothesis - the reasonable assumption of how, in what way, at the expense of what it is possible to receive the required result acts. In a way of the internal organization of the formulated hypothesis of design the real movement of thinking of the designer to new, deeper generalizations on the basis of forecasting and creative anticipation is shown. Not only careful studying of a condition of business, professional competence, but also implementation at least is necessary for a part of a predesign research for promotion of a hypothesis on the basis of polls, questionnaires, testing and other methods used in architecture and design.

The births procedure of the idea, its subsequent realization in a plan is submitted important for formation of the mechanism of city identity, and then its expansion in a hypothesis and represents the most difficult, creative element of design and design, its creative kernel making a basis of an ideation and conceptualization. And at this moment there is a need for certain recommendations about implementation of this stage, knowledge of operation of mechanisms, techniques and technologies of creative activity. It is supposed that creative process is deeply individual, peculiar and unique in the uniqueness. Also, popular belief that it's kind of does not keep within in any rules and instructions and in this sense, it is non-technological because creativity in a sense - an antipode of effective activities for the set algorithm with respect for strict technology. But it is only widespread wrong opinion of certain and far from creativity, in a sense, experts trying to justify own incompetence and small efficiency. It is characteristic that around the world during deep transformations in education two opposite trends - thirst for "rationing" and technologizing of any type of activity and at the same time - aspiration to "free thinking" in creativity accrue. However we will remember definition by Engels "freedoms as conscious need" where the last does not exclude widespread introduction of successful receptions, the approved techniques and effective procedures in the processes preceding and developing operation of mechanisms of formation of city identity in the conditions of broadcast of the best examples of international experience in city design practice. Stages of predesign researches and graphic analytics in design and research search which we considered in respect of preparation, preliminary design and planning so far should be realized practically now.

Predesign researches and the analysis - studying of a state and the possibilities of development considered processes and the phenomena precede all process of formation of city identity, accompanies it, but especially allocate an initial stage (at the beginning of the research) and the final analysis necessary for orientation of search and determination of its efficiency. As a matter of fact, predesign researches of quantitative parameters of the studied objects are under construction by the standard technique on the basis of previously developed criteria and indicators by means of known (including standardized) techniques and also received on the basis of designing of new technologies.

It is worth to remember that each design stage comprises two plans, two aspects: practical and research. At each stage the effect and at the same time - approach to clarification of design and research tasks, deepening in a subject of search and specification of design and research and constructive methods is important practical (training, bringing up, consolidating). [4]

4. Discussion

The image of the city is only one of translators of symbolical information on the city which forms city identity. Understanding of by the citizen and understanding difference˘the settlement from another, to formation of pride for the territory of accommodation are promoted as well other translators which participate in process of city identification: local myths and legends, ceremonial traditions, components of a cultural landscape (monuments, etc.), holidays and festivals, city signs and symbols. All of them serve in order that represent and keep city uniqueness which is a kernel of city identity. [5]

Also, in our opinion, various social and educational projects, cultural events, etc. belong here. It is connected with the fact that formation of city identity provides search of authenticity of the place, its originality and unique originality thanks to which there is a feeling of participation with the city, its communities and history. Thus, it is admissible to use determination of city identity as tool of symbolical and semantic "construct" of an image of the city which is defined by perception of its inhabitants and also their style of behavior in relation to the urban environment.

Assessment and interpretation of results allow to formulate preliminary theoretical and practical conclusions. These conclusions need to be comprehended in the general system of already known theoretical provisions and practical approaches, to check and specify in wider context of interpretation of experience - approbation and visualization of results and projects, to criticism and a schematization of the got experience, a problematization and programming of design thinking, its ontological analysis and the subsequent rationing of activities for formation of city identity.

The general scheme of the mechanism of formation of city identity systemically consists of 4 main units of process (see below) and 2 control units and methodological work - "The organization of a reflection and methodological providing" and "The organization of management and cooperation of activity" [2]. In turn, each of sides of process is divided into 4 stages with the logic and specifics providing normal functioning of a system of formation of city identity:

1. CRITICISM AND SCHEMATIZATION

1.1. Self-determination in a situation of target uncertainty. Formulation of a subject, purposes and problems of formation of city identity.

1.2. Collecting and processing of initial material. Principles of systematization and classification of data.

1.3. Analysis of a situation. Formulation of contradictions and problems in an initial situation.

1.4. Search of regularities of formation of city identity in a design situation. Graphical analysis.

2. PROBLEMATIZATION AND PROGRAMMING

2.1. Development of requirements to project form and content. Modeling of formal characteristics of formation of objects of city identity.

2.2. Analysis of proportional regularities of a design situation. Formulation of a composite key and schemes of a proportional system of objects of the urban environment.

2.3. Ideation and conceptualization. The principles of shaping in the conditions of formation of city identity.

2.4. Formulation of regularities of shaping. Style and factors formation of style of urban environment.

3. DESIGN AND DESIGNING

3.1. The project description in terms of a vernacular. Thinking in projections.

3.2. Alternative study of formation of city identity. Explanation, deepening and specification of sense and contents of the project.

3.3. Graphic interpretation of maintenance of model of transformation of the urban environment. Formation of the principle of "feedback"

3.4. Realization of the principles of shaping in subject to formation of city identity. Sustainable development. Dynamics, statics and reversibility of processes.

4. ONTOLOGIC ANALYSIS AND RATIONING

4.1. Detailed study of separate parts of the project of formation of city identity. Expansion of a form and subject of idea of subject to design

4.2. Development of formal language. Logic and philosophy of the description of results of process of formation of city identity.

4.3. Project presentation. Principles of the NLP and formalization of promotion of the project of formation of city identity.

4.4. Registration of materials of the project. Assessment, reflection and judgment of results of modeling and project of formation of city identity.

5. Conclusion

Thus, it is necessary to assume that for formation of city identity the projects directed to identification and judgment of unique mechanisms of formation of design of the urbanized territories, updating of significant "places of force" and "collective" memory on the designed sites can be used. Projects also should be carried to factors of active influence as they are capable to change the attitude to the city and also to symbolical factors as such projects will promote establishment or strengthening of symbolical communication between the citizen and the city. [1]

Therefore, projects of formation of city identity besides achievement of own purposes it is capable to carry out also the indirect purposes, i.e. to have more or less strong impact on identification processes. The difficult education comprising the subsystems accumulating at each other acts as subject to design and formation of city identity: society, culture and urban environment. And in this case the project of creation of the mechanism of city identity acts as a way of preservation or reconstruction social the phenomenon˘and the cultural phenomena corresponding (both quantitatively, and it is qualitative, substantial) to the developed cultural norms [2]. Design of the urban environment is very many-sided, but the purpose of the project of formation of city identity will be nevertheless growth of level and expansion of a zone of comfort for the personality that can directly affect city identity. His habitat that there was no certain disorder in vital meanings that the place of life was associated with the smallest number negative experience˘that Wednesday promoted realization of personal resources [3] has to be pleasant to the person.

It should be noted that for formation of city identity those projects of creation of city identity which are directed to updating and maintenance of meanings of city cultural and historical heritage as in this context the identification mechanism will work thanks to memory as the instrument of maintenance of cultural identity and development of a collective mention' activity will be of particular importance.

IOP Conf. Series: Materials Science and Engineering **880** (2020) 012061 doi:10.1088/1757-899X/880/1/012061

References

[1] Hokhrin E V, Smolkov S A and Hokhryakov A A 2018 *Formation of urban style of Irkutsk*: (Irkutsk: Publishing house of the Irkutsk national research technical university) p 90

[2] Hokhrin E V, Smolkov S A and Hokhryakov A A 2018 *Design of the urban environment on the example of Irkutsk*. Irkutsk: Publishing house of the Irkutsk national research technical university p 124

[3] Schedrovitsky G P 1999 Typology of situations of carrying out researches. 1986 *Programming of research and development*. From G.P. Schedrovitsky's archive **1**

[4] Potapova A V and Engel B 2012 The prospects of development of the historic center of Irkutsk taking into account the German experience of Dresden **1 (2)** pp 169-178 DOI: 10.21285/2227-2917- 2012-1-169–178.

[5] Engel B and Potapova A V 2013 The heritage is in danger – the historic center of Irkutsk *News of higher education institutions. Investments. Construction. Real estate* **1 (4)** pp 196-205. DOI:10.21285/2227-2917-2013-1-196-205.

[6] Dyagileva N S, Zhuravleva L A 2012 City identity: concept, structure, bases *City Sociology* **1** pp 46-61

[7] Dyagileva N S 2013 Theoretical aspects of city identity *Branding of the small and average cities of Russia: experience, problems, prospects pp* 54-59

[8] *The master plan of the city of Irkutsk approved by the decision of the Duma of the city of Irkutsk* of 28.06.2007 No. 004-20-390583/7;

[9] *The local standards of town-planning design of Irkutsk approved by the resolution of city admin istration of Irkutsk* of 15.07.2011 No. 031-06-1287/11

IOP Conf. Series: Materials Science and Engineering **880** (2020) 012062 doi:10.1088/1757-899X/880/1/012062

Regeneration of architectural style of the historical environment of the cultural and public downtown

E V Hokhrin [1] and S A Smolkov [2]

[1] Institute of architecture, construction and design, Irkutsk national research technical university, 83 Lermontov St., Irkutsk, 664074, Russian
[2] Institute of architecture, construction and design, Irkutsk national research technical university, 83 Lermontov St., Irkutsk, 664074, Russian

E-mail: hohrin@gmail.com

Abstract. Comprehensive consideration of problems of preservation and development of historical appearance of the city and also questions of the solution of problems of formation of the modern urban environment makes a basis of sense and contents of article. The methodology of a research of a phenomenon of architectural style in the context of the historical environment of the center of the Siberian city is formed on the basis of the analysis and criticism of specific forms of activity and functioning of city community, and ways of its organization in the form of various type of public actions, actions, movements and associations. Researches are the cornerstone various forms of ideas of the internal organization and structure of external relations, understanding and designing of mechanisms of development of own types of the activity and Wednesday providing normal development and regeneration of architectural style. The main stages of formation of procedures of creation of urban style which develops in several plans are stated: organizational and activity and subject and thematic that is provided with between structural communications in the course of cooperation with the subject circle, a way of life of people and their culture. The solution of the listed tasks allows to be objecting the choice of options that is especially necessary in a designing process and updates of the urban environment. The methods of regeneration of architectural style of the historical environment of the cultural and public downtown considered by us allow to perform the described operations over various types of basic and operational data, from formation of a range of parameters to their comparative assessment and to receive as a result indicators of compliance of the studied structure of composition to the set standard. Results of a research confirmed the existing theoretical provisions, having shown thereby efficiency of a research.

1. Introduction

Preservation and development of the cities having considerable historical and cultural potential is the purpose and sense of carrying out researches and design in the field of architecture and design [3] updates and development of architectural and spatial and art and figurative qualities of the historical environment and integration of the remained heritage into modern life of the city [1].

The modern environment of the historic town, as a rule, does not meet the requirements of modern level of comfort and therefore are in great need in protection and maintaining the unique properties, characteristics and qualities. There is a conflict between historically developed building and innovative subjects of the architectural and environmental plan. These problems of harmonization and compliance of quality of historically developed urban environment to requirements of the latest time are the cornerstone of all design problems of a modern urban planning and ensuring comfort in borders of historical settlements.

One of effective solutions of problems of interaction historically developed an image and "remake" - wide use of tools and methods of regeneration of an architectural image of the urban environment, as way of preservation and revival of the initial environment of the historic town, provoking formation of new trends in the organization of the lifestyle of citizens promoting stabilization of a valuable picture of the world and also providing support and development of traditions of a way of life of the city inhabitant that will give to his life stable comfort and will make it more dynamic.

2. Materials and Methods

In the course of formation of design of the urban environment, the system of standard restrictions and requirements performs as means of complex formation of architectural and spatial characteristics of the environment of the historic town in the course of creation of elements of complex filling and reorganization of city spaces on the basis of optimal architectural and town-planning solutions with inclusion of natural landscapes [2].

Restoration is understood, as a rule, as restoration of a monument in original form taking into account all planning, design and stylistic features while reconstruction is a reconstruction whole on the basis of the remained data (the description, the image, etc.) and the subsequent its operation in modern conditions.

Renovation assumes consideration of the available options of use of the building (esthetic, economic, functional) the architectural concept is developed for renovation of the building. For each object the individual concept on reconstruction and several interesting scenarios of use of this territory, as a rule, is specially thought over. Renovation assumes the fullest modernization of buildings, also adjacent territory in full accordance with today's policy of the capital and also, needs of business in general.

Resuscitation - revival of an architectural organism — restoration of sharply broken or lost vital functions of a system, set of actions for revival of the organism which is in a condition of "clinical death" - "the patient is rather alive, than is dead. If he is alive - he will survive, or he will not survive. If he is dead - he can be recovered, or he is impossible to treatment".

Re-cultivation – Re-cultivation of landscapes - (Latin re - the prefix expressing renewal, the return action and *cultivo* - cultivation, a processing) - a complex of the works directed to restoration of economic, medico biological and esthetic value of the landscapes broken (degraded).

Distinguish the following stages of re-cultivation: techno-sphere, biotechnological and sociocultural which realization assumes that the technical aspect of passing of all required procedures includes planning, formation of slopes, removal, transportation and drawing soils and fertile breeds, construction of roads, hydraulic engineering and other constructions.

It is necessary to mention one more direction connected with existence undeveloped, not used, ineffective and abandoned urban areas – regeneration of the urban environment including its design which is capable to update the vital functions of the city and to provide various scenarios of future use.

3. Results

The harmonious city design has to become not only esthetically subject expression of spiritual and material human life, but also the latent instrument of management and coordination of life of society by means of formation of a valuable picture of the world, support of traditions, formation of new trends of style and a way of life. [4]. Creation of the increased level of comfort of the urban environment gives to life in the city the new forms of stability that finally does it to more dynamic. A stable variety of lifestyle and activity of community in the urban environment considerably depends also on active inclusion on Wednesday of complex objects of city design and also on intensity of formation by them of shape and an image of the city.

Discomfort and disharmony of the modern environment of historic towns it is possible to call indispensable attributes and obviously expressed signs of destruction of mentality and spiritual life of modern society [5]. Studying of environmental qualities of the historic center of Irkutsk allowed to formulate the main problems which part can be solved or leveled means of design:

- loss of architectural and landscape qualities of specific points and panoramas;
- distortion of scales of the new and developed urban environment of the historic center (for example, inclusion not esthetic, in most cases standard buildings in building of the center);
- discrepancy of the actual sizes to visual parameters of functional use of city squares and streets (the large, disproportionate to the scale of surrounding building, empty areas at modern large office and public buildings);
- lack of well-planned streets – the destroyed covering, "indistinct" borders of street profiles, the started gardening;
- decay and neglect of a physical condition of facades of the majority of buildings, unreasoned coloristic decision of both separate volumes, and street in general; use of unacceptable means of reconstruction and restoration of facades;
- oversaturation and weak structure of information and advertising elements; unacceptably poor quality of advertising, imperfect graphic design;
- lack of small forms, symbols, signs and, in particular, branded, sign design.

Statement of tasks at a complete solution of the problems connected with formation and the organization of optimum performance of the comfortable environment of the historic town it is directly connected with ways of inclusion of form and content of its architectural and spatial and art characteristics in a basis of creation of next image of the projected architectural complex. Let's consider below the main methods of transformation of the architectural and art and subject-spatial environment of the city historic center.

4. Discussion

So it developed that it is accepted to carry out all operations and procedures in the existing architectural and town-planning environment with emphasis on the reversible nature of the carried-out procedures which cornerstone the plan of repeated action in value of renewal, restoration of sign or the return action – "regress", "restoration", "reproduction", "renovation", "revalorization", "resuscitation" and t is. item. In other words, everything that is implemented in the sphere of formation of architectural style of urban development of the historic center has to is under construction on knowledge, understanding and use, first of all primordial technologies, professional methods and receptions and also the historical art styles internally inherent in the place, the temporary period, specifics of the internal organization of function and a cultural originality of a designed project and also a way of life of the people inhabiting it, as makes a basis of semantic fullness of procedures of regeneration of style transformations to areas of architecture and design of the urban environment.

Other side of the problem of reconstruction in the conditions of the developed historical environment [6] the question of restoration, renovation and resuscitation of historical buildings in connection with their strong physical or obsolescence, function change, partial unauthorized reorganization and other uncontrolled actions of owners without violation of composite integrity and harmonious interaction with the environment where procedures of making decision on the fate of a separate object or even the whole complexes accept "subjects of the legal right and administration" which competences obviously do not correspond to form and content of the made decisions is represented (as it was in recent story with "A.A. Rassushin's house" in Irkutsk).

Figure 1. "A.A. Rassushin's house" in Irkutsk. The house of the end of the 19th century is located in Boulevard Gagarin, 32.

Old buildings for all the long-term history were exposed to repeated reorganizations, responding to the requests of commercial taste and change of the social relations, and, losing special unique spirit. over time fall and become useless, losing the initial shape and meaning of own existence, not to mention a condition of designs, networks and ridiculous re-planning. These trends concerning cultural heritage provide the whole complex of conservative actions for maintaining initial shape and style both the certain building, and a historical complex in general, its interrelations with the environment and to increase in sociocultural value in city life. At the same time, the main argument of "defenders of the historical environment" is that "... the buildings reconstructed today will serve future generations when the demand in an esthetics and level of comfort becomes higher also today's actions, material expenses and labor expenses will be appreciated", - but who and when simulated the system of values and criteria for evaluation of quality of the held events of "reconstruction of architectural heritage" by future generations of residents?!

With development of a way of life and the spatial and functional organization of the city the need for development of new platforms, introduction of modern design and design innovations and expansion of the existing architectural and spatial complexes increases [7]. The historical kernel is cultural, business and shopping center and the main objective maintaining integrity of historical building and harmonious interaction of the developed urban environment and the designed structures is. As a rule, the most valuable monuments of architecture are in the center of the settlement where cultural and business life of city communities is concentrated. To keep art and figurative integrity of an architectural and historical object at any type of design impact on style, structure, weight and scale of the historical environment, in a designing process it is necessary to pay special attention to creation of the mechanism of harmonization whole and separate parts of a designed project, to modeling of details and nuances of modeling of composition of all parts of the building - a socle (basement), entrance groups, a frieze, eaves, decorative frame of apertures, harmony of color and light and quality of finishing of a facade in general.

Development of improvement and application of elements of a landscape in the environment of the historic town for achievement of harmony of architectural design of the urban environment with landscape plasticity of the urbanized landscapes is a major complex of design problems of creation of comfortable life, especially in the conditions of the historic center of the city where the landscape is continuously connected with the architectural environment. A city landscape - the integral part of any settlement where it, being organically included in planning structure, supplements and develops a unique artistic image of a unique city landscape [5].

In structure of the design purposes and tasks the design of the urban environment becomes the separate projected element and the instrument of formation of city comfort and respectively – independent subject to the design [3] demanding preservation of visual communications with a surrounding landscape, an exception of the urban environment of a stylistic dissonance between the architectural and design decision of city ensemble and objects of a natural and landscape complex (green plantings, the city equipment and improvement, water constructions of park economy, the system of audiovisual registration and so forth).

Considering landscape architecture as means of interaction of city culture and the nature where it is the instrument of cultural human activity, elements of landscaping are represented as means of high-quality updating and revival of historical bases of city life and have to find active application in transformation of planning and building of the downtown for the purpose of creation of a unique city landscape, its artistic image. Preservation of historical and cultural monuments, successive development of the city and creation of art and expressive spaces has to contribute to sustainable development of the urban environment. At the same time the technological and cultural and historical aspects of activity of city communities which are not breaking integrity and keeping historical heritage of the city are considered.

Filling of city space small architectural forms makes one more complex of problems connected with the obvious shortage of city furniture and the functional equipment of public spaces. This problem has two parties – lack of the due and various nomenclature of the equipment of the corresponding quality and also ways and forms of the organization of address typology of stylistic complexes of building of the urban environment which would correspond to art quality and the level of development of modern society in general. Duping consumers visibility create a comfortable environment with use of the Small Architectural Forms (SAF) – the problem defining in the future an image of all city and its historical part. By means of SAF which an improvement component is it is possible to solve quickly problems of physical comfort and art quality of spatial characteristics of the objects of the urban environment which are harmoniously arranged in a historical context. At the same time SAF are meant as the so-called "shoddy constructions" which are situationally intended for architectural and planning and art and figurative modernization of objects of landscape architecture, creation of comfortable stay of visitors, landscape enrichment of the territory in general, esthetically and functionally supplementing the general composition of an architectural complex of historical building.

Some of the SAF elements do not bear utilitarian functions and have exclusively art and decorative or functional and spatial appointment. Being in the urban environment, objects of city design sate Wednesday emotionally, is information, artly, bring esthetic pleasure, increase comfort, bear information, necessary for fast and convenient orientation. And in this sense, it is necessary to speak about increase in information and emotional capacity of the urban environment as major lever on formation of art taste, an emotional state and mental tone of the city dweller. For this reason, the elements making a basis of SAF have to be unostentatious and innovative, plastic and convenient, simple and expressive, beautiful in a form, color and texture of material, easy, durable, ergonomic and economic, with good proportions and to correspond to the scale of the person. For our country town capital, steady forms can act as the most suitable subjects to improvement that promotes formation of pride in citizens and to feeling of stability of city life.

System approach to placement of elements of information design in the environment of the historic town has significant effect on perception of the modern city, its artistic image and architectural and decorative registration of shape. Development of forms, means and receptions of the organization of

IOP Conf. Series: Materials Science and Engineering **880** (2020) 012062 doi:10.1088/1757-899X/880/1/012062

objects of design of the urban environment is of great importance for formation of sociocultural and architectural and planning infrastructure of the city, identification of its natural-spatial components and preservation of monuments of historical and cultural heritage. It is necessary to build precisely and selectively the system of placement of elements of information design in structure of the developed urban development [8] in this connection the need for development of the comprehensive architectural and art concept which will solve all complex of problems of placement of information designs and elements of navigation in a historical part of the city increases. Within this concept the image of the urban environment of the historical settlement has to be created full architectural and art; the mechanism of creation of art composition and the style of the city equipment which are harmoniously combined with architectural, art and coloristic features of city buildings and constructions is developed; use of advanced technologies and materials on the basis of studying of domestic and foreign experience and features of local art national culture [9].

In real practice it is necessary to consider features of architectural style of a facade of the historical building on which elements of city design will be established. Again, entered designs of objects of city design have to fit in organically on the projected Wednesday and interact with it, forming new images and meanings generally a historical and art context. Means for this purpose in modern design and marketing technologies more than are enough: the designs consisting of separate elements, symbols, signs and images without substrate or with a flat substrate, flat and volume banners, designs with inclusion of effects of illumination, a design made in complex three-dimensional composition of separate elements and signs, lightboxes, poster designs, panels arms, plates, the general indexes, mobile signs on flexible material, an inscription on the sidewalk and many other things [9]. It is also often used methods of difficult configuration of volume elements and signs behind which the facade fragment and also the composition of letters and signs without substrate casting own shadow on the facade plane, etc. is visible.

Rebranding in the conditions of reconstruction of the historic town is the important moment which needs to be considered at design of the urban environment. Regardless of the size, arrangement and the administrative device, the historic town is in great need in awareness of own uniqueness, understanding of sources of originality and originality of an image, sociocultural identification and recognition of its sign elements. And to create the required relation to the city, its positive image in the foreign and domestic consumer market, it is necessary to accent historical, cultural, art and economic appeal of the city. The system complex of various type of actions for the analysis, development and deployment of a brand of the city, to maintenance of its functioning and providing conditions of successful development in which questions of formation and development of native architectural style and art image of the historical settlement act on the foreground is for this purpose formed and carried out.

City brand, in fact – his name, the strange essence lying between an impression of aboriginals - residents and a look "from outside" visitors – interventionists – an important component of city identity of image of the settlement. As a rule, it has visual expression – the sign, symbols, the motto, a logo, the branded (corporate) block and style – but has global value first of all when it is immanently connected with the strategy of development for the city, its image and semantic messages. Actually, the procedure of advance of a brand is some kind of creation of positive reputation that involves growth of direct and indirect city investments and also the subsequent its economic development and cultural prosperity [1].

City communities in any region or the settlement always hope, to attract that by means of a new brand or its part - a logo and a slogan, will be able to expand tourist appeal of the city external investments and to increase a profit share due to stimulation of a trade and tourist segment of economy. Because of the increased competition many cities are forced to position more and more actively themselves in the market of trade and tourism worldwide, to be more active in processes of attraction of new human and financial resources. However, the main problem consists that practically think of all city administrations equally and want that they were perceived as the first in many parameters — culture, arrangement, history [2]. And true sense of attraction of interest in foreign market – creation of special conditions for development of internal uniqueness of the city and support of characteristic originality, an originality

IOP Conf. Series: Materials Science and Engineering **880** (2020) 012062 doi:10.1088/1757-899X/880/1/012062

and unprecedentedness of unique city initiatives in the history of development of the market of innovative services.

5. Conclusion

Such approach assumes formation of the mechanism of regeneration of architectural style of the urban environment not only by criteria "advantage, durability and beauty", but also in terms of the prospects of its development and functional and spatial flexibility of decisions in any range of problems - from the solution of problems of transport and architectural and landscape complexes of the city before interior design of city spaces, colors-light filling and the device of systems of elements of city furniture and the equipment [9]. Methods and methods of transformation of the urban environment rely on modern engineering technologies and social requirements in which interaction there are contradictions between requirements of maintaining historical environmental values and the need for filling by the new environmental qualities corresponding to modern understanding of comfort of the urban environment [8]. Thus, creation of the mechanism of continuous increase in comfort of the living environment in the city, increases in level of its maximum safety and providing "atmosphere" of benevolent activity of all elements of the life support system also is one of important criteria for evaluation of quality of design in the conditions of regeneration of architectural style in terms of problems of modern city design.

References

[1] Hokhrin E V, Smolkov S A and Hokhryakov A A 2018 *Formation of urban style of Irkutsk*: (Irkutsk: Publishing house of the Irkutsk national research technical university) p 90

[2] Hokhrin E V, Smolkov S A and Hokhryakov A A 2018 *Design of the urban environment on the example of Irkutsk.* Irkutsk: Publishing house of the Irkutsk national research technical university p 124

[3] Schedrovitsky G P 1999 Typology of situations of carrying out researches. 1986 *Programming of research and development.* From G.P. Schedrovitsky's archive **1**

[4] Potapova A V and Engel B 2012 The prospects of development of the historic center of Irkutsk taking into account the German experience of Dresden **1 (2)** pp 169-178 DOI: 10.21285/2227-2917- 2012-1-169–178.

[5] Engel B and Potapova A V 2013 The heritage is in danger – the historic center of Irkutsk *News of higher education institutions. Investments. Construction. Real estate* **1 (4)** pp 196-205. DOI:10.21285/2227-2917-2013-1-196-205.

[6] Dyagileva N S, Zhuravleva L A 2012 City identity: concept, structure, bases *City Sociology* **1** pp 46-61

[7] Dyagileva N S 2013 Theoretical aspects of city identity *Branding of the small and average cities of Russia: experience, problems, prospects pp* 54-59

[8] *The master plan of the city of Irkutsk approved by the decision of the Duma of the city of Irkutsk* of 28.06.2007 No. 004-20-390583/7;

[9] *The local standards of town-planning design of Irkutsk approved by the resolution of city admin istration of Irkutsk* of 15.07.2011 No. 031-06-1287/11

Basic principles of development of green and open spaces in the structure of the historical city on the example of the cities of Dresden, Stuttgart

V V Kozlov[1], **K G Ivanova**[1] **and S S Belomestnyh**[1]

[1]Institute of Architecture, Construction and Design, Irkutsk National Research Technical University, 664074, Lermontov Street, 83, Irkutsk, Russia

E-mail: valery_kozlov@mail.ru

Abstract. The purpose of this study is to develop basic principles for the development of green and open spaces in the structure of the modern historical city on the example of Dresden, Stuttgart. This could serve as an important tool for the analysis of the functional and planning features of other major historical cities and would suggest the main directions of optimization and expansion of recreational zones in the design. Sources for the article were literary publications about the past and present Dresden, Stuttgart, cartographic materials, historical photographs and photographs taken during the in-kind survey. As a result, the main provisions concerning the prospects of green areas valuable for creating an enabling environment are formulated.

1. Introduction

Because of development of urbanization there is an urgent need to analyze the favorable and negative conditions for the conservation of green and open spaces in cities in general and in historical cities in particular. Analysis of the functional and planning structure of urban fabric, its features and potential is going to help to solve some problems of preserving a favorable environment for the local population. The problem of maintaining the attractiveness of the historical environment for tourists is also acute and is one of the priorities, as tourism for European countries is one of the main sources of income. Development of the main directions of development of recreational spaces will allow to clearly formulate problems and their solution, which can be used in design.

2. Materials and methods

By systematizing information from various sources (literary sources, maps, historical photographs of the late 19th and early 20th centuries, photographs taken during in-kind survey when visiting some cities of Germany) it was able to make a comparative analysis of the condition of the recreational areas and draw certain conclusions, which is set out in this article.

3. Results

Based on an analysis of the information on Dresden and Stuttgart and the main provisions of the conceptual plans for the development of these cities [1], general objectives and seven basic principles for the development of green and open spaces are proposed:

1. Preferential development of internal greening over external

The future development of the city and the architectural improvement of fragmented urban green and open spaces will be successful only if the internal development of the city gets an absolute advantage over the external one. This will require clear concepts in the placement and hierarchy of green and open spaces, both in relation to the urban center and taking into account the peculiarities of the development of the city-wide structure of greening and the function of parks.

2. Short paths as the simplest way to connect different objects and green spaces of public importance

The sustainable development of the urban centre requires as many small elements of greening as possible, mixing a variety of functions into one diversified and compact centre structure. The advantages of open and green spaces in this approach are obvious: residents and guests of the city are offered a dynamic and vibrant center with its atmosphere and aura. Trade, services and cultural institutions benefit and benefit from the effect of interaction with open and green spaces. Public and private infrastructures are being developed efficiently and economically. There is also a growing sense of security in busy and green urban areas. Proximity and comfort of greening in various objects of public service provide a high level, reduce ways of movement saving costs of transport and excluding its accumulation.

3. Impact of rational location, greening and formation of the city image on the attractiveness of the territory

The expansion and modernization of green and open spaces in the center, as well as the integration of them into a network with a city-wide system, improves the quality of life. A compact network of full open spaces creates an attractive environment for living and working. The rational location of high-quality places for play and leisure in the structure of such a network is an important factor for settlement. Attractive spaces of streets and squares, as well as richly green areas, which are closely related to each other, play an important role in the transfer of the main image of the urban landscape of Dresden and Stuttgart to small (peripheral) residential areas.

4. Separation of greening functions between the city centre and residential areas

The different efforts and capacities of individual residential neighborhoods and districts contribute to the division of greening functions between the centre and interconnected areas. In addition to housing, important intra-urban functions aimed at the further expansion of green and open spaces are urban public facilities such as theatres and museums, hotels and food retailers, government authorities, branches of economic enterprises and domestic service mills, research and development centres, and educational institutions. In the center of the city there are objects of historical and architectural value of the city with the highest degree of identification, as well as the highest value are gardens and parks. At the same time, the combination of greening with promenade, pedestrian street, urban market claim that their location is considered central among public use facilities. Thanks to the active role of green and open spaces in framing trade objects, the largest cities position themselves as the most important centers of the regions. Some mixed-use urban areas and interconnected areas create appeal by having large green areas of their own. Recognizable places that meet high requirements and conditions, with a large share of housing, maintain the balance of urban public facilities. They provide potential territory for future demand for new housing and for various projects using free spaces for experimentation and local economy. The development and temporary use of empty areas by start-up entrepreneurs with creative initiative can lead to a new perception of critical areas and thus prepare land for profitable use in the medium and long term.

5. Spatial concept of greening and step density principle

The creation of an urban centre in the traditions of European cities can be realized, for example, in Dresden only through precisely defined boundaries of spaces and a quality network of green and open spaces. The concept of most European urban development projects presupposes moderate density in the centre and in related territories, as well as continued positive trends in the inclusion of green spaces. The previously planned increase in building density going from the city center to the surrounding neighborhoods is gradually being replaced by an intense association with a network of green and open spaces, improving the quality of life. The principle of step density and concentration of housing in the city center, combined with different green forms, as a suitable response to expectations of increasing density of development and increased requirements for the quality of housing in the city.

6. High population mobility and attractiveness of urban green spaces

Meanwhile, the experience of long-term reconstruction and constant changes in historic gardens and parks with increased value and improvement of open spaces have made a sharp leap forward. Attractive built-up and green spaces create a pleasant appearance and encourage citizens to identify with their city.

IOP Conf. Series: Materials Science and Engineering **880** (2020) 012063 doi:10.1088/1757-899X/880/1/012063

The increase in visitors to public spaces extends its influence also to neighbouring areas, the value of which thus increases. As a result, attractive green spaces can initiate promising projects and investments. Active urban communities need advanced mobility. The best accessibility of the center is the accessibility of commercial places, domestic service centers, cultural institutions. All this requires an optimal division of transport. Activities to improve the culture of mobility in the future include improving the environment and reducing private vehicles. At the same time, the comparison of high mobility and the presence of attractive urban spaces does not carry any contradiction. Creative management of mobility in the city, the equipment of walking and cycling paths combined with the green network of the city, as well as the removal of motor vehicles from urban areas most affected by it are necessary.

7. Arrangement of new objects

The location of new attractive parks on promising plots and greening on unused plots is aimed at investing in high quality of life. New gardens on the river spaces, uncomfortable for the construction of relief areas, the favorable location of the center with interconnected peripheral areas contribute to the tourist popularity of the city and the comfort of the local population.

Analyzing the historical (early 20th century) [Figure 1, 2] and contemporary photographs of the cities of Dresden, Stuttgart, one can conclude that the tradition of giving importance to greening a unique historical environment continues. In certain cases, the design of various forms of plants of quays, parks and parks is even more developed in modern times [Figures 3, 4]. On other frames, the forms of greening have become simpler and more concise in favor of new trends in fashion, or thanks to the acceleration of the pace of life of modern man, the replacement of manual work with automated work [Figures 5, 6].

Figure 1. Embankment in Dresden. Photo of the early 20th century. [2]

Figure 2. Bobser is a district of Stuttgart. [3]

Figure 3. Embankment in Dresden. Photo of the early 20th century. [4]

Figure 4. Embankment in Dresden. Modern photo. [4]

Figure 5. Topiary forms of the park in Dresden. Photo of the early 20th century. [4]

Figure 6. Topiary forms of the park in Dresden. Modern photo. [4]

4. Discussion

The problems of greening in the unique environment of the historical city require an early solution. Theoretical searches and practical world experience help from numerous options to choose the necessary, applicable by certain parameters to a certain urban territory, to a certain established spatial and functional system. Working on the composition, the architect takes into account new data of sanitary and hygienic science, microclimatology, ecology. In this regard, it is necessary to identify such techniques of organization of green plantations, park improvement and urban landscape in general, which are most effective both in aesthetic and health terms, as well as create better conditions for the development of vegetation in the city environment. [5]

Each European country has its own individual features of climate, relief, culture, economy. And in each individual case we get acquainted with the unique experience of preserving the green environment in historical cities, whether it is mountain Andorra-la-Vella [6], foggy London, sunny Barcelona. In many modern cultures, the garden is an integral part of human life. Located on the roofs of houses in England or a Spanish city park, they can become open galleries for works of art and sculpture or become the basis of complex landscape architecture and design, or take on the role of conservation of rare plants [7, 8].

As an example, the functional planning organization of Stockholm can be considered. The idea of interconnecting the architectural and planning organization (morphotypes) of the neighbourhoods and larger planning elements with their key location and distribution in the urban landscape, laid down in the master plan (1997), allowed to preserve the natural component, increase the level of urbanization without perimeter expansion of the city, at the same time reduce the loads on the historical core and increase the attractiveness of peripheral areas. [9, 10,11]

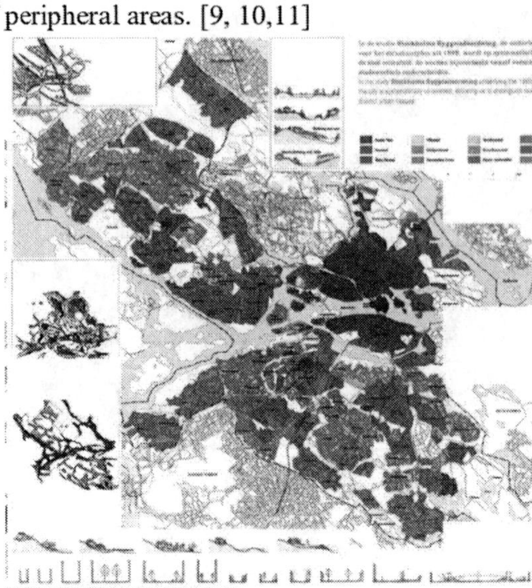

Figure 7. Urban morphology of Stockholm

Open and green spaces as part of the natural and recreational framework of the historical city are considered in the monograph «Morphogenesis of architectural and planning structure and principles of reconstruction of the historical center of Irkutsk» (Bolshakov A.G., Belomestnyh S.S.). The method of morphological zoning of the city is proposed by forming three frames (natural-recreational, urban, historical-cultural). Combined with the principles and techniques (decompaction (clearing the space of blocks), rebuild (adding elements)) of regulating the morphotypes of neighborhoods, the three frames will successfully preserve the identity of the city while developing its urban appeal. One of the most important principles is the principle of correspondence of the morphology of the quarter to its location relative to the main urban planning, natural axes and historical and cultural values [12].
For the preservation and restoration of natural territories was also proposed by Bolshakov A.G. "method of geoplasty." In the analysis of the site key points (points of intersection of two frameworks come to light (natural and functional and planning). For these points, a concept of landscape reclamation is developed, with an increase in the share of greening in its morphotype, with restoration of aquatic, soil, plant conditions. Then the issues of the combination of functions of urban and ecological (equipment and development of the territory without damage to the ecological potential of the landscape) are solved [13, 14].

5. Conclusion

Analysis of problems of development of green and open spaces in the structure of European historical cities showed that in the strategy of development of urban planning systems methods and tools of integration of green urban structures in various large-scale levels and stages are used. In the development of plans for the strategic development of historical cities, the method of layers is used, in which the greening of urban spaces is planned by a special team with the harmonization of the main provisions in the overall strategy of urban development. The benefits of green urban structures go beyond traditional urban strategies and policies to determine the competitive image and quality of living in the city.
Alternative theories and criteria have been used to develop basic concepts for the development of green urban spaces of European historical cities in the field of spatial organization, provision and management of urban greening, which in general forms the basis of the sustainability of cities, as well as the development of a critical view of its own urban environment.

Figure 8. Greening in the centre of Dresden. Modern photo. [15]

References

[1] Holger Evets 2008 Living History is the urban landscape of the city *Dresden - City Centre Perspective Plan*, **6**
[2] Dorothee Baganz 2013 Das historische Dresden p195
[3] Ulrik Barth 2014 Der Stuttgarter Bobser *Hauser – Familien–Geschichten* **47**

IOP Conf. Series: Materials Science and Engineering **880** (2020) 012063 doi:10.1088/1757-899X/880/1/012063

[4] Dr. Barbara Munch – Kienast 2011 Unser Dresden *Einekleine Zeitreise* **8, 9, 26, 27,**
[5] Under Society. Ed. A.P. Vergunov 1980 *Architecture composition of gardens and parks* **5**
[6] Ivanova K G 2012 Influence of cities forming factors on the layout of Andorra-la-Vella, its architecture and improvement *News of higher education institutions. Investments. Construction. Real estate* **1 (2)**
[7] Spencer-Jones R 2008 1001 Garden to See
[8] Parks. Gardens. Ed. Group: M. Aksenova, T. Evseeva, 5 2007
[9] Bolshakov A G 2012 Foundations of the Theory of Urban Planning and District *Planning: Educational. For Universities* **15**
[10] Case Study of the City of Stockholm and the Greater Stockholm Area: Municipal and Regional Organization, Urban Planning Tools and Issues, Planning of Transportation and the Green Network,
[11] *Housing Issues: Summary: Research Report* [Электронный ресурс]. Access mode: http://www.habitation.gouv.qc.ca/fileadmin/internet/centredoc/pubSHQ/M06301.pdf
[12] Hammarby Sjostad Case Study | CP 249 Urban Design in Planning, 2007 [Electronic Resource]. Access mode: http://www.aeg7.com/assets/publications/hammarby sjostad.pdf
[13] Bolshakov A G, Belomestnyh S S 2018 Morphogenesis of architectural and planning structure and principles of reconstruction of historical center of Irkutsk **172**
[14] Bolshakov A G *Urban planning organization of landscape as a factor of sustainable development of the territory.* Thesis for the degree of Doctor of Architecture. Abstract, 30 (2003)
[15] Bolshakov A G 2015 Restoration of disturbed and degraded territories. History of the copyright method of geoplasty *Architecture and Modern Information Technologies* **2 (31)**
[16] Katrin Nitzschke, Perk Loesch 2014 Die Prager Strasse in Dresden *Geschichte einer Prachtstrasse* **183**

IOP Conf. Series: Materials Science and Engineering **880** (2020) 012064 doi:10.1088/1757-899X/880/1/012064

Modern city: areas of modernization of the urban environment

V A Kudryavtseva[1]

[1] Irkutsk National Research Technical University, 83 Lermontov str., Irkutsk, 664074, Russia

E-mail: kudravera@mail.ru

Abstract. Modern society places new demands on the urban environment. City residents pose new challenges for the urban development which can be solved by implementing information and communication technologies and modifying the urban environment. Most cities are going through this process in an intuitive way. By structuring the transition to the smart city system, one can increase the efficiency of the urban infrastructure, ensure transparency and develop and improve the urban environment by increasing efficiency and reducing costs, contributing to the economic growth and improving living standards due to the involvement of citizens, businesses and governments in solving urban problems. In Russia, attempts are being made to implement a system of intellectual transformation of cities, which can become a developing mechanism that takes into account existing priority industry directions and make citizens believe in the future of their city based on advanced technologies. The innovative infrastructure can solve problems of technological and human capital development.

1. Introduction

The modern economic system is part of urban economies. Half of the population live in cities. More than 80% of the world GDP are produced in cities. On a global scale, 60% of the global GDP are produced in 600 largest cities (City 600). [1, 2, 3]. According to the UN and the World Health Organization, by 2050, the urban population will be 70% of the world population (about 6.4 billion people) [4]

Cities are under enormous pressure caused by complex production processes and economic relations, increased migration and socio-psychological dissatisfaction, growing environmental pressure and transport problems, terrorist threats and emergencies, excess density and "urbanization diseases" - the rapid spread of infectious diseases and viruses caused by intensive contacts. Cities do not cope with their main task being unable to meet the needs of the population.

It becomes evident that it is necessary to reorganize the urban space and improve the urban environment since large cities attract a large number of people. The urban environment should be a working area and a place for relaxation. It should provide services based on new approaches to the urban development.

At the same time, large cities are characterized by heterogeneity, a significant amount of resources, a lack of clear specialization. Digitalization of the urban environment which would help eliminate “bottlenecks”, relict technological structures and ensure the competitiveness of the entire territory for the short and long term period is required.

An analysis of Russian and foreign experience allows us to conclude that a large city does not implement its potential creating an innovative and intellectual environment based on advanced technological solutions and digitalization. It is an administrative center. [4, 5, 6, 7, 8, 9, 10]

In modern conditions, the main priority for the urban development is a safe and comfortable environment characterized by high environmental indicators and aesthetic parameters. It is necessary to develop engineering and transport infrastructures, create an effective urban space and an urban infrastructure, a high-quality and affordable housing market, i.e. a comfortable urban environment that allows us to design spaces taking into account existing buildings and infrastructure. This can be achieved through the use of innovations and high technology projects

2. Materials and methods

Intellectualization of the urban infrastructure is part of the new requirements for the efficiency of the urban environment. Therefore, a smart city is a natural condition for integration into a changing system. [9, 11]

There are different interpretations of the concept "smart city", but all of them are based on interconnected information and telecommunication technologies that allow for the most efficient provision, management of current internal processes and solution of urban problems.

Experts of the International Telecommunication Union (ITU) conducted a thorough study and analyzed more than 100 approaches to the definition of the concept "smart city". [11, 12]

According to this study, "the smart sustainable city (SSC)" is an innovative city in which information and communication technologies and other tools are used to improve the quality of life, the effectiveness of the city and urban services, and to strengthen the competitiveness. These technologies are also used to satisfy needs of present and future generations without affecting economic, social and environmental components of the city." [6] Thus, to transform a city into a smart city, it is necessary to solve the following tasks:

- creation of comfortable living conditions;
- creation of a favorable business climate;
- creation of an accessible environment and improvement of services for disabled people;
- involvement of the population in solving problems and managing territories.

Digitalization and implementation of innovative technologies can make the urban environment safer, more economical and comfortable. At the same time, this niche is becoming attractive for investors, government and local self-government bodies and local communities. According to Frost & Sullivan experts, by 2025 the technology market will have reached $ 2.4 trillion. According to the research company Markets and Markets, the market volume will have reached $ 1.2 trillion by 2022. Figure 1 shows data on the technology market financing volume for a smart city, billion US dollars. [1, 2, 3,]

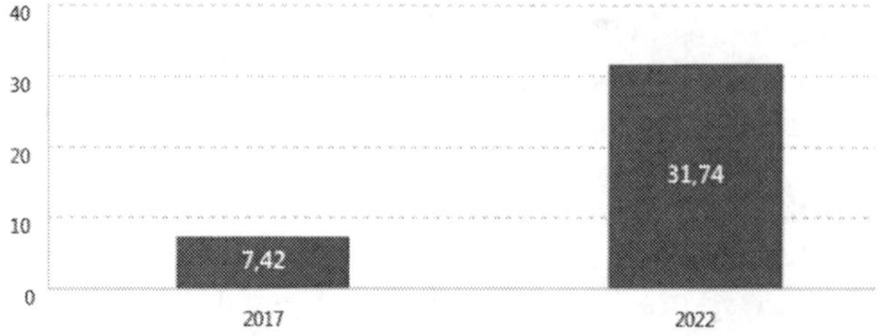

Figure 1. Volume of technology market financing for a smart city, billion US dollars.

The Markets and Market experts argue that the following elements that have a significant impact on the environmental quality are developing: urban public transport, housing and communal services, smart buildings, the Internet of things or the smart area of urban services. The inevitable growth of the urban population will be a key reference point

3. Results

In Russia, technologization began a long time ago, but only in 2015 did it get a software status. In Russia, urban development strategies have already been developed.

In 2017, the Government of the Russian Federation made an attempt to develop legal acts (e.g., the decree of the Government of the Russian Federation of July 28, 2017 N 1632-r "On approval of the program" Digital Economy of the Russian Federation" providing for a number of measures affecting creation and development of smart cities). This decree is inoperative as the decree of the Government of the Russian Federation of March 2, 2019 No. 234 "On the system for implementing the national program "Digital Economy of the Russian Federation" has come into force.

The problems of innovative and spatial development of cities and their role are solved by the Russian government. In the Address of V.V. Putin of March 1, 2018 and in the Decree of the President of May 7, 2018 No. 204 "On National Goals and Strategic Tasks of the Development of the Russian Federation for the period until 2024", the need for spatial development of Russia was emphasized [13, 14]

The Ministry of Construction, Housing and Communal Services of the Russian Federation pays attention to the smart city project. The Smart City project is aimed at improving the efficiency of city management and comfort. Modern information and communication technologies should be implemented into the city infrastructure. They are implemented through the portal "Bank of Smart City Solutions". This portal includes more than 300 projects aimed at improving various urban infrastructure areas: housing and communal services, urban public transport, communication systems, environmental safety, urban environment, urban planning, etc. All this can create a synergistic effect: the life of citizens will become more comfortable and it will save budget money.

Figure 2 presents the basic principles of implementing smart city system and their consequences.

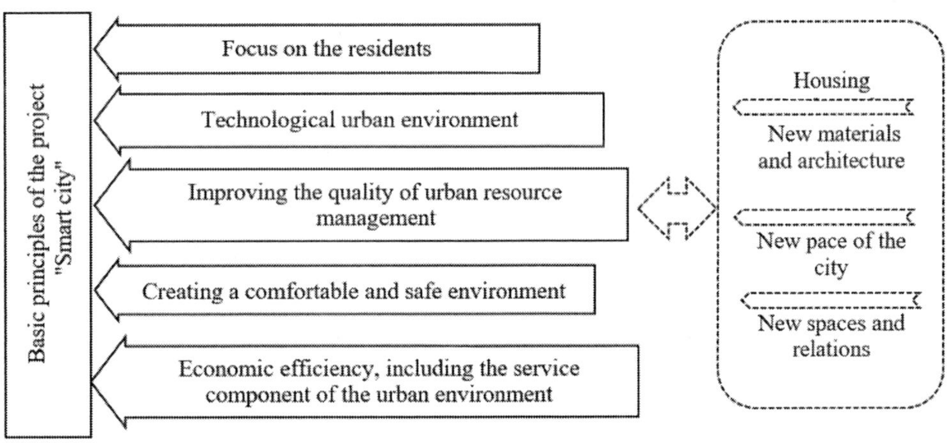

Figure 2. Basic principles of implementing a smart city system and their consequences.

Moreover, on March 4, 2019, the Ministry of Construction of the Russian Federation approved "Basic and additional requirements for smart cities (the Smart City standard)" in the following main areas:

- city management;
- smart housing and communal services;
- innovations for the urban environment;
- smart city transport;
- intelligent public safety systems;
- intelligent environmental safety systems;
- infrastructure of communication networks;
- tourism and services.

For the effective implementation of these tasks, Russian cities have to solve a number of existing problems [15]:

1. A high level of depreciation of urban public utilities systems and a constant increase in operating costs for the maintenance of old infrastructure facilities. In Russia, the average depression rate of fixed assets is 60%;
2. Deterioration of environmental indicators. For some Russian cities, the environmental situation is critical;
3. A significant decrease in the number of residents in small cities due to the outflow of the population to megacities;
4. Unreadiness of Russia to to improve comfort, quality and safety of the urban environment;
5. A lack of financial resources for improving urban conditions.

Figure 3 shows main areas of modernization of the existing urban environment.

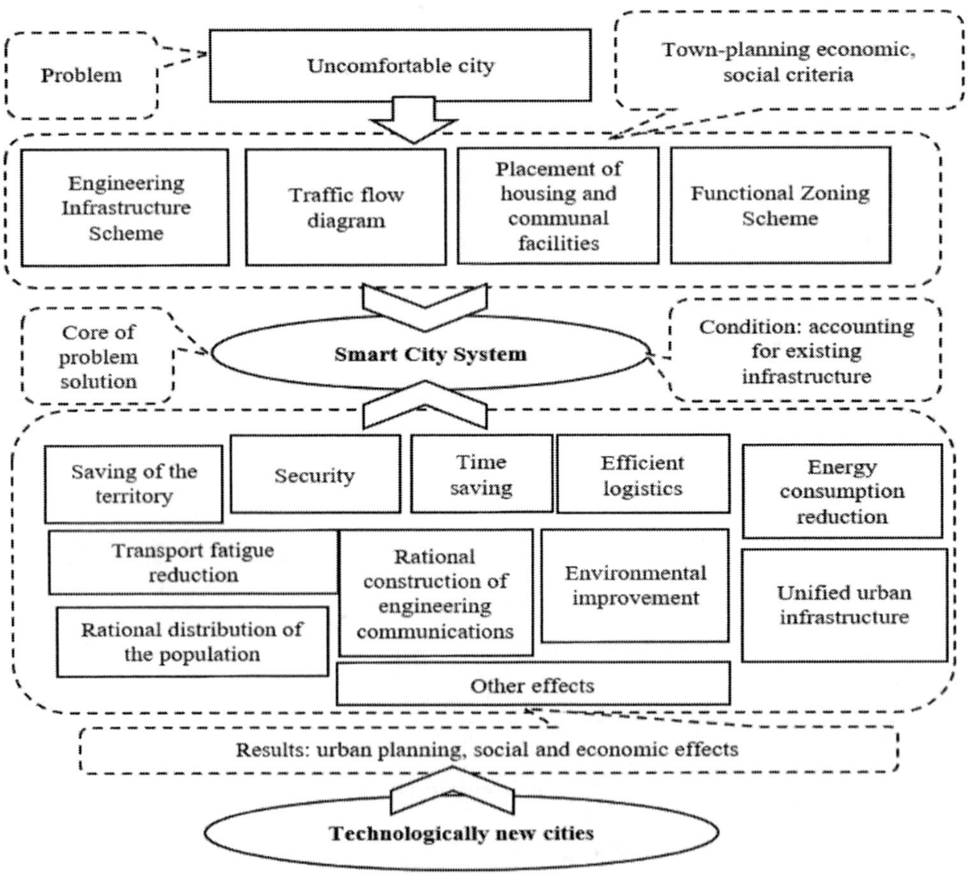

Figure 3. Directions of modernization of the urban environment.

The "smart city" system is characterized by the efficiency of smart technologies rather than their number. They have to solve existing infrastructure problems, overcome crisis situations and bring the urban environment to a new stage of development in the context of restricted budget financing.

4. Discussion

The basic concept of a smart city is availability of a "smart individual". A new generation of city residents are placing higher demands on the urban environment. Modern society is forced to save its time, use optimal routes, rationally use available resources, search for required locations, receive services, and be confident in their safety. At the same time, the residents should recognize the need for technologies being implemented to create a comfortable urban environment so that changes in the existing system do not contradict their needs and expectations. [16] Figure 4 shows possible services for creating a comfortable urban environment.

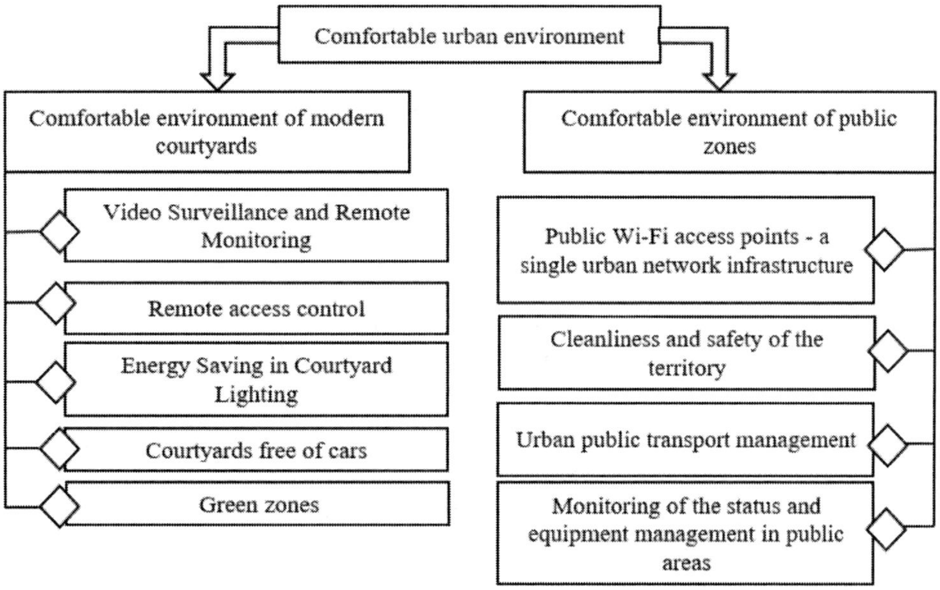

Figure 4. Services for creating a comfortable urban environment.

In other words, one can say that residents perceive their city as a system providing a wide range of services rather than a living and working area. It is in compliance with the general logic of the modern digital society and the Russian economy.

The public space has to correlate with business interests in order to stimulate the receipt and accumulation of financial resources.

According to the new model of the urban economy, money should benefit the residents. The purpose of the city is accumulation and transformation of its own experience to create a comfortable and safe environment for people. [6, 8]

The interest in smart cities caused a lot of theoretical discussions, but no final decision was made on the application of the smart city strategy. In addition, there are a number of factors that impede the implementation of the "smart city" system:

- the scale of new technologies is uncertain;
- the technologies challenge the existing urban management system;
- the technologies are understudied and their efficiency has not been tested in all sectors of the urban infrastructure

5. Conclusion

The digital transformation can solve problems of rapidly growing cities and create a safe, comfortable and economical urban environment.

This effect can be achieved by involving the population in this process and increasing the business activity of residents. Of course, technologies can change the urban architecture and structure - technologically new cities will appear along with new structural spaces, new social standards and

IOP Conf. Series: Materials Science and Engineering **880** (2020) 012064 doi:10.1088/1757-899X/880/1/012064

opportunities for involving people in environmental management. All this will increase the competitiveness of Russian cities and contribute to the human capital and innovative infrastructure development.

References

[1] Woetzel J and Kuznetsova E 2018 *Smart City Solutions:What Drives Citizen Adoption Around the Globe?* (McKinsey Center for Government) p 60

[2] Woetzel J, Remes J, Boland B and Lv K 2018 *Smart Cities: Digital Solutions for a More Livable Future* (McKinsey Center for Government) p 152

[3] Dubbeldeman R and Ward S 2015 *Smart Cities* (Netherlands: Deloitte The Netherlands) p 86

[4] Airaksinen M, Kokkala M 2015 *Smart City - Research Highlights* (Kuopio: Printed in Grano) p 140

[5] Falconer G and Mitchell S 2012 *Smart City Framework* (Cisco Internet Business Solutions Group) p 11

[6] Knyaginin V, Kuzmina A and Lipetskaya M 2018 *Priority Areas for the Implementation of Smart City Technologies in Russian Cities* (Moscow: Center for Strategic Research "North-West") p 178

[7] Vasilenko I, Lyulko A and Vasilenko E 2018 *Smart City of the XXI Century: Opportunities and Risks of Smart Technologies in City Rebranding* (Moscow: International Relations) p 256

[8] Volkov A, Sedov A and Chelyshkov P 2015 *The concept of "Smart City"* (Moscow: NRU MGSU) p 93

[9] Watson R 2014 *Smart Cities. Future. 50 Ideas You Need to Know About* (Moscow: Phantom Press) p 525

[10] Giffinger R, Fertner C and Kramar Hans 2007 *Smart Cities Ranking of European Medium-sized Cities* (Vienn: Centre of Regional Science) p 28

[11] Adcock J *Smart City* 2017 (Vancouver) p 42

[12] Borsboom-van Beurden J, Kallaos J, Gindroz B, Riegler J and Noll M 2017 *Smart City Guidance Package for Integrated Planning and Management* p 80

[13] Kurbatov M2001 *Social Designing* (Rostov on Don: Phoenix) p 416

[14] Volkov A 2014 *Knowledge is power* **3** 4-15

[15] Kudelkin V, Veredinsky S and Fomin E 2014 Creation of a Smart City (Samara: Samara State University of Economics) p 152

[16] Kupriyanovsky V, Sinyagov S and Tishchenko P 2012 ArcReview 2 (61)

IOP Conf. Series: Materials Science and Engineering **880** (2020) 012065 doi:10.1088/1757-899X/880/1/012065

Stage-by-stage construction approach in developing railway station complexes

A I Kulakov[1] and Yu V Krakovtseva[2]

[1,2] Department of Drawing, Painting, Fundamentals of Design and Historical Architectural Heritage Irkutsk National Research Technical University, 664074, Irkutsk, Lermontov Street, 83

E-mail: kulakov@istu.edu

Abstract. Developing modern-day railway stations as a rule goes along with their working. While being reconstructed, the arranged activity of railway stations has to be in-process. Railway stations have to function consistently and continue their work when providing services. To realise large-scale concepts, developing and building activities are implemented through a step-by-step approach with each building stage being autonomous. It means that the functioning of the reconstructed object is in-process when the developing works are under way. The process requires that there should be a close-cut separation of the functional zones, as well as there should be a coherent scheme to serve passengers.

When developing the project documentation, it is common for both customers and building designers to tackle the problem of identifying building process phases for staged commissioning.

This is a particularly serious issue for creating investment programs and decreasing failed construction projects. This is quite urgent either for reconstructing working railway stations, since the completing and commissioning of first stages of the planned reconstruction ensure for the proper working process of the stations and give an opportunity to continue realising the building project.

In compliance with the Urban Development Code of the Russian Federation (article 48, para.12.1), working on the project documentation can be done relating to different construction phases of improving the capital facilities by the project developer or project owner's initiative.

The project documentation for a certain phase of the construction is prepared to the extent required for putting the part of the planned building activities into practice. The requested documentation has to meet the requirements of the structure and the contents obligatory for capital construction facilities documentation formalized by the Government Decree of the Russian Federation dated 16.02.2008 № 87.

The construction stage is understood as constructing one of the capital construction facilities. According to the arrangements, the capital construction facility is to be built on one land area if such a facility can be commissioned and function independently (in other words, AUTONOMOUSLY from building additional capital construction facilities on this land area); or the construction stage is understood as constructing a part of the capital construction facility, which can be commissioned and used AUTONOMOUSLY (in other words, it can be used independently from other parts of this capital construction facility) [1].

When developing projects for railway station complexes consisting of several buildings (existing and new ones) on one land area, there appears a pre-requisite for identifying different construction stages / steps within one project. The stage-by-stage approach (staged / stepwise approach) in the construction process is tailored within the Construction method statement. However, it is required that a full set of permitting design documentation should be prepared to receive approval from the State Expert Evaluation Department of the Russian Federation (Glavgosexpertiza), to get staged investment and go through phased commissioning.

It is crucial to understand that the stage-by-stage approach in realisation of the project is impossible without having developed the complete project. The end-to-end project is of great significance for planning all the staged building activities. That is the reason why identifying the stages of the construction process is necessary through analysing all the package of measures.

Under the circumstances, when developing the project documentation, it is necessary to set the project objective correctly and allow identifying the stages / phases of the construction process in all sections of the project documentation. There are a lot of problems raised within the construction project documentation: the issues of development patterns of the building site; of the workings of technology the railway station is engaged in; of the networks of engineering infrastructure and fire safety measures should be the focus of special attention.

Let us consider the example of identifying and planning the sequent construction stages of developing the railway station complex "Irkutsk – Passazhirskii". Each of the stages requires receiving approval from the State Expert Evaluation Department of the Russian Federation (Glavgosexpertiza) and goes through phased commissioning (fig.1).

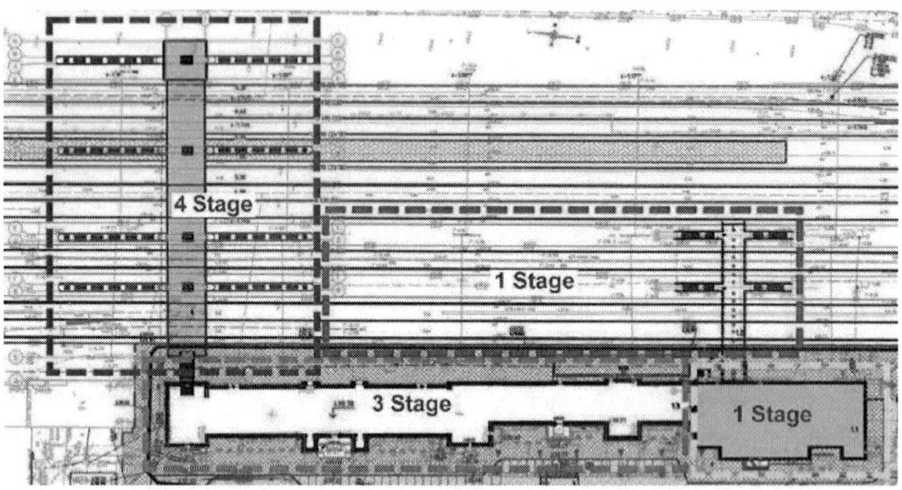

Figure 1. Example of allocation of construction stages in the design of the Irkutsk Passenger station complex

In 2010 a cooperation agreement in the area of joint implementation of the project on organising a passenger transport hub in Irkutsk and developing the surrounding land area was signed by Irkutsk Region government, Irkutsk city authorities and open joint-stock company (OJSC) "Russian Railways".

The facility at a glance. The railway station "Irkutsk-Passazhirskii" is located in Sverdlovskii administrative district of Irkutsk, along Chelnokov Street, along the embankment of the Angara River. The modern-day building of the railway station "Irkutsk – Passazhirskii" is a cultural heritage site of regional significance. Its chronologically first building was constructed in 1897-1898 upon the project of the architect Macievich. In 1904-1907 another building was constructed upon the project of the engineer V. I. Kolyanovsky. In the railway station complex there were two concourses where passengers travelling first, second and third class were served. In 1936 the railway station was reconstructed. The buildings which were constructed in 1907 and 1898 were reconstructed to be one building. The passenger in-built corridor connecting the two buildings blended with the whole concept since it was of the same architectural style. In 1964 the concourse for going through to take suburban trains was built on to the railway station building; the pedestrian tunnel was constructed; the railway station forecourt was enlarged and landscaped properly. The architectural style peculiarities of the first buildings remained intact after the reconstructions of 1998 and 2004. The railway station "Irkutsk – Passazhirskii" is a transit station of riverside type [2]. The characteristic feature of any transit station is its heavy passenger and freight traffic. At the present time the railway station "Irkutsk – Passazhirskii" is a point of departure for more than 2800 passengers travelling long-distance and 5600 commuters per day. The overall area of the railway station is two times smaller than it is required for the existing passenger traffic flow [3]. The area is not sufficient for a waiting lounge, left-luggage room, luggage offices, as well as for mother-and-child room [4]. No measures for organising service accessibility for physically challenged people is one of the most crucial downside of the railway station [5].

For developing the concept, a far-seeing urban planning project has been worked out in liaison with the Institute of Urban and Investment-driven Development "Giprogor" to improve the railway station adjacent territory within Profsojuznaya Street, Chelnokov Street, Tereshkova Street and Majakovskii Street. The development concept is aimed at reconstructing the embankment of the Angara River, at rearranging the transport infrastructure and developing the adjacent territory (fig.2) [7].

Figure 2. The concept of arrangement of the Angara embankment

Putting the concept into practice will contribute to having convenient traffic intersections, shopping centres, as well as appealing city sites for city residents and its visitors [8].

The panorama of the railway station complex is well seen from the embankment, Gagarin Avenue. The place with the railway station panorama is one of the most attractive tourist sites in the city [9].

The executive management of the company "Russian Railways" has already begun to implement the agreement reached in 2010. In 2018 the reconstruction of the railway station complex commenced. The design project is referred to as "The reconstruction of the railway station complex "Irkutsk – Passazhirskii". Taking into account the limited area intended for building and construction works, the Russian Railways authorities decided to demolish the commuter service concourse and integrate it as a part of the union railway station when constructing a new building. The building and construction works will be divided into several stages:

1. *STAGE* of the building process comprises the reconstruction of the subway crossing and platforms (fig.3).

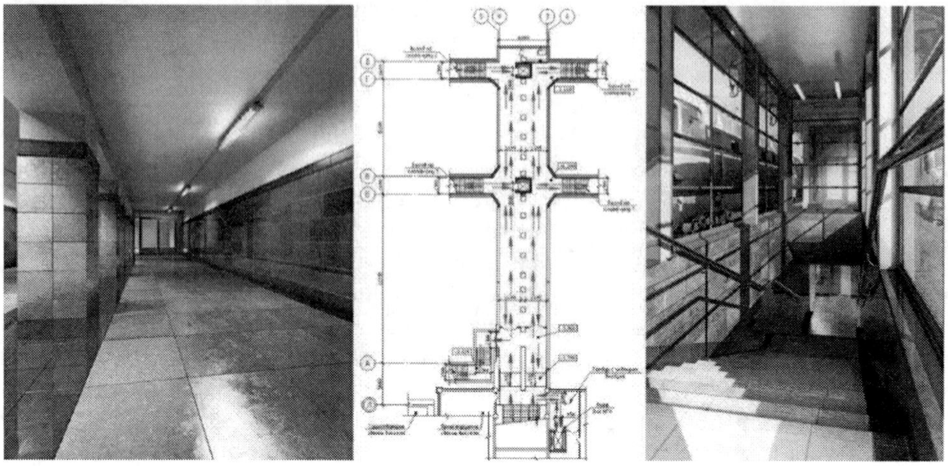

Figure 3. I stage of construction

2. *STAGE* of the building process comprises the reconstruction of the commuter service concourse. The reconstruction will change the geometry parameters of the building. The new building is supposed to be called the union railway station passenger concourse (fig.4).

Figure 4. II stage of construction

3. *STAGE* of the building process comprises the reconstruction of the modern-day railway station building to undergo restoration works (fig.5).

Figure 5. III stage of construction

4. *STAGE* of the building process comprises the construction of the concourse leading from the existing railway station building to 15 tourist route, to the embankment of the Angara River (fig.6).

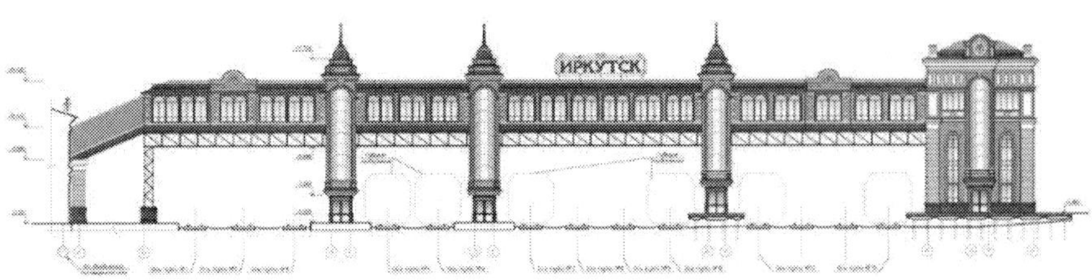

Figure 6. IV stage of construction

For the project at autonomous STAGE 1 engineering infrastructure permits were the core issue to receive approval from the State Expert Evaluation Department of the Russian Federation (Glavgosexpertiza). It was required that the connection to engineering networks should be done before moving on to the subsequent stages of the project. The project at stage 1 received the positive conclusion. The reconstruction of the the subway crossing has been completed.

For the development project at autonomous STAGE 2 the key problem was the workings of technology of the railway station. It was necessary to develop the workings of technology autonomously. The process must not be based on the subsequent stages of the development project. The building of the union passenger concourse abuts upon (southern) gable facade of the building of the railway station referred to as the cultural heritage site of regional significance "Complex of railway station buildings". The railway car shed is famous for its working people who were active participants of the revolutions and the civil war. It is located at the address: Irkutsk, Vokzalnaya Square, Chelnokov Street. Reconstructing the railway car shed is not referred to cultural heritage preservation building works. There appeared a special plan to restore this existing building at this stage of the project. But the two buildings are independent technologically. Each of the stages provides passengers with all necessary within the stage services. For the reason the development project of each building describes their different passenger capacity. Within the second stage the passenger capacity was 200 passengers for the suburban traffic service and 700 long-distance passengers.

Later on after completing STAGE 3 of the development project the passenger capacity will be equal to 1100 people in total.

Within STAGE 4 of the building process it is expected to construct the concourse leading from the existing railway station building to 15 tourist route and to the embankment of the Angara River.

Irkutsk Region government has done nothing to fulfil the obligations under the agreement so far. At this stage it is essential to develop and improve transport routes along Profsouznaya Street [10, 11]. It is necessary to analyse different variants of building a pedestrian overpass from Profsouznaya Street, across Chelnokov Street to the embankment of the Angara River as well. This pedestrian overpass will suit either passengers' and city residents' convenience. The development of the embankment along the railway station from the Glazkovskii Bridge to the existing left-luggage office. All the measures are necessary to put the development concept into existence; however, all the measures can be taken autonomously. The railway station development project including the improvement of the territory near the station is a part of the project which is aimed at improving the city environment [12].

The first thing to be done before the project commences is relocating the post office. It has to move outside the railway station adjacent territory, but still be at a walking distance from it. The authorities have already initiated the negotiations with the company Russian Post. In August 2019 an urban planning meeting with the city authorities, railway authorities, as well as the Institute "Irkutskzheldorprojekt" and Russian Post authorities was held to discuss different possible variants of relocating the Russian Post office, and creating a working advisory group affiliated with the city authorities.

The issues to tackle are ranked in the following way: the most urgent one is to improve the service conditions for people arriving to the station; then the issues of rerouteing of traffic arteries and assignment of the land parcel within the adjacent to the railway station complex streets will be paid attention to. Within the preparation process of building STAGE 2 a temporary scheme to serve passengers in the existing building of the railway station was elaborated. The scheme involves organising temporary railway ticket offices and sanitary conveniences. The ways of movement of passengers inside the railway station to get all the necessary services were assessed. According to the assessment the decision to make a revision of the development stages was taken. The reconstruction and improvement of the existing buildings of the railway station (the architectural monument) is not advantageous under the conditions of the increasing passenger throughput. The harm committed to the architectural values of the interior and facade will outweigh the potential benefits. The working space of the building will not remain functional. The servicing areas (the existing building is expected to house additional waiting lounges, sleeping rooms and a first-aid station) will remain cramped. But the most important disadvantage is that it will take passengers a lot of time to go through the pedestrian tunnel to the platforms. In some places the tunnel is not as broad as it is necessary. All the things will prevent achieving the goal to divide the passenger flow. The proposal to develop the railway station layout in order to tackle the problem has been initiated.

References

[1] Cherepanov K A 2014 Problems of choosing optimal building parameters depending on social, economic and environmental properties of the urban environment *Young scientist* **2** pp 216-232
[2] Batyrev V M 1988 Railway Stations (Moscow: Stroizdat) p 216
[3] Prikhodko A S and Fedotov S A 2016 Problems of passenger transportation quality in urban

agglomeration *Young scientist* **8** pp 281-284

[4] SP 417.1325880.2018 *Railway station buildings* Design rules. M.: 2019

[5] SP 59.13330.2012 *Accessibility of buildings and structures for people with limited mobility* M.: 2015

[6] Chuguevskaya E 2019 Ugly cities-don't live long *Architectural Bulletin* **5 (170)** pp 37–45

[7] Goncharov V V and Loginov S N 2013 Organization of pedestrian space and highways on the example of Khabarovsk *New ideas of the new century* **3** pp 87–90

[8] Prelovskaya E S, Levashev A G and Pulyevskaya E V 2014 City streets: public challenge and promising opportunities *Vestnik IrSTU* **12(95)** pp 123–127

[9] Volkova A A 2017 Small architectural forms in the space of the urban environment *Youth and science: a step to success* pp 21–23

[10] Semenikhin B A, Kuznetsova L P and Oladipupo George Olivier Oluiori 2014 Improving the organization of road traffic at the intersection of 9 Bortsov str. division-Maysky Boulevard of the city of Kursk *Modern automotive materials and technologies (SAMIT-2014)* **12(95)** pp 117–123.

[11] Kuznetsov K Yu, Semenikhin B A and Kuznetsova L P 2017 Analysis of road traffic organization at the intersection of Bortsov str., 9 division-Khrushchev Ave., Kursk *Youth and science: a step to success* pp 205–205

[12] Mikhailenko A V 2012 landscape architecture in the organization of transport flows of the city *Vestnik of Kostroma state University. N. A. Nekrasova* **3** pp 40–42

IOP Conf. Series: Materials Science and Engineering **880** (2020) 012066 doi:10.1088/1757-899X/880/1/012066

Religious buildings in developing Irkutsk city image

A I Kulakov[1] and A U Ri[2]

[1,2] Department of Drawing, Painting, Fundamentals of Design and Historical Architectural Heritage, Irkutsk National Research Technical University, 664074, Irkutsk, Lermontov Street, 83

e-mail:kulakov@istu.edu

Abstract. The article addresses some features and patterns of developing the city image. It highlights the trends to develop both natural sites and the historical heritage such as religious buildings. The research aims at looking into efficient methods to plan the urban environment when maintaining the original architectural appearance of Irkutsk.

Irkutsk at the beginning of the 21th century had more than 30 freestanding Orthodox churches, 24 of them being stone. There were also some family chapels, hospital area churches and other churches located in public buildings. Only 17 monuments of church architecture have survived so far. According to their architectural artistic and stylistic features, all the churches in Irkutsk can be divided into certain groups. Each group represents various trends in architecture: old Russian architecture, baroque, classicism and eclecticism. Baroque buildings are the quaintest in Irkutsk. Their features allow dividing them in two particular trends: "Siberian" baroque and late baroque. In the early days, in the first decades after Irkutsk Ostrog (fortress) was built, wood was the only material used in constructing churches. Their architectural appearance was similar to the buildings of the Russian North (steepled churches) and the central regions of Russia (octagonal or quadrangular structure churches). [1]

Today the cultural landscape of the city results in layering built structures of different time reference on the natural landscape. Since XVII century Irkutsk has gone through several periods of evolution. The periods are characterized by developing specific principles in planning the city environment and specific architectural-artistic approaches in the city building design. In other words, they reflect the essential features of evolution processes in developing urban territories in accordance to the necessity to increase the built-up density of the historical areas of the city to meet, on the one hand, the natural evolution requirements to change architectural styles and to follow different building density demands in certain parts of the city, on the other hand. As a result, the urban development is characterized by its being inhomogeneous both in the context of spatial planning and architectural composition, as well as by the design elements. The elements of the urban historical development of Irkutsk are represented by the following two categories. The first category was comprised of regular buildings for housing and commercial purposes. They are a contrasting featureless background for the churches, monuments, buildings intended for public use, as well as private buildings. The second group consists of some large constructions, churches, monuments and public use buildings, which are considered to be design accents for structuring the city environment. The buildings referred to the second category had to be bigger and stand out from the buildings of the first category to contribute to the city's cohesiveness and make its appearance impressive. Such a proportion between the buildings of the different categories and the buildings within the categories influenced their architectural-artistic grouping at the early stages of the city development. Mutual interaction of "monostructure", which included very tiny number of unique public use buildings, and spontaneous development of "polystructure" (residential buildings) provided architectural integrity of Irkutsk in the past. Religious architecture was popular "monostructure" in Irkutsk. Ecclesiastic buildings such as churches, being large and picturesque, contrasted with single-storey buildings and created a magnificent recognizable panorama of the city. The terrain of the city being hilly contributed to locating its most impressive constructions on the hilly sites. In this context they were felt as an architectural dominant. The analysis of locating high-altitude position buildings (religious constructions) in relation to the city terrain and the river showed that there are 3 groups of dominants in Irkutsk:

1. The group of high-altitude position dominants, which are river oriented, built along the Angara River and the Ushakovka River: Kharlampievsky Church, Roman Catholic Church, Church of the Saviour, Epiphany Cathedral, Znamenskii Convent, Kazansky Church, Church of the Transfiguration.
2. The group of high-altitude position dominants located on the terrace edge: Church of the Exaltation of the Holy Cross, Jerusalem Church.
3. The group of high-altitude position dominants located in the central part of the terrace of the Angara River. The lost pieces of religious architecture (Kazan Cathedral, Tikhvin Church, Annunciation Church).

The history of international urban planning regularities proves that a city skyline was usually constructed so that it created contrasts between vertical constructions and ribbon buildings. In Irkutsk wooden churches were four times higher than the ribbon buildings, while stone churches were five times higher than the contrasting ribbon buildings. However, there are exceptions in Irkutsk. The proportion of the height of Kharlampievsky Church to the residential buildings was the following: $h : h_1 = 5m : 48{,}040\ m = 1 : 9{,}608$. The contrast between the architectural dominants (religious buildings) and residential buildings was expressed in their architectural styles since the churches had more sophisticated architectural plastic decorative characteristics. The plain featureless silhouette of V-shaped roofs matched against heavenward steepled silhouette of the churches and their sophisticated Baroque elements and contours. All the churches had no comparable counterparts.

The contours of the tops of the religious buildings held a distinction. They were unique owing to their architectural perfectness and artistic value. In XVII century they were steepled mostly and had a lot of similarities with typical Russian North religious buildings. It is explained by the fact that Russian North and Siberian cultures are considered to have the same roots. In XVIII century it was typical of religious buildings to have pyramid-shaped silhouettes. They were constructed to have several roof layers of different size; each higher roof layer was smaller. Some churches could have stupa-shaped silhouette to vaguely resemble Buddhist stupa and Buddhist temple complexes (such as the top of the Church of the Exaltation of the Holy Cross). Later many of the religious buildings were reconstructed (for example, in the Church of the Saviour and the Epiphany Cathedral).

In contrast to the northern cities of Russia with smooth silhouettes in the soft lighting, Irkutsk religious architecture is distinguished by spacious sculptural constructions of the tops (domes) of the city architectural dominants (churches). The rich effect is shown through the perfect light-and-dark value contrast accompanied by small-scale sculpture technique. The refulgence of the light-and-dark value contrast was highlighted by the light colour of the tops of the dominants. That was whitewashing technique used first and then the surface of the tent-roofed tops of the religious buildings were covered with thin layers of gold. The range of colours on the walls and church domes was essential to convey the impression of "notability". Optical illusions were commonly used for it: light-coloured constructions seemed to be larger in contrast to the dark colours of the residential buildings in the background. In addition to the optical illusions produced by the techniques mentioned above, the impression of "notability" of the churches was seen through the contrast to people's physical characteristics. Moreover, the religious buildings were often decorated with symbols and motifs going back to a distant memory from the pagan past, inspired by the pagan rites.

This proportion between vertical dominants and residential buildings is closely related to the storey factor (the number of storeys in a building). A few cities in the world could have preserved their original appearance due to having implemented regulations for controlling the number of storeys in a building. St.Petersburg is referred to such cities where building high-rise constructions in the centre could be authorized only when meeting the requirement (1766) "The Committee on the Stone Construction of St. Petersburg", which says that no buildings bigger than 10 sazhens high (old Russian measure of distance equal to 2.13 m) are allowed on the embankment of the Neva River.

Today, when large-scale housing construction is increasing and the centre of the city is being reconstructed, the problem of regulating the number of storeys is increasingly relevant. Especially within the process of revitalization of the city landscape, it is necessary to restore its original appearance without any changes. It is highly recommended that monuments of architecture contributing to Irkutsk original historical image should be preserved. Building new high-rise constructions has to meet the need for combining with the historical architectural ensembles and monuments. The only way to preserve and develop the multifaceted image of the capital of Eastern Siberia is to consider its historical architectural composition. There is a trend in Irkutsk to develop mid-rise buildings. But it contributes to failing to have no changes in the panorama of the city. According to Evreinov Y.N. the panorama of the city will be impressive under condition of its balancing with the natural environment since buildings have to be an integral part to the landscape and adapt to it. The city space should improve the landscape when making it dynamic and expressive, or calm in terms of the urban planning situations [2]. A. G. Bolshakov believes that in order to achieve a balance of interests in developing the city centre, the new urban development policy should be based on the natural relationship of values and morphotypes. The author analyses the restoration of the centre of Irkutsk emphasizing the high quality of organizing the space. He suggests that the balance of values and their proper distribution on the territory of the historical centre of the city should be based on the urban planning rules and regulations [3]. N. N. Baranov considers that configuration of the terrain and water surfaces - seas, lakes, rivers, channels and reservoirs - as natural factors have a certain influence on the skyline of the city. The terrain being hilly contributed to locating the highest and the most impressive buildings on the hilly sites (castles, cathedrals, palaces or monuments). In this context they are felt as architectural dominant ensembles and it is the terrain that contributes to it [4].

The visual characteristics of landscapes were described by A. Vergunov [5], L. Zalesskaya, E. Mikhulina [6], J. Simonds [7]; the methodology of architectural and landscape analysis was the field of interest of A.P. Vergunov [5], A.V. Sycheva [8], M. Purvinas [9]; various aspects of adjusting the city layout with the landscape were studied by A.G. Bolshakov [10], A.P. Vergunov [5], V.R. Krogius [11], Yu. B. Khromov [12] and others; harmonizing architectural forms and the landscape was studied by Yu. I. Kurbatov [13]; identifying the dimensional structure of the landscape was analysed by V. K. Shcherban [14]; V. K. Moor [15], E. A. Erysheva studied the understanding of the city appearance through its panorama [16]

The issue of identifying and preserving the original architectural appearance of the city is related to evaluating its aesthetic qualities. So, there appears a question about the criteria for their evaluation. M. E. Weitens points out that under the conditions of the North there are two factors that make the original appearance of cities expressive and unique: nature and urban development. There is one feature that is common for such cities as Murmansk, Kirovsk and Apatity. What unites them is the visual relationships between the city environment and the natural environment. In those cities where the nature-urban development relationships are provided by the urban planning pattern, which turning the buildings to the most picturesque spaces of natural surroundings, the urbanscape is notable for being the most expressive and original. Monchegorsk and northern part of Murmansk are referred to such cities. The architectural precedent of Murmansk proves that one of the possible directions to develop a distinctive appearance of a northern city is contrasting residential buildings, where wind-protective design methods are employed, with the natural landscape. Clear functional zoning to locate the industrial port area and residential area at a distance had a positive impact on developing the urban environment of Murmansk. Its terraced spatial type allowed the observer to see clearly the terraces from top. Research of Leningrad Scientific Research Institute proves that there can be different modes of communication with the natural environment depending on the severity of the climate. There are three types of the modes: closed, nominally open and open regimes [17]. Bolshakov A.G. points out the sustainable development of the landscape as a criterion of harmony for the architectural environment, as its being disturbed cannot be aesthetically appealing [18]. Since the urban and natural spaces interact on a regular basis, they have to be considered and evaluated through one and the same criterion. Spatial structures of the landscape and the city can be taken as a criterion for such an evaluation. Y. Raninsky touches upon the consistency in developing the historical part of the city and speaks for a special "ensemble of spaces". First and foremost, the city image depends on the type of correlation these spaces have. It can be spacious or cramped, open to the light and air and, conversely, gloomy and closed. Its being picturesque and regular is primarily reflected in its spaces. People do not see the plan of the city. Spatial relationships are the first thing the observer perceives in the process of understanding and organising the city image [19]. To find these relationships, it is necessary to generalize natural forms in their hierarchical structure. It will make it possible to understand the geometric patterns of constructing the spatial structure of the landscape, to show the structural diversity and unity of the natural environment and to quantify information about the physical parameters of the space. Furthermore, it will contribute to obtaining an objective basis for architectural and spatial composition. It provides the opportunity to use common graphic symbols to denote the forms of urban and natural spaces. Z. N. Yargina claims that the spatial structure of the landscape is always a certain sign of originality. Its allocation allows for a long-term potential assessment of the development of the urban fabric, while maintaining uniqueness in the appearance of the city [20]. This is due to the fact that the urban buildings are developed to meet social requirements (sanitary and hygienic, socio-psychological, socio-demographic, socio-economic, socio-cultural). Constructive and technological needs evolve within the economic and social progress of society and are not so stable. The architectural planning stereotype of urban spaces and typical means of building make both different areas and even cities similar. The structural diversity of landscape spaces being unique due to their aesthetic characteristics creates a stable system of spatial orientation in time and can play an important role in the development of individual original features in the architectural and artistic appearance of cities.

When remembering the history of urban planning and the history of urban development, we should not forget that in many cases cities were settled on those lands that are now classified as being inconvenient. In order to be protected against enemies' attacks steep slopes and hills of the landscape were used. Settlements in the bends of the rivers were founded to make it easy to move. Various natural anomalies of certain areas contribute to developing specific individual architectural appearance of urban environment. Natural factors, both favourable and unfavourable ones, affect the original appearance of the planning structure and architectural solutions of urban development significantly" [21].

Religious buildings in the urban development were not located randomly, but they were the milestones to show the compositional elements of the landscape and make them visible in 3D pictures. So, looking for the elements of original architectural appearance of cities, it is important to start with the identification of the three-dimensional structure of the natural landscape. Selecting and organising the hierarchy of the three-dimensional structure of the landscape gives the opportunity to identify distinctive features that make the landscape unique. The main feature is a hierarchical mosaic structure of spaces that certain landforms have. The generalized structural scheme of space hierarchy will make it possible to identify the smallest module of the landscape, which visual properties have to remain intact. The scheme will demonstrate the location and the way of interaction of these dimensional units. The scheme is expected to be the tool to develop the city appearance. Aligning the built-up environment and the natural environment is the only way to make the interaction of the landscape and the urban development harmonious.

Thus, the main challenge of a professional designer when following the principle to preserve the original appearance of the city is to identify and maintain the ensemble of landscape spaces, as well

IOP Conf. Series: Materials Science and Engineering **880** (2020) 012066 doi:10.1088/1757-899X/880/1/012066

as to accentuate it by means of building city objects. All the building transformations have to be done providing for the high-rise historical dominants (religious buildings). The core role of the survived religious buildings in the city-planning policy and their architectural value are of great significance to maintain the original image of Irkutsk.

References

[1] Kalinina I V 2000 *Orthodox churches of Irkutsk diocese of XVII-XX centuries* (Moscow)
[2] Evreinov Y 1983 *Issues of form making in contemporary architecture* p 215
[3] Bolshakov A G 2012 Evaluation of building morphotypes as a reflection of the interests and values of the urban community and their balance as a principle of urban regeneration of the historical centre *Vestnik ISTU* **9** pp 89-97
[4] Baranov N N 1980 *The silhouette of the city* (Leningrad: Stroizdat) p 184
[5] Vergunov A P, Denisov M F and Ozhegov S S 1991 *Landscape design. Architecture-S* (Moscow) p 231
[6] Zalesskaya L S and Mikhulina E M 1979 *Landscape architecture (*Stroizdat. Moscow) p 240
[7] Simonds J 1965 *Landscape and architecture* (Stroizdat. Moscow) p 196
[8] Sycheva A. V. Landscape architecture. ONIKS 21st century. Moscow. 2004. 110 p.
[9] Purvinas M. 1975 Primary units of the spatial structure of the regional image and their possible characteristics Construction and architecture *Architecture and urban planning* 4 pp 5-21
[10] Bolshakov A G 2004 *Fundamentals of the theory of urban and regional planning* (Irkutsk: Irkutsk state technical university) p 214
[11] Krogius V R 1988 *Urban planning on the slopes* (Stroizdat. Moscow) p 328
[12] Khromov Yu B 1987 *Landscape architecture of cities in Siberia and European Nor*th (L.: Stroizdat) p 200
[13] Kurbatov I Yu 1988 Architectural forms and natural landscape: the compositional connection (Leningrad: Leningrad University) p 136
[14] Shcherban V K 1987 *Urban landscape and architecture* (Kiev: Budivelnik) p 87
[15] Moor V. K., Erysheva, E. A. Creating architectural and artistic appearance of seaside cities. Vladivostok: FESTU, 1997. - 88 p.
[16] Erasheva E A 1992 *Creating architectural panorama of the large seaside city (exemplified in Vladivostok)* Extended abstract of dissertation, PhD in Architecture (Moscow: Markhi)
[17] *Architectural artistic appearance of the northern cities* 1978 (Leningrad: Stroiizdat)
[18] Bolshakov A G 2003 *Developing the urban landscape as a factor of sustainable development of the territory* Extended abstract of dissertation PhD in Architecture (Moscow: Markhi)
[19] Raninskii Y 1988 Monuments of architecture and urban development (Moscow) p 25
[20] Yargina Z N 1984 Urban development analysis (Moscow: Stroyizdat) p 244
[21] Kontorovich I Y and Rivkin A B Rational use of the urban territory (Stroizdat) p 162

IOP Conf. Series: Materials Science and Engineering **880** (2020) 012067 doi:10.1088/1757-899X/880/1/012067

Development of investment zones in the renewal model for a former industrial city: Case-Study of Chattanooga

A Lyapin[1]

[1]Department of Architectural Design, Irkutsk National Research Technical University, Lermontov St. 83, Irkutsk, 664074, Russia

E-mail: a782lyapin@yandex.ru

Abstract. The article describes the advanced experience of the USA which have been developing investment zones in modernizing the economy and the planning structure of cities using the example of Chattanooga (Tennessee). The initial state of the economic and social decline of Chattanooga in the early 1980s is described. The results of the analysis of social groups that initiated the process of modernization of the urban economy in the first half of the 1980s are presented. The importance and role of urban planning and new land management policies implemented in Chattanooga are identified. The initial and strategic aspects of the economic growth which have become drivers of the urban development are identified. The role and importance of tourism, new tourist attractions and a new network of natural parks, which became the basis for the tourist flow at the national level, are described. The types of new investment zones are determined and their connection with the development of the planning framework of the city is described.

1. Introduction

The issue of conversion of the economies of former industrial cities, where one or several enterprises of the same industry formed the economic basis of the urban life, has become relevant in Europe and the USA since the early 1980s. Currently, this issue has also become relevant for Russia. The successful experience of conversion and subsequent post-industrial modernization of Chattanooga (Tennessee) is important and valuable for the theory of urban development and conversion of obsolete economic forms. Chattanooga's renewal was based on local urban resources (investment activities of Chattanooga's entrepreneurs, well-calculated municipal land management policies and activities of public organizations and volunteers).

Federal funds, state loans, and corporate sponsorship were used for the urban renewal, but did not play a decisive role. The experience of modernizing the economy and urban life of Chattanooga has become widely known and recognized. This experience has been widely discussed and used by other US cities. In Europe and Russia, the municipal experience of Chattanooga's development is much less known.

2. Methods

When conducting a study on investment zones and their effectiveness in Chattanooga, several methods were used. The author's internship in Chattanooga organizations in 2003 (the SouthEastern Center of the National Association of Recreation and Parks, the Architectural and Planning Department of Chattanooga and the architectural company "BWSK") allowed the use of the included observation method. It provided the author with knowledge of specific measures aimed at modernizing the city. The sociological analysis made it possible to determine features of the urban population and identify social groups that were initiators of the revival of the Chattanooga economy and renewal policies. The literature and press analysis identified views on the importance of the experience of Chattanooga for the theory and practice of urban planning. This experience was recognized a valuable contribution to the practice of modernizing US cities and towns at the national level. Chattanooga Magazine (1999-2003) was analyzed. Very important information was obtained using the Interview Method. 10 interviews of urban planning experts and representatives of public organizations were conducted. The analysis of urban planning documents - Master-plans and economic development schemes - determined features of urban development and the method for developing investment zones in the urban planning framework.

3. Basic stages of the renewal of the city

By the beginning of the 1980s, in developed countries, many large and medium-sized cities have experienced the influence of new trends in the global economy. During this period, the economies of developed countries began to adapt to the new level of information technology and post-industrial development. The traditional industries (metallurgy and metalworking, coal production, railways, textile and mining industries) began to play an insignificant role. Accordingly, a large number of industrial enterprises have lost their significance.

The process of restructuring priorities of the global economy affected developed countries - the United States and Western European countries. Many US cities, where coal, metallurgy, or chemical industries were developed, faced serious problems - reduced production, lower tax revenues, increased unemployment and social stress, large areas of abandoned enterprises, pollution, factors threatening the environment. In many US cities, these problems require urgent practical solutions.

The theory of urban management, like many other related disciplines, was not able to develop recommendations and formulate approaches to solving pressing urban problems. It was necessary to develop a new theoretical approach to the urban development and new models of urban processes.

Chattanooga city (Tennessee) is one of the US cities that have found ways out of the crisis and solved the problems of restructuring the urban economy in the mid-[Gratz, 1989].

It is a relatively small city with a population of 150 thousand inhabitants. It is located on both banks of the Tennessee River in the southeastern part of the state. Since the mid-19th century, the city has been a regional center for metallurgy and metalworking. Metallurgical plants and timber processing enterprises along with rail transport enterprises formed the basis of the urban economy. The metallurgy used cheap coal delivered from the Appalachian region by rail.

Wholesale business was crucial for the urban economy and budget. Since the mid-1930s, the city has hosted the Tennessee Valley Development Administration created according to the federal program developed by Roosevelt's Administration. A cascade of hydroelectric power plants was erected near the city. It became the basis for aluminum enterprises formed near Chattanooga.

In the postwar period, the economic development slowed down. Since the late 1960s, the city began to experience signs of the economic stagnation.

Since the mid-1970s, Chattanooga has been in a serious crisis. Metallurgy and timber processing enterprises were closed down or are being closed down. The railways reduced traffic and have experienced serious difficulties in competing with road transport enterprises. Wholesale trade enterprises ceased their activities.

By the pollution level, the city was one of the most disadvantaged ones in the country. According to the environmental situation, Chattanooga is the dirtiest city in the United States.

A logical consequence of the economic stagnation was a dysfunctional social situation. The levels of unemployment and urban crime were growing. The funding of the municipal education system was reduced which decreased the number of school curricula. The number of students and teachers decreased as well. The outflow of wealthy residents began. They moved to other cities with well-paid jobs and safer social environments.

The need for radical changes in Chattanooga was recognized in the early 1980s. Initiative public associations and the municipality developed and began to implement a program aimed at reviving the city [Chattanooga's R.2002].

The Chattanooga development and renewal program is one of the most successful examples of the revival of urban economies in the country. Twenty years of thoughtful and coordinated measures taken by city authorities, public associations and entrepreneurs have improved the social and environmental situation in the city. By the end of the 1990s, the success of Chattanooga in transforming its economy, restoring the ecological parameters of the urban environment and improving living standards were recognized by all urban development experts, politicians and US citizens The American environmental urban planning theorist Timothy Beatley argues that Chattanooga's practical renewal program was one of the best US programs. Its results were outstanding. [Beatley, Maning, 1997].

In the late 1990s, the city became one of the 10 best US cities by the urban environment and convenience. The city was called a "Family-Friendly City" for several years. The success of the post-industrial development of the urban economy was impressive: the city has attracted new residents and capital. It has become a center of advanced education methods and modern technologies, one of the most dynamic tourist centers in the South of the USA.

These impressive achievements were the result of a strategic model for the urban development adopted in the early 1980s.

The strategic development model for Chattanooga city has the following features:

First, public associations (real estate owners and entrepreneurs) have developed a strategic model for the urban development and initiated the city movement for improving the economic and social situation.

Second, the main efforts to improve the situation were focused on solving environmental problems and restoring the natural environment in Chattanooga. The initiators of the program made environmental issues a core of the urban revival program. The image of the Tennessee River was chosen as a symbol of the urban revival at a new stage of the history of Chattanooga. The word "River" became one of the most common in individual projects within the strategic development model [Demchenko, 2013].

Third, the initiators chose the main city revival project, the Leading Project, which fundamentally changed the role and image of the city in the region. It has become the core of the Chattanooga Revival Program for 10 years. A project on the construction of a large tourist site was developed and implemented. The Tennessee Aquarium, the largest freshwater aquarium in the United States, has become a major tourist attraction. The advertising campaign contributed to the popularity of the Tennessee River Museum of Nature in the US South. The new Aquarium made Chattanooga famous for being the city having the world's largest collection of freshwater animals and plants [Gulyaeva, 2019].

The fourth important feature of the model was constant, detailed and careful monitoring of both the implementation of the model and results of the measures taken. The public associations created at the beginning did not break up and disperse. They continued to develop individual projects and monitor the implementation of the development model. [Gulyaeva, 2019]

Formally, the beginning of the city's revival is 1984, when the Chattanooga Venture Fund business association held a session of the city seminar "Perspective Vision 2000". Based on its results, it published the model of the city's revival - "Workbook of Opportunities and Obligations". It was a new development strategy for 20 years. After a year of hard work, the main planning document of the new strategy was published and put forward for discussion. It was the General Plan of the Tennessee River Park. Economic development was initiated in the territories adjacent to the River Park, while the park itself was the result of restoration of contaminated and abandoned sections of the former industrial territory [Chattanooga's R.2002].

In May 1992, the Tennessee Aquarium Museum was opened. By the end of the year, the Aquarium received one million visitors - a large number of tourists for a city with a population of 150 thousand inhabitants. It revived the economic life - incomes of restaurants, cafes, souvenir shops located in the historical part of Chattanooga sharply increased.

In 1993, nine years after the start of the City Renaissance Program, the new citywide seminar "Perspective Vision 2000" verified the results achieved and set new city development goals.

Since 1993, travel and large private companies have invested in the construction of new hotels in Chattanooga, believing in the viability of the city. In a series of new hotels, the first object is "The Riverside Apartments" - The River Side Hotel.

During the 1990s, in Chattanooga, new hotels were built, new restaurants were opened and the system of urban public transport was renewed. Municipal authorities and public associations support the current tourist flow - festivals and events attracting tourists are held throughout the year ("Annual Brewers Festival", "Wine Festival over the Water", "Concerts of Pop Music on the River", Days of nature, parks, Harley-Davidson Motorcycle Fans, the nine-day River Ribbon music festival attracting music lovers from all over the country). The local private company "River City" plays a huge role in the revival of Chattanooga. It was created for developing the downtown.

Future development modeling was carried out by researchers, entrepreneurs and public organizations. Some researchers - geographers, ecologists, and sociologists - participated in the development of the strategic model on a voluntary basis. Other researchers - ecologists, biologists, geographers, architects, economists - were involved in the development of the model as hired specialists. Charles Landry said that public organizations had helped use advanced methods for developing a strategic model for Chattanooga [Landry, 2000]. They developed individual parts of the general model for the city development.

4. Types of investment zones

The main difference between investment zones created in Chattanooga from the usual practice of attracting investment in cities is a deep comprehensive calculation of main directions of the urban economic policy. The city authorities and representatives of public organizations defined ecology as a strategic direction for the urban revival. Each investment proposal was assessed based on the strategy of restoring ecological qualities of the urban environment. The specialization of each investment site should be part of the overall economic and planning development of the city. Taking into account interests of investors, the city planning department created favorable conditions for [Polyakova, 2010]. Cultural sites were also considered as investment zones by city planners.

The first investment zone was located along the Tennessee River Park on the river banks [The New Downtown Plan, 2003]. Some investment sites were included in the Park (e.g., the Tennessee Aquarium Museum in the city center). The Tennessee Aquarium has become a national attraction for tourists. It was built in 1992. At the beginning of the 21st century, local and national business parks were built in the investment zone of the River Park. The tourism-related investment area (hotels, restaurants, shops) was created in the historical center of Chattanooga. On the periphery of the city, on the territories of former metallurgical plants, several investment zones of modern technologies were created: enterprises designing and producing urban electric transport, solar energy, information technology. The spatial position between these enterprises and the tourist investment zone was occupied by a rapidly developing investment zone of offices and administrative buildings. In Chattanooga, investment in new housing construction is still insignificant in comparison with investment in other types of real estate.

5. Results

The study identified methods for developing investment zones on the basis of cultural and educational facilities. The interrelation of strategic urban planning strategies and the search for profitable investment sites meeting investors' interests was shown. The renewal experience, the experience of creation of investment zones in a new, post-industrial economy is relevant for many cities and towns and modernization of Russian single-industry towns.

6. Conclusion

The modernization of Chattanooga was carried out due to a thoughtfulness interdisciplinary approach applied to develop a model for a new stage in the urban development - "The Workbook of Opportunities and Obligations" of 1984. This development model was a guideline for the development of Chattanooga.

Environmental and cultural issues have become crucial for attracting investment in the urban economy. Based on the 30-year city renewal experience, one can argue that the development strategies were rational. In conclusion, it should be noted that the general strategy has been implemented by public associations of Chattanooga city. Charles Landry said that public organizations had helped use advanced methods for developing a strategic model for Chattanooga. The American environmental urban planning theorist Timothy Beatley argues that Chattanooga's practical renewal program was one of the best US programs. Its results were outstanding.

Reference lists

[1] Beatley Timothy, Manning Kristy 1997 The Ecology of Place *Planning for Environment, Economy, and Community* (Washington D.C.: Island Press)
[2] Chattanooga's Resurgence 2002 *A Decade of Downtown Development (*Chattanooga: RiverCity Company)
[3] Demchenko E A 2013 Symbolical design of the brand Social Humanistic knowledge **11 pp** 350-358
[4] Gratz Roberta Brandes 1989 *The Living City* (N.Y.: Simon and Schuster)
[5] Green Grant Sebastian 1999 An Island in the Hollow of the Mountains *Chattanooga Magazine* **8**
[6] Gulyaeva K O 2019 Architectural Program for creation of new museums at the Chattanooga Sukachev Riders **16 pp** 44-49
[7] Polyakova A G 2010 Regions for new opening up in the modernization (Tumen)
[8] Landry Charles 2000 The Creative City. *A Toolkit for Urban Innovation*
[9] Parker Lin 2003 Where The Land Reveals Its Secrets *Chattanooga Magazine* **4**
[10] Polyakova A G 2010 Regions for new opening up in the modernization. Tumen. (In Russ.)
[11] The New DowntownPlan 2003 The Master-Plan of The City of Chattanooga
[12] Kosolapova A R, Antonov A V 2019 20-th session International Baikal Winter University of urban planning: updated methodology of research and design sessions *Proceedings of Universities. Investment. Construction. Real estate* **9(1)** pp 166–173
[13] Bannoud Ghaleb 2019 Features of the historic urban living environment in the city of Aleppo *Proceedings of Universities. Investment. Construction. Real estate* 9(1) pp 158–165
[14] Makotina S A, Daggers T, Karepova A I 2019 Population mobility in a contemporary near-water city (on the example of the cities of Amsterdam and Irkutsk) *Proceedings of Universities. Investment. Construction. Real estate* **9(4)** pp 842–857

IOP Conf. Series: Materials Science and Engineering **880** (2020) 012068 doi:10.1088/1757-899X/880/1/012068

Architecture of the University campus in terms of maximum energy efficiency

Z F Nizamutdinova[1] and N E Krivchikov[1]

[1]Federal State Budget Education Institution of Higher Education Irkutsk National Research Technical University (IRNTU), Institute of Architecture, Construction and Design, Department of Drawing, Painting, Basics of Design and Historico-Cultural Heritage, 83 Lermontov str., Irkutsk, 664074, Russia

E-mail: zfartit@gmail.com

Abstract. This article presents an analysis and overview of the main energy-efficient methods in the aspect of the campus planning structure in order to preserve the maximum comfort of students in various weather periods, as well as technologies for adapting campus buildings. The purpose of the work is to determine suitable methods for applying energy-efficient technologies in Siberia, and to present possible proposals for using these technologies in relation to the architectonic appearance of University campuses in Irkutsk.

Speaking about energy efficiency in terms of space-planning solutions and the architectural appearance of the campus, it should be noted that in regions with climate conditions with sharp temperature changes, in particular, and negative in general, it remains particularly acute, because finding a balance and universal solutions for these territories is still relevant. There are many positive examples and non-trivial solutions to this problem. However, in order to proceed to concrete examples, it is necessary to note the main difficulties and obstacles dictated by certain aspects of the climate area.

To include the aspect of temperature changes inside the main campus building and, accordingly, the outdoor space, which entails a number of related problems, from the smaller - the presence of the wardrobe for students and staff of the University, the existence of a double vestibule and larger - the inability to access the campus because of extremely uncomfortable high or low temperatures, as for access to the building must interact with the street. The above problems are elegantly solved by many campus planning structures. The first is the University of Waterloo [12] in Waterloo, Canada, where all the buildings on the campus permeate the transitions between buildings and exclude contact with the adverse environment to a minimum (figure 1). This example demonstrates the persistent pedestrian connectivity of the structure of the entire University and the interpenetrating layout, which again allows students to effectively move and communicate with various specialties and directions. The second is Qasim University [7], al-Qasim province, Saudi Arabia, where we observe the same principle, but under different climatic conditions and with a different centric layout from the previous example, in the core of which is the main building of the campus, with a large covered space inside and from it there are pedestrian crossings to other buildings-dormitories, laboratories, academic buildings (figure 2). From the point of view of compositional construction of space, the main guidelines and clear structure are clearly distinguished, which again increases the efficiency of navigation and sense of space. The third example is the University of southern Denmark [10], which is dotted with transitions between buildings and sets a fairly solid structure, with all buildings of medium and low storeys (figure 3).

One of the minor problems was the entrance to the educational institution, which should be equipped with a vestibule - an air layer to preserve heat. So in the Harbin Institute of architecture in the entrance

group is a number of dense vertical fabric canopies like curtains. This is one of the ways to adapt a

building to harsh climatic conditions, if it was not designed with these in mind.

Figure 1. University of Waterloo, Canada [5]. The author analytics is N Krivchikov, 2020

Figure 2. Qasim campus structure, Saudi Arabia [5, 7]. The author analytics is N Krivchikov, 2020

Next, we can observe planning solutions that meet the challenges of wind loads and create a cascade for shelter. These include the University of Toronto Scarborough [11] and Novosibirsk State University [6]. In the first case, we observe a continuous main building that wraps around the forest and creates the same barrier against the wind (figure 4). Considering the building of the NSU, it is impossible not to note the isolation of the main building, where again the fact of wind protection of the internal space can be traced (figure 5). Thus, we can say that this type of location of the main building can create comfortable conditions of arrival both inside, due to the length or closed system, and outside, because even in winter, the recreational space remains protected from the wind.

The energy efficiency of campuses directly depends on the amount of energy that consumes and saves the entire ensemble of the campus, so it is important to use the most effective technologies for energy conservation and energy reproduction. These include the University of Queensland [9],

Queensland, Australia. On its territory there are arrays of solar panels that perform not only the function of generating electricity, but also protection from precipitation, the scorching sun. (figure 6) these panels do not look foreign and are rather a business card of an advanced University.

Figure 3. University of southern Denmark campus, Denmark [5]. The author analytics is N Krivchikov, 2020

Figure 4. Analysis of the University of Toronto Scarborough campus, USA [5]. The author analytics is N Krivchikov, 2020

Also, the use of solar panels can be observed on the territory of Stanford University, although there the panels were not integrated into civil buildings, but the array of panels located on the roof of a multi-level parking lot produces a fairly large amount of energy in relation to the California climate. This again indicates the interest of the world's leading universities in alternative energy sources and the prospects of this technology in terms of introducing students to such technologies. In addition to the roof of the Parking lot we can notice solar panels directly on the energy complex.

In addition, at the University of southern Denmark, one of the institutes was built using adaptive facade technology [4], in which the elements had a degree of freedom sufficient to change their position depending on weather conditions, thereby saving energy for cooling and heating the room. Having sensors and lights that analyze the current state of the environment allows the facade elements to change their position offline, and even if the panel is closed, the perforation on it helps natural insolation and ventilation. The elements themselves are determined by the basic shape of the building and the interiors echo this idea. Also on the roof we can note the presence of solar panels, which again emphasizes the environmental friendliness of the building (figure 7).

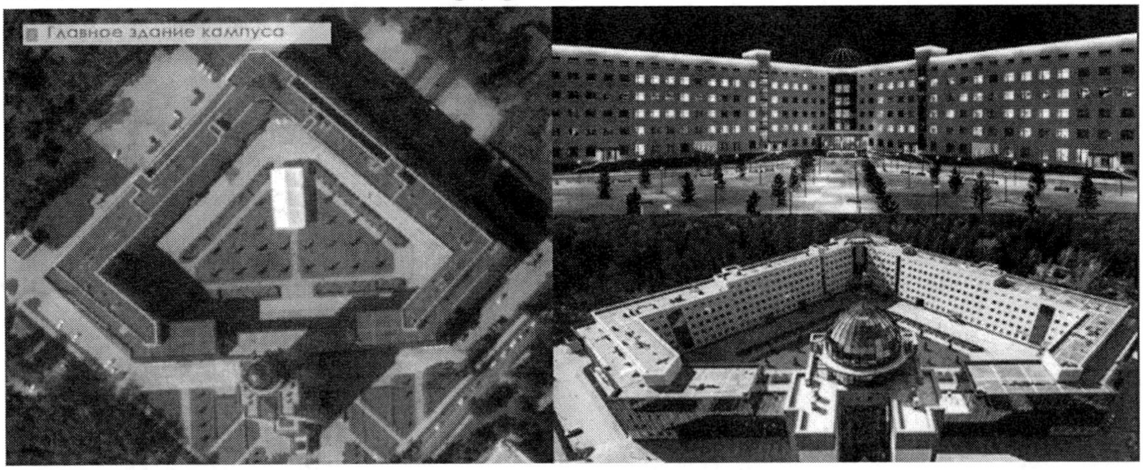

Figure 5. Novosibirsk State University, Russia [5, 6]. The author analytics is N Krivchikov, 2020

Figure 6. University of Queensland, Australia [5, 9]. The author analytics is N Krivchikov, 2020

The complex of the biomedical laboratory complex in the center of Phoenix, Arizona, USA (figure 8) is also a striking example of the use of the facade [2]. The building's cladding consists of about 5,000 recycled copper panels, which emphasize the distinctive identity of the area and protect the interior from direct sunlight [8]. The layered system of facade panels consists of rigid insulation, five centimeters of air space and a waterproofing membrane. The integrative system allows light to penetrate into interior spaces and creates a unique architectural appearance of the building, which resembles a canyon in its shape. The design used sustainable features - led lighting and chilled beams in offices and laboratories. The building is a recipient of the LEED Silver certificate from the green building council of the United States.

The Aga Khan University [1], Pakistan (figure 9), is also an example of advanced technologies of the topic under consideration. Here we can observe a symbiosis of the above techniques, including elegant facades and transitions between buildings. The architectonic appearance of the University creates a huge number of shadow spaces that hide visitors. Natural materials create a natural and environmentally oriented environment [13].

In the winter of 2018, calculations were made in the framework of the competition "vanguard of science" among 1-2 courses, pedestrian activity on the campus of the Irkutsk national research technical University, the most active pedestrian routes were determined (figure 10).

Figure 7. University of southern Denmark, Kolding [10]

Figure 8. Phoenix biomedical campus, USA [2]

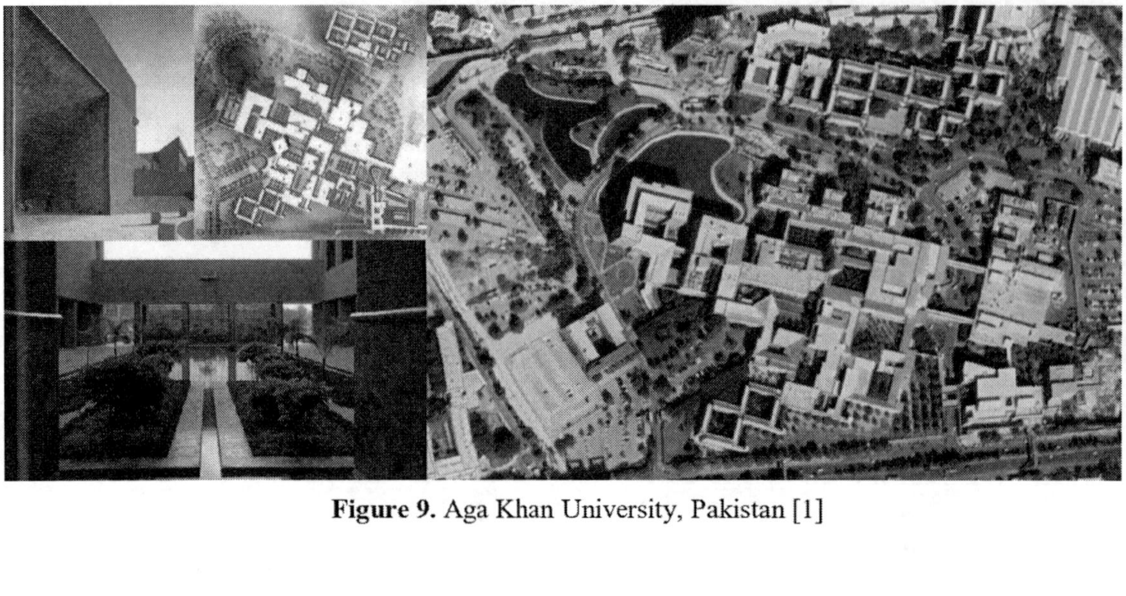

Figure 9. Aga Khan University, Pakistan [1]

In this regard, you can consider the option of comfortable organization of flows of students and employees from public stops, parks and dormitories of the University in the winter, suggesting transitions between buildings, following the example of the University of Waterloo. This will help to eliminate contact with an unfavorable environment to a minimum, and will allow you to organize a stable pedestrian connection of the structure of the entire University and the development of communication links between students and teachers of various specialties and directions.

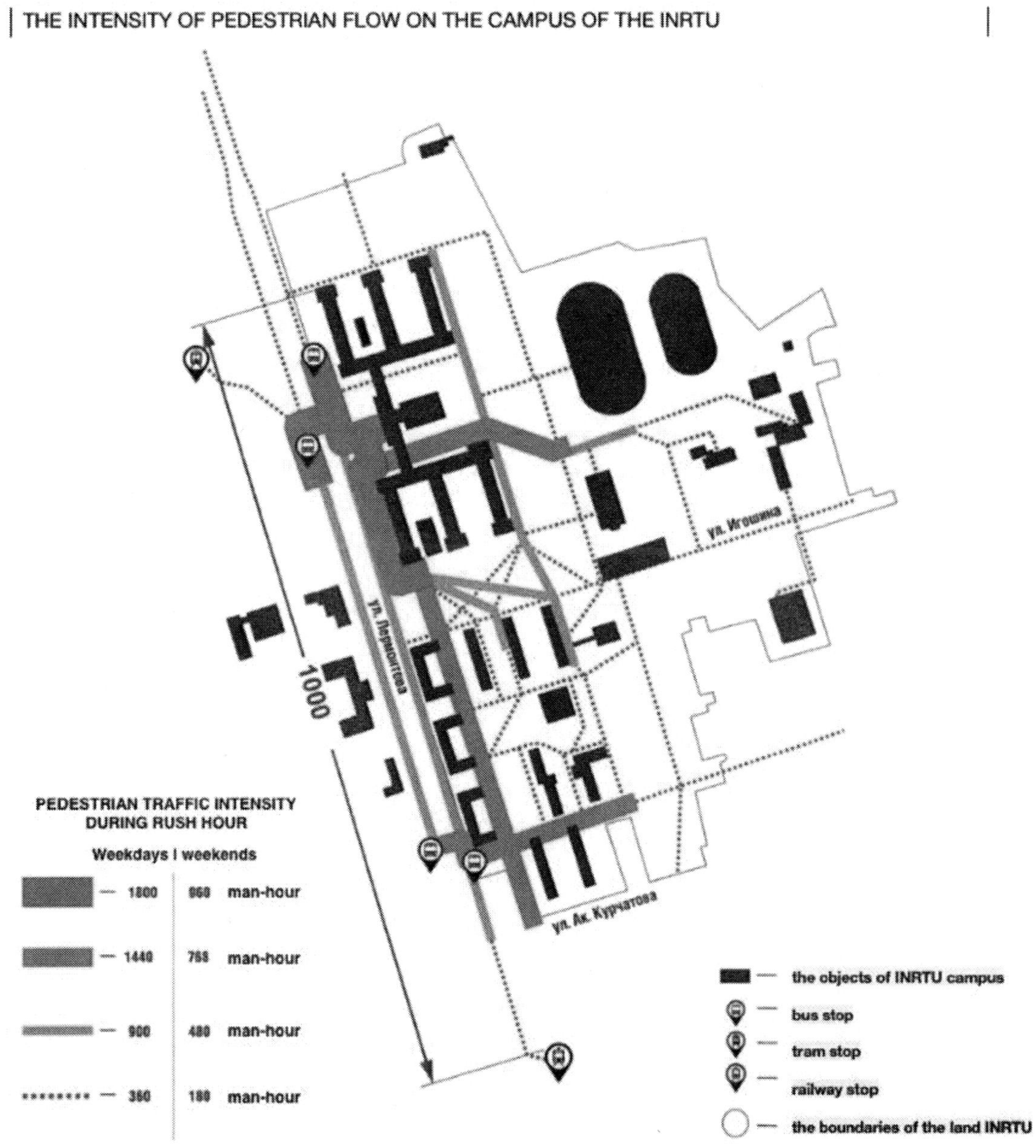

Figure 10. The intensity of pedestrian traffic on the IRNITU campus, Russia for November-December 2018, was calculated by students of the ARB-18-2 group A Vlasova and V Belova. Supervisor and author of the drawing art.: lecturer Z F Nizamutdinova [3]

IOP Conf. Series: Materials Science and Engineering **880** (2020) 012068 doi:10.1088/1757-899X/880/1/012068

The conclusion is that advanced Universities interested in the development of their territories are implementing the most energy-efficient technologies, both in terms of the purely planning structure of the entire campus, and, in particular, space-planning solutions for individual buildings. This shows that architects are more conscious of their approach to the planning of future campuses and emphasizes that the environmental adaptation of buildings is a higher priority. Taking into account the disadvantages and advantages of using these technologies, it is proposed to develop a methodology for applying energy-efficient directions on the territory of Irkutsk University campuses.

References

[1] *Aga Khan University* Available at: https://www.aku.edu/Pages/home.aspx. Accessed March, 18 2020

[2] *Biomedical Department of Arizona University* Available at: https://phoenixmed.arizona.edu/bspb. Accessed March, 17 2020

[3] Bolshakov A 2019 INRTU campus development strategy *IOP Conf. Ser.: Mater. Sci. Eng.* **667** (2019) 012015

[4] Brake A G 2015 Henning Larsen's university building has a facade that moves in response to changing heat and light *Dezeen.*, Available at: https://www.dezeen.com/2015/07/14/henning-larsen-syddansk-universitet-sdu-kolding-campus-building-denmark-green-standards-university/. Accessed March, 16 2020

[5] Google Earth. Selection of illustrations, Available at: https://earth.google.com/. Accessed March, 24 2020

[6] Novosibirsk State University, Available at: https://www.nsu.ru/n/. Accessed March, 16 2020

[7] Qasim University, Available at: https://www.qu.edu.sa/. Accessed March, 16 2020

[8] Sochalin O 2017 Innovative facade of the Biomedical campus in Phoenix – "sunscreen" made of copper panels *Architime*, Available at: https://architime.ru/news/co_architects/biomedical_sciences_partnership_building.htm - 14.jpg. Accessed March, 17 2020

[9] University of Queensland, Available at: https://www.uq.edu.au/. Accessed March, 17 2020

[10] University of southern Denmark, Available at: https://www.sdu.dk/da. Accessed March, 16 2020

[11] University of Toronto Scarborough, Available at: https://www.utsc.utoronto.ca/home/. Accessed March, 16 2020

[12] University of Waterloo, Available at: https://uwaterloo.ca/. Accessed March, 16 2020

[13] Womersley S 2003 *Payette Associates: an evolution of ideas* (Master architect series. VI. Mulgrave, Vic. : Images Publishing Group) 256

IOP Conf. Series: Materials Science and Engineering **880** (2020) 012069 doi:10.1088/1757-899X/880/1/012069

Modern aspects of the formation and implementation of environmental policy in an urban conditions

E Pesotskaya[1], L Selyutina[2] and A Chernykh[1]

[1]Department of Economics and Management in Services, Saint-Petersburg State University of Economics, 21, Sadovaya str., St. Petersburg 191023, Russia
[2]Department of Economics and Management in Construction, Emperor Alexander I St. Petersburg State Transport University, 9, Moskovsky pr., St. Petersburg 190031, Russia

E-mail: epes2@mail.ru

Abstract. The article tells about the current environmental components of a modern urban system. It describes the features of environmental policy and investigates the methodological foundations of its formation and implementation. The paper highlights the main methodological principles of environmental policy and reveals their content and the possibility of using in management tools of the environmental policy in the urban conditions. Particular attention is paid to the study of tourism development in the context of ecologization of the city's economy. The article also reveals the essence of the concept of "ecologization of tourist activity" and describes the directions of its implementation with the emphasis on the managerial aspect. It shows the necessity of developing plans and programs for the ecologization of tourism based on the analysis of the mutual influence of the functioning of the tourism sector and the environment.

1. Introduction

Today, more than half of the world's population lives in cities. The Russian Federation is a highly urbanized country, as the share of urban population is 74%. The urban environment is gradually becoming increasingly important and becoming the main condition for people's livelihoods [1-4]. Therefore, environmental indicators of the urban environment significantly affect the health status of the population of the city, life expectancy.

Favorable environmental conditions ensure the comfort of residents, harmonious socio-economic development of the city. Sustainable development of urban settlements is the development of territories and settlements during the implementation of urban development activities to provide urban development facilities with favorable living conditions for the population, including limiting the harmful effects of economic activity on the environment and its optimal use in the interests of present and future generations [5-7]. It is important to note that from the point of view of sustainable development, the formation of the urban environment is carried out in the process of urban planning, taking into account environmental priorities (city environmental organization environment).

From the point of view of state regulation, ecology as a science has practical implementation, first of all, in the form of environmental activities, the principles and requirements of which are reflected in environmental policy. However, it is only in recent years that interest in environmental policy as an important area of public administration and local self-government has increased in all countries of the world.

2. Materials and Methods

The relevance of environmental issues seems to be undeniable, and it is she who highlights the need for the formation of a scientifically based, methodologically and methodologically assured environmental policy, as well as the search for managerial mechanisms conducive to its successful practical implementation.

An even more difficult task is to form the basis of environmental policy that would create favorable conditions for achieving high rates of economic growth and increasing the level of well-being and social satisfaction of the population. As a solution to such a difficult task, many leading experts and scientists propose the greening of the economy [8-10], which is a process of systematic implementation of environmental business standards.

The *aim* of the study undertaken in this article is to determine the essence, features, and methodological basis of environmental policy in the urban conditions and to justify the need for ecologization of the tourism industry as one of the determining factors in creating an attractive and comfortable urban environment.

From the analysis of domestic and foreign experience we can see that most of the definitions for the interpretation of the term "ecologization" apply to the economic system as a whole and involve a series of actions for the evolutionary transition to a "green economy", which should become an alternative to modern post-industrial society [11-13]. At the same time, according to the authors of present article, greening should be considered in relation to specific areas of the economic life of the country and society.

The methodological basis of environmental policy should be composed of principles that reflect both the fundamental basis of the greening process and the specifics of the scientific approaches used in its study. The analysis of modern literature [14-17] allows us to formulate methodological principles, which, if integrated, can ensure the formation of a coherent, logically verified environmental policy in an urban environment.

3. Results and Discussion

The main of the methodological principles is the principle of *consistency*. This principle provides for the need to use the provisions of a systematic approach to the identification, study and resolution of environmental problems [18].

It is important to note that the provisions of the systems approach are implemented in several areas. First, environmental policy focuses on the ecosystem, i.e. on a system entity. Secondly, environmental aspects are studied in interaction and interdependence with economic and social aspects, which are aggregated into a single, structurally complex socio-ecological-economic system. Thirdly, environmental policy itself is a system of interrelated actions, and it is their systematic nature that testifies to their validity and practical feasibility.

The principle of *goal-setting* is inextricably linked with the principle of system city. The relevant sections in some of scientific papers [19, 20] are devoted to the study and formulation of environmental policy goals. Summarizing the research positions set forth there, it seems appropriate to single out three groups of environmental policy aims:

- conservation of natural diversity and the initial state of the ecosystem;
- harmonization of socio-economic development and the achievements of scientific and technological progress with the state of the environment;
- system support of phased ecologization of industries and economic sectors in order to systematically transition to their functioning, taking into account the requirements of the green economy.

The third principle of environmental policy, especially significant in modern conditions, is the principle of *harmonization of interests*. The application of this principle involves the adoption of such management decisions in the field of ecology and nature management that do not enter into confrontation with other public interests [21].

The principle of *innovation* should be highlighted as the fourth principle of modern environmental policy. Its modernity is due, first of all, to the currently implemented state concept of innovative transformation of all social processes including the processes observed in the environmental sphere and in ecology as a whole.

The content of the innovation principle comes out of the ecologization of all sectors of the economic complex, in the introduction of a green economy, in the maximum use of technologies that reduce environmental damage [22].

The fifth principle is the principle of *performance.* Following this principle requires environmental policy makers, as a prerequisite, to achieve certain results. Moreover, the results should concern not only the environmental sphere, but also be analyzed in the context of the current socio-economic situation.

The sixth principle of environmental policy can be identified as the principle of *organizational unity.* Environmental policy can only be effectively implemented if there is a rational organizational environment. Such an environment should be formed on uniform standards and approaches, on a clear distribution of responsibility for environmental violations, on organizational conditions conducive to the practical implementation of environmental activities, environmental programs at various levels, relevant strategic, tactical and operational plans [23].

The organizational support of these elements of environmental policy is combined with the creation of a special regulatory environment and mechanisms that can carry out the regulatory function in a mode of sufficient completeness, uniformity and consistency.

The seventh principle is the principle of *"open discussion"*, which implies the active involvement of public organizations, as well as individual citizens, in the discussion of topical issues related to the conservation, maintenance and development of the ecosystem in all its components.

This principle can be implemented in various forms: using information networks (Internet), social networks, mobile applications [24]. Significant effect can be achieved by holding public hearings on various events stipulated by the environmental policy, individual plans and environmental programs, as well as for any major projects, in terms of their impact on the ecosystem, and assessing the possible consequences.

Among the fastest growing market sectors in the urban environment is modern tourism, the development of which is accompanied by the emergence of a number of problems, the solution of which requires the use of special management mechanisms. The need for tourism ecologization and the implementation of a universal concept of society ecologization are among such problems.

In recent years, the term "ecologization of tourism activities" has been introduced into the terminology and methodology of tourism management [25-27]. The scientific literature presents a number of attempts to identify the essence of this concept, to determine its place in the existing conceptual and terminological apparatus. The approaches used in this case do not contain fundamental differences, but highlight individual aspects of greening, consider it generally or in structural components. Based on the analysis of existing terminological practice, it seems appropriate to focus on the following definition: greening tourism activities is a long-term, stepwise process of introducing environmentally friendly and safe forms, providing tourism services based on innovative technological support and taking into account cultural, historical and other traditions.

Currently, the ecologization of urban tourism is carried out in several areas, among which are as follows:

- formation and implementation of the concept of greening the economy, which should include strategic and tactical plans for the development of ecologization tourism, as well as determining the priorities of the ecologization process;
- introduction of resource-saving technologies in tourism, accompanied by appropriate infrastructure;

IOP Conf. Series: Materials Science and Engineering **880** (2020) 012069 doi:10.1088/1757-899X/880/1/012069

- ecologization transport networks associated primarily with the transition in tourism services to environmentally friendly transport;
- greening of the utilization of household and industrial wastes generated as a result of the functioning of tourist organizations;
- organizational support for the ecologization of the economy and tourism, which should be interpreted quite broadly with emphasis on the development of appropriate organizational schemes, norms and standards, the subject structure, and the training system for specialists capable of performing various organizational and managerial functions related to the tourism ecologization.

Plans and programs for tourism ecologization are based on an understanding of the following objective phenomenon - the mutual influence of the environment and the functioning of the tourism sector in a particular element of the territorial structure. The degree and nature of this interaction requires an adequate assessment. Such an assessment can be obtained on the basis of a sequential, progressive algorithm that includes a number of stages.

At the *first* stage, a list of indicators should be defined where the current level of ecologization is assessed. Among these indicators, as a rule, there are quantitative assessments characterizing the current state and qualitative assessments obtained using the expert method.

At the *second* stage, the collection, systematization and enlarged analysis of indicators is carried out taking into account the time dynamics.

The *third* stage is the development of a scale for measuring the dynamics of changes in each of the evaluated indicators.

The *fourth* stage is the actual measurement of the dynamics of the indicators. Here, the analysis of qualitative parameters is also carried out taking into account the dynamics of time series.

The *fifth* stage is crucial. Within its framework, a system of conditional ratings (points) is formed, which can be used to characterize the changes for each of the indicators.

At the *sixth* (final) stage, it is necessary to conduct a series of final calculations based on conditional assessments set by experts. Such calculations should take into account the significance of the analyzed indicators, their ability to reflect the intensity of the mutual influence of the environment and tourism.

4. Conclusions

In view of the above-mentioned, it seems possible to draw the following conclusions.

1. The principles considered in the study and included in the methodology for the formation of the environmental policy of urban development, are implemented at the expense of certain tools. It is customary to single out instruments of direct and indirect impact [28, 29]. Instruments of direct impact include numerous administrative measures. Indirect instruments are based on organizational and economic mechanisms. Both the one and the other group of tools are quite numerous, but, unfortunately, it is premature to talk about effective instrumental support for environmental policy.

The degree of stimulating effect of many instruments, their compatibility and complementarity is not clear. The relevance of many tools, the possibility of their application in environmental practice, raises questions. The consequences of many environmental management decisions implemented in recent years are not completely clear.

Environmental problems are not systematized, their solution and even consideration is fragmentary. A solution to such problems can only be found within the framework of complete and comprehensive studies based on a stable theoretical and methodological platform that combines the key determinants of a systematic approach.

2. The ecologization of tourism is an integral element of the ecologization of the city's economy and in the administrative aspect contains many unresolved issues. Assessing the mutual influence of the environment and the sphere of tourism development, its reflection in specific managerial decisions is only one of these issues, but it is his decision that lays the foundation for a balanced environmental policy in the part concerning tourism, which has become one of the most effective economic areas of the urban environment.

References

[1] Maleeva T V 2014 Prospects for development of agglomerations in Russia *Theory and practice of social development* **10** pp 124-126

[2] Selyutina L G 2016 Development reconstructive-building activities for the formation of investment offers on the market housing in Russia *Kant* **3_(20)** pp 126-129

[3] Frolova N et al 2019 *IOP Conf. Ser.: Mater. Sci. Eng.* **698** 077030 doi: 10.1088/1757-899X/698/7/077030

[4] Bulgakova K O et al 2018 Basics of investment projects selection for the implementation of regional investment programs in the sphere of social house building *Espacios* **39 (26)**

[5] Pesotskaya E 2011 *Management of innovative-investment processes in construction* (St. Petersburg: St. Petersburg State University of Economics) p 227

[6] Bessonov M 2016 Social housing – the strategic direction of construction in the regions of Russia *International Journal of Applied and Basic Research* **6-1** pp 142-144

[7] Selyutina L G et al 2019 *E3S Web of Conferences* **97** 06003 Retrieved from https://doi.org/10.1051/e3sconf/20199706003

[8] Maleeva T et al 2019 *IOP Conf. Ser.: Mater. Sci. Eng.* **687** 044002 doi: 10.1088/1757-899X/687/4/044002

[9] Bulgakova K 2015 Identification of risk zones and systematization of risks arising during implementation of the investment program in social housing construction *Science Review* **22** pp 366-369

[10] Vasileva N 2003 Development of housing fund's reproduction forms: a terminological aspect of the problem *Journal of the news of higher educational institutions. Construction* **1 (529)** pp 114-18

[11] Karpova G and Chernykh A 2016 Problems of ecologization of various sectors of economy *Journal of Legal and Economic Studies* **2** pp 7-13

[12] Taylor B et al 2010 Urban policy and governance in a global environment: complex systems, scale mismatches and public participation *Current Opinion in Environmental Sustainability* **2 (3)** pp 129-135

[13] Charbel José Chiappetta 2015 Environmental training and environmental management maturity of Brazilian companies with ISO14001: empirical evidence *Journal of Cleaner Production* **96** pp 331-338

[14] Pesotskaya E V et al 2019 *IOP Conf. Ser.: Mater. Sci. Eng.* **698** 077029 doi: 10.1088/1757-899X/698/7/077029

[15] Selyutina L G 2018 Innovative approach to managerial decision-making in construction business *Materials Science Forum* **931** pp 1113-17

[16] *Urban Development and Urban Policy in Germany An Overview* Retrieved from https://www.bbsr.bund.de/BBSR/EN/Publications/CompletedSeries/Berichte/2000_2009/Bd06UrbanDevelopment.html

[17] Almassy D. et al. 2018Urban Nature Atlas: A Database of Nature-Based Solutions Across 100 European Cities *Report of H2020 Project Maturation*

[18] Pesotskaya E et al 2019 *IOP Conf. Ser.: Mater. Sci. Eng.* **687** 044005 doi: 10.1088/1757-899X/687/4/044005

[19] Raven R et al 2020 Nurturing nature: Exploring socio-spatial conditions for urban experimentation *Environmental Innovation and Societal Transitions* **34** pp 7-25

[20] Udachin N 2015 Ensuring environmental safety at the municipal level *Socio-economic sciences and humanitarian research* **4** pp 132-134

[21] Rakova V et al 2015 The state's participation in solving housing problem of citizens in modern conditions *Fundamental research* **12(4)** pp 823-827

[22] Dushkova D and Haase D 2020 Not Simply Green: Nature-Based Solutions as a Concept and Practical Approach for Sustainability Studies and Planning Agendas in Cities *Land* **9 (1)** pp 19. Retrieved from https://doi.org/10.3390/land9010019
[23] Jun Bao *IOP Conf. Ser.: Earth and Environmental Science* **108** 042010 Retrieved from https://doi.org/10.1088/1755-1315/108/4/042010
[24] Irene van Kamp et al 2003 Urban environmental quality and human well-being: Towards a conceptual framework and demarcation of concepts; a literature study *Landscape and Urban Planning* **65 (1-2)** pp 5-18
[25] Chernykh A 2015 The ecologization of St. Petersburg Tourism as a Factor of Sustainable Development *Sochi Journal of Economy* **4 (38)** pp 4-11
[26] Devuyst D et al 2001 *How green is the city?: sustainability assessment and the management of urban environments* (New York: Columbia University Press) p 457
[27] Petrescu R et al 2017 Modern transportation and photovoltaic energy for urban ecotourism *Transylvanian Review of Administrative Sciences* Special Issue pp 5-20
[28] Chernykh A 2017 Analysis of consumer preferences and assessment of the level of ecologization of the tourist activities (based on the results of a survey of residents and tourists in the Leningrad region) *Journal of legal and economic studies* **2** pp 124-127
[29] *European green capitals* Retrieved from http://www.ec.europa.eu

Public spaces as a factor of the quality of the urban environment (on the example of the city of Irkutsk)

N V Polyakova [1], V V Polyakov[1] and V E Zaleshin[2]

[1] Institute of Management and Finance, Baikal State University (BSU), 11, Lenin Str., Irkutsk, 664003, Russia
[2] Independent expert, Irkutsk, 664003, Russia

E-mail: polyakovanv@bgu.ru

Abstract. The article describes the relevance of the study of public spaces as an important element of the quality of the urban environment. Successfully functioning public spaces are called upon to solve a number of urban problems, including the social isolation of citizens caused by urbanization processes. The characteristic of public spaces of a particular city is based on an assessment of a set of attributes and related criteria developed by placemaking specialists. It is proved that this complex should reflect not only the general requirements for public spaces, but also the features of the city: the diversity of the socio-ethnic structure of the population, cultural and historical heritage, and the specificity of the existing urban environment. The characteristics of the four most visited public spaces of the provincial Siberian city of Irkutsk are fulfilled: the Kirov square; historical and memorial complex "Jerusalem Mountain "; Municipal Park "Yunost Island"; shopping and entertainment complex "130th quarter". The results showed that the public spaces of the city perform a number of functions with varying degrees of success: best recreational and entertaining, partially creative and the function of storing the "collective memory" of the city. At the same time, the city does not have public spaces that fully meet the objectives of the development of urban society; this applies to both historical and newly created. Their main drawback is the inconsistency with the modern challenge for Irkutsk, namely, the solution to the problem of social inclusion of new citizens. Some measures have been proposed for City management to purposefully strengthen the elements of public spaces that implement the function of involving residents, especially migrants, in the public life of the city.

1. Introduction

In the modern post-industrial period of urban growth and transformation, the need to ensure the quality or level of comfort of the urban environment sufficient for residents is increasing. Researchers and practitioners in many countries have noted such problems caused by the global process of urbanization and the growing problems of cities, such as the physical fragmentation of urban space as a result of the industrial construction of new areas, the transport infrastructure and the underdevelopment of transport communications, the social isolation of residents of the areas and the psychological discomfort connected with such isolation.

To mitigate these problems, it is proposed to create "open and hospitable public spaces that all categories of citizens would like to regularly visit and willingly share" [1, p. 50]. And R. Rogers, one of the classics of the humanistic approach to the study of cities, directly points out: the city is primarily a public space convenient for humans [2, p. 160].

Currently, in Russian cities, the zoning model is being reproduced during the construction of new residential areas, which many cities in other countries have already abandoned. It involves the construction of "sleeping" areas, whose residents leave for work in other parts of the city every day, and then return. As a rule, there are practically no public spaces there. As a result, the urban environment breaks up into fragments that are not united by a common idea. According to O. Pachenkov, in large Russian cities "in some 5-10 years, some new buildings will turn into an absolute ghetto" [3]. Our study of the comfort of the urban environment on the example of the city of Irkutsk also showed the importance of social and psychological factors of residents' satisfaction, along with functional (general living conditions in the city, conditions for studying, working for hire, for doing business, playing sports, getting various services, etc.) [4].

These quality problems of the urban environment can be solved, at least in part, by creating and successfully functioning public spaces. Studies of the state and problems of the development of urban public spaces will provide results for a more informed development by the City management of current and future plans for improving the comfort of the city.

2. Materials and Methods

The logic of our article includes:

- a review of literary sources on the functioning and role of public spaces in the context of developing a comfortable environment in a modern city;
- a description of the state of public spaces of the city of Irkutsk;
- identification of problems of their functioning.

Researchers interpret the term "public spaces" as public places specially prepared and intended for visits by citizens for different purposes. Russian researcher A. An defines public space as a territory "free from transport" and created for "public use by an unlimited circle of people for leisure and free access to public facilities" [5, p. 175]. These include squares, parks, squares, embankments, playgrounds, places of recreation in shopping centers, stadiums, etc. Often public spaces serve as a gathering place for citizens to participate in citywide events. In addition to traditional publicly accessible public places, other locations may claim the role of public spaces. For example, a very close understanding of this concept is contained in the concept of the so-called "third place" proposed by the American sociologist R. Oldenburg [6]. "Third place" is a part of the urban space that citizens use for hanging out ("first place") and outside work ("second place"). "Transit spaces", ensuring "the movement of citizens from one object to another" should be excluded from the sphere of "third places" [7, pp. 6, 8]. The "third place" space consists of public places intended for informal meetings, where people "at any time convenient for them can freely and easily communicate" [6, p. 2]. According to the author of the concept, it is these places of recreation, entertainment, exchange of opinions, etc. "are the most important part of the everyday life of citizens and form the urban environment" [6, p. 2].

The idea of improving the urban environment, including public spaces, is largely consistent with the concept of sustainable development of R. Rogers, the goal of which is expressed in the desire to "make cities more human-friendly" [2]. The implementation of this concept is based on the complexity of using economic, environmental, social principles. Historically, beautification pursues primarily economic and environmental interests. This indicates that the improvement of the urban environment should provide, first of all, the convenience, comfort, environmental friendliness and aesthetics of urban spaces. However, this is not enough for a modern city. We agree with the opinion of O. Pachenkov, according to which traditional beautification provides residents with only "passive use" of urban space [8]. Public space is called upon and should be able to unite many interests and hear different "voices" of residents. R. Rogers defines this possible activity of people as "a new form of citizen participation in the management of urban processes" in the framework of "creative and social activities", or citizen / creative partnership [2]. O. Pachenkov, discussing the role and importance of public spaces, notes their ability not only to "turn a person into a citizen", but also to form "public" from the mass of inhabitants (population) of the city. By "public" this researcher understands “responsible citizens who are able in their thoughts, concerns, actions to rise above private, family, corporate interests and think about what is good for the public good” [8].

Thus, a public space becomes such if a “public” is formed in it, the appearance of which is associated with the culture of participation (activity) of citizens in solving the problems of the city, responsibility and care of citizens about the public good. S. McQuire believes that the traditional culture of publicity, which existed in the previous era, is now supplanted by the "mass" exodus "of people into the house, into private life." [9]. From this "obsession with oneself" a modern resident of the city can be "turn out" by organizing public spaces. The value of their reconstruction for post-Soviet cities was revealed, in particular, in the results of an empirical study of the social well-being of students in Minsk (the capital of Belarus): those who actively spend their free time in public spaces more often describe themselves as a happy person and to a lesser extent complain about bad mood [10, p. 89].

We list the functions that public spaces are called upon to perform in the context of ensuring the quality of the urban environment, based on a review of literary sources [6, 10, 11, 12]:

- communication, social, socially inclusive function – the function of the “mixer” through the interaction of the townspeople with each other: getting to know representatives of different social, cultural, ethnic, demographic groups (through personal communication and observing behavior), involving citizens in public life, forming feelings "Affiliation" and responsibility for their city;
- civic – the function of involving citizens in political life, is necessary for the formation and development of a healthy civil society, is realized through the formation of space for political discourse and the interaction of people with the government, to express the views and demands of different urban communities;
- the function of storing the “collective memory” of the townspeople, enclosed in an atmosphere of historically and culturally significant places, is necessary for the successful transmission of city-wide values and ideas of patriotism to children, youth, new residents;
- organizational, creative – involving citizens in joint activities of a different nature and forming an active life position of citizens: arranging urban facilities, providing assistance to certain categories of citizens, organizing and conducting holiday, children’s, games, sports, music, other events, etc.;
- recreational, entertaining – creating conditions for leisure activities: a pleasant and interesting vacation of various categories of citizens.

In the aggregate, the functions, thanks to such “layering”, create a synergistic effect, contributing to the preservation and improvement of the quality of the urban environment.

However, for the implementation of functions, public spaces must meet certain requirements, or, in the context of the concept of placemaking [13], have certain attributes:

1. The correct location (accessibility for different categories of citizens due to good connections – transport, pedestrian, visual – with other crowded places).
2. Comfort (safety, cleanliness, arrangement and convenience for different categories of visitors, provision of catering services, etc.).
3. Opportunities for the activities of different people (including the organization of various events).
4. Attractiveness (pleasant atmosphere for different social groups and certain categories of citizens: youth, adults and seniors, men and women, ethnic and other groups). This attribute includes the architecture and design of the space, other aesthetic elements, such as beautiful natural and artistic objects. In our opinion, this attribute should also include cultural and educational elements, which include historical, cultural, ethnic buildings, structures, ensembles, etc.

How can one determine to what extent a particular public space meets these requirements and, therefore, fulfills its functions? Placemaking experts offer a number of questions, the answers to which may reflect the success of one or another side of the space [13]. We believe that the completeness of such a complex of issues, or, more precisely, the criteria, depends not only on the general requirements that public spaces must satisfy as such, but also on the characteristics of the city: the diversity of the socio-ethnic structure of the population, cultural and historical heritage, and the specifics of the existing urban environment.

3. Results

We briefly outline the features of the city of Irkutsk that are significant for characterizing the requirements for the quality of public spaces.

Irkutsk is a provincial Russian city with a population of 623 thousand people, which has relatively strong cultural traditions. 100 years ago, it was one of the largest administrative, commercial, cultural and educational centers in the Siberian region; during the time of socialism, it became a city with developed scientific and educational potential. Currently, many visiting students of higher and secondary specialized educational institutions are studying in Irkutsk, for whom the problems of adaptation to the urban environment are relevant. In recent years, the share of students in the total population of about 11%, which, along with the city of Tomsk, is the highest indicator in Siberia [14].

Since the 17th century, Irkutsk has been forming as a multinational city, which is due to the history of its formation and development: the ethnic diversity of the indigenous people of Siberia, the colonization of Siberia through the resettlement of communities and individual families from the European regions of the Russian Empire, wide opportunities for economic activity, and finally, the status of a place of exile and hard labor. Since 1992, Irkutsk has been accepting a large number of migrants from the countries of the former USSR, as well as China, fewer from other countries: according to official data [15], in recent years, approximately 7-8 thousand people have arrived annually, according to expert estimates, even more. Despite the historical traditions of the national and religious tolerance of the inhabitants, this makes the task of assimilating the unspoken rules of communication, joint activity, coexistence in general by the new citizens of the city especially urgent.

Another feature of the city is that (unfortunately, only partially) material objects of a rich cultural and historical heritage have been preserved: unique wooden buildings, historical and architectural monuments, religious objects, parks. The city center is included in the preliminary list of UNESCO World Heritage. All this is a valuable resource for the implementation of cultural and cognitive functions: education, the transfer of knowledge about the city, the preservation of the "collective memory" of residents, as well as the creation of prerequisites for the preservation of historical and cultural monuments.

Thus, it can be argued that, on the one hand, the function of socializing new members of the urban community (children, youth, new residents) is highly relevant for public places in Irkutsk, and on the other, the city has sufficient potential for its implementation.

Currently, 18 public spaces actually operate in Irkutsk, of which the most visited are: the Kirov square with adjacent squares; historical and memorial complex "Jerusalem Mountain" (formerly Central Park of Culture and Rest); Municipal Park "Yunost Island" (located on the Angara, the main river of the city); shopping and entertainment complex "130th quarter" with an area of 6 hectares with recreated wooden buildings in the style of pre-revolutionary architecture. The first two spaces are historical, since there are more than 100 years; the last two are modern, built a few years ago.

To do this, we use a system of criteria that is based on provisions developed by placemaking specialists [13]

IOP Conf. Series: Materials Science and Engineering **880** (2020) 012070 doi:10.1088/1757-899X/880/1/012070

Table 1a. Description of the most visited public spaces in Irkutsk (according to expert judgment)

Criteria	**The Kirov Square**	**Historic and Memorial Complex "Jerusalem Mountain"**	**Island "Yunost"**	**"130th quarter"**
Location				
Location	city center	city center	city center	city center
The number of public transport routes that this space can be reached	41	26	27 (The location is far removed from public transport)	27
Walking distance	Good for all categories.	Good but not enough for people with disabilities	Good for all categories	Good but not enough for people with disabilities
External visibility	Good	Good	Very good	Satisfactory
Comfort				
Number of Seats	Sufficient	Sufficient	Sufficient	Not enough
Hiking opportunities	Satisfactory	Good	Very good	Satisfactory
Cleanliness and safety	Very good	Good	Very good	Very good
Saturation with historical and culturally significant objects	Average	High	Small	High
Food Service Facilities	Street trading	–	Cafes	Restaurants, Cafes, coffee houses, bars

Table 1b. Description of the most visited public spaces in Irkutsk (according to expert judgment)

Criteria	**The Kirov Square**	**Historic and Memorial Complex "Jerusalem Mountain"**	**Island "Yunost"**	**"130th quarter"**
Attendance and communication				
Categories of visitors by gender and age	Women and men are about the same. All age categories	There are more women than men. Mostly elderly and parents with children	There are several more women than men. All age categories	Women and men are about the same. Mostly young people, middle-aged people
Categories of visitors based on permanent residence	Mostly permanent residents, visiting students. On New Year's holidays, migrants with children	Mostly permanent residents	Mostly permanent residents, visiting students	Mostly visiting students and tourists, permanent residents
Nature of communication	Mostly in own group	Mostly in own group	Active in own group, weak with representatives of other groups	Active in own group, weak with representatives of other groups
Friendliness towards other groups	Average	Average	High	High

Table 1c. Description of the most visited public spaces in Irkutsk (according to expert judgment)

Criteria	**The Kirov Square**	**Historic and Memorial Complex "Jerusalem Mountain"**	**Island "Yunost"**	**"130th quarter"**
Activities				
Physical activity of visitors	Hiking	Hiking	Hiking, sports games (football, etc.), ice skating, skateboards, catamarans	Hiking
Participation in events	Mass celebration of the New Year (the main city tree); participation in flash mobs, concerts (as spectators)	–	Participation in sports competitions; in concerts (mainly as spectators)	Participate in the annual craft festival; in events organized by companies; in spontaneous concerts
Categories of event participants by gender and age	All categories by gender and age	–	Mostly children and youth of both sexes	Youth, middle-aged people. There are several more women than men
Categories of event participants on the basis of permanent residence	Permanent residents, visiting students. Migrants with children (in the New Year holidays)	–	Permanent residents, visiting students	Permanent residents, visiting students, tourists
Management presence	No	Yes (small office)	Yes (small office)	No

The data obtained allow us to draw some conclusions. All considered public spaces have good potential (judging by the criteria of a "physical" nature): they are located in the city center, have good transport and pedestrian access, most are perfectly visible from the outside and attract with their design, green spaces, historical and culturally significant objects. Almost all provide fairly comfortable conditions for visitors: equipped paths for walking, places for rest (benches, awnings), equipped areas for sports, festivals and other events, catering facilities. Everywhere a sufficient level of cleanliness, order and security is provided.

At the same time, the places considered only partially fulfill the functions of public space, since "physical" attributes (e.g. good arrangement) are a necessary, but not sufficient condition for its success. Attractiveness is extremely important for a wide variety of groups of citizens, and this is ensured by the quality of communication and interaction of people – both within groups and between different groups. Our initial review shows that the spaces are not attractive to all groups of citizens, even for the purpose of passive relaxation. So, migrants visit only one space - the Kirov square, and not regularly, but during one period of the year - on New Year's holidays. Most of them come with children, who are undoubtedly more adults involved in communication with the indigenous people, thanks to the school and other children's institutions. As for the nature of the communication between visitors, communication between people within their group clearly prevails, and communication with representatives of other groups is more often characteristic of youth groups, especially in the process of participation in joint activities.

Thus, it can be argued that the public spaces of the city perform a number of their functions with varying degrees of success: it is best recreational and entertaining, partially creative, and the function of storing the "collective memory" of the city. At the same time, the city does not have public spaces that fully meet modern requirements, and this applies to both historical and newly created ones. The main gap here is the mismatch between the actually performed set of functions and the modern challenge for Irkutsk, namely, the solution of the problem of socialization, social inclusion of new citizens. We believe that the problem cannot be considered acute for visiting students. It exists and has been successfully solved for over 70 years. Much more concern is the problem of social inclusion of migrants, especially those who arrived from China. Migrants from the countries of the former USSR are adapting more successfully, as they to one degree or another speak the Russian language and, at least in part, have preserved the values of internationalism. At the same time, one can note the increasing role of ethnic diasporas in the life of migrants who have arrived and have already settled, which leads to their isolation, strengthening of barriers to inclusion in the city's culture. Indeed, the domestic life of migrants (in the "first place"), life at work (in the "second place") and cultural life (participation in religious, national holidays, etc.) are usually "locked" into narrow framework of ethnic communication. The exception is the education of children in city-wide schools and other institutions.

Also, one cannot fail to note the poor performance of such functions of public spaces as civil and organizational, especially in terms of the formation of an active life position of citizens, responsibility for their city and the future of civil society.

In recent years, City management has been actively improving the quality of the urban environment of Irkutsk, in particular, the arrangement of the largest and most visited public spaces. The positive results of this activity are obvious. Of course, the success of the functioning of public spaces is determined not only by their "physical" state, but also by the degree of social maturity of the city, which is reflected in effective communication and constructive interaction of all communities and groups. Nevertheless, City management can contribute to solving this problem, as it should be guided by the strategic goal of the city's development - the formation of a prosperous and happy life for citizens. To achieve this goal, it is necessary to solve the problem of involving migrants in public life. Here public spaces can (and should!) play an important role as a "mixer". For example, their arrangement should reflect the ethnic diversity of the city's population, in particular, by placing cultural and catering facilities with national cuisines, presenting information on stands and signs in different languages, etc. The theme of mass events should be expanded by organizing national holidays, festivals, competitions of national games and types of traditional arts, educational events for children, youth, etc. To do this, we can recommend City management to work more actively with the leaders of national diasporas.

4. Conclusion

1. The creation and successful functioning of public spaces is a powerful factor of the quality of the urban environment. In this regard, in recent years, an integrated approach to the research, planning and management of public spaces of the city has been actively developing.
2. In the formation and management of public spaces, it is necessary to take into account the features and problems of the development of the city. Only in this case, the space can become a significant factor in increasing the comfort of the urban environment.
3. For the Siberian city of Irkutsk, the problem of social inclusion of new residents is relevant.
4. An expert analysis of the state of the most visited public spaces showed: despite the fact that the spaces have good potential (including cultural and historical), they do not quite meet the requirements for the development of urban society, especially in terms of solving the problem of social adaptation of new migrant citizens.

5. City management should purposefully engage in strengthening the elements of public spaces that realize a socially inclusive function – involving residents, especially migrants, in the public life of the city.

References

[1] Bauman Z 2008 City of fears, city of hopes *Logos* **no. 3** P. 24–53. (In Russian).
[2] Rogers R 1997 *Cities for a Small Planet* (London: Faber & Faber) P. 160.
[3] Kosmarsky A 2017 The absolute illiteracy of Russian urbanists: a sociologist on urbanism *Indicator.Ru* **Jan 25th** URL: https://indicator.ru/humanitarian-science/oleg-pachenkov-chast-2.htm
[4] Nina Polyakova *et al* 2019 *IOP Conf. Ser.: Mater. Sci. Eng.* **667** 012080. DOI: https://doi.org/10.1088/1757-899X/667/1/012080
[5] An A L 2012 Rol' obshchestvennogo prostranstva v munitsipal'nykh obrazovaniyakh *Public Administration Issues* no. 1. P. 174–184. (In Russian).
[6] Oldenburg R 2014 *The Great Good Place: Cafés, Coffee Shops, Bookstores, Bars, Hair Salons and Other Hangouts at the Heart of a Community* (Moscow: OOO Novoye literatrurnoye obozreniye) 456 pp.
[7] Bezrukova O *et al* 2018 A study of the development of a comfortable urban environment in Moscow and the leading cities of the world *The Boston Consulting Group.* **Jul 18th** URL:https://www.bcg.com/ru-ru/publications/2018/comfortable-urbanenvironment.aspx
[8] Gilmanov A 2018 "Oshibat'sya, delat' snova, i obyazatel'no — v komande" *TATLIN* **Apr 13th** URL: https://tatlin.ru/articles/oshibatsya_delat_snova_i_obyazatelno-v_komande
[9] McQuire S 2014 *The media city: Media, Architecture und Urban Space* (Moscow: Strelka Press) 389 pp.
[10] Lebedeva E V 2017 Publichnoye prostranstvo postsovetskogo goroda: vozmozhnosti dlya razvitiya sotsial'nosti i «krizis publichnosti» *The Journal of Sociology and Social Anthropology* **Vol. XX** № 1 (89) P. 74-92. (In Russian).
[11] Kadyrov T E 2014 Public spaces: phenomena, tendencies and processes *News of the KSUAE* **no. 4** P. 115–120. (In Russian).
[12] Zakharova E E 2017 Public spaces as factors of the socio-cultural development of local territories *Culture and Arts Herald* **no. 2** (50) P. 122–128. (In Russian).
[13] Project for Public Places 2019 *Placemaking Leadership Council* URL: https://www.pps.org/plc
[14] URL: http://vuzoteka.ru/вузы
[15] 2019 Migration from Irkutsk Region halved over the year *SIA* **May 6th** URL: https://sia.ru/?section=484&action=show_news&id=373986

IOP Conf. Series: Materials Science and Engineering **880** (2020) 012071 doi:10.1088/1757-899X/880/1/012071

Plan of reorganization of the architectural space of military campus No. 19 for the needs of the Suvorov Military School

N A Skorikova[1]

[1]Department of History and Philosophy Irkutsk National Research Technical University, 664074, 83 Lermontov str., Irkutsk, Russia

E-mail: skorikova_natalya@mail.ru

Abstract. The possibilities of implementing of a project to adapt the buildings and structures of the former military campus No. 19 in Irkutsk for the needs of the Suvorov Military School are examined in the article. The author proposes a set of specific architectural and planning solutions. The principle of compact development and rigid functional zoning has been the basis for the proposed project. Recommendations for accommodation of the personnel, arrangement of the residential and educational premises, taking into account age-related characteristics and relevant legislative norms and requirements are given.

1. Introduction

In 2009, the Irkutsk Higher Military Aviation Engineering School (IVVAIU) was disbanded. Military campus No. 19, on the basis of which IVVAIU functioned, remained ownerless. On the campus territory there were 73 facilities maintaining the operation of the school with a total area of more than 100 thousand square meters. More than 10 of them are still monuments of architecture.

More than 10 years have passed since then. Many buildings and structures have fallen into disuse. However, there is a chance of revival and salvation of a part of this architectural ensemble. One way of solving this problem may be to adapt these facilities to the activities of a new educational institution - the Suvorov Military School.

They started talking about the "reanimation" of the complex and the possibility of placing the Suvorov Military School in the buildings of the former Irkutsk Higher Military Aviation Engineering School almost immediately after the liquidation of the school. The experience of reorganizing the architectural space of the former IVVAIU to the needs of Suvorov Military School can be used in practice both in existing similar institutions and when designing new ones.

The project proposed in this article is connected not only with the problem of preserving a unique educational and residential complex on the area of military campus No. 19 and the upgrading of Irkutsk. It also has great social significance. Unfortunately, young Siberians, residents of Irkutsk and the Irkutsk region, who decided to take on a military career from an early age, have no opportunity to realize their ambitions in their native region. Today in Russia, the Suvorov Military Schools operate in Yekaterinburg, Kazan, Moscow, Vladikavkaz, St. Petersburg, Tver, and Ulyanovsk. Russian youths have the opportunity to enter the Minsk Suvorov Military School. In each instance, we are talking about military educational institutions outside our region. This involves remoteness from home, difficulties connected with leaving for vacation and the morale and psychological stability of a teenager. The emergence of the Suvorov Military School in Irkutsk will provide an opportunity for many children to start realizing in the military profession and will attract young people from our region and some other regions of the country to the city.

2. Materials and methods

The architectural and planning decisions in the process of organization of the Suvorov Military Schools (or similar institutions) rarely became the subject of scientific research. Therefore, there is, in fact, no specialized literature.

Throughout the 1990s and early 2000s, the theses devoted to various aspects of the reconstruction of educational and residential complexes of Suvorov Military School, Cadet School and universities of St. Petersburg were defended. A group of researchers, including Mukhin A.I., Klimanova S.G., Pavlova I.A., Aleksandrina A.V., published a number of articles that examined the demographic features of military educational institutions, analyzed the functional zoning of the St. Petersburg Suvorov Military School and its compliance with the moral, psychological and domestic needs of the cadets [1].

It is also possible to refer to the thesis by A.V. Alexandrin "Urban reconstruction of the educational and residential complex of the St. Petersburg Suvorov Military School" and the article "Modern foundations of the organization of the living environment of the cadet corps students" co-written by him. [2,3].

The thesis by S.G. Klimanova, which is closely related to the content of our study, concerns the architectural and planning organization of a residential complex for cadets of military universities in St. Petersburg. The author's recommendation to consider the university as a complex, i.e. a special system with strong bonds between the elements, turned out to be very usefull. This principle, according to S.G. Klimanova, should be reflected in the compactness of development [4].

The book by V.M. Polyukh and V.G. Stakhanov "Not to interrupt the thread of the traditions" was used to identify the stages of the Military camp development [5].

As the proposed project should take into consideration the specifics of the educational process in the Suvorov Military School with reference to architectural and planning decisions, the relevant regulatory acts have been studied [6,7,8,9,10,11,12,13].

The main research method is the pre-project analysis method, as well as conducting field surveys and buildings and structures photofixation.

One of the used methods is the method of interviewing. Personal conversations with the former head of the school, a PhD in engineering, an honorary worker of a higher school, and retired major general A.G. Barsukov as well as with members of the IVVAIU Veterans Council retired colonel S.N. Kuznetsov and retired colonel M.I. Laikov were held.

A retrospective method and method of comparative analysis were used to answer the question about the rationality of adapting the buildings of the former IVVAIU to the needs of the Suvorov Military School and identify possible advantages in the material and technical support of the future Suvorov Military School compared to similar military educational institutions in Russia.

3. Results

The architectural appearance of military campus No. 19 was finally formed only in the 1980s. Gradually a network of residential, service and public buildings appeared there [14]. Historically, the service area had become divided into two separate complexes. In the western part of the campus, there was a barracks area; a small part of the educational buildings; the main parade ground №1, where graduates took the oath and said goodbye to the banner; a canteen; bathroom and laundry facilities and various outbuildings. In the eastern part, there was a huge educational complex. There were also separate buildings connected by passages with the main building, in which a computer center and rooms for designing functioned. The Club building was also located there. A wide «cadet» road interconnected the western and eastern complexes. The big stadium with a football field, tracks, stands, fitness equipment, an obstacle course, tennis, volleyball and basketball courts occupied the central area of the camp. Nowadays despite military campus desolation, sports and festive events are held on its stadium. Besides a military training airfield is still located on the territory of the camp.

Of course, it would be impractical to adapt the entire huge complex to the needs of Suvorov Military School. For these purposes, we offer to use the eastern part of the campus of 7.07 hectares, the so-called «Training zone" where the former educational building No. 1 (main building) is located.

Such a choice can be justified by several circumstances:

1. The territory is located along the so-called "main road" connecting checkpoint No. 1 and checkpoint No. 3 and has three convenient access roads.

2. The most important advantage of the "educational zone" is its compactness.

All buildings are connected by convenient transitions that makes the educational and support buildings to be a monolithic architectural structure, and provides free movement of personnel within the entire complex.

3. From the economical point of view, the redevelopment of educational building No. 1 is less expensive than redevelopment of the buildings located in the western part of the campus. This is because the classrooms but not the barracks were initially located in this building.

4. The number of buildings in the designated area, their internal layout associated with the former use, can fully satisfy all the needs of the Suvorov Military School and allows adapting them to a new functional purpose without radical rebuilding and redevelopment.

5. Our calculations allow us to conclude that the redevelopment of the complex with the proper approach can fully comply with the standards specified in Code of Practice CP 118.13330.2012 «Public facilities» [15] and other regulatory documents.

In general, the chosen territory can become a full-fledged campus of the Suvorov Military School, and its structure will meet all the needs of students' daily live. In terms of the architectural space typology, the planned complex can be attributed to a multifunctional educational-residential complex, including premises for various purposes.

Particular attention should be paid to the main building, in which, according to our project, can house an educational zone, rooms of barracks, social support facilities and a library. Most likely, its building started in 1914. Due to the difficult situation in the country of that period, the construction and improvement lasted until the mid-30s of the XX century. This two-story building's layout is similar to that of shape of Russian letter "Ш". There is the basement around the perimeter. All wings of the buildings are interconnected. The main entrance is located in the center of the western facade. There is a vestibule and guardrooms. For convenience and safety, additional entrances and exits are located around the perimeter of the building. The total construction volume of the building is 126 106 cubic meters, its area is 18 609 square meters.

Classrooms, lecture halls, laboratories, a library, reading room and a book depository were located symmetrically in the north and south wings of the building. During the study time and the time allotted for self-study, more than 1,200 cadets and teachers could simultaneously be there.

The architecture of the central wing of the main building deserves special attention. The front marble staircase with a width of about 5 meters begins from the main entrance. It leads to the first and, through the corridor, to the second floor. There are columns along the perimeter of the stairwell on the second floor. Their height is 5 meters. The columns rest against a dome decorated with bas-reliefs. Unfortunately, the decoration was damaged in a number of fires, but this part of the building still looks majestic and solemn.

There is a big hall on the second floor. At the time, it was called "Greek", because there are Pilasters in each window opening that makes the hall look solemn. Award ceremonies, celebrations, military rituals were held in there. The hall is fairly spacious. Its area is 700 square meters and the ceiling height is 12 meters. The floor is covered with the tiles made of marble chips. Despite the fires, tiled floor is in good condition. The mosaic image of a red star in the center of the floor can still be seen.

On the ground floor, under the "Greek" Hall, there is another hall called the "Column" Hall. The choice of the name is not accidental: at a distance of 4.5 meters from the walls and 3 meters from each other, 8 square columns are lined up in two rows. The height of a ceiling is 3.8 meters. The columns are elements of a load-bearing frame.

Along the entire perimeter of the building, there are over 350 windows. The width of the window openings on all floors except the basement is 1.5 meters, the height is at least 2 meters, the distance between the windows is 1.5 meters. This provides natural light in some classrooms during the day. Besides such windows give some "lightness" to this monumental building.

Initially, at the beginning of the century this building was supposed to be used for the needs of the cadet school. Therefore, the corridors are wide enough (on average 6.15 meters) to parade along them.

Based on the conducted research work, we have a number of recommendations relating, first of all, to the functional zoning of the territory.

1. In order to create the most favorable living conditions for students, we propose to equip the so-called bunkrooms, with a maximum capacity of 25 students (i.e. one room for one training group), abandoning the idea of standard barracks facility.

There are at least two reasons for this:

Firstly, the construction of the building allows such layout: a wide corridor gives the opportunity to carry out platoons parades. There are premises (former classrooms) on both sides of the corridor. They are quite spacious and can be used as bunkrooms based on 4 square meters per person, which meets the standards. At the same time, all the necessary and provided by our plan auxiliary premises of the barracks zone will be kept.

Secondly, taking into account the specifics of such an educational institution as the Suvorov Military School, attention should be paid to the age of cadets. Presumably, students of 9th, 10th and 11th grades will be recruited, i.e. young people aged 15-17. It will be easier to provide an atmosphere of comfort, normal moral and psychological climate in such premises as bunkroom.

2. At the stage of construction in the courtyards of the building on the east side, two parade grounds were supposed to be built. The central wing of the building separated them. When it was the Irkutsk Higher Military Aviation Engineering School, they were turned into alleys. For the needs of the Suvorov Military School, the parade grounds should certainly be reconstructed. The separateness of the parade grounds will allow conducting military exercise of several units at the same time. A parade ground should be available at the entrance to the canteen, as the schedule of the Suvorov Military School provides for the parade of the units before and after visiting the canteen. The implementation of this plan is not difficult, because the space in front of the building chosen for the canteen is flat and building-free with a total area of more than 2,000 square meters.

3. It is necessary to use the building of Headquarters No. 1. This is a two-story building with a total area of 1,409 square meters. This building has 35 rooms, in which previously there were the offices of school commands and the reception of the head of the school. The building is connected to the main building by an arched structure, which provides access to the second floor. The internal layout of the building dictates us to preserve its functional purpose.

In addition, our calculations show that a canteen, cafe and a health center might be placed in the buildings of this architectural and construction complex in compliance with all relevant norms and requirements [16].

4. Conclusion

Having analyzed military regulations, daily routine and living conditions of cadets, peculiarities of educational and cultural processes of the Suvorov Military Schools in Chita, St. Petersburg, Astrakhan, Ussurisk and some others, we concluded that the complex of buildings for the Suvorov Military School should combine 9 main components. They are the following:

1. Classrooms to ensure the educational process according to the secondary school program.
2. A complex of specialized premises and facilities for the implementation of military training, appropriate to the age and level of knowledge of cadets.
3. A specially organized zone for sleeping and having rest, which must meet the requirements of boarding schools and the requirements of the barracks zone.
4. A catering complex, consisting of a canteen for cadets, providing five meals a day for students, according to the requirements and standards of boarding schools and an officer canteen and a cafe-shop.

IOP Conf. Series: Materials Science and Engineering **880** (2020) 012071 doi:10.1088/1757-899X/880/1/012071

5. Own health center with a sick bay and a pharmacy.
6. A complex of facilities for recreation and additional education.
7. A complex of premises related to the household services of students.
8. The complex of premises for the command staff of the school, teachers and service personnel.
9. The complex of sports facilities.

The spatial framework of the architectural environment of the object are the most suitable for the implementation of processes associated with the functioning of the Suvorov Military School. There are certainly some disadvantages. For example, a common drawback for almost all buildings are wooden beams, which nowadays does not meet the fire safety requirements.

In addition, in our opinion the compactness of buildings and premises, their maximum compliance with the functional purposes should be a prerequisite for the planning and placement of the Suvorov Military School. Cadets remain isolated from the external environment; their daily life is associated with the school that is why the convenience of moving, life safety and the aesthetic perception of the surrounding architectural environment are very important. These conditions are also achievable.

For a more accurate analysis, we had to turn both to the history of military campus No. 19 and to the experience of the placement and functioning of the different Suvorov Military Schools in Russia [14, 17, 18, 19, 20, 21]. In the course of the study, we identified the specifics of such an educational institution as the Suvorov Military School, which combines the characteristics of a secondary school, a boarding school, a military educational institution and we considered this when planning architecture.

Placing the Suvorov Military School on the territory of military camp No. 19 will certainly be a salvation for the unique architectural complex. There will be a need for additional efforts aimed at improving the whole area: the restoration of roads, the stadium and park areas.

References

[1] Mukhin V I, Klimanov S G and Alexandrin A B 1996 *Recommendations on the architectural and planning organization of military schools for the development and cadets* Final report on design and survey program "Complex" (St. Petersburg: VISI) **1**

[2] Alexandrin A V 2001 *Urban reconstruction of the educational and residential complex of the St. Petersburg Suvorov Military School* Dissertation for the Candidate of Architecture degree

[3] Mukhin V I, Alexandrin A V, Zhdankin N Yu and 0Sargsyan E R Modern principles of organization of the living environment of students of the cadet corps of the Ministry of Defense of the Russian Federation *Viviliofik Electronic Library*

[4] Klimanov S G 2000 *Architectural and planning organization of a residential complex of cadets of universities of St. Petersburg* Dissertation for the Candidate of Architecture degree

[5] Polyukh V M 1991 Not to interrupt the thread of the traditions: a brief historical outline of IVVAIU (Irkutsk: IVVAIU0

[6] *Instructions on the organization of control over the activities of the Suvorov Military Schools, the Nakhimov Naval School and Cadet Corps* Encyclopedia of Cadet Education

[7] *The concept of the federal system of preparing citizens of the Russian Federation for military service for the period until 2020* Encyclopedia of cadet upbringing and education

[8] *On amendments to the Regulation of the activities of the Suvorov Military Schools, the Nakhimov Naval School and Cadet Corps* Laws, codes and normative legal acts of the Russian Federation.

[9] *Regulations on military-patriotic youth and children's associations* Laws, codes and regulatory legal acts of the Russian Federation

[10] *The position of the Government of the Russian Federation on the Suvorov Military Schools,* the Nakhimov Naval School, music schools and cadet (naval, musical) corps of April 30, 2008

[11] *Decree of the Chief State Sanitary Doctor of the Russian Federation of December 29, 2010 N 189* Moscow "On approval of SanPiN 2.4.2.2821-10" Sanitary and epidemiological requirements for conditions and organization of training in educational institutions

IOP Conf. Series: Materials Science and Engineering **880** (2020) 012071 doi:10.1088/1757-899X/880/1/012071

[12] *Decree of the Government of the Russian Federation of April 30, 2008 N 328* On approval of the Model Regulations on the Suvorov military schools, Nakhimov naval school, military music schools and cadet (naval cadet) corps
[13] *Order of the Ministry of Defense of the Russian Federation N 337* 09/22/2003
[14] Skorikova N A 2017 The history of planning and development of military campus No. 19 in Irkutsk *News of the Laboratory of Ancient Technologies* **13** pp 132-147
[15] *Code of Practice CP 118.13330.2012* Public facilities
[16] *Departmental building codes 34-95 Planning and development of military campuses* Design Standards (M.: MO RF)
[17] Plekhanov A M, Popov A A and Heirs A V 2001 Suvorov. From the Cadet Corps of Russia to the Suvorov Military Schools of the NKVD (Ministry of Internal Affairs), KGB of the USSR (1731-1960) *Historical outline* pp 71-72
[18] Mironov V N 2002 *Cadet educational institutions in Russia: history and modernity* Dissertation for the Candidate of Historical Sciences degree
[19] Tolokolnikov G *P Suvorov, Nakhimov Cadet, Suvorovets, Nakhimovets*. URL: http://www.kadet.ru/lichno/Tolok/Suv_nah/Intr/Introduction.htm (accessed: 01/10/2020).
[20] Gurkovsky V A 2014 *The Gorky Suvorov Military School*

IOP Conf. Series: Materials Science and Engineering **880** (2020) 012072 doi:10.1088/1757-899X/880/1/012072

Mathematical model of assessing urban planning decisions based on capital construction facilities location under the city plan

A V Zedgenizov[1], A I Solodkiy[2] and K V Seliangin[3]

[1 3] The Department of Oil and Gas Engineering, Irkutsk National Research Technical University, 83 Lermontov street, Irkutsk, 664074, Russia
[2] The Department of Transport Systems, Saint Petersburg State University of Architecture and Civil Engineering, 4 Vtoraya Krasnoarmeiskaya Street, Saint Petersburg, 190005, Russia

E-mail: azedgen@gmail.com

Abstract. The article deals with the issues of placing large focal points over the city planning map. The topicality of this issue is described. The target function is determined; it takes into account the time losses of road users in the zone of the focal point influence and the amount of payments to budgets of different levels generated by the focal point functioning. The mathematical model that allows estimating time losses of road users is presented. The model has a special focus on the mechanism of transport demand development and on the assessment of crossing capacity based on the intensity of priority flows. The mathematical tools are developed; they allow calculating the financial consequences of the implementation of urban planning projects set out in the city plan, choosing focal points with the most beneficial influence on the municipal budget among the proposed ones, taking into account the minimum negative impact of the chosen focal points on the adjacent street-and-road network. Some particular cases of the transport demand development are discussed, for example, those of traffic management schemes that do not allow left-turn and through movements at uncontrolled intersections. The recommendations on the practical application of the target function when making urban transport decisions having regional and city significance are developed.

1. Introduction

Modern cities are a conglomerate of focal points (FPs) located, as a rule, in capital construction facilities (CCF) that can be different in the following characteristics: the number of floors, area, configuration, remoteness from the city center and transport communications [5, 12]. Depending on the intended use, type of ownership, degree of the target audience coverage, investment prospects, availability of engineering, transport and social infrastructure, as well as on a number of other factors, designers determine the specific location of capital construction facilities over the city map. The amount of correspondences (transport demand) made to FP will largely depend on the choice of its location. For example, if FPs are non-profit and the state authorities or public organizations define their locations unreasonably it leads to increased time losses of potential users and may create an additional load of the street-and-road network (SRN) providing these FPs functioning. When it comes to commercial organizations, transport demand is a direct factor affecting the number of visitors to such FPs and, consequently, the revenues, i.e. the commercial gains from these FPs functioning. FPs represented by

IOP Conf. Series: Materials Science and Engineering **880** (2020) 012072 doi:10.1088/1757-899X/880/1/012072

grocery stores (supermarkets) can serve as a good case of the location effect on the level of transport demand [2, 3]. Depending on its distance from the city center, i.e. proximity to apartment buildings (food is easier to buy after a work-to-home trip), the demand will increase. The demand will also depend on the distance from the public transport stop, the commuter train station or the subway: the closer the supermarket is to the stopping point, the higher the demand is. When it comes to the location in the direction of travel, the right-hand location usually provides quick easy access, therefore, the demand will be higher, while the left-hand location makes a potential visitor using public transport (PT) cross the roadway and a potential user in individual transport (IT) must make the left turn across the oncoming traffic flow.

In some cases, to increase the traffic capacity of individual sections of SRN, left turns are prohibited and, therefore, it requires significant excess mileage leading to the loss of such FPs' attractiveness for potential visitors. On the other hand, the location of FP of the CCF type, especially a large one with the area of several tens of thousands of square meters (shopping malls and entertainment centres, residential neighborhoods, sports and competition grounds, etc.) can significantly affect the functioning of the adjacent SRN as such FPs are of high attractiveness including the episodic nature of such attractiveness. Therefore, the assessment of reasonable placement of capital construction facilities included in the city plan is the most important step at various stages of urban planning and transport design from the city's general plan to traffic management projects.

2. Materials and methods

To identify the optimal location of CCF maintaining high transport demand and having minimal impact on the adjacent SRN [1,7,6], it is necessary to apply such a mathematical model that could take into account all the above-mentioned factors and the value of socially useful benefit would be considered as a resultant involving the time loss while passing the SRN section compared with the time loss absence (the basic version); besides, if this FP is a commercial one the resultant must be sensitive to the amount of taxes associated with this FP functioning. It should be noted that the time losses while passing the SRN section close to the place of the FP location should not exceed a certain reasonable limit, for example, the limit described as the service levels D or E in the special scientific literature [11]. In other words, even with the high financial significance of this FP location traffic conditions in the zone of its influence should not be appalling. Otherwise, the location of this FP should be reconsidered or some necessary changes in SRN should be introduced to reduce time losses in the zone of this FP location.

At this stage, it is worth applying the general mathematical formulation of the target function in a certain region of finite dimensional vector space limited by a set of linear and/or nonlinear equalities, as it allows finding the extremum in the discussed problem, i.e. optimal conditions within the limits of existing restrictions. The research process aim is to minimize the unit costs of FP functioning; the costs include the time of movement in the FP zone converted into the monetary equivalent (rubles) and the amount of tax payments:

$$З = \frac{1}{\mathrm{T}}\sum_{i=1}^{n} З = \frac{1}{\mathrm{T}}(З_{п} - З_{н}) \rightarrow min \qquad (1)$$

where $З$ – unit costs for the definite time period, rubles; T – estimated time, year; $З_п$ – time losses for movement, rubles; $З_н$ – costs for paying taxes to budgets of different levels induced by FP functioning, rubles.

3. Results

If it concerns uncontrolled intersections providing the definite FP functioning the traffic management scheme should be explored in order to take into account the left and right turns, the number of lanes on approach zones including additional lanes for turning maneuvers at such intersections. The transport delay should be calculated according to the following dependence [16]:

$$d = \frac{3600}{C_{res}} \qquad (2)$$

where d – an average transport delay at the intersection on the minor road, s/car; C_{res} – reserve capacity of the traffic flow on the minor road is calculated according to the formula:

$$C_{res} = C - N \tag{3}$$

where C – reserve capacity of the given traffic flow at the uncontrolled intersection, car/h; it is calculated on the basis of geometry rules, number of lanes, traffic management scheme, intensity of the priority traffic flows and other factors; the intensity of the minor road flows providing, as a rule, the FP functioning should be regarded as the transport demand of visitors using IT [4].

$$\text{N} = E_{цмт} \cdot \frac{d_{um}}{P_{um}} \cdot k_{сн}^{уб} \cdot k_d \tag{4}$$

$$\text{C} = \left(\frac{3600}{t_f} e^{-\frac{q_p}{3600}\left(t_g - \frac{t_f}{2}\right)} \right) \cdot \left(\prod_{j=1}^{m} 1 - \frac{N_{02j}}{c_{02j}} \right) \cdot \left(\prod_{j=1}^{m} 1 - \frac{N_{03j}}{c_{03j}} \right) \tag{5}$$

where $E_{цмт}$ – transport demand, persons/day; d_{um} – ratio of people using IT during the definite hour; P_{um} – average number of people in a vehicle (IT), persons; q_p – intensity of the 1st rank traffic flow intersecting the studied traffic direction, car/h; k_d – coefficient involving the ratio of the traffic flow distributed between the intersections of the given FP (if there is one intersection, $k_d = 1$); $k^{уб}_{сн}$ – coefficient of daily irregularity of departures; n – number of traffic flows in the movement direction; tg – average marginal interval, s; t_f – average following distance, s; N_{02j} – intensity of the 2nd rank traffic flow intersecting the studied traffic direction, car/h (if there is not any, it is eliminated); N_{03j} – intensity of the 3rd rank traffic flow intersecting the studied traffic direction, car/h (if there is not any, it is eliminated); c_{02j} and c_{03j} – traffic capacity of the 2nd and 3rd traffic flows intersecting the studied traffic direction, car/h (if there is not any, it is eliminated); m – number of the 2nd and 3rd traffic flows intersecting the studied traffic direction.

$$d = \frac{3600}{\left(\left(\frac{3600}{t_f} e^{-\frac{q_p}{3600}\left(t_g - \frac{t_f}{2}\right)} \right) \cdot \left(\prod_{j=1}^{m} 1 - \frac{N_{02j}}{c_{02j}} \right) \cdot \left(\prod_{j=1}^{m} 1 - \frac{N_{03j}}{c_{03j}} \right) \right) - \left(E_{цмт} \cdot \frac{d_{um}}{P_{um}} \cdot k_{сн}^{уб} \cdot k_d \right)}, \tag{6}$$

Given that the time spent on travelling should be converted and expressed in rubles, it is advisable to agree that the cost of delay $C_з$ when passing the studied section of SRN is 426 rubles/hour [9] for each IT passenger. Therefore, $З_н = d \cdot C_з$.

The amount of tax payments induced by the functioning of a commercial FP is determined under the current tax legislation and the list of taxes may vary depending on the FP type: canteens, cafes, bars, restaurants, production shops, shopping malls, administrative and office buildings, i.e. all those CCFs that are legally registered as non-residential and are used for gaining revenues. The owner calculates and makes payments under the tax laws. Terms of payments are set by regional authorities.

Thus, the final target function will look as follows:

$$З = \frac{C_з}{T} \left(\frac{3600}{\left(\left(\frac{3600}{t_f} e^{-\frac{q_p}{3600}\left(t_g - \frac{t_f}{2}\right)} \right) \cdot \left(\prod_{j=1}^{m} 1 - \frac{N_{02j}}{c_{02j}} \right) \cdot \left(\prod_{j=1}^{m} 1 - \frac{N_{03j}}{c_{03j}} \right) \right) - \left(E_{цмт} \cdot \frac{d_{um}}{P_{um}} \cdot k_{сн}^{уб} \cdot k_d \right)} - З_н \right) \to min \tag{7}$$

4. Conclusion

The developed tools are necessary for making reasonable urban and transport planning decisions aimed at identifying the maximum benefit to the State treasury when choosing the type of FP. In other words, it is possible to calculate the financial consequences of the implementation of urban development projects set out in the city plan, to find out which FP (CCF) among the proposed ones can have the most beneficial effect on the municipal budget and the minimum negative impact on the adjacent SRN.

References

[1] Belinsky A Yu *Patterns of Mobility and Resettlement of Residents of Tallinn: collected papers Urban transport* 1969
[2] Burkov D G 2017 Methodology of Assessing Transport Demand of Cultural Facilities Based on the Quantitative Characteristics of Territories *Bulletin of ISTU* **21** pp 218-224
[3] Burkov D G 2018 *Improving Traffic Management in the Maintenance of Cultural Centers Taking into Account the Forecasting of Transport Demand*: doctoral dissertation on technical sciences p 175
[4] Burkov D G, Zedgenizov A V 2016 A Mathematical Description of the Transport Demand Generated by Cultural Objects *Bulletin of ISTU* **20** No 12 pp 193-202
[5] Golts G A 1981 Transport and Accommodation p 248
[6] Gorev A E 2010 *Fundamentals of the Theory of Transport Systems* SPbGASU p 214
[7] Efremov I S 1980 *Theory of Urban Passenger Traffic* p 535
[8] Zedgenizov A V 2019 *Quality Assessment of Traffic Management Based on Transport Demand* Irkutsk: Publishing House of INRTU p 196
[9] Kichedzhi V N, Hatoyama K 2010 *Moscow: Transport Problems of the Metropolis* DPK Press
[10] Markichev V A 2016 Assessment of Losses Associated with Automobile Traffic Jams in a Megalopolis *Young scientist* **22.2.** pp 17-19
[11] *Industry Road Guidance Document. Guidelines for Assessing the Traffic Capacity of Roads* ODM 218.2.020-2012.
[12] Romm A P 2002 *Integrated Assessment and Functional Zoning of the Territory in Urban Planning*: doctoral dissertation on architecture Moscow p 206
[13] *Guidelines for the Transformation of Residential Construction in Existing Parts of Cities 1983* Central Research Institute of Urban Planning Stroyizdat p 74
[14] *Guidelines for New Cities Planning* 1982 Central Research Institute of Urban Planning Stroyizdat p 87
[15] *Guidelines for Cities Reconstruction* 1979 The Administrative Centers of Cities Stroyizdat p 159
[16] Thomas Richter 2016 Planung von Autobahnen und LandstraBen *Springer Fachmedien Wiesbaden* ISBN 978-3-658-13008-4. DOI 10.1007/978-3-658-13009-1.

IOP Conf. Series: Materials Science and Engineering **880** (2020) 012073 doi:10.1088/1757-899X/880/1/012073

Housing cost dependence on transport accessibility territory of industrial city

M I Sharov[1,2] and O A Lebedeva[3]

[1] Scientific and Research institute of motor transport, Str., Geroyev Panfilovtsev 24, Moscow, Russian Federation
[2] Moscow automobile and road construction state technical university, Leningradsky prospect, 64, Moscow, Russian Federation
[3] Angarsk State Technical University, Str. Tchaikovsky, 60, Angarsk, Russian Federation

E-mail: kravhome@mail.ru

Abstract. The study the housing cost dependence on the transport accessibility of the territory of the city will improve the efficiency of the route network and will lead to a reduction in overall transportation costs. The research algorithm consisted of the following stages: data on transport accessibility were systematized; calculation of time costs assessed transport accessibility by city zones; the dependence of the cost per square meter on the time cost of movement is obtained. When assessing the impact of transport accessibility on the housing cost it is important to pay attention to the remoteness from the central part of the city, the proximity of highways, the system of access roads; the proximity of public transport stops with a large number of routes connecting different zones. The study was conducted on the example of the city of Angarsk, the reliability of the obtained results confirmed by generally accepted statistical criteria was substantiated. The availability of tools to conduct a detailed and accurate assessment of accessibility is one of the most important scientific studies in the framework of transport planning now.

1. Introduction

As part of ensuring urban development, a comprehensive and sustainable development of territories is proposed, aimed at ensuring safety and favorable conditions for human life, limiting the negative impact of economic and other activities on the environment and ensuring the protection and rational use of natural resources in the interests of present and future generations. In this regard, the priority issue is the development of public passenger transport systems and green mobility (that is, non-mechanized movements).

The priority development of public passenger transport systems is ensured by both technical solutions and urban planning. In modern foreign urban planning theory and practice, the most common urban planning decisions that ensure the priority development of public passenger transport systems include:

• multifunctional use of the territory (mixed-use development), which reduces the need for movement over long distances;

• territorial planning focused on public passenger transport (transit oriented development), including the formation of an attractive and comfortable environment in which public passenger transport infrastructure is located;

• increasing the intensity of use of territories adjacent to the corridors of public passenger transport.

With improving the availability of public passenger transport, the attractiveness of the territories increases, which causes a concentration of service sector institutions and places of employment.

In the world practice of urban planning, the accessibility of a territory is determined by the following factors: transport accessibility of territories; availability of all necessary services; the presence of a sufficient number of public spaces.

In construction, transport accessibility is a normative indicator of the time spent on transport links between different points within the group settlement systems. When designing city streets and roads, one of the key parameters determining the quality of a city's road network is the trip time between any two points of the city [1].

On the transport side accessibility is the total time spent on travel committed for any purpose (travel to the place of work, travel on cultural and domestic purposes).

Transport security of urban and regional territories is used as one of the economic characteristics.

Improving transport infrastructure is an important mechanism for implementing regional policies. Its rational use allows reducing imbalances in the territorial development of various areas of the city and the country as a whole, and increasing the attractiveness of remote and sparsely populated territories [2, 3].

However, improving transport infrastructure affects the housing cost in this area. For example, the cost of an apartment in the city of Moscow near metro stations can differ by an order of magnitude from apartments located at a sufficient distance with a similar type of housing development. The recent opening of the Central Moscow Diameter has already led to an increase in housing prices by an average of 13%. In this regard, it is important to have tools for calculating and forecasting the cost of housing based on the transport accessibility of the territory. This article presents a study of the impact of transport accessibility on the cost per square meter of housing on the example of an average city (Angarsk).

The city of Angarsk plays a crucial role in the structure of interregional transport corridors, providing links between the European part of Russia and Western Siberia with the regions of the Far East. The territory of the Angarsk urban district is spatially heterogeneous is characterized by various socio-economic, infrastructural and environmental conditions that determine the main directions and functional priorities for the development of the territory [4]. The Angarsk city district plays an important role in the economy of the Irkutsk region and makes a significant contribution to the development of the region's competitive potential. Underestimation of the problem of determining transport demand and the compliance of local infrastructure with the socio – economic needs of society is one of the causes of economic difficulties and negative social processes. The development of the transport infrastructure of the Angarsk urban district is a necessary criterion for improving life quality of residents.

2. Theoretical basis

There are various approaches to assessing transport accessibility. One option is to define accessibility indices. Consider some of them [5-8]:

• single, arbitrary accessibility indices based on the selected landmark. These indices show the dependence of transport costs on income and take into account the minimum wage set for regions or countries. In such cases, the goal is usually to determine accessibility for low – income regions. In practice, the option is most often used when transport costs should not exceed the threshold of 10% of income. This regulatory approach, by the way, has become the standard in many countries.

The accessibility index is estimated using equations (1.2) [7]:

$$Aff = \frac{E_{pt}}{Y} * 100 \quad (1)$$

where E_{pt} – is the monthly cost of public transport; Y – is the monthly income of the family.

$$Aff = \frac{50*p}{Ymin} * 100 \quad (2)$$

where p – is fee on public transport and Y_{min} – is the minimum wages. Minimum wages are taken into account for a deeper analysis of affordability in low-income regions. The accessibility index includes 50 public transport rides relative to revenue. Based on the concept of a fixed number of monthly trips, it is proposed to take transportation costs as a third (33%) of the minimum wages.

To find out what proportion of the total transport costs Blumenberg took into account all the costs involved.

The accessibility index is estimated using equation (3):

$$Aff = \frac{E_t}{H_e} \tag{3}$$

where Et – monthly transportation costs for all modes of transport, He – represents the monthly expenses of the family.

Studies by Blumenberg (2003) – A comparison of transport accessibility in low – and high-income areas, with the aim of estimating transport costs in different countries / regions. The disadvantages of individual accessibility indices can be considered that conducting research in areas / regions with different income levels can give conflicting results. For example, in America, low-income people are more likely to use private transportation; this pattern is characteristic only for this country. In other countries, low-income people are more likely to use – public transport.

• multicriteria accessibility indicators take into account synthesized and factors capable to measure transport accessibility by alternative methods. In developing countries a fixed travel basket is proposed be used to calculate the accessibility index.

The accessibility index in this case represents the average cost of travel in public transport for a certain trip duration and the estimated minimum number of monthly trips.

The accessibility index used by Carruthers (assumes that an employee requires a package of 60 trips by public transport per month):

$$Aff = \frac{60*p}{Y_{pc}^{avg}} * 100 \tag{4}$$

где p – fee in public transport, Y_{pc}^{avg} – mean income per head.

Other tools are necessary for a qualitative assessment of transport accessibility, since the use of accessibility indices alone is insufficient.

The planning structure of the city remains relevant in terms of the formation of prospective residential buildings in the southwestern part of the city of Angarsk, as well as in the formation of new quarters of medium, low-rise and individual residential buildings in the southern and western parts of the city. First of all, construction envisages the development of territories through the construction of multi – storey residential buildings in the western part of the city in microdistricts 31 and 35 and in the eastern part of the city in blocks 256, 257, 258. The development of the Novyy – 4 microdistrict is proposed taking into account the planned location of the bus station and recreation area near Elovsky reservoir.

The central position of the main residential territory of the city of Angarsk is emphasized by established industrial zones located along its southeast, south, southwest and western periphery. At the same time, between them and the residential territory, new green-sized sanitary gaps that are significant in width are preserved and are formed, providing ecological and aesthetic comfort of the residential territory. In addition, these landscaped sanitary gaps will fulfill an important function of aeration corridors – «ventilation» corridors.

The basis of the road network of the Angarsk urban district is the need to ensure stable, maximum safe and short connections of places of resettlement, places of employment and recreation.

One of the important tasks in assessing transport accessibility is transport zoning. Transport zoning is a way of dividing an urban area according to its goals. For its implementation, the following data are required:

• population of each transport region;
• cartographic material;
• location of the main objects of the city network (industrial – production facilities, housing estates, places of employment, objects of socio-cultural purpose);
• list of streets and characteristics of the carriageway received from utilities and road organizations;
• traffic management schemes on a street-road network;

• accessibility of land use;
• transport charges;
• transport and operational indicators.

Transport zoning has two main characteristics: the number of zones into which the territory can be divided, and their dimension. The more zones, the smaller their size. During the survey, the number of zones is selected depending on the objectives of the study and the required accuracy of the results.

When performing transport zoning, the following rules should be followed:

1. It is necessary to consider the transport and pedestrian accessibility of the territory within the zone.
2. When determining the boundaries of the zone, it is necessary to combine the choice with the administrative boundaries (if any). In this case, it is possible to use the available statistical information.
3. The choice of zones should take into account the territories of one purpose: (residential, industrial, recreational).

Transport zoning is a necessary procedure for modeling a network. When choosing zones, one must focus on the rules above and find a compromise between time and level of detail.

In this study, to solve the tasks posed by assessing transport accessibility and its impact on housing cost, we select zones in such a way that development covers a sufficiently large area, for example, several residential quarters / microdistricts. When dividing into zones, we try to take into account the type of development (areas with large housing stock, areas with 2- or 3-story buildings with wooden floors, one – story buildings).

Figure 1 shows the results obtained from the study.

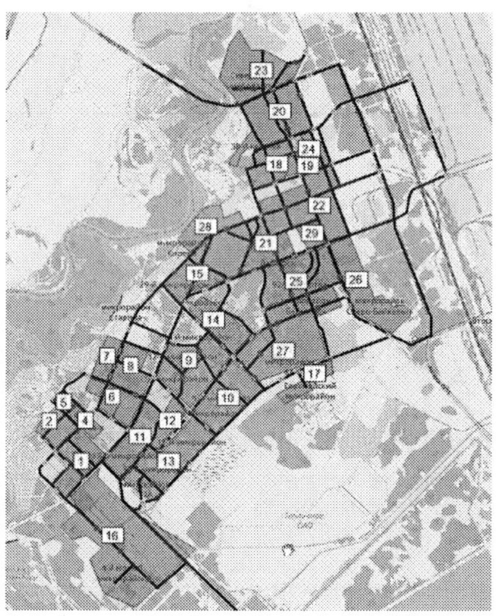

Figure 1. Transport zoning on the example of the city of Angarsk

One of the possible methods of processing the data obtained in the framework of the study is the calculation of the recovery of the time-cost matrix, which will determine the transport accessibility of the city's districts. The calculation of the values of transport accessibility was carried out using the software product PTV «VISUM» [9]

Next, we will analyze each zone separately and get the average time spent on moving around the zone (Figure 2). The following groups were identified: high availability up to 13 min.; average availability from 13 to 17 min .; low availability over 17 min.

As a result of the survey, the following conclusion can be made, the zones with the lowest availability are zones No. 1-5, 9, 11-13, 18-24, 28. The average availability at zones No. 6, 7, 26. High availability at zones No. 8, 10, 14, 15, 16, 17, 25, 27, 29.

The time spent in cities with a population of 227 thousand people for moving from places of residence to places of work in accordance with the normative indicator should not exceed 35 minutes. [10].

The data obtained allow us to assess the impact of transport accessibility on the cost of 1 square. m. of residential premises. Alternative options for determining the connection of transport accessibility with the cost of sq.m. practically does not exist, since this is one of the main parameters of the formation of housing cost. After analyzing the real estate market of the city of Angarsk, by reviewing the cost of sq. M / r on the sites avito.ru and domclick.ru, which are currently one of the most popular electronic sites for the sale of housing, the average cost of 1 sq. m for 2020. The dependence of pricing policy on zonal division is clearly shown in Figure 2.

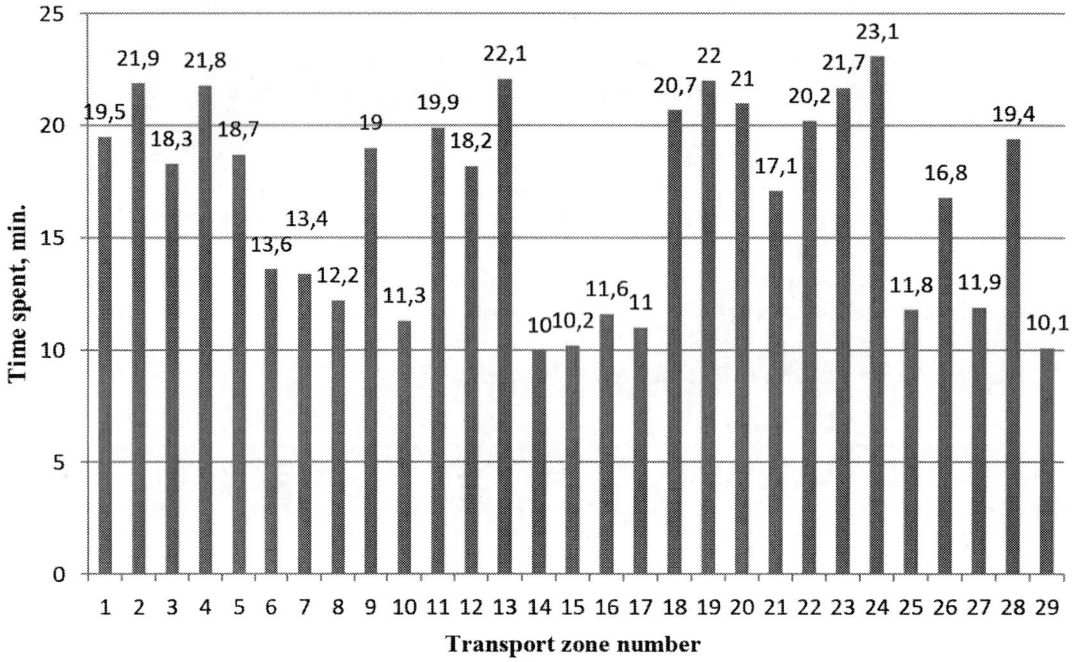

Figure 2. Distribution of transport accessibility by zones in the city of Angarsk

When assessing the impact of transport accessibility on housing costs, it is important to pay attention to a number of aspects [11-16]:

- remoteness from the central part of the city;
- proximity to highways;
- developed system of access roads;
- the proximity of public transport stops with a large number of routes connecting different zones;
- the presence of parking spaces, parking lots or garage cooperatives;
- proximity to places of cultural and social purposes (kindergartens, schools, medical centers, supermarkets);
- exposure to noise, vibration and exhaust gases.

Most often, in studies on this topic, the authors focus on assessing the distance from the city center, since all the main objects of gravity are concentrated there. Transport accessibility is measured by the time required to travel from a specific point in the city to the outer borders of the central region. Accessibility ranking allows you to create a reasonable zoning of the city.

As the level of transport availability decreases, there is an increase in the cost of travel to the central part of the city and the cost per square meter will gradually decrease.

The analysis showed that housing located in zones No. 8, 10, 14, 15, 16, 17, 25, 27, 29 will have the highest level of comfort in terms of transport accessibility, the least comfortable residential buildings located in No. 1-5, 9, 11-13, 18-24, 28 zones (Figure 3). Accordingly, the location of the object will determine their value.

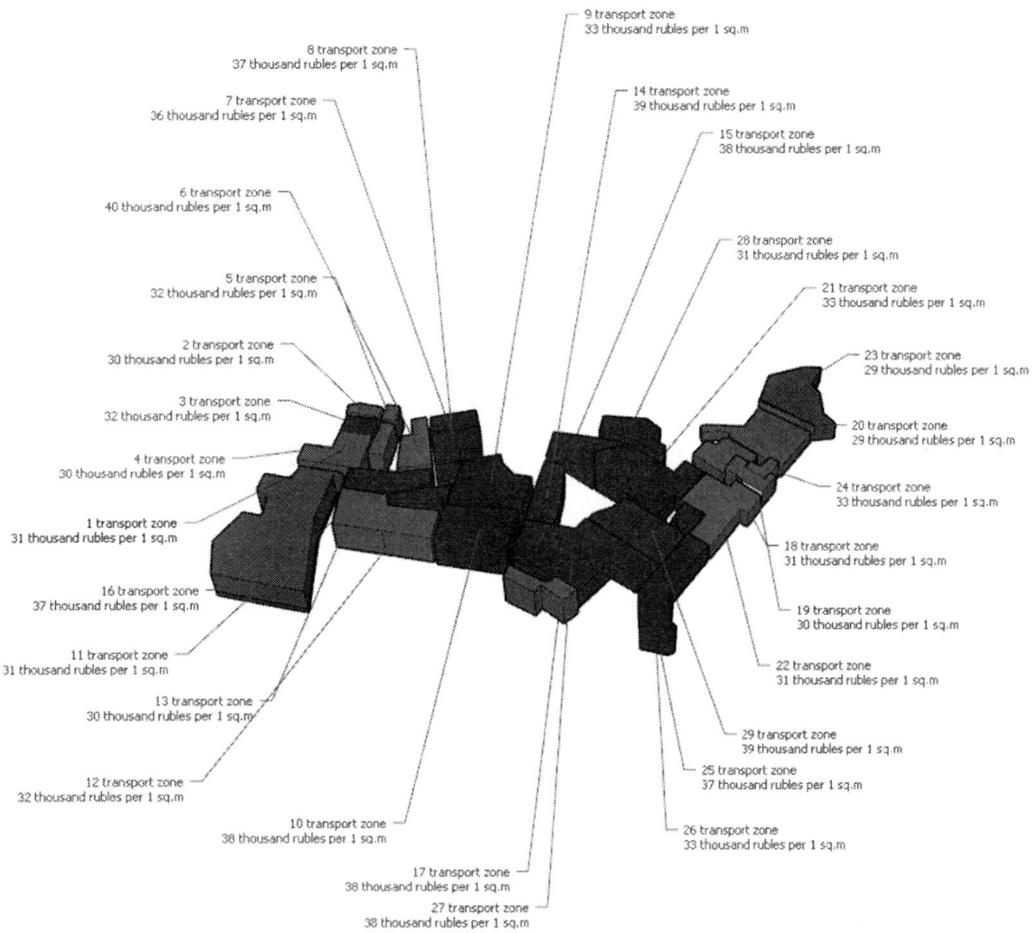

Figure 3. Per square meter price according to transport zones in Angarsk

However, it is worth noting that, despite the fact that the transport zones No. 6, 7, 26 have average transport security, the cost of one square meter is more than 33 thousand rubles, this is due to the fact that the buildings are mostly elite, and people Living in that area use private transportation.

Based on the analysis of the average market value of one square meter, we can conclude that the most attractive are residential buildings located in the central part of the city.

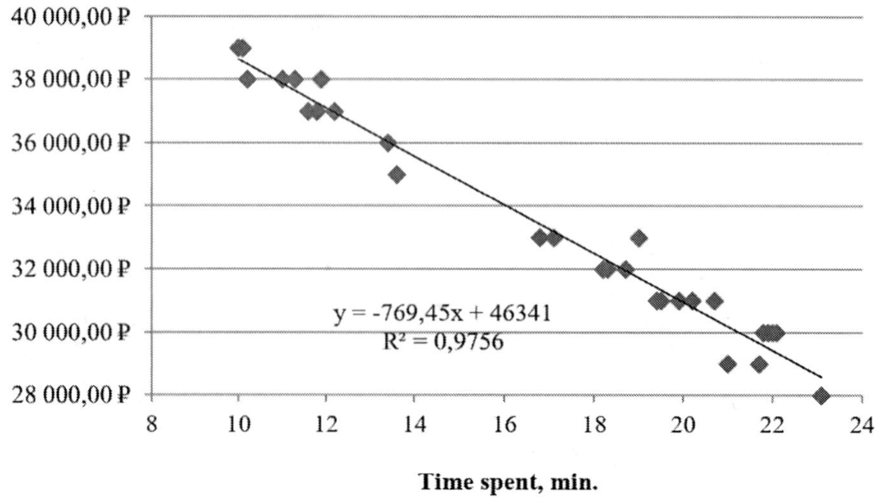

Figure 4. Dependence of the cost per square meter on the time cost of movement

Based on the analysis, the dependence of time costs for movement and the cost per square meter of residential real estate in the city of Angarsk was revealed. For a functional explanation of the dependence of the cost per square meter on time spent on moving for labor purposes, a trend line was drawn and the equation of this line was obtained (Figure 4).

With the increasing remoteness of the residential building from the geographical center of the city, the level of transport accessibility of the territory decreases, and as a result, the cost of one square meter decreases. So, in particular, the difference in the housing costs in the area with the best transport accessibility is higher by 28% than in the remote territory.

Going beyond the boundaries of the geometric center decreases the value of the property. However, the value of the object increases due to the benefits of location (transportation hubs, objects of social attraction) and causes an increase in its value.

The validity and reliability of the results can be confirmed by representative sample sizes, verification of experimental results by generally accepted statistical criteria.

3. Conclusion

A scientific study dedicated to assessing the transport accessibility of the city, which allows to improve the quality of service of the transport network, and to use this data for operational management and transport planning, is relevant.

In the process of the study, rules were identified that must be followed when conducting transport zoning, indices that must be taken into account when assessing transport accessibility. The more data available are the higher the quality of the studis. The level of detail is determined independently; more often, several microdistricts / quarters of the same type and building series are included in the zone.

The need to ensure stable functional connections of the central residential territory with peripheral zones (centripetal connections), as well as the need to organize peripheral connections, create a spatial basis for the development and transformation of the existing transport and architectural-planning structure of the city of Angarsk.

Depending on the availability of data on incomes and expenditures of the population, the relevance of information on the organization of the road network increases, the possibility of obtaining the most realistic estimate of transport accessibility identifying areas that require development of a route network in order to cover it with urban passenger transport.

As a result, we can conclude that increasing the level of transport accessibility of the territory is one of the key factors of economic growth. The availability of tools to conduct a detailed and accurate

IOP Conf. Series: Materials Science and Engineering **880** (2020) 012073 doi:10.1088/1757-899X/880/1/012073

assessment of accessibility is now one of the most important scientific studies in the framework of transport planning.

References

[1] *Guidelines for the design of city streets and roads* 1980 Under the editorship of Yu S Lanzberg Yu and A Stavnichego (Moscow: Stroyizdat) p 324

[2] Sharov M I and Mikhailov A Yu and Duchenkova A V 2013 An example of assessing transport accessibility using the software product PTV «VISUM» *University proceedings Investments Construction The property* **1 (4)**. (Irkutsk) pp 133-138

[3] Sharov M I and Lebedeva O A 2019 Modern technologies System analysis *Influence of transport zoning on the operation of the city route network* **2 (62)** (Irkutsk) pp 196-202

[4] Armstrong-Wright A and Thiriez S 1987 *Bus Services: Reducing costs, raising standards Washington*, D.C. World Bank Technical Paper 68 (Washington, DC: The World Bank)

[5] Banerjee A & Galiani S 2008 Why has unemployment risen in the New South Africa? *Economics of Transition* **16(4)** (Oxford) pp 715–740

[6] Blumenberg E 2003 Transportation Costs and Economic Opportunity Among the Poor. (Routledge Taylor @ Francis Group: New York London) **23** pp 40-41

[7] Carruthers R & Dick M 2005 Affordability of Public Transport in Developing Countries. *Transport Papers* TP-3 World Bank: (Washington DC)

[8] *Program of integrated development of transport infrastructure of the Angarsk urban district for 2017–2036 (*Angarsk) p 58

[9] PTV PARTNER, http://ptv-vision.ru/

[10] SP 42.13330 «SNIP 2.07.01-89 * URBAN PLANNING. PLANNING AND BUILDING OF URBAN AND RURAL SETTLEMENTS» (M .: Standartinform)

[11] Koryagin M 2018 Urban planning: A game theory application for the travel demand management *Periodica Polytechnica Transportation Engineering* **46(4)** (Budapesti Muszaki es Gazdasagtudomanyi Egyetem/Budapest University of Technology and Economics) pp 171–178

[12] Gil J 2016 Urban modality: Modelling and evaluating the sustainable mobility of urban areas in the city–region A+BE *Architecture and the Built Environment* **1** (Portugal) pp 1–436

[13] Lai L.W.C and Davies S.N.G 2011 Government transport land – use planning and development by implicit contract for franchised buses and ferries in Hong Kong, 1933-1972 *Planning Practice and Research* **26(4)** (United Kingdom) pp 435-466

[14] Rasay, Hasan; Golmohammadi, Amir Mohammad 2020 Modeling and analyzing incremental quantity discounts in transportation costs for a joint economic lot sizing problem *Iranian journal of management studies (Iran)* vol. **13** Issue **1** pp 23–49

[15] Arbib Claudio and Pinar Mustafa C 2020 Competitive location and pricing on a line with metric transportation costs *European journal of operational research* vol. **282** Issue **1** (Netherlands) pp 188–200

[16] Laurent Achille-B and Vallerand Steve 2020 *International journal of sustainable transportation* vol. **14** Issue **3** (United States) pp 205-214

IOP Conf. Series: Materials Science and Engineering **880** (2020) 012074 doi:10.1088/1757-899X/880/1/012074

An analysis of traffic flow distribution on roads

E V Volkova[1] and A A Stepanenko[2]

[1] Irkutsk National Research Technical University, 83 Lermontov street 664074, Irkutsk, Russia
[2] Irkutsk National Research Technical University, 83 Lermontov street 664074, Irkutsk, Russia

E-mail: ad@istu.edu

Abstract Traffic intensity is the main parameter determining the road category and all other road design and construction requirements. The aim of this article is to study patterns of traffic intensity on the example of a local highway Egorovshchina – Ryazanovshchina in Irkutsk region. The article presents results of traffic intensity measurements by hours and calculation of reduced traffic intensity taking into account the traffic flow depending on the number of axles and load capacity. Diagrams of the hourly dependence of traffic intensity and traffic flow distribution were created taking into account the traffic flow mix. Traffic intensity reduced to a passenger car was calculated. The research result is patterns of traffic intensity distribution by the traffic flow mix required for calculating reduced traffic intensity.

To ensure safe traffic on roads with permitted speed, it is necessary to know characteristics of the existing and predicted traffic intensity, taking into account the traffic flow for a specified period.

According to the Technical Regulations of the Customs Union, "traffic intensity" is the number of vehicles passing through the section of a road per unit of time (day or hour) [1]. The traffic flow composition is the share of vehicles simultaneously involved in traffic.

Existing regulatory documents [2] establish methods for calculating traffic intensity and traffic flow composition.

The methods used to calculate the traffic flow are divided into two groups: automated and visual. In terms of duration, accounting for traffic intensity is divided into long-term and short-term. With any method, the data should contain information that is a multiple of no more than 1 hour [2].

There are the following groups of vehicles in the traffic flow: cars, trucks and buses.

The trucks are divided depending on the number of axles and load capacity. The number of trucks is a decisive indicator when determining reduced traffic intensity.

By the number of axles, there are biaxial, triaxial and four-axle trucks; four-axle and five-axle road trains; three-axle, four-axle, five-axle and six-axle truck trains; cars with seven or more axles.

By load capacity, there are the following vehicles:

- trucks with a capacity of up to 2 tons, over 2-6 tons, 6-8 tons, 8-14 tons, 14 tons;
- road trains with a capacity of up to 12 tons, over 12-20 tons, 20-30 tons, 30 tons;
- buses with a small, medium or large capacity;
- articulated buses and trolleybuses [3].

The initial road design and construction stage is the substantiation of a road category. This stage is relevant both for construction and reconstruction of roads. The category of roads is determined depending on functions and the maximum traffic intensity for a passenger car. The classification by load capacity and types of vehicles is used for calculating the reduced traffic intensity (hourly or daily).

A calculation unit is equal to a passenger car. It accounts for other types of vehicles on the road, their dynamic properties and sizes in order to average averaging them to determine traffic characteristics (intensity, estimated speed, etc.) [3].

The estimated traffic in both directions is

- average annual daily traffic intensity for the last year of a 20-year period;
- hourly reduced traffic intensity achieved or exceeded for the last year of a 20-year period for 50 hours.

The category of a road is a criterion characterizing the importance of the road in the transport network of the country determined by traffic intensity. In accordance with the road category, all technical parameters of the road are set [7].

Traffic intensity for a passenger car is calculated by formula (1).

$$N_{r.} = N_{pass.} \text{ x } K_{pass.} + N_{tr.} \text{ x } K_{tr.} + N_{a.} \text{ x } K_{a.} \tag{1}$$

The factors reducing traffic intensity to a passenger car are recommended for countries of the Customs Union (Table 1).

Table 1. The recommended traffic intensity reduction coefficients for various vehicles

Types of vehicles	Reduction coefficient
Cars and motorbikes, minibuses	1,0
Trucks with load capacity, t:	
up to 2	
more than 2 " 6 "	1,3
" 6 " 8 "	1,4
" 8 " 14 "	1,6
" 14 "	1,8
	2,0
Road trains with load capacity, t:	
up to 12	1,8
more than 12 " 20 "	2,2
" 20 " 30 "	2,7
" 30 "	3,2
Small capacity buses	1,4
Medium capacity buses	2,5
High capacity buses	3,0
Articulated buses and trolleybuses	4,6

When designing a road, the type of a road surface, traffic intensity and traffic flow composition should be taken into account.

The calculation should be performed using average daily traffic intensity in both directions, calculated as the sum of intensities in each direction. When the difference between intensities is more than 10%, the calculation can be performed for each direction separately, taking into account actual traffic intensity for each direction [4].

For designing and calculating the road surface, the following characteristics reflecting intensity of the impact of a moving load are used:

- average daily load (at the end of the overhaul period), the number of driveways of all wheels located on one side of the calculated vehicle within one lane of the carriageway (traffic intensity reduced to the calculated load);
- the total number of applications of the reduced design load to a point on the surface over the overhaul period.

There are objective difficulties in determining actual intensity on a specific section of the road.

Automated methods are used for roads of high categories. They often do not capture individual sections.

The categorization of vehicles by load capacity is difficult when visually accounting for traffic intensity. It is necessary to know the car model, its load capacity or specify its characteristics. The second option is impossible without an employee of the State Inspectorate for Road Safety and leads to traffic jams, congestion.

The video surveillance and subsequent cameral processing of video files can be used as an alternative.

The authors performed recorded the traffic flow on the local Egorovschina-Ryazanovschina road in Irkutsk region. The traffic intensity was being determined during one day in September 2019. The actual traffic intensity was 74 cars per day. When determining traffic intensity, the flow composition was taken into account and trucks were categorized by load capacity and the number of axles.

Then cameral processing of field intensity data was carried out.

The hourly distribution of traffic intensity on the Egorovshchina-Ryazanovshchina road is presented in Figure 1.

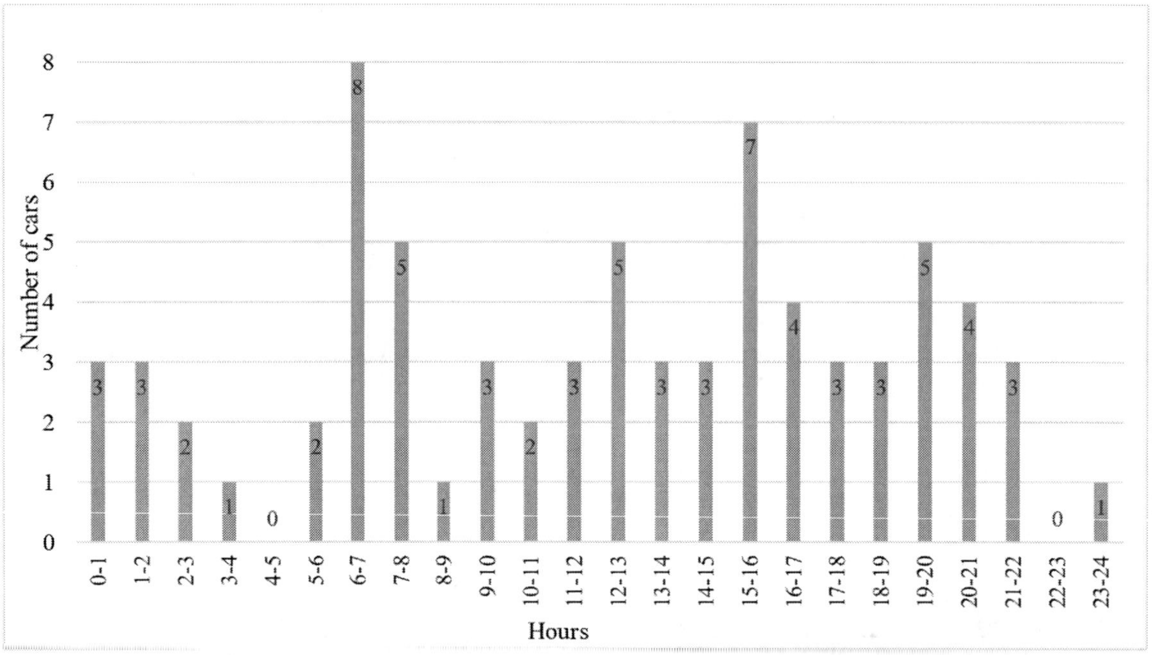

Figure 1. Distribution of traffic intensity by hours on the Egorovshchina – Ryazanovshchina road

The distribution of cars and trucks by traffic flow composition depending on the number of axles and load capacity is presented in Figure 2.

Tables 2 and 3 show the calculation of traffic intensity reduced to a passenger car. The flow contains 38% of trucks and 62% of cars.

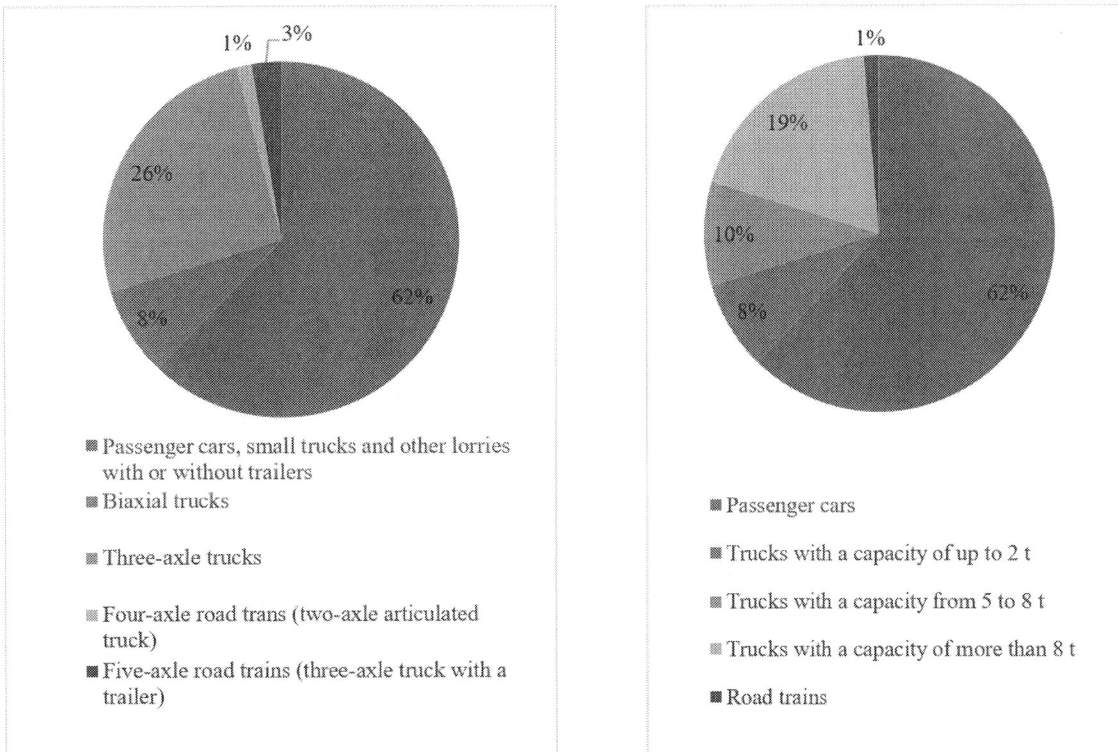

Figure 2. Distribution of cars by traffic flow composition (depending on the number of axles and load capacity of trucks)

Table 2. Traffic intensity reduced to a passenger car (depending on load capacity of trucks)

Types of vehicles	Number of cars per day	Reduction coefficient	Traffic intensity reduced to a passenger car, cars per day
Passenger cars	46	1.0	46.0
Trucks with a capacity of up to 2 t	6	1.3	7.8
Trucks with a capacity from 5 to 8 t	0	1.4	0.0
Trucks with a capacity of more than 8 t	7	1.6	11.2
Trucks with a capacity of more than 8 t	14	1.8	25.2
Road trains	1	1.8	1.8
Buses	0	2.5	0.0
Total	74		92

Table 3. Traffic intensity reduced to a passenger car (depending on the number of axles)

Types of vehicles	Number of cars per day	Reduction coefficient	Traffic intensity reduced to a passenger car, cars per day
Motorcycles	0	-	0.0
Cars, small trucks (vans) and other cars with and without a trailer	46	1.0	46.0
Biaxial trucks	6	1.5	9.0
Three-axle trucks	19	1.8	34.2
Four-axle trucks	0	2.0	0.0
Four-axle road trains (two-axle lorry with a trailer)	1	2.2	2.2
Five-axle road trains (three-axle truck with a trailer)	2	2.7	5.4
Three-axle truck trains (two-axle truck tractor with a semitrailer)	0	2.2	0.0
Four-axle truck trains (two-axle truck tractor with a semitrailer)	0	2.7	0.0
Five-axle truck trains (two-axle truck tractor with a semitrailer)	0	2.7	0.0
Five-axle truck trains (three-axle truck tractor with a semitrailer)	0	2.7	0.0
Six-axle truck trains	0	3.2	0.0
Cars with seven or more axles and others	0	3.2	0.0
Buses	0	3.0	0.0
Total	74		97

The reduced traffic intensity allows us to attribute this road section to category V.

Observations of the traffic flow composition were carried out on other sections of Irkutsk roads. The distribution of the traffic flow composition is presented in Table 4. The traffic flow includes: cars - 60-82%; trucks - 17-38%; buses (including minibuses) - 1-8%.

The results show a large number of passenger cars, which indicates a slight increase in traffic intensity compared with actual traffic intensity.

Table 4. The composition of the traffic flow in Irkutsk region

Road section	Traffic flow composition, %			Traffic intensity, vehicles per day
	Passenger cars	Trucks	Buses	
2019				
Irkutsk - Bolshoy Goloustnoye, Irkutsk - Pivovarikha section	81	16	3	8057
Irkutsk - Osa - Ust-Uda, Irkutsk - cemetery approach section (2.6 km)	76	18	6	21993
Irkutsk - Osa - Ust-Uda, Irkutsk - cemetery approach section (km 2.6) - dump cite approach section (4,7 km)	68	24	8	18606
Irkutsk - Osa - Ust-Uda, dump site approach (4,7 km) - Razdolie (8 km)	75	20	5	13 452
Irkutsk - Osa - Ust-Uda, dump site approach (4,7 km) - Razdolie (8 km) Ust-Kuda approach (17,2 km)	74	21	5	13 452
Irkutsk - Osa - Ust-Uda approach (17.2 km) - Moskovshchina (24.6 km)	76	19	5	2 737
Irkutsk - Iskra	76	22	2	727
Access to Malaya Elanka	76	21	3	3011
Access to Eloviy	62	36	2	2150
Urik - Stolbovo	60	38	2	552

Access to Karluk	76	21	3	3628
Novo-Lenino - Maksimovshchina	82	17	1	3625
Oek - Kudareyka	61	34	5	5282

The analysis allows us to draw the following conclusions.

When calculating traffic intensity reduced to a passenger car, the categorization of trucks by the number of axles or load capacity is not significant.

According to the regulatory documents [3], [6], traffic intensity reduced to a passenger car and used to determine the road category is calculated on the basis of reduction factors depending on a load capacity.

In accordance with GOST 32965-2014 "Public roads. Methods of accounting for traffic flow intensity", the calculation of traffic intensity reduced to a passenger car takes into account the division of trucks by the number of axles. Thus, traffic intensity is used to determine the throughput of roads and transport hubs.

According to PNST 265-2018 "Public automobile roads. Non-rigid pavement design" for calculating pavement, freight vehicles are divided by the traffic flow composition depending on the number of axles.

Therefore, when conducting economic surveys, a comprehensive analysis of the composition of the traffic flow is required to determine traffic intensity. It is necessary to distinguish the following groups of cars: cars, trucks and buses. Trucks should be subdivided according to the number of axles and load capacity.

References

[1] Technical regulations of the Customs Union TR CU 014/2011
[2] GOST 32965-2014 General automobile roads. Methods of accounting for traffic intensity
[3] GOST 33100-2014 General automobile roads. Road design rules
[4] PNST 265-2018 General automobile roads. Non-rigid pavement design
[5] SP 396.1325800.2018 Streets and roads of settlements. Urban design rules
[6] SP 34.13330.2012 Roads. Updated edition of SNIP 2.05.02-85*
[7] GOST 33382-2015 General automobile roads.
[8] ODM 218.4.005-2010 Recommendations for ensuring traffic safety
[9] ODM 218.4.039-2018 Recommendations for the diagnosis and assessment of technical road conditions
[10] ODM 218.2.032-2013 Guidelines for accounting for traffic on roads
[11] ODN 218.0.006-2002 Rules for the diagnosis and assessment of road condition (instead of VSN 6-90)
[12] http://www.sovstrat.ru/journals/transportnaya-strategiya-21-vek/articles/st-trans32-9.html (accessed on February 28 2019) Journal "Transport Strategy XXI Century" No 32 2016
[13] https://irkroad.ru/obsluzhivaemye-dorogi/r-255-sibir.html (accessed on March 3, 2019)
[14] E Volkova and A Stepanenko 2019 Traffic intensity on highway R-255 Siberia in Irkutsk region. *IOP Conference Series: Materials Science and Engineering*
[15] E V Volkova and A A Stepanenko 2013 *University News. Investment. Construction. Real Estate* **1** 91-95
[16] E V Volkova and A I Solonenko 2015 *University News. Investment. Construction. Real Estate* **No 2 (13)** 28-33
[17] D M Filippova, A B Chernyago and N A Slobodchikova 2013 *University News. Investment. Construction. Real Estate* **No 2** 122-125
[18] N A Slobodchikova and K V Plyuta 2018 *Collection of works of the regional scientific-practical conference "Modern problems of design, construction and maintenance of roads in the region"* 8-17
[19] V B Balabanov, A S Merkulov and V G Tatarnikov 2015 *Bulletin of Irkutsk State Technical University* **No 4 (99)** 199-205

[20] S S Shaburov and A S Kachin 2018 *Collection of works of the All-Russian scientific-practical conference "Interaction of science, education and production"* 59-74

[21] S S Shaburov and A K Fam 2017 *Bulletin of Irkutsk State Technical University* **No 5 (124)** 179-190

[22] V I Martyanov, D V Pakhomov and V V Arkhipov 2010 *Modern technology. System analysis. Modeling* **No 2 (26)** 66-72

[23] N S Kulik, V I Martyanov and D V Pakhomov 2016 *Bulletin of Irkutsk State Technical University* **No 4 (111)** 96-101

[24] A G Levashev and A Yu Mikhailov 2005 *Bulletin of Irkutsk State Technical University* **No 1 (21)** 138-143

IOP Conf. Series: Materials Science and Engineering **880** (2020) 012075 doi:10.1088/1757-899X/880/1/012075

Assessment of suburbanized areas transport demand: case study of the Irkutsk Agglomeration

A V Zedgenizov[1], A I Solodkiy[2] and I Efremenko [3]

[1 3] The Department of Oil and Gas Engineering, Irkutsk National Research Technical University, 83 Lermontov street, Irkutsk, 664074, Russia
[2] The Department of Transport Systems, Saint Petersburg State University of Architecture and Civil Engineering, 4 Vtoraya Krasnoarmeiskaya Street, Saint Petersburg, 190005, Russia

E-mail: azedgen@gmail.com

Abstract. The article deals with the issues of transport services for suburbanized areas. The significance and topicality of the research are demonstrated. Various solutions to this problem in different parts of the world are described. Currently, the wealthiest strata of society are striving to obtain a higher quality of life due to the permanent residence in the suburbs which gives rise to the demand for daily commuting between the city and its suburbs. In the article this issue is developed within the context of creating a toolbox for quantitative assessment of suburbanized areas influence on the road network load, first of all, on the load of sections providing commuting between suburbanized areas and the city. The article details the sequence of selecting individual gated communities within the transport accessibility zone of the Irkutsk agglomeration; besides, the article presents a sample study of one gated community aimed at determining the number of entrances and exits. Besides, there are results of the regression-correlation analysis that allows determining an empirical dependence making it possible to estimate suburbanized territories transport demand according to their areas.

1. Introduction

The first and subsequent technological revolutions occurred over the past few centuries have significantly changed the ratios of people living in cities and rural areas. The process of urbanization was particularly intense in the late 19th and early 20th centuries. This epoch was marked by global industrialization and required the concentration of a large number of workers living close to production sites. Later, all over the world, including Soviet Russia there appeared both large industrial centers and small towns supporting the activity of core enterprises. Such towns were called single-industry towns. Further automation and computerization of production processes, as well as development of the automotive industry and availability of individual transport (IT), allowed expanding the radius of transport accessibility around cities. The development of suburban areas involving permanent residence is called suburbanization. The fundamental difference between living in rural areas and living in suburbanized ones is that inhabitants of the latter have permanent labor and cultural ties with the city. This circumstance causes daily commuting of suburbanized areas residents to the city in the morning and back to their residence in the evening; given that 90-95% of cases involves using IT for commuting this circumstance creates significant load on the street-and-road network (SRN) especially in the sections joining an urban area and a suburb. Solving the problems associated with permanent traffic congestions requires a scientific approach to prove and predict the intensity of traffic flows at the boundaries of cities and suburban areas.

2. Topicality and Scientific Relevance of the Research

The transport demand assessment was studied and developed by many Russian and foreign researchers. In particular, I.S. Efremov [6], E.M. Lobanov [12], V.V. Silyanov [14] and others [4] [7] [8] [10] achieved the greatest progress. The urban planning issues, the basic principles of functioning and interacting of urbanized territories were described in the scientific works by Yu.N. Bocharov [1], A.O. Kudryavtsev [1], A.P. Romm [13] and others [5] [2]. Currently, it is possible to predict the intensity of traffic flows using packaged applications of micro- and macromodelling; the input data for such programs are studies based on methods of assessing transport demand which depends on the specificity of permanent residence in the suburbs. A number of parameters should be taken into account (the percentage of residents using IT, irregularity of hours, etc.); they will allow making more accurate predictions of transport demand and, subsequently, of the intensity of traffic flows in the studied sections of SRN.

3. Problem Statement

To assess transport demand of any focal points (FP) including suburban areas one should carry out transport inquiries including a complete study of correspondences made during a day. Such studies are of high labour intensity, however, they allow researchers to obtain an extensive range of data used for projections. Studies aimed at determining transport demand should begin with a preliminary estimate of the number of objects composing the general totality of the studied phenomenon; moreover, it is necessary to consider objects with different spatial characteristics, such as remoteness from the city center, area of the suburbanized territory, remoteness from the main road, availability of public transport (PT) and other factors. Each focal point must be thoroughly studied to determine the number of entrances-exits, availability of PT stopping points in the immediate proximity and of the public parking used by residents of this suburbanized area. A sample of a detailed plan of conducting a field study aimed at assessing transport demand of one of the gated communities of the Irkutsk agglomeration is presented in Fig. 1. Further, a daily study of the object is carried out to record the following information: the number of arriving and departing visitors, the number of people using IT and PT and the time of day. Moreover, it is worth recording the registration numbers of the vehicles in order to determine the duration of their parking.

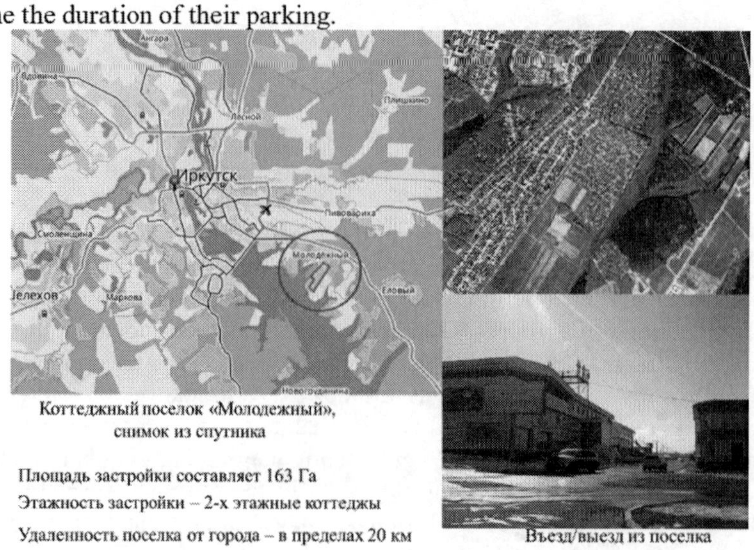

Fig. 1 The Plan of Conducting the Field Study

Molodyozhny, a gated community
satellite image

The site area is 163 ha.
The number of floors in buildings – two-storey houses.
The remoteness of the gated community from the city – about 20 km.
Entrance/exit of the gated community

4. Theoretical Prerequisites

Field experiments are necessary to determine the basic parameters of suburbanized areas functioning; these parameters in combination with additional factors should be used to calculate and predict the intensity of traffic flow generated by visitors of these areas. The mathematical model combining these parameters is presented below:

$$N_{IT} = E_{FP} \cdot \frac{d_{IT}}{P_{IT}} \cdot k_{di} \quad (1)$$

where E_{FP} stands for the transport demand of the studied FP (a suburbanized area); d_{IT} means the ratio of people using IT during the definite hour; P_{IT} – the average number of people in a vehicle (IT); k_{di} – coefficient of daily irregularity for the definite hour.

The main parameters of Formula 1 are described in Table 1.

Table 1. Basic Characteristics of Areas (FP) Functioning

Parameter	Brief Description
Transport demand, corr./day	The number of correspondences in a focal point of a definite type. As well as the concept of economic demand which is defined on the basis of comparison between the concepts of what is desired and possible, transport demand can significantly depend on the purpose of the trip (work, cultural life, etc.), on its total cost which is understood as time and money investment, on the weather and even mood.
Ratio of visitors using IT (involved in transport demand)	It shows the ratio of visitors using IT as a means of arriving at the definite FP. This characteristic depends, first of all, on the state policy in the field of public transport services, for example, in Asia there is a policy of containing the growth of automobilization, whereas in Western countries, especially in the USA and Great Britain, during the 1930s-1970s the growth of automobilization was lobbied by the state which ultimately led to disastrous consequences. Besides, the ratio of visitors depends on the remoteness of FP from the urban passenger transport lines, on the need to transport heavy (bulky) items, or on performing some sequence of correspondences one of which cannot be made using the urban passenger transport.
Daily irregularity coefficients	They reflect the ratio of correspondences (transport demand) per hour in question, usually it is a peak hour. The absolute value of transport demand, as a rule, does not present comprehensive information about the modes of FP functioning, it only shows its scale. When assessing the functioning of FP and, first of all, evaluating its impact on the adjacent SRN it is necessary to obtain information about the number of visitors during the one-hour (peak) period, since the operating modes of transport infrastructure are calculated per hour. In transport calculations the coefficient of daily irregularity having the greatest value is called the coefficient of daily maximum. In fact, it represents the relationship between daily and one-hour values of transport demand.
Average number of people in a vehicle (IT), persons/car	It reflects the average number of people in one IT when visiting FP in question. This characteristic shows the number of visitors (transport demanders) per car (IT). In fact, this is a relationship between transport demand and intensity of IT driving to/from FP.
The average duration of parking, min.	This is an average duration of staying in a parking space near FP. The average duration of parking, first of all, depends on the type of FP, on its area (size), on the duration of service and other less significant factors, for example, the

remoteness of the parking space from the entrance to FP. This characteristic is basic; it is used for calculating the required number of parking spaces. However, it should be noted that nowadays in the vast majority of cases the absence of parking spaces will lead to reluctant visits to such FPs.

5. Experimental Findings

The selection of suburbanized areas that have the most favorable location in terms of the Irkutsk agglomeration and conducting the daily studies have resulted in the fact that the team of authors obtained the data on transport demand and other parameters of 10 suburbanized territories functioning (Table 2).

Table 2. Basic Quantitative Characteristics of Suburbanized Areas of the Irkutsk Agglomeration

Name of the Area	Transport Demand, corr./day	Ratio of Visitors Using IT	Coefficient of Daily Maximum	Average number of people in a vehicle (IT), persons/car
Zui	1 218	0.768	0.175214	1.645
Khomutovo	10 743	0.894	0.119983	1.325
Molodyozhny	3 188	0.852	0.105353	1.372727
Meget	6 140	0.844	0.115122	1.533333
Kaya (inner suburb of Irkutsk)_	4 398	0.870	0.116906	1.645
Milovidy	591	0.940	0.121406	1.594286
Pivovarikha	249	0.770	0.12963	1.334667
SNT 50 Let Pobedy	160	0.740	0.063	1.355
SNT Mechta	364	0.900	0.063	1.381
Minimum	160	0.74	0.063	1.325
Maximum	10 743	0.94	0.175214	1.645
Average	3 005.66	0.842	0.1121	1.465

The regression-correlation analysis (Table 3) makes it possible to establish the dependence of transport demand on the area of the suburbanized territory.

Table 3. Regression-Correlation Statistics of Suburbanized Territories Transport Demand

Regression Statistics					
Correlation Coefficient R				0.909067	
R square				0.826403	
Observations				10	
Variance Analysis					
	The Number of Variances of System	Variance		Fisher's Variance Ratio F	
				$F_{calc.}$	$F_{95\%;2;120}$
Regression	1	91 606 065.14			
Residual	8	19 243 044.9		38.26	3.92
Total	9	110 849 110			
t – statistics					
	Regression Coefficients	t – statistics		95%[th] confidence limit	
		$t_{calc.}$	$t_{95\%;120}$	minimum	maximum
Y (corr.)	732.68	1.24	1.98	-619.73	2085.096
X_1 (m^2)	0.00118896	6.17		0.00074468	0.0016

Thus, to estimate transport demand the following regression equation should be used:

$$E = 0.00118896\ S = 11.8\ S \quad (2)$$

where S is the area of the suburbanized territory, m^2/ha. To use this relation in practice it is necessary to remember the limit intervals: 52.000 <S14 <696.000 m^2, at the distance not exceeding 20 km from the city center.

6. Conclusion

The obtained regression dependence and the accompanying quantitative characteristics of suburbanized territories functioning will allow evaluating and predicting the intensity of traffic flows as well as of the SRN load including sections of joining the urban territories and suburbs. The obtained regression equation and quantitative characteristics will improve the accuracy and quality of projections made on the basis of modern software packages for micro- and macromodelling of traffic flows. Besides, it resulted in the development of the toolbox aimed at increasing the efficiency of urban planning decisions while implementing road traffic management projects, integrated traffic management schemes and comprehensive program of transport infrastructure in the transport zones of large and metropolitan cities. This problem is especially acute for the sections of SRN connecting suburbanized territories with the city.

References

[1] Bocharov Yu N and Kudryavtsev O K 1972 Planning Structure of a Modern City (M.: Publishing House) of Literature on Construction p 160

[2] Vilner M Ya 2008 On the Town-Planning Foundations of the Innovative Development of Russia *Real Estate and Investments. Legal Regulation* **4 (37)**

[3] Volodkin P P 2010 Features of Development of Passengers' Correspondences with Spatial Self-Organization *BULLETIN OF PNU* **3 (18)** pp 123-133

[4] Vukan R and Vuchik 2011 Transport in Cities Convenient for Life

[5] Gutnov A E 1984 The Evolution of Urban Development p 256

[6] Efremov I S 1980 Theory of Urban Passenger Traffic: a manual for universities p 535

[7] Zhukovsky V S 2012 Overcoming the Investment Crisis in the Construction Sector and Creating Long-Term Investment Resources. Affordable and Comfortable Housing *Problems, Searches, Solutions: Materials from an Expert* pp 41-48

[8] Zedgenizov A V and Burkov D G 2016 Methods of Assessing Transport Demand Based on Quantitative Characteristics of Functioning of Urbanized Territories. Traffic Organization and Safety in Large Cities *A Collection of works (electronic version) of the participants of the twelfth intern. scientific-practical conf.* pp 235-242.

[9] Zedgenizov A V 2016 Correspondences Generation as the Main Quantitative Characteristic that Determines the Attractiveness of the Focal Point. Bulletin of ISTU **4** pp 187-192

[10] Zedgenizov A V 2016 Suburbanization of the City of Irkutsk: Scale and Consequences *Universities News. Investments. Construction. Real Estate* **1 (16)** pp 159-165

[11] Lobanov E M 1990 *Transport Planning of Cities* p 240

[12] Romm A P 2002 Integrated Assessment and Functional Zoning of the Territory in Urban Planning p 206

[13] Silyanov V V 1977 The Theory of Traffic Flows in Road Planning and Traffic Management p 303

[14] *Trip Generation* 8th Edition. Washington 2008

[15] Wissen L A 1989 *Model of Household Interactions in Activity Patterns*

[16] Yao L, Guan H 2008 Trip Generation Model Based on Destination *Attractive-ness Tsinghua science and technology* **13** pp 632-635

[17] Zedgenizov A V and Burkov D G 2016 Methods for the Traffic Demand Assessment Based on the Quantitative Characteristics of Urban Areas Functioning *12th Inter-national Conference Organization and Traffic Safety Management in Large Cities*, pp 28-30

[18] Zedgenizov A 2018 Location-Based Transport Demand Forecasting Methods for Suburbanized Areas *Proceedings of the international conference: aviamechanical engineering and transport (avent 2018) AER-Advances in Engineering Research* 158 pp 458-461

IOP Conf. Series: Materials Science and Engineering **880** (2020) 012076 doi:10.1088/1757-899X/880/1/012076

The current situation with the issue of modeling of flat and spatial structures based on wood

T L Dmitrieva[1], K A Podshivalova[1] and I P Pinaykin[1]

[1] Irkutsk National Research Technical University, 83 Lermontov str., Irkutsk, 664074, Russia

E-mail: dmitrievat@list.ru, 1295536@mail.ru, pinaykin@bk.ru

Abstract. The analysis of the current state of issues related to the construction of design and optimization models of flat and spatial core structures based on wood under static and seismic impacts is presented. The need to develop methods for finding the best projects for such structures, which would allow for rational technical solutions, taking into account the set criterion of optimality and all the requirements for their tense-deformed state. Particular attention is paid to the construction of adequate calculation models taking into account the anisotropic structure of wood.

1. Introduction
The key priority of the building industry is to reduce the cost of production and reduce the time of commissioning. This goal is achieved by reducing the material intensity and complexity of the manufacturing and installation of structures in buildings and structures.

One way to meet these requirements is to develop methods for the optimal design of buildings and structures. Among the most important areas of optimization theory is the justification of methodological principles for the formulation, existence and adequacy of solutions to optimality problems of typical and individual constructive solutions.

Tasks of this type are multifactorial, they include the assessment of natural and climatic conditions, taking into account the laws of stress distribution, the nature of the action of loads, the relationship of the ratio of forces with the spatial forms of elements, buildings and structures. On the other hand, these tasks should take into account economic laws governing the measure of consumption and the ratio of resources needed to implement structural solutions (cost of materials, services, operating costs, economic consequences of structural failures, reconstruction costs).

For wood-based structures, the analysis of factors affecting their dressed-deformed state is complicated by the fact that the tree as a structural material, due to its natural origin, has a large spread of mechanical characteristics, significantly larger compared to materials such as steel and concrete. Features of influence on mechanical characteristics of wood of various conditions of the work of a structural element are considered by special coefficients [1]. The calculated mechanical characteristics of wood, as well as for other structural materials, are determined by statistical methods using distribution functions.

2. Examples of optimization of individual structural elements in structures based on wood
We note some examples of the optimal design of individual elements and compounds from wood. The analysis of the behavior of natural structures of laminated wood domes and the numerous preliminary

calculations have shown the possibility of saving materials by reducing the height of cross-sections of meridional ribs. This is especially effective when you include in the design of skins, performing the role of the building shell, the collaboration with frame elements (annular and longitudinal ribs) [2].

In the works of foreign and Russian authors, numerous studies have been carried out concerning various elements of joints of wooden structures. A comparative analysis based on a series of numerical simulation experiments in the ANSYS software for determining the bearing capacity and the optimal arrangement of CFRP dowel pins in the joints of wooden structures. A comparative analysis of the stress-strain state of the compound with the results of calculations is carried out according to the procedure stipulated by the current norms and rules of the strength [3]. Experimental and numerical study of the behavior of timber beam-to-steel column connections under cyclic loads [4].

3. Features of the work of wooden structures in seismic conditions

Timber structures traditionally provided satisfactory seismic performance due to multiple known features. However, the consequences of the last major earthquakes have clearly proofed that seismic timber design must further improve. Also, at present, wooden structures are focused on high-rise construction and therefore increased seismic requirements are imposed on them. Altogether, this has led to the fact that seismic protection technologies have gained great importance in research related to woodworking, which is devoted to more than 80 publications only in the last decade [5].

In general, wooden structures can be designed to meet the life safety criterion and to deform inelastically without inducing collapse under major earthquakes. At the same time, structural and non-structural damage associated with inelastic reactions is usually very expensive to repair, which gives impetus to research in the field of increasing the seismic resistance of structures using wooden structures. In [6], a description is given of the development, experimental testing, and numerical simulation of a new hybrid joint, which was developed to improve the seismic characteristics of the frames. In [7], questions of numerical studies of the seismic characteristics of a hybrid shock-absorbing structure of wood and steel are considered. Here, friction dampers are utilized to upgrade the seismic performance of timber-steel hybrid structure, which has been proposed as an alternative structural solution for multi-story buildings. The effectiveness of various design solutions applied to high-rise wooden structures was investigated as a means of reducing seismic accelerations [8].

4. Modeling problems in finite element analysis problems

Currently, the design of complex technical objects is impossible without the use of automated software systems (PCs). About wood structures, it should be noted that in the modern design there is a need for software that allows adequate simulating the complex behavior of the wood system, accessible to practitioners.

To assess the stress-strain state of building structures among numerical methods, the finite element method has become widespread. Most computer systems and software complexes related to finite element analysis of building structures include a library of isotropic or anisotropic materials with the assumption of their isotropy. However, as shown in [9], this approach is not applicable to solving spatial problems. When modeling wood material as isotropic, there is a significant decrease in the accuracy of calculations up to complete inaccuracy. In view of this fact, it is not possible to correctly model the processes of manufacturing, installation and operation of building structures made of wood.

When calculating metal and reinforced concrete structures, shear deformations are neglected, which is quite justified, since they are negligible for metal and reinforced concrete, which cannot be said about wood. Ignoring this issue leads to inaccurate calculation results. Therefore, the determination of the stress-strain state in the calculus of wooden structures must be carried out taking into account shear deformations.

Several publications by Russian and foreign researchers are devoted to the study of various aspects of the problem of modeling wooden structures. In V. N. Volynsky's monograph [10], the main factors influencing the strength of wood are temperature, humidity, and loading speed. At the same time, Turkish scientists [11] model wood as an isotropic material. The results of studies of the elastic deformability of wood across the fibers obtained By V. N. Glukhikh [12] allow us to determine with high accuracy the relationship between the physical and mechanical characteristics of wood and environmental parameters. In [13], a finite element model of the elastic modulus of the LVL material of the Ultralam brand is described with a different number of layers and a combination of their stacking according to the mutually perpendicular and parallel arrangement of wood fibers in them. In the research, we have used numerical simulations of LVL bending tests using the FEM method according to GOST 33124–2014 in the SCAD program. The obtained data were processed in Mathcad and Excel software package. Simple analytical dependences were obtained. They allow calculating the elastic modulus of all types and thicknesses of Ultralam LVL at any given angle of wood fibers. In [14], some provisions regarding the application of non-linear static analysis of timber structures are introduced. In scientific work [15], novel building systems such as post-tensioned (PT) timber have been developed. Several numerical models of PT timber beam tests are developed and validated using general-purpose FEM software ABAQUS. This research program serves to evaluate the current capabilities or shortcomings of modeling PT timber in both ambient and fire conditions and to develop a methodology for analyzing the performance of the system.

5. Optimization of building structures in software systems

Many problems of structural engineerings, such as the rational selection of sections or the topology of structures require the use of optimization algorithms. Interest in the problem of optimal design and the development of appropriate algorithms falls on the second half of the twentieth century, where most of the tasks were formulated in the form of nonlinear programming. For the first time, the most general statement of the optimization problem was proposed by L. Schmitt [16]. He combined finite element analysis of structures with nonlinear programming methods. In 1979, a monograph was published by American scientists Edward J. Haug and Jasbir S. Arora [17], which gave a serious impetus to the development of the applied direction of optimization. It outlined general approaches to solving the problems of analysis and synthesis of mechanical systems. The 70-80s of the last century also accounted for numerous software implementations of optimization algorithms.

Since the 90s, a large number of studies have appeared in the field of engineering system optimization, which uses metaheuristic algorithms that explore the search space using random selection, combination and variation of the desired parameters based on mechanisms reminiscent of biological evolution, or physical processes occurring in nature. Here are some examples of heuristic algorithms: genetic algorithms (GA) [18], ant algorithm (ACO) [19], bee colony algorithm (ABC) [20], particle swarm method (PSO) [21], fire algorithm (FA) [22], raven search algorithm (CSA) [23], gray wolf algorithm (AGW) [24], bat algorithms (BA) [25], annealing simulation algorithm (SA) [26] and others [27-230]. In [31-36], practical problems of optimizing building structures using metaheuristic algorithms are presented.

The basis of optimization methods for building structures is research on the extremum of the target function, which is associated with the parameters of the sections of these structures, as well as with the nature of their stress-strain state, where it is possible to take into account viscous, brittle, fatigue failure, loss of stability, or destruction from other factors. The quantitative data of the target function are determined by two indicators: the unit cost of resources and the volume of resources. Genetic algorithms, in particular, have proven to be an effective basis for general tasks that model the most diverse problems of engineering practice. In [37], the TOSCA code is described, which uses genetic algorithms to solve many optimization problems that arise in the design of structures. In [38], an approach to polymorphic uncertainty modeling for timber structures is introduced. He preliminary steps aim at optimization in the design of timber structures, provided that the polymorphic uncertain design, as well as a priori parameters, are considered.

At the moment, quite a lot of research has been done related to the development of numerical optimization algorithms. However, these studies have not found practical application in Russian PCs. The most common used in Russian PC design are SCAD, LIRA-CAD, Micro-FE, and others, including developed modules of finite element analysis and structural analysis, do not contain design calculation modules that correspond to the specified optimality criterion.

6. Conclusion

The above analysis revealed the need for algorithmic and software capable of adequately simulating the complex behavior of wood-based structures, taking into account factors such as the anisotropic structure of wood and accounting for shear deformations.

Due to the fact that wooden structures are now more and more focused on high-rise construction, they are required to improve seismic protection characteristics, which, of course, should be taken into account when designing structures based on wood in seismically active areas.

Modern approaches to the automated design of structures (including wooden ones) put forward the requirements of ensuring their load-bearing capacity at a given level of reliability, as well as the fulfillment of all architectural and structural requirements, provided that the project cost reaches minimum values. Such a problem can be solved on the basis of optimization algorithms. Hence, there is a need to develop methods for optimizing wood-based building structures that would allow rational technical solutions to be obtained quickly and with minimal time and money. Further research in the direction of optimization and modeling of flat and spatial structures based on wood can be effectively implemented by constructing optimization models of not yet implemented design solutions using non-linear programming methods and metaheuristic algorithms.

References

[1] SP 64.13330.2017. Wooden structures. The updated edition of SNiP II-25-80 (with Change N 1, 2).

[2] Pyatikrestovsky K P and Sokolov B S 2018 Nonlinear analysis of statically indeterminate wooden structures and optimization of cross section dimensions of dome ribs. *International Journal for Computational Civil and Structural Engineering* **14(4)** pp 130-139 DOI:10.22337/2587-9618-2018-14-4-130-139

[3] Vodiannikov M A 2018 Results of numerical modeling of the stressed-deformed state of the joint connection of wood constructions with cfrp dowel pins. *International Journal for Com putational Civil and Structural Engineering,* **14(2)** pp 48-56 DOI:10.22337/2587-96182018-14-2-48-56

[4] Sirumbal-Zapata L F, Málaga-Chuquitaype C and Elghazoul A Y 2019 Experimental assess ment and damage modelling of hybrid timber beam-to-steel column connections under cyclic loads, *Engineering Structures,* **200**. https://doi.org/10.1016/j.engstruct.2019.109682

[5] Ugalde D, Almazán J L, Santa María H et al 2019 Seismic protection technologies for timber structures: a review. *Eur. J. Wood Prod.* **77** pp 173–194. https://doi.org/10.1007/s00107019-01389-9

[6] Gohlich Ryan, Erochko Jeffrey and Woods Joshua E 2018 Experimental testing and numerical modelling of a heavy timber moment-resisting frame with ductile steel links. *Earthquake Engineering and Structural Dynamics* **47- 6**, pp 1460-1477

[7] Zheng Li, Hanlin Dong, Xijun Wang and Minjuan He 2017 Experimental and numerical inves tigations into seismic performance of timber-steel hybrid structure with supplemental damp ers. *Engineering Structures. Elsevier* **151** pp 33-43 https://doi.org/10.1016/j.engstruct.2017.08.011

[8] Poh'sie G H , Chisari C, Rinaldin G , Amadio C and Fragiacomo M 2016 Optimal design of tuned mass dampers for a multi-storey cross laminated timber building against seismic loads . *Earthquake Engng Struct. Dyn.* **45**: 1977– 1995. doi: 10.1002/eqe.2736.

[9] Kotlov V G and Sharynin B E 2018 Development of a model of the wood material for the finite element analysis of building structures (Part 1) *Vestnik. Volga State Technological Universi ty. Ser .: Materials. Constructions. Technology* **-2 (6)** pp 58-63.UDC 624.01

[10] Volynsky V N 2012 Vzaimosvyaz' i izmenchivost' fiziko-mekhanicheskih svojstv drevesiny: monografiya [Interrelation and variability of physical and mechanical properties of wood: monograph], 2-nd ed., ISPR, *Saint-Petersburg: LAN Publishing House*, 224.

[11] Bulent Kaygin, Huseyin Yorur, Burhanettin Uysal 2016 Simulating Strength Behaviors of Cor ner Joints of Wood Constructions by Using Finite Element Method. *Drvna Industrija*. **67(2)**, 133-140. DOI: 10.5552/drind.2016.1503

[12] Gluhih V N 2007Uprugaya deformativnost' drevesiny poperek volokon [Elastic deformation of wood across the fibers], IVUZ. *"Lesnoj zhurnal" [IVUZ. "Forest journal"]* **5**.

[13] Tsiulin E Yu, and Shmidt A B 2018 Numerical Simulation of LVL Elastic Modulus with Dif ferent Combinations of Mutually Perpendicular Veneer Layers. *Lesnoy Zhurnal [Forestry Journal]* **6**, 138–148. DOI: 10.17238/issn0536-1036.2018.6.138

[14] Follesa M, Fragiacomo M, Casagrande D, Tomasi R, Piazza M, Vassallo D, Canetti D and Ros si S 2018 The new provisions for the seismic design of timber buildings in Europe. *Engi neering Structures* **168** pp 736-747

[15] Quiquero H, Gales J, Abu A et al. 2020 Finite Element Modelling of Post-tensioned Timber Beams at Ambient and Fire Conditions. *Fire Technol* **56** 737–767. https://doi.org/10.1007/s10694-019-00901-0

[16] Schmit L A. Structural design by systematic synthesis. *Proceedings of second ASCE confer ence of electronic computation*. 1960 pp 105-122.

[17] Haug Edward J and Jasbir S Arora. Applied Optimal Design: Mechanical and Structural Sys tems. New York: *Wiley* 1979 pp 487-498

[18] Dao S D, Abhary K, and Marian R 2017 An innovative framework for designing genetic algo rithm structures *Expert Systems with Application* **90** pp 196-208. https://doi.org/10.1016/j.eswa.2017.08.018.

[19] Kaveh A and M H Ghafari 2015 Plastic analysis of planar frames using CBO and ECBO algo rithms. *Inter-national journal of optimization in civil engineering* **4(5)** pp 479-492.

[20] Dogan A, Gürcan Y, Thomas S 2017ABC-X: a generalized, automatically configurable artificial bee colony framework . *Swarm Intelligence* **11(1)** pp 1-38. https://doi.org/10.1007/s11721-017-
0131-z.

[21] Bonyadi M R and Michalewicz Z 2017 Particle Swarm Optimization for Single Objective Con tinuous Space Problems. *A Review. Evolutionary Computation* **25(1)** pp 1-54 https://doi.org/10.1162/evco_r_00180

[22] Nekouie N and Yaghoobi M. 2016 A new method in multimodal optimization based on firefly algorithm . *Artificial Intelligence Review* **46(2)** pp 267-287. https://doi.org/10.1007/s10462-016-9463-0.

[23] Alireza A A 2016 Novel metaheuristic method for solving constrained engineering optimization problems: Crow search algorithm . *Computers & Structures* **169** pp 1-13 https://doi.org/10.1016/j.compstruc.2016.03.001.

[24] Mirjalili S, Mirjalili S M and Lewis A 2014 Grey Wolf optimizer. *Advances in Engineering Software* **69** pp 46-61. https://doi.org/10.1016/j. advengsoft.2013.12.007.

[25] Cai X, Wang H, Cui Z, Cai J, Xue Y and Wang L 2017 Bat algorithm with triangle-flipping strategy for numerical optimization . *International Journal of Machine Learning and Cyber netics* **9(2)** pp 199-215. https://doi.org/10.1007/s13042-017-0739-8.

[26] Samma H, Mohamad-Saleh J, Suandi S A and Lahasan B 2019 Q-learning-based simulated an nealing algorithm for constrained engineering design problems . *Neural Computing and Ap plications*.1-15. https://doi.org/10.1007/s00521-019-04008-z.

[27] Ho-Huu V, Nguyen-Thoi T, Vo-Duy T and Nguyen-Trang T 2016 An adaptive elitist differen tial evolution for optimization of truss structures with discrete design variables. *Computers & Structures* **165** pp 59–75. https://doi.org/10.1016/j.compstruc.2015.11.014.

[28] Cheng M -Y, Prayogo D, Wu Y-W and Lukito M M 2016 A Hybrid Harmony Search algorithm

for discrete sizing optimization of truss structure. *Automation in Construction* **69** pp 21-33 https://doi.org/10.1016/j.autcon.2016.05.023.
[29] Jalili S and Talatahari S 2017 Optimum Design of Truss Structures Under Frequency Con straints using Hybrid CSS-MBLS Algorithm. *KSCE Journal of Civil Engineering* **22(5)** pp 1840-1853. https://doi.org/10.1007/s12205-017-1407-y.
[30] Khatibinia M and Yazdani H 2018 Accelerated multi-gravitational search algorithm for size op timization of truss structures. *Swarm and Evolutionary Computation* **38** pp 109-119. https://doi.org/10.1016/j.swevo.2017.07.001.
[31] Rao R V and Saroj A 2018 Multi-objective design optimization of heat exchangers using elitist-Jaya algorithm. *Energy Systems* **9** pp 305-341. http://dx.doi.org/10.1007/s12667-016-0221-9.
[32] Lieu Q X, Do D T T and Lee J 2018 An adaptive hybrid evolutionary firefly algorithm for shape and size optimization of truss structures with frequency constraints. *Computers & Structures.* **195**, 99-112. https://doi.org/10.1016/j.compstruc. 2017.06.016.
[33] Degertekin S O, Hayalioglu M S 2009 Optimum design of steel space frames: tabu search vs. simulated annealing and genetic algorithms . *International Journal of Engineering and Ap plied Sciences (IJEAS)* **1(2)** pp 34-45
[34] Kaveh A and Zakian P 2017 Improved GWO algorithm for optimal design of truss structures, *Engineering with Computers* **34(4)** pp 685-707. https://doi.org/10.1007/s00366-017-0567-1.
[35] Jalili S and Hosseinzadeh Y 2018 Combining Migration and Differential Evolution Strategies for Optimum Design of Truss Structures with Dynamic Constraints. *Transactions of Civil Engineering* **43(1)** pp 289-312. doi:10.1007/s40996-018-0165-5.
[36] Venkata Rao R, Ankit Saroj. 2017 A self-adaptive multi-population based Jaya algorithm for engineering optimization. *Swarm and Evolutionary Computation*. **37**, 1-26. http://dx.doi.org/10.1016/j.swevo.2017.04.008.
[37] Chisari C and Amadio C 2018 TOSCA: a Tool for Optimisation in Structural and Civil engi neering Analyses. *Int J Adv Struct Eng* **10** pp 401–419. https://doi.org/10.1007/s40091-018-0205-1
[38] Schietzold F N, Graf W and Kaliske M 2018, Polymorphic Uncertainty Modeling for Optimiza tion of Timber Structures. *Proc. Appl. Math. Mech*. **18**: e201800426. doi:10.1002/pamm.201800426

IOP Conf. Series: Materials Science and Engineering **880** (2020) 012077 doi:10.1088/1757-899X/880/1/012077

Corporate information systems in construction

S V Fedorova[1]

[1]Institute of Natural Resources Management, Department of Industrial Ecology and Life Safety, Irkutsk National Research Technical University, 83 Lermontov st., Irkutsk, 664074, Russia

E-mail: fsta65@yandex.ru

Abstract. Under continuous increasing competition, only companies that use the most optimal methods can survive and succeed. One of the most relevant and successful business models is outsourcing. It was developed by the Electronic Data System (EDS) Company in 1963. In the 1980s, the outsourcing became an effective way to optimize costs and increase profitability. In the 1990s, it was revealed that the outsourcing contributes to the rationalization of company structures and their activities focusing on the main subject and transferring peripheral, official functions to external specialists. The transition to the post-industrial economy contributed to the implementation of this method in data processing and company management systems.

1. Introduction

Among the types of outsourcing, IT outsourcing is especially interesting for an analysis. This type has appeared recently, but it has become one of the most popular outsourcing services. Many companies are implementing IT outsourcing schemes. The IT outsourcing is a relevant and interesting research issue. The research subject is IT outsourcing models implemented in various economic sectors. The aim is to analyze outsourcing models applied in various professional fields. The novelty of the study is due to the attempt to analyze the IT outsourcing market by studying models applied by various economic sectors.

2. Materials and methods

In the late 1990s, there was a sharp increase in the number of companies - providers of outsourcing services. With the growing number of outsourcing companies, outsourcing models (ASP, MSP, SSP, MSSP, WASP and AAA) were being developed.

The ASP model (online applications) [1] provides customers with software products, software packages and IT elements, maintains these programs on the basis of a fixed monthly fee via the Internet [2]. The ASP company provides the following services: management of the hardware and software infrastructure; system integration; customization of applications to the requirements of various industries; web hosting; application hosting; administrative support; industry knowledge and experience; business consulting [3]. The ASP model allows enterprises to reduce the total cost of IT ownership, application implementation risks, speed up the application implementation. The model allows companies to apply the latest enterprise management technologies.

The MSP model is used to provide management services [4], services for remote management of equipment, software and network resources: management of updating and backup procedures; monitoring and performance reporting; stress testing; real-time management of client system information on availability, security and performance of the IT infrastructure; analysis of system

IOP Conf. Series: Materials Science and Engineering **880** (2020) 012077 doi:10.1088/1757-899X/880/1/012077

performance and its optimization. When enterprises are abandoning e-commerce projects aimed at developing web sites and online directories and using web technology to implement business processes, the MSP industry is growing rapidly. The infrastructure for supporting these systems is complex and requires the qualified staff.

The SSP model is used to provide remote data storage services: leasing of the storage infrastructure; storage process management: downloading of client data, creation of archival copies; centralized management of distributed data warehouses.

The MSSP model is used to provide Internet security services: vulnerability assessment that identifies potential weaknesses, security policies and procedures; security audit that meets various industry standards to monitor firewalls and other security devices; intrusion detection services that alert customers; investigation of security incidents; security policy preparation; training of the security personnel [5].

The WASP model is used to provide wireless online application management services: modification of applications for using by wireless devices; sale of general-purpose wireless applications to other service providers and portals; development and leasing of specialized applications developed for organizations that want to provide wireless services [6].

The AAA model is used to provide services related to the integration of various applications into a common interface through which customers access these applications: ensuring functions of an integrator of offers of various application providers: concluding contracts with suppliers and negotiating service agreements; development of a single interface; payment for suppliers' services - customers receive an invoice; application hosting; integration of new applications purchased by the client with existing ones; consulting [7].

3. Results and analysis

The main advantages of IT outsourcing services are as follows:

- saving financial resources. Attracting experts requires large monthly expenses, while IT outsourcing is much less expensive. The company can increase or decrease the volume of outsourcing services, optimizing its own costs;
- tax savings. All costs of outsourcing services can be taken into account as company expenses reducing the tax burden.
- professionalism and competence. The accumulated experience of professionals from outsourcing organizations is more comprehensive and extensive than that of a person who works for one company. A variety of problems and situations contribute to the solution of the most complex and non-standard tasks;
- availability of specialists. Maintenance services are provided on an ongoing basis;
- high quality of services. A detailed description of all types of services, their quality parameters and control methods are fixed in the contract between the customer and the supplier. Outsourcing companies regularly provide detailed reports;
- narrow specialization of specialists. When looking for a full-time employee, it can be difficult to find a professional having comprehensive experience.
- no staff training costs. The constant development of technology requires regular training which increases costs. Using IT outsourcing services, the company saves on training and receives a full range of professional services;
- the ability to concentrate on own business. If the IT sphere is non-core for the company, it takes a lot of time. Outsourcing companies allow employees to focus on urgent problems;
- rapid implementation of new technologies. Outsourcing engineers are aware of new IT trends and can immediately begin to implement IT projects;
- the ability to use services of highly qualified specialists: small companies can afford only one system administrator, but IT outsourcing changes this situation.
- security and privacy. The agreement between the customer and the service provider stipulates conditions for confidentiality of commercial information.

Let us compare quality of service (professionalism) of a full-time IT administrator and an IT outsourcing specialist. The results are presented in Table 1.

Table 1. The comparative analysis of quality of services provided by the full-time IT administrator and an IT outsourcing specialist

Quality of services, professionalism	Staff IT specialist		IT outsourcing specialist	
	1 full-time specialist	2 full-time specialists	1 specialist	Group of specialists
Security	+	+	+	+
Valuable information backing-up	+	+	+	+
Documentation	-	-	+	+
ATS and other communication systems	-	-	+	+
Ability to use various OS	-	-	+	+
Ways to reduce IT costs	-	-	+	+
Ability to solve any IT problem	-	-	+	+
Holiday interchangeability	-	+	+	+

We cannot say that IT outsourcing is always more profitable than traditional information services.
If the company has a larger share of typical IT processes, it is better to choose IT outsourcing. If there are more unique IT processes, it is better to use insourcing (full-time employment). Both models can be used as well.

Table 2. The comparative analysis of financial indicators for services provided by the full-time IT administrator and an IT outsourcing specialist

Financial indicator, rub.	Staff specialist		IT-outsourcing specialist	
	1 server Full-time IT administrator, 1 full-time unit	30 servers Full-time IT administrator, 2 full-time unit2	1 server 1 specialist	30 servers Group of specialists
The average salary in Russia (net of taxes)	25000	50000	no	no
Workplace rent (6 sq.m per person)	4000	8000	no	no
Depreciation of office equipment (1 person / month)	500	1000	no	no
Administrative	1500 (5%)	1500 (5%)	no	no

expenses, rub. (%)				
Personal income tax, rub (%)	3250 (13%)	6500 (13%)	no	no
Social tax deductions, rub (%)	8500 (34%)	17000 (34%)	no	no
Leave allowance (100%)	yes	yes	no	no
Sick payment (50-100%)	yes	yes	no	no
Total, month	42750	84000	3000	15000
Total, year	513000	1008000	36000	180000

4. Discussion

As a result, full-time specialist costs are about 42,750 rubles per month. This does not include sick payments, bonuses, salary for replacing an employee on vacation, processing fees and other unforeseen expenses [8].

This is a salary for 8 hours a day, 40 hours a week, 160 hours a month. In addition, managers have to monitor the performance of workers. The above example shows that IT outsourcing is more advantageous for the company.

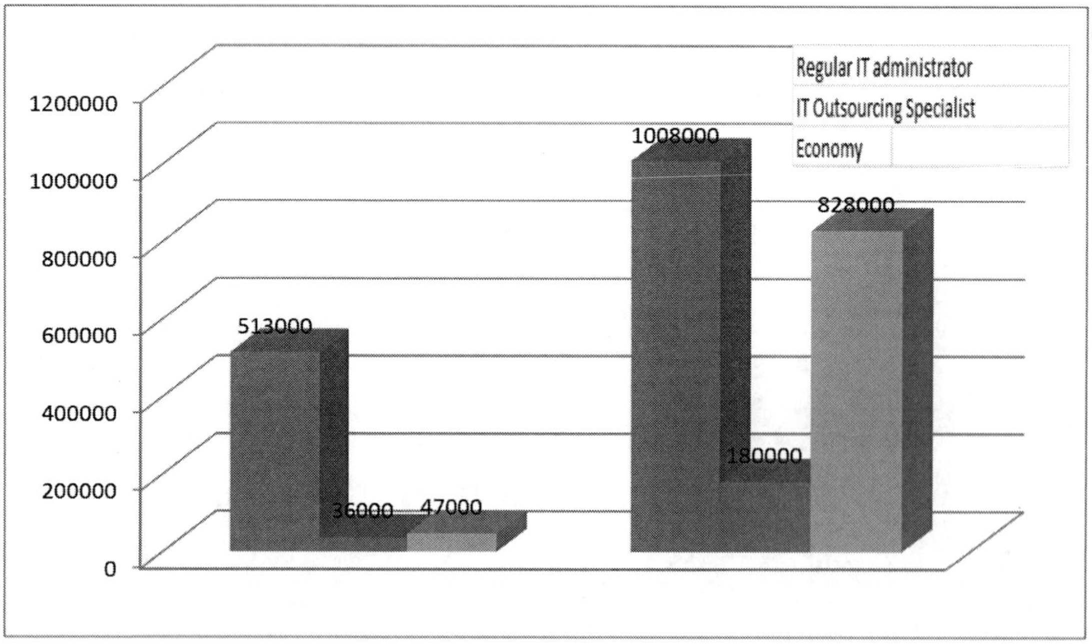

1 server 30 servers Full-time IT administrator IT-outsourcing specialist savings

Figure 1. Savings due to tje IT outsourcing per year, rub.

The relations of the customer and the IT outsourcer are regulated by contracts and service provision agreements [9]. According to the latter, the contractor gives financial guarantees that a quality service will be provided. Agreements regulate the speed of software functions or the absence of repeated bugs

IOP Conf. Series: Materials Science and Engineering **880** (2020) 012077 doi:10.1088/1757-899X/880/1/012077

when correcting system errors. In addition, the outsourcing company is responsible for competencies and motivation of its employees [10].

5. Conclusion

IT outsourcing is one of the new types of outsourcing. Companies are switching to IT outsourcing earlier than to other types of outsourcing. IT outsourcing involves the maintenance of the information infrastructure of the company by a specialized company [11]. The company does not need to hire IT specialists [12]. There is no unified classification of IT outsourcing. Several types of IT outsourcing can be distinguished: external and internal; resource and functional; outsourcing of IT infrastructure, IT processes or full outsourcing; one-time and long-term; primary and secondary [13]. There are several IT outsourcing models: ASP model (provides online applications); MSP model (provides management services); SSP model (provides remote data storage services); MSSP model (provides managed security services on the Internet); WASP model (provides wireless online application management services); AAA model (provides services related to the integration of various suppliers' applications into a single interface through which customers access these applications); IT outsourcing has a number of advantages: financial savings, the ability to engage in core activities, professional services, and the speed of implementing new technologies [14]. IT outsourcing has the following shortcomings: the problem of data security, trade secrets, low-quality services. Before using IT outsourcing services, the company should analyze business processes and determine the economic efficiency [15].

References

[1] Gelfand A O 2015 The history of development of IT outsourcing *Notes of the scientist* **3** pp 26-28

[2] Gribanov Yu I 2013 *Improving the efficiency of system outsourcing services for IT infrastructure* p 185

[3] Guseva E N 2015 Evaluation of effectiveness of IT outsourcing Modern equipment and technologies **6 (46)** pp 131-135

[4] Dubitsky L G 2013 Outsourcing and quality of products and services *View of the problem* p 296 p

[5] Izotova Yu D 2016 IT outsourcing as an effective and economical *technology New science: experience, traditions, innovations* **4-1 (77)** pp 74-77

[6] IT outsourcing. Practical guidance p 254

[7] Kashaev R A 2015 The IT outsourcing market in Russia and its development *Young scientist* **18** pp 273-274

[8] Kurbanov A Kh 2014 Outsourcing: history, methodology, practice p 110

[9] Kurbanova F D 2016 IT-Outsourcing in Russia Improving accounting, analysis and control as mechanisms of information support for sustainable economic development **2** pp 199-203

[10] Kurylenk Yu V 2014 IT outsourcing as a tool to ensure global competitiveness Regional Economics and Management **2 (02)** pp 147-150

[11] Lisnevskaya O S 2016 Features of development of IT outsourcing in Russia *Scientific almanac* **2-1 (16)** pp 218-220

[12] Lukyanov I V 2015 IT outsourcing in Russia: problems and prospects *Modern studies of social problems* **4 (48)** pp 379-388

[13] Mansurova N A 2014 Analysis of the effectiveness of IT outsourcing *Bulletin of TvSU Series Economics and Management* **19** pp 129

[14] Timofeev M A 2017 *Development of tools for implementing an innovative project using process - control outsourcing* p 188

[15] Shalamay A V 2017 IT outsourcing: essence, development trends in Russia *New science : current status and development paths* 3 pp 173-176

[16] Ergardt N O 2011 IT outsourcing as a tool to reduce the cost of information *Bulletin of the Institute of Economics of the Russian Academy of Sciences* **1** pp 263-272

IOP Conf. Series: Materials Science and Engineering **880** (2020) 012078 doi:10.1088/1757-899X/880/1/012078

State of the problem of numerical modeling of optimal design of load bearing structures of linear pipeline systems

A V Kablukov[1] and T L Dmitrieva[1]

[1] Irkutsk National Research Technical University, 83 Lermontov str., Irkutsk, 664074, Russia

E-mail: andrewkablukov@mail.ru

Abstract. The features of the design and construction of oil and gas facilities are indicated. Metal flyovers were selected from the general list of capital construction projects in the industry for further study. Special requirements of regulatory documents for flyovers are considered. An example of reducing the cost of design and construction by creating unified design solutions is given, the disadvantages of this approach are identified. The problem of optimal design of flyovers as spatial metal structures is indicated. Brief results of monitoring of scientific papers describing modern optimization methods are presented.

1. Introduction

The oil and gas complex is the basis of energy supply in Russia and one of its most important national economic sectors, since the products manufactured by this complex are raw materials for the functioning of a number of other industries, and the output of their final products depends on this.

Design and construction of oil and gas industry facilities has a special specificity, which differs from industrial construction as a whole in a number of features, of which the following can be distinguished:

– significant volumes of construction and its duration;

– the dispersion of oil and gas field facilities on a large territory of the country's regions, which requires the construction of high-tech oil and gas pipelines and technological facilities for their maintenance;

– inaccessibility of areas of oil and gas fields;

– construction in severe climatic conditions of the Far North;

– fire and explosion hazard of work at the facilities.

All of the above features lead to a significant increase in the cost of production and transportation of oil and gas in comparison with foreign practice.

An important component of oil and gas production in Russia is the construction of oil and gas pipelines. One of the constituent parts of the pipeline system is flyovers (Figure 1) of various types, made primarily of metal structures. When designing space-planning and structural solutions of these capital construction projects, it is also necessary to pay attention to the industry features: permafrost conditions and increased requirements for the safety of facilities [1, 2].

An additional difficulty in creating design models of flyovers creates the need to take into account the special requirements of regulatory documents [3, 4]. We list some of these requirements:

- calculation of building structures of free-standing supports and flyovers should be performed as flat structures. If it is necessary to make more accurate calculations and take into account additional factors,

IOP Conf. Series: Materials Science and Engineering **880** (2020) 012078 doi:10.1088/1757-899X/880/1/012078

the calculation of building structures of free-standing supports and flyovers should be performed as the calculation of spatial systems, taking into account their joint work with pipelines;

- when laying pipelines on an flyover, the longitudinal horizontal load from the friction forces in the movable supporting parts of the pipes is perceived by the span and anchor supports and is not transferred to the intermediate supports;
- normative vertical load from pipelines on supports and flyovers should be taken as the sum of vertical loads from all pipelines.

Figure 1. Image of a flyover

One of the measures aimed at reducing the costs of design and construction is the creation of standard unified projects of buildings and structures for the arrangement of oil and gas fields. An effective solution to such a problem can be realized with the help of a mathematical optimization apparatus, introduced in the design of the load-bearing elements of these structures.

An example of the unification of design decisions is the creation of CSDS (Company Standard Design System) developed by NIPI, which are part of Rosneft, PJSC. This system makes it possible to accumulate optimal technical and economic design solutions for key technological facilities and unified requirements for the purchase of equipment and materials for the arrangement of oil and gas fields in a common database, which is used in design production. At the same time, it should be noted that optimization of construction projects is not widely used in real design yet. Such software systems as: LIRA-CAD, SCAD, MicroFE, which are used in the practice of modern Russian design, allow solving verification calculation, as well as perform constructive calculations in accordance with regulatory requirements, but they are not provided optimization modules.

World-class software systems (MSC Nastran, ANSYS and ABAQUS), in addition to the developed FE analysis apparatus, contain optimization modules. However, the main obstacle to the effective introduction of this software for optimal design tasks is that programs do not have a verification module based on Russian regulatory documents, accordingly there are not have the ability to change sections according to grade. For this reason, the development of effective optimization algorithms and programs based on them is one of the urgent problems of the entire industry.

2. Approaches to modeling overpass optimization tasks

Let us consider a variant of an integrated approach to the optimal design of structures associated with oil and gas production using the example of flyovers as spatial metal structures.

The optimality criterion for these structures is the minimum weight or volume of the structure. Limitations are set in the form of regulatory requirements in accordance with SP 16.13330.2017 Steel Structures. The state parameters of structures (internal forces, nodal displacements, stress, etc.) are determined by solving the problem of statistical analysis.

In general, the research of many Russian and foreign scientists is devoted to questions of the optimal design of building structures. We will name only a few authors who have made a significant contribution to this direction: N.P. Abovsky, N.V. Banichuk, A.I. Vinogradov, T.L. Dmitrieva, V.A. Kiselev, V.A. Komarov, L.S. Lyakhovich, V.P. Malkov, D.A. Matsuliavičius, Yu.M. Pochtman, I.M. Rabinovich, Yu.A. Radzig, A.R. Rzhanitsyn, N.D. Sergeev, N.N. Skladnev, A.F. Smirnov, A.P. Fippippov, A.A. Chiras, L. Schmitt, J. Arora, N. Adele, Z. Vasyutinsky, D. Keller, M. Levy, Z. Moruz, F. Niordson, N. Olhoff, V. Prager, D. Rozhvany, D. Taylor, M. Turner, E. Hog, R. Shield and many others.

Let us consider in more detail some works affecting the problem of optimal design of spatial structures.

The general statement of the optimization problem was first proposed by L. Schmit [7]. In his work, he pointed out the admissibility of the application of structural analysis using the finite element model and the nonlinear programming method in the presence of various forms of constraints.

In 1991, Hazela and Berke presented optimization algorithms based on a neural dynamic model. When solving the optimization problem, scientists used neural networks that replaced the force-displacement ratio in a static analysis [9].

In 2002, an article was published that proposed a solution to the problem of optimal design of spatial structures by the method of practical nonlinear inelastic analysis [10].

The book [8] analyzes various problems associated with the design of steel structures of industrial facilities, including multi-tiered spatial core structures. Particular attention is paid to saving and calculating costs, which allows you to compare costs and significantly save on the design stages.

Alpatov's article is devoted to the search for the optimal geometric shape of structures, as well as the optimal distribution of material in them with unchanged geometry. The article [12] describes the methodology for the optimal design of spatially-rod structures based on random search and directional search algorithms.

The ANSYS PC has been widely used in solving optimization problems. Examples of its use for solving optimal design problems with a description of the approximation method of a subtask and a first-order method are described in the verification report, performed by the scientific staff of CJSC SIC StadiO [13].

The work [15] is devoted to solving the problem of structural optimization of wireframe steel structures using a special genetic coding algorithm to establish a strategy for detecting an ideal group of elements and using the penalty method developed by the authors to take into account restrictions.

The application of the "harmony search" algorithm in optimizing the design of steel structures was considered in [17].

In 2015, a monograph was published by IRNITU researchers, which addressed the optimization problem in the form of a nonlinear mathematical programming problem and described the application of the author's development: ROSK PC - a software package for calculating and optimizing building structures [18].

In the dissertation, Le Chan Minh Date described in detail the solution to the problem of optimizing spatial metal structures, with the possibility of varying both the parameters of the cross section and the coordinates of the nodes on a continuous and discrete range [21].

It should be noted that over the past two decades, special attention has been paid to the description of metaheuristic algorithms for solving the optimal design problem. The use of metaheuristics in the optimization of various kinds of engineering processes is an extremely popular and discussed topic in recent years. In general, metaheuristic algorithms for optimizing structural design were considered by scientists from three countries in the article [22] in 2016. Also, a book [25] is devoted to developments in the field of metaheuristics and their application in construction and design. This publication describes in detail various metaheuristic algorithms and provides practical examples of demonstrating their real application in optimizing the design process.

We give a brief presentation of the works describing the use of metaheuristic algorithms. The use of optimization algorithms based on the method of colony of ants and harmonic search in the optimal design of spatial steel frames with a discrete change in the parameters of the sections according to the criterion of minimum weight with restrictions on strength and stability is given in [14]. In [16], the upper bound strategy for reducing the computational cost of the Big Bang – Big Crunch metaheuristic algorithm and its modifications used in the optimization of spatial steel structures is reflected. The development of a metaheuristic optimization algorithm based on eagle hunting at the junction of MATLAB and SAP2000 software systems (Figure 2) to reduce the metal consumption of steel spatial frames was discussed in [19].

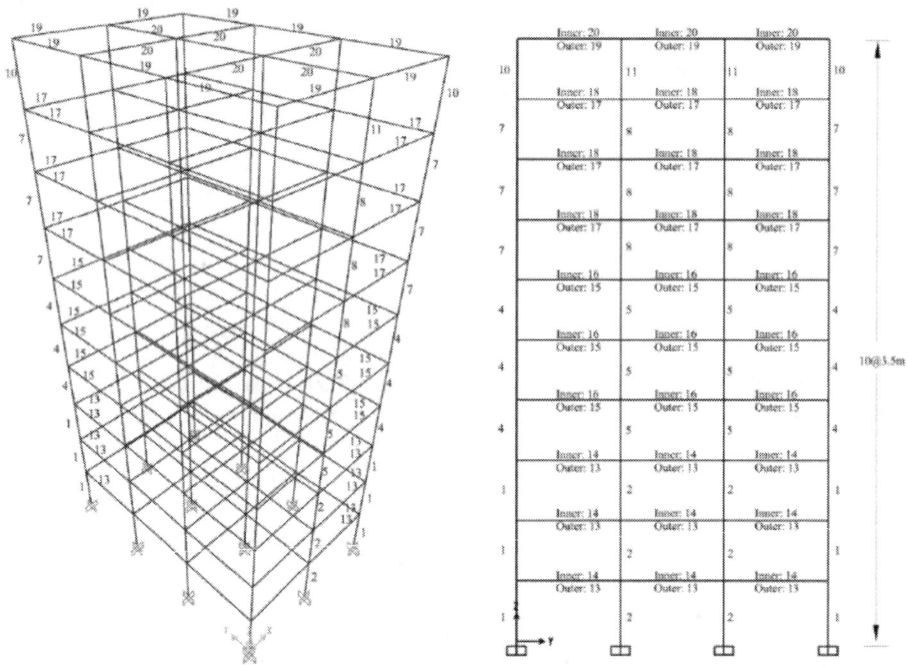

Figure 2. Spatial metal structure model in SAP2000

Optimization in the design of spatial steel structures under seismic loading based on the cuckoo metaheuristic algorithm was studied in [20]. Using the "Productive Design" approach and metaheuristic algorithms to solve the problem of optimizing eccentrically fixed frame structures (Figure 3) is given in [23].

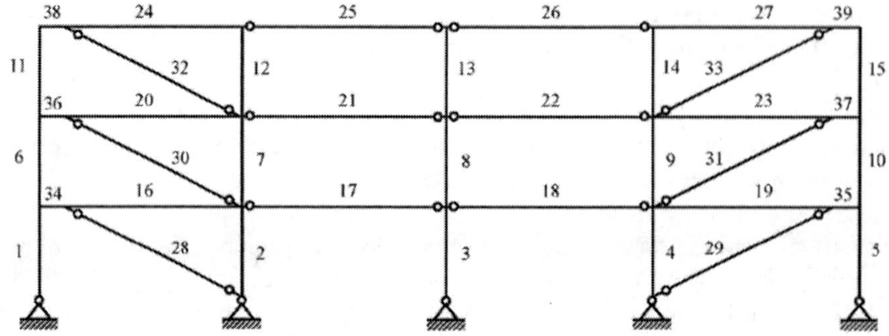

Figure 3. Eccentrically braced frame structure

Optimal design of structures is presented in great detail in Arora's book "Introduction to Optimal Design" [24]. This publication illustrates various concepts and procedures with simple examples and demonstrates their applicability to engineering design tasks using Excel and MATLAB [11].

3. Conclusion

Based on the analysis of the sources cited, the following conclusions can be drawn.

Calculation and optimization of flyovers as typical unified parts of the piping system is an urgent and demanded task. At the same time, their modeling as spatial rod systems is proposed.

Despite a large number of studies in the field of optimization of spatial metal structures, none of these approaches can be used as universal for solving the problem of optimal design of technical flyovers for oil pipelines in accordance with Russian regulatory documents in design and construction.

The main areas of work on this issue will be:

1. Construction of a model for the problem of optimizing linear load-bearing structures of piping systems taking into account regulatory requirements.
2. Conducting a comparative analysis of the effectiveness of algorithms of different classes (gradient, metaheuristic, zero-order algorithms, etc.) in relation to solving conditional-extreme problems of optimization of flat and spatial rod structures.
3. Investigation of the possibility of embedding neural networks in the optimization process of spatial core systems of large dimension.

References

[1] Mustafin A M, Bykov L I and Mokhov V N 2008 *Building structures of oil and gas facilities* (SPb: Nedra) p 780
[2] Belyaeva V Ya, Mikhaylichenko A M, Baraz A N and Gabelaya R D 2005 *Oil and gas construction* (-M, Omega-L) p 774
[3] A guide for designing stand-alone supports and flyovers for process pipelines (to SNiP 2.09.03-85) 1989 p 75
[4] SP 43.13330.2012 Construction of industrial enterprises. Updated edition of SNiP 2.09.03-85 (with Changes № 1, 2).
[5] SP 16.13330.2017 Steel structures. Updated version of SNiP II-23-81 * "(with amendments, with Changes № 1).
[6] Garina S V 2005 *Mathematical modeling of the process of constructing estimates of the optimality of building structures* (Saransk).
[7] Schmit L A 1960 Structural design by systematic synthesis *Proceedings of second ASCE Conference of Electronic Computation* pp 105-122
[8] J Farkas and K Jarmai 2008 *Design and optimization of metal structures* (Woodhead Publishing) p 328
[9] Hajela P and Berke L 1991 Neurobiological Computational Models in Structural Analysis and Design *Computers and Structures* **41 (4)** pp 657-667
[10] Se-Hyu Choi and Seung-Eock Kim 2002 Optimal design of steel frame using practical nonlinear inelastic analysis *Engineering Structures* **24** pp 1189-1201
[11] Ketkov Yu L, Ketkov A Yu and Schulz M M 2005 *MATLAB. Programming, numerical methods* (SPb, BHV-Petersburg) p 752
[12] Alpatov V Yu and Kholopov I S 2009 Optimization of the geometric shape of spatial-rod structures *Metal constructions* **1** pp 47-57
[13] *Verification report on the ANSYS Mechanical software package* (CJSC SIC StaDio) 2009
[14] Ibrahim A. 2010 *Optimum design of 3-D irregular steel frames using ant colony optimization and harmony search algorithms* (Monograph) pp 185
[15] Afonso C C, Helio J C, Alvaro L G A, Carlos C H Borges 2011 Multiple cardinality constraints and automatic member grouping in the optimal design of steel framed structures *Engineering Structures* **33** pp 433-444

[16] Kazemzadeh Azad S, Hasançebi O, Kazemzadeh Azad S 2013 Upper bound strategy for metaheuristic-based design optimization of steel frames *Advances in Engineering Software* **57** pp 19-32
[17] Patrick Murren, Kapil Khandelwal 2014 Design-driven harmony search (DDHS) in steel frame optimization *Engineering Structures* **59** pp 798-808
[18] Dmitrieva T L, Le Chan Minh Dat and Nguyen Van Thou 2015 *Implementation of numerical optimization algorithms in modern software systems* (Irkutsk National Research Technical University) p 160
[19] Siamak Talatahari, Amir Hossein Gandomi, Xin-She Yang, Suash Deb 2015 Optimum design of frame structures using the Eagle Strategy with Differential Evolution *Engineering Structures* **91** pp 16-25
[20] Kaveh A, Bakhshpoori T, Azimi M 2015 Seismic optimal design of 3D steel frames using cuckoo search algorithm *Struct. Design Tall Spec. Build* **24** pp 210-227
[21] Le Chan Min Dat 2016 *Numerical models and their software implementation in the optimization of spatial metal structures* (Irkutsk National Research Technical University) 173 p
[22] Saka M P, Hasançebi O and Geem Z W 2016 Metaheuristics in structural optimization and discussions on harmony search algorithm *Swarm and Evolutionary Computation* **28** pp 88-97
[23] Mohammad Ali Fathali, Seyed Rohollah Hoseini Vaez 2020 Optimum performance-based design of eccentrically braced frames *Engineering Structures* **202** pp 1-15
[24] Jasbir Singh Arora 2011 *Introduction to optimum design, Third edition* (Academic Press) p 896
[25] Amir Hossein Gandomi, Xin-She Yang, Siamak Talatahari and Amir Hossein Alavi 2013 *Metaheuristic Applications in Structures and Infrastructures* (Elsevier) p 568

IOP Conf. Series: Materials Science and Engineering **880** (2020) 012079 doi:10.1088/1757-899X/880/1/012079

On the development of a united digital platform in the construction industry

V A Kudryavtseva[1] and N V Vasileva[2]

[1] Irkutsk National Research Technical University, 83 Lermontov str., Irkutsk, 664074, Russia

[2] Saint-Petersburg State University of Economics, 21 Sadovaya str., St. Petersburg, 191023, Russia

E-mail kudravera@mail.ru

Abstract. Currently, in the the world economic system, company development models are transforming. One of such transformations is active development of digitalization, which contributes to the system development and implementation of digital technologies. At the same time, digitalization is being implemented into different spheres of human life - government, economy, industry, education, construction, etc. Today the construction industry has a sufficient number of information and communication technologies which are the basis for creating digital (volumetric) models of buildings, correcting projects and managing the processes of construction and maintenance of buildings during their life cycles, including managing the engineering infrastructure of capital construction objects. The main task of digitalization of the construction industry is to create a united digital platform.

1. Introduction

Currently, the world is undergoing continuous changes in the economic system. The transformation increases the influence of innovation, advanced technology and digitalization on various aspects of human, social and business activities. At the same time, digitalization is an integral part of the national economies.

Currently, digitalization is a global process that involves a huge number of countries focused on the development of human capital, a competitive and efficient economy, and improvement of living standards. It contributes to the transformation of the world economy in general and national economies in particular. In fact, historically developed economies are changing within the new technological, digital vector, and there are examples of the digital economic systems in various industries and enterprises.

From a historical point of view, the concept “digital economy” was introduced into circulation by the American computer scientist Nicholas Negroponte in 1995. He compared the process of transforming the old economy into a new one by formulating the concept of electronic (digital) economy. [1, 5, 13]

Today the digital economy can lead to a noticeable growth of the global economic system, increase labor productivity, and create new areas. Professor of St. Gallen University (Switzerland) W. Brenner argues that “the aggressive use of data transforms business models, contributes to the production of new products and provision of new services, creates new processes, generates great utility and introduces a new management culture” [2, 3].

IOP Conf. Series: Materials Science and Engineering **880** (2020) 012079 doi:10.1088/1757-899X/880/1/012079

Therefore, the transition to the digital economy is one of the main directions in most countries, allowing them to remain competitive and diversify the economy.

2. Materials and methods

According to the study conducted by the international consulting company Accenture, the digital economy accounted for 22.8% (or $ 17 trillion) in the global GDP in 2016, and by 2021 it is planned to increase the GDP by 1.9% (Figure 1). In 2010-2016, the share of the digital economy in the GDP of developed countries increased by 1.2% (from 4.3% to 5.5%), and in the GDP of developing countries, it increased from 3.6% to 4.9%. The Boston Consulting Group predicts that the volume of the digital economy will reach $ 16 trillion by 2035. [4, 8, 9]

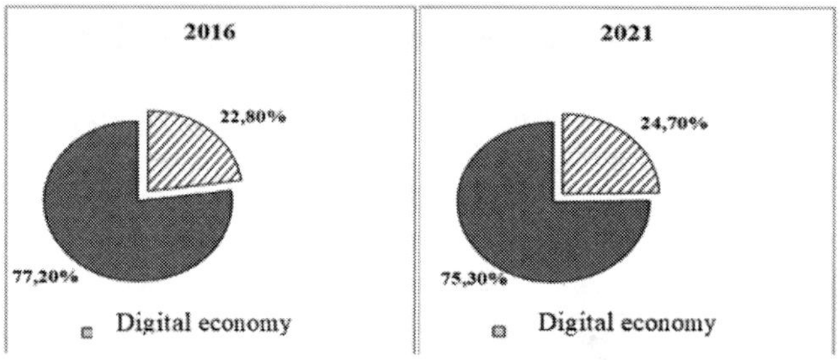

Figure 1. The share of the digital economy in the world GDP.

Thus, the digital economy exists in all developed countries, including Russia, for which it is one of the priorities of the Strategy for scientific and technological development, allowing them to develop industries and markets based on breakthrough innovative technologies.

In Russia, the digital economy is rapidly developing. According to the studies conducted by the international consulting company McKinsey, in 2011-2015, the share of the digital economy increased by 59% (nine times faster than the national GDP) and reached 3.9% of GDP in 2015 [6].

The national course is directed towards the implementation of the digital model proposed by V.V. Putin in his message to the Federal Assembly in 2016. In 2017, the Government of the Russian Federation developed and approved a program aimed at creating conditions for the transition to the digital economy, according to which the digital economy is a key area of development of the Russian Federation until 2030. Initially, the program “The Digital Economy of the Russian Federation” included five key areas: regulation, personnel training and education, development of research competencies and technical backlogs, information infrastructure and information security.

In February 2019, the Prime Minister of the Russian Federation Dmitry Medvedev signed a decree that cancels the program “The Digital Economy of the Russian Federation”. The national project “The Digital Economy of the Russian Federation” was developed. It entered into force in December 2018.

The Decree of the President of the Russian Federation of May 7, 2018 No. 204 “On national goals and strategic objectives of the development of the Russian Federation for the period up to 2024” (paragraph 11) contributed to the implementation of the national program “Digital Economy of the Russian Federation”.

The main objectives of this project are as follows:

- creating the stable and secure information and telecommunications infrastructure for high-speed transmission, processing and storage of large amounts of data accessible to all organizations and households,

- using domestic software by federal, regional and local authorities and organizations.

The digitalization will affect most sectors of the economy: education, medicine, energy, agriculture, construction, etc.

Figure 2 shows the intensity of using digital technologies by organizations in 2018 (the percentage of organizations using digital technologies in the total number of organizations).

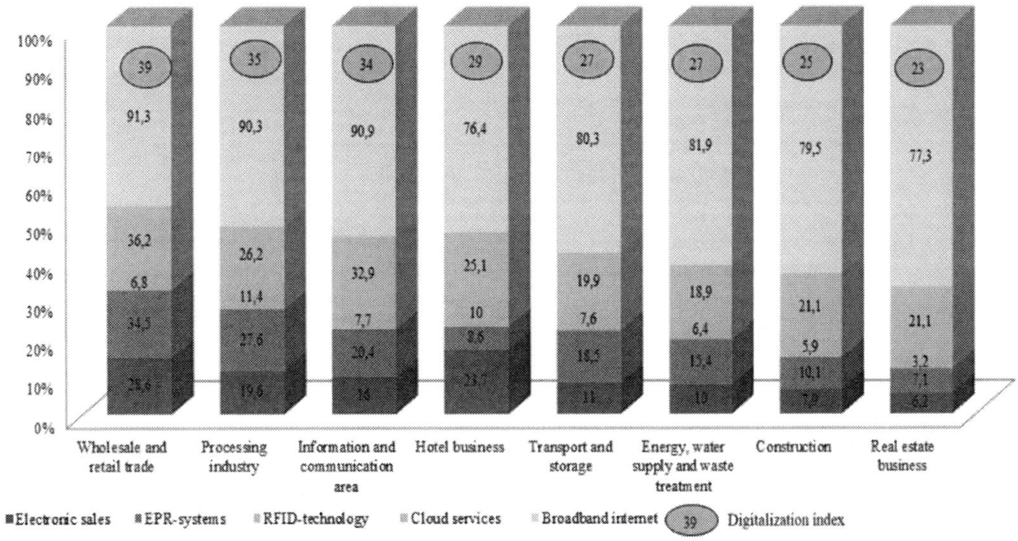

Figure 2. Intensity of the use of digital technologies by Russian organizations in 2018.

Government authorities pay attention to the issues of digitalization, since the digitalization of the Russian economy will become a platform for qualitative changes in its structure and long-term capabilities. In addition, the strategic orientation of the development of the digital economy is important. It will be the basis for the development and implementation of relevant documents of sectoral strategic planning. [10, 11]

3. Results

The construction industry is one of the key industries of the Russian economy. With the development of science and technology, construction projects have become complex. They involve ongoing technological processes and meet increased human needs. A modern construction object is a multi-level system with a large number of internal and external relations. Moreover, due to the changed guidelines caused by the development of the digital economy, increasing market requirements for technical and economic offers in construction, increased possibilities of architectural and construction practices and human needs, it is necessary to implement practices of digital construction. The modernization of the design and construction industry is active. For example, the section "Digitalization of the construction industry" is included in the draft strategy for the Development of the Construction Industry until 2030, and digital construction is another federal project within the Digital Economy program. Thanks to this project and achievements in the development of information technology in 2022, all federal objects will be built using digital modeling technologies (BIM, Building Information Model).

In general, the government has specified areas of digitalization in the construction industry:

- using an electronic format for implementing urban planning procedures;
- using an electronic database for storing documents of urban planning activities;
- collecting digital statistical data making it possible to process and publish data by extracting them from information systems;
- developing search and reference platforms;
- implementing information modeling technologies for capital construction projects at all stages of their life cycles;
- registering lands in the form of electronic documents (unless this scheme was registered by a citizen), etc.

The main directions of digitalization of the construction industry are presented in Figure 3.

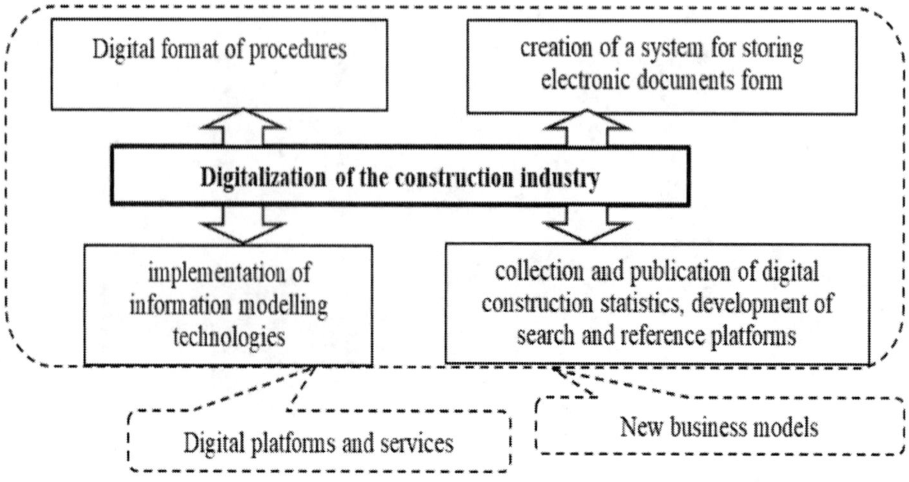

Figure 3. Directions of digitalization of the construction industry.

Thus, the construction industry cannot be not involved in the large-scale digitalization process, which is the strategic goal of the government, businesses and society as a whole.

Modernization of the construction industry aimed at improving construction quality is one of the main tasks of the government reflected in the Order of the President of July 19, 2018. No. Pr-1235. It suggests using BIM technologies. The main features of BIM technologies are shown in Figure 4. [8, 12]

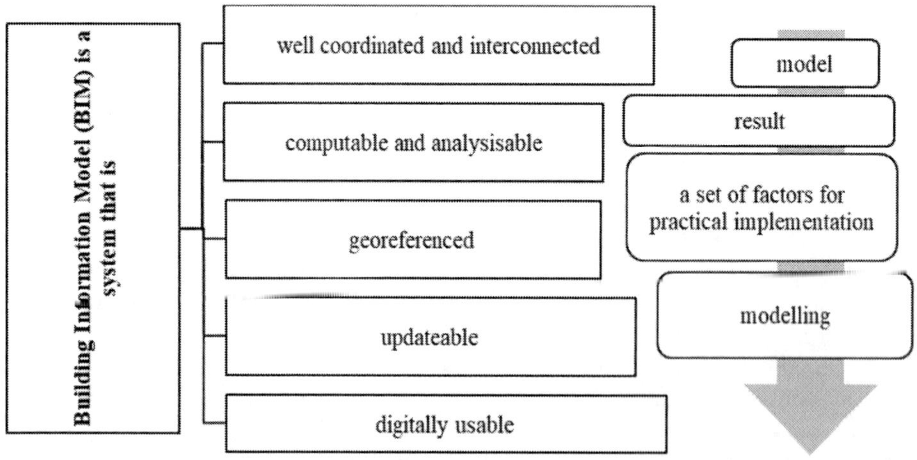

Figure 4. Key features of BIM technology.

Digitalization of the construction industry is aimed at automating all the stages of the life cycle of an object, including management of the life cycle of capital construction objects. According to experts, by 2024, when BIM-technologies will have been implemented, construction costs will decrease by 20%, the amount of design errors will decrease by 40%, and the project implementation time will decrease by by 30%. In addition to the above advantages, information modeling technologies will contribute to the maximum openness of the construction business to all its participants. [7, 14]

Figure 5 presents target results of digitalization of the construction industry.

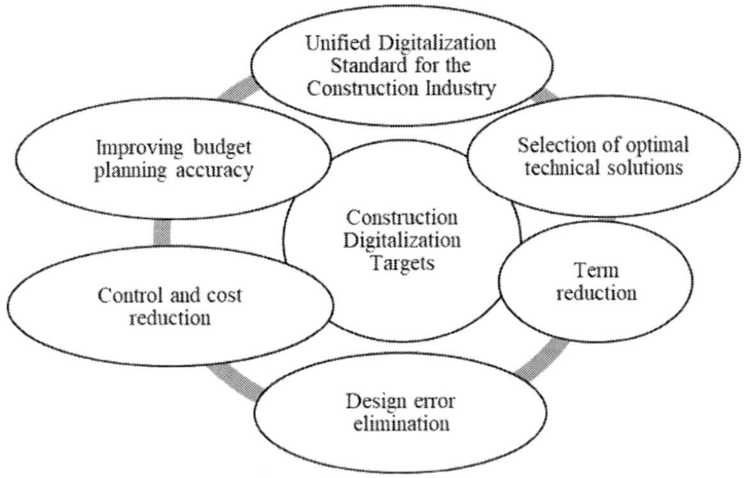

Figure 5. Targeted results of digitalization of the construction industry.

Figure 5 shows that digital technologies will serve as one of the main factors in the innovative development of the construction industry, and the use of modern software and digital technologies will contribute to the more effective monitoring and regulation of the entire construction process.

4. Conclusion

The digital technologies are being implemented in the construction industry. Therefore, most construction enterprises have already designed objects using information modeling technologies. For other enterprises, the only rational solution is implementation of the BIM technologies.

To digitalize the construction industry, economic incentives to implement new technologies are required. There should be no regulatory restrictions on individual solutions.

References

[1] Negroponte N 1995 *Being Digital* (New York: Alfred A. Knopf) p 245
[2] Tremblay S, Iturria-Medina Y, Mateos-Pérez J, Evans A and De Beaumont L 2017 *Eur J Neurosci* **46** 1956-967
[3] Stone B 2014 *The Everything Store* (Moscow: ABC Atticus) p 136
[4] Banke B 2016 Russia online? *Catching up can't be left behind* (Boston: The Boston Consulting Group) p 56
[5] Berman S 2012 *Strategy and Leadership* **40** p 16-24/
[6] Aptekman A 2017 *Digital Russia: a new reality* (Moscow: report of McKinsey) p 133.
[7] Vasilyeva N, Bachurinskaya I 2018 *Bulletin of the Altai Academy of Economics and Law* **7** p 39-46
[8] Abdrakhmanova G, Vishnevsky K, Volkova G and Gokhberg L 2018 *Indicators of the digital economy:* 2018 (Moscow: HSE) p 268
[9] Abdrakhmanova G, Vishnevsky K and Gokhberg L 2019 *What is the digital economy? Trends, Competencies, Measurement* (Moscow: HSE) p 82
[10] Dobrynin A, Chernikh K and Kupriyanovsky V 2016 *International Journal of Open Information Technologies* **1 (4)** p 4-10.
[11] Evtyanova D 2017 *Economic systems* **3 (38)** p 54–57
[12] Kudryavtseva V and Mironova M 2018 Transport Infrastructure of the Siberian Region **10** p 538-43
[13] Tabunchikov U 2018 *Energy Saving* **7** p 1-13

[14] Travush V 2018 *Construction Sciences* **3** p 107-11

IOP Conf. Series: Materials Science and Engineering **880** (2020) 012080 doi:10.1088/1757-899X/880/1/012080

Features of digital transformation in the construction industry

V A Kudryavtseva[1], O V Nikishina[1] and N V Vasileva[2]

[1] Irkutsk National Research Technical University, 83 Lermontov str., Irkutsk, 664074, Russia

[2] Saint-Petersburg State University of Economics, 21 Sadovaya str., St. Petersburg, 191023, Russia

E-mail: kudravera@mail.ru

Abstract. The introduction and development of digital technologies and innovations is one of the necessary elements for the long-term and effective functioning of most sectors of the economy, not only in Russia, but also in the world. The construction sector accounts for a significant portion of Russia's GDP. Therefore, the issue of introducing digital and innovative technologies is quite acute. In order to develop a mechanism for digital transformation of the construction sector, special attention should be paid to the possible maximum coverage of business processes existing in construction. Since the synergistic effect of the introduction of digital technologies in the construction sector will be to increase the competitiveness of construction organizations in the construction market. The formation and development of the concept of digital transformation of the construction industry is one of the necessary elements for the full implementation of the digital economy of the state.

1. Introduction

The construction sector of any state belongs to one of the main production complexes that occupy a significant place in the world or within the state economic system. In connection with the global transformation of the world economic system, the construction industry already needs to apply current scientific achievements in the field of technology, new technologies and effective methods of production organization for effective functioning and formation of competitive advantages. The creation of prerequisites for the introduction of new construction organization schemes should be based on the implementation of design solutions in the context of the entire life cycle of construction objects.

Currently, the effective implementation of construction projects is largely determined by and at the same time depends on the level and quality of the technological connection of administrative, managerial and construction processes. Since modern methods used in the design, management, technology and organization of construction determine and have a greater impact on the performance of construction and installation works and processes. The construction of buildings and structures is carried out using modern requirements of building codes and rules used in the design and construction of the object. But we cannot limit ourselves to this aspect only, because today it is important to comprehensively introduce elements of digital technologies into the production process. Another impetus for the development of digitalization in the construction industry is the large-scale growth of the population of cities – on average, the number of citizens in the world increases by 200 thousand people every day and this number will grow year by year. Thus, according to forecasts of the world health organization, by 2050, about 70% of the World's population, or about 6.4 billion people, will live in urban areas. At the same time, most of the population will live in large cities, making very different demands on their places of residence, which will lead to the need to provide the population not only with affordable and high-quality housing, but also with infrastructure facilities.

2. Materials and methods

To date, the global construction industry has begun to undergo large-scale changes, primarily due to the use of innovative ways to implement construction projects instead of the established methods of design and construction. Increased requirements for modern project documentation, increasing pressure on deadlines, problems with communication between participants, a large amount of information or lack of it at the right time – all this contributes to the poor performance of the industry as a whole.

Digital representation of information has much greater capabilities than traditional methods, allowing the formation of complete technological environments (ecosystems, platforms, platforms), within which the user can create the necessary friendly environment for himself (technological, instrumental, methodological, documentary, partner, etc.) in order to solve entire classes of problems [11, 12].

For Figure 1, it is presented what aspects of society's life are affected by the digital transformation of the investment and construction sector.

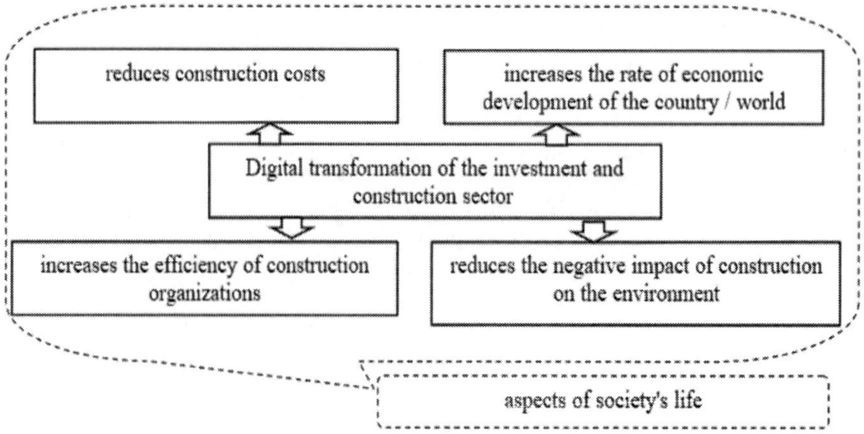

Figure 1. The impact of digital transformation of the investment and construction sector on the life of society.

It should be noted that the introduction of digital technologies in business at the basic level has already taken place several years ago, but the management of the transition to digitalization, in its essence, is an endless process, because the introduction of new and new technologies in business will continue and the process will go deeper and deeper into the relevant industry. Digital transformation is being implemented in different areas of business in waves and they are moving with a shift in time. The baton passes from one industry to another (Figure 2).

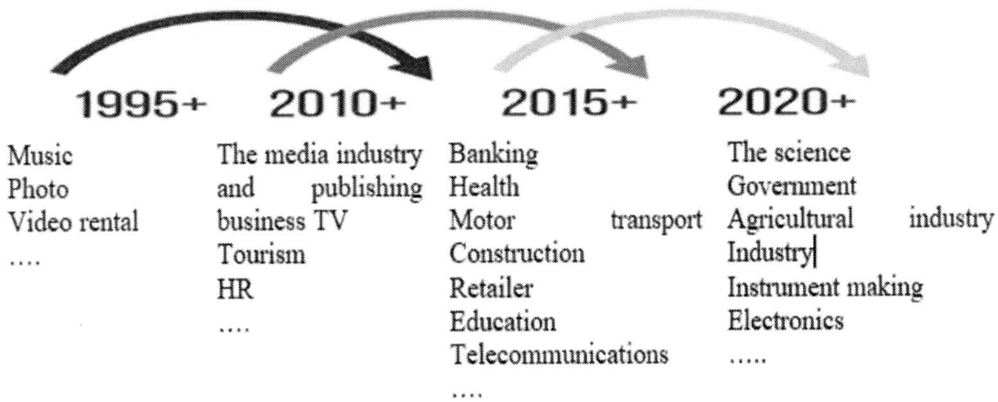

Figure 2. Waves of digital transformation. [10]

In general, scientists pay a lot of attention to the development of digitalization in organizations. For example, foreign scientists, Burman S.J. and Bel R. of the IBM Institute for business value, identified three stages in the evolution of digital transformation of organizations (Figure. 3). [1, 9]

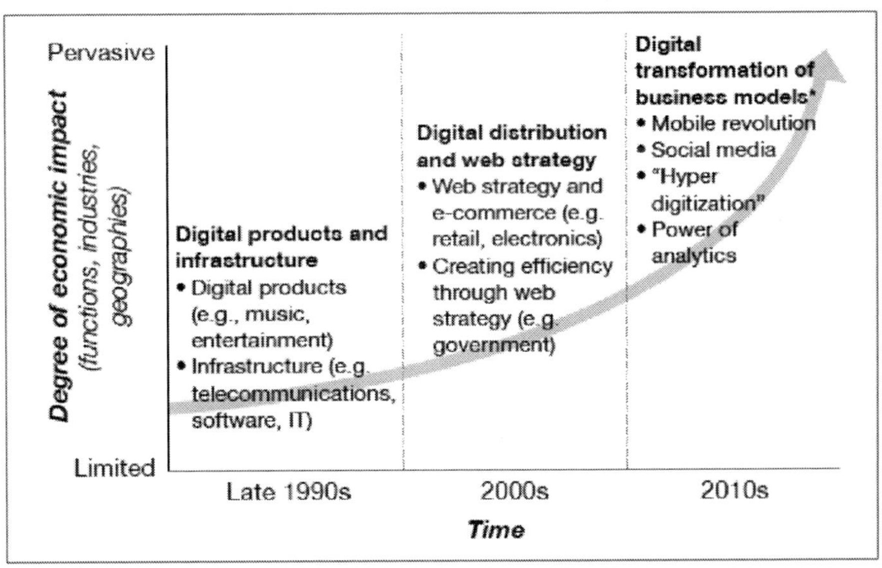

Figure 3. Evolution of digital transformation of organizations. [9]

A significant part of the existing business models and ecosystems used by western companies are an obstacle to successful competition in the digital economy. Therefore, the only option is digital transformation, which can work on the basis of continuous improvements in the entire process of product value formation. [5, 6, 7, 8]

The Table 1 illustrates the nature of business changes when switching to digital platforms.

Table 1. The nature of business changes in the transition to digital platforms. [10]

Business before switching to digital platforms	Business based on digital platforms
Supply chain (a linear addition of values)	Ecosystem based on a digital platform (nonlinear value addition)
Ability to control the supply chain	Ability to control the ecosystem
Economies of scale on the supply side	Economies of scale on the demand side
Increase in value as the number of suppliers increases	Increase in value with an increase in the number of consumers
Reduction of return on tangible assets due to their depreciation	Increasing returns as the network effect of the digital ecosystem develops
	Ability to quickly search for partners – effective cooperation with other participants
	Moving production of products and services closer to the consumer
	Avoiding intermediaries, direct access to resources, savings due to collective (crowd) projects (organized collective purchases, construction in a warehouse
	Avoiding long supply chains and eliminating a succession of intermediaries
	Accelerate the introduction of new products
	Reduce the cost of manufactured products and significantly reduce the time of its release

To create a successful ecosystem, an organization must have the maximum number of relevant competencies, such as:

- management of integration processes;
- flexible organization management;
- customer relationship management;
- organization of interaction between the center for innovative solutions and existing divisions of the company;
- innovation management.

3. Results

Trying to assess digital transformation in various industries, including construction, it should be noted that the essence of it is that the organization must have the ability or final competence to regularly transform its processes using digital technologies.

If we consider the impact of a particular industry on the size of GDP, we should highlight the construction industry. According to research and analytical bases, construction accounts for no more than 6% of world GDP [1, 2]. At the same time, in developed countries, the construction industry accounts for 5% of GDP, in developing countries-up to 8% of GDP. By 2025, it is planned to increase the total annual income of construction organizations to 15 trillion dollars. 10 trillion dollars USA in 2015 [3]. Already, more than 100 million people in the world are employed in the construction sector [4].

In Russia, the indicator for total investment in fixed assets is 50% and it is accounted for the construction of buildings (residential and non-residential) and structures. The indicator of profitability of construction organizations for goods and services sold for the period 2010-2018 was in the range of 4.8–7%, while the average economic indicator was 7.7-12.3%. Currently, the construction industry in Russia employs about 6.3 million people. Calculations by the Analytical center based on OECD data showed that the GVA per person employed in the industry in 2017 was 41.6 thousand dollars. (in current prices), which is significantly lower than in developed countries- Sweden (96.7 thousand dollars), France (93.4 thousand dollars), Great Britain (77.4 thousand dollars), USA (76.7 thousand dollars), Germany (75 thousand dollars), Japan (58.8 thousand dollars)

The issues of the development prospects of the construction industry are dealt with at the highest state level. To date, the work on the formation and approval of a Strategy for the development of the construction industry until 2030 is being carried out by the Ministry of construction of Russia. According to the decree of the Government of the RF dated August 16, 2018 № 1697-r, the strategy under consideration is included in the action plan ("road map") for the development of competition in the country's economic sectors.

This strategy is based on a number of principles that will eventually lead to a competitive, efficient, and innovatively viable construction industry (Figure 4).

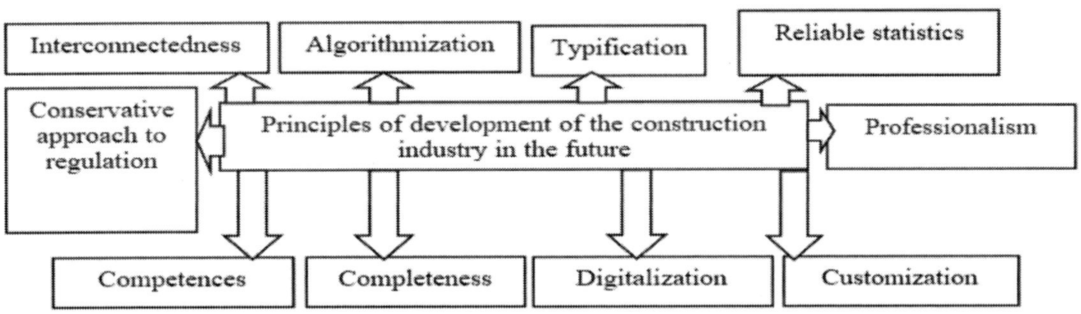

Figure 4. Principles underlying the development of the construction industry until 2030.

For effective implementation of digital technologies in the construction organization, changes must affect all areas of the organization's activities. It is not possible to implement the technology separately only in the production department, at the level of a separate project or a separate specialty. This approach gives some results, but in the end does not lead to significant changes, bringing only a small share of the benefits that are possible with the full implementation of digitalization. The entire process requires a number of tools or systems that the organization will operate on (Figure 5).

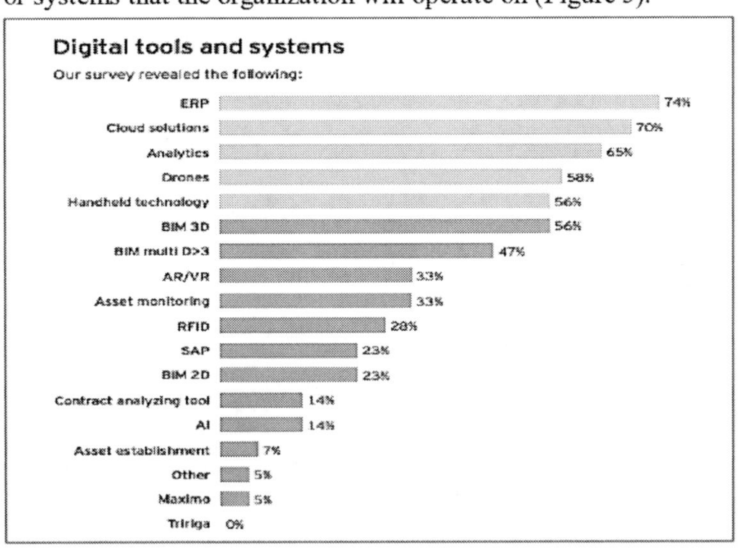

Figure 5. Digital tools for construction organizations. [5]

Without a clear concept and good management at the top level, attempts to implement digital platforms will lead to a waste of resources, so organizations need their own strategy that takes into account the specific features of their activities.

4. Conclusion

Having a certain specificity, digital technologies have their own characteristics and their own set of technologies that are suitable for a specific industry and that define the essence of such a phenomenon as digital transformation. In relation to investment and construction, digital transformation has an impact on many aspects of society's life, for example, it reduces construction costs, increases the efficiency of construction organizations, reduces the negative impact of construction on the environment, and increases the pace of economic development both at the level of individual regions and countries, and at the global level.

References

[1] Berman Sand Bell R2011 "*Digital transformation: Creating new business models where digital meets physical"* (USA: IBM Institute for Business Value) pp 1-1
[2] World Development Indicators. World Bank 2015. – URL: http://data.worldbank.org/data-catalog/ world-development-indicators (date of request: 13.11.2018).
[3] Berger R 2016 "*Digitization in the construction industry"* (Germany: Civil Economics, Energy and Infrastructure Competence Center) pp 1-156
[4] Banke B 2016 Russia online *Catching up can't be left behind* (Boston: The Boston Consulting Group) pp 56
[5] Grinis M and Buisman A 2018 *How are engineering and construction companies adapting digital to their businesses* (London: Ernst & Young) pp 16
[6] Aptekman A 2017 *Digital Russia: a new reality* (Moscow: report of McKinsey) pp 133
[7] Abdrakhmanova G, Vishnevsky K, Volkova G and Gokhberg L 2018 "*Indicators of the digital economy:* 2018" (Moscow: HSE) pp 268
[8] Vasilyeva N, Bachurinskaya I 2018 *Bulletin of the Altai Academy of Economics and Law* **7** pp 39-46
[9] Vishnevetskaya A and Ablyazov T 2019*Bulletin of the Altai Academy of Economics and law* **3(2)** pp 28-37
[10] Prokhorov and Konik L 2019 *Digital transformation. Analysis, trends, international experience* (Moscow: LLC " Alyans Print») p 368
[11] Khalin V and Chernova G 2018 *Management consulting* **10** pp 46-63
[12] Accept the challenge of the digital economy [Electronic resource]. URL: http://expert.ru/siberia/2017/48/prinyat-vyizov-tsifrovoj-ekonomiki/ (date of request: 10.02.2020).

IOP Conf. Series: Materials Science and Engineering **880** (2020) 012081 doi:10.1088/1757-899X/880/1/012081

Smart technologies as a type of intangible assets of a construction organization and mechanisms for their implementation

A V Peshkov

[1] Irkutsk National Research Technical University, 83 Lermontov str., Irkutsk, 664074, Russia

E-mail: expertiza@istu.edu

Abstract. The company's assets include intangible assets, a distinctive feature of which is the absence of physical assets. Today, it is becoming more obvious that the so-called tangible assets are not the only factor in ensuring the profitability of the organization, and that there are other types of them that do not have such a classic feature as a material substance, but can play a crucial role in the process of making a profit for the enterprise. The relevance of this topic is obvious, because in modern conditions, the formation of complete information about economic processes is almost impossible without information about intangible assets. Moreover, with the development of innovative technologies, organizations in the construction industry need to step forward and use such type of intangible assets as Smart platforms. This type of intangible assets allows construction companies not only to meet a high innovative level of development, but also provides a reduction in costs and generate more revenue.

The efficiency of any construction project depends not only on the level of technical competence and responsibility of the implementation agencies, but also from the quality control a huge number of interrelated and interdependent processes, as well as coordination of all project participants. In accordance with the development trends in all sectors of business, digital technologies are beginning to take an increasing place, the possession of which will eventually become a mandatory requirement for the implementation of each project. The construction industry is the most conservative with regard to the use of digital technologies. At the same time, it has a high potential for applying digital and innovative technologies.

In 2018, the president of the RF issued an instruction "on modernizing the construction industry and improving the quality of construction", which is aimed at introducing digital technologies into the construction industry [1]. This project should ensure the digital transformation of the industry by 2024. Within the framework of this order, a set of measures "digital construction" is being developed, which are aimed at fulfilling this order. In the process of digitalization of construction, it is expected to reduce the cost and time for the construction of objects by up to 20% [2].

Digitalization of the construction industry is developing in many directions. Participants in the construction market are actively implementing digital information technologies that cover almost all business processes: recruitment, accounting, internal document management, planning, development and placement of advertising, customer search and support, procurement, production, work, services, monitoring of contracts, and many others.

Prospects for the development of digitalization consist in a radical transformation of industrial relations, the creation of a digital ecosystem, which is characterized by the following:

– all elements of the investment and construction system are present simultaneously in the form of physical objects, products and processes, as well as in the form of their digital copies (mathematical models);

– all physical objects, products and processes become part of an integrated it system due to the presence of a digital copy and the "connectivity" element;

– through the presence of digital copies (mathematical models) and being part of a single system, all elements of the investment and construction system continuously interact with each other in a mode close to real time, simulate real processes and predicted States, and ensure constant self-optimization of the entire system.

The key advantage of digital transformation is the implementation of the ability to automatically manage the entire system (or individual components), as well as its almost unlimited scaling without loss of efficiency, which allows you to significantly improve the efficiency of economic management (economic activities and resources of the country in various industries) at the micro and macro levels.

However, the construction industry does not fully realize the potential of implementing digital technologies. For example, project planning often remains inconsistent between the office and the field office and is often done on paper, without the use of digital devices. Also, contracts do not include incentives for risk sharing and innovation. The industry has not yet introduced new digital technologies that require prior investment, even if the long-term benefits are significant. RDA spending in construction is significantly lower than spending in other industries: less than 1% of revenue compared to 3.5–4.5% for the automotive and aerospace industries [3,4].

However, investment and construction projects are becoming more complex and large-scale, both in terms of construction volumes and investment. This means that traditional methods must change. It is obvious that the deep problems of the construction industry, such as lack of personnel and resources, will also require new ways of thinking and working. Traditionally, this sector has focused on incremental improvements, in part because many believe that each project is unique, that it is impossible to scale up new ideas, and that the introduction of new technologies is impractical. The McKinsey global institute estimates that the world as a whole will need to spend $ 57 trillion on infrastructure by 2030 to keep up with global GDP growth [5]. This is a huge incentive for construction industry actors to find solutions to transform productivity and implement projects using new technologies and improved methods.

In the RF, the greatest attention when implementing digitalization technologies in the construction industry is given exclusively to building information modeling. This approach is not correct, because digitalization of construction is not only an information modeling of buildings, it is also the formation of electronic information systems for urban planning, digital libraries of standard elements, the creation of machine-readable regulatory documents, the introduction of Smart platforms in all business processes of a construction company.

The author of the article identified five areas of the Smart platform that will be practical and relevant for the subjects of the construction sector:

1. The implementation of surveying and geo-high resolution
2. next-generation 5-D BEAM Information modeling
3. Digital collaboration and mobility
4. Internet of things and advanced analytics
5. Modern construction materials

Let's look at each element of the Smart platform in detail. The occurrence of errors during geodetic works is the main reason for the delay of construction projects and excess of estimated costs. New technologies that combine high-definition photos, three-dimensional laser scanning and geographic information systems, thanks to recent improvements in unmanned aerial vehicle and unmanned aerial vehicle technology, can significantly improve the accuracy and speed of geodetic work. Advanced shooting techniques are complemented by geographic information systems that allow you to overlay maps, images, distance measurements, and GPS positions. This information can then be uploaded to other analysis and visualization systems for use in project planning and construction.

The use of BIM technologies in construction has been supported and promoted by the government of the RF since 2010, but so far it has not been possible to achieve high results in this area. Today, the world's leading construction companies use a new generation of information modeling 5-D BIM. This is a five-dimensional representation of the physical and functional characteristics of any project. It takes into account the cost and schedule of the project in addition to the standard spatial design parameters in 3-D [6,7]. This also includes details such as geometry, technical characteristics, aesthetics, thermal and acoustic properties. The 5-D BIM platform allows owners and contractors to identify, analyze, and record the impact of changes on project cost and planning. To take full advantage of BIM technology, project owners and contractors must enable its use right at the design stage, and all stakeholders must adopt standardized design and data presentation formats that are compatible with BIM. In addition, owners and contractors should allocate resources to implement BIM and invest in capacity building.

The digitization process means moving from paper to real-time online information sharing to ensure transparency and collaboration, timely progress and risk assessment, quality control, and ultimately better and more reliable results.

One of the system-forming elements of the Smart platform is the joint work of all project performers. The main reason for low productivity in the industry is that it still relies primarily on paper to manage its processes and results, such as drawings, project drawings, purchase orders and supply chains, equipment logs, daily progress reports, and stamp lists. Due to the lack of digitization, information exchange is delayed and cannot be universal. This is why owners and contractors often work with different versions of reality. Using paper makes it difficult to collect and analyze data; this is important because in purchasing and contracting, historical performance analytics can lead to better results and risk management. Errors in the paper also usually cause disagreements between owners and contractors on issues such as construction progress, change orders, and claims management. Finally, paper trails just take longer. Based on this, contractors need to develop software within the company's intangible assets that runs in a cloud-based mobile field surveillance platform that integrates project planning, design, physical control, budgeting, and document management for large projects (Picture 1) [8,9].

Design management	Scheduling	Materials management	Crew tracking
• Visualize drawings and 3-D models on-site, using mobile platforms • Update blueprints in the field with markups, annotations, and hyperlinks	• Create, assign, and prioritize tasks in real time • Track progress online • Immediately push work plan and schedule to all workers • Issue mobile notifications to all subcontractors	• Identify, track, and locate materials, spools, and equipment across the entire supply chain, stores, and work front	• Provide real-time status updates on total crew deployed across work fronts, number of active working hours, entry into unauthorized areas, and so on
Quality control	**Contract management**	**Performance management**	**Document management**
• Offer remote site inspection using pictures and tags shared through app • Update and track live punch lists across projects to expedite project closure	• Update and track contract-compliance checklists • Maintain standardized communication checklists • Provide updated record of all client and contractor communications	• Monitor progress and performance across teams and work areas • Provide automated dashboards created from field data • Offer staffing updates and past reports generated on handheld devices	• Upload and distribute documents for reviewing, editing, and recording all decisions • Allow universal project search across any phase

Picture 1. Software aspects

In fact, the digital collaboration and mobile solutions segment has attracted almost 60% of all venture capital funding in the construction technology sector [10]. One of the startups has developed apps for tablets and smartphones that allow you to make changes to design drawings and plan real-time transmission to local crews; photos of the site can be linked to construction plans. This solution supports a core set of documents with automatic version control and cloud access. Other companies offer mobile timekeeping, real-time cost coding, employee location detection, and logging and problem tracking.

As advanced users such as project managers, merchants, and operators implement real-time crew mobility applications, they can change the way the industry does everything: managing work and change orders, tracking time and materials, dispatching, scheduling, measuring performance, and reporting an incident.

The digital solutions segment described above is based on the Internet of things. On the construction site, the internet of things allows construction machinery, equipment, materials, structures, and even formwork to “communicate” with a central data platform to collect critical performance parameters [11,12]. The Internet of things is the concept of a computer network of physical objects equipped with built-in technologies to interact with each other or with the external environment, which allows them to collect, analyze and transmit data among themselves using software and technical devices. Let's look at the main applications of the Internet of things in the construction industry:

1. Monitoring and repair of equipment. Advanced sensors allow equipment to detect and report maintenance requirements, send automatic preventive maintenance alerts, and collect usage and maintenance data [13,14].
2. Inventory management and ordering. Connected systems can predict and notify site managers when inventory runs out and when orders need to be placed. Marking and tracking materials using NFC can also accurately determine their location and movement, and help coordinate physical and electronic inventory.
3. Quality assessment. "Smart structures" that use vibration sensors to test the strength and reliability of the structure during the construction phase can detect flaws and then correct them in advance.
4. Energy efficiency. Sensors that monitor environmental conditions and fuel consumption for assets and equipment can help improve energy efficiency on site.
5. Security. Wearable lanes can send alerts if drivers and operators fall asleep, or if a vehicle or asset is stationary or inactive for a set amount of time during shift hours.

All of the above applications are combined in a single platform, where devices and sensors that track and analyze all construction processes work in real time. This approach allows you to reduce not only the time of work and their labor intensity, but also their cost. Special attention should be paid to control. Already, large companies are using drones, GPS devices and all sorts of scanners in order to comply with plans, speed of construction of houses and the appropriate quality. End-to-end analytics will also affect the physical work of staff on the construction site [15].

The use of the above programs and tools is aimed at increasing the role of the organization's intangible assets. Competent and timely action on the implementation and use of IA significantly transforms the economic and industrial image of the organization [16]. By implementing the IA management cycle, an organization can increase the efficiency of its activities by making the most complete and rational use of information, reputation, and the knowledge that it has rights to use. It should be noted that IA is quite complete in relation to the economic turnover of the organization: they are used in sales and purchase transactions, they are subject to property claims, they are made as a contribution of legal entities and individuals to the share capital when purchasing shares of the company, they include their cost in fixed assets and amortize, including costs [17].

In modern realities, the use of Smart space in construction organizations allows you to reduce production costs, get additional income, increase the level of business reputation after its severe decline due to economic difficulties and get an investor or an important partner.

References

[1] Biglova A A 2018 State of the public procurement system in Russia *Science, education, innovation: approbation of research results Materials of the International (correspondence) scientific and practical conference* pp 153-156

[2] Avagimyan Yu 2017 System of management of current assets of the enterprise in modern conditions: (on the example of JSC "Tandem") *Scientific Bulletin* **1 (13)**

[3] Matveeva M 2018 Synchronization of the activity of participants during the implementation of housing construction projects *Proceedings of Universities. Investment. Construction. Real estate* **8** pp 31-41 DOI: 10.21285/2227-2917-2018-2-31-41.

[4] Gaskova Yu.I. "Integrative approach to the formation of a mechanism for the coordinated development of industry and infrastructure in the region" // Proceedings of the II International scientific and practical conference "Economics and modern management: theory, methodology and practice". - Penza: 2018. - p.358-362.

[5] Kalyuzhnova N, Violin S and Matveeva M 2019 Smart Specialization of Russian Regions: Prospects and Limitations *Investments. Construction. Real Estate: New Technologies and Targeted Development Priorities IOP Conference Series: Materials Science and Engineering*

667 012036
[6] Sayfutdinova R V 2016 *Modern Fundamental and Applied Researchers* **1(20)** pp 167-173
[7] Peshkov V V and Gamayunova E V 2018 Specific features of investment support for the repair and reconstruction of buildings and facilities *Proceedings of Universities. Investment. Construction. Real estate* **8** pp 56-72 DOI: 10.21285/2227-2917-2018-2-56-72
[8] Biglova A A 2018 State of the public procurement system in Russia *Science, education, innovation: approbation of research results Materials of the International (correspondence) scientific and practical conference* pp 153-156
[9] Yaskova N Yu and Sarchenko V I 2018Development of strategic approaches to the creation of aqua-territorial industrial complexes in the Arctic zone *Proceedings of Universities. Investment. Construction. Real estate* **8** pp. 84-93 DOI: 10.21285/2227-2917-2018-2-84-93
[10] Gertsekovich D, Gorbachevskaya L, Grigorova L and Peshkov V 2019 Return on investment in REIT real estate funds *Investments. Construction. Real Estate: New Technologies and Targeted Development Priorities IOP Conference Series: Materials Science and Engineering* **667** 012063
[11] Vovseenko E A 2018 *Financial Instruments and Hedging: New Rules of IFRS* European Proceedings of Social and Behavioural Sciences **L** DOI: https://dx.doi.org /10.15405/epsbs.2018.12.167
[12] Khokhlova G., Kretova N., Burov V. 2019 *The problems of investment activity of entrepreneurship and methodological aspects of credit risks assessment,* ICRE IOP Conference series: materials science and engineering **667** DOI:10.1088/1757-899X/667/1/012038
[13] Matveeva M, Kalyuzhnova N, Skripka S 2019 Intellectual specialization of Russian regions: prospects and limitations *IOP Conf. Series: materials Science and engineering* **667**
[14] Simon French 2017 Cynefin: uncertainty, small worlds and scenarios *Operational Research Society* **66 (10)** pp 1635-1645
[15] Morozko N I 2017 Financial management in small business p 217
[16] Matveeva M, Kalyuzhnova N 2019 Industrial construction: investment support for entrepreneurship *IOP Conf. Series: materials Science and engineering* 667
[17] Rasulev A F, Gorbachevskaya E Yu, Nikityuk L G, Timchuk O G 2019 Increasing the level of innovative potential of construction industry enterprises. *University news. Investment. Construction. Realty* **9** pp. 260-273

Difficulties in transforming the construction industry under the digital economy

O G Timchuk[1] and M V Evloeva[2]

[1,2] Irkutsk National Research Technical University, 83 Lermontov st., Irkutsk, 664074, Russia

E-mail: oksana-timchuk@yandex.ru

Abstract. Solving key tasks of economic growth and increasing the level of social responsibility of the construction industry require the use of the scientific and technical potential, implementation of long-term innovative development plans and programs, transformation of external conditions of investment and construction activities, adaptation of the construction management system to the digital economy. In this regard, management of structural elements of the construction industry and digitalization of the construction industry are crucial. The relevance of the research issue is due to the fact that in modern conditions, the innovative potential of construction companies is a key factor contributing to competitiveness in the market and transition to the digital economy. The object of research is construction companies operating within the market economy. The subject is organizational and economic relations taking shape in the process of development of the construction industry under the digital economy. The authors considered the problems of transition of the construction industry to the digital economy, analyzed the construction industry, and proposed their own definition of the digital economy in the construction industry. The factors that impede the development of the construction industry were described taking into account modern aspects of economic development, which makes it possible to identify areas of construction business management under the digital economy. The competences of an employee of the construction company in the context of the digital transformation of the construction industry were identified. The authors drew conclusions proving the need for the transition of the construction industry to the digital economy

One of the key sectors of the national economy is construction. Today, there is a need for the transition of all sectors of the economy to the digital economy - the construction industry is no exception. The construction industry is a locomotive for the development of the Russian regions, as it improves the infrastructure. The success of a construction company depends on the successful transformation of the industry, as the digital economy dictates its own rules for doing business.

According to I.A. Chekhovskaya, the private sector dominates in the Russian construction market, and the number of public organizations is decreasing [1].

A. V. Tumashova and E. O. Proshkina believe that the construction business is the riskiest one, as it depends on market changes [2].

The current realities make every entrepreneur take into account economic changes. The most important task is digitalization of business processes.

"Information" has become an engine of the economy, it caused the transformation of the economy as a whole, namely the transformation of the economy into a digital one.

The digital economy allows the implementation of new technologies in order to create a single digital construction platform [3].

The digital economy has become a ubiquitous element of everyday life in most countries. The rapid spread of the Internet has changed the way businesses operate and negotiate with consumers, citizens get public services, and public bodies work. Digitization gives rise to a new business model, a new cross-border supply chain and new risks. Products and services of the online marketing, digital content and analytics data are rapidly becoming globally traded consumer goods.

Like the Internet, the digital economy is global; it has no boundaries, and anyone can immediately gain access to markets around the world [4]. Traditional consumer protection approaches, fair contracts and information should be reevaluated. Moreover, laws, regulations, and policies governing the digital economy must work along with efforts to improve the operational environment, including IT infrastructure, services, platforms, ecosystems, and devices.

Although many factors influence the digital economy, this guide focuses on four areas: consumer protection; data protection; cybersecurity; electronic transactions, in particular electronic payments (e-payments) and electronic signatures [5]. These topics constitute a favorable environment for the digital economy and affect almost all aspects of Internet business. These issues can act as a multiplier of forces for wider reforms, and they are central to business opportunities.

Digitalization of the construction industry is a global process in which all countries aimed at developing human capital and creating a competitive economy are involved [6].

A systematic feature of the modern development of the construction business is acceleration of transformation processes that require the readiness of capital-creating industries for efficient and expedient activities [7]. The responsibility of the construction industry as a whole, as well as large construction companies in solving development problems is increasing. Figure 1 shows the dynamics of growth in the number of construction companies in Russia from 2010 to 2017. The dynamics proves the significant role and relevance of construction companies in the market.

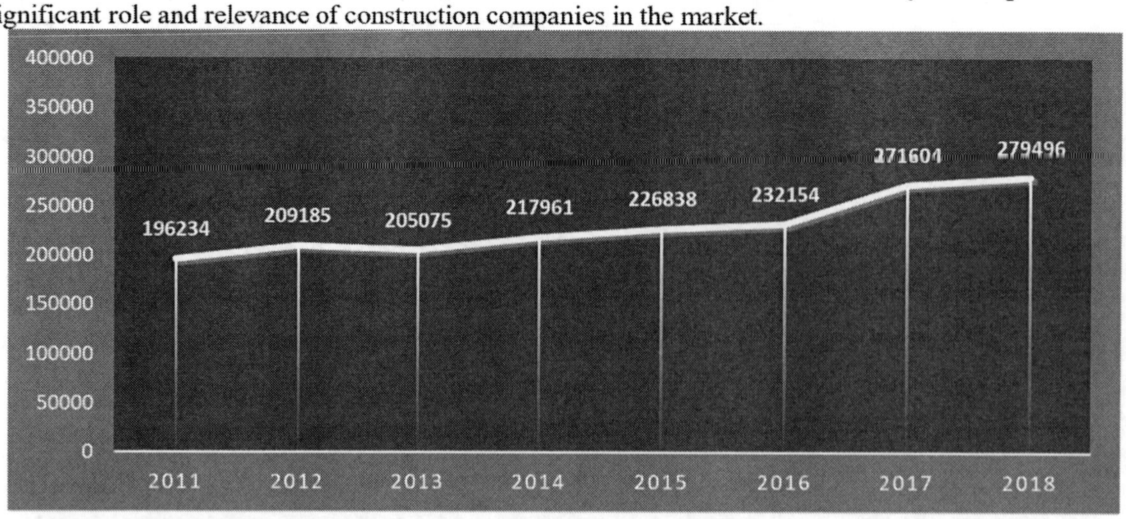

Figure 1. Dynamics of the number of construction companies in Russia in 2010-2018, units

One of the indicators of socio-economic development of the country is the rate of construction volumes. First of all, the construction industry aims at satisfying housing needs [8]. The dynamics of the volume of works performed by the construction industry from 2010 to 2018 is presented in Figure 2.

Figure 2 shows that the volume of works performed by construction companies from 2011 to 2018 increased by 3118.9, i.e. 1.7 time, which indicates that the industry is in a stable state.

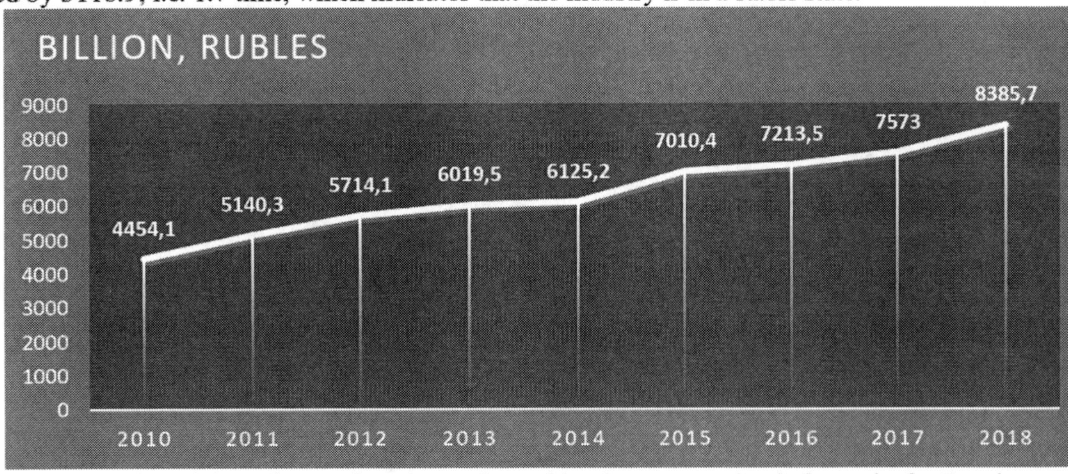

Figure 2. Dynamics of the volume of works performed by the construction industry in the Russian Federation from 2010 to 2018, billion rubles.

After the collapse of the USSR, a shortage of investment funds was characteristic of the Russian construction industry. The situation changed dramatically in the early 2000s, and the government maintained a positive trend. But there is no significant foreign investment in the industry. Figure 4 shows the dynamics of commissioning of residential and non-residential buildings from 2010 to 2018, m^3

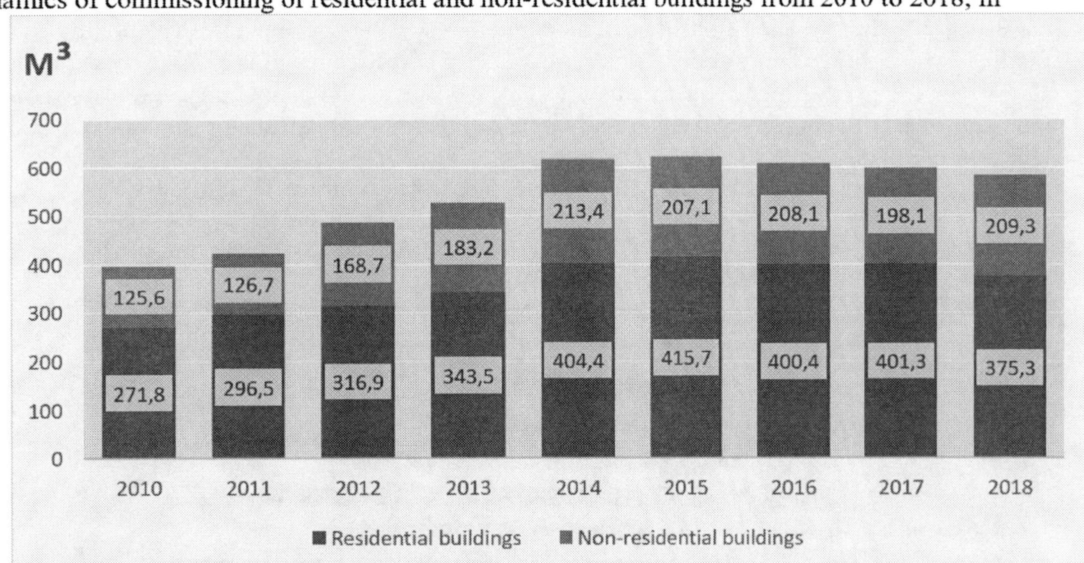

Figure 4. Dynamics of commissioning of residential and non-residential buildings in the Russian Federation from 2010 to 2018, m^3

An analysis of the data showed that more than 65% of buildings are residential. This sector of construction accounts for about 80% of all investments in the construction industry [9]. Despite this fact, the level of provision of the population with high-quality and comfortable housing is low, which indicates a high demand for this product. Figure 5 shows the dynamics of the housing stock in Russia from 2010 to 2018.

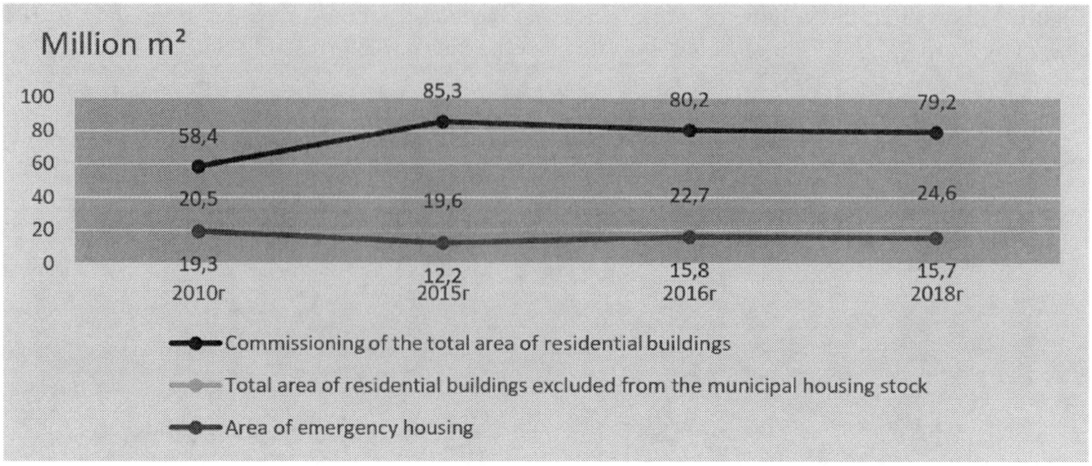

Figure 5. Housing stock dynamics in Russia in 2010, 2015-2018, million m²

Figure 5 shows positive results of the housing stock development. The total area of the housing stock increased 1.4 times, the volume of disposal of emergency housing increased 1.2 times in the period from 2010 to 2018.

Despite the positive growth dynamics, there are problems of transition to the digital economy.

The most important problem is corruption and dishonesty in tendering [10]. A number of points are not taken into account: the cheapest project wins, it is used by organizations reselling tenders; sometimes companies do not calculate the project, do not justify prices, do not take into account the specifics of the construction industry. The result of such tenders is prosperity of companies which aim to raise more funding, which makes construction a long process or a defective result [11]. One more problem is the staff issue: the lack of highly qualified staff [12]; managers who lack knowledge of the construction industry; "cheap" labor of illegal emigrants which entails fines and arrests for violation of the legislation [13].

Figure 6 presents factors hindering the development of the construction industry.

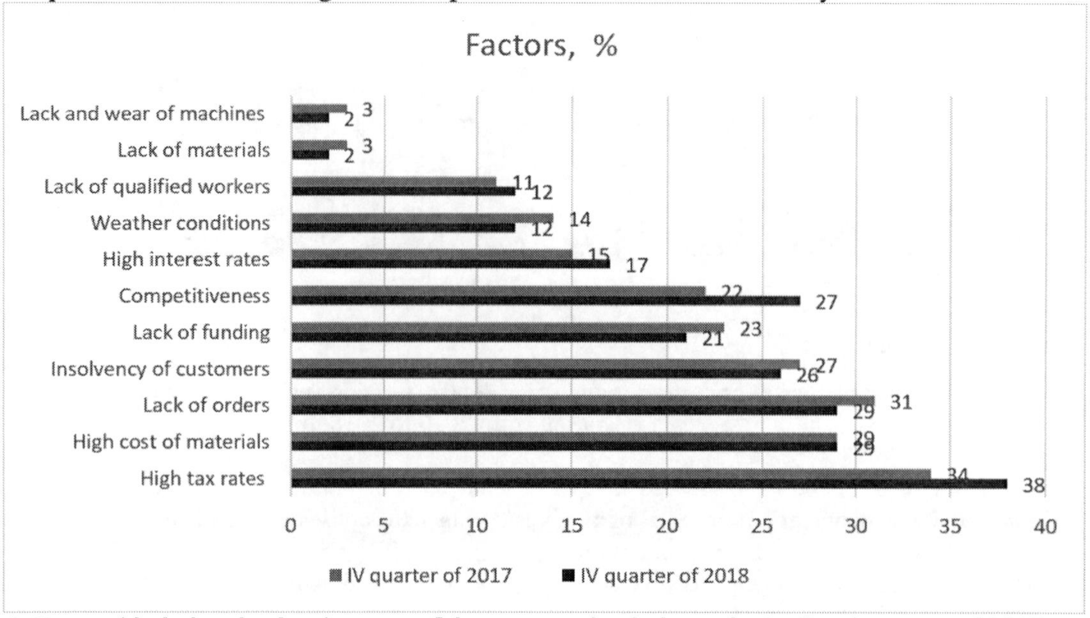

Figure 6. Factors hindering the development of the construction industry in the fourth quarter of 2017 and 2018 (based on a sample survey), % of respondents

IOP Conf. Series: Materials Science and Engineering **880** (2020) 012082 doi:10.1088/1757-899X/880/1/012082

An analysis of the factors indicates that the main problem is a high level of taxes and competition.

After analyzing the current situation in the construction business, the authors presented advantages of the transition of the construction industry to the digital economy (Figure 7).

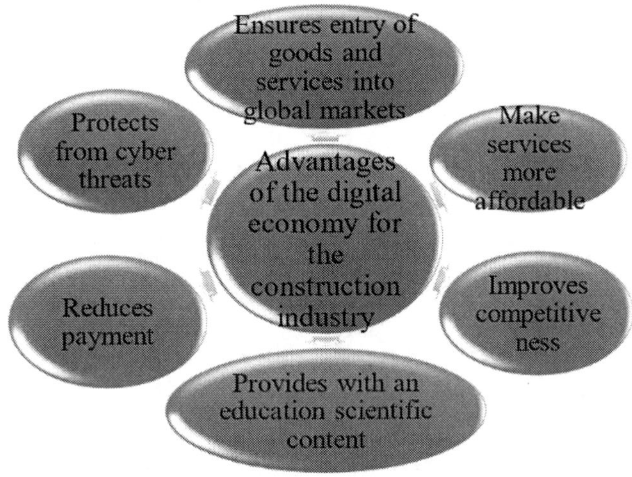

Figure 7. The main advantages of the digital economy for the construction industry

The digital economy allows the implementation of new technologies in order to create a single digital construction platform [14].

According to the authors, the digital economy in the construction industry is a combination of information, organizational and economic methods, forms, levers of influence on the construction business, based on an integrated approach to the large-scale implementation of digital technologies for regulating management processes in the construction industry [15].

Construction companies are aware of the importance of involving marketing as a way to adapt to changes in the industry, meet the needs of customers, remain competitive and improve their business strategy [16]. This article presents results of a quantitative and qualitative analysis of the current state of marketing in the construction industry. The study was carried out by searching for literature, processing and analyzing documents. Despite the efforts made in the last decade, it is difficult to generalize how to apply marketing in construction companies [17]. Nevertheless, there are strategies that can be used by construction companies. Knowing the needs of their customers, they will be able to choose relevant strategies.

The digital economy requires qualified workers [18]. An employee of the construction company should possess qualities presented in Figure 8.

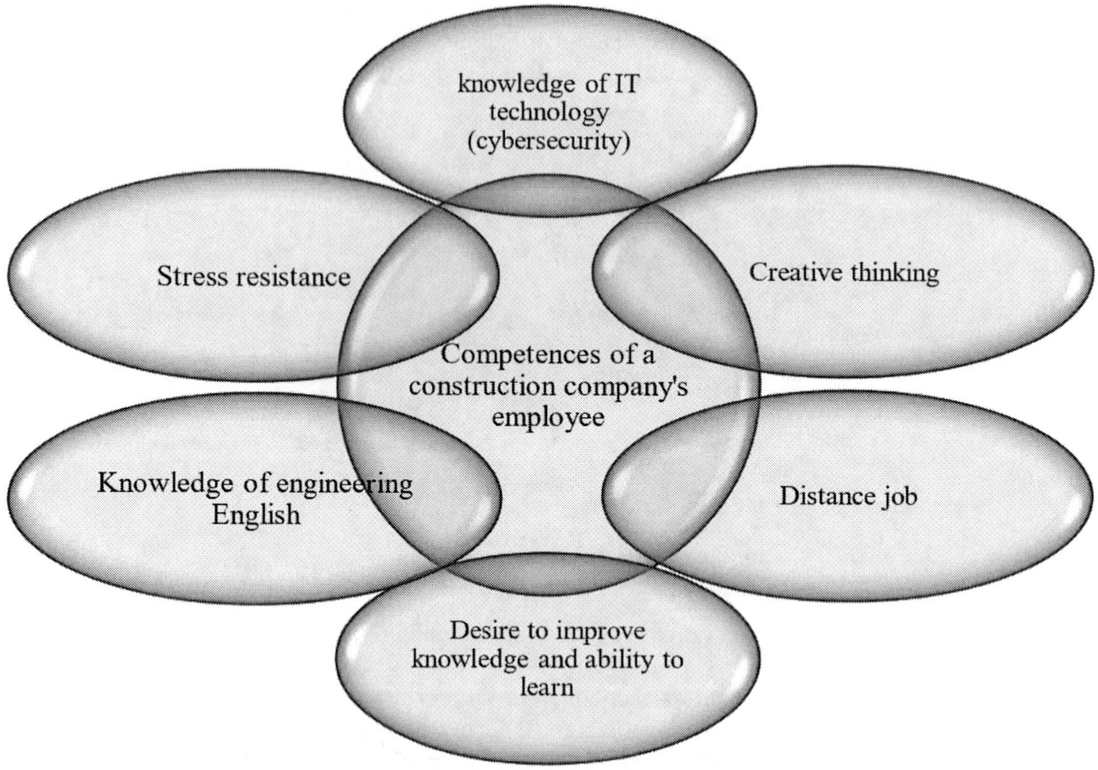

Figure 8. Basic requirements for a construction company's employee in the digital economy

The lack of qualified specialists may hinder the rapid transition of the construction business to the digital economy. The Government of the Russian Federation has already raised issues of IT staff shortages.

The current state of the industry was analyzed and main problems and factors that impede the development of the construction business were identified. Figure 9 presents the author's methodology for the transition of the construction industry to the digital economy.

Figure 9. The methodology for the transition of the construction industry to the digital economy

Given that a significant part of works is carried out by subcontractors, the creation of a digital platform will ensure the interaction of small businesses (formal and informal) with each other and large companies in order to establish contractual relationships, use experience and special skills.

Thus, the digital transformation will give positive results; new markets and businesses can be created. Reliability of control systems can improve.

It is important to note that for digital transformation, it is not sufficient to develop information technologies (digitalization); the restructuring of business processes, data processing and development of new business models and strategies, training of qualified personnel, development of digital culture and digital society are required. These processes should be interconnected, and the company's potential should be considered in the context of the ecosystem.

References

[1] Jennings R and Marques A 2013 Amortized cost for operating lease assets. *Accounting Horizons* 27 pp 51–74

[2] Kolontaevskaya I F and Isabekova O A 2015 Digital culture of engineers: problems and Solutions *Science 2014: results, prospects: Proceedings of international scientific-practical conference* pp 72-76

[3] Krishnan S V and Moyer C R 1994 Bankruptcy Costs and the Financial Leasing Decision *Financial Management* **23** pp 31–42

[4] Lewis C M and Schallheim J 1992 Are Debt and Leases Substitutes? *Journal of Financial*

and Quantitative Analysis **27** pp 497–511
[5] Medvedev D 2019 Russia-2024: Strategy of social and economic development
[6] Maloletko A N and Maloletko N E 2016 Philanthropy of big business and the pitfalls of "social entrepreneurship": a new threat to the economic security *Innovative development of the economy* **3-1(33)** pp 293-297
[7] Medvedev, D. The government of the Russian Federation from July 28, 2017 № 1632-p http://www.tadviser.ru/images/a/af/9gFM4FHj4PsB79I5v7yLVuPgu4bvR7M0.pdf
[8] Miller M and Upton 1976 Leasing, Buying, and the Cost of Capital Services *Journal of Finance* **31** pp 761–786
[9] Myers S C and Majluf N 1984 Corporate financing and investment decisions when firms have information that investors do not have *Journal of Financial Economics* **13** pp 187–224
[10] Nagornova T S and Chikin A A 2014 Digital culture as a factor of social and vocational Mobility *Socio-professional mobility in the XXI century: collection of materials and reports of International conference. The collection of materials and reports of International conference* pp 182 -186. http://elar.rsvpu.ru/handle/123456789/668
[11] Nikityuk L G, Grigorova L E, Timchuc O G and Ogloblin V A 2018 Encouragement of Innovation Activity of Industrial Enterprises. *The European Proceedings of Social & Behavioural Sciences, International Conference on Research Paradigms Transformation in Social Sciences* **50** pp 900-910
[12] Putin V V 2017 The message to the federal assembly Vladimir Putin *Bulletin of the Russian nation* **53** pp 9-24
[13] Putin V V 2017 The speech at the plenary session of the St. Petersburg international economic forum
[14] Putin V V 2017 *100 and 1 quotes. Compiled by S. M. Henkin, I. A. Istomin, S. M. Kretov* **8. 1(26)** p 161
[15] Shchegolev I *Speech at the opening of the College*
[16] Timchuk O G, Gorbachevskaya E Yu, Nikityuk L G and Evseeva G G 2018 Role of e-commerce in development of innovation economy *The European Proceedings of Social Behavioural Sciences, International Conference on Research Paradigms Transformation & in Social Sciences* 2018 **50** pp 1179-1187
[17] Shchegolev I *Speech at the opening of the College*
[18] Veretekhina S V, Shinkareva O V, Kozhaev J P, Telepchenkova N V, Kuznetsova E A and Zaitseva N A 2017 Evaluation methodology of the multiplier effect for the region as the result of the cluster formation *Eurasian Journal of Analytical Chemistry* **12(1)** pp 1-2

IOP Conf. Series: Materials Science and Engineering **880** (2020) 012083 doi:10.1088/1757-899X/880/1/012083

Digital transformation of the Russian construction industry

V N Turkova[1], A N Archipova[2] and Z G Fedorovna[3]

[1]Irkutsk National Research Technical University, 83 Lermontov st., Irkutsk, 664074, Russia
[2]Irkutsk National Research Technical University, 83 Lermontov st., Irkutsk, 664074, Russia
[3]Ph.D., Associate Professor, Department of Investment and Real Estate Management, East Siberian State University of Technology and Management, Ulan-Ude, Russia

E-mail: nikolaevna_ur@mail.ru

Abstract. Construction is an impactful industry and like everything impactful, it has to evolve with the needs and demands of the modern day consumers. The article is devoted to the analysis of the current state and prospects of development of digital technologies in construction. The fundamental concepts of digital economy, digital production and digital technologies are introduced; a peculiar "evolution" of the design process in construction is described (transition from so-called "paper-based" technologies to computer aided design systems and further to information modeling of construction objects). Construction technologies solutions and apps for the construction industry have significant growth potential, and we can nurture a cohort of new digital technology firms by supporting the development of connections with the construction industry, the broader IT sector, incubators and sources of capital. However, the largest scale impact is likely to be realised through effective implementation of BIM with the support of digital technology. Among the very topical issues raised in the article we should also highlight BIM - a fundamentally new approach in architectural and construction design, which includes creation of a computer model of a new building (structure), which carries all the information about the future of the object and is an instrument of control over its life cycle. Given the rapid pace at which new technologies and innovations are being developed, the specialization in Digital Technologies in Construction prepares graduates for the imminent, exciting and necessary, wide-scale digital transformation of the construction industry. Courses include Digital Technologies for Civil Engineering, BIM and 5D Planning, Technology and Innovation in Road Construction, Simulation and Optimization of Construction Processes and Construction Supply Chains and Digitization.

1. Introduction

The digital economy has changed the relationship between customers, workers and employers, as the digitalization of Russia is fast. The digital transformation is determined by the general technology, i.e., the ability to constantly transform the economy, gradually expanding and increasing productivity in all sectors and industries. The modern digital technological revolution depends on computers, the technical basis of the Internet, search engines and digital platforms. Due to delays in adapting to new processes, time is needed to accelerate the growth. At the early stages of such revolutions, more and more resources are used for innovation purposes. Digitalization will change jobs. The expanding digital economy, which includes cross-border services and e-commerce, can be an important engine for democratic and economic development, creating new market channels for local businesses, and raising tax revenues for governments to expand the list of basic services. As digital innovations spread around the world, local business communities face obstacles when eliminating technological and digital gaps. National policies, laws, and regulations governing this new space have a significant impact on development outcomes. The digital economy has become a ubiquitous element of everyday life in most countries. The proliferation of the Internet has changed the ways how businesses operate and transact with consumers, how citizens receive public services, and how regulators work at the national and international levels. Digitalization is generating new business models, new cross-border supply chains, and new risks. Online goods and services, digital content and data analytics are rapidly becoming global commerce products. Local enterprises and organizations representing their interests should be equipped with everything required to create favorable conditions contributing to the economic growth in the digital future. In industries, the construction sector can achieve significant efficiency gains due to new technologies.

2. Despite the fact that many executives recognize the potential of new technologies, they often hesitate to implement multi-billion dollar projects

Technologies used to increase construction productivity can have a profound effect on public and private expenses. Institutions and organizations use information and communication technologies to reduce their operating costs and serve customers more quickly. [1] The most significant areas of uncertainty that create risks for the innovative development of the region and can have a serious impact on its prospects include: strategic Federal policy in relation to the region; the volume and structure of demand for innovative products and services on the Russian and international markets. [2] The public sector accounts for a significant share of these expenses. Excluding residential and private projects, the costs of building public infrastructure facilities (e.g., healthcare, education and transport) reached a significant amount in 2018. Many companies are reluctant to experiment with their digitalization construction projects. Government spending accounts for a significant share of non-residential and public construction projects, and government agencies work closely with private companies to implement these complex infrastructure projects. Such projects cover a wide range of infrastructure facilities (roads, buildings and sewer systems). The purchasing power of the government affects each aspect of the construction industry, while its regulatory power allows it to set standards that are most easily implemented using new technologies or even authorize their use. The construction industry has not adopted an integrated platform that covers project planning, design, construction, operation and maintenance. Instead, the industry continues to rely on customized software. In addition, project owners and contractors often use different platforms that are not synchronized with each other. As a result, there is no single source that ensures an integrated real-time representation of the project, its cost and schedule. The next generation of 5-D BIM is a five-dimensional representation of physical and functional characteristics of any project. It takes into account the cost and schedule of the project, 3-D standard design parameters. It also includes such details as geometry, specifications, aesthetic, thermal and acoustic properties. The 5-D BIM platform allows owners and contractors to identify, analyze and record the impact of changes on project costs and planning. The visual and intuitive nature of 5-D BIM gives contractors a better chance to identify risks and make better decisions. For example, designers can visualize and evaluate the impact of a design change on a project's cost and schedule. One approach is to develop a digital construction strategy that encourages the use of new tools to reduce the time and cost of community service projects. For example, clear goals could be set for using off-the-shelf or modular components supported by digital collaboration tools such as BIM, which would reduce the number of rework and order changes. Some countries have already taken steps in this direction. In Ireland, for example, the BIM National Council published the National Strategy for the Construction

Industry in 2017, which included clear digital goals. As part of its vision, the council seeks to reduce project implementation time by 20 percent, increase the export of building materials by 20 percent and reduce the capital costs by 20 percent by 2021. According to the order of the President of the Russian Federation of July 19, 2018, it is necessary to gradually implement information modeling technologies in the construction industry. A very short deadline for the implementation of this order was set. It is one year. [3] The government intends to make the transition to BIM-designing comfortably and smoothly for the domestic construction market. A roadmap developed by the Ministry of Construction for the implementation of information modeling technologies (BIM) at all stages of the life cycle of a capital construction project has been signed for the Russian construction industry. In Russia, at the federal level, a decision on the long overdue transition to industry digitalization standards was made. It has proven their effectiveness. According to the Ministry of Construction, the use of information modeling technologies in the design and construction process will reduce costs by 20%. In addition, the use of BIM should reduce administrative barriers and shorten the construction time. It is worth noting that the concept of BIM technologies will be introduced in the Town Planning Code of the Russian Federation in the near future. The life cycle management system of the capital construction project, including the phased digitalization of the industry, will have been implemented by 2021. In general, the concept will be implemented during three stages: Stage - 2019-2021; Stage 2 - 2022-2024; Stage 3 - 2025-2030. [4, 5] Currently, the government can use its budget and tender procedures to create meaningful incentives for construction companies. For example, government grants may be offered to assist companies in the development of technologies that facilitate the development and implementation of projects. National and regional competitions can provide "new builders" with additional financial support, as well as publicly recognize the importance of technologies used to accelerate and reduce construction costs. Similarly, the government may consider creating incubators that will be used to test new low-risk applications. In addition, government contractors may insist that successful bidders use digital collaboration tools in government projects. In addition to creating real incentives for the implementation process, the government should help reduce barriers and risks inherent only in these new technologies. For example, procurement provisions often place a great emphasis on the past performance of the contractor when choosing a source of funding. However, contractors who want to try out new technologies will not have as much experience or proven examples as those who offer traditional solutions. If this is seen as a serious flaw, it may make it difficult to use public procurement processes to stimulate the implementation of new technologies. Rethinking these guidelines in order to take into account new technologies can be crucial for accelerated implementation. At the same time, the government should consider accepting some of the contractor's risks associated with testing new digital technologies or attracting additional resources in the form of experts from construction institutes (faculties) who can analyze risks of a construction company. For example, in individual projects, the subject of the federation (represented by the government) should offer contractors to reimburse costs if new technologies do not bring expected savings. Such guarantees may seem bold, but they can be successful if they focus on the target components of the project, phases or solutions with a significant potential for long-term savings. Measures can also be taken to increase transparency of costs and implement government projects. This transparency is supported by digital technologies that provide information on the progress of large projects in real time. In turn, increased transparency creates pressure to complete projects on time, which becomes easier when new digital technologies are implemented. It is necessary to create more affordable digital platforms that will track costs and schedules throughout the project implementation period, as well as the possibility for continuous monitoring by customers. In particular, the system monitors construction projects and evaluates the effectiveness and conditions of the agreement concluded by the parties. These indicators can be presented on online panels that allow the public and other interested parties to monitor the implementation of the project in a region of the Russian Federation. As in most industries, the construction sector is struggling to find talents for the effective use of new digital technologies. Currently, it is impossible to get rid of the paper version. Russian companies cannot do this, because, according to current standards, most of the workflow in the construction industry should be on paper. These are journals of works and executive documents. Builders refuse any thoughts about automation, otherwise documents are filled in electronic and paper forms. It is easier for interested consumers to solve the problem by calling a right person. It is also worth highlighting one more problem - stereotypical thinking. In IT, this is manifested most clearly. Many managers are opposed to any project data being uploaded to external servers and going beyond the office and employees' computers. In this case, being legal entities they enter into contractual relationships and guarantee data security. All the files and data uploaded to the digital platforms are stored on Russian servers. In Irkutsk region, there is a regional designer using advanced BIM technologies.

These technologies save 10-30% of financial resources at the stage of project implementation, reduce the time required to commission facilities by 25%, save up to 50% of the costs of operating facilities, allow full control of all stages of construction, provide reliable data on any object, identify violations in the construction technology, organize full financial and legal control for each object, make it possible to view objects under construction in a 3-D model. The company designs social facilities in the region (schools, hospitals, kindergartens, fitness centers, feldsher-obstetric centers, etc.). Thus, in Russia the construction industry has a good potential for digitalization. Despite many adverse factors, a significant number of the most progressive construction companies are implementing new technologies, seeing their high potential and efficiency. Moreover (the example of "Mobile Construction Solutions" confirms this), there are developers who are able to develop world-class software that proves quality of their products with a high implementation effect. Therefore, federal measures aimed to "legalize" information modeling technologies have a great success. According to the government plans, the Russian construction industry will enter the digital age in the next two years. Transparency in the construction market (digitizing all stages of construction in a single information system) allows you to control a construction project at all the stages. Higher quality of the finished product of the construction market reduces the number of construction defects. Better maintenance of buildings reduces their wear. The renovation of buildings becomes cheaper, as the 5D, 6D-, 7D-models contain all the information on replacing worn-out parts.

3. Conclusion

Trillions of dollars are spent on building skyscrapers, highways, pipelines, schools and other structures, and the resources that could be saved with advanced analytics, automation, machine learning and other technologies are staggering. Being the main investor, the supplier of infrastructure projects and the customer of national economies, the government has an incentive to accelerate decision-making in the construction industry. New quality and competitiveness depend on breakthrough scientific and technological development, permanent technological renewal, which will be beneficial to business and ensure the growth of national welfare. For this, it is not enough to intensify innovative activities or implement universal standard solutions.

References

[1] Lane N 1999 Advancing the Digital Economy into the 21st Century. *Information Systems Frontiers* **1:3** pp 317-320

[2] Belostotsky A 2017 Contemporary problems of numerical modelling of unique structures and buildings *International Journal for Computational Civil and Structural Engineering* **13** pp 9-34

[3] Klymchuk M and Chernyshev D 2018 The organization of biosphere compatibility construction: justification of the predictors of building development and the implementation prospect *International Journal of Engineering & Technology* **7 (3.2)** pp 584-586

[4] Kupriyanovsky V and Namit D 2016 Digital economy - Smart way to work *International Journal of Open Information Technologies* **2** pp 26-33

[5] Yudina T 2016 Understanding the digital economy *Theoretical Economics* **3** pp 12-16

[6] Kuzior A 2007 Internet jako narzędzie budowania świadomości zrównoważonego rozwoju *Problemy Ekorozwoju* **2** pp 95-101

[7] Kwilinski A, Dalevska N, Kravchenko S, Hroznyi I, and Kovalenko I 2019 Formation of the entrepreneurship model of e-business in the context of the introduction of information and communication technologies *Journal of Entrepreneurship Education* **22** pp 1-7

[8] Czernich N, Falck O, Kretschmer T and Woessmann L 2009 Broadband Infrastructure and Economic Growth *CESIFO Working Paper* **2861**

[9] Adamtsevich A, Ginzburg A, Shilova L and Shilov L 2016 Implementation of BIM-technologies in Russian construction industry according to the international experience *Journal of Applied Engineering Science* **14 (4)** pp 457-460

[10] Sharmanov V V, Simankina T L and Mamaev A E 2017 BIM in the assessment of labor protection *Magazine of Civil Engineering* **69(1)** pp 77–88

[11] Hugo Consciência Silvestre, Rui Cunha Marques, Ricardo Corrêa Gomes 2018 Joined-up Government of utilities: a meta-review on a public–public partnership and inter-municipal cooperation in the water and wastewater industries *Public Management Review* **20:4** pp 607-631

[12] Gendarz P, Rabsztyn D 2014 Finite element based optimization in the process of selecting the geometric design features constructions *International Journal of Modern Manufacturing Technologies* ISSN 2067-3604 **VI** pp 15-22

[13] Popov E V, Semjachkov K A, Simonova V L 2016 Evaluation of the influence of information and communication technologies on innovation activity of regions *Finance and credit* **46** pp 46-60

[14] Stefanova N A, Sedova A P 2017 Model of digital economy *Karelian scientific journal.* **6** pp 91-93

[15] Popov E V, Semjachkov K A and Simonova V L 2016 Evaluation of the influence of information and communication technologies on innovation activity of regions *Finance and credit* **46** pp 46-60

IOP Conf. Series: Materials Science and Engineering **880** (2020) 012084 doi:10.1088/1757-899X/880/1/012084

Development of a system for monitoring environmental parameters of cities and open areas

R Belousov[1], A Nikolaev[2], E Fiskin[2] and M Fiskina[2]

[1]LLC "Scientific and Technical Center" KUMIR ", 83a Lermontov street, office 201, Irkutsk, 664074, Russia
[2]Irkutsk National Research Technical University, 83 Lermontov street, Irkutsk, 664074, Russia

E-mail: belousov@ntckumir.ru

Abstract. This article is devoted to the development of a system for monitoring the state of the ecological environment. The system under development should provide ongoing monitoring of environmental air parameters, both in open areas and in the city, and is designed to timely detect excess concentrations of hazardous substances in the air. A distinctive feature of the created system is high productivity at low economic costs per accounting point. An appropriate selection of the component base and software optimization of the operation algorithms will allow to achieve the maximum data transmission distance over a distance of up to 100 km in an open area. Interim results on the development of a monitoring system are presented, planned tasks for this year are indicated.

Today, the environmental situation in many cities and even nearby open areas is subject to increased pollution due to harmful emissions into the atmosphere by factories and factories, automobiles and other equipment that uses petroleum products for work.

The most dangerous substance resulting from human activities is formaldehyde. The main sources of formaldehyde in the outside air are motor vehicles, chemical plants, wood processing plants, waste incinerators, and it is also contained in tobacco smoke. This gas is colorless with a pungent odor, soluble in water, in our country it is assigned the first hazard class. As a result of many years of research on formaldehyde, it is possible to compile a table with the reaction of the human body to various concentrations of this substance in the air.

Table 1. Reaction of the human body to the concentration of formaldehyde.

Formaldehyde concentration, mg / m3	Reaction of the human body
0,03	10% of the subjects felt the smell
0,04	starts to decrease expiratory flow rate in children (bronchospasm begins - a linear relationship according to the WHO Research)

0,05	change in the EEG electroencephalogram (brain waves become unnatural) - this concentration is accepted as the maximum permissible in Russia
0,07	in such homes there is an increase in the number of diseases in children with asthma and chronic bronchitis (WHO statistics)
0,1	irritation of the mucous membranes begins (generally accepted concentration at which the smell is felt)
0,2	50% of the subjects felt the smell
0,6	90% of the subjects felt the smell
0,7	all subjects felt eye irritation
1	30% of subjects experienced breathing difficulties (maximum permissible concentration in the air)
2	as a result of a single exposure - eye and nose irritation, headache, nausea for the next 24 hours
4	difficulty breathing, severe lacrimation
40	pulmonary edema, pneumonia, danger to life
60	Lethal concentration

To date, there are several systems for environmental monitoring, in European countries and the United States is actively using the "Worldwide Network of Automated Monitors" uRADMonitor.
This system allows real-time monitoring of environmental parameters such as temperature, pressure, humidity, tracking volatile organic substances, noise level in the city, formaldehyde, radiation level, and the content of ultrafine particles.
This manufacturer offers several options for device models for organizing a monitoring system. Models of monitoring stations are summarized in table 2.

Table 2. Comparison of models of monitoring stations.

Model	A	KIT1	D	A3
Used sensors	Dallas DS18B20 (Temperature) SBM20 (γ, X-rays)	SBM20 (γ, , X-rays) Hackable (For an additional sensor)	Bosch BME680 (Temperature, Pressure, Humidity, Volatile Organic Compounds) GP2Y1010AU0F (Ultrafine particles PM 2.5) LND LND712 (α, β, γ, X-rays)	Bosch BME280 (Temperature, Pressure, Humidity) Winsen ZH03A (Ultrafine particles PM1.0, PM2.5, PM10) Winsen ZE25–O3 (Ozone) Winsen MH-Z19B

				(Carbon dioxide) Winsen MP503 (Volatile organic matter) SPU414/MAX4466 (Noise level)
Case size, mm	110×80×24	110×60×12	110×70×24	110×80×24
Supply voltage	External 5V–9V	External 6V–28V	Built-in battery 1500mAh	External 6V–28V
Battery life, years	–	–	–	–
Operating frequency, MHz	No built-in modem (Ethernet data transfer)	No built-in modem (Ethernet data transfer)	Wi-Fi data transfer	Data transmission via Wi-Fi, Ethernet, GSM, LoRaWAN
Cost, rub.	21.987	9.387	37.737	34.587

The disadvantages of these systems include the economic factor, focus on an external power source, which gives its limitations when monitoring in open areas, the ability to transmit data over the air only with the premium model, the closed device architecture without the possibility of replacing sensitive elements, or adding new ones.

Thus, in order to solve existing problems, it was decided to develop our own monitoring system for the water area of Lake Baikal, consisting of sensors for analyzing hydrophysical and meteorological characteristics, a radio modem, a concentrator, and a GSM module.

To constantly monitor the environmental situation and prevent the increase in the concentration of hazardous substances in the air, we have developed a special monitoring system. This system includes a sensitive element (sensor), a radio modem, a hub and a server for storing received data. The data transmission scheme is presented in Figure 1.

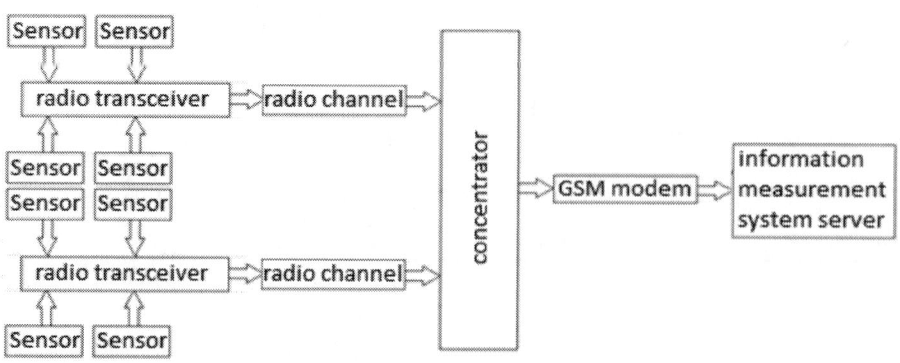

Figure 1. The transmission scheme of the measured data.

As a sensitive element, we use the air quality sensor ZP07-MP503, which has proven itself in various air quality monitoring systems. Appearance is shown in Figure 2.

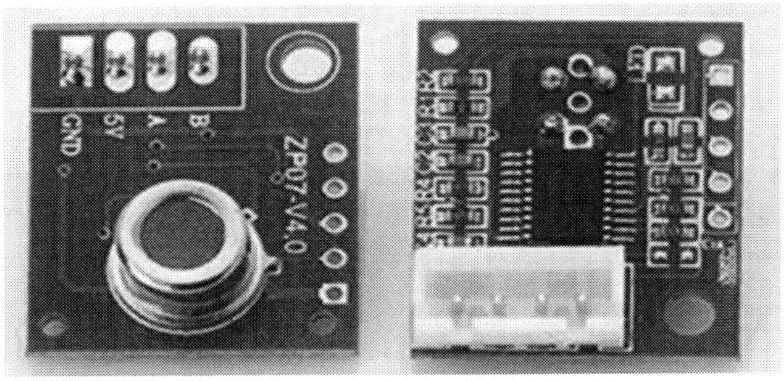

Figure 2. Appearance of the ZP07-MP503 sensor.

This sensor has an output UART data interface, operating current less than 60mA, which will create a system with low power consumption. The used sensor for recording formaldehyde level is semiconductor with a resolution of 10 ppb (0.012458 mg / m3) and an error of ± 5%. Measurement range from 0 ppm to 5 ppm (6.22884 mg / m3).
The use of this sensor will detect an increase in the concentration of formaldehyde in the air at levels that are safe for human health, and high sensitivity will help to determine even a small deviation of the concentration of formaldehyde from the norm. A radio modem is used to collect, process and transmit data. The main characteristics are summarized in table 3.

Table 3. The main characteristics of the prototype radio modem in comparison with the premium model A3.

Parameters	Prototype	A3
Number of channels	RS-232 to 4 RS-485 to 32 CAN to 125	Closed architecture Factory set of sensing elements.
Data transfer protocol	Radio channel, GSM 900/1800 MHz	Optional data transfer: Wi-Fi, Ethernet, LoRaWAN
Transmitter power	radio channel – 100mw	LoRaWAN – 25mw
Distance	To 100 km	n / a
Source of power	Built in: 3V External: 9-12V	External: 6-28V
Cost, rub.	5.000-10.000	34.587

The presence of a built-in GSM transmitter allows you to transfer the received data directly to the information-measuring server without the need to install a hub and additional equipment for data transfer. It is very convenient for use in urban environments. For data transmission in areas with poor coverage of the GSM network or its absence at all, the radio modem has an integrated data transmitter over the radio channel at a frequency of 868 MHz with a data transfer rate of 1.2 Kbps.
To date, communication range tests have been conducted in urban environments. As a result, the range of sustainable data transmission was 6 km. Using the CC1120 transceiver in conjunction with the low-noise power amplifier CC1190 will theoretically allow data to be transmitted in an open area over a distance of about 100 km. An experimental confirmation of the data transmission range in the open area

IOP Conf. Series: Materials Science and Engineering **880** (2020) 012084 doi:10.1088/1757-899X/880/1/012084

is planned in the near future. Thanks to the upgraded power supply system, the radio modem can operate independently, from the built-in power supply, and from an external voltage of 9-12V.

This radio modem supports several data exchange interfaces, which allows you to work with a wide range of sensors designed to monitor the environmental situation of the environment, hydrophysical characteristics of water areas and other applications. The connection diagram for several sensors is shown in figure 3.

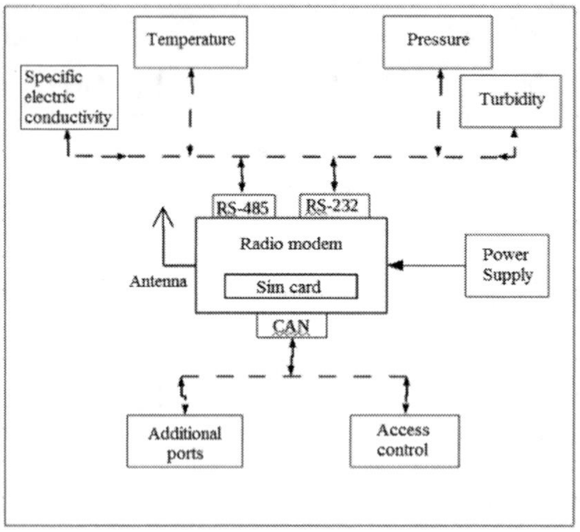

Figure 3. Diagram of connecting different types of sensors to a radio modem.

A short list of supported sensors is given below:

1) CO_2 Gas Sensor
2) CO sensor
3) Flammability sensor
4) Dust Sensor
5) H_2 Sensor
6) Toxic Gas Sensor
7) Air Quality Sensor (VOC)
8) Alcohol Sensor
9) Humidity Sensor
10) Gas Flow Sensor
11) NH_3 Gas Sensor
12) CH_2O Gas sensor
13) O_2 sensor
14) O_3 Gas Sensor
15) SO_2 Gas Sensor
16) H_2S Sensor
17) C_2H_4 Sensor
18) PIR Sensor
19) MEMS Gas Sensor
20) Pressure Sensor

If it is necessary to change the measured air parameter, this radio modem has an open architecture and at any time it is possible to replace one sensor with another without the need for parsing or soldering.

I would like to note that for the work in the conditions of the city of Irkutsk, the entire infrastructure has been prepared, consisting of hubs, GSM modems and an information-measuring system server, which will allow to deploy an environmental monitoring system for air quality in the city in the shortest possible time and at minimal cost.

Thus, this system is applicable for use both in urban conditions and in open areas for the current assessment of the ecological environment.

References

[1] Sizykh M S, Nikolaev A A and Pinkin A A 2016 Radio modem for apartment-based resource accounting: *Materials of the All-Russian scientific-practical conf. "Improving the efficiency of production and use of energy in the conditions of Siberia" ISTU* **2** pp 450-453

[2] Belousov R A, Pinkin A A, Fiskin E M and Fiskina M M 2015 The possibilities of "Kumir-Resource" system for SMART GRID and apartment level energy accounting *EURO-ECO 2015. Das Internationale Symp. «Ökologische, Technologische und Rechtliche Aspekte der Lebensversorgung»* pp 15-16

[3] Belousov R A, Pinkin A A, Fiskin E M and Fiskina M M 2014 Data collection system for SMART GRID *EURO-ECO 2014. Das Internationale Symp. «Ökologische, Technologische und Rechtliche Aspekte der Lebensversorgung»* pp 28-29

[4] Belousov R, Fiskin E and Fiskina M 2018 Integrated Systems for Commercial and Technological Control of Energy Resources Consumption in Urban Areas and Businesses in Russia *E3S Web of Conf. Scopus* **39**

[5] CC1120 High-Performance RF Transceiver for Narrowband Systems [Electronic resource] Article: [site]. [2020]. URL: http://www.ti.com/lit/ds/symlink/cc1120.pdf

[6] CC112x-CC1190 BoosterPack [Electronic resource] Article: [site]. [2018]. URL: http://www.ti.com/lit/an/swra492/swra492.pdf

[7] Matched Integrated Passive Component for 868 / 915 MHz operation with the CC112x, CC117x & CC12xx high performance radio series [Electronic resource] Article: [site]. [2020]. URL: http://www.ti.com/lit/an/swra407/swra407.pdf (date of the application 28.01.2020).

[8] 850 – 950 MHz RF Front End [Electronic resource] Article: [site]. [2020]. URL: http://www.ti.com/lit/ds/symlink/cc1190.pdf

[9] Usmanova L M, Safarova V I 2015 Sources of Formaldehyde and its Effect on the Living Organism: Scientific community of students of the XXI century. *NATURAL SCIENCES: Electronic collection of articles based on the materials of the XXXI student international scientific and practical conf.* - Novosibirsk: Ed. "Sibak" **5 (30)** [Electronic resource] - access Mode. - URL: http://www.sibac.info/archive/nature/5 (30).pdf

[10] Nikolaev A, Belousov R, Fiskin E and Fiskina M 2019 Development of equipment for monitoring the lake Baikal water area *E3S Web of Conferences Scopus* **102**

[11] Nikolaev A A, Fiskina M M and Melnikov M S 2019 Creating a compact radio modem with ultra-low power consumption *Materials of the all-Russian scientific and practical conf. with international participation " Improving the efficiency of energy production and use in Siberia* **2** pp 408-411

[12] Dorogova V B, Taranenko N A and Rychagova O A Formaldehyde in the environment and its effect on the body.: *Angarsk branch of East Siberian scientific center ORAMN ECH SB RAMS research Institute of occupational medicine and human ecology in 2010.* [Electronic resource] - Access Mode. - URL: https://cyberleninka.ru/article/n/formaldegid-v-okruzhayuschey-srede-i-ego-vliyanie-na-organizm-obzor/viewer.

IOP Conf. Series: Materials Science and Engineering **880** (2020) 012085 doi:10.1088/1757-899X/880/1/012085

Urban reorganization and development of Baikal macro-region

A Kolesnikov[1]

[1] Russian Institute of urban planning and investment development "Giprogor", 119311, Moscow, a/ya 93, "Giprogor"

E-mail: Arka-92@mail.ru

Abstract. One of the most important areas of spatial development of Russia is the innovative development of the territories of the constituent entities of the Russian Federation, which is reflected in the main documents of strategic planning - the forecast of socio-economic development of the Russian Federation for the period until 2030 [1]; Strategies for the innovative development of the Russian Federation for the period until 2020 [2]; Federal Law "On Territories of Advancement of Socio-Economic Development in the Russian Federation" [3].With the formation of the Baikal macro-region, which includes the Irkutsk region, the Republic of Buryatia and the Trans-Baikal Territory, the role of the natural territory of Lake Baikal becomes the cornerstone. Without understanding the resettlement tasks in this area of Lake Baikal, it is impossible to implement the package of measures provided for by the Law on the Protection of Lake Baikal of May 1, 1992 No. 94-ФЗ [4] and the Federal Target Program "Protection of Lake Baikal and the Socio-Economic Development of the Baikal Natural Territory for 2012-2020 " [5]. Determination of the basic principles of the organization of urban planning systems in the Baikal and Trans-Baikal regions, and, above all, in their systemic core, the Baikal Natural Territory, is an extremely urgent urban planning problem..

1. Introduction

The purpose of the work is the justification of an integrated approach to the study, planning and forecasting of the Baikal natural territory, as the most important component of the three constituent entities of the Russian Federation.

Issues related to the deployment of the country's productive forces and the provision of a sound spatial development strategy by 2018 have come to the fore in government resettlement programs in the Russian Federation. The concept of the country's spatial development strategy up to 2035 with the prospect up to 2050 was adopted [6]. According to the proposals of the Ministry of Economic Development, Russia is divided into 14 macro-regions. Each of them reflects and even contains in the names the essence of its territorial and, possibly, resettlement aspect.

20 years ago, academician of the RAABS I.M.Smolyar noted that in the constitution of the Russian Federation, urban planning is not included in the list of subjects of competence at the federal level of state power, as well as in the list of subjects of joint jurisdiction of the federal government and subjects of the Federation [7].

Today, the situation in the resettlement of the country at the state level is recognized and, finally, an important document on the spatial development strategy of Russia has been adopted.

Irkutsk region, the Republic of Buryatia and the Trans-Baikal Territory became part of the sub-territorial block - the Baikal macro-region, united by proximity to Lake Baikal, and have a special specificity of socio economic and ethno-cultural development and mutual influence. The regions included in the Baikal macro-region have all the signs of landscape-ecological and economic-cultural integrity, which allows us to consider the existing formation as a conditionally uniform territorial entity.

The factors uniting the territory:

— The presence of significant natural resource potential;
— General transport and energy infrastructure;
— Significant environmental requirements and restrictions on doing business in the zone of influence on the ecosystems of Lake Baikal;
— Proximity to the rapidly developing countries of East Asia (Mongolia, China, Korea) - the main promising sales markets, as well as future sources of large investments.

Picture 1. Baikal macro-region

2. Subjects of the Russian Federation in the structure of the Baikal macro- region

The unification of the Irkutsk region, the Republic of Buryatia and the Trans-Baikal Territory into a single territorial community, the Baikal macro-region is justified on the one hand by the necessity and possibility of joint participation of subjects in solving the national task: the revival of the industrial and economic complex in the east of the country, since further east and north only separate, mostly sectoral, industrial units are located. Therefore, the subjects of the Baikal region rightfully acted and are a powerful base for the development of the Far Eastern and northern regions of Russia.

On the other hand, programs and joint actions of state and regional institutions are required to overcome [8]:

— territorial disunity and the elimination of infrastructure "gaps";
— inconsistencies in the preparation of sectoral territorial planning schemes for individual entities and their parts (agglomerations) among themselves;
— unevenness and low density of settlement;
— a "peripheral" approach to the problems of protection and rehabilitation of the coastal territories of Baikal and the lake as a whole in the territorial planning schemes of the Irkutsk region, Trans-Baikal Territory and the Republic of Buryatia, that is, important strategic documents for the development of the constituent entities of the Russian Federation.

The materials of territorial planning developed in modern conditions for all three subjects of the Baikal macro-region are distinguished by the presence of common shortcomings:

An "intraregional" approach to the development of territorial planning schemes, underfunding of measures laid down in federal and regional target programs to improve the socio-economic and natural-ecological situation of the Baikal natural territory, inefficiency and uncoordinated measures, aggravate the unfavorable situation in each of the constituent entities of the Russian Federation individually.

STP of the Irkutsk region is an illustration of a "peripheral" approach to resolving issues related to the rehabilitation and development of the coastal territories of Lake Baikal. The administrative boundaries of the STP of a constituent entity of the Russian Federation limit the scope of consideration, exclude the necessary accounting for the influence of Lake Baikal and neighboring constituent entities of the Russian Federation on the urban planning system as a whole [9].

Uncoordinated activities exacerbate the adverse situation in each of the constituent entities of the Russian Federation individually [9].

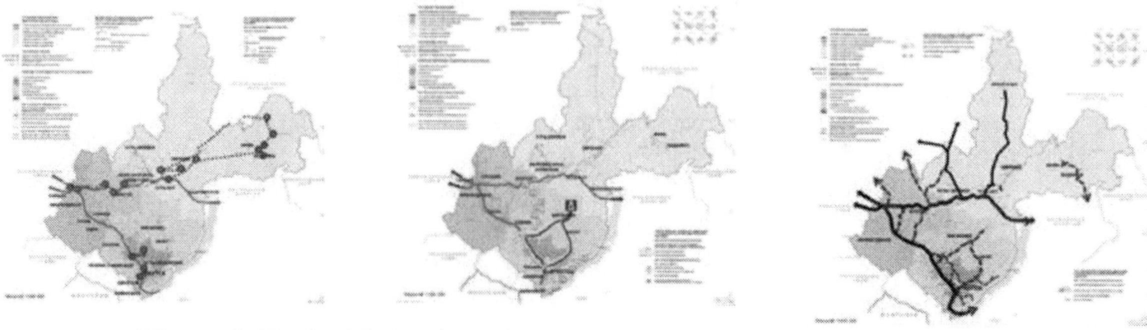

Picture 2. Territorial planning scheme of the Irkutsk region. GIPROGOR 2018

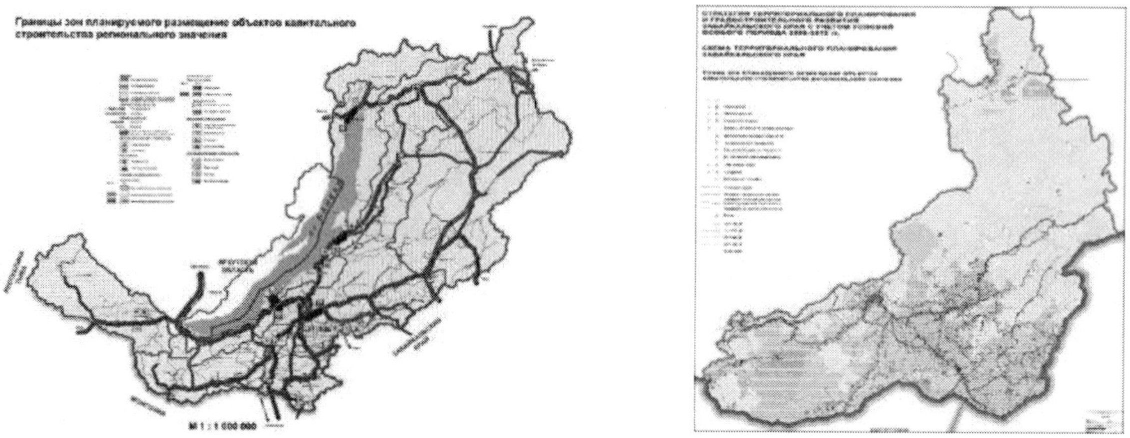

Picture 3-4. Territorial planning scheme of Republic of Buryatia and Zabaykalsky Krai. GIPROGOR 2018

3. Baikal natural territory as a whole object of research and territorial planning

In the framework of the current Urban Code of the Russian Federation, the development of planning strategic documentation of an intermediate nature is a territorial planning scheme for a part of the subject of the federation.

This regulation is used by Russian urban planners - designers in the development of large urban agglomerations. In our case, the existing territorial planning schemes of the Irkutsk, Chita and Ulan-Ude city agglomerations are considered.

Specialists - urban planners confirm the need to develop intermediate, transitional planning documents for the development of urban planning systems from the territorial planning scheme - the upper level of consideration, to the lower level - the general plan of the settlement and, further, to the planning projects of the territories of settlements.

The design of the Baikal natural territory on the basis of the Law on the Protection of Lake Baikal, as an independent object of socio-economic regulation, is necessary for implementing the provisions of the Federal Target Program "Protection of Lake Baikal and the program for the socio-economic development of the Baikal natural territory for 2012-2020."

The design of the Baikal natural territory on the basis of the Law on the Protection of Lake Baikal, as an independent object of socio-economic regulation, is fixed by the provisions of the Federal Target

Program "Protection of Lake Baikal and the Socio-Economic Development of the Baikal Natural Territory for 2012-2020".

This suggests, in our case, the formulation of urban planning, settlement approaches as a self-sufficient object of study. Only on the shores of Lake Baikal live more than 200,000 people. The volume of negative impact on the Baikal natural territory per year (2016-2018) by the subjects of the Russian Federation was characterized by the following indicators:

Irkutsk region: Emissions - 391.5 thousand tons Discharges -30.1 mln.m3 Waste - 275.7 thousand tons The Republic of Buryatia: Emissions - 119.6 thousand tons Discharges -495.4 million m3 Waste - 11712 thousand tons Transbaikal region: Emissions - 10.9 thousand tons Discharges - 1.3 million m3 Waste - 366.6 thousand tons

The growth rates of these indicators are growing and, according to expert estimates, average 15-17% per year, both in terms of emissions and discharges, and 4.5% - waste generation [10].

At the same time, it becomes obvious that the development of individual subjective measures with the existing methodology for their coordination is doomed to receive a low-effective or even negative result.

The Baikal natural territory is considered as an integral and independent object of research and territorial planning.

Three zones of the Baikal natural territory form three types of urban development approaches.

Central zone:

- Environmental measures;
- Stabilization of resettlement and improving the quality of socio-economic and environmental living conditions of the population: transport links, jobs (fish farming, fishing, tourism and recreation).
- Atmospheric impact zone: Introduction and development of all types of modern technologies in production, excluding harmful emissions into the atmosphere;
- Transition to gas supply and gasification of public and personal vehicles.

Buffer Zone: The development of technologies and production in industry and agriculture sharply reducing the discharge of harmful substances.

Earlier, experts noted the need to take into account the processes of transformations and changes in trans-boundary territories and adjacent foreign ones.

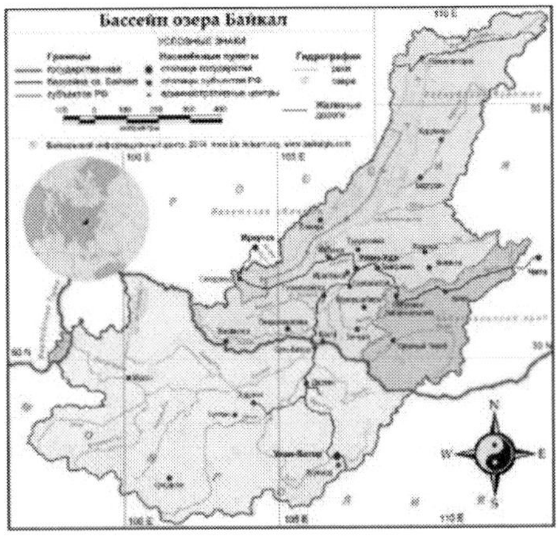

Picture 5. Territory of Mongolia in the zone of influence on the Baikal Natural Territory

The drains of the river systems of the border regions of the Mongolian People's Republic are of decisive importance for the further formation of the buffer zone of Lake Baikal - 70% of the length of the Selenga River falls on the territory of Mongolia and 50% of the volume of water income of Lake Baikal.

The river remains the main supplier of controlled substances to Lake Baikal. Through the "closing gate" of the Selenga, 87.6% of suspended solids, difficult-oxidizable and easily-oxidizable organic substances come in each year - 78.0% of the total amount of these substances with water. The main source of air pollution are energy and motor transport enterprises both in Mongolia and in Russia, as well as Selenginsky Pulp and Cardboard Mill located in the immediate vicinity of Lake Baikal.

4. Infrastructural background of rehabilitation and development of the Baikal natural territory

The Ministry of Transport of the Russian Federation today formulated its priorities: "The implementation of the Strategy will ensure the rapid development of the transport network in the territories of the new development of the Far East and the Baikal region, including in the north-eastern part of the Far East" [1]. This position is shared by other departments and state corporations that have their own "sectoral" and targeted programs, oriented, as a rule, to profit.

Academician S.D. Valentai notes that current economic problems are caused by the fallacy of the policy of minimizing state participation in the technological renewal of domestic industry; in regulating the system of labor relations; in training in accordance with the challenges of the XXI century.

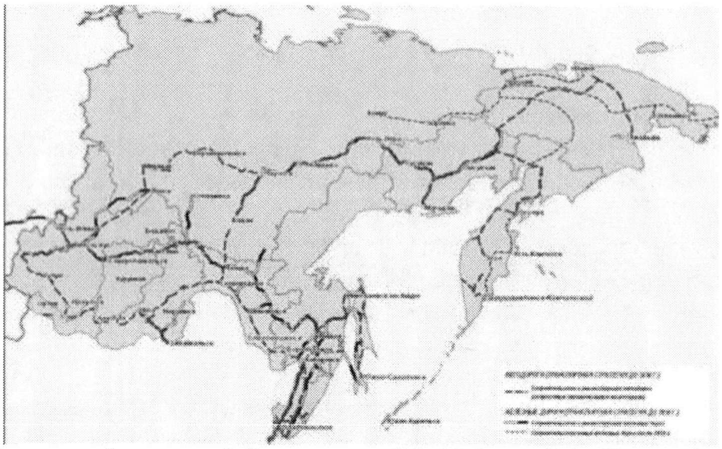

Picture 6. Development of transport infrastructure of the Baku and Far East regions (scheme of the Ministry of Transport of the Russian Federation)

Leading researcher at the Central Scientific Research Institute of Industrial Property of the RF Ministry of Construction E.O. Tovmasyan continues this idea, and in the urban planning aspect notes that improving the socio-economic status of society based on a nationwide policy is impossible without active state regulation of the spatial features of this process. Otherwise, in the development of the transport infrastructure of the Baikal macro-region, in general, and the Baikal natural territory in particular, the following shortcomings arise:

Priority development of the north-east and far east directions;

The "transit" role of the territories of the Baikal macro-region;

Until 2035, targeted programs of the Russian Federation do not provide for the development of high-speed lines and SM in the Baikal macro-region;

The prospects of "vertical" entrances to the regions of Mongolia and China were not taken into account. An almost critical situation has developed in the electrification of the territories adjacent to and affecting the ecosystem of Lake Baikal.

IOP Conf. Series: Materials Science and Engineering **880** (2020) 012085 doi:10.1088/1757-899X/880/1/012085

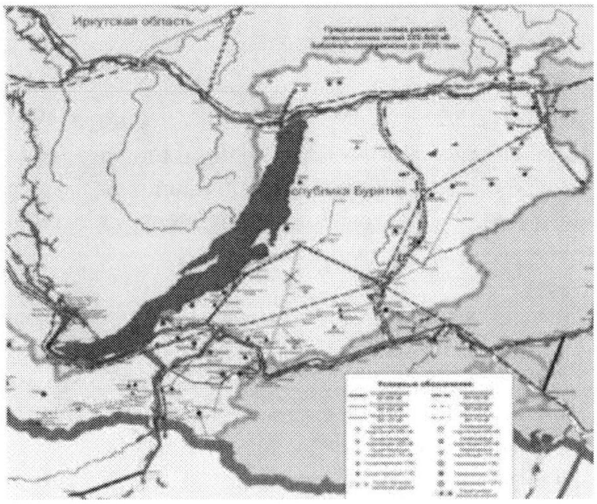

Picture 7. Proposed scheme for the development of electric networks of the Transbaikal region

The backbone of the main directions of electric networks, the orientation toward the export of energy resources lead to a shortage of electricity in the macro-region, the underdevelopment of the general energy supply network;

The distortions in pricing in the industry and the monopolism of state corporations in foreign markets led to the idea of developing in Mongolia the concept of creating a hydropower complex in the buffer territory of Lake Baikal, consisting of three hydroelectric stations on the Selenga River and its two tributaries. According to preliminary estimates, the total volume of export of energy resources is 5-6 times higher than the region's own energy consumption.

The industry's lack of interest in the relevant ministries and the Gazprom corporation in the urgent need for gasification of the Baikal natural territory, the insufficient volume and number of consumers, in their opinion, make gasification of the Baikal and Trans-Baikal areas unprofitable until 2035. At the same time, the main sources of energy (TPPs and large boiler houses operate on brown coal and fuel oil).

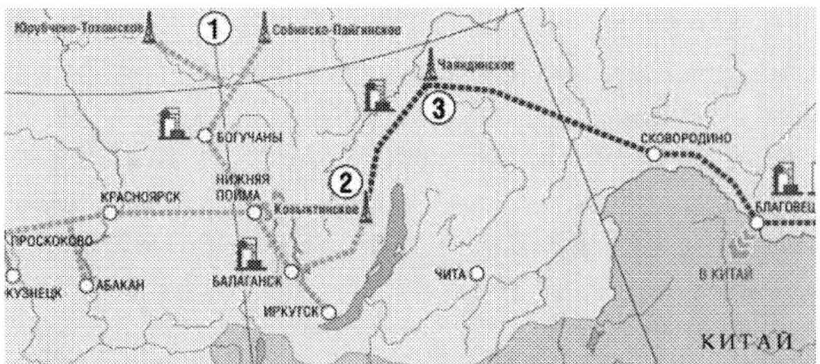

Picture 8. The proposed gas supply scheme in the Baikal macro-region (Power of Siberia)

There was a problem of the domestic price of local (Russian) raw materials and energy systems 30-50% higher than world prices with the same understatement when moving abroad, which hinders the development of domestic production and is the greatest economic diversion against our country in the interests of geopolitical competitors.

It is the unified state strategy of own development and processing of resources of the Baikal region that should be of primary geopolitical importance.

IOP Conf. Series: Materials Science and Engineering **880** (2020) 012085 doi:10.1088/1757-899X/880/1/012085

5. Conclusion

As a result, we can reasonably suggest that the Baikal natural territory is endowed with all the signs of a holistic object of study of territorial planning and urban planning. Lake Baikal is the core of a complex of urban planning prerequisites for the development, improvement and regulation of the settlement system.

References

[1] *Forecast of the long-term socio-economic development of the Russian Federation for the period until 2030* [Electronic resource]: developed by the Ministry of Economic Development of the Russian Federation-URL: http // base.garant.ru (accessed: 11/21/2018)

[2] *The strategy of innovative development of the Russian Federation for the period until 2020* [Electronic resource]: Order of the Government of the Russian Federation of 08.12.2011 N 2227-p. On the Strategy for Innovative Development of the Russian Federation for the period until 2020 - URL: http: consultant.ru/document/cons_doc_LAW_123444/ (Date of access 03.19.2018)

[3] Decree of the President of the Russian Federation "On national goals and strategic objectives of the development of the Russian Federation for the period until 2024 of 05/07/2018.

[4] Law on the Protection of Lake Baikal of May 1, 1999 N94-FL

[5] Federal Target Program "Protection of Lake Baikal and Socio-Economic Development of the Baikal Natural Territory for 2012–2020

[6] The concept of the Strategy for spatial development until 2035.

[7] *Town Planning Code of the Russian Federation* [Electronic resource]: Federal Law of December 29, 2004 No. 190-ФЗ (as amended of July 3, 2016) (as amended and supplemented, entered into force on September 1, 2016): Adopted by the State Duma on December 22. 2004 Approved by the Federation Council on December 24, 2004. - URL: http://consultant.ru/document/cons_doc_LAW_51040/ (accessed: 08/20/2017)

[8] Vilner M Ya 2018 On the aspects of spatial development in the methodological basis for the preparation of a long-term forecast for the socio-economic development of the Russian Federation *Management of the development of the territory (Anniversary issue)* **1** pp 29-32

[9] The list of activities prohibited in the central economic zone of the Baikal Natural Territory (approved by Decree of the Government of the Russian Federation of August 30, 2001 No. 649)

[10] Analytical review of environmental and natural topics: Pulse Radio LLC; Report on the state of the determining environment of the Baikal basin. 2015 p 136

IOP Conf. Series: Materials Science and Engineering **880** (2020) 012086 doi:10.1088/1757-899X/880/1/012086

Development of technologies for improving the water resources of contaminated areas using plants

E Vasilevich[1]

[1]Department of Engineering Communications and Life Support Systems, IRNITU, Lermontova 83, Irkutsk, Russian Federation

E-mail: elvira.vasilevich@yandex.ru

Abstract. The article provides an overview of relevant phytotechnologies for improving water resources and wastewater, and indicates the regulatory framework for the best available technologies. Attention is paid to the significance of the natural processes of biodegradation of water and wastewater pollution, the most studied plants and trees used for the decontamination of pollution in water and soils are highlighted. The emphasis is placed on the correct selection of plants for the effect of reducing the toxic effects of individual or total pollutants, as well as phytostimulation of the local soil microflora by plants to intensify the processes of self-cleaning of polluted environments. The experience of studying the work of plants in various climatic conditions, including sharp continental ones, helps to expand the capabilities of this method and solve problems of improving reservoirs and territories at a minimum cost.

The development of modern industries invariably leads to an increase in the amount of waste, emissions and pollution of the water used, respectively, to an increase in the load on soil, air and water resources. Even with the wisest natural legislation and law-abiding producers and the population, this burden is present. From year to year, pollution becomes more complex in composition and accumulate in the environment until their destruction occurs. Separate complex substances remain in the natural environment for many decades.

Technology purification from contaminants is constantly evolving, summarized, and as a result of cooperation of experts can work at the international level. Due in large part to the European Bureau on integrated prevention and pollution control-European Integrated Pollution Prevention And Control Bureau (EIPPCB), which over 20 years accumulating the experience of professionals of Europe to prevent pollution and manage the production of BREFs, Best available techniques Reference document— (BAT) [1,2].

In 2012, the Russian Federation also came standard on the application of BAT. [3]. Since 2015, 50 BAT reference books have been published in Russia for various industries.

Special attention in the treatment of contaminated water currently being paid to bio - and phytotechnologies in which to maximize the use of natural mechanisms of destruction of contaminants. They are also reflected in the reference document «Information and technical reference on the best available technologies for wastewater treatment using centralized water disposal systems in settlements and urban districts» [4]. The document identifies the most frequently used plants and technologies and types of systems for purification (e.g.,type of bio plateau). Russian scientists actively study and apply various methods of rehabilitation of water bodies and sewage. The choice of health measures depends on the correct estimation of the geoecological state of the hydraulic, features of hydraulic zoning, water and hydrochemical regime and the degree of contamination. A more accurate picture of the state of the reservoir can give the geoecological characteristic of bottom sediments. [5].

Climate characteristics of the region when you need to use the outdoor area sewage treatment plants or surface waters.

For example, "application Eichhornia is possible only at water temperatures above 20 °C for the Central part of Russia restricts the period of use is 2-3 months, for the South of Russia — 4 months. In the Northern and North-Western regions, the process is not applicable. To start the culture Eichhornia part of the plant you want to save in the winter while maintaining the required temperature, light and nutrition "[4].

In addition Eichhornia crassipes for FOS (photoacustic systems) we recommend using bulrush, cattail, reeds and other hygrophilous plants. It is known that the reed lake absorbs water from phenols. One plant cane weight of 100 g capable of removing from water to 4 mg of phenol. In addition to the absorbed phenol and its derivatives (catechol, resorcinol, xylene, etc.) [6].

Degradation of organic compounds in photosysteme, both in aerobic and anaerobic conditions, is actively carried out by bacteria. The microorganisms used for the development of underground or submerged parts of plants and the surface of the mineral load. Oxygen as the most important pokazateli comes from the atmosphere through diffusion from the roots and rhizomes of plants. The use of artificial aeration, in comparison with natural, certainly, accelerates the process but is more costly (requires the use of a blower pump).

It is very important that every higher plant root system forms a kind of community of bacteria, feeding him enzymes, oxygen and bio-available organic compounds. This natural mechanism is used in the generated phytosystems. Also used the ability of plants to consume the nitrogen and phosphorus.

Nitrogen from the water is removed through nitrification/denitrification and assimilation by plants (during the growth of biomass).

Phosphorus is primarily removed through exchange and sorption reactions on the surface of the mineral load, forming insoluble and slightly soluble salts.

Along with these plants, the researchers scrutinized the use of barley straw (Hordeum vulgare) as an inhibitor in the development of algae phytoplankton. Evidence that barley straw can inhibit algal blooms in freshwater systems, accumulated since the late 1970-ies. These are examples of the processing ponds of all sizes, canals, reservoirs and rivers in the British Isles and the United States. Many of the published works say about the success of such measures to prevent the flowering in places where it was observed regularly. The method was successful for 6 years without any information about the resistance of the algae or a noticeable change in their species composition) [7].

The theoretical basis of the method is that the decomposition of straw into the water allocated substances inhibiting the development of algae – different oxidized phenolic compounds (decomposition products of lignin) and some others. A necessary condition for the application of the method is the free access of the oxygen necessary for the bacterial degradation of plant material [8].

Barley straw is preferable to make in the warmer time of the year immediately before the beginning of mass development of algae, in the amount of 25-400 g/m^3 or 25-50 g/m^2 [6]. These values of the dose of the straw in the calculation of reservoir volume and per unit water area is justified by the fact that intensive "blooming" of water, the depth of the photic zone, where the processes of photosynthesis is about one meter. Both laboratory and field experiments have shown that straw secretes phenolic compounds, which, before and after its decay, inhibit the development of algae

A positive result of barley straw obtained in the harsh continental climatic conditions of Siberia by researchers at the national research Irkutsk state technical University and laboratory of toxicology of Irkutsk University, where he estimated a reduction of toxicants heavy metals in the presence of barley straw. Straw that has undergone preliminary fermentation has shown the greatest effectiveness in reducing the toxicity of the studied substances.

Phytoremediation measures allow you to:

- clean up toxins at the site of contamination (for example, by placing bags of fermented straw directly into the water body to be cleaned);
- Make the process of cleaning affordable in comparison with traditional methods;
- Use the most natural mechanisms;
- Monitor the cleaning process using bioassay methods;
- The degree of splitting of polluting components is comparable to the results of using traditional methods, especially with a small amount of waste water. [10,11]

Financial and technological accessibility of using barley straw to restore the purity of natural water and sewage emphasizes its importance and the possibility of using in different regions, mainly in the warm period of the year.

From the limitation of phytoremediation, the duration of the process should be noted (cleansing a site can take from several agricultural seasons to several years. Thus, for sites with a very high concentration of pollutants, phytoremediation cannot be selected by the recovery method. Some authors note that this technology is better most suitable for remote areas where human contact is limited or where pollution does not require an immediate response.

Significantly affects the physiological activity of plants, their age, especially the roots. It is important to use healthy young plants to remove contaminants more efficiently.

Also, a high concentration of pollutants can inhibit plant growth. The main limitation of the concentration of toxic elements is the maximum level that can accumulate in plants. [12]

Phytodegradation technology proved effective in cases of soil and water contamination with high concentrations of aliphatic, aromatic, and polycyclic hydrocarbons, phenols, pesticides, etc. Usually all plants have the property of splitting toxicants with the help of enzymes. However, the highest phytodegradation characteristics are shown by such annual herbaceous plants as fescue (Festuca L.), horseradish (Armoracia rusticana L.), alfalfa (Medicago sativa L.) and woody plants: oak (Quercus robur L.), poplar (Populus L.), willow(Salix alba L.), cypress (Cuprḗssus L.). By breaking down toxins in the soil, they prevent stormwater runoff from polluting nearby reservoirs and groundwater.

Bright representative of phytoncide tree is the poplar (lat. Pópulus) and silver poplar (lat. Pópulus álba) which is recommended to be planted around the perimeter of industrial areas, utilities, and polygons. Separate studies were established seasonal dynamics of change of phytoncide activity of woody plants in relation to the test cultures of gram-positive Bacillus subtilis IMB B - 7018 and gram-negative Esherichia coli UKM In-926 microorganisms.[13]

Frost resistance, ruggedness and isolation of large amounts of oxygen puts it almost on the first place in the greening of cities of Russia. [14]

Search new methods of cleaning soil and water scientists discovered an amazing plant southern latitudes - the tree Moringa (Morínga oleifera), which grows in tropical and subtropical climates. Since ancient times the plant was used for household, culinary and medicinal needs.

Left over from making butter oil cake, local people use for water purification. Water, passed through filters from Moringa oil cake, suitable for use within 24 hours.

It is established that Moringa seeds have the property of natural coagulants / flocculants, which allows to grow a tree for the use of treatment plants for drinking water purification from turbidity and sewage sludge [15].

Staff from the Universities of Pennsylvania and Oklahoma (USA), proposed a new method of water purification with seeds Morínga oleifera .Scientists proposed the "F-sand" ("functionalized sand"). By isolating eight seed proteins and adsorbing them on the surface of silicon dioxide particles, scientists have obtained a sand filter for water purification. The seed proteins that have become part of the filter destroy microorganisms and significantly reduce turbidity. Water passed through such a filter is ready for use in everyday life and for drinking. After special treatment, the F-sand filter can be used again to purify water.

It is proposed to use the F-sand method for water purification in regions lacking this valuable resource (residents of India, Mexico, Sri Lanka, Vietnam, Africa and other countries in which Moringa oleifera grows) [16].

Phytotechnology has clear advantages in the treatment of wastewater and natural waters during the rehabilitation of water bodies.

With the right selection of plants, phytosystems can provide deep wastewater treatment, where the inherent natural destruction mechanism reduces energy consumption, eliminates the ingress of chemicals and uses a minimum of equipment. No waste or emissions.

The possibility of widespread implementation is reduced by factors such as a long period of negative temperatures and the availability of free areas (under the bioplate), since natural mechanisms require up to several days. But, despite these shortcomings, the direction of phytotechnology is very in demand and is constantly developing.

References

[1] Directive 96/61/EC of the European Parliament and of the Council of 24 September 1996 concerning integrated pollution prevention and control *Official Journal L 257 , 10/10/1996 pp. 0026 – 0040* https://eur-lex.europa.eu/legal-content/EN/TXT/?uri=CELEX%3A31996L0061) accessed date: 05.03.2020

[2] Mezentseva O, Volosatova M 2014 The introduction of BAT in the countries of the European and customs unions.., *The quality Control* N°6 date of the application 05.03.2020stk.ru/mas https://ciscenter.org/upload/iblock/e88/vnedrenie_ndt_v_stranax_evropejskogo_i_tamozhennogo_soyuzov.pdf

[3] GOST R 54097-2010 Group T58 NATIONAL standard of the RUSSIAN FEDERATION

[4] RESOURCE conservation BEST AVAILABLE TECHNOLOGIES a Methodology for identifying Resources saving. Best available techniques. Identification methodology OKS 13.020

[5] ITS 10-2019 Information technology reference best available technology wastewater treatment with the use of centralized Sewerage systems for settlements and urban districts. Moscow, BAT Bureau, 2019

[6] Nikitin O, Latypova V, Pozdnyakov S 2015 Ecotechnologies of water reservoir restoration: a training manual p 139

[7] elhttps://portaleco.ru/ekologii-vysshih-vodnyh-rastenij/vodnye-rastenija-i-fenolnoe-zagrjaznenie.html accessed date: 10.03.2020

[8] *http://aquavitro.org/2011/06/22/yachmennaya-solomka-protiv-vodoroslej/* accessed date:.03.2020

[9] Rusanov A, Stanislavskaya E 2007 Evaluation of the effectiveness of barley straw as an inhibitor filamentous algae in laboratory conditions Theory and practice of restoration of internal reservoirs pp 328-334

[10] Butler R Use of barley straw for algal control in Prairie Dugouts: final report p 43

[11] Vasilevich E, Bezryadina I, Konovalov A 2014 Reducing toxicity model solutions of the toxicants in the presence of barley straw. *The CLIVET Solution for achieving a high level of comfort and energy savings a Collection of papers* pp 28-31

[12] Ostroumov S, Shestakova T 2009 Ecological safety: Decreased measured concentrations of Cu, Zn, Cd, Pb in the water of experimental systems with Ceratophyllum demersum: phytoremediation potential *REPORTS OF THE ACADEMY OF SCIENCES* **428** pp 282–285

[13] Bryuzgina A 2013 Phytoremediation as a method for cleaning soils contaminated with cu, ni, zn *Scientific community of students of the XXI century. NATURAL SCIENCES* **13**. URL: http://sibac.info/archive/nature/Natur07.11.2013.pdf (accessed date: 03/25/2020)

[14] Glukhov A, Volodarets S 2013 Phytoncide activity of woody plants in urban environment by example the city of Donetsk *Proceedings of the Samara scientific center,* Russian Academy of Sciences**15** pp 2122-2125

[15] SP 82.13330.2016. Landscaping.

[16] Ayerza R 2012 Seed and oil yields of moringa oleifera variety periyakalum-1 introduced for oil production in four ecosystems of south America *Ind. Crops Prod* **36 (1)** pp 70–73

IOP Conf. Series: Materials Science and Engineering **880** (2020) 012086 doi:10.1088/1757-899X/880/1/012086

[17] https://www.researchgate.net/publication/ 321230927_Moringa_oleifera_f-sand_Filters_for_Sustainable_Water_Purification Boya Xiong, Bethany Piechowicz, Ziyuhan Wang, Rose Marinaro, Emma Clement, Taylor Carlin, Adam Uliana, Manish Kumar, * and Stephanie Butler 2017 Velegol Moringa oleifera f-sand Filters for Sustainable Water Purification / accessed date:25.03.2020

Overcoming information asymmetry in the construction industry based on open and closed contractor activity data

I Anokhov[1]

[1] Department of Enterprise Economics and Entrepreneurship, Baikal State University, 664003, г. Irkutsk, Lenin st, 11, Russia

E-mail: i.v.anokhov@yandex.ru

Abstract. The purpose of the research is to find measures to reduce information asymmetry which significantly affects the relationship between customer and contractor, leading to higher prices for construction work and enhanced opportunism. Information asymmetry can make contractor and customer move away from direct contractual relations to more and more costly ways of interaction: legal, law enforcement and ethical. In the long run, the most effective level of interaction is ethical, as it significantly reduces information asymmetry as well as attendant costs. The article proposes two approaches to analyse asymmetry of contractor's information: in the situation of limited access to internal information of a construction company and in the situation of maximum access to information of the activities of construction company's individual departments. In the first approach, an assessment of price discrimination and changes in the cost of construction due to a deliberate deterioration in its quality is used. In the second approach, a set of indicators is considered, grouped by the functional levels of the contractor: physical, procurement, economic, technological and design.

1. Introduction

Key players of the construction industry are construction company (contractor) and investor (customer) who are interested in bilateral relations [1]. The building owner bears significant costs for the construction of industrial, residential and infrastructure facilities, as well as their maintenance over the life of the product. Therefore, it is very important for him to evaluate the reliability of a construction company in the long run. To assess reliability, the customer has to overcome information asymmetry in his relations with the developer. Such information asymmetry (hereinafter - IA) is naturally overcome in the course of long-term cooperation. From this point of view, a long-term construction contract not only brings the contractor cash income, but also maintains its reputation and trust among customers. As a result, IA and associated risks are reduced.

If the contractor and the customer lack experience in interacting with each other, information asymmetry and attendant risks are at their maximum. It forces the contractor to use market signals to overcome it:

- ratings and returning client testimonials
- construction objects guarantees
- quality system standardization (for example, standards ISO 9000 series);
- guarantee of third parties (professional construction organizations, government agencies, etc.);
- disclosure of construction technology and materials used
- video recording of the construction process

IOP Conf. Series: Materials Science and Engineering **880** (2020) 012087 doi:10.1088/1757-899X/880/1/012087

- testing by the customer (including individual stages of construction);
- phased delivery of construction projects to the customer.

Nevertheless, in the construction industry, information asymmetry and risks still remain even in the relations with long-term partners. It was clearly demonstrated by the crisis of 2007, which caused a wave of bankruptcies in the world, including bankruptcies of construction companies. Thus, additional mechanisms are needed to reduce IA and attendant risks.

2. Features of information asymmetry in construction transactions

In any construction transaction, the relationship between customer and contractor take place not only in the form of a contract, final occupancy and its payment. If either party behaved opportunistically (i.e. has not fully complied with the terms of the contract), cooperation could move to other levels (Figure 1).

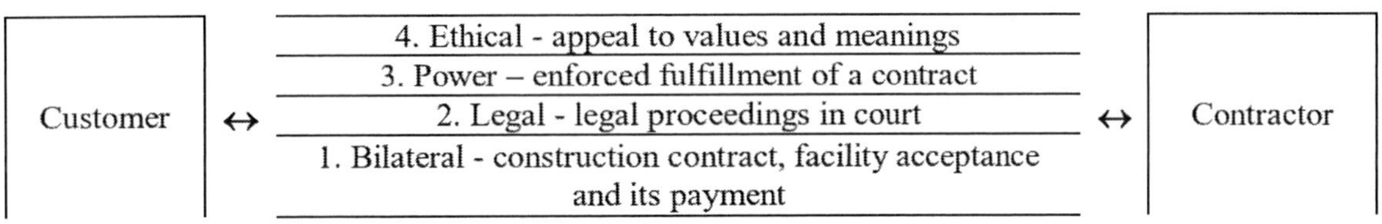

Figure 1. The levels of interaction between contractor and customer

If for some reason the construction contract has not been fulfilled, the parties move to the next legal level. As a result of this, the parties will incur additional costs for resolving legal issues, and the transaction itself will become more costly and less profitable for both parties.

If, at the legal level, it was not possible to reach a settlement that works for both parties, they proceed to the next power level, which involves coercion to fulfill the terms of the contract in various forms: providing unflattering media coverage, putting pressure on banks and suppliers, informing other contractor's customers about the situation, initiating state inspections, filing an appeal to the parent company, etc. It further increases the costs of each of the parties [2].

If a compromise has not been reached at the power level, then the parties could try to settle it at the ethical level, which is a system of rules and values of society that aims to overcome the limitations of society, to achieve inter-individual harmonization and personal growth through increased awareness. This level is most high-maintenance because it requires investing time and energy over a prolonged period. At the same time, this level is the most effective because it automatically harmonizes all the levels below. Thus, it is to the benefit of all conscientious participants in the construction industry to establish a unified ethical standard in the industry, which will help them reduce long-term costs. If such a single ethical standard is not available, customers can use indirect indicators that can reduce IA in relation to the contractor.

3. Indirect evaluation of information asymmetry

Information asymmetry allows the contractor to raise prices for construction work (+ ΔP) and / or to lower the cost of construction work (-ΔC) with the help of a conscious deterioration of the quality of these works. It is relatively easy to identify such facts if the contractor builds standard objects for different customers. In this case, if prices of such objects at fixed costs differ, then there is price discrimination caused by information asymmetry (Figure 2).

IOP Conf. Series: Materials Science and Engineering **880** (2020) 012087 doi:10.1088/1757-899X/880/1/012087

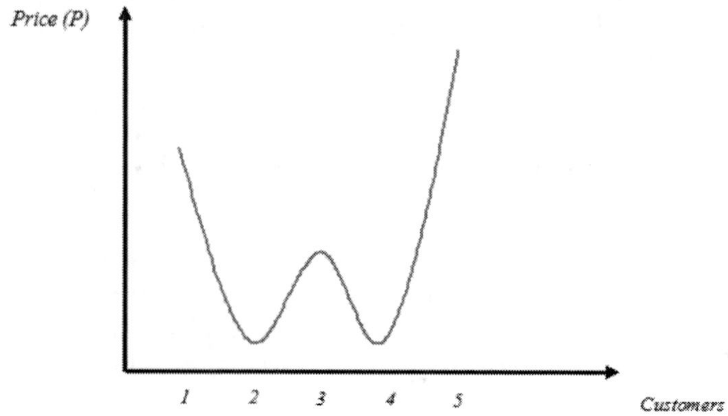

Figure 2. Price discrimination of customers as a result of information asymmetry

If, at a fixed price for identical construction projects, their cost is significantly different due to deliberate quality deterioration, this phenomenon can be called discrimination of customers on the quality of construction work (quality discrimination) (Figure 3).

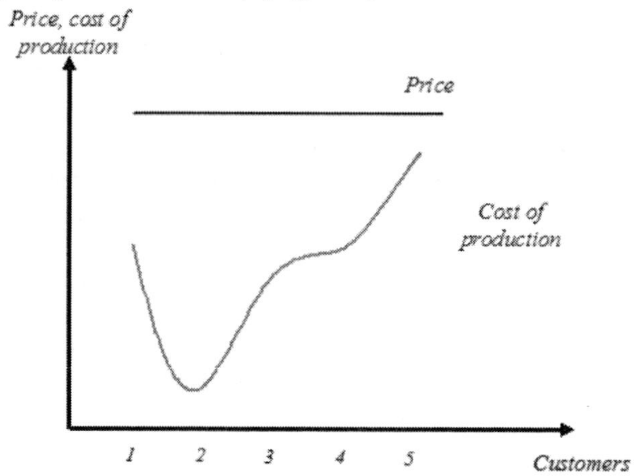

Figure 3. Quality discrimination of customers as a result of information asymmetry

Based on this, a change in information asymmetry (ΔAI) can be expressed through the equation:

$$\Delta AI = \sum_{i=1}^{n}(\Delta Pi + |\Delta Ci|) \qquad (1)$$

where

n - the number of customers

i - a specific customer of construction products.

The above formula allows us to represent information asymmetry in monetary value. Practical application of this formula is complicated by the problem of separating information asymmetry factor from other factors that may affect price and cost.

If the contractor performs non-standard construction orders or the customer has access to the contractor's internal information environment, the customer can conduct a comprehensive analysis of contractor's activities to reduce information asymmetry.

4. Comprehensive analysis of the contractor activities to reduce information asymmetry

Comprehensive analysis performs a set of following functions:

1. Physical function – creation of construction projects.

2. Supply function – organization of interaction between the individual construction sites of the contractor, logistic operations, storage facilities, energy supply, provision of labor resources (including a training system), etc.

3. Economic function – support for construction work using cash flow (income generation, payment for construction materials, wages for workers, taxes, etc.).

4. Technological function – development and implementation of construction, administrative, human resources and other technologies.

5. Project function – identification of strategy, relationships with external actors, relations with state agencies, work with key customers, etc.

For each of these levels we can apply their own indicators to reduce information asymmetry regarding a specific aspect of the contractor's activities. Consider them below.

Physical level

The following indicators can be applied at this level:

– increase of the number of participants in the construction process (suppliers and subcontractors). It indicates an increase in specialization and a subsequent increase in construction efficiency

– share of work for long-established customers out of the total volume of construction work. It indicates stability of the construction quality and confidence in the contractor;

– steady increase in the average load of the main equipment

– reduction of the share of manual labor and construction labor.

Supply level

At the supply level, the quality of construction work should be evaluated first. It is the most difficult aspect of external analysis, because it is not always possible to evaluate it in a professional manner. However, if the contractor is interested in getting a correct evaluation of his activities, he will provide all the evidence of quality stability of construction, including acts of expert examination from specialized organizations.

The risks of deterioration in construction quality are strategic. If the contractor consciously takes this step, then, firstly, he is not able to maintain his position by other means, i.e. its competitive performance is questionable. Secondly, he destroys his long-term competitive performance by deliberately deceiving customers and compromising his reputation as a credible partner. Of course, all of the above does not apply to cases where the cost reduction is caused by positive effects of the economy of scale, new technologies implementation or reduction of prices for any of the resources.

To analyze the supply level of the contractor, the following indicators can be applied:

1. Changes in a number of reclamations and consumer complaints about the construction quality. Such information is fairly easy to collect if you consult former and installed base customers. In this case, you should take into account not only changes in a number of complaints, but also changes in the volume of construction. Therefore, a more reliable indicator would be a number of reclamations and consumer complaints about product quality per unit of construction products.

2. The share in the total volume of construction work, which quality is beyond dispute. Declining quality of construction can not always be recorded even by users of fully commissioned construction objects. Therefore, a more objective evidence of products quality could be quality certificates (ISO, etc.), certificates of expert examination, etc. A positive development could be the availability of voluntary Quality Conformance Certificates.

3. Industry awards for quality, ranking among contractors, overall reputation. Such indirect indicators are created by independent parties and therefore objectively reflect a change in the construction quality, although sometimes with some delay.

4. Duration of contracts with key customers. In this case, what is measured is how successful the contractor has been in maintaining relationships with key customers, whether the work of the contractor is in demand, and whether the clients order work regularly and in sufficient quantity.

5. The share of expenses for the payment of the accounts of the main sub-suppliers, sub-contractors, etc. in the construction cost. This indicator reflects the relations between the enterprise and external partners who provide the contractor with productive resources, work and services. It helps to determine

manufacturing reliance of the enterprise and direction of its outgoing financial flows. It also allows to identify a whole set of other risks such as regional, political, risk of currency deprecation, transport and others.

Economic level

At the economic level, changes in prices, profit margins, market volumes, and terms of credit are thoroughly examined. In addition, an analysis of price discrimination is necessary. Such analysis aims to increase transparency of methodology of setting individual prices for identical or the same construction project at fixed costs. It means that all information about price structure (including the percentage of profit) should be considered. In fact, this aspect of assessing information asymmetry reveals the degree of monopoly market power of the contractor, which allows him to determine the price of his work and group customers according to the degree of their elasticity of demand. In addition, it allows one to find out whether the contractor is inclined to outsource some of the construction activities to third-party contractors and appropriate a brand name.

The most obvious indicators that allow us to identify the implicit risks of the contractor in terms of price policy analysis are the following:

1. Seasonal and business cycle price fluctuations at fixed production costs. The difference in the construction price during the period of decline in demand and during the period of its maximum value allows us to indirectly identify threshold limit value of the price and surplus the contractor receives in case of speculative demand [3, 4]. An important condition for correct evaluation is rigidity of production costs, which allows us to exclude from the analysis cases that are not the subject of evaluation (for example, cases of purchase cost decline of raw materials and consumables).

2. Difference in price of an identical or similar construction object for different consumers at fixed production costs. The higher the value of this indicator is, the greater the difference in the elasticity of customer demand. Here, we study cases in which a unit of construction work is sold at a bid price, i.e. at the highest possible price that the customer is willing to pay. This policy is called price discrimination based on the income of the buyer.

3. Price discrimination based on market segmentation. In this case, customers are divided into separate groups according to the elasticity of their demand in terms of price and other qualitative parameters: industry affiliation, form of ownership, geographical location, etc. Different price levels for these customer groups allow us to estimate the stability of the contractor's position and the possibility of sales diversification.

4. Assessment of middlemen's profit.

In this case, we evaluate the role of the contractor as an intermediary between customers and a third-party subcontractor. In the real world, it is not rare that the construction is indeed carried out by third-party contractors, and the contractor under evaluation only resells the project to its customers. This type of situation should be evaluated as highly risky because both the actual producer and end-use consumer are objectively interested in eliminating unnecessary intermediaries with corresponding financial losses. In this regard, sales revenue should be adjusted to the difference between purchase price and sales price. To do this, the difference between product price received from customers and product price paid to third-party contractors is summed up.

Economic level assessment should also include an analysis of the contractor's main activities in the context of their profitability. This aspect of analysis is focused on increasing the transparency of sources of income and profits. As a result, we should have clear and complete understanding of what types of contractor's activities are most likely to affect overall profits.

Indicators to identify the risks of the contractor's income and profits may include:

1. Cost effectiveness and profit-making capacity of contractor's activities. Due to the availability of financial statements of most companies, it is possible to establish facts of changes in the economic efficiency of the main production profile of the contractor, as well as identify new, financially interesting segments of the market.

2. Cost effectiveness and profit-making capacity of the contractor in his relations with the clients. In most cases, each company has both key customers and less important customers. While key customers

do not always deliver the highest profitability, their impact on revenue is generally high. If we know the difference between cost effectiveness and profit-making capacity of these two customer groups, we can estimate the degree of reliance of the enterprise on specific customers, as well as the risks that may arise from the loss of these customers.

3. Share of income from non-basic activities in revenue from sales of products. The concept of "core business" for a particular enterprise can transform as it develops, captures adjacent markets and acquires non-core assets. An increase in the share of income from non-core activities may signal qualitative shifts in the competitiveness of the enterprise in the "home" market and in the "new" market. Such facts should be recorded by an external analyst and evaluated in terms of new risks and opportunities for the enterprise.

Technological level

At the technological level, the following aspects should be evaluated:

– compliance of construction methods with market requirements;

– achievement of target indicators of cost effectiveness and profit-making capacity

– flexibility to change production capabilities of the enterprise according to business activity in the economy;

– setup costs and staff retraining.

– compliance of construction technology with quality standards and environmental requirements

Design level

At the design level, the following aspects should be evaluated:

- compliance of the enterprise with community demand in the long term;
- information openness of a contractor
- participation of the contractor in environmental and humanitarian projects aimed at harmonization of social relations;
- promotion of long-term survival and development of society.

As a result of the conducted analysis, a set of indicators will be created to comprehensively evaluate long-term sustainability and predictability of the contractor, which substantially reduces information asymmetry in relations with a contractor. This, in turn, reduces the cost of monitoring the progress of construction and the fulfillment of the terms of the contract. As a result, the price level for construction work can be significantly reduced in the interests of all participants in the construction industry.

5. Interdependence between functional levels

There is interdependence between different functional levels. So, for one phase of the construction activity cycle at the physical level (construction phase of the facility) there are at least two phases of the supply cycle (at least transportation of workers and materials to the construction site, as well as preparation of the territory for construction and cleaning of the construction site after construction is completed) [5, 6]. In turn, for each phase of supply there are at least two phases of the economic cycle (at least execution of a contract with suppliers and subcontractors, payment for materials, work and services, monitoring the implementation of the contract by suppliers and receiving payment from the construction customer). Similarly, for each phase of the economic cycle there are at least two phases of the technological cycle (at least the development of project documentation, control over the activities of units, adjustment of internal technological processes and commissioning of the facility). Similarly, the design level cycle is more intense than the technological one. The project cycle includes continuous monitoring of demand for construction work, development of business processes for each type of work and services, coordination of construction activities with authorities, trade unions and other external entities).

The intensity of transactions at each of the levels can be calculated using the formula:

$$Int = \frac{Q_{trans}}{\mathrm{T}}$$

where

Int – transaction intensity at functional level.

IOP Conf. Series: Materials Science and Engineering **880** (2020) 012087 doi:10.1088/1757-899X/880/1/012087

Q_{trans} – the number of transactions of units of a separate functional level of the contractor with other construction entities (internal and external).

T – time period.

If the contractor is in the growth stage, then the intensity of the processes at the project level is much greater than at other levels, because at this level, the company operates with many external entities in a constantly changing external environment, which requires immediate and constant adjustment of construction technology at all other functional levels [7]. In this case, we can observe the following ratio of transaction intensities of functional levels:

$$I_{project} > I_{technological} > I_{economic} > I_{supply} > I_{physical}$$

Between the indicated levels, quantitative relationships can also be observed. For example, if we know the planned volume of construction (physical level) it is quite easy to calculate the complexity of the work, the number of workers, the required amount of materials, storage space, transport units, special equipment, etc. It determines the quantitative relationship between the *physical and supply* levels.

Based on the data of the *supply* level, you can determine the main indicators of the *economic* level: payroll, transportation and storage costs, taxes, profits, payment to suppliers, etc.

The parameters of the physical, supply and economic levels are directly related to the parameters of the *technological* level, primarily with the volume of development of construction technology and construction documentation.

Thus, if we know the quantitative parameters of one of the levels, it is possible to determine the intensity of processes at all other levels. To accomplish this, it is also necessary to know the features of the construction industry (road, housing, industrial, etc.).

Any change in the phase of the contractor' cycle first becomes apparent in a change of the intensity of transactions at the project level, which activates all the levels below correspondingly (Figure 4).

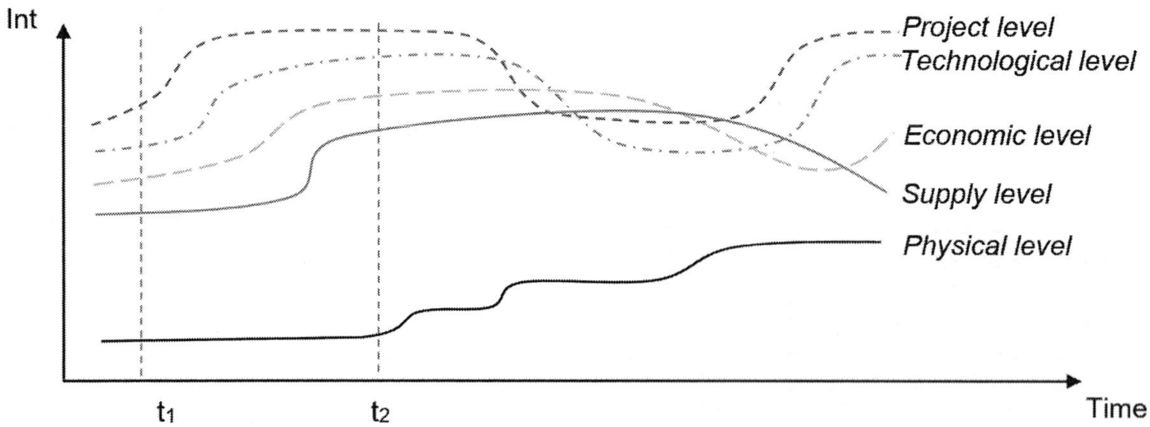

Figure 4. The relationship of the phases of the cycle at different functional levels of the contractor

The size of the interval t_1t_2 shows the degree of flexibility of the contractor's organizational structure to change in the market demand for construction work. It largely determines its competitiveness in the long run.

The exact quantitative relationships of different functional levels for a particular contractor can only be determined empirically. It will allow the customer to predict in advance the change in the state of the contractor and thereby significantly reduce the information asymmetry.

6. Conclusion

Thus, according to the study, we can conclude that the customer can significantly reduce the information asymmetry in their relations with the contractor. Moreover, the more detailed internal information about the contractor you can get, the more reliably you can predict his behavior in the long term. The proposed methods to reduce information asymmetry can also reduce the risks of all construction participants and facilitate long-term cooperation.

References

[1] Astafiev S, Dombrovskaya K 2012 Process management of investment-construction complex in conditions of self-regulation. *Bulletin of Baikal State University* **3** 104

[2] Tagarov B Consequences of information asymmetry for labor market subjects 2014 *Bulletin of Baikal State University* **2** 1

[3] Yakimova E 2018 On Issue of Fundamental Nature of Business Activity as a Scientific Category. *Baikal Research Journal* **9** 1. DOI: 10.17150/2411-6262.2018.9(1).1.

[4] Kubasova T 2016 Organizational designing of logistic mortgage and construction systems of coalition type. *Baikal Research Journal*, **6** 1 DOI: 10.17150/2411-6262.2016.7(6).10.

[5] Grushina O, Shelomentseva N 2018 Construction of standard and comfort class housing: Functionality of costs and their use in the interest of developers. *Economic Analysis: Theory and Practice* **7** 1318

[6] Steinle C, Schiele H, Ernst T 2014 Information Asymmetries as Antecedents of Opportunism in Buyer-Supplier Relationships: Testing Principal-Agent Theory. *Journal of business-to-business marketing* **21** 123

[7] Anokhov I 2019 Theoretical Aspects of Investment. The Number of Individuals Involved as a Criterion of Payback. *Journal of Economic History & History of Economics* **3** 506–529. DOI: 10.17150/2308-2488.2019.20(3).506-529

[8] Pesek A, Smithwick J, Saseendran A, Sullivan K. 2019 Information Asymmetry on Heavy Civil Projects: Deficiency Identification by Contractors and Owners. *Journal of management in engineering* **35** 4

IOP Conf. Series: Materials Science and Engineering **880** (2020) 012088 doi:10.1088/1757-899X/880/1/012088

A model for managing the development of low-rise housing construction using the method of discursive design

O A Bezrukikh[1], T O Pinchuk[1] and S V Latysheva[1]

[1] Irkutsk National Research Technical University, 83 Lermontov str., Irkutsk, 664074, Russia

E-mail: olga.bezrukikh11@gmail.com

Abstract. In the Russian Federation, the housing problem is one of the strategic and top-priority goals of the state's social and economic policy. The indicators of housing supply prove that low-rise buildings created by means of large-panel housing construction technology can become a solution to the housing problem in the Russian Federation. Today, the technology of large-panel housing construction has opened up completely new prospects for the construction sector, combining all the advantages of industrial mass technology and providing originality and individuality of architectural solutions. In this context, switching from institutional to personal discursive design will help to increase the investment attractiveness of low-rise construction projects.

Both a massive growth rate of digital economy and an increase in innovations in all industries lead to new development trends. Despite the extension of innovations, the housing problem is still urgent and has national-scale relevance. It is commonly accepted that the issue of providing housing is not only a social problem, but also a determining factor for the health of citizens, demographic situation, and attitude to the government. Based on this, the solution of the housing problem is one of the ways to increase the indicators of population's living standard and the effectiveness of public administration.

When implementing the tasks of providing citizens of the Russian Federation with affordable and comfortable housing, as well as overcoming the trends of population concentration in megacities, a key question is to develop low-rise housing construction, which includes the principles of energy efficiency, environmental friendliness, comfort and safety of living [1]. This trend goes beyond the national construction industry, and arises in the global construction discourse. Being a kind of institutional discourse, construction discourse encompasses corpuses of documentation having common content criteria: the same discourse address the same subject, theme, and concept. This forms a network of semantic and pragmatic characteristics acting as a common system of utterances, unified by communicative and functional-target relations [2]. The study focuses on a corpus of Russian-language and English-language specialized documentation that represent the practices of low-rise construction.

Modern construction often applies industrial methods, in particular the method of large-panel housing construction. The main reason of this is the former economic crisis, which forced developers to review construction costs in order to reduce cost price by means of low-cost and timesaving technologies providing high work performance, high percentage of precasted elements used in construction, and high consumer qualities. The above efficiency indicators make it obvious that there is still no alternative to industrial housing construction [3,4]. Large-panel housing construction demonstrates extensive experience and huge growth prospects in the future, due to the deep modernization of production processes. The new projects of large-panel housing construction will change the existing views on industrial construction and create a brand new living environment to meet the needs of consumers implementing the main tasks of housing policy of the state.

Based on a comparison of national and English-language discursive practices of the construction industry, a general trend arises to show the contrast of low-rise and industrial housing construction, with a prevailing positive assessment of the development of low-rise construction. At the same time, English-language texts, along with institutional relations, represent variable strategies of personal discourse that activate an individualized perception of the issues discussed. For example, it represents both the description of lower energy efficiency or greater environmental impact of high-rise buildings, and the personal-oriented meanings of an individual involved in interaction with the urban environment [5]. However, in the Russian-language construction discourse, there is a predominance of institutional factors. They are the representations of systemic, but nor personal relations, for example, the management system of a low-rise construction complex in Russia.

The construction complex is an important industrial sector of each region of Russia, providing affordable, high quality, safe and comfortable housing, improving settlements and their infrastructure. To fulfill these tasks, the construction complex needs an effective management system of organizational and economic mechanisms that allow achieving the strategic goal of improving the quality of life of all groups of the region's population [6,7]. To achieve this goal it is necessary to form multi-facet managing mechanisms bolstering the development of management system, as well as by means of discursive design.

In the Russian Federation, there are three major and commonly used mechanisms of the low-rise construction management system:

1.Centralized target programme approach. As centralized state planning disappeared, and private independent enterprises arose, one of the most effective approaches is the development of targeted programmes, in particular, a system of coordinated measures aimed at solving specific problems of the regions of the Russian Federation. In this context, the government defines priority issues based on regional specifics and the needs of each territory of the Russian Federation. Accordingly, this approach applies general perception, so personal discursive practices are minimal [8].

2.Strategic planning. Since 2014, the development of strategic planning documents has become a mandatory element for the design of the country's development concept at the federal level, at the level of regions and municipalities [9]. At the legislative level, there are seven stages of the strategic planning for the development of the construction complex, which allow both identifying the problems of the industry and defining the direction of development of construction enterprises, designing their structure and management system. In terms of discursive models, strategic planning is completely institutional.

3.Management system based on development indicators. To create a mechanism coordinating the interests of state and non-state actors involved in construction activities, it is effective to apply indicators of management and economic development of enterprises in the Russian Federation. These indicators include the design and implementation of state support measures contained in strategic development plans [10]. This process ensures consistency of decisions at all levels through the adoption of regional priorities, forecasting, and development of state and regional support measures. In this case, the degree of discourse impersonality reduces, but institutional content remains predominant.

This research reviews the analysis of several interrelated factors, including management system mechanisms, a model of organizational and economic relations of construction companies in Russia, and statistics on the activities of construction companies in the Russian regions [11]. It reveals a set of key inconsistencies in the theory of low-rise construction management that currently exist:

- an executive chain of command known or "vertical" management structure, in which the object of management is an organization or its division;
- infill construction of low-rise buildings lacking necessary infrastructure;
- insufficient development of low-income and middle-income housing in the sector of low-rise construction;

— lack of a mechanism to improve the efficiency of organizations' cooperation in the construction sector, by establishing general management system to make construction businesses and organizations collaborate;
— loopholes in the laws, regulating purchase contracts in construction, as well as participation in federal target programmes. This leads to the lack of competitiveness as the sector is under regulation of unofficial groups whose interests are to control decision-making authorities and cash flows in order to take advantage of this;
— the lack of free access to sufficient data on the activities of organizations that perform construction works, which does not allow developers to compete and does not allow the population, objects and subjects of management of the construction industry to have information about the performance of the construction industry as a whole;
— depreciation of fixed assets, refusal to use innovative technologies, which negatively affects the quality and timing of construction.

The theoretical framework described above and the discussed issues of low-rise construction management do not fully correspond with modern market relations, current technology, and they lack information and legislative support [12]. This results in the lack of information management, which prevents the sector from effective development showing reduced growth in comparison with developed countries. To optimize information exchange within the framework of construction discourse in the field of low-rise construction, it is necessary to balance institutional and personal strategies, using the method of discursive design to level excessive institutionality and bring social responsibility and individual needs to the fore.

In order to minimize the effects of negative factors in the functioning of low-rise construction management system in Russia, it is necessary to develop a set of recommendations aimed at establishing a managing and economic mechanism for regulating construction sectors in the regions of the Russian Federation. In particular, it should include the following:

1.To define low-rise construction and individual housing construction as the primary areas of urban planning regulation and development and as a sustainable development of the country's territories. To do this, it is advisable to design a "Road Map" for the development of low-rise construction, which will act as an operational and strategic plan. The Road Map should arrange the formation of legislative and technical regulations by including low-rise construction in federal, regional and municipal state target programmes. Sufficient financial support of low-rise construction projects should also be a key question, and industrial construction method can provide this.

2.At the municipal level of the Russian regions, low-rise construction and individual housing construction should become the basis of development policy that is in accordance with the needs of citizens. The regional executive authorities should transfer effective foreign practices of housing construction, taking into account not only institutional, but personal responsibility.

3.For the implementation of The Development Strategy of Construction Materials Industry (in 2020-2030 perspective) approved by The Government Executive Order No. 668-p dated 10.05.2016 it is necessary to provide budget for the modernization of industrial housing and the development of territories' infrastructure.

4.In order to improve interaction between investing and constructing companies, such organizational forms as cartel, consortium, or cluster are preferable for the implementation of large low-rise development projects.

5.In order to provide Russian citizens with affordable housing, it is recommendable to establish the transition to industrial housing construction, in particular to the use of large-panel housing construction technologies, which will reduce the cost of construction, while ensuring the necessary requirements for environmental friendliness and energy efficiency.

To ensure the implementation of the above recommendations, state programmes should promote large-panel housing construction, which will generate demand for this type of housing. This will be facilitated by a personalized discursive strategy focused on the needs of residents of a particular region. It is necessary to create a ready-made construction site with engineering communications, produce high-quality construction materials at the local level of each region, provide the construction industry with highly qualified personnel, and prepare general layout for the development of territories. For a total transition to large-panel construction technology, it is necessary to create a competent form of interaction between all subjects of the construction process. The main organization forms of business interaction are holdings, trusts, alliances, cartels, consortiums. One of the most preferable forms may be a consortium. For projects of large-panel low-rise building, it is recommendable to create a consortium that is a temporary association of two or more companies with the objective of participating in a common activity or pooling their resources for constructing large-scale projects. This form of organization will provide the main criteria and parameters, design solutions, as well as mechanisms for the production of large-panel residential buildings. At the same time, it will create comfortable, safe, diverse and visually attractive living environment, designed to provide conditions for harmonious lives of citizens, most fully taking into account the individual characteristics of the regions.

In recent decades, consortiums have become more widespread in the construction industry. First, this is due to the appearance of large, unique engineering and industrial facilities, and the development of new energy-efficient technologies. Secondly, as the level of competition in the contract market increases, construction companies are forced to increase the economic efficiency and reliability of their products. Based on this, in most cases, consortiums are effective in implementing investment projects of different scale and nature, with a high degree of cooperation and collaboration between the consortium members [13]. Based on this, the prospects for the existence of a consortium of low-rise large-panel housing construction are favourable, since this organization form, due to economic integration, is able to significantly reduce the cost of construction and installation work and, consequently, increase both its competitiveness and profitability of final products, both for consumers and for consortium members. Improving and implementing a new management method, in this case, the form of organizing a consortium, will allow construction organizations to ensure effective interaction with elements of the macro environment, ensure the implementation of the strategy for the development of the housing sector, efficiently allocate the resources of organizations and achieve their goals, both in the long and short term.

References

[1] Budanov I A and Ustinov V S 2016 Processes and mechanisms of perspective development of the complex of structural materials in Russia *Problems of Forecasting* **1** pp 22-37

[2] Kirillova E D and Kazantsev I O 2016 Technologies that improve the energy efficiency of construction projects: accumulation of heat *Young scientist* **28** pp 95-98

[3] Trushin G Low-rise housing: pros and cons *[Electronic resource]* URL: http://realty.vesti.ru/zhile/maloetazhnoe-zhile-plyusy-i-minusy (accessed date: 16.01.2020)

[4] Alekseev S A 2019 Spatial restructuring. New meanings and rules of investment and construction activity p 454

[5] Yaskova N Yu and Sarchenko V I 2018 Formation of strategic approaches to the creation of aqua-territorial production complexes of the Arctic zone *News of Higher Educational Institutions. Investment. Construction. Realty* **2** pp 84-93

[6] Chernyavskaya V E 2002 Discourse and discursive analysis: traditions, goals and directions *Text stereotyping and creativity* pp 122-135

[7] Skolubovich A Yu and Matveeva M V 2019 Development of a methodology for assessing the readiness of enterprises for public-private partnership in the field of public water supply *Proceedings of Universities. Investment. Construction. Real estate* **9** pp 274–283 DOI: 10.21285/2227-2917-2019-2-274-283

[8] Kazimirov I A and Peshkov V V 2019 Determination of price behaviour in the secondary residential real estate market using a multidimensional regression model *Proceedings of Universities. Investment. Construction. Real estate* **9(3)** pp 476–487 DOI: 10.21285/2227-2917-2019-3-476-487

[9] Matveeva M 2018 Synchronization of the activity of participants during the implementation of

housing construction projects *Proceedings of Universities. Investment. Construction. Real estate* **8** pp 31-41 DOI: 10.21285/2227-2917-2018-2-31-41.

[10] Yaskova N Yu and Sarchenko V I 2018 Development of strategic approaches to the creation of aqua-territorial industrial complexes in the Arctic zone *Proceedings of Universities. Investment. Construction. Real estate* **8** pp. 84-93 DOI: 10.21285/2227-2917-2018-2-84-93

[11] Yaskova N Yu 2018 Construction in the context of new views on the goals of economic development. *Proceedings of Universities. Investment. Construction. Real estate* **8** pp 133-142 DOI: 10.21285/2227-2917- 2018-1-133-142

[12] Kudryavtseva V A, Zaslavsky A M and Larina O P 2018 Multicriterial comparison at the inclusion of investment projects of manufacture of building materials into the state program of the region (at the example of Irkutsk region) *Proceedings of Universities. Investment. Construction. Real estate* **8** pp 60–72 DOI: 10.21285/2227-2917-2018-1-60-72

[13] Gertsekovich D, Gorbachevskaya L, Grigorova L and Peshkov V 2019 Return on investment in REIT real estate funds *Investments. Construction. Real Estate: New Technologies and Targeted Development Priorities IOP Conference Series: Materials Science and Engineering* **667** 012063

[14] Peshkov A V 2019 Remote mortgage lending as a resource for housing development *Investments. Construction. Real Estate: New Technologies and Targeted Development Priorities IOP Conference Series: Materials Science and Engineering* **667** 012073

Contemporary perspective of energy efficiency: essentials, content, challenges, implementation in the Russian Federation

O A Bezrukikh[1], E Yu Gorbachevskaya[1] and N G Sivtseva[1]

[1] Irkutsk National Research Technical University, 83 Lermontov str., Irkutsk, 664074, Russia

E-mail: olga.bezrukikh11@gmail.com

Abstract. Natural resources conservation and sustainable living have become important issues worldwide. Among natural resources, energy is one of the important prerequisites for sustainable development. Therefore, in order to meet the high demand for energy, it is necessary to bring its sustainable consumption to the forefront of the development strategy of construction industry. The construction sector development in the Russian Federation is largely determined by national projects whose main goal is to provide citizens with comfortable and affordable housing. The national program announced by the government can be successfully implemented by the means of energy-efficient technologies usage, in particular, while building low-rise housing.

Currently, the Russian economy is moving towards the energy-saving path of development. This trend originated in 1990 when there appeared a doctrine involving the rationing of buildings; the main purpose of this doctrine was to minimize energy consumption. The government developed regulatory documents that made the construction complex of the Russian Federation apply energy-efficient technologies in the construction and management of buildings at the state level. As a result, the concept of "an energy-efficient building" was introduced in the construction industry and in the country's economy.

The conceptual framework of the energy-efficient sphere is described in the Russian Building Norms and Rules (SNiP) 23-02-03 Thermal Protection of Buildings. However, it should be noted that the norms of this SNiP explain energy efficiency only as a property of an object and its engineering systems that provides the necessary level of heat energy consumption in order to maintain optimum microclimate parameters of housing [1].

In the process of energy-efficient technologies development the term "energy efficiency" has changed a lot and the boundaries of its influence in the construction sector have expanded. Due to the fact that energy efficiency technologies include not only the use of thermal energy, but also the use of other types of energy resources and energy, it is necessary to give a more precise definition of this concept.

In the process of realizing the entire technological potential of energy-efficient technologies, three categories of measures can be applied; they are shown in Figure 1 (figure 1).

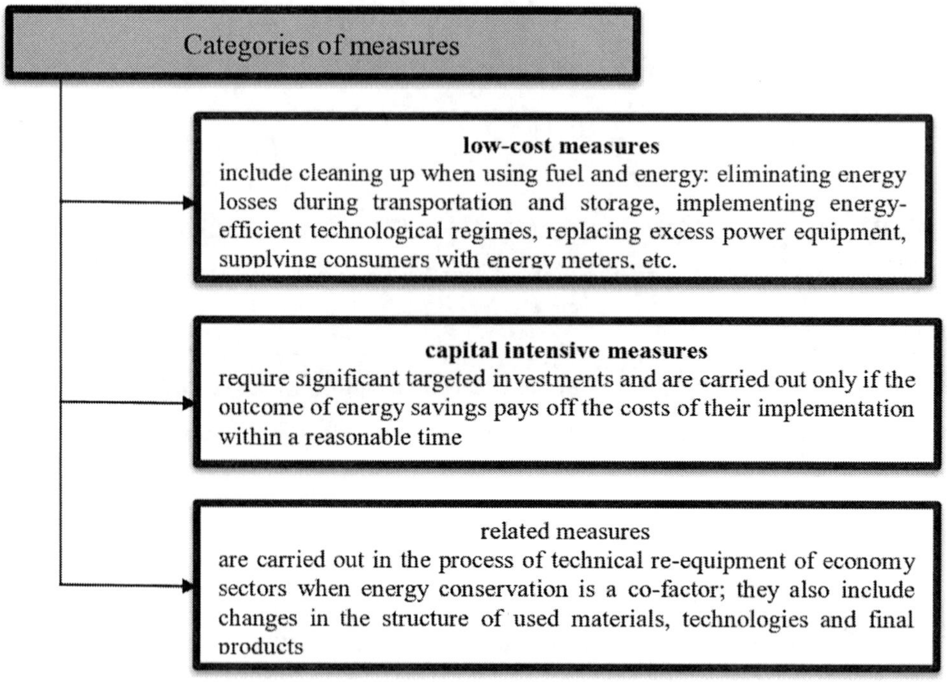

Figure 1. Categories of measures used for realizing the technological potential of energy-efficient technologies

The energy-efficient technologies development in a country is determined by numerous measures and factors. Energy efficiency factors include the structure of the country's economy, the type of the industrial base used, the exchange rate of the national currency, the area and climate of the country, as well as affordability of energy services [2]. In the Russian Federation electricity consumption is increasing every year, which undoubtedly causes both the environmental degradation and increased citizens' expenses associated with paying electricity bills (figure 2) [3].

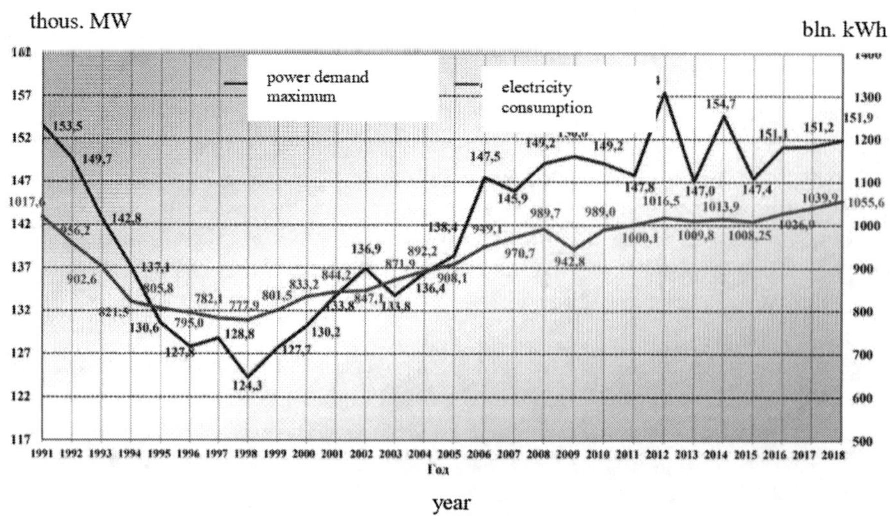

Figure 2. Dynamics of demand for electricity from 1991 to 2018

Thus, in 2019, there was another increase in electricity tariffs by an average rate of 4%. Since the territory of the Russian Federation is rather large in comparison with other countries, in Russia the electricity tariff rate varies depending on the region. In 2019 the lowest tariff was recorded in the Irkutsk region – 1.06, the highest tariff in large cities was recorded in Moscow – 5.38. If one considers all the regions of the Russian Federation, then the highest tariff is in the Chukotka region – 8.2 [3]. These data indicate a fairly high cost of electricity for the Russian citizens. According to the officials, high tariffs should stimulate citizens to use energy-saving technologies. However, the problem of increased cost of energy resources is one of the main state problems. If the problem of higher cost of energy resources is solved, it will both strengthen the position of the Russian Federation in the list of developed countries of the world and provide an increase in the standard of citizens' living. Russia is a resource base for the majority of countries in the world; however, the huge base of fuel and energy resources in our country does not involve energy waste, since only in an open market economy the energy efficient management can be the most important factor in Russian goods and services competitiveness [4, 5]. Currently, the state has a strategic goal in the economic sphere – to double the level of gross domestic product by 2025. This goal can be achieved only by changing the understanding of energy conservation, and, as a result, by introducing energy efficiency technologies.

The Russian Federation as well as all the countries around the world recognize the urgent need to change existing approaches to energy usage. Ensuring energy security, reducing energy prices are of primary importance in the development of energy efficiency policies.

The focus on the development of energy-efficient technologies in the Russian Federation can be traced in The Energy Strategy of Russia for the Period up to 2030, which treats energy conservation and energy efficiency as the main issues. Besides, various state programs involve grants of at least 50 billion rubles [6, 7] to the budgets of constituent entities of the Russian Federation allowing them to implement regional programs. Nevertheless, the state budget is not enough to popularize energy-efficient technologies; therefore, the issue of raising extrabudgetary funds for energy-efficient projects implementation is still relevant. The Government of the Russian Federation is developing and approving regulatory acts, decrees and laws that contribute to reduced energy consumption while constructing new buildings, reconstructing and overhauling. For example, in order to stimulate businesses to switch to the energy-efficient policy, the state developed the economic levers including tax incentives and potential reimbursement of interest on loans whose funds were used for energy saving projects or projects aimed at improved energy efficiency of buildings.

Based on the experience of foreign countries, the Support Fund for the Reform of Housing and Utilities Sector of the Russian Federation together with 42 constituent entities of the Russian Federation implements pilot projects of energy-efficient blocks of flats as part of programs that allow relocating citizens from emergency housing. At the end of 2018, 40 residential buildings were commissioned, the remaining 17 houses are at the design and construction stages [8]. These pilot projects contribute to the development of sustainable construction in Russia, which aims at reducing the impact of housing development on the environment and human health.

One of the major projects implemented in Russia is Green Balance private houses located in the Moscow region. Initially, their peculiarity lies in the fact that they demonstrate the usage of energy-efficient construction technologies in various designs: traditional and original [9, 10].

When designing a Green Balance residential building, the architects' aim was to reduce the total costs during construction and subsequently during the operation of the house, as well as to build a house with low-energy heating. The project was implemented by the aid of investments and application of technologies of Rockwool company, which is a major supplier of energy-efficient technologies not only in Russia but worldwide.

A Green Balance residential building has an area of 208 m2; the living space is 132 m2. The project implementation period was 15 months (figure 3).

IOP Conf. Series: Materials Science and Engineering **880** (2020) 012089 doi:10.1088/1757-899X/880/1/012089

Figure 3. House architecture

The architectural solutions promote energy efficiency of the house – due to the compact shape of the recessed semi-basement floor (which is also used as a residential floor), the area of building envelopes to the usable area of the house is less than in traditional houses. This fact made it possible to ensure energy efficiency of the house and reduce the number of building envelopes that allow heat to pass through [11]. Besides, due to these architectural solutions natural resources can be used for the operation of the building all year round, as there are no energy costs for cooling in summer and in winter the house is heated by solar energy. The window openings provide the maximum flow of daylight into the house throughout the year, which allows saving on artificial lighting.

Moreover, a number of measures have been taken to increase the level of livability in the house. One of such measures is creation of sound comfort as building envelopes are made of stone wool which absorbs noise from outside.

As for the utilities technologies in the *Green Balance* house they are as follows: heating and hot water supply of the house are based on the individual boiler station; the boiler is convertible, namely, it is a solid fuel boiler allowing installing a high efficiency gas or diesel burner. Brick fuel is used [12].

Traditional radiators are used for water heating system. The ground floor space is additionally warmed by heated water floor. For heating thermal radiation is partly used, it is obtained due to the built-in heating of the wall surface.

The usage of energy-efficient technologies allowed the *Green Balance* house to reach energy consumption of 63 kWh per m^2 per year, which is 60% less than the standard indicator. Thus, the house can get *A* energy efficiency class (see figure 4) [13].

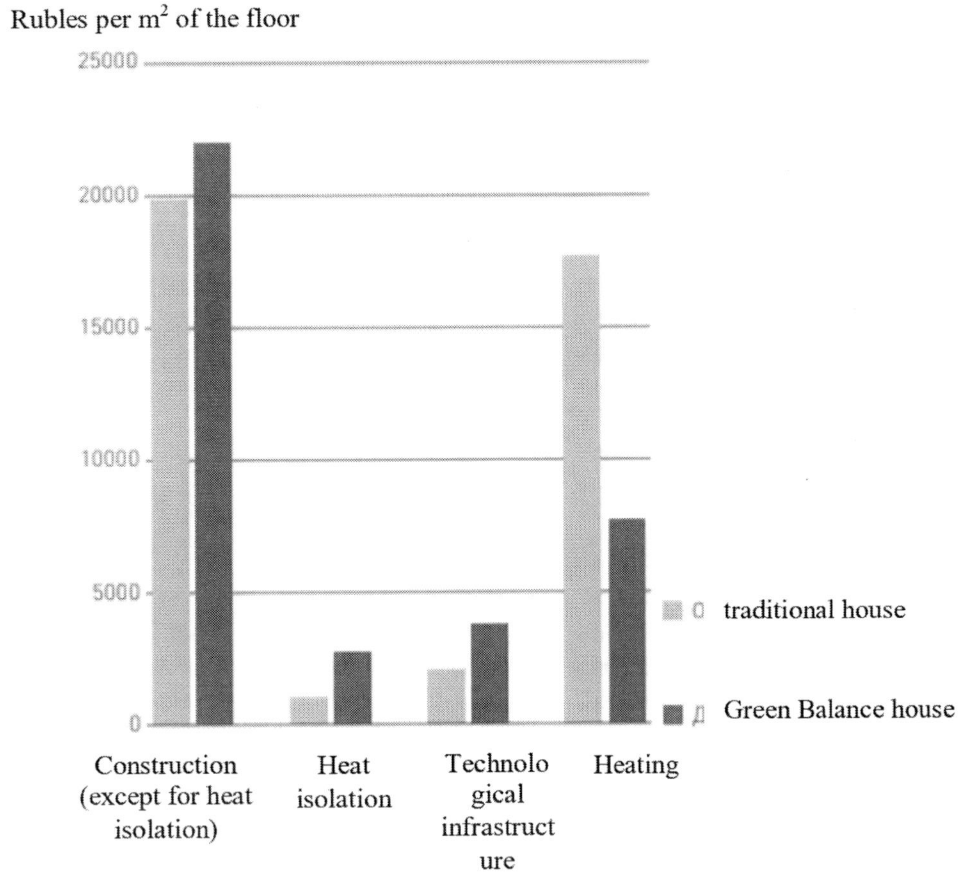

Figure 3. Indices of the house energy consumption

Considering investment attractiveness, we can draw the following conclusion: at the design and construction stages investments in energy-efficient technologies increased the cost of construction by 14.5%, however, due to reduced energy costs for heating residents can save up to 32 850 rubles annually [14].

Thus, the project of *Green Balance* houses has the following characteristics:

• Energy efficiency class of *A*;
• Comfortable microclimate;
• Stylish and non-standard design;
• Good soundproofing;
• Reduced energy consumption by 60% compared with current standards;
• Annual savings – 32 850 rubles.

In this regard, it can be argued that the *Green Balance* house indicates that there are no barriers to the development of energy-efficient construction in Russia, as it confirms its simplicity, efficiency and environmental friendliness.

The implemented projects allow stating that initially energy efficient technologies should be used in the type of construction that has promising developments. Low-rise housing construction can be referred to such type of construction. In 2018, 75.3 million of m^2 were built in the Russian Federation, and the share of low-rise housing construction reached the highest level for the entire period of statistical survey – 42% of the total volume of housing commissioned in 2018. Thus, low-rise housing construction can be defined as one of the most comfortable, resource-saving and cost-effective types of housing allowing people to live in harmony with nature.

The relevance and diversity of the issue of improving the energy-efficient technologies performance in low-rise building construction determine the need to improve the methodological foundations of management and to introduce energy-efficient technologies in production and economic processes. Improving and implementing relevant methodological foundations and principles will allow developing a systematic approach to the issue of energy-efficient technologies mass use in the Russian Federation.

References

[1] Sibikin Yu D 2017 Power Supply of Industrial and Civil Buildings p 89
[2] Kudrin B I 2016 Power Supply p 160
[3] Statistical Review of World Energy *[Electronic Resource]* URL: https://web.archive.org/web/20150813113139/http://www.bp.com/en/global/ corporate/about-bp/energy-economics/energy-outlook.html (access date: 15.02.2020)
[4] Makarov A A and Voropay N I 2018 Systemic Research in the Energy Sector: Methodology and Results p 309
[5] Plakitkina L S and Plakitkin Yu A 2017 Coal Industry of the World and Russia. Analysis, Trends and Development Prospects p 374
[6] Statistical Yearbook of World Energy *[Electronic resource]* URL: https://yeabook.ennerdata.ru/ (access date: 02.16.2020)
[7] Aran Shankar V K, Umashankar S, Paramasivam S and Hanigovszki N A 2016 Comprehensive Review on Energy Efficiency Enhancement Initiatives in Centrifugal Pumping System *Applied Energy* **181** pp 495-513
[8] Lontsikh P A, Moskvitina V A, Golovina E Yu and Kibirev Yu V 2019 Relevant Trends in Construction Industry of the Irkutsk Region *University News. Investments. Construction. Real Estate* **9** pp 10–25
[9] Kazakbaev V, Prakht V, Dmitrievskii V, Ibrahim M, Oshurbekov S and Sarapulov S 2019 Efficiency Analysis of Low Electric Power Drives Employing Induction and Synchronous Reluctance Motors in Pump Applications *Energies* **12** doi: 10.3390/en12061144
[10] Topchiy D V, Yurgaitis A Yu, Kravchuk A S and Shevchuk D A 2019 Energy Audit of Capital Construction, Reconstruction and Reprofiling Facilities before Commissioning *University News. Investments. Construction. Real Estate* **9** pp 134–143
[11] Yanovsky A B 2019 Results of structural reorganization and technological re-equipment of the coal industry of the Russian federation and objectives for prospective development *Russian Coal Journal* **8** pp 8-16
[12] Volokitina V M and Gedich T G 2017 Investment Component in the Current Situation and Development of the Coal Industry in Russia *Economic Analysis: Theory and Practice* **16** p 1262
[13] Chupin V R and Dushin A S 2019Assessment of the Reliability of Providing Consumers with Water. Development of Reliability Indicators of Water Supply to Consumers. *University News. Investments. Construction. Real Estate* **9** pp 578-593 DOI: 10.21285 / 2227-2917-2019-3-578-593
[14] Topchiy D V, Klimina V V, Kodzoev and Khaliullin I M 2019 Research of the Stress-Strain State of the Structure by Tensometric Method in the Renovation Implementation *University News. Investments. Construction. Real Estate* **9** pp 372–383 DOI: 10.21285 / 2227-2917-2019-2-372-383
[15] Matveeva M 2018 Synchronization of the activity of participants during the implementation of housing construction projects *Proceedings of Universities. Investment. Construction. Real estate* **8** pp 31-41 DOI: 10.21285/2227-2917-2018-2-31-41

IOP Conf. Series: Materials Science and Engineering **880** (2020) 012090 doi:10.1088/1757-899X/880/1/012090

Problematic aspects of the management system for large-panel low-rise construction

O Bezrukikh[1]

[1] Irkutsk National Research Technical University, 664074, 83 Lermontova str., Irkutsk, Russia

E-mail: olga.bezrukikh11@gmail.com

Abstract. Solving the housing problem in the Russian Federation is one of the strategic and priority directions of the state's social and economic policy. The government of the Russian Federation is taking consistent steps to activate low-rise housing construction in the construction market. In 2018, the share of low-rise housing construction in the Russian Federation was 42% of the total volume of housing commissioned. It should be noted that in order to solve the housing problem in Russia, when building low-rise buildings, it is necessary to give priority to technologies that will ensure low cost of housing for consumers, fast construction times and the required design solutions. Based on these requirements, the progressive technology of low-rise housing construction is large-panel housing construction. The relevance and versatility of the problem of increasing the efficiency of large-panel housing construction technology in low-rise construction determines the need to improve the entire economic mechanism and management of production and economic processes. Improving and implementing new management methods, in this case, the form of organizing a consortium, will allow construction organizations to ensure effective interaction with elements of the macro environment, ensure the implementation of the strategy for the development of the housing sector, efficiently allocate the resources of organizations and achieve their goals, both in the long and short term.

To date, Russia has seen a period of multiple increases in the share of low-rise housing construction in the housing stock and a period of changing accents in domestic urban planning. Many researchers and scientists, based on the growth rate of the housing stock and global trends, argue that in the near future low-rise housing construction will replace the construction of multistory building construction and will become the foundation of the city of the future and housing for citizens. This statement is confirmed by the fact that the growth of low-rise construction occurs with a general fall in the volume of housing being built on the territory of Russia, so since 2010 there has been a steady trend of an annual increase in the volume of low-rise housing being built with an index of 0.85 [1]. Construction of low-rise housing today is one of the most promising areas of the construction industry, a key factor in the integrated development of territories and the most important direction of state policy in the Russian Federation. Thus, state authorities have developed measures of program-targeted support to stimulate low-rise housing construction using federal and regional programs, such as “Housing”, “Providing housing for young families” and others [2]. Moreover, bank support for consumers of low-rise construction is widespread through the use of special banking products for lending and mortgages.

The analysis of foreign experience of management and organization of low-rise construction in such countries as Germany, Great Britain, and the United States revealed a rich experience and identified the main components of the management system that can also be applied in our country [3].

However, before implementing innovative components of the management system, it is necessary to solve a number of problems that hinder the development of low-rise construction in Russia.

The implementation of projects in low-rise housing construction largely depends on both external conditions and the investment attractiveness of the projects themselves, which is influenced by a large number of various factors (figure. 1). Let's consider in detail the micro-factors of low-rise construction projects, since it is these factors that the subject of the low-rise construction management system has a direct impact on [4].

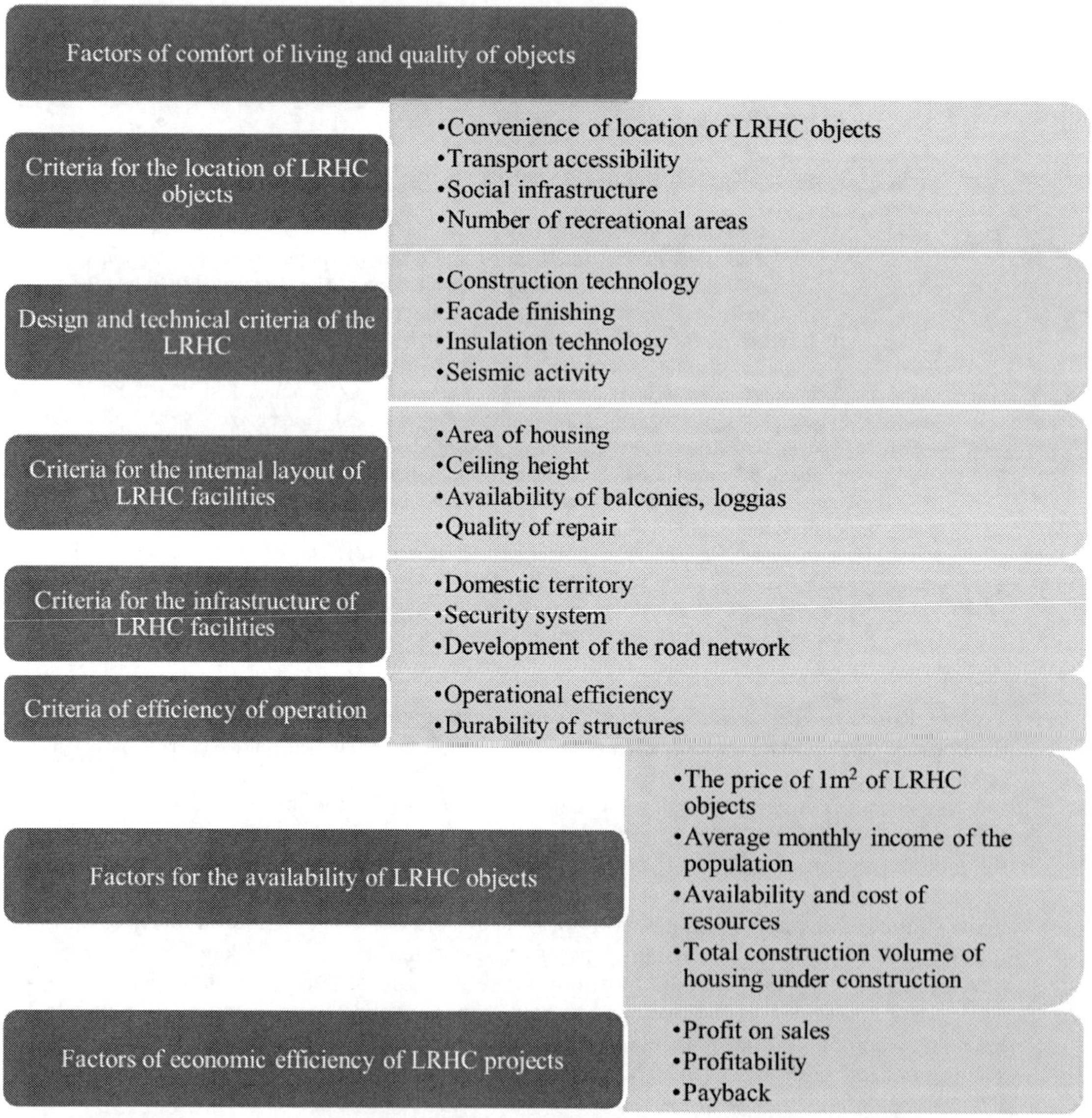

Figure 1. System of factors affecting the implementation of low-rise construction projects

In order to bridge the gap between the Russian management system and foreign country and ensure the implementation of urban planning tasks, it is necessary to solve a number of problems that at this stage negatively affect the system of factors that affect the implementation of low-rise large-panel construction projects.

The construction industry is a complex dynamic system that requires competent management influence directed by the subject of management. In the process of organizing mass low-rise housing construction in the subjects of the Russian Federation, it is necessary to solve a number of issues presented in table 1.

Table 1. Identification of problems of management of low-rise large-panel housing construction

Type of problem	Characteristic of the problem
urban planning problem	lack of territorial planning documents for municipalities in the regions, which has a deterrent effect on the timely preparation of land plots suitable for housing construction
resource problem	acquisition of land or the receipt of rent is an issue for the individual developer and for the developer-builder; land plots intended for integrated development for housing construction from the lands being in state or municipal property, are provided by results of auction on selling rights· at the conclusion of the contract of lease of a land plot without preliminary agreement on the object location, however, to participate in the auction of land must pass state cadastral registration; as a result, the share of land plots granted under auction conditions in the total area of plots occupied for construction is less than 15% [5]
technological problem	cheap when choosing contracting organization for construction of comfortable, aesthetically pleasing home that meets optimum technical solutions and costs; need to move on the most effective system with a wide step of supporting structures and the use of prefabricated wall systems allow to reduce production costs by about 10-15%; the most promising types of residential buildings for low-rise development are: one-, two story with attics, detached or blocked buildings, built on the basis of effective light structures using local building materials
social problem	lack of developed transport and social infrastructure and insufficient number of places of employment
financial problem	lack of financial opportunities for local self-government bodies of municipalities to perform land management preparation of territories for the purpose of organizing auctions for land plots, their arrangement with engineering and social infrastructure

In order to solve the above problems, it is necessary to ensure the development and creation of functional and territorial zones in the general plans for the development of settlements with predominantly low-rise buildings. At the same time, measures should be taken to update the rules of land use and development during urban development activities. These measures will eliminate administrative barriers and create a market for affordable low-rise housing.

The problem of insufficient development and provision of engineering and transport infrastructure for land plots allocated for low-rise buildings is quite acute. In most cases, the sites are not attractive for investment and require high costs to ensure the connection of external engineering networks. That is why the formation of the land market provided with the required engineering infrastructure is a particularly urgent problem of large-panel low-rise construction. It is necessary to make qualitative changes to the urban planning and land codes, aimed at the possibility of providing engineering-secured and fully ready-to-build land plots. In the future, it will be possible to organize land development on the basis of this project, which will fully provide a design approach to managing low-rise construction [6].

The high cost of land plots in most cases is more than half of the cost of construction. Land markets are focused on large developers, which is why when conducting auctions for land plots, those developers who adhere to the trend of increasing prices win, which leads to the monopolization of the market for

IOP Conf. Series: Materials Science and Engineering 880 (2020) 012090 doi:10.1088/1757-899X/880/1/012090

low-rise large-panel housing construction. It will be very promising to hold an auction to lower the price, where the initial price of the auction will be the market price of the sale of finished housing.

To ensure proper management of large-panel low-rise construction, it is necessary to segment consumers in this category. Large-panel low-rise construction belongs to the economy class segment, which in turn makes up more than 54% of the population able to purchase this housing. In connection with segmentation, it will be possible to create investment and construction processes focused on efficiency, energy efficiency and environmental friendliness. Since large-panel low-rise construction is more of an economy-class segment, it is necessary to use innovative energy-efficient technologies when implementing projects [7]. The use of these types of technologies will increase the competitive advantages of large-panel housing construction over other technologies by reducing the cost and time of construction.

The main problem is the lack of an organizational mechanism for regulating and managing large-panel low-rise housing construction at the federal, regional, municipal and project levels [8]. It is necessary to determine the method of cooperation of participants at all levels, which will increase the competitiveness and profitability of large-panel low-rise construction projects.

In order to develop large-panel low-rise construction, it is necessary to ensure the design of the management system of the subjects, which will be aimed at popularizing large-panel low-rise construction and creating demand. It is appropriate to define conceptual approaches aimed at technical, technological, organizational and managerial solutions that cover all levels of management in the management system of low-rise housing construction (figure 2).

The proposed concept is implemented at four levels. The first level is federal, which includes a system for managing the development of large-panel low-rise construction at the level of the Russian Federation, implying the development of federal laws, target programs and federal standards. The second level is regional, which includes a system for managing the development of large-panel low-rise construction at the level of the subject of the Russian federation, which implies the development of a regional concept and strategy.

The third level is municipal, which includes evaluating the strengths and weaknesses of the current state of large-panel low-rise housing construction in each municipality, as well as identifying needs to change the existing characteristics of the participants in the investment and construction process that hinder the effective development of this area in the region. The fourth level is the project level, which includes direct management of the design, construction and commissioning of individual buildings or groups of large-panel low-rise housing construction.

These levels form the main subsystems of development through interaction with each other. As the main functional directions of regulation of the management mechanism, formation and development of large-panel low-rise housing construction, it is proposed to use the active subsystems presented in the figure (figure 3).

At the same time, the effectiveness of the system management will be tracked by the following indicators:

— Criterion for managing housing affordability for consumers. As an indicator of affordability, a coefficient is used that is mandatory for use in all federal and regional housing programs. The coefficient is calculated as the ratio of the cost of a typical house with an area of about 90 m^2 to the annual income of a family of three people. By its content in the final value, the coefficient shows the number of years for which a family can buy a house into ownership, if all the income will be used to pay the cost of the purchased house [9-10];

— Criteria for managing the effectiveness of investment projects and programs. To evaluate this criterion, economic performance indicators are used, which include: net present income, internal rate, payback period, and other accepted indicators.

When implementing large-panel low-rise housing construction programs, it is advisable to use management methods that contain operational and technological functions and economic and mathematical methods that are implemented when making management decisions. A key role in the implementation of large-panel low-rise housing construction programs is played by participants, i.e. subjects of the investment and construction process. Participants activities are tracked when implementing management decisions made at various phases of the program's life cycle.

It is important to note that the quality of management will consist in encouraging the regional management body of project-level participants to choose certain actions related to technological and organizational solutions in accordance with the developed management concept. Based on this, the effectiveness of the territorial level depends directly on the results of the activities of the project level participants [11]. Thus, the territorial management body is interested in providing a set of control actions on the participants of the project level that has the maximum efficiency

At the same time, one of the ways to reduce the degree of uncertainty in management decisions when implementing large-panel low-rise construction projects is to create integrated organizational and production structures that seek to find a balance between centralization and decentralization of management. It should be noted that this kind of cooperation does not slow down the development of existing types of integration of companies, but rather complements them. Due to this, the nature of relationships between companies in the construction sector becomes more complex and diverse, due to the possibility of cooperation of integrated structures.

The limitations and insufficiency of common methodological tools for organizing low-rise housing construction predetermines the need to create an association of construction companies, manufacturers of construction materials, suppliers of material, labor, scientific and investment resources, organized in the main technological chain of creating added value to ensure the technological development of large-panel low-rise construction [12].

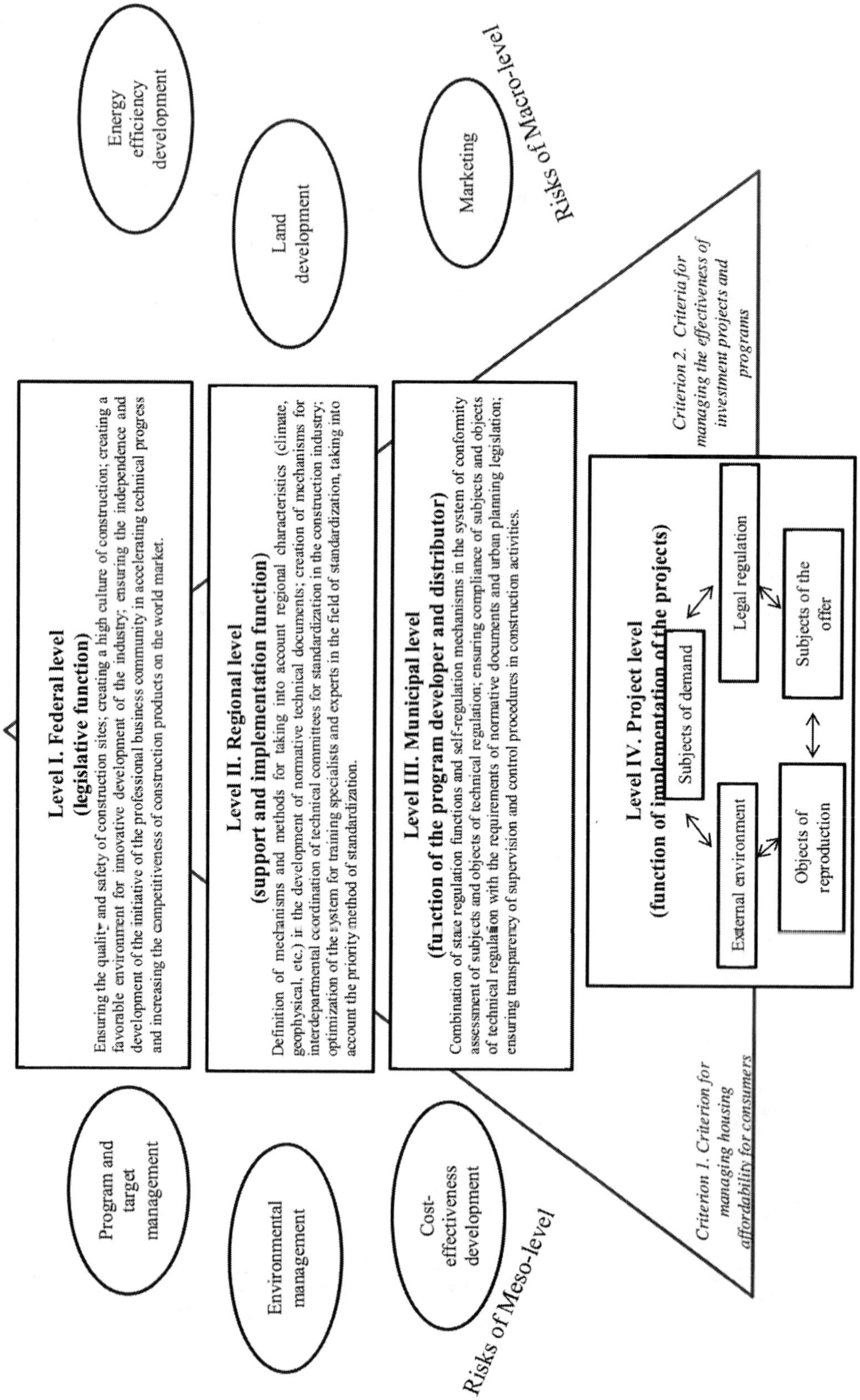

Figure 2. Large panel low rise construction management system

Development of cost-effectiveness of objects, which includes monitoring and managing the price level of real estate objects in the territorial markets for various segments of demand

Development of energy efficiency of objects, which includes monitoring compliance with requirements in construction for the class of buildings for energy efficiency and increasing requirements for this category of management in construction for all types of residential buildings.

Environmental development, which includes the priority direction of the transition to environmental friendliness of construction and the formation of ecological systems of the human environment.

Marketing, including analysis of federal, regional and municipal territorial markets for low-rise housing construction by various management entities.

Land development, which includes territorial and urban planning, the formation and presentation of land plots for large-panel type of development by all branches of government with the organization of open land auctions for large-panel low-rise construction.

Program and target management aimed at creating and developing specialized programs for demand management

Figure 3. Active subsystems of the low-rise housing construction management system

This type of association is called a consortium in the scientific literature. A consortium is a temporary association of several independent companies that are united to implement a specific project. The main goal of the consortium is to combine all the resources of each organization to improve the effectiveness of the project. As practice shows, the main advantage of the consortium is the ability to reduce the cost of construction and installation work by integrating the material, labor and financial resources of organizations in the construction industry.

References

[1] Belyaev M K and Lysenko A V 2013 Development of low-rise housing construction in Russian cities. *Sociology of the city* **2** pp 3-8
[2] Biryukov D A 2010 Legal regulation of consortia activities. *Economy and law* **1** pp 118-123
[3] Vetrova G A 2014 Panel houses can be energy efficient *Roofing and insulation materials* **4**
[4] Gaskova Y I 2018 Integrative approach to the formation of a mechanism for the coordinated development of industry and infrastructure in the region *Proceedings of the II International scientific and practical conference "Economics and modern management: theory, methodology and practice* pp 358-362
[5] Zenin S A, Sharipov R Sh, Kudinov O V, Shapiro G I and Hasanov A A 2016 Calculations of large-panel buildings for stability against progressive collapse by methods of ultimate equilibrium and finite element *Academia. Architecture and construction* **4** pp 48-53
[6] Klavina J N 2018 Economic efficiency of applying new materials and technologies in low-rise Construction. *Economy and management* **1** pp 89-92
[7] Lukmanova I G 2015 Methodological bases of formation and development of the innovative basis of the construction industry *Economics and entrepreneurship* **6-3** pp 985-988
[8] Minaev N N 2007 Scenario options for managing the city's housing and communal complex *Regional economy: theory and practice* **5** pp 178-181
[9] Sapacheva L V and Yumasheva E.I. 2014. Large-panel housing construction remains the fastest

and most economical *Housing construction* **10** pp 3-10
[10] Khmyzov A S 2015 Study of the state and prospects of development of the construction materials industry in the regions of the Siberian federal district *Modern problems of science and education* **1-1** pp 734-738
[11] Official website of the Ministry of Economic development of the Russian Federation. Recommendations for implementing public-private partnership projects. Best practices [Electronic resource]. - Access mode: http://economy.gov.ru/minec/activity/ sections/privgovpartnerdev/support/20160829. (date accessed: 15.11.2019).
[12] Official website of the federal state statistics service [Electronic resource]. – Mode of access: http://www.gks.ru/. (date accessed: 11.10.2019)
[13] Gertsekovich D, Gorbachevskaya L, Grigorova L and Peshkov V 2019 Return on investment in REIT real estate funds *Investments. Construction. Real Estate: New Technologies and Targeted Development Priorities IOP Conference Series: Materials Science and Engineering* **667** 012063
[14] Peshkov A V 2019 Remote mortgage lending as a resource for housing development *Investments. Construction. Real Estate: New Technologies and Targeted Development Priorities IOP Conference Series: Materials Science and Engineering* **667** 012073

IOP Conf. Series: Materials Science and Engineering **880** (2020) 012091 doi:10.1088/1757-899X/880/1/012091

Innovative system of organization of capital repairs of common property of apartment buildings: problems and solutions

E Dediukhina[1]

[1]Irkutsk National Research Technical University, 664074, Irkutsk, Lermontov st. 83, Russia

E-mail: ded_es@mail.ru

Abstract. The analysis and evaluation of the effectiveness of the results of the first stage of the regional capital repair program of the Irkutsk region was carried out. The problems of the innovative system of organization of capital repairs of common property in apartment buildings of the Irkutsk region are considered and ways of their solution are proposed.

The question of carrying out major repairs of the housing stock is very relevant, since it concerns every person who has their own housing. The issue of capital repairs is particularly acute for residents of apartment buildings (hereinafter- AB), due to the presence of common property in such houses and the common responsibility of owners for the safety of this property. It is worth noting that Russia is characterized by a high proportion of multi-apartment housing in the existing housing stock, much more than in Europe and the United States [1].

In accordance with article 178 of the housing code of the RF, the subjects of the country have created regional operators, called the capital repair fund for apartment buildings (hereinafter- the fund), to implement the mechanism for accumulating funds and organizing capital repairs of the common property of the AB. The functions of the funds include the accumulation of

IOP Conf. Series: Materials Science and Engineering **880** (2020) 012091 doi:10.1088/1757-899X/880/1/012091

contributions for major repairs, which are paid monthly by owners in apartment buildings, as well as ensuring the organization of repairs [2].

The minimum contribution that owners of premises in the AB are required to make for major repairs is set by the subject of the RF. It depends on the climatic conditions and geographical location, the type and number of floors of a residential building, the availability of engineering systems, elevator equipment and garbage disposal in the AB [3].

The dynamics of collection of contributions in the Irkutsk region for the period from 2016 to 2019 has a growing trend and is 85% for 2019, 79.53 % for 2018, 73.76% for 2017, and 71.78% for 2016 [4]. However, this indicator is lower than in the whole of the RF, which is about 95% [5].

Owners of premises in an apartment building have the right to choose to transfer contributions for major repairs to a special account or to the account of a regional operator. A special account is intended for accumulating funds for the repair of a particular AB, the owners of which have chosen this method of accumulation. It should be noted that in this case, the entire burden of organizing major repairs, acceptance of completed works falls on the shoulders of the owners [6].

The regional operator finances capital repairs at the expense of the owners, provides capital repairs of common property in the AB, and is responsible for the quality of work. It should be noted that the regional operator's account is shared by all AB's [1].

In 2019, in the Irkutsk region, 95.8% of owners of premises in the AB to accumulate funds for major repairs chose the regional operator's account, 0.7% a special account of the regional operator, 1.1% a special account of the management company, 2.4% a special account of the HOA, HC, HCC [7].

The program of capital repairs of common property in the AB on the territory of the Irkutsk region for 2014-2043 (hereinafter- the program) was approved by The resolution of the government of the Irkutsk region of March 20, 2014 № 138-pp, the purpose of which is to plan and organize capital repairs [8].

Within the framework of the program, a list of the Irkutsk region's AB is formed, where major repairs are planned for each object. The program is planned to be implemented in six stages, each lasting five years. The distribution of houses by stages is approximately the same.

At the moment, the first stage of the program implementation, which took place from 2014 to 2018, has been completed. As a result of the analysis of the results of this stage, the effectiveness of the program is estimated as low. For the specified period, the percentage of plan completion is 36. The effectiveness of the program is recognized as high, if the degree of implementation of the program's activities is more than or equal to 80% [8].

At the same time, there is a tendency to reduce the level of implementation of short-term plans. The dynamics of implementation of the plan for the period of the first stage of the program by year is shown in Figure 1 [4].

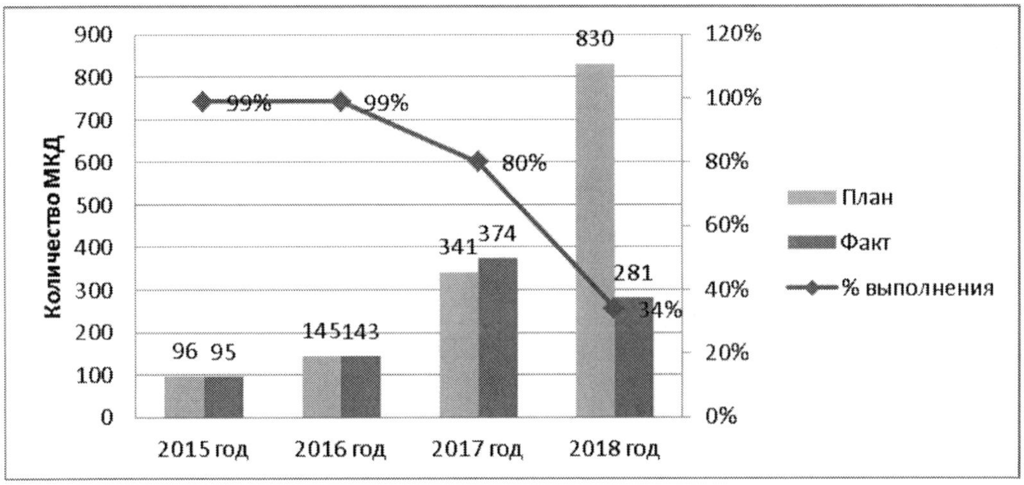

Figure 1. "Dynamics of implementation of the regional program plan of the Irkutsk region"

The reasons for unsatisfactory results in the implementation

In addition, there are problems in planning capital repairs, one of the reasons for which is the lack of a methodology for determining the cost of capital repairs at this stage. Determining the estimated cost using the element-by-element calculation method requires a large amount of input data and is very time-consuming for the investment planning stage, so it is necessary to use methods for calculating the aggregated estimated cost.

In foreign practice, consolidated indicators of the cost of structural elements and types of work per unit of physical volume of buildings and structures are used [9]. There are no consolidated indicators of the cost of repairs in the regulatory framework of pricing in the RF. At the same time, it is necessary to use such standards at the stage of investment planning [10].

When planning the volume of investments for major repairs of the AB, the same accuracy is given by the costs calculated per m2 of the total area or per m3 of the building volume. At the same time, special attention should be paid to determining the actual, rather than the estimated amount of wear and tear of the building, and also the time of operation of the object before major repairs [11]. The accuracy of determining the cost plays an important role, as this determines the effectiveness of investments for major repairs [12]. Minimization of calculation errors when evaluating the cost of repair and construction works will allow for more reliable planning of funds for major repairs of apartment buildings and reduce investment risks [13].

The reasons for unsatisfactory results in the implementation of the first stage of the program and suggestions for their elimination are given in table 1.

Table 1. Recommendations on elimination

Reason	Recommendations on elimination
insufficient awareness and involvement of owners in the process of organizing capital repairs	providing owners with documents for approval that indicate the state of the AB and the cost of its repair
lack of available credit programs to finance capital repairs of the AB	development of state measures to attract credit organizations, including the use of preferential lending rates
disadvantages in ways of accumulating funds for major repairs	implementation of a payment system that recalculates the amount of contributions after major repairs, in accordance with its actual cost
the method of calculating the amount of contributions is too generalized	the calculation of the cost contribution based on the actual wear of the AB, time elapsed since last overhaul, regulatory overhaul of the timing equipment, systems and building structures AB, list of necessary works on capital repair of concrete AB
accumulated funds of citizens are not protected from depreciation due to inflation	provide a state guarantee of annual indexation of accumulated funds, not lower than the percentage of inflation
citizens funds they accumulate on the accounts of regional operators and, as a result, are devalued	optimization of the process of drawing up short-term plans and introduction of mandatory requirements for regional operators regarding the need to use the collected funds
there are no requirements for a thorough examination of the AB before including it in the program	amendments to the housing code of the RF
poor quality of repairs performed	optimization of the system of distribution of contract works for major repairs,

	selection of the contractor not only on the principle of minimizing costs, but also on reliability; inclusion in the list of works performed, those works that affect the safety of further operation of the object [14]
the presence of "extra links" when concluding contracts and performing repairs	when organizing auctions, exclude the possibility of placing lots with a high cost, so that small organizations can participate in the auction, and not work on a subcontract
the use of false information in the formation of regional programs of capital repairs of	updating and adjusting regional programs based on real data obtained from the survey of buildings on the deterioration of houses, and as a result, the cost of repairs

As a result of the unsatisfactory performance of the capital repair fund, the losses are borne by residents who make contributions for capital repairs. A persistent trend of non-fulfillment of short-term plans leads to an increase in the fund's account balances, as a result of the depreciation of funds and a decrease in the efficiency of their use. At the moment, the fund's accounts in the Irkutsk region have accumulated about 4.5 billion rubles. At the same time, in 2019, repairs were planned at a cost of 3.320 billion rubles, and work was performed at 1.43 billion rubles, which is 43% of the planned [15].

Despite the active implementation, the innovative system of organization of capital repairs of the AB common property cannot be called effective. For the successful implementation of the program, it is necessary to increase the involvement of both the owners of the premises and the state as a guarantor of the interests of the country's citizens in the organization of capital repairs of the housing stock. It is necessary to immediately make an inventory of the AB included in the capital repair program in order to identify their actual technical condition and determine the physical deterioration of buildings.

It is very important to optimize the process of accumulating funds for major repairs, by recalculating the contributions of citizens, depending on the actual cost of repairing the AB in which they are the owners of the premises. An equally important point is the problem of the quality of work performed, which will not be solved without mechanisms for regulating the process of distribution of contract works for major repairs.

IOP Conf. Series: Materials Science and Engineering **880** (2020) 012091 doi:10.1088/1757-899X/880/1/012091

From all the above, we can conclude that it is necessary to adjust the existing mechanisms for capital repairs of apartment buildings, as well as to develop innovative solutions to this problem.

References

[1] Ovsiannikova T Y, Rabtsevich O V, Karpova E A 2015 *Tomsk State University of Architecture and Building* **5** pp 235
[2] ConsultantPlus, *Housing Code of the Russian Federation: Federal Law of the Russian Federation No. 188-FZ of December 29, 2004* (http://www.consultant.ru/document/cons_doc_LAW_51057/).
[3] Government of the Irkutsk region, *On the establishment of a minimum contribution for capital repairs of common property in apartment buildings located on the territory of the Irkutsk region for 2020: Decree of the Government of the Irkutsk region dated 27.11.2019 N 1007-pp* (Irkutsk, 2019).
[4] Report on the activities of the capital repair Fund of apartment buildings in the Irkutsk region for 2018 [Electronic resource] // (http://www.fkr38.ru/images/pdf/Godovoy_otchet_2018.pdf) pp.7, 33 (Irkutsk, 2019)
[5] https://www.reformagkh.ru/overhaul [Electronic resource].
[6] Fund for Assistance to Housing and Utilities Reform, *Capital repairs in apartment buildings: questions and answers. Comments and explanations of experts of the state corporation* (CJSC Bibliotechka RG, Moscow, 2013)
[7] https://www.reformagkh.ru/analytics#overhaul [Electronic resource].
[8] Government of the Irkutsk region, *Regional program of capital repairs of common property in apartment buildings on the territory of the Irkutsk region for 2014-2043* (Irkutsk, 2014)
[9] Technical Services Division, *2017 cost schedule* (http://sfdbi.org/sites/default/files/ Cost%20Schedule_0.pdf, SanFrancisco, 2017)
[10] Sayfutdinova R V, *Modern Fundamental and Applied Researchers*, **1(20)**, pp 167-173 (2016)
[11] Dedukhina E, Torgashina I 2018 *MATEC Web of Conferences* ***212***
[12] Dobysheva T V 2015 *Izvestiya Vuzov. Investments. Building. Real Estate* **1(12)** pp 33-38
[13] Yamshchikova I V, Sayfutdinova R V 2015 *Proceedings of the Irkutsk State Technical University* **3(98)** pp 307-312
[14] Konnov N S 2013 *Construction and industrial safety* **4** pp 80-85

[15] http://www.ogirk.ru/2020/02/07/neudovletvoritelnoj-priznana-rabota-fonda-kapremonta-mkd-irkutskoj-oblasti/ [Electronic resource]

[16] Peshkov A V 2019 Digitalization of the economy, implementation of national projects and development of the construction industry *Investments. Construction. Real Estate: New Technologies and Targeted Development Priorities IOP Conference Series: Materials Science and Engineering* 667 012074

[17] Gertsekovich D, Gorbachevskaya L, Grigorova L and Peshkov V 2019 Return on investment in REIT real estate funds *Investments. Construction. Real Estate: New Technologies and Targeted Development Priorities IOP Conference Series: Materials Science and Engineering* **667** 012063

[18] Yaskova N Yu and Sarchenko V I 2018 Development of strategic approaches to the creation of aqua-territorial industrial complexes in the Arctic zone *Proceedings of Universities. Investment. Construction. Real estate* **8** pp. 84-93 DOI: 10.21285/2227-2917-2018-2-84-93

IOP Conf. Series: Materials Science and Engineering **880** (2020) 012092 doi:10.1088/1757-899X/880/1/012092

Analysis of the DuPont model of the construction industry in the Baikal region

E A Filatov[1] and V I Bunkovsky[2]

[1] Irkutsk scientific center, Siberian branch of the Russian Academy of Sciences, 134, Lermontov street, Irkutsk, Russian Federation
[2] Department of management, Irkutsk national research technical university, 83, Lermontov street, Irkutsk, Russian Federation

E-mail: johnru3000@rambler.ru

Abstract. In the context of globalization, the role of the investment component in the dynamics of economic development is significantly increased. The DuPont model allows you to determine which factors caused the change in profitability or to make a factor analysis of the return on equity. The object of the study is construction companies (construction of residential and non-residential buildings) in the Baikal region of the Russian Federation (Irkutsk region, Republic of Buryatia, Zabaikalsky Krai). The article describes the methods of factor analysis developed by Filatov E. A., which make it possible to make an accessible and relatively simple conclusion about changes in the financial position of the enterprise, as well as to assess the degree of influence of factors on changes in the studied indicator. The purpose of this article is to generalize and systematize the theoretical and methodological foundations of factor analysis in a form that is accessible to specialists in the field of economics and management. The article should contribute to the formation of a basis for managers and economists to develop economic thinking, understanding the essence of economic processes and phenomena occurring within the framework of economic entities, and most importantly – the development of skills for their quantitative assessment, economic interpretation and search for reserves to improve the efficiency of functioning. The article was carried out within the framework of the scientific project of the Inc SB RAS no. XI.174.1.4 «Activation of the internal development potential of regions of resource specialization (on the example of the Baikal region)».

1. Introduction

Construction is one of the key fund – forming industries, on which the formation of the national economy largely depends. In the structure of the gross value added of the Russian Federation for 2017, the construction sector occupies 6.2%.

Construction is a fund – forming industry, on which the formation of the national economy depends, and that is, the formation of a complex economic, social, organizational, and scientific and technical system depends.

The construction industry sets the pace of economic development and is a very significant sector in the economy of many countries of the world. The relevance of the article is due to the dynamic development of construction services in the last few decades in the world economy. At the same time, this industry is very sensitive to even minor changes in the global market environment.

2. Research questions

Improving the methods of factor analysis is a key task of economic analysis. The more detailed the influence of factors on the value of the indicator is considered, the more accurate the results of the analysis and the assessment of the quality of the decision made. In some situations, without a deep and comprehensive study of the direct impact of factors, it is impossible to draw reasonable conclusions about the company's performance.

One of the most important tasks of financial analysis of any economic phenomenon is to identify factors whose level and dimensions have a decisive influence on the formation and change of the level of the phenomenon, considered as effective in relation to these factors (Hakstian, A. R., Roqers, W. D., & Cattell, R. B., 1982).

One of the main tasks of financial management in the company is to ensure financial stability, which could contribute to the growth of return on equity.

In order to evaluate the performance of any company, you need to use some methods of analysis. The analysis methodology is understood as a certain sequence of operations, techniques, actions, and rules for the most appropriate execution of analytical work (Kim, J. O., & Mueller, C. W., 1978a).

Deterministic factor analysis is a method of studying the influence of factors that are related to the performance indicator is functional, that is, the performance indicator can be represented as the product of various factors (Kim, J. O., & Mueller, C. W., 1978b).

The profitability of the company and the growth of its value for shareholders is reflected by the profitability coefficients. Managing the profitability of an enterprise becomes a key task for all levels of management: strategic, tactical, and operational. DuPont has proposed a simple way to manage profitability by decomposing the profitability coefficient into factors that reflect various aspects of the company's activities.

In practical analytical work, the DuPont method was used by DuPont specialists in the 1920s.

The DuPont equation (also the DuPont Model or the DuPont Formula) is a modified factor analysis that allows you to determine which factors caused the change in profitability. The basis of the factor model in the form of a tree structure is the indicator of return on equity (**ROE**), and the signs are characterizing the factors of production and financial activity of the enterprise (Ivanov, P. I., 2019; Kazakova, N. A., 2018; Rozanova, N. M., & Zoroastrova, I. V., 2018). Simply put, the factors that affect **ROE** are broken down to find out which factors more or less affect the return on equity. The main three factors:

1. return on sales (this coefficient is calculated as the ratio of net profit to revenue (net) from all types of sales);
2. asset turnover (the asset turnover ratio or total capital return is calculated as the ratio of revenue from all types of sales to the total amount of assets);
3. the equity multiplier (the inverse of the financial independence coefficient or the autonomy coefficient characterizes the ratio of total assets to the organization's equity).

The purpose of the study is to analyze the DuPont model of the construction industry in the Baikal region using new author's methods of factor-based deterministic analysis (Filatov methods), which more reliably and argumentatively evaluated its results.

3. Materials and methods

The initial data for the analysis of the DuPont model of the construction industry in the Baikal region using Filatov's methods (Filatov, E. A., 2018) are presented in table 1.

The object of the study is construction companies (construction of residential and non-residential buildings) in the Baikal region.

The sources of the information base on the financial results of construction companies in the Baikal region were financial statements according to Russian accounting standards: the Balance sheet as of December 31, 2017 and 2018; the Report on financial results for 2017 and 2018 (IS «BIR-Analyst», 2019).

Table 1. Initial data for factor analysis of the Baikal region by Federal Subjects

No.	Indicators	№ fac-tor's	2017 year Plan (0) *	2018 year Fact (I) **	Deviation (Δ) ***
1	**V** – The net revenue, thousand $		1 194 149	1 086 591	-107 559
	including:				
	- Irkutsk region		883 524	806 341	-77 183
	- Republic of Buryatia		139 895	133 197	-6 697
	- Zabaikalsky Krai		170 731	147 052	-23 678
2	**P** – Net profit, thousand $		33 980	-21 583	-55 562
	including:				
	- Irkutsk region		16 198	24 524	8 326
	- Republic of Buryatia		8 313	-51 781	-60 094
	- Zabaikalsky Krai		9 469	5 675	-3 794
3	**A** – Value of assets, thousand $		1 633 701	1 459 641	-174 060
	including:				
	- Irkutsk region		1 109 751	1 054 510	-55 241
	- Republic of Buryatia		294 550	232 175	-62 375
	- Zabaikalsky Krai		229 400	172 956	-56 444
4	**SK** – The average cost of equity, thousand $		126 426	180 520	54 094
	including:				
	- Irkutsk region		49 299	112 796	63 497
	- Republic of Buryatia		28 202	25 530	-2 673
	- Zabaikalsky Krai		48 924	42 194	-6 730
5	**ROS – Return On Sales** (2/1)	F_1	0.028455	-0.019863	-0.048318
	including:				
	- Irkutsk region		0.018334	0.030414	0.012080
	- Republic of Buryatia		0.059420	-0.388756	-0.448176
	- Zabaikalsky Krai		0.055460	0.038590	-0.016870
6	**AT – Asset Turnover** (1/3)	F_2	0.730947	0.744423	0.013476

	including:				
	- Irkutsk region		0.796146	0.764659	-0.031487
	- Republic of Buryatia		0.474944	0.573694	0.098751
	- Zabaikalsky Krai		0.744249	0.850228	0.105979
7	**EM – Equity multiple** (3/4)	F_3	12.922192	8.085758	-4.836434
	including:				
	- Irkutsk region		22.510630	9.348788	-13.161842
	- Republic of Buryatia		10.444132	9.094342	-1.349791
	- Zabaikalsky Krai		4.688876	4.099078	-0.589798
8	**ROE – Return On Equity** (2/4) = (5 * 6 * 7)		0.268774	-0.119560	**-0.388334**
	including:				
	- Irkutsk region		0.328572	0.217417	-0.111155
	- Republic of Buryatia		0.294745	-2.028284	-2.323030
	- Zabaikalsky Krai		0.193538	0.134493	-0.059045

where: * 0 – past (base) period (year) taken as a basis for comparison; ** I – reporting (current) year; *** Δ – change for the period is calculated as the difference between the fact and the plan (I – 0).

The decrease in such basic absolute financial indicators as revenue by -9.01% and total assets by 10.65% is due to the depreciation of the ruble against the dollar during 2018 by 17.09%.

In this case, the original formula for factor analysis will have the following form (formula 1):

$$ROE = \frac{P}{V} * \frac{V}{A} * \frac{A}{SK} = ROS * AT * \mathrm{EM} = F_1 * F_2 * F_3 \quad (1)$$

The deviation balance is found as follows (formula 2):

$$\Delta ROE = \sum_{n=1}^{3} ROE\,(F_n) = \Delta ROE\,(F_1) + \Delta ROE\,(F_2) + \Delta ROE\,(F_3) \quad (2)$$

The algebraic sum of the influence of factors must necessarily be equal to the total increase in the effective indicator. The absence of such equality indicates mistakes in the calculations.

Next, we will consider 10 methods of deterministic factor analysis developed by Filatov E. A.

Auxiliary data on the author's comparative coefficients for factor analysis are presented in tables 2, 3.

Table 2. Multiple comparative coefficients for one factor in the whole Baikal region

Comparison of factors	The designation of comparative coefficient	Value	The product of coefficients (value)
$F_{1(I)} / F_{1(0)}$	A_1	-0.698041	1.00
$F_{1(0)} / F_{1(I)}$	A_2	-1.432581	
$F_{2(I)} / F_{2(0)}$	A_3	1.018437	1.00
$F_{2(0)} / F_{2(I)}$	A_4	0.981897	
$F_{3(I)} / F_{3(0)}$	A_5	0.625727	1.00
$F_{3(0)} / F_{3(I)}$	A_6	1.598142	

Table 3. Multiplicative comparative coefficients for two factors in the Baikal region as a whole

Comparison of factors	The designation of comparative coefficient	Factor factors	Value
$(F_{1(I)} * F_{2(I)}) / (F_{1(0)} * F_{2(0)})$	B_1	$A_1 * A_3$	-0.710911
$(F_{2(0)} * F_{3(0)}) / (F_{2(I)} * F_{3(I)})$	B_2	$A_4 * A_6$	1.569211

Author's (alternative) methods of factor analysis are presented in table 4.

Method No.1.1 (formulas 1.1 – 1.3 in table 4) is based on the difference between the effective targets, which is adjusted by comparative coefficients (A_1, B_1).

Method No. 1.2 (formulas 2.1 – 2.3 in table 4) is based on the difference between the effective actual indicators, which is adjusted by comparative coefficients (A_6, B_2).

Method No. 2.1 (formulas 3.1 – 3.3 in table 4) is based on the relation of deviation of the initial factor to the initial planned factor multiplied by the planned effective indicator which is corrected by the comparative coefficient (A_1, B_1).

Method No. 2.2 (formulas 4.1 – 4.3, in table 4) is based on the relation of deviation of the initial factor to the initial actual factor multiplied by the actual effective indicator which is corrected by the comparative coefficient (A_6, B_2).

Method No. 3.1 (formulas 5.1 – 5.3 in table 4) is based on the difference between the effective actual and planned indicators, which are adjusted by comparative coefficients (A_1, B_1).

Method No. 3.2 (formulas 6.1 – 6.3 in table 4) is based on the difference between the effective actual and planned indicators, which are adjusted by comparative coefficients (A_6, B_2).

Method No. 4.1 (formulas 7.1 – 7.3 in table 4) is based on the ratio of the deviation of the effective factor to the difference between the effective actual and planned factors, which is adjusted by a comparative factor (A_1, B_1).

Method No. 4.2 (formulas 8.1 – 8.3 in table 4) is based on the relation of deviation of the effective factor to the difference between the effective actual and planned factors, which is adjusted by comparative coefficients (A_6, B_2).

Method No. 5.1 (formulas 9.1 – 9.3 in table 4) is based on the ratio of the deviation of the effective factor to the difference between the actual effective factors, which is adjusted by comparative coefficients (A_1, B_1).

Method No. 5.2 (formulas 10.1 – 10.3 in table 4) is based on the ratio of the deviation of the effective factor to the difference between the planned effective factors, which is adjusted by comparative coefficients (A_6, B_2).

Table 4. Methods of alternative factor analysis using comparative coefficients

№ formulae	formulas / calculations	
	the main part of the formula	adjustment factors
1.1	Δ ROE (F_1) = ROE $_0$*(A_1) – ROE $_0$	–
1.2	Δ ROE (F_2) = (ROE $_0$*(A_3) – ROE $_0$)*	A_1
1.3	Δ ROE (F_3) = (ROE $_0$*(A_5) – ROE $_0$)*	(A_1 * A_3) or B_1
2.1	Δ ROE (F_1) = (ROE $_I$ – ROE $_I$ * (A_2))*	(A_6 * A_4) or B_2
2.2	Δ ROE (F_2) = (ROE $_I$ – ROE $_I$ * (A_4))*	A_6
2.3	Δ ROE (F_3) = ROE $_I$ – ROE $_I$ * (A_6)	–
3.1	Δ ROE (F_1) = (ΔF_1/ $F_{1(0)}$) * ROE $_0$	–
3.2	Δ ROE (F_2) = (ΔF_2/ $F_{2(0)}$) * ROE $_0$)*	A_1
3.3	Δ ROE (F_3) = ((ΔF_3/ $F_{3(0)}$) * ROE $_0$)*	(A_1 * A_3) or B_1
4.1	Δ ROE (F_1) = ((ΔF_1/ $F_{1(I)}$) * ROE $_I$)*	(A_6 * A_4) or B_2
4.2	Δ ROE (F_2) = ((ΔF_2/ $F_{2(I)}$) * ROE $_I$)*	A_6
4.3	Δ ROE (F_3) = ((ΔF_3/ $F_{3(I)}$) * ROE $_I$	–
5.1	Δ ROE (F_1) = (ROE $_I$*A_4*A_6) – ROE $_0$	–
5.2	Δ ROE (F_2) = ((ROE $_I$*A_2*A_6) – ROE $_0$)*	A_1
5.3	Δ ROE (F_3) = ((ROE $_I$*A_2*A_4) – ROE $_0$)*	(A_1 * A_3) or B_1
6.1	Δ ROE (F_1) = (ROE $_I$ – (ROE $_0$*A_3*A_5))*	(A_6 * A_4) or B_2
6.2	Δ ROE (F_2) = (ROE $_I$ – (ROE $_0$*A_1*A_5))*	A_6
6.3	Δ ROE (F_3) = ROE $_I$ – (ROE $_0$*A_1*A_3)	–
7.1	Δ ROE (F_1) = Δ ROE – (ROE $_I$ – (ROE $_0$*A_1)	–

7.2	**Δ ROE (F_2) = Δ ROE – (ROE $_I$ – (ROE $_0$*A_3))***	**A_1**
7.3	**Δ ROE (F_3) = Δ ROE – (ROE $_I$ – (ROE $_0$*A_5))***	**(A_1 * A_3) or B_1**
8.1	**Δ ROE (F_1) = Δ ROE – ((ROE $_I$ *A_2) – ROE $_0$)***	**(A_6 * A_4) or B_2**
8.2	**Δ ROE (F_2) = Δ ROE – ((ROE $_I$ *A_4) – ROE $_0$)***	**A_6**
8.3	**Δ ROE (F_3) = Δ ROE – ((ROE $_I$ *A_6) – ROE $_0$)**	–
9.1	**Δ ROE (F_1) = Δ ROE – (ROE $_I$ – (ROE $_I$*A_4*A_6))**	–
9.2	**Δ ROE (F_2) = Δ ROE – (ROE $_I$ – (ROE $_I$*A_2*A_6))**	**A_1**
9.3	**Δ ROE (F_3) = Δ ROE – (ROE $_I$ – (ROE $_I$*A_2*A_4))***	**(A_1 * A_3) or B_1**
10.1	**Δ ROE (F_1) = Δ ROE – (ROE $_0$*A_5*A_3) – ROE $_0$)***	**(A_6 * A_4) or B_2**
10.2	**Δ ROE (F_2) = Δ ROE – ((ROE $_0$*A_5*A_1) – ROE$_0$)***	**A_6**
10.3	**Δ ROE (F_3) = Δ ROE – ((ROE $_0$*A_3*A_1) – ROE $_0$)**	–

4. Results

The result for methods 1.1, 2.1, 3.1, 4.1, 5.1 is shown in table 5, the result for methods 1.2, 2.2, 3.2, 4.2, 5.2 is shown in table 6.

Table 5. Results for methods 1.1, 2.1, 3.1, 4.1, 5.1 for the Baikal region

No.	**the main part of the formula**	**adjustment factors**		**result**
1	**Δ ROE (F_1)** = -0.456389	–		-0.456389
2	**Δ ROE (F_2)** = 0.004955	-0.698041	**A_1**	-0.003459
3	**Δ ROE (F_3)** = -0.100595	-0.710911	**A_1 * A_3**	0.071514
	-0.552029			**-0.388334**

Table 6. Results for methods 1.2, 2.2, 3.2, 4.2, 5.2 for the Baikal region

No.	**the main part of the formula**	**adjustment factors**		**result**
1	**Δ ROE (F_1)** = -0.290840	1.569211	**A_6 * A_4**	-0.456389
2	**Δ ROE (F_2)** = -0.002164	1.598142	**A_6**	-0.003459
3	**Δ ROE (F_3)** = 0.071514	–		0.071514
	-0.221490			**-0.388334**

As can be seen from the final result of tables 1, 5, 6, the purpose of the analysis is achieved – the determination of the influence of factors is disclosed without deviations.

The final change in the return on equity of the construction industry in the Baikal region had a positive impact:

- decrease in the equity multiplier by -483.64%, caused an increase in return on equity by **7.15%**.

The final change in the return on equity of the construction industry in the Baikal region had a negative impact:

- decrease in return on sales by -4.83%, caused a decrease in return on equity by **-45.64%**;
- an increase in asset turnover by 1.35% caused a decrease in return on equity by **-0.34%**.

The combined effect of three factors led to a decrease in return on equity by **-38.83%**.

The main decrease in the return on equity of the construction industry in the Baikal region is due to a decrease in the return on sales of the construction industry in the Baikal region. In turn, the decrease in the profitability of sales of the construction industry of the Baikal region is mainly due to a decrease in the net profit of the construction industry of the Republic of Buryatia by -60.1 million dollars.

The turnover of assets of the construction industry in the Baikal region has not changed much (a change of about 1%), which has almost no effect on the change in the return on equity of the construction industry in the Baikal region. The turnover of the construction industry of the main economy of the Baikal region – Irkutsk region decreased by 3%, which was offset by a 10% increase in the turnover of the construction industry of the Republic of Buryatia and the Zabaikalsky Krai.

The main increase in the return on equity of the construction industry in the Baikal region is due to changes in the equity multiplier of the construction industry in the Baikal region. In turn, the change in the equity multiplier of the construction industry in the Baikal region was mainly influenced by an increase in the equity of the construction industry in the Irkutsk region by +63.5 million dollars or 2.3 times.

5. Discussion

The investment attractiveness of the region depends on the level of return on equity. From the point of view of owners, profitability is most reliably displayed in the form of return on equity. This indicator is important for the company's shareholders, as it characterizes the profit that the owner will receive from one monetary unit of investment invested in the company. Return on equity provides an idea of the revenue that a company earns for shareholders. Return on equity is compared with possible alternative investments in shares of other companies, bonds, bank deposits, and so on.

The minimum (standard) level of profitability of an entrepreneurial business is the level of bank deposit interest. The minimum regulatory value of the return on equity (**ROE_N**) is determined using the following formula 3:

$$\mathbf{ROEn} = \boldsymbol{Sd} * (\mathbf{1} - \boldsymbol{Sp}) \qquad \mathbf{(3)}$$

where: $\mathbf{ROE_N}$ – the Minimum standard value of the return on equity indicator; **ROE** – return on equity; **Sd** – the average rate on Bank deposits for the reporting period; **Sp** – the income tax rate.

Let's calculate $\mathbf{ROE_N}$ for 2017 in Russia. Weighted average interest rates on deposits (deposits) of individuals attracted by credit institutions in rubles for the whole of the Russian Federation up to 1 year, including «on demand» in 2017, based on data from the Central Bank of the Russian Federation, amounted to 5.86%. The income tax rate is 20%. As a result, $\mathbf{ROE_N}$ = 0.0586 * (1 – 0,2) = 0.0469 or **4.69%**.

Let's calculate $\mathbf{ROE_N}$ for 2018 in Russia. Weighted average interest rates on deposits (deposits) of individuals attracted by credit institutions in rubles for the whole of the Russian Federation up to 1 year, including «on demand» in 2018, based on data from the Central Bank of the Russian Federation, amounted to 5.35%. The income tax rate is 20%. As a result, $\mathbf{ROE_N}$ = 0.0535 * (1 – 0,2) = 0.0428 or **4.28%**.

Count $\mathbf{ROE_N}$ for 9 months of the year 2019 in the Russia. Weighted average interest rates on deposits (deposits) of individuals attracted by credit institutions in rubles for the whole of the Russian Federation up to 1 year, including «on demand» for 9 months of 2019, based on data from the Central Bank of the Russian Federation, amounted to 5.63%. The income tax rate is 20%. As a result, $\mathbf{ROE_N}$ = 0.0563 * (1 – 0,2) = 0.045 or **4.5%**.

In 2017, the **ROE** of the construction industry: Irkutsk region 32.86%, Zabaikalsky Krai 19.35%, the Republic of Buryatia 29.47%.

In 2018, the **ROE** of the construction industry: Irkutsk region 21.74%, Zabaikalsky Krai 13.45%, the Republic of Buryatia -202.83%.

6. Conclusion

Analysis based on the DuPont method allows you to determine the strengths and weaknesses of economic entities. Each weak financial ratio may be subject to a deeper decomposition to get an accurate idea of the cause of the weakness. When the sources of problems are identified, the company's management can develop measures to improve return on equity, for example, improve cost control, improve asset management, improve marketing, and so on (Boddy, D., 2002; Dolan, E. G., & Lindsey, D. E., 1988; Farnham, P. G., 2005; Glen, A., 2002; Naylor, J., 2004; Zikmund, W. G., 2003).

As you can see, the return on equity under the DuPont model depends on three factors: the level of return on sales, the speed of asset turnover, and the equity multiplier. Thus, the directions for improving the return on equity are directly indicated. DuPont's three-factor model shows the impact of operating activities (sales), investment and financial activities on the firm's return on equity.

Factor analysis is aimed at identifying the impact of individual factors on the performance indicator, so deterministic modeling of factor systems is a simple and effective means of formalizing the relationship of economic indicators, which serves as a basis for quantifying the role of individual factors in the dynamics of changes in the generalizing indicator (Bartholomew, D.J., 1984). Due to the fact that factor deterministic factor analysis is aimed at identifying the influence of factors on the value of the effective indicator of interest excluding error, it is most relevant for practical use in market relations (Joreskog, K. G., & Sorbom, D., 1979).

References

[1] Bartholomew D J 1984 The foundations of factor analysis. *Biometrika* **71** pp 221-232
[2] Boddy D 2002 *Management An Introduction* (Prentice Hall) p 598
[3] Dolan E G and Lindsey D E 1988 *Economics (*The Dryden Press) p 921
[4] Farnham P G 2005 *Economics for managers* (Prentice Hall) p 558
[5] Filatov E A 2018 Deterministic factor analysis of three-factor DuPont model using Filatov methods *The European Proceedings of Social & Behavioural Sciences* ***45*** pp 364-372
[6] Glen A 2002 *Corporate financial management* (Prentice Hall) 1098
[7] Hakstian A R, Rogers W D and Cattell R B 1982) The behavior of numbers of factors rules with simulated data *Multivariate Behavioral Research* **17** pp 193-219
[8] Ivanov P I 2019 *Management, financial and investment analysis* p 160
[9] *Information system «BIR-Analyst»* 2019
[10] Joreskog K G and Sorbom D1979 *Advances in factor analysis and structural equation models* (Cambridge, MA: Abt Books)
[11] Kazakova N A 2018 *Economic analysis in business assessment and management of the company's investment attractiveness* (Moscow: Finance and statistics) p 240
[12] Kim J O and Mueller C W 1978a *Factor analysis: Statistical methods and practical issues*
[13] Kim J O and Mueller C W 1978b *Introduction to factor analysis: What it is and how to do it (*Beverly Hills, CA: Sage Publications)
[14] Naylor J 2004 *Management* (Prentice Hall) p 668
[15] Rozanova N M and Zoroastrova I V 2018 *Economic analysis of the firm and the market* p 280
[16] Zikmund W G 2003 *Business research methods* (THOMSON) p 748

IOP Conf. Series: Materials Science and Engineering **880** (2020) 012093 doi:10.1088/1757-899X/880/1/012093

Analysis of the Filatov model of the construction industry in the Baikal region

E A Filatov[1] **and L G Rudykh**[2]

[1]Irkutsk scientific center, Siberian branch of the Russian Academy of Sciences, 134, Lermontov street, Irkutsk, Russian Federation
[2]Department of history and philosophy, Irkutsk national research technical university, 83, Lermontov street, Irkutsk, Russian Federation

E-mail: johnru3000@rambler.ru

Abstract. In recent years, due to the slowdown in industrial development and the decline in the economic growth of the Russian Federation in trade and services, many companies from these activities in order to minimize their costs stop expanding production through new construction and at the same time freeze what they have started. In these conditions, managing the profitability of construction companies is one of the most important tasks for all levels of management: operational, tactical and strategic. Return on assets is the most important indicator of the efficiency of a commercial company, the main standard or average value in a market economy, with which individual indicators of companies are correlated to justify their competitiveness. The article suggests a way to manage the return on assets by decomposing the Filatov model (return on assets) into factors that reflect various aspects of the company's activities. The object of the study is construction companies (construction of residential and non-residential buildings) in the Baikal region of the Russian Federation (Irkutsk region, Republic of Buryatia, Zabaikalsky Krai). The article presents the method No. 2 of integral analysis developed by Filatov E. A. The article was carried out within the framework of the scientific project of the Inc SB RAS no. XI.174.1.4 «Activation of the internal development potential of regions of resource specialization (on the example of the Baikal region)».

1. Introduction

Profitability is a relative indicator of the company's level of profitability, it characterizes the efficiency of work. Profitability, in contrast to profit, more fully reflects the final results of management, as it shows the ratio of the effect with cash or consumed resources. The efficiency of any commercial company's business activities and the economic feasibility of their operation are directly related to their profitability (Alexandrov, O. A., 2016; Kazakova, N. A., 2016; Kireeva, N. V., 2019; Kogdenko, V. G., 2015; Lysenko, D. V., 2019; Markin, Yu. P., 2018; Rusakova, E. V., 2018).

Profitability indicators are used to evaluate the company's performance and as a tool in investment policy and pricing. The construction complex of the Russian Federation is one of the largest and most important sectors of the economy, which largely determines the socio-economic development of the country. The profitability of the construction industry is slightly higher than the average for the entire Russian economy as a whole.

2. Research questions

The Filatov model is a modified factor analysis that allows you to determine which factors caused the change in profitability. The basis of the factor model in the form of a tree structure is the indicator of return on assets (**ROA**), and the signs are the characterizing factors of the enterprise's activity. Simply put, the factors that affect the **ROA** are broken down to find out which factors more or less affect the return on assets. The main three factors:

1. return on equity (this ratio is calculated as the ratio of net profit to equity);

2. the author introduced the coefficient of income coverage with equity (this coefficient is calculated as the ratio of equity to revenue (net) from all types of sales);

3. asset turnover (the asset turnover ratio or total capital return is calculated as the ratio of revenue from all types of sales to the total amount of assets).

The difference between return on assets (**ROA**) and return on equity (**ROE**) is that **ROA** shows the effectiveness of all assets of the enterprise, and **ROE** only those that belong to the owners.

The advantage of the Filatov model is the «splitting» of complex indicators into factors that make up them. This allows you to determine the root causes and the relationship of changes in complex performance indicators of the

IOP Conf. Series: Materials Science and Engineering **880** (2020) 012093 doi:10.1088/1757-899X/880/1/012093

company. With the help of a factor analysis scheme for key indicators, you can clearly trace the influence of primary factors on the formation of complex indicators, give their comparative characteristics and determine the reasons for their change.

Return on equity, the ratio of return on equity and asset turnover characterize simultaneously three types of company's activities: investment, financial, and operating.

3. Materials and methods

The initial data for the analysis of the Filatov model (return on assets) of the construction industry in the Baikal region using the Filatov integral method No. 2 (Filatov, E. A., 2019) are presented in table 1.

The object of the study is construction companies (construction of residential and non-residential buildings) in the Baikal region.

The sources of the information base on the financial results of construction companies in the Baikal region were financial statements according to Russian accounting standards: the Balance sheet as of December 31, 2017 and 2018; the Report on financial results for 2017 and 2018 (IS «BIR-Analyst», 2019).

Table 1. Initial data for factor analysis

No.	**Indicators**	**No. factor's**	**2017 year Plan (0)** *	**2018 year Fact (I)** **	**Deviation** (Δ) ***
1	**V** – Revenue from sales, thousand $		1 194 149	1 086 591	-107 559
2	**P** – Net profit, thousand $		33 980	-21 583	-55 562
3	**A** – Total assets, thousand $		1 633 701	1 459 641	-174 060
4	**SK** – Own capital, thousand $		126 426	180 520	54 094
5	**ROE – Return On Equity** (2/4)	F_1	0.268774	-0.119560	-0.388334
6	**EICR – Equity Income Coverage Ratio** (4/1)	F_2	0.105871	0.166134	0.060263
7	**AT – Asset Turnover** (1/3)	F_3	0.730947	0.744423	0.013476
8	**ROA – Return On Assets** 2/3 = (5 * 6 * 7)		0.020799	-0.014787	**-0.035586**

where: * **0** - past (basic) period (year), taken as reference base; ** **I** – reported (current) period (year); *** **Δ** – change for the period, calculated as the difference between the fact and the plan (**I – 0**).

To formalize the integral method No. 2, a multiplicative dependence is chosen – this is a dependence in which all factors are multiplied among themselves. In this case, the original formula for factor analysis will have the following form (formula 1):

$$\mathbf{ROA} = \boldsymbol{ROE} * \boldsymbol{EICR} * \boldsymbol{AT} = \boldsymbol{F_1} * \boldsymbol{F_2} * \boldsymbol{F_3} = \prod_{n=1}^{3} \boldsymbol{F_n} \quad (1)$$

The total deviation of the resulting indicator (**ΔROA**) is determined by the formula 2:

$$\mathbf{\Delta ROA} = \sum_{n=1}^{3} \mathbf{\Delta ROA}(\boldsymbol{F_n}) = \mathbf{\Delta ROA}(\boldsymbol{F_1}) + \mathbf{\Delta ROA}(\boldsymbol{F_2}) + \mathbf{\Delta ROA}(\boldsymbol{F_3}), \quad (2)$$

The calculation of the influence of factors on the change of the effective indicator in the author's integral method No. 2 is presented in formulas 3.1–3.3:

$$\Delta \text{ ROA } (F_1) = (\Delta F_1 * \text{FO№2}_1) + \text{Z№2} \quad (3.1)$$
$$\Delta \text{ ROA } (F_2) = (\Delta F_2 * \text{FO№2}_2) + \text{Z№2} \quad (3.2)$$
$$\Delta \text{ ROA } (F_3) = (\Delta F_3 * \text{FO№2}_3) + \text{Z№2} \quad (3.3)$$

When using the traditional integral method, the additional increase in the effective indicator («non-decomposable remainder» – Z) formed as a result of the interaction of factors is distributed equally between them (formula 4).

$$\text{Z№2} = \Delta\text{ROA} - \sum ((\Delta F_n * \text{FO№2}_n) / n \quad (4)$$

where: FO№2_n – the main part of the formula of the author's integral method is calculated using formulas 5.1–5.3:

$$\text{FO№2}_1 = ((F_{2(0)} * F_{3(I)}) + (F_{2(I)} * F_{3(0)})) / 2 \quad (5.1)$$
$$\text{FO№2}_2 = ((F_{1(0)} * F_{3(I)}) + (F_{1(I)} * F_{3(0)})) / 2 \quad (5.2)$$
$$\text{FO№2}_3 = ((F_{1(0)} * F_{2(I)}) + (F_{1(I)} * F_{2(0)})) / 2 \quad (5.3)$$

To form the main part of the formula (FO№2_n), it is necessary to use the principle of selecting factors disclosed in table No. 2.

Table 2. Selection of factors for the main part of the formula (FO_n) by the author's integral methods

Under influence No. factors	the sum of the factors			
	The 1st factor		The 2nd factor	
	0	I	I	0
1	2	3	2	3
2	1	3	1	3
3	1	2	1	2

где: **m** – number of indicators in the main part of the formula (table 2). **m** is defined by the formula 6:

$$\mathbf{m = n * (2*(n - 1))} \quad (6)$$

With 3 factors in the model (**n** = 3), **m** will be 12 (m = 3 * (2 * 2) = 3 * 4).

4. Results

Testing of the author's method of factor integral analysis No. 2 presented above is presented in tables No. 3, 4.

Table 3. Components of the formula according to the author's integral method No. 2

No. formula e	part of the formula		
	ΔF_n	the main part of the formula (FO№2_n)	Z
1	ΔROA (F_1) = ΔF_1 *	(($F_{2(0)}$* $F_{3(I)}$) + ($F_{2(I)}$* $F_{3(0)}$)) / 2	Z
2	ΔROA (F_2) = ΔF_2 *	(($F_{1(0)}$* $F_{3(I)}$) + ($F_{1(I)}$* $F_{3(0)}$)) / 2	Z
3	ΔROA (F_3) = ΔF_3 *	(($F_{1(0)}$* $F_{2(I)}$) + ($F_{1(I)}$* $F_{2(0)}$)) / 2	Z

Table 4. The result of the author's integral method No. 2

No. factor's	part of the formula			
	ΔF_n	the main part of the formula (FO№2_n)	Z№2	Final result

1	ΔROA (F_1) = -0.388334	0.100124	-0.000105	-0.038986
2	ΔROA (F_2) = 0.060263	0.056345	-0.000105	0.003290
3	ΔROA (F_3) = 0.013476	0.015997	-0.000105	0.000110
Total	-0.035271		-0.000315	**-0.035586**

Factor analysis allows you to obtain a quantitative assessment of the influence of factor deviations on the deviation of the value of the studied indicator. As can be seen from the final result of tables No. 1, No. 4, the goal of the author's method has been achieved – the determination of the influence of factors is disclosed without deviations.

Thus, the change in the return on assets of construction companies in the Baikal region was influenced by:

- decrease in return on equity by 38.83 %, caused a decrease in return on assets by **3.90 %**;
- an increase in the equity income coverage ratio by 6.03 % caused a decrease in the return on assets by **0.33 %**;
- increase in asset turnover by 1.35 %, caused a decrease in return on assets by **0.01 %**.

The combined effect of three factors led to a decrease in the return on assets of construction companies in the Baikal region by -3.56 %.

The return on assets for the construction industry in the Baikal region in 2017 was 2.08 %, that is, 2 cents of net profit was received from $ 1 of the property of construction companies in the Baikal region. In 2018, the return on assets of the construction industry in the Baikal region decreased to -1.48 %, meaning that 1.5 cents of loss was received from $ 1 of the property of construction companies in the Baikal region.

5. Discussion

Construction as an economic sector participates in the creation of fixed assets for all sectors of the national economy. This is one of the leading industries where the vital tasks of restructuring the material base of the country's entire production potential and developing the non-production sector are being solved. Both the rate of recovery from the crisis and the competitiveness of the domestic economy largely depend on the efficiency of the construction complex. It is not for nothing that it is said that if the economic situation in the country worsens, its construction industry first of all suffers.

According to the state and development of the construction industry in

Russia, we can judge our realities and aspirations. The global financial crisis affected all industries in Russia, as well as in other countries. First of all, banks were affected, and construction companies also felt the impact of the crisis.

If you take, in recent years, construction has developed at a high rate, thanks to lending and favorable investment conditions in the industry. A significant outflow of investment as a result of the crisis had a significant impact on construction.

6. Conclusion

Each weak financial ratio can be subjected to a deeper decomposition to get an accurate idea of the cause of the weakness. When the sources of problems are identified, the company's management can develop measures to improve the return on assets, for example, improve the efficiency of asset management, improve marketing, and so on (Ali, M., Boulden, G., & Brake, T., 2002; Corrado, C. J., & Yordan, B. D., 2002; Hill, C. W. L., & Jones, G. R., 2004; Kenneth, A. J., 1974; Ross, S. A., Westerfield R. W., & Jordan, B. D., 2006; Saunders, A., & Cornett, M. M., 2006; Teece, D. J., 1984).

Filatov's three-factor model shows the impact of investment, financial and operating activities (business activity) on the company's return on assets. Therefore, the return on the company's total capital (assets) directly determines both the overall return and the investment attractiveness.

Analysis based on the Filatov model allows you to determine the strengths and weaknesses of the company. The return on assets under the Filatov model depends on three factors: the level of return on equity, income coverage with own funds, and the speed of asset turnover. Thus, the directions for improving the return on assets are directly indicated.

References

[1] Alexandrov O A 2016) *Economic analysis* p 334
[2] Ali M, Boulden G. and Brake T 2002 *Successful manager's Handbook* (Dorling Kindersley) p 864
[3] Corrado C J and Yordan B D 2002 *Fundamentals of investments. Valution and management* p 604
[4] Filatov E A 2019 Author's integrated methods of analysis of the model of return on assets of construction companies *IOP Conf. Series: Materials Science and Engineering* ***667*** 012024. DOI: 10.1088/1757-899X/667/1/012024
[5] Hill C W L and Jones G R 2004 *Strategic management. An integrated approach* (Houghton Mifflin Company)
[6] *Information system «BIR-Analyst»* (2019).
[7] Kazakova N A 2016 *Economic analysis* (Moscow: Infra-M) p 256

IOP Conf. Series: Materials Science and Engineering **880** (2020) 012093 doi:10.1088/1757-899X/880/1/012093

[8] Kenneth A J 1974 Limited knowledge and economic analysis. *American Economic Review* **64** pp 1-10

[9] Kireeva N V 2019 *Economic and financial analysis* (Moscow: Infra-M) p 368

[10] Kogdenko V G 2015 *Economic analysis* (Moscow: Unity) p 392

[11] Lysenko D V 2019 *Comprehensive economic analysis of economic Activity* (Moscow: Infra-M) p 320

[12] Markin Yu P 2018 *Economic analysis* (Moscow: Omega-L) p 864

[13] Ross S A, Westerfield R W and Jordan B D 2006 *Corporate finance fundamentals* (McGraw-Hill Companies) p 844

[14] Rusakova E V 2018 *Comprehensive economic analysis of the company's activities* (Saint-Petersburg: Peter) p 208

[15] Saunders A and Cornett M M 2006 *Financial institutions management. A risk management approach* (McGraw-Hill Companies) p 856

[16] Teece D J 1984 Economic analysis and strategic management *California Management Review* **26(3)** pp 87–110

IOP Conf. Series: Materials Science and Engineering **880** (2020) 012094 doi:10.1088/1757-899X/880/1/012094

Optimization of trading and purchasing policies in the market of the construction material

D A Gercekovich[1], E Yu Gorbachevskaya[2], E A Abramov[3] and E V Rybakova[4]

[1] Department of natural disciplines, Irkutsk State University, Lenin st., 1, Irkutsk, 664074, Russia
[2] Department of expertize and management of real estate, Irkutsk National Research Technical University, 83, Lermontov st., Irkutsk, 664074, Russia
[3]Department of commerce and trading policy, Irkutsk State University, Lenin st., 1, Irkutsk, 664074, Russia
[4]Department of Foreign Languages for Technical Specialties №1, Irkutsk National Research Technical University, Lermontov st., 83, Irkutsk, 664074, Russia

E-mail: eugorbachevskaya@mail.ru

Abstract. In the present research, the authors' attention is concentrated on the methodological fundamentals of surveying the situation on the construction material market using mathematical analysis methods, the model of "Profitability – Risk" in particular. The significance of the paper under consideration is conditioned by a great value share of construction materials in the integrated capital of an investment construction project and by the necessity to make the dependence on imported construction materials lower. The construction material market regulation and its conditioning oriented at its more effective functioning are to affect the construction field favourably. The information received in the process of analysis with regard to a retrospective information is to become a reference basis in managerial decision making. The objectives to be achieved on managerial decisions making depend on the quality of the reference information. A mathematical algorithm for studying a retrospective information on constructive material value is offered in the paper under consideration along with the analysis of other factors. The algorithm has been worked out basing on the portfolio theory concepts. While gestating the offered algorithm practically, construction materials were ranged to form groups, and following that, several types of materials were defined for each group. In the process of the analysis, a number of items were calculated: yearly price changes, expected average price changes and root-square deviations as risk level indicators and an average values ratio in a price movement to the corresponding risks. Besides, some groups of materials were identified as having considerable value fluctuations. The reasons of their fluctuations having been revealed, the most cost-effective and mostly exposed to risk groups of construction materials were indicated as being mostly on demand in the home markets. The offered recommendations are the result of the investigation and are meant for main participants of the construction material market, such as the government, construction material producers and developers engaged in organizing and planning their activities in the sphere of investment-construction project realization.

1.Introduction

Construction is an integral part of a country existence. Housing, enterprise, infrastructure construction provides working places for the people, takes important socio-economic decisions and improves living standards. The subjects of the investment-construction process are represented by a considerable number of enterprises orientation: ordering parties, project owners, construction material distributors, dealers, design engineers, etc.

A high capital capacity is characteristic for an investment-construction sphere, this fact determining the orientation of construction production management efforts to reduce the expenditures with regard to realization of the projects. Depending on the type of an investment-construction project, the region of its realization and a number of other factors the material expenses in the total project expenditures fluctuate within 40% to 60% . Therefore, the control of material resources value will make it possible to solve a number of problems that are essential for the branch under consideration.

The construction material market had naturally been formed as a result of an economic intercourse developed in the process of production followed by purchase and sale of materials. The price of a certain material depends on a great number of factors including a demand for the material, its accessibility, the availability of deterrent factors such as sanctions, production conditions, facilities of access for a raw material resource, etc. The crisis that the Russian Federation has met with lately has affected all the spheres of social life and has certainly had an impact on the construction material market. The analysis of changing prices on the market makes it possible for the construction enterprise managements and the contacting officer-suppliers to realize different solutions connected with their main activities and in particular allows:

1) to follow promising trends on the construction material market;

2) to optimize the realization strategies that are common for investment-construction companies with regard to the companies engaged in construction;

3) to estimate the future profits and losses of companies producing construction materials and to diversify the production basing on the results obtained;

4) to adapt the lines of activities concerning the trading and purchasing companies choosing more profitable ones with regard to the analysis data;

5) to model the government policy regarding one or another aspect of the market under consideration and to develop a subsidization strategy to support the construction field.

The given paper is aimed at an algorithm development oriented at an increased force of price attraction with regard to different kinds of construction materials on the part of all the market members in addition to its practical approbation.

To achieve the purposes some objectives have been set:

1) using the initial data and the mathematical analysis methods it is necessary to follow the parameters of price changes calculating the mean values of price changes and the risk (a standard deviation), followed by a scatterogram in correspondence with the axes of risk and an average mean value change and to drop out the duplicates which are the materials that have close values of risk and profitability parameters, their conditions being inferior, and to separate groups characterized by some dependence that can be followed;

2) to study the causes of changing different construction material prices and price dependence in separate groups. To analyze the data obtained from the point of view of a producer, supplier, stock object acquirer and organs of power(that have the ability to make decisions at the federal and regional level);

3) to work out the recommendations concerning the force of investment attractiveness with regard to this or that kind of materials, to determine the most commercial kinds of construction materials and enterprises that do not need any implementation of special instruments as well as to identify the kinds of materials that are least commercial but are necessary as the ones to be used in the building process.

2. The algorithm used to analyze construction material price changes.

In practical work, the analysis results obtained facilitate developing effective action strategies meant for each of the market members that in their turn contribute to an affluence growing.

A financial analysis is performed basing on the portfolio analysis, developed by G. Markowitz (1952), U. Sharp (1970). A model designed with regard to an investment risk and expected returns is meant by the methodology. These two notions are interconnected in the model. The objective is to compile a bundle of assets contributing either to a minimum risk achieved at the target level of profitability or to a maximum income at a target risk level. [9].

The process of a model designing anticipates:

1) A risk calculation meant as an average square deviation of returnes and the calculation of total returns meant to be a simple average of returns for equal periods of time [10].

2)A diagram plotting with the horizontal axis being marked with risk measurements while its vertical one being marked with returns and the subsequent dropping out. Among the pairs of double points under consideration, there is a point which is in a worse position in the diagram. This point is to be excluded from the further examination. Two points are considered to be double points, their risk or returns being correspondingly different from each other by no more than 0.1 %. In such a pair, the point positioned to the right is not taken into consideration if dropping out takes place in correspondence with a risk index, it being even lower if dropping out happens in correspondence with the index of returns.

3) On having been dropped out, the points are removed and the trend line is added followed by its equation.

4)Selecting groups with regard to some common characteristic [9].

The assets you see in the right lower corner are most unprofitable ones because they combine the greatest risk and the lowest returns. The assets in the left lower corner are most reliable, though not the most profitable ones and they are used to save the funds, whereas the assets in the right upper corner are most risk-related and suitable for the ones seeking for a quick profit, whereas the most attractive assets are located in the upper left corner and combine a high profitability and a low risk relation [4].

To approbate a price changing analysis algorithm with regard to construction materials, the data on the value of 18 kinds of construction materials were subjected to the analysis, the materials being divided into several groups. The paper covers the period of 2010 – 2016 [7]. Value changes on an annual basis, expected average price changes and mean square deviations as the measure of risk relation were calculated in the process in addition to the mean values of price changes in the relation with the corresponding risks (table 1).

Table 1. Average price changes, risk level and their relation with construction materials

Construction material	Average (%)	Standard deviation (%)	relation of profitability to risk
Concrete which is ready for filling (custom concrete), m3	2,9	5,8	0,5
Construction mortars, m3	3,9	6,9	0,6
Ceramic refractory building brick, thousand relative units	6,2	8,2	0,8
Brick of silicate, thousand relative units	5,6	4,8	**1,2 (min)**
Insulating building fiberboards, sheeting, panel boards, paving tile, and similar products made of gypsum or compounds on its basis that are not covered or reinforced with paper or card board with the exception of ornamented products or the ones agglomerated with gypsum,m2	**-5,1 (max)**	21,3	**-0,2 (min)**
Broken natural stone, м3	5,7	12,5	0,5
Sundry natural sands, м3	8,3	24,3	0,3
Sundry expanded mineral materials (gravel ceramsite), m3	6,3	8,5	0,7
Sundry pre-fabricated construction elements of steel, tons	10,3	16,5	0,6

Hot-rolled and shaped rolled products, hot drawn, extruded and forged from unalloyed steel, tons	7,0	9,2	0,8
Converted timber common that does not belong to any other groups, and some other products, m3	5,8	**5,2 (min)**	1,1
Wood chipboards of continuous pressing, m2	**10,9 (max)**	22,0	0,5
Compounds and products from heat insulating materials that are not covered by any classification, m3	6,8	24,4	0,3
Ceramic glazed tiles for the inside wall cladding, m2	7,3	16,3	0,4
Cast and profiled rolling glass sheets, m2	2,2	9,5	0,2
Cement, tons	6,0	12,2	0,5
Dry oil, tons	9,3	**30,8 (макс)**	0,3
Ready-mix paint, т	2,6	9,8	0,3

The materials classified as stable one and belong to the group of "Wooden materials and products" have been marked for the greatest price growth. Purpose-made materials (molding) are the most unstable ones and some kinds belonging to this group, on the contrary, would reduce in price. "Sand-lime brick" is particularly marked for its returns to risk relationship indicator -1,2 which

means that in fact returns grow more quickly than risks. This is quite contrary to "insulating building fiberboards, sheeting, panel boards, paving tile, and similar products made of gypsum or compounds on its basis that are not covered or reinforced with paper or card board with the exception of ornamented products or the ones agglomerated with gypsum" with the value of -0,2.

To make the reasons for a price leap regarding a number of materials well founded, the legislation and competitive environment were studied. Some other factors were revealed that could have led to construction material price changing. For example, since January 1, 2015 the general State Standard of 32389 – 2013 Drying oils. General technical conditions. has come into force. This general state standard was introduced on the territory of RF and the neighboring countries and considerably tightened the regulations of the drying oils, the process of their production and its control. A price leap is a natural market reaction resulting from tightened standards followed by expenses. The countries that have accumulated state standards have got the markets that are closely interconnected. Therefore, the growing prices cannot be compensated by an import trade for cheaper products from neighboring countries. A similar situation has taken place with the item of "Compounds and products from heat insulating materials that are not covered by any classification".

On the other hand, unlike the above materials, fiberboards, sheeting, panel boards, paving tiles and similar products made of gypsum or its compounds but neither covered nor reinforced with paper or cardboard became considerably cheaper in 2013, ornamented products or gypsum sintered ones being the exception. This was facilitated by:

1) The establishment of works to produce partition blocks of gypsum on the territory of "Samara Gypsum Center" in 2013. This caused the appearance of a new market member and an increased volume of products, conditioned the competitive struggle for the consumer and lower prices (www.samaragips.ru).

2) The growth of inner production resulted in the growth of a monthly export volume by 30 %: 682 thousand dollars in January 2013 as compared with 904 thousand dollars in December 2013 [3,4] along

with a decreased import volume of expensive products from Europe by 20 % monthly: 2.1 million dollars in January 2013 as compared with 1.7 million dollars in December 2013 [3,4].

For the further analysis to be performed, it is reasonable to plot the data on a diagram with the axes of risk-returns and leave the outsiders with no further consideration (Fig.1). All the construction materials included into the group of leaders can be divided into several groups, though conventionally to some degree. The materials that are used in a rough-edged construction of the most objects in the RF are to be marked first. They are "concrete which is ready for filling (custom concrete)" which is a basic compound of any building foundation, "converted timber common that does not belong to any other groups, and some other products" such as sawn timber and other materials which are popular with private construction along with "brick of silicate" which is the basic material for the walls in rental apartments as well as "pre-fabricated construction elements of steel" that are used in the frames of great buildings," the expanded mineral materials (such as gravel ceramsite") should also be mentioned as the ones used to facilitate the thermoinsulation of the foundation, floor, roof, utility systems, they being added to concrete. All the materials that belong to this group with the exception of concrete are positioned above the trend line (Figure 1) and consequently possess the best indices of returns-risk relation.

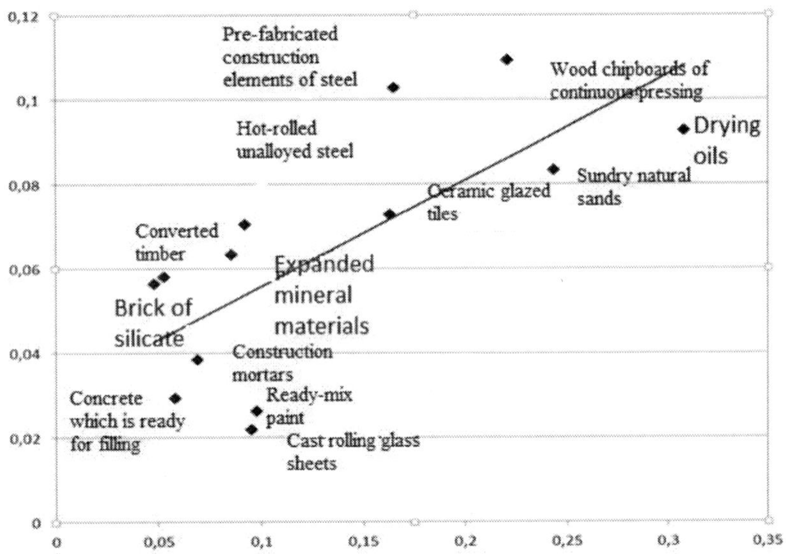

Figure 1. The diagram of dispersal with regard to construction materials possessing the best indices.

A correlation analysis has been made basing on the materials forming the group of leaders. It resulted in revealing that there is no correlation between average price changes developed by leading materials.

Conclusion

There are some recommendations for the management of companies engaged in the production of materials or planning to start their production in the near future:

1) The production of materials for rough-edged construction is preferable, i.e. for walls, the foundation and the roof. Artificial materials of stone and wooden demonstrating a steady price growth for a long period are the most profitable ones.

2) The production of materials for interior finish is an extremely risk-related business and calls for detailed studies of quickly changing trends on the market of home decoration construction materials, a more narrow segmentation and an assessment of a potential market volume.

At the federal level:

1) The examples dealing with drying oils and heat insulating compounds prove of the great influence that the state regulation exerts on the market. The introduction of many standards for one occasion only results in a price leap. Besides, the goods bought before new standard introduction become unsaleable, so a customer has to buy the product he needs once again which is sure to cause higher prices resulting from a higher demand conditioned by a limited stock-in-trade.

2) The analysis performed allows to identify the regions that need a state support for their development. The situation with trowel firms has been unsteady and their assets are at risk. At the same time, this portion of the construction market is as important as the "rough" one. It is in this market segment that a high dependence on import is registered in creating additional threats for economic stability.

For a building owner:

1) For a building owner, whether we speak of a physical or a legal entity, it is important to understand that there are materials with a value growing steadily, and these are the materials that are used for a foundation, load-bearing wall panels and a roof erection and that a delay with the rough work orders results in a greater value of a "case" construction. A forecastable rise in price should have been made provisions for in the capital expenditures of an investment project.

References

[1] Blasi J R, Kruse D L and Markowitz H M 2008. Risk and Lack of Diversification under Employee Ownership and Shared Capitalism *NBER Working Papers* **14229** pp 95 – 107

[2] CJSC "Samara gypsum plant" opened a new factory for the production of tongue-and groove plates of the European standarts. URL: https://samaragips.ru/company/news/zao-samarskiy-gipsovyy-kombinat-otkryl-novyy-zavod-po-proizvodstv/ (27.12.2019).

[3] Export from Russia "stone, plaster, cement, asbestos, mica and etc." URL: https://rstat.com/date-M201301-201401/RU/export/world/1368 (28.12.2019).

[4] Gercekovich D A 2017 Formation of an optimal investment portfolio for a complex of effective portfolios *Bulletin of Moscow University Ser.: Economics* **5** pp 86-101

[5] Gercekovich, D A and Babushkin R V 2019 Dynamic portfolio analyses of global stock Indices *The world of economy and management* **19 (4)** pp 14-30

[6] Import in Russia "stone, plaster, cement, asbestos, mica and etc." URL:https://ru-stat.com/date-M201301-201401/RU/import/world/1368 (28.12.2019).

[7] *Investment in Russia* 2017

[8] Markovitz H M 1952 Portfolio selection. *J. of Finance* **(1)** pp 77 – 91

[9] Markowitz H M 1959 *Portfolio Selection: Efficient Diversification of Investment* New York: Wiley p 368

[10] Markowitz H M 1991 Foundations of Portfolio Theory *Journal of Finance* 46 pp 469 – 477

[11] Sharpe W F 1970 *Portfolio Theory and Capital Markets* New York: Mcgraw-Hill p 316

[12] *GOST 32389 – 2013 Dry oils. General technical specifications* 2014 (M.: Standartinform) p 10

[13] *GOST R 54358 – 2011 Dry cement base decorative coats for facade thermo insulation composite systems with external plaster lays* 2011 Standartinform p 10

IOP Conf. Series: Materials Science and Engineering **880** (2020) 012095 doi:10.1088/1757-899X/880/1/012095

Legal regulation of construction quality and control

I A Guseva[1]

[1]Irkutsk National Research Technical University, Department of Theory of Law, Constitutional and Administrative Law, 83 Lermontov st., 664074, Irkutsk, Russian Federation

E-mail: mif-leg@mail.ru

Abstract: The Institute of State Construction Supervision of urban planning legislation system occupies a key place in the matter of quality assurance of construction, that is a necessary condition for the safety of buildings under construction. The article gives a brief analysis of state architectural-building supervision mechanism of the RSFSR/USSR of Soviet period and developed processes of that period of time. By Decree of the Council of People's Commissars of August 16, 1944 No. 1110 "On the State architectural-building Supervision", the authorities (organizations) of architectural-building supervision were created. The main activity of GASK authorities was control functions in construction process, the Construction Norms and Regulations compliance, quality assurance, project solutions relevance, the correctness of capital investments. Analyzing the legal regulation of construction quality and control of Soviet period, the author notes that the political regime of the state does not affect the essence of the state construction supervision, but defines forms and implementation methods of it.

Currently, developers are violating building codes, rules and technical standards. The violations are caused by their desire to gain more profit, reduce the construction period and turnover terms, win tenders, etc. These illegal initiatives can be implemented due to corruption practice. As a result, customers and consumers cannot enjoy quality facilities and there is a risk of technological disasters. To eliminate causes and conditions that give rise to negative phenomena, it is necessary to strengthen federal and regional government control in the construction sector.

Significant efforts aimed at improving state inspection and enforcement activities are underway in Russia. The final beneficiaries of such public functions are usually citizens, as inspection and enforcement affect public health, the safety of property as well as other public values such as environment or cultural heritage [1].

Federal laws provide requirements to the planning of supervision activities, to the order of organization and holding supervision, to documenting and arranging the results of such activities. The facts of ignoring the requirements, which brings about infringing the rights and legitimate interests of the citizens in entrepreneurship, legal persons, remain a significant negative factor due to their mass character, and hamper the economic development of the Russian Federation as a whole [2].

The relevance of this issue is due to the fact that in the field of legal regulation, problems require improvement of the legislation. Serious changes in the Russian legislation on state construction supervision were made in 2005-2006. In recent years, this institution has been reformed. At the same time, proposals for its further improvement are being made by the expert community and the government. Sometimes these proposals are diametrically opposed. During the period of development of market relations, it is necessary to create an effective mechanism to ensure quality and safety of facilities under construction. Therefore, to understand in which direction the state construction supervision system should be improved, let us analyze the history of this institution in the Soviet period.

The purpose of this article is a historical and legal study of development of the state construction supervision system in the Soviet period.

This study was carried out using various general scientific and private scientific methods: historical-genetic, logical, system-structural, etc.

In pre-revolutionary Russia, architectural and construction supervision was carried out by provincial and local authorities. According to I.V. Kropacheva, "this scheme existed for a long time and was implemented in modern Russia" [3]. In Soviet Russia, the construction supervision system was administered by other state bodies created by the new state.

In the 20s of the XX century, supervision of the construction industry and prosecution of violators were carried out by local executive committees. The militia regulated construction works. Then, the situation changed and the militia began to carry out their immediate task of protecting law and order and public safety. Local authorities, for example, the department of district engineers, carried out construction supervision after the revolution [3].

On September 27, 1923, the Council of People's Commissars of the RSFSR adopted the decree "Regulation on the Directorates of Provincial Engineers." Under the provincial executive committees, departments which consisted of engineers (higher technical education and construction experience) appointed by the provincial executive committee were created. The staff was formed in accordance with local needs and plans of the provincial executive committee.

The provincial engineer performed his duties both directly and through district engineers appointed by the provincial executive committee on the recommendation of the local executive committees. Their activities were carried out on the basis of a special provision approved by the provincial executive committee.

The provincial engineers performed construction control, in particular, "general technical control over compliance with national building codes and local mandatory regulations when issuing permits and performing local construction works", organization and performance of construction works; accounting and distribution of building materials; preparation of opinions on construction and planning projects and various technical conditions, local government decisions regarding local building codes; designing; inspection of facilities for reconstruction, etc. [4]. According to paragraph 1 of Section II of the Regulation, provincial engineers were able to punish people violating national construction standards, suspend works in case of a threat to public safety.

The departments of provincial engineers were replaced by the departments of construction control in accordance with the decree of the All-Russian Central Executive Committee and the Council of People's Commissars of the RSFSR "On the Approval of the Regulation on the Departments of Construction Control in the Territories, Regions and Provinces of the RSFSR of February 27, 1928. The Departments of Construction Control were subordinate to the local executive committees, as well as higher building control authorities [5].

The Construction Commission of the RSFSR was a central governing body in relation to the Building Control Departments. It supervised activities of building control bodies according to Russian laws.

Local engineers were appointed by the local executive committee by special orders. Local construction authorities were appointed by special decrees. In districts, there were district engineers who were subordinate to higher construction control departments.

In addition to construction control, the functions of construction control departments included:

accounting;

technical and economic regulation of local construction;

implementation of public safety measures.

The Regulation defined the concept of technical and construction control as follows: "consideration and approval of construction, reconstruction and overhaul projects and ... control at the construction sites ... supervision of works performed in compliance with construction projects and permits" [5].

Officials were empowered to visit construction works and inspect objects, give orders to suspend works "if there are significant deviations from the approved project or when performing works not agreed with the Construction Control Department, or works performed carelessly and threatening public safety" [5].

In addition, the construction control authorities were empowered to liquidate construction organizations and terminate activities of private contractors.

In the resolution "On improving housing construction" of April 23, 1934, the Council of People's Commissars of the USSR said that "... the existing practice of housing construction does not correspond to the cultural level and needs of workers" [6]. A number of drawbacks were emphasized: poor quality of construction and finishing works, and the design of mass housing.

As a result of reorganization of the building control authorities in 1935, functions of the state construction supervision were entrusted to the RSFSR Chief State Inspector for Construction according to the decree of the All-Russian Central Executive Committee and the Council of People's Commissars of the RSFSR "On Reorganization of Construction Control". It was subordinate to the Council of People's Commissars of the RSFSR; its duties were assigned to the People's Commissar of Communal Services of the RSFSR.

To carry out functions assigned to the Chief State Controller for Construction, the department of the Main State Construction Control of the RSFSR was created. The staff and funding were approved by the Council of People's Commissariats of the RSFSR.

The staff of the State Construction Control under the heads of regional, communal administrations and people's commissars of communal services of the Autonomous Soviet Socialist Republic and presidiums of city councils were approved by the regional, executive committees and councils of people's commissariats of the Autonomous Soviet Socialist Republic.

The Chief State Inspector for Construction had the following rights: to issue general instructions and specific instructions on the operation of the State Construction Control System of the RSFSR; arrange meetings and conferences of employees of the State Construction Control in order to instruct workers and share experience; require the submission of information and reports on the work of state inspectors for construction and supervise their activities.

In 1938, the Committee for Construction Affairs under the Council of People's Commissars of the USSR was created to coordinate activities of construction organizations. The main tasks of the Committee were as follows: project and construction management, development of production standards and technical regulations. In 1939, its functions were transferred to the People's Commissariat for the Construction of the USSR which was created by a special Law. More than hundred departmental building trusts as well as territorial and specialized heads and functional departments were subordinate to the People's Commissariat. The general construction control was carried out by the People's Commissariat, which was divided according to the industrial principle in 1946: the People's Commissariat for the construction of fuel enterprises, the People's Commissariat for the construction of military and naval enterprises. In March 1946, in the USSR, there were 29 all-Union and 19 Union-Republican People's Commissariats. On March 15, 1946, all People's Commissariats were transformed into the ministries according to the law "On the Transformation of the Council of People's Commissars of the USSR into the Council of Ministers of the USSR".

The main document which regulated bodies of architectural and construction supervision was the Resolution of the Council of People's Commissariat of August 16, 1944 No. 1110 signed by Molotov. In accordance with the decree, the Main Directorate of the State Architectural and Construction Control of the USSR was created as part of the Committee for Architecture of the Council of People's Commissars of the USSR. There were departments of architecture under the Council of People's Commissars of the Union Republics, inspections of the State Architectural and Construction Control of the Union Republics; departments of architecture under the CPC of the Autonomous Soviet Socialist Republic and executive committees of the Councils of Deputies - the Inspectorate of the State Architectural and Construction Control of the Autonomous Republics, territories, regions of Moscow, Leningrad and Kiev; in the offices of the chief architects (according to a special list approved by the

USSR Council of People's Commissars) - state inspections of the State Architectural and Construction Control.

The State Architectural and Construction Supervision Committee of the Committee for Architecture controlled development of large cities and urban settlements, quality of construction of the most important buildings and structures, architectural designs of facades of industrial buildings, city highways and driveways."

These bodies monitored quality of building materials and products, did not allow the use of building materials and products that did not comply with approved projects and technical specifications. They had the right to suspend construction works performed in violation of the legislation and in case of emergencies. They demanded the removal of persons who did not have the right to perform construction works, brought persons who had violated construction laws, building rules and norms to disciplinary and criminal liability and demanded production expertise and testing of materials, parts and individual structures.

In 1972, a new Construction Regulation was approved by the order of the State Architecture Committee. It was valid until the collapse of the USSR.

In accordance with the decree, the state architecture control bodies carried out the following tasks: "...control of construction of housing and civil facilities in cities, towns, rural and other settlements, regardless of their departmental subordination; control of building materials and products used in the construction of housing and civil facilities" [7]. This control was carried out at construction sites and plants. The authorities were entrusted with monitoring functions: implementation of approved projects and compliance with technical requirements for industrial buildings and structures in terms of architectural decisions and quality of façade works "included in city highways and passages; planning and development of rural settlements in accordance with approved projects; compliance with standards and technical conditions for the transportation of warehousing and storage of building materials, products, sanitary and other engineering equipment, buildings and structures at construction sites and manufacturing plants, etc"[7].

The bodies were entrusted to give binding instructions to correct poorly completed construction and installation works, replace low-quality products and materials.

The Regulation established the right of the State Architectural Control bodies to retain the cost of poorly performed construction and installation works. The deducted amounts were recovered only after defects were eliminated. It was an effective mechanism. As a result, the legislation did not establish administrative responsibility.

The system of architectural and town-planning control changed only in 1991. As a result of the federal policy, the private construction system was eliminated, since the state monopolized distribution of building materials, developed construction plans and established a complicated procedure for obtaining construction permits and land plots.

Undoubtedly, government control was carried out by those ministries and departments which performed construction works. According to the Decree of the Council of Ministers of the RSFSR dated December 25, 1987 No. 513 "On Further Development of Architecture and Urban Planning in the RSFSR", ministries and departments, councils of ministers of autonomous republics, executive committees of local Councils of People's Deputies were responsible for quality of designing and construction works, development of urban plans, activities of design and construction organizations. According to I.V. Kropacheva "... the bodies of internal affairs bring persons who violated construction norms to liability" [3]. Thus, these bodies were not supervisory.

The situation changed as a result of the decree of the President of the RSFSR of November 27, 1991 No. 105-r "On the of bodies of the State Architectural and Construction Supervision of the RSFSR".

The inspections of the State Architecture and Construction Supervision of the RSFSR had to ensure the compliance of buildings and structures, building materials and products with standard technical and design documents; organizational and legal procedures. The inspections were empowered to take administrative measures (penalties) for violation of regulatory acts and state standards for the successful completion of tasks. Consequently, their activities can be characterized as supervisory.

In 1991-1992, state departmental control over quality of construction works was eliminated.

In post-Soviet Russia, the reform of the state construction supervision system began. Currently, in accordance with the legislation, the state construction supervision system includes a number of authorities (see Table 1).

Table 1. The structure of state construction supervision in the Russian Federation

State Construction Supervision	
Federal executive body	Executive authorities of constituent entities of the Russian Federation (independently determine the structure of authorized bodies of constituent entities of the Russian Federation)
Supervision during the construction, reconstruction, overhaul of nuclear facilities, hazardous production facilities, communication lines, defense and security facilities, facilities that are state secret, especially dangerous, technically complex and unique facilities	Supervision during the construction, reconstruction, overhaul of other facilities subject to state construction supervision
Federal Service for Ecological, Technological and Nuclear Supervision (during construction, reconstruction, overhaul of facilities specified in clause 5.1. Article 6 of the Town Planning Code of the Russian Federation)	
Ministry of Defense of the Russian Federation (during the construction, reconstruction and overhaul of military infrastructure facilities of the Armed Forces of the Russian Federation)	
Federal Security Service (during construction, reconstruction and overhaul of facilities under the jurisdiction of the federal security agencies (Decree of March 6, 2008 No. 325)	
Federal Security Service of the Russian Federation (during the construction, reconstruction and overhaul of facilities under the jurisdiction of security agencies) (Decree of November 28, 2007 No. 1594)	
The procedure for the implementation of state construction supervision in the Russian Federation is established by a decree of the Government of the Russian Federation of 1.02.2006. No. 54 (as amended by Decree of the Government of the Russian Federation of July 18, 2019 No. 926)	

A. Lapidus and D. Topchiy come to the conclusion that"...a modern approach to construction supervision based on the criteria of a risk-based approach should be the basis for improving the quality and safety of construction products at all stages of their formation" [8].

In the situation of the current construction boom, the existing construction control procedures fail to prevent growing accidents at buildings and structures or construction product defects [9]. One of the solutions to this problem is to develop an automated design system for the quality management process of construction products, which will be used to improve construction control [10].

Thus, in the history of development of the state construction supervision system, there are several stages:

1. 1917-1938 - The bodies of architectural and construction supervision (administration of provincial engineers, construction control) which operated in difficult political and legal conditions;

2. 1939-1991 - development of Soviet architecture and urban planning legislation, creation of the People's Commissariat for Construction, state architectural and construction control bodies.

There were no differences between the concepts "control" and "supervision", the main functions were assigned to the executive authorities, which can be observed in modern Russia.

During the Soviet period, the viable structure of architectural and construction supervision bodies was created. It solves problems of quality of design materials and urban development. Being a part of Soviet society, the state construction supervision system was influenced by administrative bureaucratic methods, the totalitarian regime, but the essence of the state construction supervision system did not change: it ensures compliance with the legislation on urban planning, technical regulations and design documents.

References

[1] Yuzhakov V N, Dobrolyubova E I, Pokida A N and Zybunovskaya N V 2019 Effectivenessof public inspection and enforcement activity for final beneficiaries. *Ekonomicheskaya Politika* **14(1)** pp 92-115
[2] Zyryanov S M 2016 The typology of errors and violations by state authorities and their officials in exercising state control (Supervision) *Public Administration* **1** pp 149-159
[3] Kropacheva A V 2010 The history of the development of administrative regulation of construction activities *Modern Law* **2** pp 156-162
[4] Knyazeva L K 2009 History of state construction supervision in Russia *Labor safety in Industry* **8** pp 10-20
[5] On approval of the Regulation on building control departments in the territories, regions and provinces of RSFSR: Resolution of the All-Russian Central Executive Committee, SNK of the RSFSR of 02.27.1928.
[6] On improving housing construction: Decree of the Council of People's Commissars of the USSR of 04.23.1934, No. 945.
[7] State architectural control [Electronic resource]. Access mode: https://www.ktovdome.ru/74/482/17791.html (accessed February 1, 2020).
[8] Lapidus A, Topchiy D 2019 Construction supervision at the facilities renovation. *Topical Problems of Architecture, Civil Engineering and Environmental Economics* **91** pp. 1-6
[9] Dobrolyubova E I and Yuzhakov V N 2015 Evaluating Eff ectiveness and Efficiency of the Control (Supervision) Activities *Public Administration* **4** pp 41-64
[10] Topchiy D, Tokarskiy A 2019 Formation of hierarchies in the organization system of the state construction supervision during reshaping of city territories *International Journal of Engineering and Advanced Technology* **8** pp 44-46
[11] Peshkov A V 2019 Digitalization of the economy, implementation of national projects and development of the construction industry *Investments. Construction. Real Estate: New Technologies and Targeted Development Priorities IOP Conference Series: Materials Science and Engineering* **667** 012074
[12] Gertsekovich D, Gorbachevskaya L, Grigorova L and Peshkov V 2019 Return on investment in REIT real estate funds *Investments. Construction. Real Estate: New Technologies and Targeted Development Priorities IOP Conference Series: Materials Science and Engineering* **667** 012063

IOP Conf. Series: Materials Science and Engineering **880** (2020) 012096 doi:10.1088/1757-899X/880/1/012096

Specificity assessment and management of financial stability of construction companies

N V Kretova[1], G I Khokhlova[1], A A Kretova[2] and A Yu Khokhlova[3]

[1] Irkutsk National Research Technical University, 83 Lermontov str., Irkutsk, 664074, Russia
[2] Saint Petersburg state University of civil aviation, 38 Pilotov str., Saint Petersburg, 196210, Russia
[3] Russian-Asian College of Economics and law, 258a Baikal str., Irkutsk, 664050, Russia

E-mail: vip.kretova@inbox.ru

Abstract. The purpose of the scientific article is to identify the features of assessing and managing the financial stability of construction companies. The object of this study is entrepreneurial activity in the construction sector. The subject of the study is specific aspects of analysis and management of financial stability in construction organizations. To achieve this goal, the following tasks are set: 1) determining the features, problems and trends in the development of the construction industry in Russia; 2) identifying the specifics of the functioning of finance in the construction sector; 3) systematization of factors that affect the activity of enterprises in the construction sector; 4) assessment of the financial stability of the construction company, taking into account the regulatory and industry average values. A classification of factors affecting financial stability by a number of system-forming features is proposed. The most significant indicators for assessing the financial stability of construction companies are determined, taking into account the industry specifics for making management decisions.

1. Introduction
Construction is "the creation of buildings, structures, and structures (including on the site of demolished capital construction projects)" [1]. The construction industry has a significant impact on the competitiveness of the Russian economy, as the construction sector is one of the top five Russian segments that make up more than 40% of the gross value added structure, for example, in 2018 this figure was 5.7% compared to 6% in 2017. [2].

The study of the specifics of the financial condition and identification of problems in the construction sector is important for making management decisions at the enterprise. The competitiveness of an enterprise depends on the rational management of financial resources [3]. Modern conditions for the construction of buildings and structures, increased competition, and its imperfection have dramatically increased the need to effectively manage the financial stability of construction organizations.

2. Materials and methods
The concept of financial status abroad is mainly identified with financial position and financial sustainability. Important indicators are solvency, cost recovery and repayment of obligations, the

ability to create an innovative product from existing resources and receive income at the same time [4]. Foreign practice of assessing the financial condition is sometimes identified with the analysis of financial statements or the activities of the enterprise as a whole [5], [6], [7]:

In Russian practice, the interpretation of the financial state and the essence of the analysis of the financial state differ depending on the approach of a particular author. Some authors define the financial condition of an enterprise as "an economic category determined by the composition and placement of funds, the structure of their sources, the speed of capital turnover, the ability of the enterprise to repay its obligations on time and in full, as well as other factors" [8]. Other authors identify the financial state with "the state of capital in the process of its circulation and the ability of the subject to self-development at a fixed time" [9]. Despite different approaches to determining the financial condition and methods for determining it, the main aspect of the analysis is to assess the effectiveness of using financial resources to make decisions about the need to regulate their volume.

3. Results

Financial stability of a construction company is a characteristic of its activity and financial well-being, the result of its current, financial and investment development, contains the necessary information for the investor, reflects the company's ability to fulfill its obligations and is a forecast of further development. A financially stable company has advantages in attracting investments, obtaining loans, and selecting suppliers and consumers [10]. Such a company is more independent from unexpected changes in market conditions, therefore, the risks of becoming insolvent and being on the verge of bankruptcy are minimized.

In general, the objective conditions for the functioning of the construction "industry" led to the fact that recently in Russia there were contradictory trends in its development, which generated the appearance of risk situations and affected the financial stability of the industry's enterprises:

- despite a 2.3% increase in the profitability of goods, products, works and services sold, 852 organizations were unprofitable in 2018;
- accounts payable exceeded accounts receivable by 911.8 billion rubles or 26.6%;
- the share of overdue accounts receivable in the total volume of accounts receivable was 4.7% and increased in 2018 compared to 2017 by 0.7%, overdue accounts payable amounted to 118.4 billion rubles in 2018 and decreased by 6.8% compared to 2017, overdue accounts receivable – 160.7 billion rubles and increased by 29.3%, respectively;
- due to insufficient own financial resources, the share of overdue wage arrears in construction amounted to 13% of the total amount of overdue debt in the economy;
- there was a decrease in the volume of investments in the construction of buildings (including housing) and structures, in particular, in 2018, 56.0% of investments in fixed assets were directed (compared to 57.4 in 2017%);
- the degree of depreciation of fixed assets is still higher than in the economy as a whole by 1.1% (48.4%), the coefficient of renewal of fixed assets in 2018 was 5.7% and compared to 2017 increased slightly (by 0.6%), the coefficient of disposal of fixed assets in construction has not changed and amounted to 1.1%;
- the average level of capacity utilization in the fourth quarter of 2019 decreased from 63% to 62% compared to the third quarter of 2019. At the same time, 16% of organizations had a load level of no more than 30%, 16% of organizations - more than 90%. Among small businesses, the average level of capacity utilization was 54% [11].

Today, among the leaders of the Russian construction business, the third place, according to Forbs, is occupied by the St. Petersburg construction company PC "LRS Group" with revenue (2016-2018) of 305.1 billion rubles and commissioning of 2976,000 m^2 [12].

As for assessing the financial stability of construction companies, the most important indicators that characterize their liquidity, solvency and financial independence are the absolute and current liquidity ratios, the autonomy coefficient and the investment coverage coefficient. These indicators are calculated by the authors on the basis of the public financial statements investment coverage coefficient of the company "LSR Group" [13]. The calculation results are shown on figure 1.

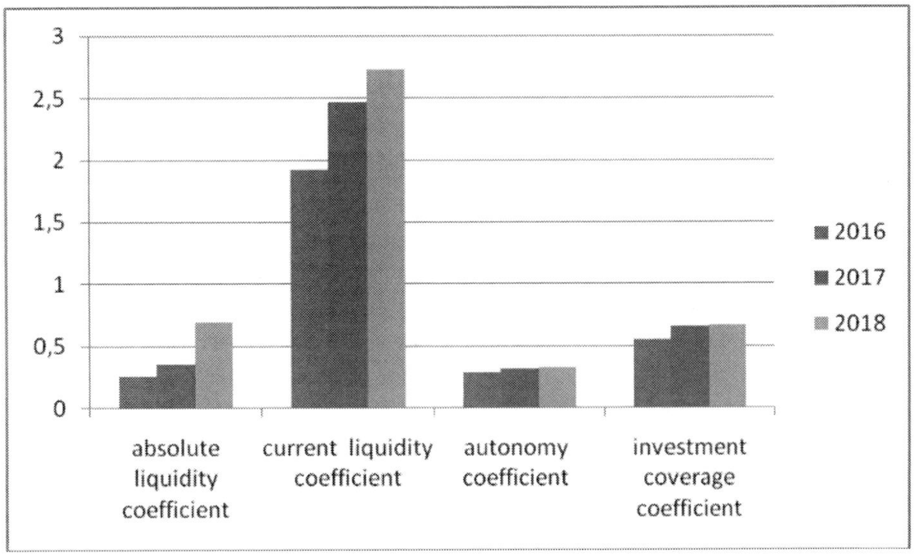

Figure 1. Financial coefficients of the company "Group LSR"

As shown by the dynamics of figure 1, all indicators for the study period increased, while:

- the absolute liquidity ratio (corresponds to the standard > 0.1–0.2) and shows a steady growth trend of the company from 0.25 in 2016 to 0.69 in 2018, which indicates a balance and synchronization of the inflow and outflow of funds of the enterprise;
- the current liquidity ratio of PC "LSR Group" (corresponding to the standard 1.5-2.5) has a positive trend and increased from 1.92 in 2016 to 2.73 in 2018, and in 2018 exceeded the recommended value, which indicates the possibility of more active use of current assets;
- the coefficient of autonomy, which characterizes the financial independence of the enterprise, increased from 0.28 in 2016 to 0.32 in 2018, while it was less than the standard 0.5, but higher than the industry average of 0.11;
- the investment coverage ratio, which is directly related to the company's solvency, increased from 0.55 in 2016 to 0.66 in 2018 and more than 4 times exceeded the industry average value of this coefficient of 0.16.

Thus, the calculated indicators indicate the effectiveness of the LSR group's activities, its financial stability and the tendency to increase this stability.

Cash flow, sales revenue, profit and profitability are the most important financial indicators of the company [14].

The final financial result of the company is net profit and profitability. Thus, the profit of the construction company "LSR Group" in 2016 amounted to 9163 million rubles, an increase in 2018 by 177% and reached 16230 million rubles, which indicates the effectiveness of management activities and an increase in the sources of formation of its financial resources. According to the federal tax service's assessment, in the construction sector, the average industry returns on sales in 2017 is set at 7.2%, in 2018 at 6.1%, and the average industry return on assets in 2017 at 1.8%, in 2018 at 1.7% [15]. In PC "LSR Group", the sales margin increased from 16.2% in 2017 to 16.9% in 2018, exceeded the industry average by more than twice, and was characterized by positive growth dynamics, which indicates that the company's pricing policy was correctly chosen and that costs were monitored and controlled accordingly. The return on assets of the "LSR Group" construction company reached 6.3% in 2017 and 6.25% in 2018, which is more than 3.5 times higher than the industry average and is currently an indicator of the company's profitability and efficiency, as well as its investment attractiveness.

Trends in the studied indicators characterize the performance of construction companies and reflect the functioning of finance in the construction sector, which has its own specific industry features, namely [16],[17],[18]: a longer production cycle, the need to build new temporary premises and change the scheme of transportation of materials; estimated funding, a large share of the article "construction in progress" in the structure of working capital; uneven revenue from the delivery of completed works; close relationship between the need for working capital and the amount of material and labor intensity of work in different periods of construction, etc.

4. Finding

The complexity of studying financial aspects in the construction industry lies in the variety of organizational and economic forms of the construction production process, a large number of participants who perform various functions. In the process of construction production (construction of any object), as a rule, an investor, a customer, a designer, a contractor, specialized subcontractors, as well as credit organizations and insurance companies can be involved. In addition to these direct participants in the construction process, dozens of manufacturers of technological equipment, construction machines and construction materials, etc., participate in the construction process, each of which, first of all, seeks to maximize its own income – profit.

The long construction cycle of objects affects the nature of cost management and financial results of the organization. Due to the fact that during the construction period, large funds and material assets of the organization are excluded from circulation, any measures aimed at reducing the duration of construction are profitable, but this contradicts the organization's desire to expand the "front" of work on the object and increase the number of objects under construction.

5. Discussion

There are different approaches to understanding the essence and structure of financial analysis, its specifics in the conditions of innovative development, which directs users of economic information to carefully choose methods of analysis, since this determines the quality of conclusions made with their help for making management decisions.

We believe that the analysis of sustainability requires a systematic approach that allows us to determine the "nature" of the integrity of this characteristic of economic activity. In this regard, it is advisable to classify factors that affect financial stability by a number of system-forming features, which include the following groups:

– Group 1: basic, external and internal environmental factors. Internal ones include: industry affiliation of the business entity, product structure, its share in the total effective demand, the amount of paid-up authorized capital, the amount and structure of costs and their dynamics, the state of property and financial resources, including reserves, their composition and structure. External factors are: political and legal factors, economic, social, cultural and technological factors.

– Group 2: a time continuum that includes short-term and long-term assets and liabilities. This group contributes to the formation of indicators of short-term and long-term financial stability, which is associated with the tasks of financial and economic analysis. The main focus in the short term is to assess the structure, size and dynamics of stocks, as well as the rhythm of work. In the long term, the investment attractiveness of projects should be considered taking into account their impact on various indicators that form the overall efficiency of the enterprise.

– Group 3: self-sufficiency, ranging from highly profitable to pre-bankruptcy and forming the following types of financial stability: absolute, normal, unstable and crisis.

So, the specificity of financial analysis in construction based on the study of peculiarities of functioning of enterprises, the versatility of providing construction services, the adequacy and optimality of own funds and liquid assets for the implementation of activities, assessment of investment payback.

6. Conclusion

Qualified management of financial resources is a guarantee of survival and the basis of the company's stability. it should be aimed not only at maximizing profits in the short term, but also at maintaining a state of financial stability.

The problem of assessing financial stability is one of the most significant not only financial, but also general economic problems, since low financial stability can lead to the lack of funds for enterprises not only to develop production, but also to resume current activities, as well as to insolvency and even bankruptcy.

We believe that choosing the most appropriate indicators for the company's financial analysis and identifying factors that objectively reflect the conditions for providing construction services is a key issue in managing the financial stability of organizations in this industry. Thus, the difference in approaches to understanding the essence and structure of financial analysis directs users of economic information to carefully select analysis methods, since the quality of their conclusions depends on them.

References

[1] Urban planning code of the Russian Federation from 29.12.2004 N 190-FZ
[2] Official site of Rosstat [electronic resource]. - Access mode: . https://gks.ru/accounts/ Ver. From the screen
[3] Khokhlova G, Kretova N and Burov V 2019 The problems of investment activity of entrepreneurship and methodological aspects of credit risks assessment *ICRE IOP Conference series: materials science and engineering* **667** doi:10.1088/1757-899X/667/1/012038
[4] Brigham E F and Gapenski L C 1991 Financial management: theory and practice (Dryden Press)
[5] Keynes J M 1936 *The General Theory of Emloyment, Interest and Money* (London Macmillan)
[6] Schumpeter J M 1993 *Business cycle. A Theoretical, Historical and Statistical Analysis of the Capitalist Process* (New York Toronto London)
[7] Drucker P F 2002 Management challenges for the 21st century (United Kingdom Harper Collins Publishers Ltd London)
[8] Sheremet A D 2017 *Analysis and diagnostics of financial and economic activity of the Enterprise* (Moscow: INFRA-M) p 374
[9] Savitskaya G V 2016 *Analysis of economic activity of the enterprise* p 378
[10] Kretova N V, Tsaregorodtseva E Y and Khohlova G I 2017 Applying advanced methods of evaluation of level of innovative development for industrial sector *Advances in Economics, Business and Management Research, TTIESS* **38** doi: org/10.2991/ttiess-17.2017.116
[11] Official site of Rosstat [electronic resource]. - Access mode https://gks.ru/folder/210/document/13227/ Ver. From the screen
[12] Forbes for the first time evaluated the reliability of Russian developers [electronic resource]. – Access mode: https://www.forbes.ru/biznes-photogallery/386153-20-samyh-nadezhnyh-zastroyshchikov-strany-reyting-forbes?photo=1 /Ver. From the screen
[13] Consolidated financial statements of PJSC "LSR Group". Access mode: https://lsrgroup.ru/investors-and-shareholders/portfel-investora/financial-statements. Ver. From the screen
[14] Vovseenko E A 2018 Financial Instruments and Hedging: New Rules of IFRS *European Proceedings of Social and Behavioural Sciences*
[15] Appendix # 4 to the Order of the Federal tax service of Russia dated 30.05.07 # MM-3-06/333@ Profitability of goods, products, works, services sold and return on assets of organizations by type of economic activity, as a percentage http: //www.consultant.ru/ document/cons_doc_LAW_ 55729/01bf784784f784b06ae280a3d45a7f9a8f594989/ Ver. From the screen

[16] Peshkov V 2018 The building complex of the region in the aspect of implementing the main directions of the Strategy for the Development of the Housing Sector *MATEC Web Conf.* **212**
[17] Peshkova M V 2011 Assessment of investment preferences and investor competencies for the implementation of investment construction projects *Kazan Science* **9** pp 89–93
[18] Peshkov A V 2018 The economic efficiency of the use of intangible resources in investment and construction activities *University News. Investments. Construction. The property* doi: 10.21285 / 2227-2917-2018-4-44-51

IOP Conf. Series: Materials Science and Engineering **880** (2020) 012097 doi:10.1088/1757-899X/880/1/012097

On legal regulation of investment of financial resources of the compensation fundscreated by self-regulatory construction organizations

I V Kuryshova[1] and A A Timofeeva[2]

[1]Irkutsk National Research Technical University, Department of Theory of Law, Constitutional and Administrative Law, 83 Lermontov st., 664074, Irkutsk, Russian Federation
[2]Baikal State University, Department of Finance and Financial Institutions, 11 Lenin st., 664003, Irkutsk, Russian Federation

E-mail: kiw09@mail.ru

Abstract. The article deals with the issues of legal regulation of investment offinancial resources of compensation fundscreated by self-regulatory organizations, identifies features of this process in the construction industry of the Russian Federation, analyzes methods of investing financial resources of compensation funds in modern Russian reality. The shortcomings of the investment policy are identified, conclusions about the need to improve legal regulation in the Russian legislationare drawn.

1. Introduction

In the economic area, self-regulation is one of the main factors of innovative development of the government, its implementation indicates a qualitatively new stage in the system of interaction between the government and society. The development of self-regulation leads to a redistribution of management functions between public structures and the self-regulation system. It is aimed at increasing efficiency of economic development. Self-regulation creates a favorable environment for doing business, which is characterized by a decrease in administrative barriers, as well as an increase in the role of the state and business in the economy) [1].

In foreign countries, self-regulatory structures have been developing for a long time, have extensive experience and are used as an alternative to public administration. In Russia, this institute has been developing since 2007, when Federal Law No. 315 On Self-Regulating Organizations was adopted. According to this law, in various industrial sectors, self-regulatory organizations (SROs) were created. One of the most striking examples is the construction industry.

Since 2007, the regulation of SROs in the construction sector has beendeveloping and improving. The amendments were made by Federal Law No. 315 on SROs, special legislation and by-laws and regulations governing the construction sector. One of the main documents defining the specifics of self-regulatory organizations is the Urban Planning Code of the Russian Federation [2]. According to N. Seraya, "the most significant innovations were implemented by Federal Laws No. 372 and No. 340 7 (in particular, the concept ofa legal status of SROs established by Article 55.2 of the Town Planning Code of the Russian Federation, which allowed transferring the procedure for obtaining SRO permits to

IOP Conf. Series: Materials Science and Engineering **880** (2020) 012097 doi:10.1088/1757-899X/880/1/012097

the regions providing for the compensation funds (compensation funds limit the list of persons, types of work for which registration of SROs is required, replacement of the admission mechanism by mandatory membership, etc.) [3].

Compensation funds are one of the basic mechanisms of self-regulation. They embody the basic idea of self-regulation - the collective responsibility of members of self-regulatory associations for the results of their work. In world practice, there are various forms of compensation mechanisms, but their main element is the compensation fund created by intermediaries with active participation of the public and market infrastructure institutions.

2. Compensation funds

According to Russian law, design, research and construction organizations have to be members of one of the SROs. A self-regulatory organization, according to B. Bedin, is one of the institutions of the real estate market [4]. As of January 30, 2020, 120 SROs were included in the state register. They can be of the following types:

1) self-regulatory organizations including members performing engineering surveys;
2) self-regulatory organizations including members developing project documents;
3) self-regulatory organizations including members performing construction works.

At the same time, regardless of the membership, they have to create two funds - a fund for damage compensation and a fund to secure contractual obligations of a self-regulatory organization. Both funds are created to ensure property liability of SRO members: obligations as a result of damage to the person or property destroyed fully or partially. The need to create a compensation fund for securing contractual obligations is caused by the non-fulfillment or improper performance of obligations by members of self-regulatory organizations under concluded contracts. Thus, there is a two-level legal capacity for members of the SRO. On the one hand, the emergence of the legal capacity of members arising in connection with the payment of a contribution to the compensation fund, giving rise to the right to conclude construction contracts by members for the performance of works that affect the safety of capital construction projects, and secondly, special legal capacity arising in connection with payment of a contribution to an additional compensation fund for securing contractual obligations, which is mandatory only for members who intend to conclude construction contracts for you completing works affecting the safety of capital construction projects using competitive contracting methods [5].

The resources of each compensation fund are placed into special bank accounts opened on an unlimited basis. Further movement of resources is regulated by article 55-16 of the Town Planning Code of the Russian Federation. This article was introduced in 2016 by the Federal Law "On Amendments to the Town Planning Code of the Russian Federation and Legislative Acts of the Russian Federation" and other rules were in force until its adoption. Until 2016, the resources of the SRO compensation fund could be deposited in any assets, except for bank accounts of members of the self-regulatory organization; it is not allowed to invest resources placed into these accounts in securities issued by members of the SRO. The amendments adopted in 2016 tightened the rules for investing resources of the compensation funds of the SRO.

According to paragraph 8 of Article 55.16-1 of the Town Planning Code of the Russian Federation, resources of only one fund can be invested (resources of the damage compensation fund). Maintaining and increasing the size of the compensation fund is a goal of its investment.

A further analysis of legislation showed that SROs are vested with the right to choose a financial market entity that will facilitate the transformation of resources of the compensation fund into investment resources. It is possible to transfer the funds of the fund to trust management of a management company that has a license for the management of investment funds, mutual funds and private pension funds, or to a professional participant in the securities market - trust manager. (Clause 9 of the Town Planning Code). In addition, it is permissible to place resources into deposit accounts.

3. Investment of resources of the compensationfund

IOP Conf. Series: Materials Science and Engineering **880** (2020) 012097 doi:10.1088/1757-899X/880/1/012097

Today, investing resources of the compensation funds into securities is a common practice of the SROs. This is due to the development of self-regulation in the construction and housing and utilities sectors, the arbitration management and insurance sectors, the securities market, the audit sector, etc [6]. SROs make decisions in favor of management companies. The possibility of transferring compensation funds to trust management is provided for by paragraph 5 of article 13 of Federal Law "On self-regulatory organizations" [7]. Due to the fact that there are no additional legislative acts regulating the interaction between management companies and SROs, general practice can be used as an example. One of the main aspects of the interaction between the management company and the SRO is development, coordination and adoption of an investment declaration. This document should contain a list of assets available for placement of resources, as well as the structure of the investment portfolio. The management company must take into account the risk acceptable to the founder of trust management, urgency, expectations regarding future portfolio profitability and other parameters. Federal Law No. 315 imposes the following restrictions on the management company:

- more than 10% of resources of the compensation fund cannot be invested in real estate;
- at least 10% of resources of the compensation fund should be in federal government securities.

Consequently, additional requirements for the composition and structure of the investment portfolio formed at the expense of the compensation fund are determined by the investment declaration adopted by the management company in conjunction with a self-regulatory organization.

Based on a service agreement, a specialized depository monitors compliance with restrictions on compensation fund assets, rules for placing such resources and investment requirements.

Both the management company and the specialized depository are selected as a result of a competition, which allows the SRO to maximize goals pursued when investing in the compensation fund.

Data on the number of companies licensed to manage securities are presented in Table 1.

Table 1. Number of companies licensed to manage securities*

Parameter	unitofmeasurement	I quarter 2017	I quarter 2018	I quarter 2019	III quarter 2019
Number of companies having securities management licenses	unit	321	264	227	208
Clients	thousandpersons	39,7	92,6	168,.9	259,4
includingcustomers - individuals	thousandpersons	39,1	92,0	168,4	258,9
includingcustomers - companies	thousandpersons	0,6	0,6	0,5	0,5
includingactive customers	thousandpersons	12,0	27,0	32.0	54,1
Shareofactivecustomers	%	30,2%	29,2%	19,0%	20,9%
Number of companies with 75% of customers	unit	3	3	3	2
Volumeofinvestmentportfolios	billion rubles	-	801	914	1 051

* Source: Bank of Russia [Electronic resource]. Access mode: http://www.cbr.ru (accessed February 14, 2020) [8]

According to Table 1,since the first quarter of 2017, there was a decrease in the total number of companies engaged in securities management activities. A decrease occurred against the background of increased requirements for their activities set by the Bank of Russia. We can assess this trend as positive, since measures taken by the Bank of Russia contribute to the recovery of this sector and should improve quality and reliability of the compensation funds.

There is one more method of investing resources of compensation funds - placement of resources into deposit accounts. In order to regulate the mechanism of placement of resources of the damage compensation fund, the “Rules for the placement and (or) investment of resources of the compensation fund of a self-regulatory organization dealing with engineering surveys, architectural and construction designing, construction, reconstruction, and overhaul of capital construction objects” were adopted;they were approved by the Decree of the Government of the Russian Federation of April 19, 2017 No. 469 [9]. According to this document, the resources of the compensation fund can be placed only into a deposit account in the currency of the Russian Federation. Requirements for credit organizations are set by the Decree of the Government of the Russian Federation "On requirements for credit organizations in which it is allowed to place resources of compensation funds of self-regulatory organizations dealing with engineering surveys, architectural and construction designing, construction, reconstruction, capital repair of capital construction projects" of September 27, 2016 No. 970 [10]. According to this decision, the credit institution has to meet two criteria:

• have a universal license of the Bank of Russia for banking operations;
• the amount of own resources should be more than 100 billion rubles.

As of February 1, 2020, there are three hundred and ninety-six credit organizations in Russia. The number of banks with equity of more than 100 billion rubles is significantly less (Table 2). Thus, approximately 3% of credit institutions operating in the Russian Federation meet the requirements of the Decree of the Government of the Russian Federation No. 970.

Table 2. Banks with equity of more than 100 billion rubles. as of December 1, 2019, thousand rubles *

No	Name	Thousandrubles
1	Sberbank	4 510 352 502
2	VTBBank	1 652 738 873
3	BankGPB	772 413 091
4	Rosselkhozbank	488 656 516
5	Alpha-Bank	469 673 252
6	FCOtkrytie	325 126 053
7	MOSCOW CREDIT BANK	275 597 790
8	UniCreditBank	218 958 900
9	Raiffeisenbank	188 361 613
10	Promsvyazbank	153 690 045
11	Rosbank	150 905 627
12	Bank «VBRR»	133 840 277
13	Sovcombank	118 136 665

* Source: Bank of Russia [Electronic resource]. Access mode: http://www.cbr.ru (accessed February 14, 2020) [8]

In addition to the fact that resources of the compensation fund can be placed into the deposit accounts of the same credit organization where a special bank account is opened, Rules No. 469 provides for several more restrictions.

First, the placement of resources becomes possible if there is a decision of the general meeting of members of the self-regulatory organization.

Secondly, Rules No. 469 established a limit on the resources (no more than 75% of the compensation fund of the self-regulatory organization). Thus, 100% of the resources cannot be placed indeposit accounts. At the same time, the placement of resourcesinto a fixed term deposit account is a more attractive option. In terms of risks, the term deposit and the special bank account have no differences, because if a license is withdrawn, the credit institution can lose these resources in the same amount. The deposit insurance system does not apply to legal entities. Given the level of interest rates and the level of inflation, the resources are placed into deposit accounts to be preserved rather than increased.

Thirdly, it is forbidden to purchase deposit certificates issued by this credit institution.

The SRO can terminate the bank deposit agreement unilaterally for a number of reasons, after which the credit institution is obliged to transfer the resources placed into the deposit account to a special bank account. The obligation should be fulfilled during ten days; the term of the bank deposit agreement (one year) makes it impossible to invest resources of the compensation fund in the long-term active operations of banks.Despite large amounts accumulated by the construction sector, it does not solve the problem of attracting " long "money in the Russian economy.

IOP Conf. Series: Materials Science and Engineering **880** (2020) 012097 doi:10.1088/1757-899X/880/1/012097

4. Conclusion

Monitoring of judicial practice demonstrates a large number of litigations on the issue of compensation funds created by SROs. An analysis of the existing legislation and law enforcement practice allows us to identify problems of compensation fund regulation. The issues requiring timely resolution include civil liability of SROs in the field of construction, the legal regime of the SRO's compensation funds and the legal regime for investing resources of SRO's compensation funds.

References

[1] Romanova A I 2016 Improving the Quality of Construction Works in Terms of the Self-Regulation. *2 nd International Conference on Industrial Engineering*

[2] Town Planning Code of the Russian Federation of December 29, 2004 No. 190-FZ (as amended on December 27, 2019) Collection of legislation of the Russian Federation 2005

[3] Seraya N A 2018 Actual problems of regulating the civil status of self-regulatory organizations in the construction industry based on the analysis of judicial practice (2016-2018) *Russian Law Online* **4** pp 38-47

[4] Bedin B M 2017 Economics and Real Estate Management: [Electronic resources]: Textbook: BSU, 2017. Access mode: http: // lib-catalog@bgu.ru. (date of the application February 14, 2020)

[5] Dautia T V 2018 Features of the legal status of a member of a self-regulatory organization in the field of construction, Journal of Russian Law **11** pp 76-84

[6] Astafiev S A 2011 The Impact of the Accession of the Russian Federation to the WTO on the Formation of Self-Regulation in the Investment and Construction Sphere *Bulletin of the Pacific State University* **3 (22)** pp 111-118

[7] On self-regulatory organizations: Federal Law of December 1, 2007 No. 315-FZ (as amended on August 3, 2018) Collection of the legislation of the Russian Federation 2007 49 Art. 6076.

[8] Bank of Russia [Electronic resource]. Access mode: http://www.cbr.ru (date of the application February 14, 2020)

[9] On approval of the Rules for investing funds of the compensation fund for damages to a self regulatory organization in the field of engineering surveys, architectural design, construction, reconstruction, overhaul of capital construction projects: Decree of the Government of the Russian Federation of April 19 *Collection of the Russia legislation* 18 2781

[10] On requirements for credit organizations which can invest compensation funds for engineering surveys, architectural and construction design, construction, reconstruction, overhaul of capital construction objects: Decree of the Government of the Russian Federation of September 27, 2016 *Collection of the legislation of the Russian Federation* 41. Art. 5824

IOP Conf. Series: Materials Science and Engineering **880** (2020) 012098 doi:10.1088/1757-899X/880/1/012098

Improving construction administration technologies in emergency response: targeted focus

M V Matveeva

[1] Irkutsk National Research Technical University, 83 Lermontov str., Irkutsk, 664074, Russia

E-mail: expertiza@istu.edu

Abstract. The housing development strategy of the Russian Federation for the period up to 2025 sets the goal of providing the population with comfortable and affordable housing. In this regard, the construction management system requires changes in accordance with the tasks of the government of the Russian Federation. Implementation of investment and construction projects in housing construction is quite a complex and multifaceted process. Solving the housing problem requires improving the mechanisms of interaction between participants in the implementation of investment and construction projects in housing construction, especially in the event of emergencies, the number of which increases exponentially every year. In this regard, when determining the possible scenarios and relationship of participants of investment and construction activities in liquidation of consequences of emergency situations it is necessary to consolidate and focus the resources of all participants in the Investment and Construction Projects, which will allow not only to concentrate all resources on realization of the investment construction project within the given time but will also ensure the achievement of the maximum effect to the principle of synchronization of economic interests.

The socio-economic development of the Russian Federation is largely determined by indicators of housing provision. In 2019, 80 million m^2 were built in the Russian Federation, and the share of low-rise housing construction reached the highest indicator for the entire period of statistical observations of 73% of the total volume of housing commissioned in 2019. Such a significant increase in low-rise construction was observed in many regions of Russia. This significant growth was largely facilitated by the declared end of the terms of the country amnesty, as well as the popularization of low-rise buildings on the part of the state [1]. Despite the relatively high rates of housing commissioning, demand in the primary and secondary markets is declining, which causes each region to identify the housing problem as a strategic and priority areas of housing policy.

Within some years, investment and construction activities have been considered as an important component of the country's economic potential. Investment and construction activities, being difficult to manage, high-risk and subject to the influence of a variety of market, structural, institutional, communication, economic and other factors, to a greater extent than other types of activities, need to develop fundamental scientific research aimed at improving the management system.

The current state of the construction industry, in general, requires adjustment of management systems. Currently, there are clearly visible shortcomings of the management system of investment and construction activities associated with a comprehensive assessment of resources, costs and results of production, economic and financial activities of construction organizations at various levels of management. The most detailed shortcomings of the management system are shown in emergency situations [2,3].

The implementation of investment and construction activities, including, in response to emergency situations, makes sense if the potential for economic efficiency is growing faster. Otherwise, there will be processes of recapitalization of real estate in which investment and construction activities will lose economic/social/ environmental and other meanings [4]. Thus, investment and construction activity a priori must have certain properties that determine its effectiveness, both by type of construction, stages of the project cycle, forms of reproduction, and territorial affiliation. Based on this, investment and construction activities in all cases should focus on the creation of construction products which obviously must meet not only the strategic development goals of the country and its regions but also consider emerging emergencies. Thus, it is advisable to supplement the strategic focusing of investment and construction activities with interactivity, prevention, mobility and proportionality. Outside of the above properties, it is impossible to ensure the economic efficiency of investment and construction activities in the event of emergencies.

The problem of ensuring the safety and survivability of buildings and structures considering various natural disasters is very relevant, since buildings and structures are among the most vulnerable objects to such impacts. Significant importance is given to the safety and survivability of buildings in the tasks specified in the National project "Housing and a comfortable urban environment " (approved by decree of the President of the Russian Federation dated May 7, 2018 No. 204 on national goals and strategic objectives of the development of the Russian Federation for the period up to 2024), in particular in the Federal project "Housing" [5]. So, for example, Irkutsk oblast, like other Russian regions are region with specific climatic conditions like possible flooded areas, where are apartment buildings, long winter period and the challenges this autumn-winter and winter-spring periods. It is in the aspects of these features that the most effective way to use the methodology of targeted focusing of investment and construction activities, which involves a balanced increase in the construction, reconstruction and renovation of facilities, based on compliance with the principles of integrated development, taking into account the requirements of energy and resource conservation, environmental safety, durability, technical and social reliability [6, 7]. Based on this, the importance of ensuring the target focus of investment and construction activities is to justify the project management format in the event of emergencies and the need to ensure the target focus of investment and construction activities aimed at improving the reliability of residential buildings and improving the safety of people living in specific conditions of regions of Russia with possible flooding of territories. In the end, the result of targeted focus reduces the economic and social costs associated with dealing with the consequences of natural emergencies.

This is the situation that has developed, for example, after the large-scale floods in the summer of 2019 in the Irkutsk region, which led to the death of people, breakthroughs of protective dams and flooding of a significant number of residential buildings. Official information about flood victims is constantly updated, and the final report shows the number of victims is more than 4,000. Irkutsk region is in the difficult natural and climatic conditions. The region is characterized by a long cold winter (in the Northern regions of the region, winter temperatures are extreme), a long ice age and warm, with heavy precipitation in the summer, which leads to severe floods. Also, the territory of the Irkutsk region is one of the most seismically active areas in the Russian Federation. According to the Ministry of emergency situations, the Irkutsk region belongs to the subjects of the 1st degree of danger from the possible impact of natural and related man-made disasters. The number of emergencies in the Irkutsk region tends to increase every year [18]. Based on this, the process of providing housing to the population and creating favorable conditions in the event of emergencies (mass disasters) becomes particularly important.

Similar emergencies also occur in such regions of the country as the Primorsky krai, Krasnodar krai, the Russian far East, and others [9]. According to the Ministry of Emergency Situations, natural disasters affected about 500 thousand km^2 floods with catastrophic consequences - 150 thousand sq km, where are located some 300 cities and tens of thousands of settlements, a large number of economic objects, more than 7 million hectares of farmland. Analyzing the causes and nature of the prerequisites and occurrence of emergencies over a long period of time, it is easy to see that any emergency that occurs once has a cyclical nature of repetition [10]. In this regard, the relevance of modifying the management system of the investment and construction sector increases.

In such circumstances, investment and construction activities should be focused on solving the housing problem for the population affected by an emergency. Moreover, it is important not only to focus investment and construction activities on the construction of new facilities, but also to direct resources to the development of recommendations for ensuring the safety of living in residential buildings that are in the zone of emergency situations and have been flooded by flood waters [11]. An important issue is the development of recommendations of a compensating nature to ensure the operability of residential buildings for the subsequent period of operation considering natural and climatic conditions and the consequences of flooding. In this regard, for reducing damage and ensuring the most rapid and high-quality restoration work, the problem of improving the management systems of investment and construction activities in the aftermath of emergencies comes to the fore.

Based on the above-mentioned problems, one of the strategic goals of investment and construction management should be the development of methodological foundations and methodological recommendations for improving the management system of investment and construction activities in emergency situations, as necessary conditions and processes aimed at providing affordable housing to the population. Based on a specific goal, the following tasks are defined:

1. Systematizing the challenges and problems of national economic development at the present stage, as the basis for restructuring the target space for the development of investment and construction activities in emergency response.
2. Forming a concept of target focus of investment and construction activity management at various stages of the emergency response process.
3. Determining the main directions of development of the target focus of the investment and construction complex, as the basis for spatial and temporal focus and synthesis of resources to provide housing for the population in emergency situations.
4. Developing methods of targeted focus of investment and construction projects in the context of managing the process of providing housing for the population in emergency situations.
5. Developing a paradigm for the target focus of the management system of investment and construction activities in emergency response
6. Analyzing the submitted technical documentation based on the results of the survey of the technical condition of residential buildings that have fallen into the flood zone, their classification according to the degree of danger to the life and health of residents, taking into account the natural and climatic conditions and the likelihood of progressive deformations and damage
7. Developing recommendations of a compensating nature in order to ensure the operability of residential buildings for the subsequent period of operation, taking into account natural and climatic conditions and the consequences of flooding.
8. Justifying the effectiveness of using large-panel housing construction technology to solve the housing problem, including in the aftermath of emergencies in the Irkutsk region
9. Developing an organizational and functional process for implementing projects using large-panel housing construction technology, including in response to emergency situations [12].
10. Developing a model for a comprehensive assessment of the economic efficiency of the target focus of management implemented in the main projections of investment and construction activities in response to emergency situations.

The proposed goal and objectives are theoretical and based on a complex of scientific methods of knowledge that allow a comprehensive analysis of methodological tools to select appropriate subject area of research General scientific methods (deduction, induction, inversion, analysis, synthesis, etc.), special economic methods (comparison, functional statistical, graphical, analytical groupings, SWOT analysis, PEST analysis, TAIDA method) methods of economic and mathematical modeling (simulation, game theory, optimization, etc.), methods of selective observation, methods of expert assessments [13].

The underlying tasks of implementing the target focus of investment and construction activities have scientific novelty, as they consist in the development of methodological foundations and methodological recommendations for the target focus of investment and construction management in emergency response, which provides an afterburner-front, balanced in the industry and spatio-temporal sections, the type of development implemented in the project format to achieve equifinal efficiency when using large-panel housing construction technology.

The result of completing the research tasks will be reflected in some forms:

- The author's model of interpretation of the process of management of investment and construction activities in emergency response
- Developing recommendations and technical solutions to ensure the safety of living in residential buildings exposed to flood waters and showing signs of reduced load-bearing capacity and stability of structural elements, including ground conditions.

The practical significance of the research is expressed in a wide range of possible use of methodological developments on:

- Environmental and economic assessment of the effectiveness of the management process in emergency response.
- Modeling the organizational and economic mechanism of interaction between construction participants in the process of emergency response [14].
- Development and justification of the effectiveness of construction technologies in emergency response.
- Use of environmental and economic methods of construction management to transform the management concept and bring it in line with priority tasks in emergency response [15].

Emergencies caused by natural phenomena cause huge damage to the entire industry structure of the economy, including the construction industry. Implementation of investment and construction projects in housing construction in response to emergency situations is quite a complex and multi-faceted process that requires, in addition to the implementation of effective economic activities of a single economic entity, also effective joint work of all participants in the project to achieve the final goal [16]. Therefore, it is necessary to focus resources on achieving the final goal. From the point of view of focusing the resources of economic activity, it is necessary to analyze and take into account the influence of extensive and intensive factors of a technogenic nature on the final result of activity. Targeted focus will allow you to concentrate all resources on the implementation of an investment and construction project within the specified time frame. This fully applies to the activities of each participant individually as an independent business entity, and to their Association for the purpose of implementing an investment and construction project (ICP) as a whole.

References

[1] Peshkov A V 2019 Digitalization of the economy, implementation of national projects and development of the construction industry *Investments. Construction. Real Estate: New Technologies and Targeted Development Priorities IOP Conference Series: Materials Science and Engineering* **667** 012074

[2] Peshkov A V 2019 Remote mortgage lending as a resource for housing development *Investments. Construction. Real Estate: New Technologies and Targeted Development Priorities IOP Conference Series: Materials Science and Engineering* **667** 012073

[3] Gertsekovich D, Gorbachevskaya L, Grigorova L and Peshkov V 2019 Return on investment in REIT real estate funds *Investments. Construction. Real Estate: New Technologies and Targeted Development Priorities IOP Conference Series: Materials Science and Engineering* **667** 012063

[4] Peshkov V, Gertsekovich A and Gorbachevskaya L 2019 Dilapidated and dilapidated housing in the aspect of the Federal project "Ensuring sustainable reduction of uninhabitable housing"

Investments. Construction. Real Estate: New Technologies and Targeted Development Priorities IOP Conference Series: Materials Science and Engineering **667** 012075

[5] Kudryavtseva V A, Zaslavsky A M and Larina O P 2018 Multicriterial comparison at the inclusion of investment projects of manufacture of building materials into the state program of the region (at the example of Irkutsk region) *Proceedings of Universities. Investment. Construction. Real estate* **8** pp 60–72 DOI: 10.21285/2227-2917-2018-1-60-72

[6] Abdrakhmanova G., Vishnevsky K., Volkova G. and Gokhberg L. 2018 *Indicators of the digital economy:* 2018 p 268

[7] Yaskova N Yu 2018 Construction in the context of new views on the goals of economic development. *Proceedings of Universities. Investment. Construction. Real estate* **8** pp 133-142 DOI: 10.21285/2227-2917- 2018-1-133-142

[8] Olga Lavygina, Oksana Grebneva, and Irina Maizel 2019 *IOP Conf. Series: Materials Science and Engineering* 667 012056

[9] Peshkov V V and Gamayunova E V 2018 Specific features of investment support for the repair and reconstruction of buildings and facilities *Proceedings of Universities. Investment. Construction. Real estate* **8** pp 56-72 DOI: 10.21285/2227-2917-2018-2-56-72

[10] Komarov A K, Ivanov I A and Lundenbazar B 2019 Theory and practice of the use of gabions for forming protective structures *Proceedings of Universities. Investment. Construction. Real estate* **9(1)** pp 78–89 DOI: 10.21285/2227-2917-2019-1-78-89

[11] Peshkov V V 2008 *Potential of development as a basis for the formation of management strategy of investment and construction sphere.*Bulletin of Irkutsk state technical University № 2(34) pp 57-61

[12] Khokhlova G.I. and Okladnikova D.R. 2014 *Classification of factors as grounds for analysis of innovative–investment activity of small and medium businesses*. Middle East Journal of Scientific Research Vol 20 № 8 pp 900–904

[13] Kazimirov I A, Peshkov V V 2019 Determining price dynamics in the secondary market of

[14] residential real estate using a multidimensional regression model *University news. Investment. Construction. Realty* **9(3)** pp 476–487

[15] Peshkov V V 2018 Some aspects of the development of the construction industry in 2018 *IOP conference series: materials science and engineering* **667**

[16] Yaskova N Yu 2019 Imitation of shared-equity construction reforms as a deterrent to the inevitable systemic transformation of the industry *University news. Investment. Construction. Realty* **9** pp. 296-303

IOP Conf. Series: Materials Science and Engineering **880** (2020) 012099 doi:10.1088/1757-899X/880/1/012099

A concession agreement as an effective mechanism for the development of innovative activities in the housing and utilities sector

L G Nikityuk[1]

[1]Irkutsk National Research Technical University, 83 Lermontov st., Irkutsk, 664074

E-mail: Love175@yandex.ru

Abstract. The state of the housing and utilities sector is one of the most significant indicators of the level of development of the national economy. In many countries, the state of the housing and utilities sector does not fully satisfy needs of the population. Under the slow economic development, an innovative development path can ensure high quality and reliability of services provided. The concession mechanism of cooperation between government and businesses should become a key tool which can attract investment in the development of innovative activities in the housing and utilities sector. The author has developed a support scheme for the investment and innovation project implemented in the housing and utilities sector using a concession agreement within the PPP which ensures stability of the industry and improves quality of services provided. The scheme of concession tools can be used to achieve a social and economic effect. The article presents a method for assessing the economic effect of concession agreements used when implementing the investment and innovation projects in the housing and utilities sector taking into account risks. It can balance interests of the government and housing and utilities companies. The concession mechanism will contribute to the development of the most promising areas of the housing and utilities sector.

1. Introduction

The process of implementing reforms in the housing and utilities system of the Russian Federation can be attributed to one of the most protracted modern reforms. Currently, special attention is paid to problems of the housing and utilities sector: the increasing share of fixed assets depreciation, low quality of services and high capital intensity.

In the market economic conditions, in order to provide high-quality housing and utilities services and ensure financial stability of housing and utilities companies, competitiveness is crucial. It should be based on innovative approaches to the management of the housing and utilities sector. However, the lack of effective methods and mechanisms for managing investment and innovation activities is a problem [7, 12, 18, 19]. The concession mechanism within the public-private partnership can be an effective tool for the development of investment and innovation activities in the housing and utilities sector.

Concession agreements are widely used to attract private capital in order to modernize the housing and utilities sector [1, 3]. In Russia, the need for investment is 25.9 trillion rubles, including 17.8 trillion - for the modernization and construction of transport infrastructure. The energy sector requires 5.5 trillion rubles, the information and communications sector needs2 trillion rubles, the water supply and sanitation system needs about 0.5 trillion rubles [14].

The concession agreement in the housing and utilities sector is a contract concluded between public agencies represented by federal, regional, local authorities and housing and utilities companies aimed at developing investment and innovative activities of the housing and utilities sector and ensuring the socio-economic effect.

2. The investment and innovation project support scheme in the housing and utilities sector using a concession agreement within the PPP

The author has developed a support scheme for the investment and innovation project using a concession agreement (Figure 1).

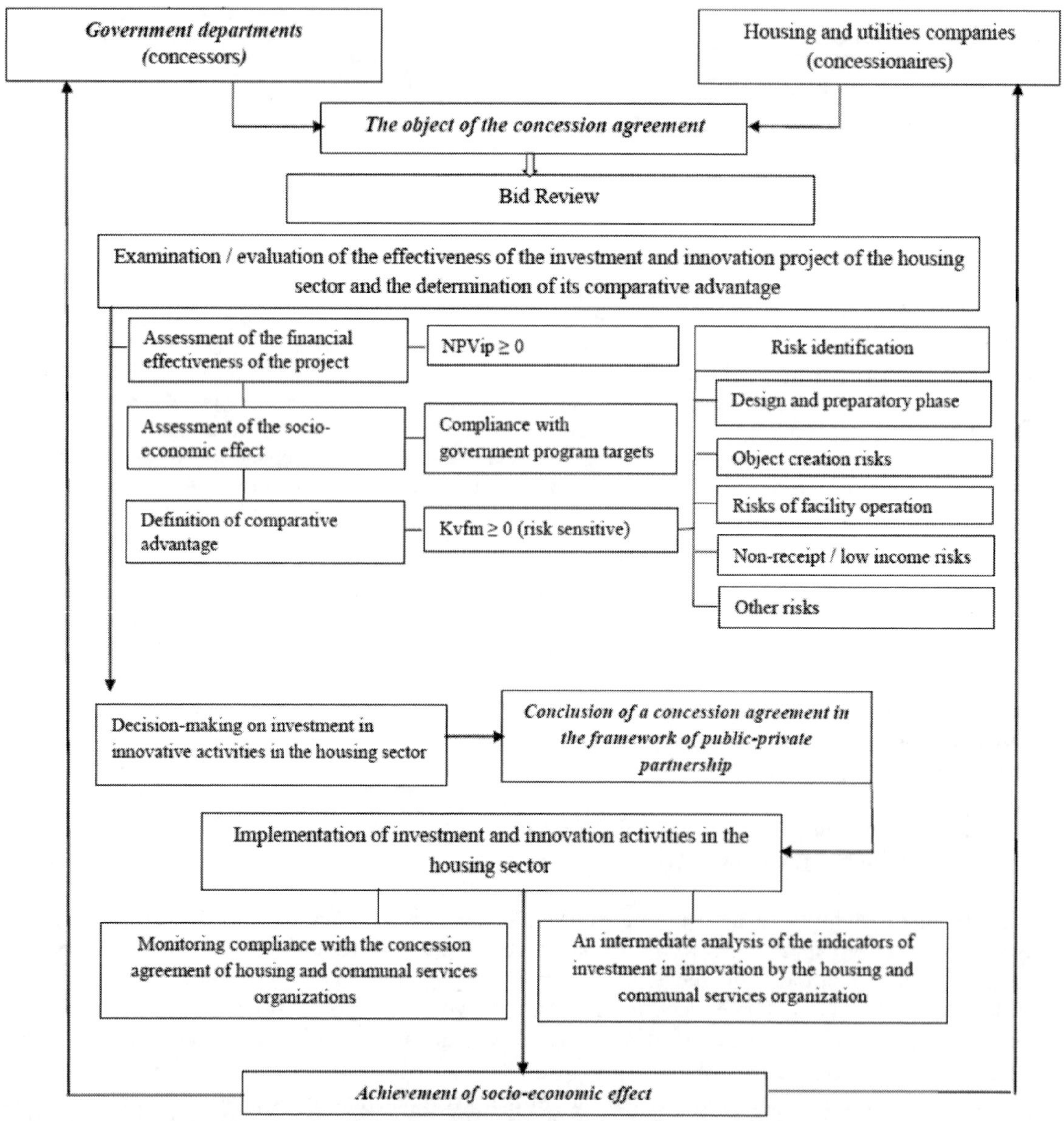

Figure 1. Scheme of support for investment and innovation projects in the housing sector using a concession agreement

The effectiveness of innovative and investment projects depends on the coordination of actions of all participants in the investment process.

IOP Conf. Series: Materials Science and Engineering **880** (2020) 012099 doi:10.1088/1757-899X/880/1/012099

The scheme of concession tools makes it possible to support the innovation and investment project in the housing and utilities sector to achieve a social and economic effect.

The review of a competitive proposal of housing and utilities companies by government agencies. A concession agreement is concluded as a result of open / closed tenders.

Housing and utilities companies (concessionaires) submit proposals to the public authorities (concessors) [7, 8].They attach an investment and innovation project to the documents. The authorized bodies evaluate the effectiveness of the project (financial efficiency; socio-economic effect) and determine comparative advantages [17].

3. The financial and economic model for assessing the effectiveness of the innovation and investment project in the housing and utilities sector

The effectiveness of concession results involves an assessment of each stage of the concession relationship: preparatory, organizational, and implementation ones.

At the preparatory stage, the effectiveness of the investment-innovative housing and utilities project is assessed. It is preceded by the determination of its comparative advantage.

To assess the financial effectiveness of the investment and innovation project, it is necessary to calculate the net present value:

$$NPV_{ip} = \sum_{t=1}^{T} \frac{CF_t^d - K_t}{(1+R)^t} + \frac{V_t}{(1+R)^T} \quad (1)$$

whereNPV_{ip} – net value of the investment and innovation project;

CF_t^d – total discounted cash flow balance;

K_t – concession payment;

R - annual return on capital;

V_t - value of assets created;

T –implementation term;

t – implementation time.

The innovation and investment project is effective if NPVip ≥ 0 [16].

An assessment of the socio-economic effect of the investment and innovation project is carried out by correlating goals of federal / municipal programs.

If the project is recognized effective according to each of the specified criteria, its comparative advantage is determined.

Evaluation of the effectiveness of the federal project and determination of its comparative advantage are carried out by the federal authorized body (the Ministry of Economic Development of Russia); the authorized body of the region of the Russian Federation evaluates regional and municipal projects.

Determination of the comparative advantage of the project. To determine the effectiveness of the concession agreement, it is necessary to determine comparative advantages for the federal / municipal budgets.

The comparative advantage of the investment and innovation project is recognized provided that:

$$\begin{cases} K_{vfm} = 1 - \frac{PBV_{ppp} + PRV_{ppp}}{PBV_{cp} + PRV_{cp}} \geq 0 \\ PBV_{cp} + PRV_{cp} \geq 0 \end{cases} \quad (2)$$

$$\begin{cases} K_{vfm} = \frac{PBV_{ppp} + PRV_{ppp}}{PBV_{cp} + PRV_{cp}} - 1 \geq 0 \\ PBV_{cp} + PRV_{cp} \leq 0 \end{cases} \quad (3)$$

whereK_{vfm} – coefficient of comparison of benefits of the project when implementing the federal / municipal concession agreement.

PBV_{ppp} - net discounted expenses of the innovation concession project

PBV_{cp} - net value of the federal / municipal contract.

PRV_{ppp} – the total amount of obligations of the public partner in case of risks of the innovative concession project

PRV_{cp} - the total amount of obligations of the public legal entity in case of risks of the federal / municipal contract [16].

In addition, it is necessary to assess risks that may arise between the parties to the investment and innovation project in the HUS.

The distribution of risks between the parties to the concession agreement is carried out by including terms on special obligations (table 1). Special circumstances are various events that may occur during the execution of the concession agreement. They affect investment and innovation activities of the housing and utilities companies.

Table 1. The list of risks of investment and innovation activities in the housing and utilities sector

Name of a risk
Risks at the preparatory stage
provision of land, engineering infrastructure; land preparation; failure to meet the deadline for facility designing; failure to meet the deadline for preparatory activities
Construction / reconstruction risks
elimination of consequences of actions of third parties; elimination of natural disasters; elimination of environmental consequences; failure to meet the deadlines for construction / reconstruction of the housing and utilities facility; failure to meet the deadlines for commissioning the housing and utilities facility; an increase in construction costs due to the increased exchange rate; an increase in construction costs due to the increased inflation rate; an increase in construction costs due to the increased interest rate
Maintenance risks
an increase in maintenance costs for the facility transferred to the public partner; an increase in maintenance costs due to an increase in the exchange rate; increase in maintenance costs due to an increase in the inflation rate; increase in maintenance costs due to an increase in the interest rate; increase in maintenance costs due to an increase in the tax rate
Income Risks
non-receipt of payments ensuring minimum profitability; a decrease in revenue due to a decrease in the volume of services; a decrease in revenue due to lower prices (tariffs) for housing services; a decrease in revenue due to non-payment
Other risks
termination of the agreement by public bodies; termination of the agreement by the housing and utilities company; loss of the facility; force majeure circumstances; other risks

When determining the financial efficiency and comparative advantages, the negative result of the evaluation is the basis for finalizing the investment and innovation project or refusing to implement it.

If the authorized body makes a decision on the possibility of concluding the concession agreement, the public bodies conduct negotiations with housing and utilities companies in order to discuss the terms of the concession agreement [4, 14].

The efficiency of the concession project can be evaluated in the following areas:

1. Economic efficiency is determined by assessing financial consequences of the concessionaire's participation in the investment and innovation project [2, 13].

2. Social efficiency is determined by evaluating results of the project that can improve quality of life of the population, including quality of infrastructure facilities, roads, transport infrastructure, health and education infrastructure, increase the number of jobs in the region, improve working conditions [5]. When calculating social efficiency, it is necessary to take into account indirect results of the concession project, such as changes in the income of enterprises and citizens, the market value of real estate, loss of natural resources as a result of accidents.

Thus, the main goal of housing and utilities companies is to maximize profits, and the main goal of public bodies is to protect public interests.

When implementing the investment-innovative project, the parties are interconnected by the terms of the concession agreement and other relations [10, 13]. The authorized bodies exercise control over the concession facility and activities of housing and utilities companies [11, 20]. The fact that housing and utilities companies interact with consumers determines their special relations with society. In this regard, the control over activities of housing and utilities companies and the concession facility should be carried out by consumers and specialized public organizations. Information on compliance / non-compliance with public interests should be received by both housing and utilities companies and public authorities [9].

An important stage is the monitoring of changes in the concession facility when implementing the investment-innovative project, which involves an assessment of socio-economic efficiency of the concession project. The compliance of the results achieved with the conditions of the concession agreement and the goals set by the concessor is assessed.

The implementation of a new investment and innovative concession mechanism in the housing and utilities sector will increase the level of competitiveness by improving the efficiency of innovative processes. All changes will positively affect organizational and economic relations that contribute to the effective management of the housing and utilities sector.

4. Conclusion

Given the lack of budgetary funds, the public system of management of the housing and utilities sector focuses on the use of a concession agreement within the public-private partnership as the main way to overcome large-scale infrastructure imbalances. The government (concessor) transfers real estate objects and related movable property to a private person (concessionaire) for temporary possession and use, and the housing and utilities company undertakes to improve them (reconstruct, modernize, etc.) for the development of innovative activities. Thus, the use of the concession mechanism is aimed at enhancing innovation activities of the housing and utilities sector, developing the industry of innovative technologies for the production of housing and utilities services. The application of innovative investment methods by housing and utilities companies will contribute to the implementation of modern highly efficient technologies and equipment, provision of high-quality services and efficiency of housing and utilities companies. As a result, costs of housing and utilities services and housing and utilities fees will be reduced.

References

[1] Airapetyan M S 2009 Foreign experience of public-private partnership application. *State power and local government* **2** pp 35-42

[2] Critical performance factors of public sector organizations in concession-based public-private partnership projects *Property Management* **37:1** pp 17-37.

[3] Alklychev A M 2019 Foreign experience in the effectiveness of concessions and infrastructure investments in public-private partnerships *Regional problems of economic transformation* **1** pp 74-81

[4] Burganov R A and Yarullina A G 2015 Institute of Public-Private Partnership in the Electric Power Industry: Content and Forms *University News. Problems of Energy* **5-6** pp 44-52

[5] Bychkov A I 2011 Concession agreement as a mixed agreement. Property relations in the Russian Federation **10** pp 102-110
[6] Chernyakhovskaya Y V and Korolkov D L 2017 Public-Private Partnerships in Nuclear Energy: The USA experience *Finance: Theory and Practice* **21(1)** pp 91-105
[7] Federal Law of July 13, 2015 No. 224-FZ "About public-private partnership, municipal-private partnership in the Russian Federation and amendment of certain legislative acts of the Russian Federation".
[8] Federal Law of July 21, 2005. No. 115-FZ "On Concession Agreements".
[9] Gorovenko V V and Samolovov D A 2014 Concession agreement and public-private partnership agreement as contractual forms of public-private partnership *Academic Bulletin.* **4 (30)** pp 69-75
[10] Hugo Consciência Silvestre, Rui Cunha Marques, Ricardo Corrêa Gomes 2018 Joined-up Government of utilities: a meta-review on a public–public partnership and inter-municipal cooperation in the water and wastewater industries *Public Management Review* **20:4** pp 607-631
[11] Kameneva E A 2010 The specifics of housing and communal services as an object of financial relations in the housing and communal services *Finance and credit* **22**
[12] Krupina N N 2014 An innovative approach to enhancing the potential of regional enterprises of housing and communal services - water utilities *Regional Economics: theory and practice* **14** pp 51–59
[13] Leus D I 2013 Investment support of housing and communal services in the region. *Innovations and Investments* **1** pp 130–135
[14] Malikova O I and Maksimova A M 2017 Concession agreement as a form of involving small private companies in the effective management of state property *Management Sciences* **4** pp 6-15
[15] Nikityuk L G and Timchuk O G 2015 Investment of innovative activity in the housing and communal services sector on the basis of public-private partnership. Bulletin of the SSGUTU **5** pp101-108
[16] Order of the Ministry of Economic Development of Russia dated November 30, 2015 N 894 "On approval of the Methodology for assessing the effectiveness of a public-private partnership project, a municipal-private partnership project and determination of their comparative advantage" (Registered in the Ministry of Justice of Russia December 30, 2015 No. **40375)**
[17] Petyukov S E 2017 Opportunities for financing investment projects in the electric power industry using various models of public-private partnerships *Investment and Innovation* **4** pp 31-34
[18] Prachitha John, Ashwin Mahalingam, Akash Deep and Thillairajan A 2015 Impact of Private Sector Participation on access and quality of services: systematic review of evidence from the electricity, telecommunications and water supply sectors *Journal of Development Effectiveness* **7:1** pp 64-89
[19] Timchuk O G, Nikityuk L G and Gorbachevskaya E Yu 2019 Public-private partnership as an effective mechanism for financing innovation in the Russian Federation *Bulletin of NSUU* **3** pp 43-55
[20] Zhukova N S 2011 Concession mechanism for the implementation of state innovation policy *Problems of the modern economy* **4** pp 25-2

IOP Conf. Series: Materials Science and Engineering **880** (2020) 012100 doi:10.1088/1757-899X/880/1/012100

Organization of work on the development of measures to ensure the survivability of buildings exposed to flooding

V V Peshkov

[1] Irkutsk National Research Technical University, 83 Lermontov str., Irkutsk, 664074, Russia

E-mail: pvv@ex.istu.edu

Abstract. Nowadays the issue of ensuring the survivability of buildings in emergency situations is relevant. Floods are one of the most frequent and serious types of natural disasters. Therefore, flood protection becomes an important task for cities, regions and countries that are exposed to the threat of flooding. Because along with human tragedies and damage to the national economy, floods cause the failure of Central supply points, on which the population is particularly dependent in an emergency. A methodological analysis of the results of flood response in the Irkutsk region in the affected areas has shown that there are number of issues related to the design construction and restoration of residential buildings in areas subject to natural emergencies. In order to solve these problems, theoretical generalizations of domestic and foreign experience are required, as well as the development of principles techniques and models of urban development placement in conjunction with technological solutions. Together, emergency response measures should have the properties of mobility, urbanization, and efficient use of the environment and resources in the implementation of investment and construction projects.

In the vast territory of the Russian Federation, natural and geographical conditions are very different. The availability of certain building materials, climate and soil that differ significantly in certain geographical areas, the variety of natural resources, minerals and associated agriculture and industry – all these factors have had a direct impact on the formation and development of various types of housing. The features of housing depend to a large extent not only on the direction of human economic activity, the historical process, the level of development of society property and class differences, but also on natural and geographical conditions [1]. The Russian Federation, due to natural and climatic factors, namely, the complexity of the terrain geographical location and features of air mass circulation, is periodically affected by natural phenomena. Depending on the strength of their manifestation, the country's economy and its population suffer significant material damage, sometimes it is associated with human casualties.

The maximum number of natural emergencies is caused by floods, which account for about 40% of the total number of emergencies. According to the Federal service for Hydrometeorology and environmental monitoring of Russia (Roshydromet) floods in the Russian Federation-prone areas a total area about 500 thousand square km; at the same time, floods can have catastrophic consequences – 150 thousand square km, which houses more than 300 cities, tens of thousands of settlements, more than 7 million hectares of agricultural land [2,3].

One of the most dangerous is flooding, caused by the breaking of a dam, dam or other hydraulic structure, or the overflow of water over the dam due to the overflow of the reservoir. Flooding of the area located below the structure is carried out suddenly, with the arrival of the so-called break wave (displacement, skipping), the height of which can reach several meters, and the speed of movement - several tens of m/sec. This type of flooding occurred in the Irkutsk region in 2019.

In connection with the extraordinary situation on the territory of Irkutsk region as a result of flooding of settlements, deterioration of hydrometeorological conditions in the territory of the Irkutsk region and the threat of disruption of life-support systems of the population of the Irkutsk region, and economic facilities on the territory of Irkutsk region introduced a state of emergency.

In order to assess the damage and develop measures to minimize the damage consider the preliminary results of the assessment of damage from the first wave of flooding [4]:

1. The total number of flooded houses was 10,890, the number of residential premises-17,370 with a total area of 1,122. 8 thousand square meters;
2. The total number of affected socio-cultural objects is 97;
3. The total length of flooded roads of regional, inter-municipal and municipal significance is 730.8 km, the number of bridges affected is 35 bridges, of which 11 bridges were completely destroyed the remaining 24 partially.

Based on the results of surveys of housing and social facilities, it was found that the total number of residential premises requiring resettlement or major repairs was 9,867, including 7,242 premises deemed unfit for habitation, and 2,625 premises deemed to be subject to major repairs [5].

The survey showed that the remaining flooded low-rise residential buildings are losing their capital particularly wooden buildings are damaged by rot. The destruction of wooden buildings and structures is mainly due to insufficient strength of the foundations, except of pile foundations. In brick buildings, there is a gradual destruction of masonry with the loss of bricks. Metal structures and reinforced concrete reinforcement are subject to corrosion.

In panel low-rise residential buildings with enclosing structures made of two-layer panels made of non-autoclaved reinforced concrete, the foam concrete insulation is peeled off, and a layer of light concrete is destroyed in solid wall panels. The durability of concrete and reinforced concrete elements, Foundation blocks, pile heads and grillage are reduced under the influence of water, which leads to a decrease in the capital of buildings. If there is insufficient concrete density in the protective layer of reinforced concrete elements, rebar steel intensely corrodes. Especially intensively corrode embedded parts and welds of external wall panels. Low-quality manufacturing of large-size structures, as well as damage during their heat and humidity treatment, contributes to the reduction of durability and capital during flooding.

As a result, a set of measures to protect residential buildings, structures, utilities, etc.in flood-prone areas, as well as in flood-prone areas, should include forecasting, planning and implementation of work before the onset of flooding, as well as during its passage and after the end of a natural disaster.

After a natural disaster, it is advisable to analyze, summarize and develop recommendations for ensuring the safety of living in multistorey residential buildings that have been flooded by flood waters. To develop recommendations, it is necessary to have these conclusions on visual inspection of residential multistorey and low-rise buildings and technical solutions to ensure the safety of living in residential multi-apartment and low-rise buildings that have been flooded by flood waters

Based on this, the methodology for developing recommendations includes the following activities:

1. Analyze the submitted technical documentation based on the results of a survey of the technical condition of residential apartment buildings that have fallen into the flood zone, perform their classification according to the degree of danger to the life and health of residents, taking into account natural and climatic conditions and the likelihood of progressive deformations and damage;
2. Prepare the final report on the results of the analysis of the submitted deliverables, develop recommendations and technical solutions to ensure safe living in residential homes exposed to flood waters and having signs of a decrease in bearing capacity and stability of structural elements [6,7];
3. taking into account typical failures of load-bearing and enclosing structures of buildings, develop recommendations of a compensating nature in order to ensure their operability for the subsequent period of operation, taking into account natural and climatic conditions and the consequences of flooding;

4. carry out a selective (at least 5 representative buildings) assessment of the technical condition of buildings by means of their microdynamic survey using contactless laser equipment with computer processing of the results.

At the municipal level, to organize the above activities, it is advisable to organize a separate service, the main activity is the provision of public services, execution of works and execution of the state functions to ensure the implementation of statutory powers of public authorities of the Irkutsk region in the part of realization of measures on construction, reconstruction, capital repair of capital construction objects, which are lost or damaged as a result of an emergency situation.

In the event of an emergency caused by flooding of residential buildings, there was a problem in the regulatory and legal protection of the population, the economy and the national heritage of the country as a whole from natural emergencies [8]. In many respects, this problem is caused by the complete absence or imperfection of legislative and regulatory documents that should clearly fix the legal status, content, timing of drafting and mechanisms for implementing support measures. On this basis, at the level of the government of the Russian Federation, special attention should be paid to the relations regulated by legislation in the field of architectural activity, in particular, the placement and construction of low-rise residential buildings on territories subject to natural emergencies.

Considering the high probability of emergency situations on the territory of the Russian Federation, the formation of housing in special natural conditions must be carried out in accordance with the approved urban planning documentation, namely, the settlement scheme. The type of settlement is determined by the peculiarities of the structure of the economic base and size of settlements and its shape is determined by density of the network of settlements, their relative placement within a particular area, as well as the level of development of different kinds of functional relations between them. It should be remembered that the implementation of a new model of economic and social development of the country is possible only if favorable conditions are created for the reproduction of its main resource – human potential, which in turn is directly related to the reconstruction of the settlement system and a cardinal improvement in the quality of the urban environment.

It follows that the solution of problems of effective housing formation in special natural conditions is possible only on the basis of joint work of all levels of government, scientists, architects, designers, engineers, economists, technologists, designers, ecologists and other interested specialists. When designing and constructing residential buildings, it is necessary to consider not only the current regulatory and methodological framework of the Russian Federation, but also the world's promising trends in optimizing both spatial planning and design solutions.

According to the results of surveys conducted in July 2019, by August 1, 2019, it was found that about 47% of families who lost their homes wanted to stay in their localities. Considering the survey data, the government of the Irkutsk region decided to build new housing. The Ministry is taking measures to organize integrated development of territories in the city of Tulun and the city of Nizhneudinsk. 3 territories for complex development have been identified.

Housing construction is provided under 3 mechanisms:

1. Three-way contracts (payment is made at the expense of certificates for the purchase (construction) of residential premises).
2. Formation of the state housing Fund at the expense of the Federal budget for citizens who refused or did not receive a certificate.
3. Formation of a specialized housing Fund at the expense of the regional budget, with subsequent compensation of costs from the Federal budget.

For the complex development of land plots from the regional budget of 2019, the customer Service of the Irkutsk region has mastered 621,192. 3 thousand rubles (49% of the plan) [9]. These funds are used to prepare urban planning documentation for land plots, design and construction of residential premises engineering and transport infrastructure facilities.

According to the results of 2019, the transfer of funds for major repairs of residential premises was carried out on 1,721 certificates (65.6% of the total volume of residential premises subject to major repairs) for a total of 327,433,2 thousand rubles. According to 1,126 certificates, the first part of the social payment was transferred in the amount of 50% in the amount of 173,276. 1 thousand rubles. 337.7 thousand rubles were transferred to advance work on 4 residential premises where major repairs were carried out within the framework of trilateral agreements. According to 491 certificates (18.7% of the total volume of residential premises subject to major repairs), payment was made in full for 153,819,4 thousand rubles [10,11]. Major repairs in residential premises have been completed.

A comprehensive analysis of the current situation in the Irkutsk region allowed us to determine the main elements of the formation of low-rise and multistorey housing in areas subject to natural emergencies (Picture 1).

The types of activities in the territories subject to emergency situations of natural character

Urban development actions

1. Engineering protection of coastal territories from flooding
2. Measures to lower the water table
3. Measures for drainage of wetlands
4. Construction of residential buildings on high places

Architectural and planning actions

1. Designing the height of the basement of a residential building at least 1.5 m from the level of the planned ground level
2. Design of the floor level of a residential building at least 2 m from the level of the planned land mark
3. Creating attic rooms in the attic space

Engineering and constructive actions

1. Ensuring the necessary rigidity and stability of residential buildings
2. Increasing the rigidity of the coupling structures
3. Consideration of the possibility of hinge joints that do not violate the operational reliability of buildings

Constructive and technological actions

1. Consideration of the possibility of using steel reinforcement in the construction of monolithic concrete foundations
2. The use of major materials in the construction of walling
3. Effective strengthening of a roof overlappings
4. Development of measures to ensure the necessary rigidity and stability of residential buildings
5. Increasing the rigidity of the interface of building structures
6. the device of hinge joints that do not violate the operational reliability of buildings

Picture 1. List of activities in territories subject to natural emergencies

Initially, on the territory of the Irkutsk region, traditional technologies of building houses were used to quickly ensure housing construction. However, it is advisable to use alternative technologies, for example, the development of an original creative idea that can serve as a scientific and practical basis for solving the topic of architecture of residential buildings located on territories subject to natural emergencies. There is the experience of Western countries, where such a type of house as an amphibious house is common. This is a system that rises when flooding occurs and remains on the surface. The Foundation used in it is hollow and filled with foam, due to which, such a structure is almost impossible to flood: fixed to the front and back on special piles, it is able to painlessly transfer the increase in water level to 5.5 m. At the same time, residents are not left without light, gas and other benefits of civilization, which are supplied using flexible PVC-pipes, and easily float along with the house [12, 13]. The second type of buildings, floating houses, unlike amphibian houses, will never touch the ground, and will be in the river all year round. Scientists are constantly developing a way to integrate their technology with existing buildings, which is the best solution.

This approach to the practice of housing construction in areas subject to natural emergencies shows that under the influence of scientific and technological progress, changes occur in building structures and materials, engineering equipment and operating systems in the field of individual residential buildings [14, 15]. These trends give rise to new housing typologies that require the adaptation of old residential buildings to new conditions, especially to extreme natural conditions, in the event of which special attention should be paid to the creation of flexible structures of residential buildings and housing complexes.

References

[1] Chupin V R and Maizel D I 2020 Aggregation and decomposition in routing oil and gas pipeline systems *Proceedings of Universities. Investment. Construction. Real estate* **10(1)** pp 108–115 https://doi.org/10.21285/2227-2917-2020-1-108-115

[2] Pinus B I, Pinus J N and Khomyakova I V 2015 Changing of structural properties of concrete during cooling and freezing. *Bulletin of ISTU* **2** (Irkutsk) pp 111-115

[3] Matveeva M 2018 Synchronization of the activity of participants during the implementation of housing construction projects *Proceedings of Universities. Investment. Construction. Real estate* **8** pp 31-41 DOI: 10.21285/2227-2917-2018-2-31-41.

[4] Trofimov B Ya and Pogorelov S N 1986 The study of concrete deformation during cyclic freezing. InterVUZ Bulletin *Performance of building materials under the impact of various operational factors* (Kazan) pp 79-82

[5] Wang H L and Song Y P Fatigue capacity of plain concrete under fatigue loading with constant confined stress *Mater Struct* 2011 **44** pp 253–62

[6] Lee M K and Barr B I G An overview of the fatigue behaviour of plain and fibre reinforced concrete *Cem Concr Comp* 2004 **26** pp 299–305

[7] Melekhov E S and Shlepnev O K 2018 Development and reconstruction of district water supply systems *Proceedings of Universities. Investment. Construction. Real estate* **8(4)** pp 114–127 DOI: 10.21285/2227-2917-2018-4-114-127

[8] Peshkov A V 2019 Remote mortgage lending as a resource for housing development *Investments. Construction. Real Estate: New Technologies and Targeted Development Priorities IOP Conference Series: Materials Science and Engineering* **667** 012073

[9] Pinus B I 1985 *Ensuring the durability of reinforced concrete structures at low temperature impacts.* Dissertation Doctor of Technical Sciences (Moscow) p 367

[10] Korneeva I G 2019 Extensibility of fibre reinforced concrete *IOP Conf. series: materials science and engineering* (2019) **667** 012044

[11] Matveeva M V, Doroshenko T G and Sukhanova I A 2019 On the issue of cost and timing of contract tenders for industrial and civil construction *Investments. Construction. Real Estate: New Technologies and Targeted Development Priorities IOP Conference Series: Materials Science and Engineering* **667** 012063

[12] Kalyuzhnova N, Violin S and Matveeva M 2019 Smart Specialization of Russian Regions: Prospects and Limitations *Investments. Construction. Real Estate: New Technologies and Targeted Development Priorities IOP Conference Series: Materials Science and*

Engineering **667** 012036

[13] Khan V V, Dekanova N P and Khan P V 2017 Heat Supply of Villages and Towns of Eastern Siberia. Problems and Ways of Development *International Conference on Construction, Architecture and Technosphere Safety* **262** DOI: 10.1088/1757 899X/262/1/012081.

[14] Chupin V R, Pham N M and Chupin R V 2019 Optimization of developing district water supply systems taking into account variability of perspective water consumption *IOP Conference Series: Materials Science and Engineering* **667** 012018 doi: 10.1088/1757 899X/667/1/012018/

[15] Shcherbakov V, Akulshin A, Chizhik K and Tolstoy M 2018 Design of interactingwells for optimization of investments and operating costs while constructing water-diverting structures *6-th International Scientific Conference on Integration, Partnership and Innovation in Construction Science and Education* **251** *03037*

IOP Conf. Series: Materials Science and Engineering **880** (2020) 012101 doi:10.1088/1757-899X/880/1/012101

Winter city: staff training features

V V Peshkov

[1] Irkutsk National Research Technical University, 83 Lermontov str., Irkutsk, 664074, Russia

E-mail: pvv@ex.istu.edu

Abstract. The education system is one of the most important social institutions, the qualitative state and the result of which determines the level of development of the economy, science and culture of the country as a whole. The problem of training and retraining of workers for the construction industry and housing and communal services is becoming increasingly important for the economy of the country. It is obvious that the necessary volume of housing can be introduced only with appropriate quantitative and qualitative provision of human resources to the construction industry, especially in regions with harsh climatic conditions. Construction organizations need highly qualified personnel. Investment in construction has always been profitable, as this sphere is growing rapidly. Meanwhile, the increase in construction does not automatically entail a tide of human resources into this sector of the economy. However, the analysis shows that the shortage of specialized personnel reduces the pace of construction, especially in regions with severe climates, because for such regions it is necessary to provide a different approach to training, which should provide the necessary set of competences for specialists.

The climate has a significant impact on the lives of people around the world as a determining factor for the functioning of the economy. To a greater extent, it is the harsh climate and the prolonged winter period that cause the organization of the territory of the Russian Federation, which has a huge infrastructure importance for the entire economy. Russia is a northern country, 2/3 of which is in the area of permafrost [1,2]. The climate in the territory of the Russian Federation is characterized by the lowest temperatures among all countries of the world, at which 80% of the territory of the country is northern. Russia as a whole is a northern country, so the concept of "areas of the Far North and similar areas" is quite extensive and includes not only polar and circumpolar territories [3,4]. The 65% territories of the country are directly classified as areas of the Far North and similar areas. In this regard, the long winter period affecting most of the country puts Russia in an economic imbalance with the West.

Strong winds, high relative humidity, low natural illumination in autumn-winter period, in general, long winter period generates high transport and infrastructure costs, higher cost of production and construction. In this regard, the most negative factor affecting the quality of life in areas with a long winter period is the state and development of housing stock and housing and communal services. The following factors should be taken into account in the design of northern buildings[5,6.7]:

• Development of general plans of enterprises and settlements taking into account the patterns of wind conditions and snow traffic in winter;

• Maximum possible blocking of buildings and structures;

• Inadmissibility of complex configurations of buildings in plan and height differences;

IOP Conf. Series: Materials Science and Engineering **880** (2020) 012101 doi:10.1088/1757-899X/880/1/012101

• Use of buildings with minimum perimeters of external fences per unit area;
• Reduction of glazing surfaces;
• Reliable sealing of windows, gates and doors;
• Improvement of aesthetic level of buildings and structures;
• Creation of comfortable working conditions in working rooms.

These features and climate characteristics of the Russian Federation dictate a new trend of staff training for the construction industry. The construction complex of Russia is currently more than 130 thousand organizations and enterprises of various forms of ownership, 9.3 thousand enterprises of industry of construction materials, more than 10 thousand design and survey organizations, with the official number of employees 1,683,000 people [8]. Today, this market for the construction complex can be defined as labour-intensive, i.e. there are a large number of unemployed and, accordingly, an excess of labour supply (unskilled or underskilled in terms of work professions).

During the meeting of the Presidium of the Public Council under the Ministry of Construction in October 2019, Volkov Andrei, Rector of the Moscow State University of Civil Engineering or MGSU noted: "The shortage of engineering personnel, economists, managers and other specialists with the highest specialized education in the field of construction in Russia exceeds 100 thousand people. The number of specialists with higher education in construction organizations decreased 1.5 times, the share of staff of retirement age increased, while the share of staff of the economically active age category under 40 years decreased [9,10,11]. There is a steady increase in the outflow of staff, and the number of vacant seats exceeds 10%. Statistical indicators confirm the fact that the personnel potential of the industry construction does not meet the requirements of the market. Many companies solve the personnel issue in various ways. One of the most common and low-cost is the assistance of the "labor force" from the countries of the near abroad. But now the market dictates high demands on workers and specialists, such as knowledge and possession of new materials and technologies. Therefore, large construction companies and companies involved in construction materials pay great attention to the topic of training existing staff and recruiting staff with distinctive knowledge, including knowledge related to the construction, design, architecture, and urban planning of populated areas in winter conditions.

Today, the education system of Russia consists of 741 universities, 65% of their state educational organizations make up the state educational organizations (Picture 1) [12].

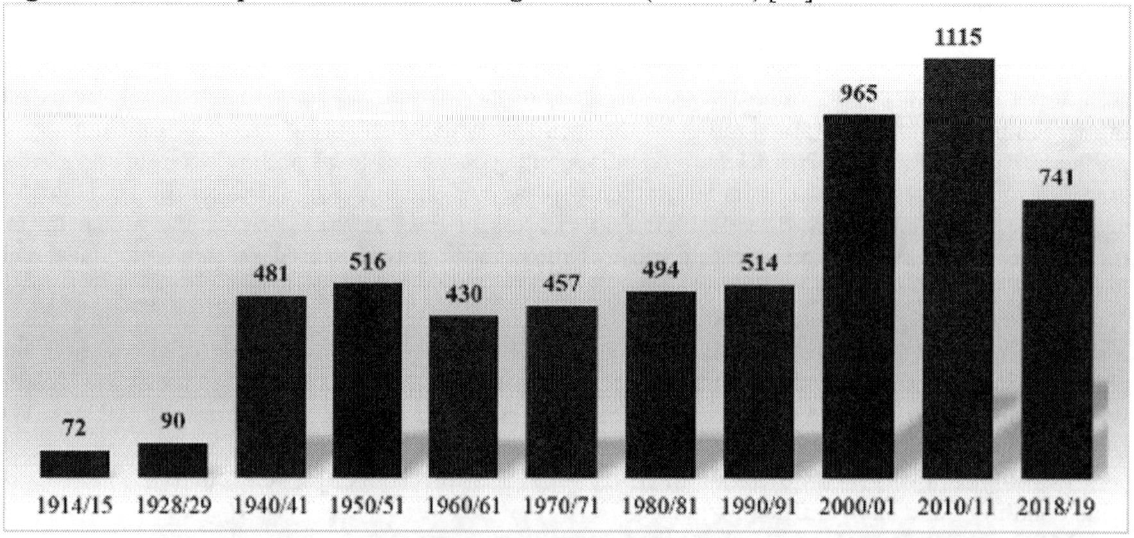

Picture 1. Dynamics of the number of universities of the Russian Federation

Of the fairly large number of universities, the critically low proportion of educational institutions pays attention to the training of students based on the climatic characteristics of the region. Teachers of the Irkutsk National Research Technical University (INRTU) investigate problems of design, construction and operation of cities in cold climatic conditions, In particular, issues of increased cost, low density of interaction of people (short light day, negative temperatures, snow, ice), seasonality of many activities, reduced emotional background, unreliable engineering systems and low transport availability [13,14]. The researchers assume that the cost of living in winter cities should not exceed the average of the cities of the southern strip. Students, teachers and investors are invited to create joint start-ups that will allow to produce products demanded in winter cities. Moreover, the study of architecture, urban planning, construction and design in winter conditions is taught to students in a number of disciplines:

The direction of preparation of bachelors 07.03.04 - "Town planning":

• Basics of urban zoning;
• Planning organization of land.

Bachelor Training Direction 07.03.01 - "Architecture":

• Design of thermal protection of buildings;
• Housing energy efficiency concepts;
• Architecture of buildings in northern climatic conditions.

Bachelor's training direction 08.03.01 - "Construction," profile "Roads and airfields":

• Research and design in difficult conditions;
• Research and design of roads in permafrost zone;
• Features of road design in permafrost zone.

Direction of bachelor's training 08.03.01 - "Construction," profile "Town planning and economy":

• Energy saving in urban construction and agriculture;
• Resource conservation in urban construction and agriculture;
• Engineering and environmental survey in construction;
• Environmental safety of buildings and structures during construction and operation in northern climatic conditions.

Direction of bachelor's training 08.03.01 - "Construction," profile "Industrial and civil construction":

• Construction in the northern construction and climatic zone;
• Special technologies in construction;
• Modern materials in construction.

Bachelor's training direction 08.03.01 - "Construction," profile "Expertise and real estate management":

• Real estate economy;
• Pricing in construction (taking into account winter conditions);
• Basic organization of operations in harsh climatic conditions;
• Tariff regulation in Housing and communal services.

In the course of training of specialists the purpose to give to students the corresponding competences in the directions of preparation "Construction", "Architecture", "Town planning", "Design" and to focus graduates on activities for design, designing, construction and reconstruction of buildings and constructions in the conditions of the winter city is set. In this regard, the following competences of the student can be distinguished:

Competence of students in the direction: Town planning

To be able:

• Apply the compact principle in the formation of residential buildings and complexes in cold climates [15].

Know:

• Peculiarities of urban planning systems formation taking into account climate;
• Influence of social, historical, economic and natural-climatic factors is a planned solution;
• Techniques of protection of the dwelling against adverse climatic factors.

To own:

• Classification of typological features of residential buildings, ensuring their performance, corresponding to local climate;

• Characterization of climate studies and climate typology of buildings.

Competence of students in the direction: Architecture

To be able:

• Determine the dependence of architecture on climatic factors;

• Is able to take into account and evaluate the role of the main components of urban ecosystems in the formation of architectural objects in different climatic conditions;

• Explore and create new architectural and urban objects with maximum understanding of the degree and nature of environmental factors.

Know:

• Rational design of cities taking into account climatic conditions;

• The impact of climatic zones on space-planning and architectural and design decisions;

• Methods and means of designing a comfortable and safe artificial environment and its components, taking into account environmental factors.

To own:

• Methods of assessing the degree and nature of environmental factors in the design.

Competence of students in the direction: Construction

To be able:

• To calculate the scope of works and the required labour intensity for construction of structures taking into account climatic phenomena and their impact on construction;

• To solve issues of preparation of construction territory, snow-clearing, construction sites.

Know:

• General information and conditions of construction in the Northern construction and climatic zone;

• Physical and geographical conditions of construction;

• Organization of frost-soil monitoring;

• Basis of construction organization design during spring and winter periods;

• Principles of using permafrost soils as building innovations;

• Operation of construction machines and mechanisms at low temperatures;

• Production of earthen, concrete and enclosing structures in harsh climates.

To own:

• Normative and instructional documents, manuals reflecting the specifics of construction in the Northern construction and climatic zone and in the conditions of the Far North;

• Peculiarities of planning of construction and installation works, taking into account the impact of climatic phenomena on construction production;

• Drawing up calendar and network schedules for the provision of preparatory and basic works on construction.

Having learned all competences, the student is aware of theoretical and practical methods of sustainable development of winter cities, which solve problems of snow, cold and strengthen economic potential of the region.

References

[1] Skolubovich A Yu and Matveeva M V 2019 Development of a methodology for assessing the readiness of enterprises for public-private partnership in the field of public water supply Proceedings of Universities. Investment. Construction. Real estate 9 pp 274–283 DOI: 10.21285/2227-2917-2019-2-274-283

[2] Kalyuzhnova N, Violin S and Matveeva M 2019 Smart Specialization of Russian Regions:

Prospects and Limitations *Investments. Construction. Real Estate: New Technologies and Targeted Development Priorities IOP Conference Series: Materials Science and Engineering* 667 012036

[3] Chupin V R and Maizel D I 2020 Aggregation and decomposition in routing oil and gas pipeline systems *Proceedings of Universities. Investment. Construction. Real estate* **10(1)** pp 108–115 https://doi.org/10.21285/2227-2917-2020-1-108-115

[4] Matveeva M 2018 Synchronization of the activity of participants during the implementation of housing construction projects *Proceedings of Universities. Investment. Construction. Real estate* **8** pp 31-41 DOI: 10.21285/2227-2917-2018-2-31-41.

[5] Peshkov A V 2019 Remote mortgage lending as a resource for housing development *Investments. Construction. Real Estate: New Technologies and Targeted Development Priorities IOP Conference Series: Materials Science and Engineering* **667** 012073

[6] Kalyuzhnova N, Violin S and Matveeva M 2019 Smart Specialization of Russian Regions: Prospects and Limitations *Investments. Construction. Real Estate: New Technologies and Targeted Development Priorities IOP Conference Series: Materials Science and Engineering* **667** 012036

[7] Khan V V, Dekanova N P and Khan P V 2017 Heat Supply of Villages and Towns of Eastern Siberia. Problems and Ways of Development *International Conference on Construction, Architecture and Technosphere Safety* **262** DOI: 10.1088/1757 899X/262/1/012081.

[8] Chupin V R, Pham N M and Chupin R V 2019 Optimization of developing district water supply systems taking into account variability of perspective water consumption *IOP Conference Series: Materials Science and Engineering* **667** 012018 doi: 10.1088/1757 899X/667/1/012018/

[9] Shcherbakov V, Akulshin A, Chizhik K and Tolstoy M 2018 Design of interactingwells for optimization of investments and operating costs while constructing water-diverting structures *6-th International Scientific Conference on Integration, Partnership and Innovation in Construction Science and Education* **251** *03037*

[10] Selezneva O I, Radaev S S and Baranova A A The impact of thermal treatment on the granulometric and chemical composition of diatomite of the Kamyshlov field when obtaining liquid-glass binder *Investment. Construction. Real estate* 9(3) pp 550–555

[11] Kazimirov I A and Peshkov V V 2019 Determination of price behaviour in the secondary residential real estate market using a multidimensional regression model *Proceedings of Universities. Investment. Construction. Real estate* **9(3)** pp 476–487 DOI: 10.21285/2227 2917-2019-3-476-487

[12] Melekhov E S and Shlepnev O K 2018 Development and reconstruction of district water supply systems *Proceedings of Universities. Investment. Construction. Real estate* **8(4)** pp 114–127 DOI: 10.21285/2227-2917-2018-4-114

[13] Tyuryukhanov K Yu and Pugin K G 2019 Impact of the surface of particles of moulding sand on the structural formation of asphalt *concrete Proceedings of Universities. Investment. Construction. Real estate* **9(3)** pp 566–577

[14] Dobruskina M A, Petrov A V and Bat-Erdene Z 2019 Improving the technology of retaining walls in the Irkutsk region using gabion baskets *Proceedings of Universities. Investment. Construction. Real estate* **9 (2)** pp 312–323

[15] Moskvitin V A., Emelyanova N A. and Mashovich A Y 2019 Experimental studies of air permeability indicators of composite "Poroplast CF" *Investment. Construction, Real estate* **9 (2)** pp 342–353

IOP Conf. Series: Materials Science and Engineering **880** (2020) 012102 doi:10.1088/1757-899X/880/1/012102

Organization of investment support for entrepreneurial activity in energy construction

V S Ryutin[1] and M V Matveeva[2]

[1] EvrosibEnergo, 22, Rabochaya str., Irkutsk 664052, Russia
[2] Irkutsk national research technical University, 83, Lermontov str., Irkutsk 664074, Russia

E-mail: vasilii.ryutin@gmail.com

Abstract. Process of reforming and forming a new model of the heat energy market in Russia is associated with the increasing role of interaction between private business and the state in the development of Russian energy sector and achieving energy security. Due to the long overdue need to attract private investment to the industry, the issues of investing activities financing of heat supply organizations are relevant and of great research interest. The article analyzes the effectiveness of the state policy on attracting private investment in the heat supply industry and builds a model for optimizing sources of financing for investment activities of heat supply enterprises.

1. Introduction

Problems of infrastructure projects implementing in the field of heat power have been forming for a long period, and are systemic in nature. Today, to modernize the communal infrastructure in Russia, a course has been taken to attract private investment [1].

In this area, infrastructure modernization is carried out through the implementation of public-private partnership infrastructure projects. Increase in total number of public-private partnership projects is primarily due to the development of legislation on public-private partnerships and the interest of public partners to applying this mechanism for the infrastructure development and modernization [2]. It should be noted that until recently, the legislation has seen serious shortcomings in part of stimulating the attraction of private investment to modernization of communal infrastructure, including the heat supply field [3].

The main prerequisites for attracting private investment to the field of heat supply are: limited state revenues for large-scale modernization; emergence of private owners in this sector of the economy associated with the process of the energy industry liberalization; lack of effective management and strategic planning in the country regions.

Solving the issues of attracting affordable sources of financing investing activities of heat supply organizations plays a significant role to ensuring the necessary pace of industry modernization.

2. Materials and methods

Currently, large-scale processes are taking place in Russian market of infrastructure utilities. All inefficient municipal unitary enterprises in housing and communal services are transferred to private business through concession agreements [1].

The current trend is the emergence of active dialogue regimes between business and government, international organizations and national business entities. Since 2016, long-term tariffs have been introduced for heat, hot and cold water, water disposal for a period of at least 3 years at first use and at least 5 years thereafter. In 2014–2015, regions could voluntarily apply long-term tariff regulation, since 2016, long-term tariffs have become mandatory for all subjects of the Russian Federation [2].

On July 31, 2017, the federal law "On heat supply" and certain legislative acts of the Russian Federation on improving the system of relations in the field of heat supply №279-FZ were published. The document was signed on July 29, 2017 by the President of the Russian Federation Vladimir Putin. The law establishes new rules for regulating tariff formation, with the transition from state regulation of tariffs to the establishment of a maximum level of the price of thermal energy for the consumer - the level of "alternative boiler room". It is expected that such changes will continue to effectively reform the industry and create better conditions for attracting investment to heat supply modernization.

To assess the effectiveness of public policy and identify role of private investment in Russia heat supply sector, it is advisable to analyze the dynamics of financing concessions sources in heat supply over the past 5 years.

A striking example of a concession for heat supply is the city of Irkutsk. Heat supply systems were concessioned to Baikalenergo JSC (a subsidiary of Irkutskenergo OJSC) in accordance with the concession agreement dated 26.02.2007 No. 010-64-83 / 7 in 2007 even before the approval of the model concession agreement by the Government of the Russian Federation. A comparative analysis of the activities of the Irkutskteploenergo MPTP until 2007 and Baikalenergo JSC (PJSC Irkutskenergo) in the period 2007–2015, one can observe a significant improvement in the indicators of municipal property management shown in the table. From 2007 to 2018, 15 inefficient municipal boiler houses (11 coal; 2 fuel oil; 1 electric boiler; 1 diesel boiler) were closed (transferred to the reserve), which led to a significant reduction in the volume of pollutant emissions into the atmosphere and a decrease in the fuel component in tariffs. The total economic effect for consumers from the implementation of measures aimed at saving resources, increasing energy efficiency and reducing operating expenses of Baikalenergo JSC amounted to 116 million rubles [2].

For the period 2018-2022, an investment program with an investment volume of 854 million rubles, financed by profit and depreciation, is operating in the city of Irkutsk under the concession agreement. Based on this investment program, it is proposed to build a model for optimizing the available sources of investment financing.

3. Results

To assess the effectiveness of public policy and identify role of private investment in financing of heat supply in Russia, a schedule was constructed reflecting the share of private investment in the total investment in the heat supply industry on the basis of concession agreements. The dynamics of the number concluded concession agreements is also analyzed.

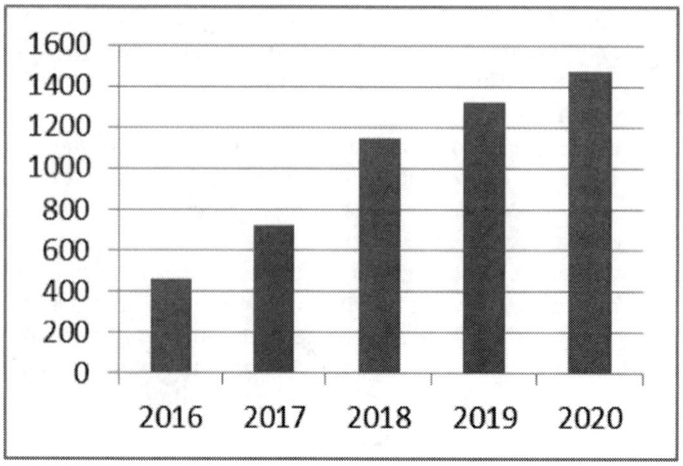

Figure 1. Number of concluded concessions in the heat supply field

Analyzing the dynamics of the number of concluded concession agreements in the field of heat supply in figure 1, we can say about the constant increase in the volume of such agreements. Nevertheless, the increase in concessions since 2018 decreased from 60% per year to 10%. Considering that the volume of investments in the heat supply sector is growing at a faster rate than the number of concessions, it can be concluded that the number of large investors is growing, which has a positive effect on the development of the industry.

Figure 2 shows that over the past 5 years, the volume of investments under concession agreements in heat supply has increased 5.5 times and amounted to 193 billion rubles in 2020 or 92% of the total investment. Basically, this growth is achieved due to an increase in private investment in the industry. For the period from 2016 to 2020, the share of public investment decreased from 16% to 8%. The volume of public investment in 2020 amounted to 15 billion rubles.

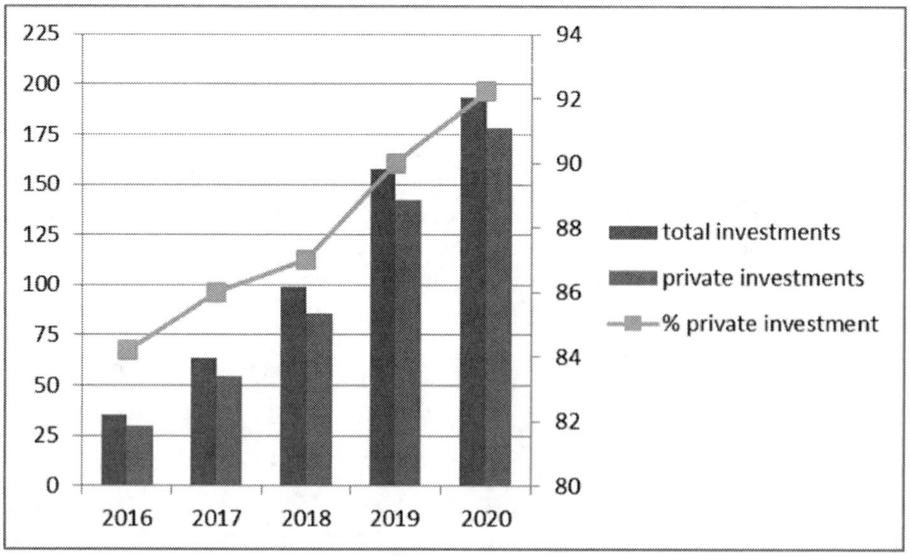

Figure 2. Share of private investment in total investment to heat supply industry

Thus, we can say that the main problems of the implementation of infrastructure projects in the field of thermal energy with the use of extra-budgetary funds are solved using:

1. implementation of concession agreements that reduce the political and legal risks of non-fulfillment by a state of obligations under an agreement with a private partner;

2. establishing long-term and understandable to a private partner rules of tariff regulation and tariff setting in the field of heat supply.

Profit and depreciation are traditional sources of financing investment programs aimed at increasing equity. Considering the growing role of private investments in the heat supply sector, in addition to own funds, as part of off-budget sources, the following sources can be distinguished: loans, bond loans, organization loans, funds of external investors, the use of leasing from which investment loans and loans are widely used in practice. Being the most traditional source of financing investment programs in the heat power industry, bank lending is distinguished by such characteristic features as the possibility of refinancing it in the future, quick terms for issuing loans with significant volumes and their relatively low cost [3].

In order to build a model for optimizing the structure of financing investing sources in the heat supply industry, it is necessary to solve the task of finding the best ratio of own and borrowed funds to finance investments.

Required for search variable is the share of loan funds for financing investments x in the field of definition from 0 to 1. The main optimization criterion will be the maximization of borrowed investment capital:

$$f_1(x) = LCx \rightarrow \max \tag{1}$$

LC, borrowed investment capital.

The second criterion is selected based on the need to minimize the weighted average cost of capital when attracting borrowed funds to finance investments:

$$f_2(x) = \%x(1 - Tax) + ROE(1 - x) \rightarrow \min \tag{2}$$

%, borrowed funds cost; Tax, profit tax; ROE, return of equity.

Given the specifics of the heat supply industry, which is characterized by continuous operation, heat supply organizations are trying to minimize the time to attract financing to investment. Thus, one more optimization criterion can be designated:

$$f_3(x) = (1 - x)t \rightarrow \min \tag{3}$$

t, term of mobilization own sources financing.

The last used criterion Debt burden ratio is selected based on the need to ensure the normative level of solvency of the company:

$$f_4(x) = \left\{ \frac{LC + ICx}{EBITDA + (IC - ICx)} - K_{norm} \right\}^{1/2} \rightarrow \min \tag{4}$$

IC, necessary investment capital; K_{norm}, recommended value Debt burden ratio.

Based on the list of selected additional criteria and a given focus, the required level of achievement for them is 0.

Thus, to solve the problem, there is a possible to use the decision-making algorithm of choosing the best alternative to additional optimization criteria that meets the necessary levels (Figure 3).

In order to simultaneously bring the selected optimization criteria to the established levels, it is necessary to minimize the following additive function:

$$f_{ad} = \sqrt{t_i^2 + w_i^2 + (d_i - d_{opt})^2} \rightarrow \min \tag{5}$$

t, term of mobilization own sources financing; w, WACC; d, Debt burden ratio.

According to calculations results based on the investment program until 2022, the optimal share of borrowed funds (credit) for the organization of Baikalenergo JSC is 60%. The additional effect of the optimization can be estimated using the financial leverage indicator.

Table 3 Key indicators of the model

Leverage ratio	WACC	DBR	t	DFL,%
Actual value, 0%	7.31	6.08	8	0
Optimized value, 60%	7.95	7.82	3.2	27.64

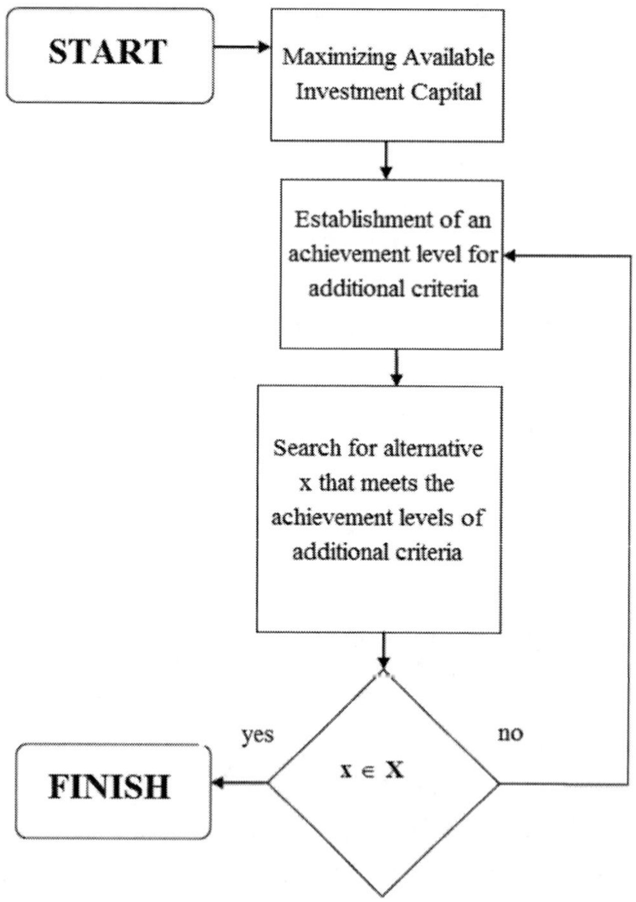

Figure 3. Decision-making algorithm for choosing the best alternative to additional optimization criteria

With the optimized value and minimization of the additive function, the weighted average cost of capital and the ratio of borrowed capital to EBITDA increased insignificantly with a significant reduction in the term for attracting financing. The obtained value of borrowed capital for financing investments allows to significantly increase the return on equity.

4. Conclusion

Analysis of the state policy effectiveness to attract private investment in the field of heat supply showed that the use of concession agreements practice for the modernization of the heat supply systems infrastructure is fully justified. Implementation of concession agreements allows to reduce political and legal risks of non-fulfillment by a state of obligations under an agreement with a private partner. In addition, in private investment growth is due to the establishment of long-term rules for tariff regulation and tariff setting in the field of heat supply that are understandable to a private partner.

For the most effective investment programs implementation, private investors in the field of heat supply are invited to use as sources of financing investments a reasonable amount of borrowed sources of financing calculated on the presented model basis.

Calculations presented in this article for example of Irkutsk city heat supply organization suggest that the proposed model can be used by heat supply organizations to make a decision about choosing optimal structure of sources of financing investment activity.

References

[1] Peshkov V., Matveeva M., Ryutin V. Concession as a tool for modernization of infrastructure facilities. Vestnik Volgogradskogo gosudarstvennogo universiteta. Ekonomika. 2016. № 4 (37). p. 130-139.

[2] Peshkov V., Matveeva M., Borisova S., Ryutin V. Concession agreements in the field of region's industrial investment programs implementation. Izvestiya vuzov. Investitsii. Stroitelstvo. Nedvizhimost. 2016. № 4 (19). p. 48-56

[3] Skolubovich A., Matveeva M. On the issue of public-private partnerships in the field of public water supply. Izvestiya vuzov. Investitsii. Stroitelstvo. Nedvizhimost. 2018. № 3 (26). p. 45-53

[4] Nikolaev M. Investment activity: Textbook. 2007. – p. 307

[5] Kudryavtseva V A, Zaslavsky A M and Larina O P 2018 Multicriterial comparison at the inclusion of investment projects of manufacture of building materials into the state program of the region (at the example of Irkutsk region) *Proceedings of Universities. Investment. Construction. Real estate* **8** pp 60–72 DOI: 10.21285/2227-2917-2018-1-60

[6] Lebedeva K Yu 2013 Research of physical and mechanical characteristics of bituminous-rubber composite binders *JSRP* **2**

[7] Gokhman L M 2017 Substantiation of regulatory requirements for polymerasphaltconcrete according to GOST 9128-2013 *Bulletin of the KNARU* **79**

[8] Yaskova N Yu and Sarchenko V I 2018Development of strategic approaches to the creation of aqua-territorial industrial complexes in the Arctic zone *Proceedings of Universities. Investment. Construction. Real estate* **8** pp. 84-93 DOI: 10.21285/2227-2917-2018-2-84-93

[9] Peshkov V V and Gamayunova E V 2018 Specific features of investment support for the repair and reconstruction of buildings and facilities *Proceedings of Universities. Investment. Construction. Real estate* **8** pp 56-72 DOI: 10.21285/2227-2917-2018-2-56-7

[10] Kazimirov I A and Peshkov V V 2019 Determination of price behaviour in the secondary residential real estate market using a multidimensional regression model *Proceedings of Universities. Investment. Construction. Real estate* **9(3)** pp 476–487 DOI: 10.21285/2227-2917-2019-3-476-487

[11] Chupin V R and Maizel D I 2020 Aggregation and decomposition in routing oil and gas pipeline systems *Proceedings of Universities. Investment. Construction. Real estate* **10(1)** pp 108–115https://doi.org/10.21285/2227-2917-2020-1-108-115

[12] Peshkov A.V. 2018 *The economic efficiency of the use of intangible resources in investment and construction activities*. University News. Investments. Construction. The property doi: 10.21285 / 2227-2917-2018-4-44-51

[13] Khokhlova G.I. and Okladnikova D.R. 2014 *Classification of factors as grounds for analysis of innovative–investment activity of small and medium businesses*. Middle East Journal of

Scientific Research Vol 20 № 8 pp 900–904
[14] Official site of Rosstat [electronic resource]. - Access mode: http:// www.gks.ru/storage/ mediabank/str6.xlsx Ver. From the screen
[15] Yaskova N Yu 2018 Construction in the context of new views on the goals of economic development. *Proceedings of Universities. Investment. Construction. Real estate* **8** pp 133-142 DOI: 10.21285/2227-2917- 2018-1-133-142

The role and place of small innovative enterprises in the construction industry in the modern economy of single-industry towns

Y A Skorobogatova[1], A S Bovkun[2], M Yu Ivanov[3] and O S Shilova[4]

[1] Baikal State University, 11 Lenin St., Irkutsk, 664003, Russian Federation
[2] Irkutsk National Research Technical University, 83, Lermontov str., Irkutsk, 664074, Russian Federation
[3] Bratsk State University, 40, Makarenko St, Bratsk, 665709, Russian Federation
[4] Irkutsk State University,1, Karl Marx St., Irkutsk, 664003, Russian Federation,

E-mail: Bovas87@yandex.ru

Abstract. This article gives an author's interpretation of a Russian single-industry town, in which the key factor in the socio-economic development of a territory is the successful development of a town-forming enterprise. It is proved that the development of single-industry towns can be promoted by the development of small innovative entrepreneurship. One of the promising forms of small business development may be the construction industry, which is one of the most promising industries in Russia. The main social functions of small business in the construction industry are highlighted.

1. Introduction

Currently, the Russian economy is aimed at solving a large number of problems, one of which is the stabilization of the situation single-industry towns and identifying possible scenarios for their development. Nowadays, single-industry towns are experiencing a lot of difficulties. **[1]**.

Currently, the Russian economy is undergoing changes, namely, it is moving on to the path of sustainable development and innovation. In these conditions, the most important task is the development of small business, without which it becomes impossible to ensure the successful socio-economic development of Russia.

2. Theory

Consider the concept of "single-industry town" and its characteristics for a better understanding of the current state of the economy of single-industry towns. At present, a clear and statutory definition of "single-industry town" does not exist. This indicates insufficient regulation of the economy of single-industry municipalities and the weakness of their institutional status. The essence of such a phenomenon as "single-industry" consists in a strict relationship between the financial and economic situation of the town-forming enterprise and the state of all spheres of the town's life [2].

A single-industry town is a complex structure in which a town is inextricably linked to an enterprise. Moreover, the company performs not only economic, but also social functions in the form of ensuring living conditions. In this case, the costs of maintaining social infrastructure are included in the cost of production of the enterprise, because of which these products often become uncompetitive.

After analyzing the scientific literature [3,4,5]., we can give the following definition of the concept of "single-industry town". A single-industry municipality or a single-industry town is a municipality with a large enterprise or group of enterprises united by one technological cycle, the number of employees is more than 20% of the economically active population of the town and the share of manufactured products is more than 50% of the total volume of products shipped.

In the Decree of the Government of the Russian Federation of July 29, 2014 No. 709, the following criteria were identified for classifying municipalities as single-industry towns [6]:

- the resident population of the town is more than 3 thousand people;
- the municipality has the status of an urban settlement or urban district (the exception is the municipalities in which the legislative authority of the constituent entity of the Russian Federation is located);
- the implementation by one of the enterprises of activities for the production, processing of industrial products and (or) for the extraction of minerals (excluding gas and oil);
- the number of employees of one of the enterprises is 20% of the number of employees of all enterprises located in the town.

Consider the main quality characteristics of single-industry towns:

- the presence in the municipality of technologically interconnected organizations producing goods for one market, with the exception of firms that serve the internal needs of the town;
- the presence in the municipality of one or more companies that serve one segment of the industry market or belong to the same industry, and the remaining organizations of the municipality serve only the internal needs of the town and the population living there;
- a strong dependence of the revenue side of the budget of the municipality on the work of a large enterprise located in this town;
- the lack of developed transport infrastructure connecting the town with the outside world or the greater remoteness of the municipality from larger settlements, which reduces the possibility of mobility of town residents and causes difficulties in transporting goods;
- the population of the town is equally professional.

Today in Russia, according to the estimates of the Ministry of Economic Development of the Russian Federation, there are 313 single-industry towns in 61 constituent entities of the Russian Federation, which cover about 15.5 million people. In the Irkutsk region there are 8 single-industry municipalities with a population of more than 380 thousand people [7,8]..

Town-forming enterprises form the socio-economic environment of single-industry towns, the qualitative characteristics and level of development of which determine the living standards of the population of this town.

In turn, the investment attractiveness of this territory, the development of social, economic, engineering infrastructure, government policy and other conditions make up the external environment of the enterprise. Thus, the development strategy of single-industry towns depends on the activities of town-forming enterprises.

The main problem of the development of single-industry towns is the narrow specialization of the economies of such towns. Due to the high degree of mono-profile, the development of the municipality becomes dependent on environmental changes and unstable to negative external influences. Therefore, a decrease in demand and prices for products manufactured at town-forming enterprises and a change in the external environment are introducing the economy of the entire town into a long crisis [9]..

The formations of a market economy, the formation of a new external environment and the change of ownership of most companies have made the company experiencing great difficulties, the corresponding single-industry towns and their residents. It has become even more difficult with the onset of the global financial and economic crisis of recent years [10].. This crisis has greatly changed the socio-economic situation of single-industry towns. The closure of investment programs, the fall in the value and demand for export products, technological backwardness and insufficient resources lead to a significant decrease in the profitability of production and the profit of town-forming enterprises.

In this regard, enterprises introduce anti-crisis measures: they reduce investment programs and staff costs, many enterprises completely stop their work. Because of this, social problems are exacerbated. [11]. Residents of many single-industry municipalities are on the verge of survival due to a strong reduction in wages and lack of work. The difficult situation is aggravated by the increase in tariffs for housing and communal services, the rapid growth of inflation, unpaid loans and an increase in prices for goods and services.

Monotowns are now in conditions of ever-increasing unemployment, declining incomes of the population, migration of able-bodied young people to other cities, the town budget is heavily dependent on taxes paid by the town-forming enterprise, insufficient job opportunities and low birth rates. All this suggests that single-industry towns are now at risk of transition to a group of depressed towns. Depressiveness can also be characterized by low investment, hidden unemployment and poor public health. [12]..

According to the Ministry of Economic Development, today 89 Russian single-industry municipalities have a relatively favorable situation. The unemployment rate does not exceed the national average of 5.2% (data for 2017) [13]. 149 single-industry towns currently have more unemployed than the national average. And 75 single-industry towns have unemployed 2-3 times higher than the average for the Russian Federation. These factors are the reason for the mass migration of residents from single-industry towns. [14,15]..

Thus, the problem of the functioning and development of single-industry towns has become very acute and has acquired strategic importance not only for individual regions and municipalities, but also for the Russian economy as a whole. The solution to this problem requires government intervention.

3. Experiment

In order to develop the economy of single-industry towns, the Government of the Russian Federation has developed a set of measures to support them using the selected single-industry development strategy (see table 1). This set of measures gives particular importance to supporting small businesses as a factor in the possibility of diversifying the economy of single-industry municipalities. Small business primarily contributes to the formation of a competitive environment and provides residents of single-industry towns with goods and services. It also performs social functions at the level of single-industry towns, such as creating new jobs for the population of the municipality and ensuring the growth of tax revenues in town budgets.

Table 1. Choosing a development strategy for single-industry towns based on their typological characteristics

Characteristic of the economic condition single-industry towns	Development strategy
Single industry towns that are temporarily in economic decline according to industry dynamics	Investment support, diversification of the economy
Single-industry towns in a gradually worsening situation	Diversification of the economy
Economically promising single-industry towns, but with a mistakenly chosen line of business	Diversification of the economy
Economically promising single-industry towns	Maintaining a single-industry economy
Subsidized single-industry towns as a result of tax policy by	Adjustment of tax policy

the state	
Single-industry towns developing as part of the investment and economic trend	Development maximization
Diversifying single industry towns	Identification of promising areas of development

From the data in table 1.6 it is seen that the development of single-industry towns in a difficult socio-economic state or in a risk zone needs a strategy for differentiating the economy through the development of small and medium-sized enterprises.

At the moment, single-industry towns of Russia have poorly developed small businesses. The share of small enterprises in most single-industry Russian municipalities is from 1% to 13%, which is much lower than the average for Russia [16].. The development of small business in single-industry cities cannot take place without impressive support from the state due to the crisis situation in single-industry towns, outdated infrastructure, the lack of personnel of different professional backgrounds (most of the population has been working in a city-forming enterprise all its life and has the same professional tasks) and other factors limiting the activity of small enterprises in single-industry towns [17].

The administrations of most single-industry towns with a crisis state of the economy have modernization plans aimed at developing small and medium-sized businesses through the implementation of a range of incentive measures. For example, the state contributes to the development of small business in the city of Baikalsk, Irkutsk Region (a single-industry town with a difficult socio-economic situation) through the implementation of the municipal program "Development of Small and Medium Enterprises in the Baikal Municipal Formation for the Period 2016–2022". The following activities are planned under this program:

- creation of a favorable external environment for the development of small and medium enterprises;
- financial support for small and medium-sized businesses, as well as support for organizations that form the infrastructure for supporting small and medium-sized enterprises;
- support for small and medium enterprises in the field of innovation, industrial production, agriculture, social entrepreneurship, folk art crafts, handicrafts and eco-tourism.

The role of small business is not only its economic functions. Small business is also presented as a social phenomenon that implements entrepreneurial initiatives, provides not only employment but also self-employment, introduces innovations and provides many other effects in society.

It should be noted the political role of small business. Small business has the greatest propensity for political stability, economic freedom and the principles of democracy in the presence of conditions of sustainable political and socio-economic development in society. The layer of small owners is desperately ready to protect their property in case of a threat, because for small entrepreneurs, unlike large and medium-sized entrepreneurs, their property is often an important way of expression and the only way to make money.

The middle class, represented by small business, defending their rights can form their own political movements or can become the object of the struggle for the votes of voters of various political forces in the country. Therefore, small business has an inextricable link with the interests of regional and local political movements. This fact determines the significance of the national basis for small business and forms its patriotic moods.

Small business has a strong political influence in various countries, because on the basis of this social group the middle class has long been formed. This is the largest class, which expresses the political mood and preferences of a significant part of society. In Russia, this class does not exert such a strong influence on politics due to the insufficient development of small business.

Small business is becoming a very important factor in solving social and economic problems with a developed infrastructure, a high number of small businesses and state support. Let us single out the main social functions of small business:

- creation of a social layer of owners;
- formation of additional jobs;
- disclosure of people's talents, realization of their creative abilities, development of various types of folk crafts;
- reduction of social tension in the country and unemployment;
- involvement in labor activity of certain groups of the population, the employment of which is limited in large firms (pensioners, housewives, students, disabled people);
- solving problems of municipal and regional development.

Social and economic tension is especially felt in regions with a raw material orientation of the main industries, poorly developed infrastructure and agriculture, remote from trade and industrial centers and roads, with a monostructure of towns. The development of such cities is impossible without the participation of small enterprises. Small business helps improve the functioning of the economy of small towns, reduces the differences between central and provincial towns and improves the intellectual level and lifestyle of the population.

Small business because of its features has a pronounced municipal and regional focus. Small enterprises usually conduct their activities based on the needs of the local market and the possibilities of saturation, as well as on the basis of the structure and volume of local demand. In the conditions of economic independence of municipalities and regions, small business should be considered as a necessary means of improving the structure and overall state of the municipal and regional economy, including the economy of single-industry towns.

4. Conclusions

Thus, small business plays not only an economic role in the development of the entire state, including single-industry towns, but it can also fulfill political and social functions. Small business provides the population with goods and services, creates a competitive environment, provides employment and the implementation of entrepreneurial initiatives, which is so important for the development of single-industry towns, which are now in a difficult situation. Small business is able to act as a factor in the socio-economic development of single-industry towns if the authorities implement an effective policy to support and develop small business, especially with regard to the competent implementation of tax and financial policies.

Small business is a sector of a market economy that includes enterprises that have restrictions on the number of employees, the size of revenue and the structure of the authorized capital called quantitative criteria. Small business also has qualitative criteria (qualitative characteristics) such as independent management of the company by the entrepreneur, ownership by the entrepreneur of the largest share in the authorized capital of the company, direct and constant contact of the entrepreneur with staff, and much more.

The development of small business in single-industry towns should be accompanied by state support, which in turn requires constant improvement. At the moment, the state helps the development of small businesses through measures such as the creation of a support infrastructure, the provision of financial and informational assistance, and legal regulation of the activities of small businesses.

Thus, the development of small business is an important criterion for the effectiveness of the economy and a condition for a stable social and economic situation in the country, in the regions and in such special municipalities with an unstable position as single-industry towns. The problems of single-industry towns are currently associated with the difficult state of city-forming enterprises, with high unemployment, and increasing population migration.

References

[1] Vilensky A 2014 State support for small and medium-sized businesses in modern Russia on the second wave of global finance *Economic Issues* **11** pp 95–106

[2] Glebova I S 2008 *State regulation of small and medium-sized businesses* p 326

[3] Davydova I 2011 *Sh. Single-industry towns: problems and development prospects* p 220
[4] Ryakhovskaya A N 2013 Evaluation of the effectiveness of state programs implemented in single-industry towns *Problems of the modern economy* **1** pp 126–132
[5] Tkachuk S N 2010 Small business is an important reserve for the development of a single-industry town *Economic and social changes: facts, trends, forecast* **3** pp 34-43
[6] Antipina N V 2019 Optimization of investments in fixed assets of an oil company. Proceedings of the Baikal state University **29** pp 262-272
[7] Antipina N V 2019 Regression analysis of oil export dynamics in the Russian Federation *Intellectual and resource potentials of regions: activation and increase of efficiency of use Materials of the V all-Russian scientific and practical conference. Under the table* pp 15-21
[8] Baldanova E V 2019 Forecasting of the financial state of oil companies *State adviser* **1 (25)** pp 5-9
[9] Zakharov S V 2014 Analysis of innovation activity in the Russian Federation *Innovation Management: theory, methodology, practice* **8** pp 21-26
[10] Kutergina G V 2015 Management of development of single-industry towns: domestic and foreign approaches to modeling *Bulletin of the Perm University* **3** pp 69-78
[11] Maslova A N 2011 Single-industry Towns in Russia: problems and solutions *Problem analysis and state-management design* **5** pp16-28
[12] Manaeva I V 2016 Methodological recommendations for choosing a scenario for the development of a single-industry city *Problems of territory development* **6** pp 196-208
[13] Novikova I 2018 Development of a methodology for evaluating the investment potential of an oil company *Methodology for the development of the economy, industry and services in terms of digitalization* pp 534-559
[14] Popovicheva N E 2015 Problems of development of single-industry towns in modern conditions *Central Russian Bulletin of social Sciences* **4** pp 186-193
[15] Skorobogatova Yu. A., Shagunov A. P. (2018) The role of small enterprises in the socio-economic development of the region. In the collection: Problems of management of production and innovative systems: materials of articles of the all-Russian scientific and practical conference 2018 pp 178-183.
[16] Skorobogatova Yu A 2018 The significance of the "production functions" tools in the management of the investment component in the socio-economic development of the region *Economic almanac materials of the I International scientific and practical conference "economy of infrastructure transformations: problems and prospects of development* pp 344-346
[17] Titov M B 2011 Development of single-industry towns in Russia *Bulletin of the Nizhny Novgorod University* **5** pp 207-210
[18] Kazimirov I A and Peshkov V V 2019 Determination of price behaviour in the secondary residential real estate market using a multidimensional regression model *Proceedings of Universities. Investment. Construction. Real estate* **9(3)** pp 476–487 DOI: 10.21285/2227-2917-2019-3-476-487
[19] Peshkov V, Gertsekovich A and Gorbachevskaya L 2019 Dilapidated and dilapidated housing in the aspect of the Federal project "Ensuring sustainable reduction of uninhabitable housing" *Investments. Construction. Real Estate: New Technologies and Targeted Development Priorities IOP Conference Series: Materials Science and Engineering* **667** 012075

IOP Conf. Series: Materials Science and Engineering **880** (2020) 012104 doi:10.1088/1757-899X/880/1/012104

The factors that influence the dynamic development of the construction industry in a remote region

O V Tayurskaya[1], D R Okladnikova[2] and K R Bibarsov[3]

[1] Irkutsk National Research Technical University, 83 Lermontov str., Irkutsk, 664074, Russia
[2] Irkutsk, Regional Guarantee Fund, 2 Rabochaya str., Irkutsk, 664003, Russia
[3] Irkutsk National Research Technical University, 83 Lermontov str., Irkutsk, 664074, Russia

E-mail: ovt58@mail.ru

Abstract. This article discusses the development of the construction industry on the example of a remote region. The construction complex has always played a leading role in the development of the Russian economy as a whole and in some of its regions. The construction industry accounts for 6 % of global GDP. moreover, the construction sector is interconnected with almost all other sectors of the economy. Construction is one of the main types of economic activity, as it provides for the creation of new, expansion and reconstruction of existing fixed assets. Due to the developed construction industry, the most important tasks of structural adjustment of the material base of the entire production complex in the country are being solved, as well as conditions for the development of non-production sectors are being created. Construction is a branch of material production where objects of production and non-production infrastructure (buildings, structures, highways, etc.) are created. The state of the construction market is always an indicator of the successful development of the country's economy as a whole. [1]

1. Introduction

The construction industry today and the past period of time retains the status of a significant and at the same time problematic industry among the main sectors of the country's economy.

The construction industry is experiencing a powerful impact of the economic crisis. As a result, many construction and design projects are suspended or their budgets are subject to significant changes which destabilizes the implementation of the planned programs.

The main problems for construction companies (especially in remote northern regions) are:

- seasonal nature of the work performed (for example, in the winter time the works on the construction of objects are suspended), as a result-a slow pace of completion of work, therefore a decrease in the amount of revenue received for commissioned objects.

Also the main reason is the high dependence of the construction industry on borrowed funds, that is, there is a problem of insufficient level of development of credit and financial mechanisms in the construction industry today.

IOP Conf. Series: Materials Science and Engineering **880** (2020) 012104 doi:10.1088/1757-899X/880/1/012104

2. Materials and methods

Special attention should be paid to the remote northern regions of the country where the problems in the construction industry are much greater than in the central regions, each region has its own characteristics. Despite this, according to the results of the first nine months of 2019, the Siberian Federal district has a positive dynamics of disbursed funds in the construction industry, the volume of work performed increased by 6.9% compared to the same period of 2018.

During the analyzed period the funds were disbursed:

- Kemerovo region – 119 billion rubles.
- Irkutsk region – 116 billion rubles.
- Krasnoyarsk territory-98 billion rubles.
- Novosibirsk region – 91 billion rubles.

The least developed is the Republic of Altai – only 4 billion rubles.

Some regions showed a significant over fulfillment of funds disbursements by 9 months of 2018, for example:

- in the Republic of Tyva mastered by 48.1% more than in 2018,
- in the Irkutsk region-by 43.7%,
- in the Omsk region - by 25.9%.

There was a significant decrease in some regions:

- in the Republic of Altai-reduced by 40.6%,
- in the Krasnoyarsk territory-by 20.7% . **[2]**

The main reason for the slowdown in the volume of work performed remains the high dependence of the construction industry on borrowed funds, i.e. today there is a problem of insufficient level of development of credit and financial mechanisms in the construction industry.

The conducted Express analysis of the Irkutsk region allows us to draw conclusions:

- high interest rates on loans for legal entities and individuals-significantly reduce-the amount of work performed,
- the dynamics of the volume of loans granted to legal entities-residents and individual entrepreneurs by type of activity "construction" for 2012-2016 is shown in figure 01. **[3]**

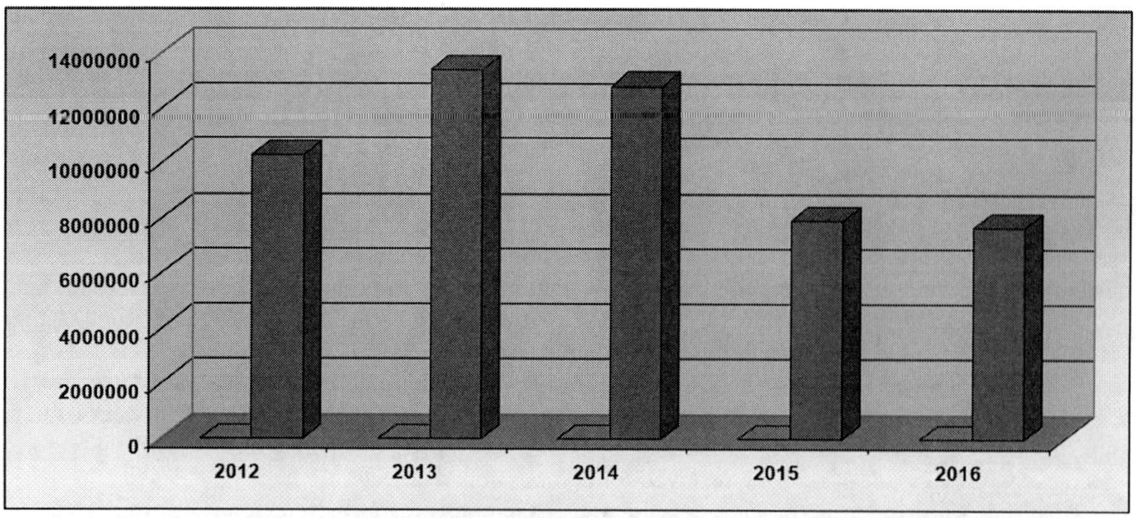

Figure 1. Dynamics of the loans volume granted to construction companies for 2012-2016 (Irkutsk region).

The volume of loans in 2013 increased by 30 % compared to 2012 and reached the maximum value.

There has been a decrease in the volume of the loans taken in the following years:

- in 2014-by 4.54 %;
- in 2015-by 38 % compared to 2014.
- in 2016 the volume of the loans decreased by 4.7 % compared to 2015.

This situation can be explained by the fact that the construction companies refuse the services of credit institutions and use their services less often due to the high interest rates.
[4,5]

An important indicator is also the housing commissioning indicator. The figure 02 shows data for the Irkutsk region for 2012-2016:

- it can be noted that the region reached its maximum value in 2013 but there was a significant decrease in 2014, as we can see this indicator does not have a certain stability.

The decline in consumer demand also negatively affects the dynamic development of the industry, for example, mortgage lending to individuals is not a panacea for purchasing the desired housing space due to the general increase in prices for all types of goods and services in the region.

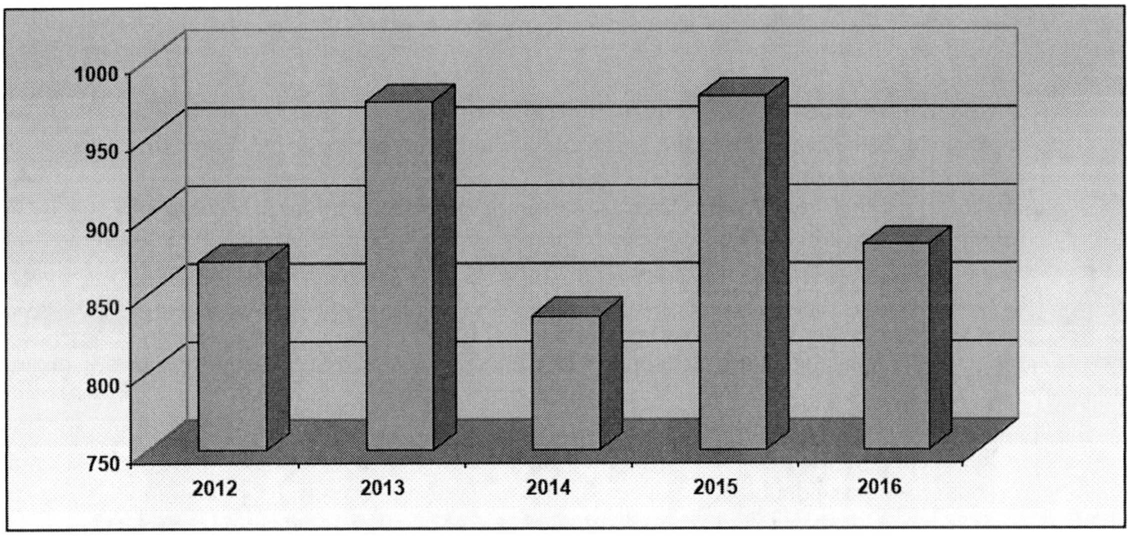

Figure 2. Commissioning of housing in the Irkutsk region.

Also it is worth noting the fact that the heads of construction companies specialize primarily in the construction of commercial real estate (office space), this type of work brings a large profit and a quick payback period.

The price often becomes much higher due to the presence of many intermediary firms on the construction market also.

Another factor that reduces the dynamic development of the industry is the increase in certain taxes, the receipt of fines and penalties for certain inaccuracies or errors.

In the first half of 2019 there were no significant changes in the construction industry as a whole, investment activity did not increase and some major construction projects are underfunded, all this is the basis for the introduction of new measures to activate the construction market of the country and regions. In times of crisis, effective measures are always: first, reducing the tax burden for industrial structures, and second, reducing loan rates and concluding joint activity agreements with similar firms, both in the surrounding regions and with foreign partners. The development of joint industrial facilities will also help to activate economic processes in the domestic market of the country and regions. all these measures should be implemented in a comprehensive manner, with the control and support of Federal management services, since today remote regions need significant support in the form of the concessional loans, effective activities of leasing companies which should create more comfortable conditions for lessees, for example, by increasing the repayment period of debt [6].

Noting the problems in the construction industry, you cannot ignore many economic indicators, such as:

- the level of income of the population in each region (unfortunately, this indicator has significant differences, despite many existing state programs to support residents of the country),
- another important indicator is the volume of retail turnover, on the basis of which many social and economic processes in the region can be activated. Unfortunately, both indicators in the last three years tend to decline. Thus, in the first half of 2019, the real incomes of the population decreased by 1.3% compared to the same period last year.

An important condition for activation in the construction industry is mortgage lending to the population. This system of lending should stimulate consumer activity in the housing market, however, in recent years, the volume of mortgage lending is not developing as actively as it was 7-8 years ago.

Regarding equity participation, there may be difficulties due to changes in the law on shared construction. Previously, the population could buy apartments at an early stage of construction, so they saved up to 30% of the final cost of housing, now you can buy housing on other terms and at a higher price.

You can add to the problems of the construction complex also:

- the presence of high seismicity in the region which undoubtedly contributes to an increase in the price per meter of housing-about 15 %;
- due to the territorial remoteness from the central industrial regions, to the problems in the Siberian regions, you can add quite high prices for imported the building materials which also does not allow you to get a larger amount of profit and direct it to your own development and modernization.

The problems include the fact that many developers do not meet the established requirements during inspections and many must leave the construction market. This may lead to the emergence of the monopolistic firms in the region which will be able to inflate prices in the absence of healthy competition.

The banks have a significant impact on the dynamics of the construction industry and the provision of housing for the population also, setting inflated rates on loans. As a result, the developer will shift them to the buyer which undoubtedly reduces the purchasing power and a vicious circle is formed, since all participants in the process of construction and sale of housing are closely interrelated.

According to many experts despite these problems the construction market in the Irkutsk region has a good chance for successful development in spite of many existing problems, since housing is actively purchased by citizens coming from the northern territories or the Far East.

It is necessary to develop the domestic market (for each region) to solve many problems developing its own production of building materials, achieving the main goal to align demand and supply in the regional market.

Special attention should be paid to the topic of qualified personnel in construction for the implementation of many construction projects, there is a shortage of personnel in the Irkutsk region. Therefore, there is a need to quickly develop the system of training and retraining of specialists in the region and neighboring regions. [7]

In General, the prospects for the development of the construction industry in the region are quite optimistic, since there is a significant amount of old housing, and the need to improve their living conditions among the population is high.

However, another problem is the high cost of housing for many years there has been a noticeable increase in real estate prices, the reason is that Irkutsk is an attractive city for residents of more remote regions.

IOP Conf. Series: Materials Science and Engineering **880** (2020) 012104 doi:10.1088/1757-899X/880/1/012104

According to the statistics of 2018, almost 30% of the realized housing stock fell on nonresident buyers.

The prices of 1m2 of housing stock are also affected by the high cost of survey work.

3. Results

The rapid analysis revealed two negative trends in the housing sector of the Irkutsk region:

1) The high unmet need of the population for housing is accompanied by a low level of income and a constant increase in the cost of construction and housing prices. At the same time, the current volume of housing construction does not cover even the existing effective demand of the population. The problem of providing citizens with housing can be solved with mass housing construction.

According to the indicator of the entered volume of housing for each resident, the region ranks 55th in the country. Among the Siberian regions:

- Buryatia is on the 71st place,
- The TRANS-Baikal territory is in 80th place .

The maximum increase in the rate of housing construction compared to the previous year-over the past five years was observed?

- in 2014-by 18.2%,
- the total housing area increased by only 1.4% in 2015,
- in 2016 decreased by 6.0%,
- in 2017-by 1.3%,
- in 2018 decreased by 4.5%.

Figure 03 shows data for:

- entering the volume of housing stock,
- the volumes of financing of the industrial enterprises,
- indicator of the average per capita income of the residents of the Irkutsk region for the years 2000-2017. [2, 6].

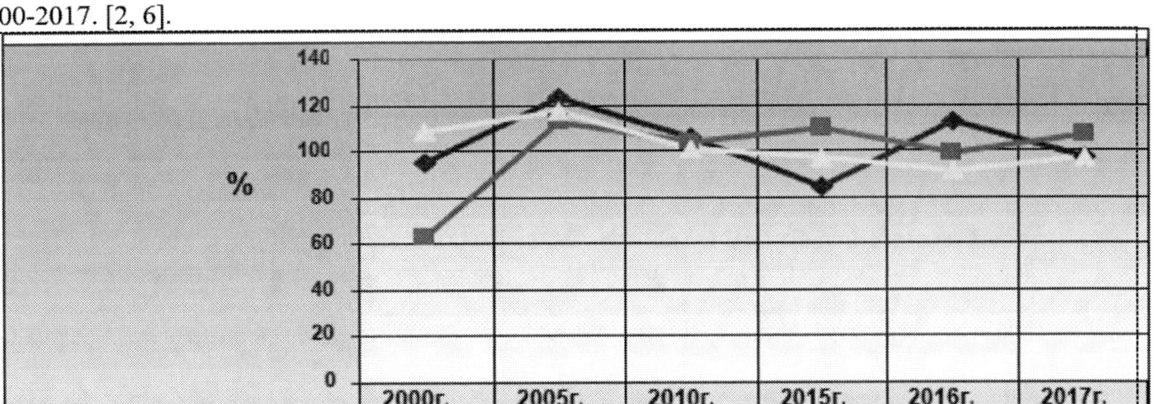

	2000г.	2005г.	2010г.	2015г.	2016г.	2017г.
Investments in the main capital	95	123	106	85	113	97
the housing	63	113	104	110	99	107
capita income	109	118	101	97	91	97

Figure 3. Dynamics for 2000-2017.

As we can see, the investments in fixed assets, growth of average per capita income of the population and the dynamics of housing construction in the region do not have a constant positive dynamic.

IOP Conf. Series: Materials Science and Engineering **880** (2020) 012104 doi:10.1088/1757-899X/880/1/012104

The significant factors that reduce the volume of housing stock input today are:
- still - high credit rates;
- administrative barrier;
- lack of an effective housing credit system,
- high cost-engineering infrastructure.

That is, for the dynamic development of the construction industry in remote regions, a set of the measures and programs is necessary, both at the Federal and regional levels, such programs are available but at present they cannot provide the necessary level of development of the housing and construction industry and solve the housing problem in the region.

The issues of housing security are particularly important because the availability of housing is an indicator that characterizes many others, for example:
- standard of living of the population,
- the level of prices in the housing market,
- the level of income and wages,
- the demographic situation in the region, etc., and these indicators, in turn, affect others the investment climate and investment attractiveness of the regions, the importance of the latter is especially important in the context of the ongoing crisis. [9, 10]

4. Conclusions

The express analysis allows us to draw certain conclusions:
- the problem of development of the construction industry is still important, the investors are currently not ready to actively invest in construction, production of building materials as they are not satisfied with the payback period which is difficult to predict.

The main problem is the slow implementation of many national projects, and weak control over the timing of their implementation. [11, 12, 13]

1. It is necessary to revive (as it was 30 years ago) the close relationship between educational institutions and industrial enterprises to solve the problem of qualified personnel in the construction industry.

2. Rapid implementation of innovative technologies will help to reduce the time of commissioning of objects and their quality characteristics.

3. It is necessary to actively implement methods of investment attractiveness to attract investors (domestic and foreign), for example:

- provision of the benefits and tax holidays for major taxes: income, property, VAT.

The regional government has been using and implementing a set of the measures in recent years but tax incentives could have been used more actively.

4. Special attention should be paid to the productivity indicator and the effective use of available resources.

5. Reducing the cost of work and materials used should be one of the priorities, since the constant increase in prices for the objects sold by the builders leads to a decrease in consumer demand.

6. The development of partnerships with neighboring regions will increase competition in the construction market, resolve issues of import substitution, etc.

In general, the development of the construction industry in remote regions is a rather complex task that needs to be solved at all levels of government: Federal, regional, municipal, important support measures can be a system of state orders and industrial structures with the allocation of priority industrial sectors for each region which can become the basis for the development of non-production sectors. [14, 15,16, 17]

All these measures should ultimately ensure economic growth and investment activity in the region.

References

[1] Dynamics of the volume of work performed by the type of activity "Construction" [Electronic resource]//Irkutskstat:website.URL.http://irkutskstat.gks.ru/wps/wcm/connect/rosstat_ts/irkutskstat
[2] Information Service of IA "USI", https://stroyrf.com/news/
[3] Kaverzina L A, Kubasova T I, Belsky O K and Dobrya A V 2017State of the construction market of the Irkutsk region and Bratsk Problems of socio-economic development of Siberia pp 9-15
[4] Volume of mortgage housing loans granted to individuals-residents in rubles for Federal districts [Electronic resource] Central Bank of the Russian Federation: ofic. website. URL. [http://www. cbr.ru/statistics/
[5] The Volume of loans granted to legal entities-residents and individual entrepreneurs in rubles, by type of economic activity and certain areas of use of funds [Electronic resource] Central Bank of the Russian Federation: ofic. website. URL. [http://www.cbr.ru/ statistics/UDStat.aspx?TblID=302-01]
[6] Dynamics of the volume of work performed by the type of activity "Construction" [Electronic resource] / / Irkutskstat: website. URL. http://irkutskstat.gks.ru/wps/wcm/connect/rosstat_ts/irkutskstat
[7] Sladkova E A and Dukhovnikova K A 2015 personnel management System as a factor for improving the performance of the investment and construction complex pp 329-335
[8] Federal state statistics service. [Electronic resource]. Mode of access: https://irkutskstat.gks.ru/
[9] Yu Zhou 2008 Synchronizing export orientation with import substitution: creating competitive domestic high-tech companies in China *Global developments* **36** pp 2353-2370
[10] Adewale A R 2017 Import substitution industrialization and economic growth - data from the BRICS group of countries. Future Business Magazine **3** pp 138-158
[11] Kotlyarova S N 2016 Mechanisms for stimulating import substitution in the construction industry **12** pp1146-1156
[12] The Concept of development of the construction industry in the Irkutsk region for the period up to 2020 [Electronic resource]. URL: http://irkstroy.media/content/articles/3079 (11.09.2017).
[13] Demyanov K V 2015 Methodological solution of the problem of housing reproduction in the country *Izvestiya vuzov. Investment. Construction. Realty* **2 (13)** pp 11-21
[14] Yershova I and Yershov A 2016 Development of import substitution strategy *Economics and Finance theory* **39** pp 620-624
[15] Volkov, Sedov and Chelyshkov 2015 the concept of " Smart cities "(Moscow: higher school of Economics MGSU) p 93
[16] Rodrigues M 2017 Import substitution and economic growth *Journal of Monetary Economics* **57** pp175-188
[17] Skolubovich A Yu and Matveeva M V 2019 Development of a methodology for assessing the readiness of enterprises for public-private partnership in the field of public water supply *Proceedings of Universities. Investment. Construction. Real estate* **9** pp 274–283 DOI: 10.21285/2227-2917-2019-2-274-283
[18] Larina O P and Khalabuda E I 2019 Accident prevention during construction and installation works at height *Proceedings of Universities. Investment. Construction. Real estate* **9(4)** pp 734–741 https://doi.org/10.21285/2227-2917-2019-4-734-741
[19] Tayurskaya O V, Okladnikova D R and Solodova N G 2018 Investment in the regional economy: challenges and prospects *Proceedings of International Conference on Research Paradigms Transformation in Social Sciences The European Proceedings of Social & Behavioural Sciences EpSBS* pp 1161-1168

IOP Conf. Series: Materials Science and Engineering **880** (2020) 012105 doi:10.1088/1757-899X/880/1/012105

Infrastructure project planning

N G Urazova[1], N V Kotelnikov[1] and A V Martynyuk[2]

[1] Irkutsk National Research Technical University, Department of Industrial Enterprises Management, 83 Lermontov st., 664074, Irkutsk, Russian Federation

[2] Irkutsk National Research Technical University, Department of Chemical Technology, 83 Lermontov st., 664074, Irkutsk, Russian Federation

E-mail: c12@ex.istu.edu

Abstract: This article analyzes problems arising when implementing large infrastructure projects whose number is increasing due to the transition of companies to project management. The study focuses on significant budget overruns and longtime delays in project implementation. The authors developed an approach that allows for effective project stratum decisions. The works of foreign scientists who studied experience of large infrastructure project implementation were analyzed. The analysis helped identify main causes of cost overruns and construction delays (lack of high-quality analysis of data used for project budgeting and scheduling).

1. Introduction

The current stage of economic development characterized by enhanced competition, high risks and uncertainty makes it imperative for companies to implement new building technologies and apply effective methods of corporate and project management. Companies implementing large construction projects are currently moving from the extensive development path based on the use of additional production factors to the intensive development path based on investments in large construction projects, scientific and technological achievements and innovation.

This requires a new management approach to planning of investment projects. At the planning stage, the key parameters (performance time, cost, resources) of the project are identified, and the decision on project launching is made. At this stage, it is important to analyze information. The key aspect of implementation of any investment project is a pre-design stage which involves planning, developing schedules and budgeting. Then the decision on project launching is made. Specialists of the planning or design department of the construction company analyze information obtained by the departments involved in the project implementation process.

2. Materials and Methods

When planning the construction of large objects, in addition to the data gathered by specialists of the project department, corporate project management regulations are used as well. They contain a list of typical works and typical performance time. In budgeting, prices of similar objects or estimates used for the construction of other objects with regard to correction factors (e.g., the inflation coefficient) are used.

The data on performance time are summarized in the schedule, while the data on costs are summarized in the budget scheme. If the documents are approved, a target plan and a project budget are developed. Schedules are often developed taking into account only the launching date determined by the top management within the limits of a strictly stipulated budget. This practice of corporate management causes the need for schedule revision which reduces the performance time that lies on a critical path and limits the budget.

This approach is possible if the decision to set a deadline and allocate a strictly limited budget concerns investment projects in a field in which the company has already had successful construction experience. The approach can be justified if the company analyzes managerial decisions made before launching the project and after completing it. There is a correlation between retrospective decisions and the result. Second, there are no changes in the professional project team over the long period, or its structure changes slightly.

For a new project, the approach to the performance time and project cost should be flexible. First, probability of unaccounted expenses and events affecting the performance time is extremely high. As a result, the size of the allocated budget and available reserves will not be sufficient which can "freeze" the project. Second, the performance time and costs often become known to the public; third parties are involved in these processes, and the company strives to implement the project on time to strengthen its position in the market (to obtain the government and banking support, to cooperate with reputable contractors).

If the deadline has been violated and the cost has increased, the company has a reputational risk that could undermine trust of "stakeholders" - government authorities, creditors, potential consumers, and other organizations.

Thus, in scheduling and budgeting, it is not enough to rely only on experience of the company management and expert opinions. Errors made at the planning stage can have a significant impact on both financial and temporary effectiveness of the project.

The most common causes of failures of large construction projects are as follows.

IOP Conf. Series: Materials Science and Engineering **880** (2020) 012105 doi:10.1088/1757-899X/880/1/012105

First, the company relies on previous projects. Balancing on the "time edge" is an important principle of successful project management. The rapid flow of changes in the external and internal environment of construction organizations forces the company's management to think about a variety of time horizons. The principle "balancing on the edge of time" prevents the company from "slipping" into the past. If too much attention is paid to previous experience, the company's strategies turn out to be enclosed in outdated competitive models. However, it is impossible to ignore previous experience. If we ignore the accumulated potential of implemented projects, it is impossible to achieve positive results. The company always starts everything from the beginning, repeating past mistakes, and the result is a slowdown of its development. Second, it is inaccurate determination of risk probability. Third, it is "rapid" solutions.

There are a lot of examples of inefficient financial planning. Table 1 presents the most famous projects in the field of large-scale infrastructure construction, with the highest excess of planned costs of construction and installation works (CIW).

Table 1. Projects with the most significant excess of costs of construction and installation works

Project	Excess of planned costs of construction and installation works, %
Suez Canal	1900
Sydney Opera House (Australia)	1400
Scottish Parliament (UK)	1075
Calcutta Metro (India)	500
Panama Canal	200
Denver International Airport (USA)	200
Boston-Washington-New York Railroad (USA)	130
Brooklyn Bridge (USA)	100
Shinkansen Railway (Japan)	100
Washington Metropolitan (US)	85
Mexico Metropolitan (Mexico)	60
Baltimore Metropolitan (US)	60
Metropolitan Tyne and Wear (UK)	55

Based on [1].

In the 2000s, the expert community analyzed railway construction projects implemented in 1970-2000. In most cases, the planned number of passengers was overestimated, and the cost of projects increased by more than 50%. Despite the thirty years of experience, managerial conclusions have not been drawn. The construction of the Scottish Parliament building is a well-known example. The

IOP Conf. Series: Materials Science and Engineering **880** (2020) 012105 doi:10.1088/1757-899X/880/1/012105

planned construction costs were £ 40 million. As a result, they exceeded £430 million.

The construction of Denver International Airport is an example of inaccurate planning of the performance time. It was planned to open the airport in 1993. The decision to build the airport was made in 1988. The construction was completed on October 31, 1993, except for the automatic baggage handling system, which was on a critical path. Due to the completion of construction of the tunnels designed for servicing the automated system of remotely controlled trolleys, there were no alternative baggage handling methods (reconstruction of the tunnels would take up more time than construction of the baggage handling system). Owing to the commissioning delay, the airport was launched only in 1995. [2]

The reason for the launching delay was ignorance of the development time for a similar system (Baggage handling system "Crisbag") at Franz Josef Strauss Airport in Munich. The The developers of Munich airport spent 2.5 years developing, testing and debugging a similar system. [3] These dates were ignored by the DIA Board.

As can be seen from Table 1, errors occur in projects implemented in different countries at different time and by different companies. According to the International Association of Engineering Insurers IMIA, disruption of construction deadlines resulting in a significant increase in the project cost was y influenced by management errors, including material and equipment delivery delays (33%). Other reasons are low quality of CIW (23%) and designing errors (12%).

The study on planned and actual costs of large infrastructure projects shows that the lack of in-depth analysis of factors that may affect project costs significantly affects its budget. Overruns for railway projects amounted to 45%, tunnel and bridge construction projects – 34%, road construction projects - over 20%. [4]

Table 2 presents data on the most famous projects.

Table 2. Excessive costs of large infrastructure projects

Project	Cost overruns, %
Large Boston Tunnel (USA)	196
Humber Bridge (UK)	175
Boston-Washington-New York Railroad (USA)	130
Railway tunnel Great Belt (Denmark)	110
Motorway A6 (UK)	100
Shinkansen Railway (Japan)	100
Washington Metropolitan (US)	85
Mexico Metropolitan (Mexico)	60
Metropolitan Tyne and Wear (UK)	55

Based on [4].

The study on the effectiveness of large infrastructure projects conducted by the University of Aalborg in Denmark revealed deviations of actual costs from the

IOP Conf. Series: Materials Science and Engineering **880** (2020) 012105 doi:10.1088/1757-899X/880/1/012105

designed ones by 28%. The conclusions drawn from this study are as follows:

- in 9 out of 10 transport infrastructure projects, the costs were underestimated which caused the excess of costs;
- underestimation of costs and cost overruns have not decreased over the last 90 years; there are planning errors;
- underestimation of costs and cost overruns are due to planning errors and strategic distortion of data, deception [5].

The findings of these scientists are complemented by in-depth studies [6] carried out by B. Flivbjorg who believes that "the main cause of cost overruns is the lack of realism in the original estimate." He formulated the following causes of unsuccessful planning:

- a method: integrated production planning and logistics (econometric software packages). Probability of using an inappropriate model is high due to specific requirements.
- insufficient database;
- biased assessment: the desire to include political preferences into the forecast, experience of implementation of large infrastructure projects in the country of his origin.
- a non-permanent model of behavior and the influence of additional factors (attractiveness of stations; their location near bus stops, shops and parking facilities; consumer safety; navigation data; compliance with hygiene requirements in transport; availability of access roads; flexibility of urban development and housing policy, etc. [7]
- unexpected changes in exogenous factors (social and political upheavals, changes in energy prices);
- unexpected political situation: changes in government decisions on urban infrastructure, taxation, etc.
- biased assessment by the project manager: distortion of data when the manager wants to launch the project regardless of possible unaccounted over-costs. [8]

Thus, the most significant factor influencing the performance time and project costs is poor-quality data. In the conditions of shortage of time and competition, the project management is more likely to decide to launch the project.

3. Results

A conclusion drawn by D. Kahneman, the founder of modern behavioral economics and hedonistic psychology, who received the Nobel Prize for applying psychological techniques in economics, can confirm our viewpoint. In his well-known work “Think slowly, decide quickly,” he points out: “The main source of forecasting errors might be a trend to underestimate or ignore distribution information. Therefore, planners should work hard to identify the problem of forecasting and gather and analyze distribution information.” [9]

Distribution information are data on similar projects. According to Kahneman, information collection eliminates planning errors." [10]

To reduce the impact of this factor (since full elimination of the factor of uncertainty is impossible), it is necessary to use the methods which can solve this problem.

Considering the root cause of the problem of incomplete collection of initial data for making decisions on implementation of investment and innovative projects, let us analyze the theory of decision making developed by D. Kahneman and A. Tversky, according to which "optimistic distortion is an important source of risk acceptance". This theory says that in standard rational economic models, people take risks in favorable conditions, i.e. people accept probability of losses with sufficient probability of success.

In project planning, under the influence of deceptive optimism, the developers overestimate project benefits and underestimate project costs analyzing successful scenarios and skipping tricky places where you can make errors. It causes budget overruns, temporary delays, non-return of funds, etc. [11]

Having eliminated the planning error, one can increase probability of a right decision. Any investment project is implemented under high uncertainty and risks, and elimination of planning miscalculations enhances project certainty.

To reduce planning errors, the methods presented in Table 3 can be applied.

Table 3. Methods for eliminating planning errors

Method	SWOT-analysis	Network Planning Analysis	Critical Path Method
Essene	Description of advantages, weaknesses, opportunities and threats of the project	Assessment and analysis of the project, performance time, labor and other resources	Identification of the critical path of the project, i.e. a temporary shift in one of the works which increases or decreases the construction period
Advantages	Analysis of advantages and disadvantages of the organization which allows for general assessment of the company capabilities and limitations regarding the construction of the facility	Applicable for large construction projects. The method takes into account uncertainty of the project through simulation of three project development options: pessimistic, optimistic and	Simplicity. The method identifies works that are on the critical path

Method	SWOT-analysis	Network Planning Analysis	Critical Path Method
		realistic	
Disadvantages	There are no clear instructions that require additional analytics. In assessing the weights, errors that are difficult to identify can be made.	It underestimates the duration of the project; does not take into account resource constraints and scheduler actions.	It is suitable for understandable projects in which execution time is easy to predict. It does not take into account uncertainty of execution of works

The analysis of the methods presented in Table 3 shows that the use of one of them is not enough for effective planning; the company will need to take additional measures to eliminate planning shortcomings.

For leveling planning deficiencies, it was proposed to use an extended approach, which is similar to the method for eliminating a priori probability ignorance errors developed by Kahneman (see Table 4). It consists of three successive planning stages.

Table 4. Description of planning stages

Stage	Description	Example
Identification of the source category	Identification of the object for which source data should be gathered	Railway construction
Data capturing	It is necessary to gather all data on previous projects carried out by the company, as well as on similar objects.	The cost of the railroad track construction depending on the season
Using specific information to improve the baseline forecast	Each object of the internal and external environment, which can affect the project is analyzed. Expert organizations are involved to assess company capabilities, availability of conditions (internal and	For example, for a small-scale, relatively short-term facility, it is necessary to foresee the use of wooden sleepers impregnated with creosote instead of reinforced concrete. Otherwise, there may be disproportionate goals of the project: - high cost of sleepers due to the

Stage	Description	Example
	external) for the project. All data are analyzed and consolidated to make a decision on implementation / cancellation of the construction project as well as on application of information on promising projects	use of concrete and pre-tension reinforcement; - high cost of fasteners; - high cost of transportation; - difficult installation may increase the duration of project implementation

At the first stage, it is necessary to determine the nature and "bottlenecks" of the project, and take into account its basic parameters: goal, planned deadlines, budget, etc. This information is included in the project charter. It is necessary to consolidate all the project documents: minutes, reports, meeting materials, materials of video and audio conferences, decisions.

According to experts, [12] in order to successfully plan a project and achieve target results, it is necessary to set a goal using the SMART ideology:

- Specific: a target result;
- Measurable: quantitative expression of the result (absolute or relative indicators);
- Achievable: a realistic goal (the company need to have resources);
- Relevant: checking the goal for relevance (whether the project is significant now and will be significant in the future);
- Time-bound: the goal statement contains the deadline.

After the project documents have been prepared, the second planning stage begins. After the main project parameters have been identified, the project management needs to develop a Work Breakdown Structure - this is a graphic representation of the project, i.e. a set of interrelated project elements [13]

Depending on the type of a project - a new original project, or a project similar to the previous one, different methods are used for developing the WBS:

1) the bottom-up method is used when developers lack information about the upcoming works. In practice, the method of brainstorming and grouping of works and tasks are applied. The volume of works is estimated. This grouping is carried out before elements reach the top level.

2) the top-down method assumes that developers are aware of the upcoming works. Decomposition is carried out from the topmost elements to indivisible units of work, resources, etc.

In developing the WBS, special attention should be paid to the statistical data on similar works. It is necessary to use all possible sources of information:

- information on previous projects, if these projects have already been implemented by the company;
- analysis of the current state of the market and forecasts for the future;
- gathering of information from all specialists involved in the project ("experience gathering");
- research carried out by professional organizations;
- involvement of companies having benchmarking experience;
- statistics, schedules and other information about similar projects implemented by other companies.

The information is accumulated and structured for further analysis.

4. Conclusion

At the final stage, specific information is used to improve the baseline forecast. All the information is analyzed by the project management. The company develops an approximate plan (Plan A). It is necessary to adjust this plan to the results of analysis of the information gathered. This is the second variant of the schedule (Plan B) which must be saved and compared with Plan A

In addition to the standard quantitative methods, the critical chain method (CPM), the Monte Carlo and others methods used for analyzing the performance time, the authors suggest analysis of qualitative parameters (climatic conditions, possible inclusion of the land leased for the project into the sanitary protection zone, companies operating in the area, attitude of local people to facilities constructed on their land, etc.).

It is necessary to take into account all factors influencing the project. Based on the in-depth analysis, plan B is adjusted and plan C is developed. The latter reflects real performance time and project costs.

When "bottlenecks" are identified, it is easier to develop appropriate measures to minimize or eliminate project uncertainty. The final stage is comparison of plans. As a rule, at this stage, the project team has already had an idea of the upcoming works, their challenges and opportunities. The comparison makes it possible to identify differences in the superficial "optimistic" planning and in-depth preliminary project analysis. After the comparison has been carried out, a final document is developed. It is presented to the top managers. Based on this report, the management makes a decision on the project.

Thus, in order to make a balanced decision on the project, it is necessary to eliminate planning errors which entail too optimistic forecasts of the project costs and performance time. To eliminate planning errors, it is necessary to carry out in-depth and high-quality gathering and analysis of data on the project.

References

[1] Fourace, Allport and Thompson. *The Perfomance and Impact of Rail Mass Transit in Developing Countries, TRRL Research Report*
[2] De Marco T 2005 Waltz with bears *New Delhi Prime Minister's Office* pp 12-15

[3] Description of the automatic baggage handling system at the Munich International Airport // Munich International Airport (MUC/EDDM) – Airport Technology. Baggage handling system. URL: http://www.airport-technology.com/projects/munich-international/ (дата обращения: 09.09.2015).

[4] Gref G O 2013 Megaprojects - investments or ...? Where is the border*? Report of the President, Chairman of the Board of Sberbank Gref G.O. at the XII International Investment Forum "Sochi-2013"*

[5] Flyvbjerg B, Holm M and Buhl S 2002 Underestimating Costs in Public Works Projects: Error or Lie? *Journal of the American Planning Association* 68 **3** pp 290

[6] Flivbyorg B, Bruzelius N, Rotengatter V 2014 Megaprojects and risks. Anatomy of ambition p 10

[7] *Department for Transport. Variable Demand Modelling. Transport analysis guidance (TAG)* TAG UNIT M2. January 2014 p 84

[8] Flivbyorg B, Bruzelius N and Rotengatter V 2014 Megaprojects and risks. Anatomy of ambition (Moscow) pp 64-69

[9] Kahneman D *Think slowly ... decide quickly* p184

[10] Ibid. P. 188.

[11] Kaston M. *Why do not All Startups Become Successful* Startup Market Research. UBR.ua URL: http://ubr.ua/business-practice/own-business/pan-ili-propal-pochemu-ne-vse-startapy-stanoviatsia-uspeshnymi-301929 (accessed on 18.09.2015).

[12] *Project management* 2010 p 312

[13] Kalyuzhnova N, Violin S and Matveeva M 2019 Smart Specialization of Russian Regions: Prospects and Limitations *Investments. Construction. Real Estate: New Technologies and Targeted Development Priorities IOP Conference Series: Materials Science and Engineering* 667 012036

[14] Kazimirov I A and Peshkov V V 2019 Determination of price behaviour in the secondary residential real estate market using a multidimensional regression model *Proceedings of Universities. Investment. Construction. Real estate* **9(3)** pp 476–487 DOI: 10.21285/2227-2917-2019-3-476-487

Formation of the fuel and energy complex of Eastern Siberia in the 1960s–2010s: construction of energy enterprises

O A Uvarova[1] and O E Naumova[2]

[1,2]Department of History and Philosophy, Irkutsk National Research Technical University, 83 Lermontov Str., Irkutsk, 664074, Russia

E-mail: uvarova.o.a.@gmail.com

Abstract. The article discusses the formation experience of the fuel and energy complex of Eastern Siberia. Based on the analysis of a large number of archival materials, it is shown how enterprises of oil and gas production, coal industry, and hydropower were built under the conditions of the controlled and planned economy. The geography of Eastern Siberia contemplated the construction of powerful industries, which became natural monopolies under the new economic conditions. The coal industry was mostly privatized. Today, the enterprises producing and processing oil, gas, and hydropower operate in the frame of private-public partnership. Siberia has never experienced a shortage of electricity. Its cost was low; therefore, the enterprises were highly profitable. The market economy identified new trends: strengthening of the export-orientedness of the enterprises in the energy sector; attraction of foreign investment into organizing the construction of the newest enterprises; rapid introduction of new technologies; and increase in the uneven development of the region. Summary of historical experience can serve as a source of understanding of modern economic realities.

1. Introduction
The article discusses the formation experience of the fuel and energy complex of Eastern Siberia. Based on the analysis of a large number of archival materials, it is shown how enterprises of oil and gas production, coal industry, and hydropower were built under the conditions of the controlled and planned economy. The geography of Eastern Siberia contemplated the construction of powerful industries, which became natural monopolies under the new economic conditions. The coal industry was mostly privatized. Today, the enterprises producing and processing oil, gas, and hydropower operate in the frame of private-public partnership. Siberia has never experienced a shortage of electricity. Its cost was low; therefore, the enterprises were highly profitable. The market economy identified new trends: strengthening of the export-orientedness of the enterprises in the energy sector; attraction of foreign investment into organizing the construction of the newest enterprises; rapid introduction of new technologies; and increase in the uneven development of the region. Summary of historical experience can serve as a source of understanding of modern economic realities.

2. Problem Statement
By the early 2020s, it became apparent that the Western model of globalization, relying on the strength and dominance of one center in the world, had become tarnished. Gradually, regional centers of "installation" of a new economic reality began to form. Russia became one of such centers. Its past Soviet experience in industrialization, including the experience in the formation of the fuel and energy complex, is used today by China. The growing economies of Asian countries suffer a shortage of energy;

therefore, they need it. To some extent, they push the development of new technologies in Russia. The political stability of our country, development trend, lack of desire to weaken our neighbors, and equal partnership attract more and more Asian economies to cooperate with Russia. In this case, studying the economic experience of the past is always relevant. That will contribute both to a new understanding of the economic realities of our country and to the formation of prospects for its development (international, too). This fact determines the relevance of new research on this issue.

3. Purpose of the Study

The studies on the development of the energy sector in Eastern Siberia were mainly completed quite a long time ago, and they concerned the Soviet period. One of the first studies was the book by Ya.A. Mazover **[3]**, where he briefly dealt with Eastern Siberia as well. Of great interest are the parallel studies of V.V. Alekseev **[4]** and A.N. Zykov **[5]**. They describe the initial period of large-scale electrification of Siberia. Cheap electricity allowed building large enterprises of heavy industry in tens. Siberia developed very dynamically in the 1960s–1970s. Therefore, such material was reflected in I.I. Komogortsev, G.N. Harin, V.S. Balakin, G.S. Prazdnov, V.V. Chernyh and others. The development of the national republics of Siberia, including Buryatia, was especially closely studied **[6]**. However, there were some blank gaps even in avalanchine publications. With the collapse of the USSR and the destruction of the Soviet economic system, many enterprises disappeared. However, this hardly affected the energy sector, especially large enterprises. Enterprises were updated, sales markets were optimized, and innovations began to be actively introduced. Scientific research was also adjusted by the formation of market mechanisms: lots of publications of narrowly focused economic content appeared; the authors used one or several enterprises of a similar profile as a factual base. That did not allow evaluating the development of industries as a whole. The purpose of this article is to summarize the historical experience of the formation of the fuel and energy complex of Eastern Siberia over the past half century using the more comprehensive systematic approach. An attempt was made to determine trends in the energy sector development, as well as changes in the configuration of its sub-sectors. The study is based on the wide use of data from state archives, many of which are first-ever to be introduced into scientific field. That will close some gaps in the history of the region.

4. Research Methods

The study used both general scientific methods (comparison, description, induction, deduction, generalization, etc.) and methods of historical research (chronological, historical genetic, and comparative historical). A large array of analyzed data entailed the use of statistical and historical sociological methods. The whole set of possibilities, classification, and the systemic method allowed recreating an objective scientific picture of the issue under consideration.

5. Research Questions

The fuel and energy complex includes industries related to the extraction and production of energy resources, their processing into other types of fuel, conversion to other types of energy, as well as industries related to the transportation of this energy to consumers. Its formation in the USSR was an urgent need in the harsh climate and vast territory. Siberia played a special role with its rich and various natural resources: brown and fossil coal, water, and hydrocarbons. The country's government relied on a scientific approach and state planning. As far back as 1947, a conference was held to study the productive forces of the Irkutsk region. In the report of academician A.V. Vinter, Eastern Siberia was noted to be unique in terms of energy resources: 37% of potential reserves of water energy, high-quality coals of various grades. The necessity for the construction of the cascade of hydroelectric power stations on the Angara River was expressed in the report. Energy-intensive industries on local raw materials were to become the main energy consumers **[7]**. Six hydropower plants were planned to be built;

however, there were only four to be eventually built. The Irkutsk Hydroelectric Power Station became the first of the 4 built stations. Many pilot solutions were tested in it **[8]**. Then the Bratsk Hydroelectric Power Station (main construction 1954–1961, and then until 1966) and the Ust-Ilimsk Hydroelectric Power Plant (main construction 1962–1974, and then until 1979) were built and launched. They constituted a powerful core around which territorial-industrial complexes began to form **[9]**.

The Krasnoyarsk Hydroelectric Station on the Yenisei River (main construction 1956–1967, and then until 1971) remained the most powerful in the country for a long time. To meet the demand for electricity in the north of the Krasnoyarsk Territory, the Ust-Khantai Hydroelectric Power Station was built (1970–1972). That made it possible to increase the electric power generation in the Krasnoyarsk Territory from 33.8 billion kWh in 1970 to 44.7 billion kWh in 1974 **[10]**. It amounted to 57 billion kWh in 1980 **[11]**. The eighth and ninth five-year plans were periods of powerful investments in the energy sector of the Krasnoyarsk Territory. They amounted to 361.1 and 928.6 million rubles in the eighth and ninth five-year plans, respectively **[12]**. This allowed not only expanding a number of thermal power plants, but also directing money to reconstruction of facilities and equipment modernization. Sayano-Shushenskaya Hydroelectric Power Station on the Yenisei River was gradually being built. It is unique in many respects. However, eliminating design errors took a lot of time and money. Its construction entered its most active phase in 1978 and was finally completed in 2014 with the liquidation of the consequences of the 2009 accident.

The change in the economic structure and forms of ownership intensified the construction of the fourth hydroelectric power station on the Angara (the Boguchanskaya hydroelectric power station). The first preparatory work started in 1974. However, its construction completed only in 2012–2014 due to the lobbying of the project by Oleg Deripaska, RUSAL (Russian Aluminum). The share of hydropower in Siberia has always been high, and in 2017 it reached 46% **[13]**. In Russia, only the Volga and Angara Rivers have cascades of large hydroelectric power stations, the first five most powerful ones of which are located in Eastern Siberia. Thus, it became possible to preserve and build new energy-intensive industries in the region. The development of older enterprises is carried out in line with the equipment modernization and the production reconstruction. Newer enterprises expand and introduce new hydraulic units.

As compared to hydropower enterprises which were located in the Irkutsk Region and the Krasnoyarsk Territory, coal mining was carried out in all the four regions of Eastern Siberia. This is the most traditional form of natural resources, and its active (albeit small) development was carried out since the beginning of the twentieth century. Large-scale post-war development plans for the eastern regions of the country (especially since the early 1960s) gave a new impetus to the development of the coal industry. In the Irkutsk Region, coal mining is conducted by Vostsibugol, a company established by the state in 1945 as a regional coal mining enterprise. The peak of its development occurred in the 1970s **[14]**. During Perestroika, it changed its ownership several times and found Irkutskenergo as a parent company. Today, work is carried out at seven open-pit mines (mines have died), including the Cheremkhovsky and Tulunsky ones.

In the Krasnoyarsk Territory in the 1970s, coal was taken both by mines (mines of Abakan, Yenisei, Kotuy) and by open-pit mining. From the 1970s, huge deposits of brown coal were actively developed at the Berezovsky, Nazarovsky, Irsha-Borodinsky open-pit mines. The equipment modernization, introduction of new machines, use of inventions and efficiency proposals, as well as environmental events became permanent **[15]**. Brown coals became the basis for the heat generation of several areas. The Berezovskaya state district power station became the largest enterprise in the Krasnoyarsk Territory. In February 2016, the billionth ton of coal from the date of launching was mined at the Borodinsky open-pit mine. Today it is a huge field measuring seven by two kilometers, where the square of stripping operations is more than two thousand hectares. Two most powerful excavators in Russia (ERP-2500) work at this open-pit mine. In 2013, the open-pit mine was named after the legendary USSR Minister of Coal Industry M.I. Shchadov.

Fossil coal is mined at the Izykhsky and Chernogorsky open-pit mines in Khakassia (the Krasnoyarsk Territory). The Chernogorsky open-pit mine dates back to coal works of 1907 but it began to develop actively in the 1960s and 1970s. The staff of the enterprise worked tirelessly to increase production efficiency and introduce best practices. During the tenth five-year plan, more than 10 million tons of coal was mined. The average monthly labor productivity per worker reached 308 tons. The prime cost of one ton of high-quality coal was 6.09 rubles **[16]**. Today, the open-pit mine is part of the Siberian Coal Energy Company (SUEK) and is actively being developed. Records are still setting at this open-pit mine. In January 2016, the team of V. Yarosh set a world record for the shipment of stripping soils over the month (1075 thousand cubic meters) **[17]**.

The coal industry of Transbaikalia developed in the 1970s at a more modest pace. The Gusinoozerskaya mine in Buryatia and the Vostochnaya mine in the Chita region experienced some difficulties. They were connected with roadways in coal seams. The Bukachacha mine was considered more successful. But it also could not be called trouble-free. In the 1970s, the party committee regularly reviewed issues related to the elimination of accidents and fires in the mine, whereas the issues of scientific and technological progress were never discussed (!). Despite the reconstruction, the share of manual labor was 40% in 1985 **[18]**. Due to its unprofitability, the mine was closed in 1998.

Since the late 1970s, the government relied on open-pit coal mining as cheaper and safer one. In 1978, the Holboldzhinsky open-pit mine was launched. The Kharanorsky open-pit mine (launched in 1970) began to work more actively. Its design capacity is 5 million tons of brown coal per year **[19]**. After technical re-equipment in 1990, 9 million tons of coal was mined. At the turn of the century, in 2002, in the course of industry optimization, the open-pit mine became part of SUEK (Siberian Coal Energy Company). Serious investments and competent management have made the enterprise performance sustainable and profitable. Old traditions are supported and filled with new content at this open-pit mine. In Soviet times, there was a "club of millionaires" at the Kharanorsky open-pit mine. It was a competition between teams for the extraction of one million tons of coal. Nowadays, the financial side has added to moral incentives: such a team receives a prestigious car and one million rubles for all team members.

In the 1970s, the rich Tugnuisky coal-bearing area on the border of Buryatia and the Chita region was considered promising. It began to be developed in 1989. But almost immediately, the Tugnuysky open-pit mine fell into a zone of instability along with the country and survived due to barter for several years. In 2001, he became part of SUEK. Sustainable development of this open-pit mine is associated with the appointment of Valery Kuletsky as the Director General in 2008. In recent years, coal mining has tripled and amounted to 14 million tons in 2018 (totally, 439.3 million tons of coal was mined in Russia in 2018) **[20]**.

Considering energy in general, we mean first of all traditional energy based on coal, hydrocarbons, water. For Eastern Siberia, there was no need to build nuclear power plants and develop unconventional sources of energy using the sun and wind. Another peculiarity of Eastern Siberia was due to the vast territory and underpopulation. Industrial enterprises are grouped in a very limited area and are always tied to powerful energy sources. Territorial-production complexes arose and developed this way. Energy enterprises are almost always very large. That is why almost all of them survived the difficult perestroika period. Over the past thirty years, private property has developed and strengthened in Russia. There are public-private partnerships in the hydropower field. Almost all coal enterprises have become private and are part of large holdings. Thus, almost 60% of the coal mined in 2018 accounted for seven companies. The leader is SUEK, which accounts for a quarter of Russian coal mining. The enlargement allowed reducing costs, increasing profitability, and implementing investment projects. The industry is developing dynamically. Russia is on trend since over the past forty years, coal in the world's electricity production has permanently been 38–39% **[21]**. In the 21st century, Russia became a reliable exporter of coal.

The development of oil and gas production concentrated in Western Siberia for a long time. However, geological exploration yielded positive results both in the Irkutsk region and the Krasnoyarsk Territory. Thus, in the Krasnoyarsk Territory, by the mid-1970s, the Nordvik oil deposit and such gas deposits as Pelyatkinskoye, Kazantsevskoye, Nizhne-Khetskoye, Sukhotungusskoye, and Kuyumbinskoye were explored **[22]**. The exploitation of the Messoyakha and Soloninsky gas deposits was launched. However, the region economy could not be called energy-deficient; therefore, there was no special economic sense to develop oil and gas deposits.

Everything changed with the advent of a new economic reality, shift or change in world development trends, gradual acquisition of strength and new opportunities by Russia. Having surveyed its Siberian resources **[23]**, in the 2000s, Russia started its turn to the East, to Asia. China met this turn positively since economic interests of both countries coincided **[24]**. The Eastern Siberia-Pacific Ocean oil pipeline (ESPO) was the first major interstate initiative implemented. Oil from Western and Eastern Siberia began to be destined for Japan and China. In 2019, the Power of Siberia gas pipeline was launched. Gas from the Kovyktinskoe and Chayandinskoe fields of the Irkutsk region went to China. This project confirmed the farsightedness of China, which, in the context of the unfolding trade war with the United States, relied on Russia **[25]**. The game on foreign oil trade markets, building positive relations and equal partnership with Asian neighbors open up new prospects for our country. The construction of the second stage of the ESPO and the Power of Siberia-2 gas pipeline is planned. Thus, Russia is actively involved in Eurasian integration processes **[26]**. Expanding production in Eastern Siberia, such companies as Gazprom, Transneft, Rosneft, and NOVATEK are creating new jobs. A number of technical and technological solutions used in the projects have become innovative.

The development of the North is very important today. In 2009, the commercial operation of the Vankorsky oil and gas field was launched. It was opened in 1988, 140 kilometers far from Igarka (the Krasnoyarsk Territory). Together with the Lodochny, Tagulsky, and Suzunsky deposits, it forms the so-called Vankorsky block. Today, Russia attracts foreign investment (in particular Indian) to further develop this cluster. To develop the natural resources of the Arctic **[27]**, Russia attracts China since it will also get preferences from the development of Siberian hydrocarbons **[28]**. The new century has opened up new opportunities in the development of hydrocarbon production in Eastern Siberia. On the one hand, the rapid development of Asia has helped diversify markets towards the Asia-Pacific region (APR). On the other hand, with the increase in foreign trade and the growth of Russian own economy, it has become possible to invest heavily in the construction of ultramodern industries. To some extent, investor interest has shifted to the rapidly growing oil and gas industry. The development of the energy sector leads to the development of infrastructure and electric power industry. All that serves as a stable basis for economic growth. It can be stated that the adaptation to new realities and the formation of new development trends have occurred.

6. Conclusion

The authors can summarize all the foregoing as follows:

1. For more than half a century, Eastern Siberia has been and remains a place where traditional types of energy based on the rich natural resources available are developed. Large enterprises remain stable and have the potential for their further development in case of high and modern managerial competence.
2. According to the past experience, private, joint-stock form of ownership is more effective in a number of cases: when the value of the enterprise is not prohibitively high and when the capital stock is available both financially and technologically. This conclusion is confirmed by the rapid development of the coal industry in the 21st century.
3. More complex and high priced hydropower and hydrocarbon industries require public-private partnerships since neither large investment nor interstate initiatives are possible without the organizing role of the state. The market economy stimulates the exploration of the new but does not impede the development of the old, economically sound and cost-effective.
4. Restructuring of the economy under the new economic and old geographical conditions in Siberia led to the formation of natural monopolies. Under the state supervision, such companies as SUEK, Gazprom, and others serve as a driver of economic growth and territorial development. Large enterprises are trying to support traditions, develop the social component, and take care of the environment.

5. Strengthening of the export-orientedness has become a new trend in the 21st century. In conditions of relative overproduction, enterprises trade and cooperate with foreign partners successfully.
6. In our opinion, there has been some reconfiguration of the subsectors. The share of coal remains stable, the share of hydropower decreased slightly, and the share of oil and gas increased. Moreover, the course towards deeper hydrocarbon processing is well-established.
7. Under the conditions of the controlled Soviet economy, the government tried to develop territories more or less evenly. Under the conditions of several decades of the market economy, the gap widened between the Krasnoyarsk Territory and the Irkutsk Region on the one hand, and Transbaikalia on the other hand. In the first case, we have a complete set of energy resources and a complete set of enterprises, accordingly. In the second case, the lack of oil and gas leads to a “downfall” of the economy, as evidenced by numerous indicators.

It is possible to assume that some generalization of historical experience allows us not to get lost in the present. The power economy development can be assumed to serve as an indicator of the development of the economy as a whole.

References

[1] Alekseev V V. 2018 The energy challenge for imperial Russia *Ural Hist. Bulletin* **59** 6–12
[2] Bazarov V B 2008 Transboundary East-Asian regions in Eurasian interaction *Ural Hist. Bulletin* **19** 131–136
[3] Koksharov V A 2013 Trans-border cooperation in Central Asia: post-soviet syndrome *Ural Hist. Bulletin* **39** 96–101
[4] Mazover Ya A 1967 *Fuel and Energy Centres of the East of the USSR* (Moscow: Nauka) 207
[5] Alekseev V V 1973 *Electrification of Siberia. Historical Research. 1961–1970* (Novosibirsk: Nauka) 272
[6] Zykov A N 1967 *The CPSU is the Organizer of the Construction of Hydropower Plants in Eastern Siberia (1950–1967)* (Irkutsk: East Siberian Book Publishing House) 574
[7] Baldano M N 2001 *Industrial Development of Buryatia (1923–1991): Achievements, Costs, Lessons* (Ulan-Ude: Publishing and printing complex of East Siberian state institute of culture) 432
[8] Vinter A V and Veitz V I 2012 Power resources of Eastern Siberia and the development of the energy sector of the Baikal-Cheremkhovsky complex *Vostocho-Sibirskaya Pravda* **26603** 2
[9] State governmental institution The State Archive of the Irkutsk Region f. 2918p, op.1, dd. 48,97,220,348,890,1041,1837
[10] State governmental institution The State Archive of the Irkutsk Region f. 2879p, op.1, dd. 312,346,408,490
[11] Regional State public institution State Archive of the Krasnoyarsk Territory f. 26, op.8, d. 606, l. 59
[12] Regional State public institution State Archive of the Krasnoyarsk Territory f. 26, op.10, d. 10, l. 37
[13] Regional State public institution State Archive of the Krasnoyarsk Territory f. 26, op.8, d. 63, ll. 148–149
[14] Antonov N V 2018 *Electricity production and consumption in the Russian Federation in 2017* (Electronic resource. Access Mode: https://1prime.ru/sience/20181115/829538943.html)
[15] State governmental institution The State Archive of the Irkutsk Region f. 2690p, op.1, dd. 1245,1353,1460,1474,1508,1639
[16] Regional State public institution State Archive of the Krasnoyarsk Territory f. 2303p, op.1, dd. 536,637,638,747,858,859,960,963,975
[17] State public institution of the Republic of Khakassia National Archive f. 199, op.1, d. 22, l. 11
[18] Literaturnaya Gazeta 2018 **6636** 17

[19] State governmental institution The State Archive of Zabaykalsky Krai, f. 4182p, op.1, dd. 38,42,43,46,47,56
[20] State governmental institution The State Archive of Zabaykalsky Krai, f. 6263p, op.1, dd. 1–26
[21] The Tugnuisky open-pit mine (Electronic resource. Access Mode:http://www.oaort.ru/about/news/352)
[22] Regional State public institution State Archive of the Krasnoyarsk Territory f. 26, op.8, d. 697, ll. 36–40
[23] Alekseev V V 1986 The discovery of Siberian oil *Questions of History* **5** 68–80
[24] Borokh O N 2019 The impact of ideology and politics upon modern economics in China *World Economy and International Relations* **10** 30–41
[25] Mikheev V V and Lukonin S A 2019 China-USA: multiple vector of "trade war" *World Economy and International Relations* **5** 57–66
[26] Marchenko O V, Podkovalnikov S V, Savel'ev V A, Solomin S V and Chudinova AYu 2018 Russia in Eurasian electric power integration *World Economy and International Relations* **6** 18–29
[27] Gutenev M Yu 2019 Northern sea route in Arctic policy of Russia *World Economy and International Relations* **1** 83–87
[28] Zagorsky A V 2019 China Accepts Rules in the Arctic *World Economy and International Relations* **7** 76–83

IOP Conf. Series: Materials Science and Engineering **880** (2020) 012107 doi:10.1088/1757-899X/880/1/012107

Cost optimization for overhauling the apartment blocks in Irkutsk, Russia

I V Yamshchikova[1] and I A Lisetskij[1]

[1]Department of real estate expertise and management, Irkutsk National Research Technical University, Lermontov Street 83, 664038 Irkutsk, Russia

E-mail: yamsirina@yandex.ru

Abstract. This paper analyzes the specifics and major shortcomings of the following documents: the Irkutsk Oblast Act No. 167-OZ of December 27, 2013 On the Arrangement of Common Property Overhauls in Irkutsk Oblast and the pertaining regulations, programs, and Short-Term Plans Under the Irkutsk Oblast Regional Apartment Block Overhaul Program (hereinafter, common property stands for common facilities used by the residents of a single apartment block, e.g. rooftops, utilities, etc.). It further analyzes the financial reports of the Irkutsk Oblast Apartment Block Overhaul Fund. The analysis demonstrates the costs of overhauling the apartment blocks in Irkutsk Oblast with breakdown by sources, the amount and source of funds coming from the region's municipalities as shown in the general funding structure of the Plan, and the specific jobs arranged under the City of Irkutsk Short-Term Plan Under the Regional Apartment Block Overhaul Program in 2015-2016. To simplify the drafting of overhaul programs, the authors hereof present the estimates of roof overhaul costs per square meter with breakdown by overhaul-related jobs; the estimates assume the use of novel materials: MasterSeal*501 and MasterEmaco*S 48. Such analysis reveals the roof job-specific coefficients (multipliers) one needs to apply to find the actual overhaul costs for Series 135 panel buildings in Irkutsk. The authors propose an scenario-based approach to design when drafting time-adjusted short-term overhaul plans. Time adjustment helps optimize such plans and minimize the change of a building being classified as a dilapidated apartment block.

1. Introduction

Housing must be in good condition and meet some minimum requirements to be suitable for comfortable and safe residential use. So far, the Russian Federation's attempt to compile apartment block (AB) data sheets have shed light on the situation. Based on such data sheets, they drafted a 30-year overhaul program. Irkutsk Oblast's Ministry of Housing Policy and Energy issued its Order No. 73-mpr of September 1, 2014, which approved the *Irkutsk Oblast Regional Common Property Overhaul Program for 2014-2043* [1]. The overhaul queue is adjusted every year for how long this or that building has been in use, how worn it is, and what its specifications are.

Pursuant to Article 178 of the Housing Code of the Russian Federation and the Irkutsk Oblast Act No. 167-OZ of December 27, 2013 *On the Arrangement of Common Property Overhauls in Irkutsk Oblast* , the region now has a regional operator, the Irkutsk Oblast Apartment Block Overhaul Fund.[2] The mentioned documents specify the following core functions of this company:
(1) accumulation of fees collected from landlords or allocated from various budgets specifically for the purpose of overhaul;
(2) arranging the overhaul of common property in apartment blocks in Irkutsk Oblast.

2. Materials and methods

Data for this research was collected from the reports on carrying out the *Irkutsk Oblast Regional Common Property Overhaul Program for 2014-2043*, as well as from the financial reports of the Irkutsk Oblast Apartment Block Overhaul Fund.

Methods included economic statistical analysis, systematic method, and systematization of theoretical and empirical data.

AB overhaul valuation and cost optimization were based on finding the costs of construction products relevant in the Russian Federation as of the timeframe of this research.

3. Results

Pursuant to the *2014 Short-Term Plan Under the Irkutsk Oblast Regional Common Property Overhaul Program for 2014-2043*, the Plan shall be funded by the Irkutsk Oblast Apartment Block Overhaul Fund as well as from the regional budget, the municipalities' budges, and AB landlords' funds collected and stored on the Overhaul Accounts of the Regional Operator. [2]

The Program sets forth the following breakdown of overhaul funding sources.

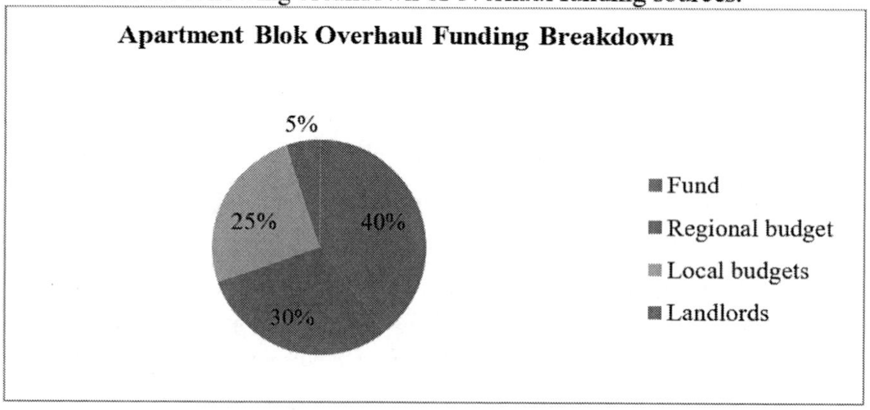

Figure 1. Apartment Block Overhaul Funding Breakdown

Funding under the Plan totals 338,368,762 rubles that come from the following sources:

Table 1. Amount and sources of funds in the total funding under the Plan: breakdown by municipalities

Municipality of Irkutsk Oblast	total funds for AB overhauls, rubles				
	total	including funds from			
		the Fund	regional budget	the local budgets	landlords
City of Irkutsk	139,390,494	55,016,956	42.223,000	35,000,000	7,150,538
total for Irkutsk Oblast	338,368,762	133,554,956	102,500,000	84,957,000	17,356,806

For this Plan, the compiled general data sheets of apartment blocks in Irkutsk Oblast. Data provided by the region's municipalities revealed 1287 apartment blocks to be covered by the Program; these have a deterioration of >70% [3]. The short-term plan for 2015 covered 100 blocks commissioned in the 1950-1980s, of which only 8 had seen any overhaul. 34 of the scheduled-for-overhaul buildings were in the Municipality of Irkutsk, none had ever been overhauled. The oldest one was 80 years old (commissioned in 1935), and the newest one was 35. When making the short-term plan for 2015, its authors adjusted the amount of investment, which resulted in the following figures.

Table 2. Short-Term Plan Under the Irkutsk Oblast Regional Apartment Block Overhaul Program for 2015-2016, rubles

AB overhaul costs in 2015-2016, Irkutsk	total	landlords' fees	the local budgets	regional budget
	266,632,812	266,632,812	0	0

Breakdown by overhaul types.

Table 3. Short-Term Plan Under the Irkutsk Oblast Regional Apartment Block Overhaul Program for 2015-2016, thousand rubles (breakdown by jobs)

indoor wiring	indoor heat piping	indoor cold water supply systems	indoor hot water supply systems	indoor sewerage systems	roof	basements and scaffolding	facades	construction monitoring
45,991	78,851	9,919	12,220	8,694	29,890	15,585	61,810	3,669

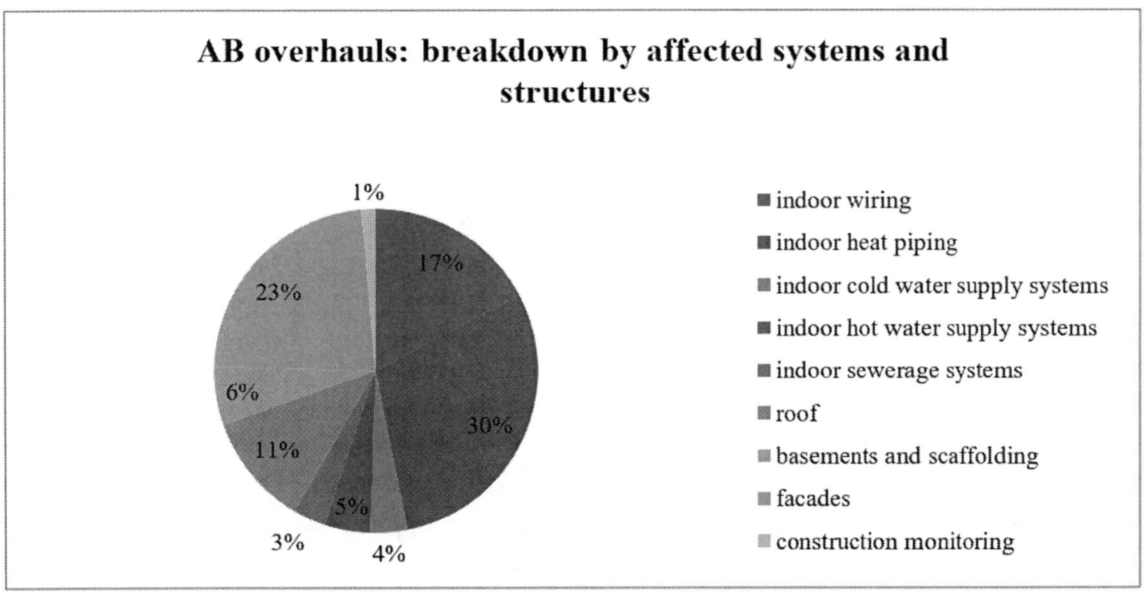

Figure 2. Short-Term Plan Under the Irkutsk Oblast Regional Apartment Block Overhaul Program for 2015-2016, thousand rubles (breakdown by jobs)

As shown in the diagram, it is the indoor engineering systems, roofs, and facades that cost the most to overhaul

4. Discussion

Pursuant to the *Regulations on Common Property Overhauls in Irkutsk Oblast for 2014-2043* [ ], drafting the designs and estimates is done by the following roadmap:

- Inspect the apartment block;
- Run the calculations;
- Draft the designs and estimates;
- Overhaul the apartment block.

Apparently, none of these steps implies scenario-based design, which could help reduce the costs of this or that job. When drafting the original plan under the Program for 2014-2015, scenario-based approach was not an option. The Program covered mainly nearly-dilapidated apartment blocks and sought mainly to preserve the existing housing. However, the today's adjustments to further plans for until 2043 must be made with optimization in mind.

For instance, consider overhauling the roofs, which accounted for 11% of the total overhaul costs, or 29,890 thousand rubles in 2015-2016. When overhauling a roof, they can redesign it to make a non-ventilated roof into a ventilated one with air outlets in the rooftop; such overhauls can also be associated with the repair or replacement of roofing elements and drains.

The City of Irkutsk is facing challenges pertaining to the overhaul of flat reinforced-concrete rooftops of Series 135 panel buildings. The first stage of the ICD survey revealed the possibility of using different methods of roof repair, depending on the degree of wear and tear. For repairs, the authors hereof propose novel materials: MasterSeal*501 and MasterEmaco*S 488.

Table 4. Overhauls of flat reinforced-concrete rooftops in Series 135 panel buildings in the City of Irkutsk

No.	Repair job	Scope of work
Job 1	Surface restoration	Prepare the slab surface (brush with metal brushes, remove the dust, moisturize) Apply MasterSeal*501 to the surface (in two applications)
Job 2	Upper layer restoration	Dismantle the protective concrete layer Prepare the slab surface (brush with metal brushes, remove the dust, moisturize) Apply MasterSeal*501 to the surface (in two applications). Apply MasterEmaco*S 488. Apply MasterSeal*501 to the restored surface (in two applications).
Job 3	Coating restoration	Dismantle the protective concrete layer Prepare the slab surface (brush with metal brushes, remove the dust, moisturize) Seal the opened seams with a high-grade solution. Mount the reinforcement grids. Arrange the timbering. Pour the concrete mix. Apply MasterSeal*501 to the surface (in two applications). Apply MasterEmaco*S 488. Apply MasterSeal*501 to the restored surface (in two applications).

Job 4	Roof slab replacement	Place safety strain poles and bindings. Dismantle the roof slab. Arrange the timbering. Mount the reinforcement grids. Pour a mixture of high-grade modified concrete. Dismantle the timbering. Prepare the slab surface (brush with metal brushes, remove the dust, moisturize) Apply MasterSeal*501 to the surface (in one application).

To find repair job-specific costs, the authors hereof compiled process maps to estimate the repair costs by sourcing.

Calculations returned job-specific costs per square meter applicably to overhauling the reinforced-concrete rooftops in Series 135 panel buildings in Irkutsk.

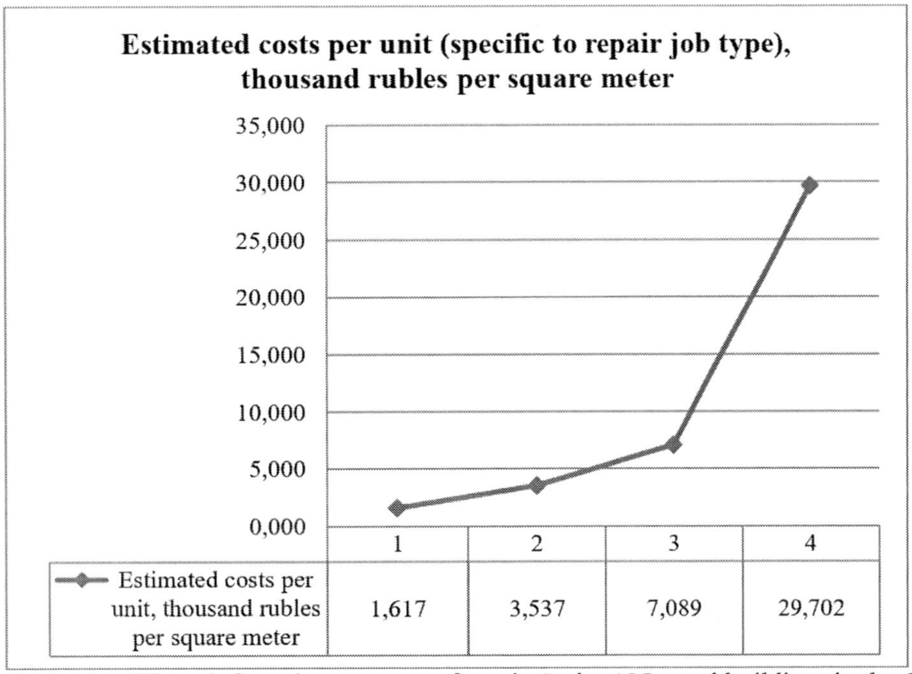

Figure 3. Overhauls of flat reinforced-concrete rooftops in Series 135 panel buildings in the City of Irkutsk: job-specific cost per square meter

The estimates demonstrate the job-specific multipliers that can help adjust the short-term AB overhaul plans in the context of repair jobs.

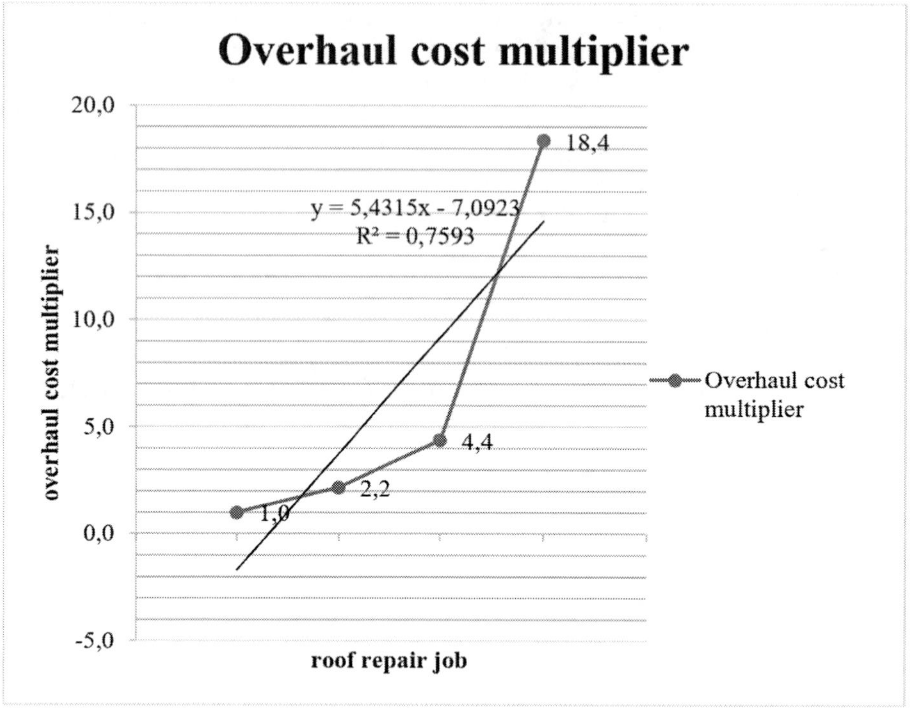

Figure 4. Repair job-specific coefficients (multipliers) one needs to apply to find the actual overhaul costs for Series 135 panel buildings in Irkutsk

5. Conclusion

Analysis of the Regulations on Common Property Overhauls in Irkutsk Oblast and *Short-Term Planы Under the Irkutsk Oblast Regional Apartment Block Overhaul Program for 2015-2016* showed that such plans were often drafted and approved with no due consideration of the choice of design solutions. Applying the scenario-based approach to repair planning can reduce the roof repairs (rooftop element replacements) by a factor of up to 18. However, such reduction will require adjustments for time to exclude nearly-dilapidated buildings.

References

[1] Approval of a Short-Term Plan for 2014 Under the Irkutsk Oblast Regional Common Property Overhaul Program for 2014-2043 [Electronic source]: Irkutsk Oblast's Ministry of Housing Policy and Energy Order No. 73-mpr of Setmber 1, 2014 (rev. November 17, 2014). Accessed from ConsultantPlus.

[2] On the Arrangement of Common Property Overhauls in Irkutsk Oblast [Electronic source]: Act of Irkutsk Oblast No. 167-OZ of December 27, 2014. Accessed from ConsultantPlus.

[3] Approval of a Short-Term Plan for 2015 Under the Irkutsk Oblast Regional Common Property

[4] Overhaul Program for 2014-2043 [Electronic source]: Irkutsk Oblast's Ministry of Housing Policy and Energy Order No. 45mpr of May 25, 2015. Accessed from ConsultantPlus.

[5] Yamschikova I V and Naumov E I 2018 Innovative Design Product Cost *Optimization MATEC Web of Conferences*

[6] Yamschikova I V and Lisetskij I A 2018 Analysis of costs for land development in price of construction products *MATEC Web of Conferences*

[7] Yamschikova I V and Alecsandrova A C 2019 Analysis of the program "Relocation of citizens Russian Federation from a dilapidated and emergency housing background» in the Irkutsk region *ISTU Bulletin of Youth* **9**

[8] Mashoshina T V 2013.Appraisal of Design Work with Account of Indirect Costs as the Main Way to Improve the Efficiency of the Development and Release of Design Products in the Railway Industry *Bulleting of Research Outcomes* **4(9)**

[9] Borovsky O N 2017 Questions and Development Prospects of Design Work in Russia *Russian Entrepreneurship* **22**

[10] Kalyuzhnova N, Violin S and Matveeva M 2019 Smart Specialization of Russian Regions: Prospects and Limitations *Investments. Construction. Real Estate: New Technologies and Targeted Development Priorities IOP Conference Series: Materials Science and Engineering* **667** 012036

[11] Matveeva M V and Kalyuzhnova N Y 2019 Industrial Construction: Investment Support in Entrepreneurship *Investments. Construction. Real Estate: New Technologies and Targeted Development Priorities IOP Conference Series: Materials Science and Engineering* **667** 012062

[12] Peshkov A V 2019 Remote mortgage lending as a resource for housing development *Investments. Construction. Real Estate: New Technologies and Targeted Development Priorities IOP Conference Series: Materials Science and Engineering* **667** 012073

IOP Conf. Series: Materials Science and Engineering **880** (2020) 012108 doi:10.1088/1757-899X/880/1/012108

The importance of the feasibility study of investments in the development of the Russian oil and gas industry

T V Dobysheva[1] and A M Gidaspova[2]

[1]Irkutsk National Research Technical University, chair of real estate expertise and management, Lermontov str. 83, Irkutsk, 664074, Russia
[2] LLC Irkutsk oil company, 664007, Irkutsk, Bolshoy Liteyny prospect 4, Russia

E-mail: DobishevaTV@mail.ru

Abstract. The article considers the issues of determining the cost of construction of oil and gas facilities at the pre-project stage. The great importance of investing in the construction of oil and gas facilities for the industry and the country's economy as a whole is analyzed. Shortcomings were identified when calculating the cost of construction at the pre-project stage. Ways of improving the accuracy of the calculations the cost of construction of industrial facilities in the development of feasibility study of investments, including a proposed supplement to its existing base of estimated regulatory documents in large-size collectors of the rates of industrial construction. Also, as part of the method of determining the estimated cost based on enlarged collections and a data bank of previously constructed and designed objects, it is proposed to make adjustments to the register of cost-effective project documentation for reuse. The authors are convinced that increasing the accuracy of calculating the cost of construction products will improve the efficiency of investment in industrial construction.

1. Introduction

Today, the Russian oil and gas industry is the largest structure that exerts its influence not only within the country, but also abroad, and also occupies a key position in the economy of the RF.

Today, oil is a powerful argument in conducting foreign policy affairs. Russia has great potential on the world market of fuel and energy resources. according to estimates, about 13% of the world's oil reserves and 34% of natural gas reserves are concentrated on the territory of Russia, which shows the country's definitely strong position in this area [1].

Having large reserves of hydrocarbon raw materials and supplying them to world markets, the RF has a significant influence on the formation of the international energy security system [2].

The feasibility study is a key document that reflects the effectiveness of project financing, and is a reference when deciding on the implementation of investment projects in construction. Capital construction plays a key role in the implementation of scientific and technical achievements, thereby creating and updating the production base. The oil and gas industry is the main beneficiary on the russian market. According to the federal state statistics service, investments in fixed assets in the RF by type of economic activity: production of crude oil and natural gas have the following dynamics: in 2014- 1 494,9 billion rubles, in 2015- 1 582,7 billion rubles, in 2016- 1 597,1 billion rubles, in 2017- 1 807,1 billion rubles, in 2018- 1 851,7 billion rubles [3].

The above factors justify the relevance of the chosen topic, as the development of the industry requires the establishment of efficient production and improvement of the previous activity of the enterprise, and given that the construction of such plants consumes high investment needs proper planning not only investment, but also obtained income.

2. Problem Statement

Despite the drop in oil and gas prices in recent years, and the general problems of the russian economy, the country's oil and gas companies continue to invest billions of dollars in new projects for the development and production of oil and gas resources. In this regard, the efficiency of individual regions and territories depends directly on the performance of individual enterprises that are part of the production potential of the region. The gas industry is one of the fastest growing in the fuel and energy sector.

The main condition for expanding economic reproduction in Russia is the efficiency of investment in real sectors of the economy. An increase in the volume of capital construction, as a rule, leads to an increase in demand for the purchase of construction materials, which increases the investment attractiveness of projects for the development of production at domestic enterprises of the construction industry.

The oil and gas industry consistently ranks first in the number of capital construction projects. The share of these objects in the total volume of conclusions prepared by the main state expertise of the RF at the end of 2017 was about 85%. The construction process in the oil and gas industry is characterized by a long implementation time, capital and material intensity. The products are individual in nature, as they are intended for a separate customer [4]. The average life of an oil and gas project is 25-35 years, which is due to the geological, technological and design features of the industry. In this regard, a well-founded, thorough evaluation of the project is of high importance.

In this regard, we can identify a number of problems and issues that need to be addressed immediately. The analysis of trends and identifying the dominant structural elements in the field of construction investment in the oil and gas industry are essential for further development of not only the industry but also the economy as a whole.

3. Research Questions

The greatest importance in considering issues related to the development of a feasibility study for investment in industrial facilities is the inability to determine the cost of construction at the pre-project stage using enlarged collections. Based on the data obtained in the course of the study, it is possible to form measures aimed at supplementing the existing budget and regulatory framework.

4. Purpose of the Study

The purpose of the scientific article is to determine the significance of the accuracy of the development of a feasibility study of investments in oil and gas facilities for the development of this industry, as well as to identify trends and directions that contribute to the development of this area, based on investment and innovative approaches. The results obtained will provide recommendations for improving the accuracy of calculating the cost of construction at the pre-project stage when developing a feasibility study for investments.

5. Research Methods

This study should present different points of view that give the concept and structure of the feasibility study of investment in construction.

Today, the main task of construction activities of oil and gas companies is to develop the most effective investment and construction projects. Each implemented investment project must be effective under any conditions of its implementation [5].

At all stages from development to implementation, a feasibility study of the decisions made is carried out, which differs in the methods of carrying out, due to the required degree of accuracy, resulting from the tasks facing the justification.

The main regulatory document regulating investment activities in the RF is Federal Law № 39-FL "On investment activities in the RF carried out in the form of capital investments".

In accordance with article 1 of this law, an investment project is understood as "Justification of the economic feasibility, volume and timing of capital investments, including the necessary project documentation developed in accordance with the legislation of the RF, as well as a description of practical actions for the implementation of investments (business plan)" [6].

Before implementing an investment and construction project, it is necessary to make sure (without forgetting the risks and uncertainties) that the benefits (usually expressed in monetary terms) resulting from the successful completion of this project and the achievement of its goals should exceed the costs of its implementation. The document confirming the usefulness of the investment and construction project is a feasibility study.

Thus, the feasibility study (feasibility study) is the main document that is being developed for investment projects in order to determine their effectiveness and feasibility of implementation [7].

The feasibility study of the project is one of the main project documents for the construction of facilities. On the basis of the feasibility study, approved in the prescribed manner, preparing tender documents and bidding of the contract for construction of the facility, a draft of the construction documentation, the agreement (contract) for the construction, offer financing.

When the feasibility study of investment and construction projects identifies the main technological, constructive and space-planning decisions; evaluates operational, sanitary-epidemiological and ecological safety of the project; determined the economic efficiency of the project and its social effect for companies [7].

It is impractical for oil and gas companies to implement a project if it is technically feasible but economically inefficient. Although, often, when evaluating the effectiveness of some projects, the efficiency obtained after the completion of the project (with a posteriori evaluation of efficiency) and the planned efficiency calculated at the pre-project stage may not coincide [8].

In practice, there is no universal approach to the feasibility study and its universal model covering industrial projects of any type, size or category. We use a wide range of recommendations for most investment projects.

In the "Guidelines for evaluating the effectiveness of investment projects and their selection for financing" approved by the ministry of economy of the RF on 21.06.1999, the approach to evaluating the effectiveness of investment projects is adapted to the market conditions of the economy. The main principles of evaluation are: multi-criteria evaluation, accounting for the impact of inflationary processes, bringing different-time costs and revenues to the conditions of their commensurability in terms of economic value to the initial period.

The feasibility study usually provides a project financing scheme, investment plan, and sales plan; describes production, organizational, and financial plans; and calculates coefficients that reflect the financial condition of the enterprise, as well as indicators that characterize the economic efficiency of the project [8].

The effectiveness of an investment project is a category that reflects the compliance of the project that generates this investment project with the goals and interests of its participants [9].

Setting requirements for the feasibility study helps to improve the quality of investment decisions, allows you to maximize the effectiveness of investments, and establish a system of evaluation criteria for preliminary investment decisions [7].

As of 01.01.2019, the production of oil and gas condensate (oil raw materials) on the territory of the RF was carried out by 290 organizations that have licenses for the right to use the subsoil, and 251 enterprises that produce natural and associated petroleum gas.

The main volume of national oil production (85.2% of the national figure) is still formed by the largest vertically integrated companies. At the same time, the share of independent oil companies is increasing, whose total contribution to the total oil production in the country reached 14,8% in 2018.

Based on the data from the website of the ministry of energy of the RF in picture 1 we consider the dynamics of oil and gas condensate production and natural and associated petroleum gas.

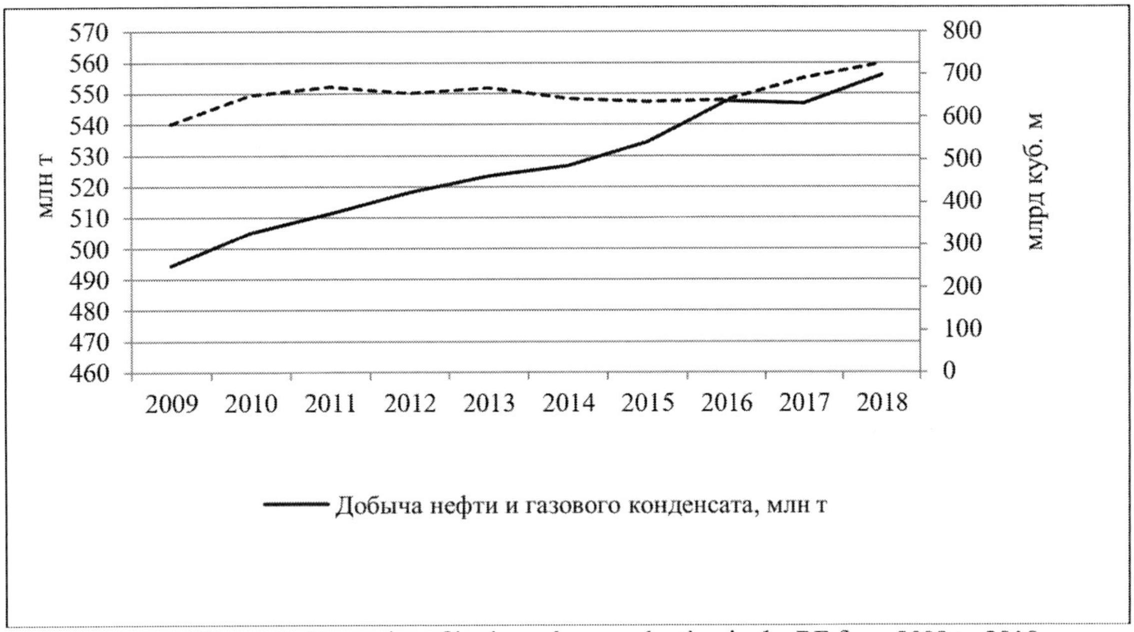

Figure 1. Dynamics of hydrocarbon production in the RF from 2009 to 2018

According to the results of 2018, the volume of national production of crude oil increased by 9,1 million tons compared to 2017 and amounted to 555,9 million tons in absolute terms. Of the resulting increase, more than half- 5 million tons- was obtained from new fields with a service life of up to 5 years.

In 2018, total gas production (natural and associated oil) in the RF increased by 5% and amounted to 34,3 billion m3 by 2017, reaching a new record level for the entire period of Russian gas production- 725.4 billion m3.

In the structure of russian gas production, the obvious trend in recent years is the decline in the share of Gazprom. If in 2009 the group's share in gas production in Russia was 80%, in 2018 it was 68%.

In 2017, the industry accounted for 9.3% of the country's GDP, 52.8% of exports, and 38.4% of revenues in the federal budget. In 2017, Russia exported 253 million tons of oil worth $ 93 billion.

The main part used in Russia natural gas consumed from the heat - and electricity- 37%, about 11% of the population, 9%- company FEC, 8%- household sector, a 6% metallurgy, 29%- other consumers. Most of russia's natural gas is consumed for heat production (29% in 2015), followed by the electricity sector (24%).

In comparison with other gas-producing countries, Russia is in a unique position, since it simultaneously meets the needs of a huge domestic market, as well as supplies volumes of gas that are unique for world practice for export.

The RF is firmly ranked second in the world in terms of oil exports. The leading Saudi Arabia's oil exports in 2017 were estimated at 360-370 million tons, while the "catching up" Iraq and Canada- at170-180 million tons.

Gas exports from Russia increased by 14 billion m3 in 2018 to 213 billion m3, including deliveries of pipeline gas abroad increased by 4,1%, and LNG by 70%.

Let's look at Figure 1 and 2 for the dynamics of hydrocarbon exports from Russia from 2012 to 2017. [10].

Figure 2. Dynamics of oil exports from Russia from 2012 to 2017

Figure 3. Dynamics of gas exports from Russia from 2012 to 2017

MEASURES TO INCREASE THE CALCULATION OF CONSTRUCTION COSTS OIL AND GAS FACILITIES AT THE PRE-PROJECT STAGE

Oil and gas companies are committed to developing new areas of activity, in particular oil and gas chemistry, power generation, and renewable energy. The stable growth of the share of renewable energy sources is an attractive fact for investing in this segment, which allows oil and gas companies to participate in the development of new approaches to energy production [11].

Each construction object is implemented on an individual project and is tied to a specific territory, so the means of labor and labor force are constantly moving from one object to another. The nature of the objects being constructed and the work being performed is very diverse, the cost of which is determined by a specific indicator-the estimated cost. The final result of construction is construction products that represent the main production assets put into operation [5].

If the investment project involves the construction of an object, then all stages of construction and installation work should be displayed in the investment plan of the feasibility study, designed to systematize information about the costs of the investment stage. The construction stages of an object can be performed either sequentially or in parallel. A GANT chart is usually used to visualize the project's network graph. The diagram shows the time dependencies of the construction stages and the relationships between them, which helps to plan the construction stages [8].

Also, when describing the project, it is necessary to plan possible costs for reengineering and redevelopment, maintenance costs and property management. Another additional condition for making an effective decision may be several variations of the same project. It is obvious that in this way, it is possible to optimize the decision-making process for the project implementation [9].

For a more in-depth and detailed analysis of the investment plan of the feasibility study, participants can study the project documentation of the object under construction in accordance with the decree of the government of the RF of 16.02.2008 № 87 "On the composition of sections of project documentation requirements for their design" [8].

However, the availability of all this information with sufficient accuracy is possible only when the project documentation (PD), including the construction organization project (COP) and the feasibility study are developed. This means that the design is already underway, and the collection of project information and design is estimated at 3-5% of CAPEX [9].

In the practice of oil and gas companies, for accounting purposes, real investments are divided into two groups- CAPEX and REVEX.

CAPEX represents capital investments in fixed assets, while REVEX represents non-capital expenditures for the current period that are included in the cost price. CAPEX investments can be represented as investments in the construction of wells, in the construction of production and transport infrastructure, and in the purchase of equipment.

Examples of REVEX include all exploration costs, except for the construction of wells, the cost of drilling exploration and evaluation wells that did not produce an industrial flow of oil, the cost of performing certain geological and technical activities, the cost of non-capitalized research work, the cost of maintaining roads, the cost of managing the implementation of a business project or investment program at an early stage.

Oil and gas companies consider investment areas by type of activity: exploration, production, processing, transport, sales and logistics. When analyzing the structure of real investments of oil and gas companies, it can be found that more than half of the funds are spent on exploration and production, and the main part of them is spent on the development of fields.

Thus, we understand that the investor needs a valuation method that would allow him to provide all the necessary data with sufficient accuracy even before the start of the design. In this regard, the following problems may occur at the pre-project stage:

It is quite difficult to determine the cost of construction of CAPEX facilities at the pre-project stage, when the project documentation, including the estimate documentation, has not been developed. In the method of determining the estimated cost of construction products in Russia is one of the methods of determining cost: based on aggregated estimate standards, including the bank data on the cost of previously built or designed analogues [12]. Currently, the budget and regulatory framework of the RF has developed a group of enlarged collections of construction price standards (CPS), which are recommended to be used to determine the estimated cost at the pre-project stage [13]. The calculation of the cost of aggregated indicators is usually used at the preliminary stages of design or at the stage of investment planning, when information about design solutions is not available in full, or to simplify the estimated calculations. The calculation of the cost of aggregated indicators is based on the premise that the necessary costs to obtain a unit of useful final effect in the same state of general economic conditions and the ratio of supply and demand in a particular market should be equal. That is, the unit costs per unit of useful effect (output, installed capacity, area, volume, etc.) on the market are determined not by the individual costs of each individual contractor, but by the average level of costs that occur at a given time [14].

Industrial objects are not included in the enlarged collections of the CPS. Then, for the purposes of preliminary calculation of the estimated cost of construction in the oil industry, you can only use a data bank on the cost of previously built or designed analogs. The website of the ministry of construction of the RF presents a register of cost-effective project documentation for reuse [15]. The above-mentioned register does not include industrial objects. Given all of the above, the developers of the feasibility study face a difficult task-to determine the preliminary cost of construction of objects, without having sufficient information for this in the existing budget and regulatory framework.

6. Findings

Taking into account the results obtained in the course of the study, it is possible to form the main measures for improving the calculation of the cost of construction at the pre-project stage as part of the feasibility study. To do this, you must:

1. Develop collections of consolidated construction price standards for production facilities at the Federal level.
2. To expand the range of existing collections of the CPS is to include an additional representative projects.
3. Include in the register of cost-effective project documentation for reuse not only residential and civil objects, but also industrial objects.

7. Conclusion

Investment is one of the foundations of social and economic development at the macroeconomic level. The feasibility study is a key document that reflects the effectiveness of project financing, and is a reference when deciding on the implementation of investment projects in construction.

Our proposed ways to improve the accuracy of construction cost calculations at the pre-project stage will allow us to determine the customer's investment costs in the framework of developing a feasibility study of investments. This in turn optimizes the volume of investment in the construction of oil and gas facilities.

References

[1] Karkhov V A 2015 *Investment Economy. Right. Management* **1** pp 1–5
[2] https://inecon.org/docs/2018/Voloshin_Sokolov_20181127.pdf [Electronic resource]
[3] https://www.gks.ru [Electronic resource]
[4] Nor Azmi Ahmad Bari, Rosnah Yusuff, Napsiah Ismail, Aini Jaapar, Rizan Ahmad 2012 *Procedia - Social and Behavioral Sciences* **35** pp 689-696
[5] Generalova N M, *Modern economy* 2017 **1** pp 16–18
[6] ConsultantPlus, *About investment activity in the Russian Federation carried out in the form of capital investments: Federal Law of the Russian Federation No. 39-FZ of february 25, 1999* (http://www.consultant.ru/document/cons_doc_LAW_22142/)
[7] Zharikov I S, Motunova M S 2017 *Science of the young* **4** pp 129–131
[8] Chugunov V I, Prokaeva I. G 2017 *Questions of economics* **27** pp 64–69
[9] Sherbakova N S 2016 *Economic and management issues* **5** pp 119–128
[10] https://minenergo.gov.ru/node/1209 [Electronic resource]
[11] ConsultantPlus, *About strategic planning in the Russian Federation: Federal Law of the Russian Federation No. 172-FZ of june 28, 2014* (http://www.consultant.ru/document/cons_doc_LAW_164841/)/
[12] ConsultantPlus, *Methods for determining the value of construction products in the territory of the Russian Federation, (MDS 81-35. 2004)* (http://www.consultant.ru/cons/cgi/online.cgi?req=doc&base=LAW&n=164458&fld=134&dst=1000000001,0&rnd=0.07496834998232427#024570607944855982)
[13] ConsultantPlus, *Methodological recommendations for the use of state construction budget standards - consolidated construction price standards various types of capital construction projects for non-production purposes and engineering infrastructure (MDS 81-02-12-2011)* (http://www.consultant.ru/cons/cgi/online.cgi?req=doc&base=LAW&n=124896&fld=134&dst=1000000001,0&rnd=0.08819234265125786#06795515104030775)

[14] Dedyukhina E S 2015 *Izvestiya Vuzov. Investments. Building. Real Estate.* **4(15)** pp 26-32
[15] https://www.minstroyrf.ru/trades/gradostroitelnaya-deyatelnost-i-arhitektura/17/ [Electronic resource]
[16] Skolubovich A Yu and Matveeva M V 2019 Development of a methodology for assessing the readiness of enterprises for public-private partnership in the field of public water supply *Proceedings of Universities. Investment. Construction. Real estate* **9** pp 274–283 DOI: 10.21285/2227-2917-2019-2-274-283
[17] Matveeva M 2018 Synchronization of the activity of participants during the implementation of housing construction projects *Proceedings of Universities. Investment. Construction. Real estate* **8** pp 31-41 DOI: 10.21285/2227-2917-2018-2-31-41.

IOP Conf. Series: Materials Science and Engineering **880** (2020) 012109 doi:10.1088/1757-899X/880/1/012109

Some aspects of risk management in the investment and construction sector

G I Khokhlova[1], J A Yushina[1] and V Y Burov[2]

[1] Irkutsk National Research Technical University, 83 Lermontov str., Irkutsk, 664074, Russia
[2] Transbaikal state University, 30 Alexandro-Zavodskaya str., Chita, 672039, Russia

E-mail: hgi08@yandex.ru

Abstract. Risk is an important factor that results from the execution of any economic decision and occurs in the conditions of uncertainty and cyclical development inherent in a market economy. Classification and assessment of risks specific to construction organizations is a complex and strategic task based on the application of various models, taking into account the industry's risk management features. The purpose of this work is to study the methodological and practical aspects of risk assessment for making management decisions in the investment and construction sector. The study focuses on such interrelated concepts as "uncertainty" and "risk" and their impact on business performance in the market conditions. The authors prove that the risks typical for the construction industry in Russia have their own specific features and depend on such significant factors as the level of taxes, the volume of orders, the cost of materials and structures, the availability of sufficient financial resources, and the payment capabilities of customers. Depending on the categories of construction companies, the most significant risks are identified (technological, commercial, investment and innovation risks). The authors studied the essence and features of risk assessment taking into account international standards and practical aspects in the investment and construction complex. The scientific article reveals the role of risk management in risk management in the investment and construction sector, the main problems of using methods to minimize risks for various categories of construction organizations in Russian practice.

1. Introduction

The investment and construction complex is "a socio-economic system in which industry enterprises in the region realize their economic interests by accumulating investment resources and participating in the process of creating and selling the final product of construction" [1].

Despite the important role of the investment and construction complex in solving the problems of economic growth of the state, we can note the contradictory trends that characterize its state in Russia in recent years, namely:

first, the number of construction organizations has recently increased by 80% (from 155 thousand to 278 thousand), while at the same time there is a reduction in the number of state and municipal construction organizations and an increase in private companies, small businesses account for one-fourth of the volume of construction work performed by organizations of private ownership;

secondly, after the decline in construction production in the period from 2014 to 2017, there was a tendency for their growth, in 2018, the volume of work performed by construction organizations is higher than the previous year by 5.3%;

third, there is an uneven volume of construction in the country's regions in 2018. in 34 regions, the volume of construction production increased several times, on the contrary, in 51 regions, construction production decreased, for example, in Mordovia (by 46.5%), Udmurtia (by 30.8%), Novgorod region (by 40.3%), Arkhangelsk (except for the Nenets Autonomous district) (by 40.2%) [2].

In addition, the socio-economic situation in Russia is characterized by an insufficiently high level of stability, and it is difficult to predict investment processes that are influenced by financial (7.25%), economic (4.6%), managerial (3.1%), criminal (2.8%), and social (1.9%) risks [3]. As a result, in the investment and construction sector, risk situations are created due to rising prices for the main types of materials and structures, an increase in accounts receivable, including overdue debts, an increase in debt on loans and loans received, etc.

In these conditions, there is a need to improve the quality of design and competitiveness of enterprises in the investment and construction sector, taking into account risks and uncertainties.

2. Materials and methods

The risks of economic development in a competitive environment and its impact on the business activity of economic entities were considered in the works of well-known scientists F.Night focused on the fact that the concepts of "uncertainty" and "risk" are not equivalent, risk is "the probability of a decision in the conditions of information uncertainty", and also analyzed the formation of business profits taking into account risk and uncertainty [4].

G. Markowitz proved the relationship between the yield of various assets and the risk, which increased the efficiency of managing the total portfolio through diversification, reducing its risk in the conditions of changing market conditions [5].The construction of the capital asset pricing Model, called the Capital Asset Pricing Model (CAPM), was based on the provisions of the Markowitz portfolio theory W.Sharp justified the assumption that in a competitive environment, the expected value of the risk premium will be directly proportional to the "beta" coefficient [6]. The concept of evaluating the effectiveness of investment projects by G.Birman and S.Schmidt is based on the application of the current value method adjusted for risk [7].

Each stage of the project life cycle corresponds to a specific focus of risk management (table 1)

Table 1. Stage of the life cycle of the project and the focus of risk management

The name of the stage	The focus of risk management
The concept and definition of the project	The strategic threats and opportunities of the project
Preliminary feasibility study of the project	Choosing a preferred project option based on risk analysis
The project's design and development	Analysis of the design and delivery strategy to determine the final application area
The installation an commissioning. Release, implementation and implementation of the project	Execution, testing and transfer project in accordance with the agreed scope of application
Operation and maintenance of the project	Analysis of the operating results and software maintenance ensuring the operation of the project

We believe that the most significant provisions of the standard in the field of risk analysis which may be important for the development of the measures to minimize the consequences of the risk situations are the results of qualitative and/or quantitative risk analysis at different stages of design, as well as data on the reliability and cost of the life cycle of project options and further design detail.

In this case, each organization can determine the magnitude of the consequences of risks using the most appropriate methods and criteria for it, in particular, in the scale of consequences of risk situations, such parameters as "people", "environment", "finances", "reputation of the organization" are accepted.

3. Results

Construction is one of the activities where increased risks are generated. Risk factors in the investment and construction sector are presented in figure 1.

.

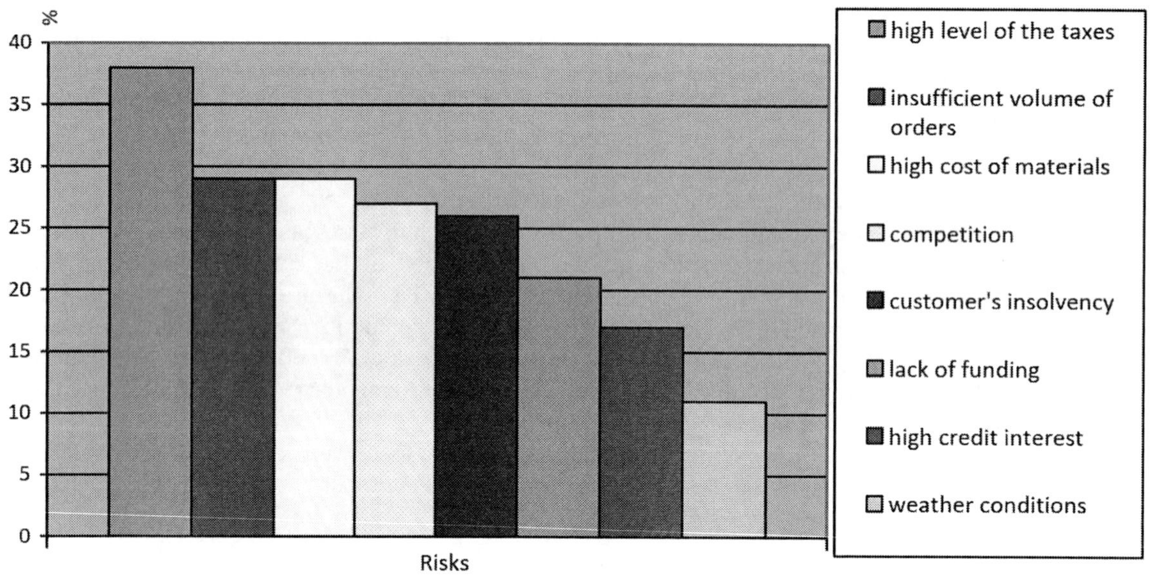

Figure 1. The risk factors specific to the construction activity in Russia

The analysis of statistical data showed that according to the estimates of construction companies managers, the most significant risk factors in Russia are: a high level of tax burden, insufficient volume of orders and high cost of materials and structures, insufficient amount of own resources to Finance projects and customers' insolvency [2].

According to experts, most of the profit losses of construction organizations (from 47 to 50%) are due to internal factors related to the risks of using labor resources, to a lesser extent (about 20%) with the organization and technology of construction, with the use of materials (4%), external factors account for 28% of the profit losses [10].

The construction companies can be divided into two main categories today: construction companies that build typical houses and cottages, the companies that build large, unique objects. For each of these construction companies, the risk situations have their own specifics and management features.

For the first group of the companies, the technological risks can be reduced to zero thanks to the flow method. Cottage construction, for example, is carried out using positively proven technologies. At the same time, the issues of commercial risks in the field of sales, deliveries and investment risks are very relevant. Risk management related to the external infrastructure of such construction companies is often replaced by diversification of activities. In practice, construction companies prefer to work with their own designers, suppliers, create divisions (firms) that are engaged in transportation, perform the functions of object protection.

However, investment in these activities does not allow you to develop the main (construction) business. This not only reduces investment opportunities, but also increases the risks of newly created firms.

For the second group of the companies which usually have stable financing (for example, budget financing), the technological and innovative risks prevail. Here, the commercial and investment risks are minimized but technological risks increase due to the uniqueness of construction projects. These risks are minimized by observing technology and safety practices, using years of experience accumulated since Soviet times, creating a contingency reserve that is traditionally included in estimates. Meanwhile, it is impossible to completely eliminate the possibility of major risks in the construction of complex and unique objects, and if a risk situation occurs, it may be difficult to cover the damage at the expense of budget funds. In this case, insurance of construction risks plays an important role in the insurance protection of property interests of organizations (insurance of property, personnel, liability of construction companies to third parties). For example, as a result of an accident during the construction of the Tsarev Garden business center (Moscow), it was necessary to repair the load-bearing structures again. The damage amounted to about $ 2 million and was paid by Ingosstrakh which provided comprehensive insurance for the construction of the center.

Currently, the rates for insurance of construction and installation works, depending on the characteristics of the object and the risks, vary from 0.12 to 1.15% of the insured amount. The average insurance rate in 2019 was 0.39%, the average insurance premium was 37.6 thousand rubles, and the average limit was close to 9.6 million rubles.

4. Finding

In practice, the risk assessment techniques based on expert assessment and a point system are usually used to make investment decisions. The algorithm of their application assumes the following sequence:

- the expert method determines the number of points for each risk group or for a specific risk;
- the overall risk assessment of the project is derived based on the weighting factors and the conclusion about the project's risk group is formulated;
- all parameters of the project adjusted with the adjustment for risk.

When assessing the risks, it is necessary to calculate not only the probability of risk, but also the amount of losses from its impact. For each risk indicators such as the "amount of possible losses" and "probability of occurrence" are determined in points which are entered in the corresponding rows and columns of the special matrix. Multiplying the received points (risk probability indicators (P) and losses from it (I)) gives the risk index or risk rank (Ir).

Scales for assessing the risks of investment projects can be developed by organizations themselves, taking into account the scope of the risks. At the same time, risk classification plays a key role in identifying and analyzing project risks.

5. Discussion

For risk analysis in construction, it is possible to base on the classification of factors that affect the innovative and investment activity of entrepreneurship which in the conditions of instability of the Russian economy determine: "organizational and managerial level, human resources, production and technical, financial and economic levels of enterprise development and scientific and technological capabilities of enterprises" [11].

As for the specifics of managing investment activities in construction, domestic experts attach paramount importance to "economic, market-related, and communication risks that the construction industry enterprises are exposed to a greater extent than other types of activities. An urgent task of strategic importance is the automation of the production management process, which takes into account not only profit and expenses, but also the avoidance of possible damage when evaluating the effectiveness of the process" [12,13,14].

In order to assess the effectiveness of investment projects taking into the account risks, attention should be paid to the classification which provides for the division of risks into such two types as:

- systematic (clean), these include factors such as the natural and geographical conditions, environmental, political and a number of commercial risks (property, production and trade);
- non-systematic (speculative) which are determined by management decisions [15].

Risk research in the context of investment project management is inextricably linked to the introduction of innovations [16]. At the same time, identification and management of such risks is subject to difficulties in quantifying them, so it is possible to conduct a qualitative assessment from the position of "minimum, acceptable and critical levels of risk" based on comparison of losses and income and / or the value of property, namely:

- the minimum risks are risks that result in losses that do not exceed the amount of profit;
- the acceptable risks are risks associated with losses within the amount of planned income;
- the critical risks are risks when a full or partial loss of invested funds is possible [17].

It should be noted that organizations perception of specific risks is of high importance at the project level for making decisions about ways to minimize them[17].We believe that the most significant types of risks for construction companies can be identified as production and technological, socio-economic, commercial, investment and innovation risks, as well as operational and organizational risks.

6. Conclusion

In the process of risk management in the investment and construction sector, it is necessary to take into account such methods of risk minimization as diversification (distribution of the total risk by combining with other participants), risk compensation (by creating mechanisms for hazard prevention) and insurance.

Insurance plays a key role in risk management and compensation in construction. However, the "stagnation" in the construction market associated with the instability of the Russian economy and the exclusion from the cost estimates of insurance costs for construction and installation works, significantly weakened the protection of the construction market subjects, reducing its investment attractiveness. In this regard, when managing risks in the investment and construction sector, it is necessary to take into account and improve the procedure for selecting contractors, optimize the cost and speed of work; improving their quality; develop information and telecommunication technologies in construction, green construction; improve working conditions and the quality of training of specialists; increase the level of construction safety.

References

[1] Asaul A N The directions for the development of regional investment and construction complexes in the Russian *Federation Successes in modern natural sciences* **2** pp 124-127
[2] Official site of Rosstat [electronic resource]. - Access mode: http:// www.gks.ru/storage/ mediabank/str6.xlsx Ver. From the screen
[3] Official site of the Russian regional investment resource [electronic resource]. - Access mode: https://raex-a.ru/ratings/regions/2018/att1Ver. From the screen
[4] Night F.H. 2003 *Risk, uncertainty and profit* (Moscow: Business) p 359
[5] Markowitz H 1952 *Portfolio Selection* https://assetallocation.ru/portfolio-selection/
[6] Sharp W F, Alexander G J and Bailey J V 2012 *The investments*. (Moscow: INFRA) p 1028
[7] Birman G and Schmidt S 2003 *Investment Economic analysis of investment projects* (Moscow: Unity-Dana) p 632

IOP Conf. Series: Materials Science and Engineering **880** (2020) 012109 doi:10.1088/1757-899X/880/1/012109

[8] National Standard of the Russian Federation GOST RF 2011 / ISO Guide 73: 2009 [electronic resource]. - Access mode: https://files.stroyinf.ru/ Data2/1/ 4293792/4293792106.pdf Ver. From the screen
[9] National standard of the Russian Federation GOST R IEC 62198-2015 Project management. Guidance on the application of risk management in design [electronic resource. - Access mode: https://docviewer.yandex.ru/ Ver. From the screen
[10] Abakumov R G 2014 *Reproduction management of fixed assets of the organization: theory, methodology, concept* (Belgorod: Publishing house of BSTU) p 506
[11] Khokhlova G I and Okladnikova D R 2014 Classification of factors as grounds for analysis of innovative–investment activity of small and medium businesses *Middle East Journal of Scientific Research* **8 (20)** pp 900–904
[12] Peshkov V V 2008 Potential of development as a basis for the formation of management strategy of investment and construction sphere *Bulletin of Irkutsk state technical University* **2(34)** pp 57-61
[13] Peshkova M V 2011 Assessment of investment preferences and investor competencies for the implementation of investment construction projects *Kazan Science* **9** pp 89–93
[14] Peshkov A V 2018 The economic efficiency of the use of intangible resources in investment and construction activities *University News. Investments. Construction. The property* doi: 10.21285 / 2227-2917-2018-4-44-51
[15] Balabanov I T 1996 *Risk management* (Moscow: Finance and statistics) p 192
[16] Kretova N V, Tsaregorodtseva E Y and Khohlova G I 2017 Applying advanced methods of evaluation of level of innovative development for industrial sector *Advances in Economics, Business and Management Research, Proceeding of the International Conference on Trends of Technologies and Innovations in Economic and Social Studies* **38** doi: org/10.2991/ttiess-17.2017.116
[17] Agafonova I P 2003 Building an effective risk management system at an enterprise during the implementation of an innovative project *Management in Russia and Abroad* **4** pp 106-107
[18] Vovseenko E A 2018 Financial Instruments and Hedging: New Rules of IFRS European *Proceedings of Social and Behavioural Sciences* **L** doi: https://dx.doi.org/10.15405/epsbs.2018.12.167

IOP Conf. Series: Materials Science and Engineering **880** (2020) 012110 doi:10.1088/1757-899X/880/1/012110

Determinants of the housing engineering entrepreneurship

A B Kogan[1,2]

[1]Economics and Finance of Development Department, Novosibirsk State University of Architecture and Civil Engineering (SIBSTRIN), 113, Leningradskaya, Novosibirsk 630008, Russia
[2]Innovations and Entrepreneurship Department, Novosibirsk State University of Economics and Management, 56, Kamenskaya, Novosibirsk 630099, Russia

E-mail: KogAnt@mail.ru.

Abstract. Current specifics of entrepreneurship in the field of apartment buildings construction (AB) and the impact of modern financing rules set on it, written in the relevant Federal Law are highlighted in the article. The purpose of the discussion is to develop the method foundations for analyzing the financial boundaries of housing (MFAFBH). The author's opinion is based on the results of monitoring facts and expert opinions, texts' analysis of normative acts, averaged requirements of banks for AB projects, methods of equivalent mathematical transformations, calculations according to special dynamic financial models of AB projects (MFPAB), principles of financial decision-making, laws of the primary market real estate (PRMRE), conjuncture of the Novosibirsk PRMRE. The fundamentals of MFAFBH were developed, the MFPAB was compiled for one of the typical apartment buildings in Novosibirsk, the boundary price of housing under construction was calculated, below this prize any AB project could not receive funding, i.e. construction would be impossible. The MFAFBH proposed by the author, can be applied to any region of the country, to forecast housing construction volumes on its basis, to develop and take measures, to manage this field of activity.

1. Introduction

Traditionally, there are planning volumes of Housing Engineering in Russia. Therefore, by the Presidium decision of the Presidential Council on Strategic Development and National Projects the National Project "Housing and Urban Environment" was approved, including the Federal Project "Housing" (dated by December 24, 2018). This document establishes (among other things) plans for the construction (commissioning) of housing in Russia in the period of 2019-2024. In general, it is planned to increase the commissioning of housing to 120 million square meters in 2024 (per year) from 79.2 million sq.m. in 2017 [3]. So, this market growth is planned at a rate of 6% per year (compared to the previous year).

From the standpoint of classical economic theory, such plans are somewhat paradoxical with the task, which is posed by the Regional Authorities that do not have leverage to influence the macroeconomic situation. Meanwhile, there may be a limitation of the residential real estate market development at the macro level. However, any plan is rather debatable. Nevertheless, this national project plays a role, influencing entrepreneurship in the investment and construction sector. Further, we will consider the modern determinants of entrepreneurship (factors that have a positive or negative impact on).

First, we should describe the terms and the dominant business model in the field of residential real estate development. The bulk of housing is created according to the tenement houses construction. In clause 6 of the provision "On recognition of premises as residential premises, residential premises of those, that are unsuitable for living, an apartment building, which are emergency and subject to demolition or reconstruction, a garden house of a residential one and a residential house of a garden" (approved by the Decree of the Government of the Russian Federation (dated by January 28, 2006 N 47) contains the following definition. "An apartment building is a combination of two or more apartments having independent exits either an adjoining territory to a residential building or to common premises in such a house." However, the majority of apartment buildings (AB), created in Russia, are multi-storey buildings consisting of tens of apartments and thousands of square meters.

Shared housing construction is practiced in Russia unlike many countries. It means that developers can sell an unbuilt property to the end customer. We have to clarify the terminology. "Developer" is an entity that owns a land plot for AB construction. At the construction stage, he can sign an agreement on participation in shared construction (SCC). The second side of such agreements is participants in shared construction (persons entering into agreements in order to obtain an apartment or other real estate after AB readiness). We emphasize: at the construction stage only the developer is entitled to conclude SCC, and there can be no other agreements. This is regulated by the basic norm in the field of apartment buildings creating, it's the Federal Law "On participation in shared construction of apartment buildings and other real estate and on amending certain legislative acts of the Russian Federation" dated 12.30.2004 No. 214-FZ (hereinafter - 214- Federal Law).

The shared construction rules are established at the Federal level and can be periodically changed. Normative innovations are mainly aimed at reducing risks for share holders. This is largely accomplished by tightening the regulation of a developer's activities, which entails his spends increase for the AB projects implementation. The first proposal, in such situation, is rising costs will push developers to increase prices. If this really happens, share holders will suffer losses. This leads to a dilemma: in order to increase the attractiveness of housing under construction, additional functions and guarantees (as a result of spending) of the developer are necessary, which entails a negative economic effect (price increase), reduces the attractiveness of housing under construction. For this reason, all legislative changes are faced with business objections.

2. Materials and Methods

Please follow these instructions as carefully as possible so all articles within a conference have the same style to the title page. This paragraph follows a section title so it should not be indented. Shared housing construction did not always give the desired result. There are quite a few unfinished AB constructions in Russia. As the result, there are "deceived share holders" - people who signed SCC, but did not receive either an apartment or money paid for it. The 214-FL included various mechanisms for protecting the investments of participants in shared construction. Initially, it was a bank guarantee, then an agreement with an insurance company. However, these tools did not exclude the appearance of problematic objects.

In order to eliminate this acute social problem, the amendments were made to the 214-FL and the Federal Law, dated July 29, 2017 N 218-FL "On a public law company which protects the rights of citizens, participating in shared construction in case of insolvency (bankruptcy) of developers and on making amendments was adopted into separate legislative acts of the Russian Federation." The law determined the procedure for the creation and operation of the "Fund for the Rights Protection of Citizens-Participants in Shared Construction", which completes 174 "problematic" AB, but there are significantly more such constructions in Russia (see table).

The Unified Housing Information System (UHIS) was created in Russia (source) in order to increase the transparency of this economy sector. In this work's author opinion, UHIS is a really useful resource. It includes, in particular, registers of the following objects and entities: distressed objects, housing under construction, developers, housing and construction cooperatives. Along with, there is a service for checking housing under construction (for compliance with 214-FL) and information on a programs number for housing preferential lending. This allows you to get market information to participants in shared construction and other entities involved in AB construction.

Table 1. Summary data of problem projects for AB creation

Status of the AB projects	Developers	Houses	Housing area, thousand square meters
Bankruptcy proceedings was started in accordance with Federal Law No. 127-FL	**486**	**1 699**	**10 141**
Another bankruptcy procedure was introduced in accordance with Federal Law No. 127-FL	**84**	**377**	**1 537**
Deadlines for construction completion exceeded for more than 6 months	**145**	**258**	**1 546**
Deadlines for shared construction commissioning exceeded for more than 6 months	**360**	**647**	**4 428**
Total	**1 053**	**2 981**	**17 651**

Note. Data were taken from a single registry of problematic objects, dated February 2020 [13].

In order to exclude the problem of "deceived share holders" it was proposed to carry out "a transition from housing construction using citizens' funds attracted to create apartment buildings and other real estate to bank lending and other forms of financing that minimize risk to citizens " in the National project "Housing and Urban Environment". Revisions were again made in 214-FL, which made it mandatory (from July 1, 2019) to open escrow accounts to store money from participants in shared construction paid on SCC. This national project contains the following target characteristics of the number and type of contracts (without escrow accounts and with escrow accounts, respectively): in 2019, 653,000 and 34,000; in 2021, 71,000 and 635,000; in 2024, 0 and 1 014 000. [3]

The current norms of 214-FL include the mandatory participation of a bank in a project in the event that the developer intends to sign SCC share holders. It should be noted that SCC are signed at the construction stage, other agreements of real estate purchase and sale (REPSA) will be signed after getting permission for AB commissioning. Now three entities are interacting: a participant in shared construction transfers funds according to REPSA, they come to the escrow account, but not to the developer. The developer will be able to receive this money only after presenting the bank with confirmation of the AB commissioning.

The Bank also performs another function. It carries out compulsory project lending to the developer. A loan is issued in the amount of the work cost, performed to build AB, based on certain accounting documents. It's rather important to note, that the money is transferred not to the developer account, but to the accounts of his contractors, who performed these works. At the same time, neither the bank nor the developer can refuse to participate in the project. In the case of a delay in AB commissioning and in a number of other cases, the bank returns the amount of money to the participant in the shared construction, he paid according to SCC. Escrow accounts are insured by the State Corporation "Deposit Insurance Agency" (DIA) in the amount of 10 million rubles. This means that in case of bankruptcy, the DIA will return the funds from escrow accounts within 10 million rubles to the participant in the shared construction.

The risks financing of the share holder is close to zero in this scheme. This is a way to earn 20-40 million rubles for the bank on each AB project with its specific risks. In this case, the bank must take part in a new building construction. The introduction of escrow accounts was led banks to play a role in the housing process. In this sense, it is quite reasonable to call "financial permission for construction."

Therefore, it turns out to be a big problem for developers not only to obtain a loan, but also to apply for it. One of the reasons is that the application includes a financial model of the project, which must be formed according to the requirements of the bank. The developer's employees simply do not have such

IOP Conf. Series: Materials Science and Engineering **880** (2020) 012110 doi:10.1088/1757-899X/880/1/012110

knowledge, and banks do not provide training. The big question is what will happen to those facilities which are refused to grant loans for? The concerns about a massive stoppage of housing construction are being expressed. This problem is described by the Heads of the Ministry of Construction of the Novosibirsk Region, developers and representatives of their associations [11].

Certainly, here is a discussion in scientific circles about the specifics of escrow accounts, as well as whether it makes sense to use escrow accounts in development or not. An analysis of this problem is proposed by T.I. Kubasova, N.G. Novikova, A.P. Sukhodolov [10] A.R. Kirsanov [4], N.Ya. Konyakhin [9], and in a number of other author's works. We should note that publications on escrow accounts appeared long before the described changes in 214-FL. The first article about this tool, contained in the National E- Library is the work of R.S. Churakov [1].

3. Results

The transition to escrow financing seems irreversible. It has positive aspects that will manifest themselves in the medium and long term. The fact is that some of the projects for AB construction were "torn" from the market and were implemented mainly through barter transactions. The problem of "deceived share holders" arose in many respects precisely on such objects. Escrow financing will eliminate this phenomenon and bring development to a higher business level.

Now, the developer has two ways to attract citizens' finances:

1) sell apartments after commissioning the building (concluding real estate purchase and sale agreements);

2) sign an arrangement at the stage of the building construction (participation contracts in shared construction using escrow accounts).

The first method requires the developer to make large investments of his own funds. The second method allows him to make less investment, but the developer has the above-described dependence on the bank. The first method of financing has never dominated in modern Russia. It is on the specifics of the second financing method, the volumes of commissioning housing and the investment and construction business depend. For brevity, we will call the second method "escrow financing".

Most developers implement projects using SCC. We describe an important property of prices and the sale dynamics of real estate under construction. As a rule, participants in shared construction are ready to sign SCC no earlier than the underground work completion of AB. The farther the sign SCC moment is from the expected moment of AB commissioning, the greater the discount, i.e. unit price of real estate is lower (price of one square meter of real estate sold). AB is weakening the negative impact of a long period of waiting for paid housing with an increase in the degree of preparedness, and unit prices of real estate increase.

Another factor contributing to the growth of real estate prices with the AB readiness is the reduction in the cost of shared construction participant to the transaction financing. We should clarify this situation. A shared construction participant can use either only his own funds, or combine them with credit funds. Consider the first case, the sooner a share holder signs SCC, the longer it will freeze funds. If he had the opportunity to complete a deal for the same apartment after AB commissioning, then this money could be put into the bank to receive interest. Hence the early execution of the transaction leads to this kind of lost profit; it is one of the reasons for reducing the unit price of real estate. In the second case, when a loan is used, the early SCC will be singed (as compared to the transaction close to AB commissioning) it leads to an increase in the interest paid by the share holder, which also influences on prices.

These price properties and transaction dynamics are the law of the primary real estate market, which can be formulated as follows. The unit price of real estate and the number of transactions are growing with an increase in AB readiness. Moreover, the typical dynamics of transactions is as follows: the most transactions (90%) occur before AB commissioning; apartments with the smallest area are sold earlier than others. Here, of course, there is a connection between prices and dynamics: if prices are too high, the number of transactions will decrease. The appropriate price level can be determined by the average discount for the market (in relation to the various stages of construction) in relation to the above financing costs. In this case, we are taking under consideration prices, cleared of inflation. The manifestation of this law in some cases can be veiled by random factors that will lead to one or another change in the ratio of prices and transaction dynamics.

The development business is also specific to its tax liabilities arising. Developers can use one of two taxation systems: general or simplified. At the same time, the company can choose one of two objects of taxation on a simplified system: "income" or "income, reduced by the amount of expenses". Often, a second taxable item is chosen because it provides lower tax payments.

The moment of tax liabilities occurrence depends on the type of contracts transactions are carried out. If SCC is used, then their revenues are targeted financing, i.e. do not fall under the tax in accordance with the Tax Code of the Russian Federation (on profit). Only after AB has been commissioned, so-called developer savings will be obtained (difference between the income from the SCC and the costs of AB Construction), which the tax obligations of the developer will be calculated from. For this reason, developers allocate payment to the developer's services in SCC. These amounts are subject to taxation upon receipt, and the availability of this money allows the developer to spend on promising tasks.

Now, we should consider the concerns of escrow financing will impede the implementation of the National Project "Housing and a Comfortable Urban Environment". Implementation of development plans is possible only with the simultaneous presence of the following conditions:

1) the urban planning situation allows developers to obtain permits for the planned volume of construction;

2) the characteristics of the projects allow them to obtain the consent of banks to escrow financing;

3) the characteristics of developers allow them to obtain the consent of banks to escrow financing or there are developers who can construct AB without escrow financing;

4) there is solvent demand (the population has income sufficient to buy an apartment without a loan or with a loan).

Developers are private companies that act at their own risk in order to make a profit. This means they are implementing projects that "promise" to be effective. The principles and problems of project evaluation are described by Vlasova O.N., Vorbochuk V.A., Doroshenko T.G., Dymchenko V.A., Kazimirov I.A., Kudryavtseva V.A., Matveeva M.V., Naumov I. .V., Nikishina O.V., Peshkov A.V., Peshkov V.V., Pinchuk T.O., Polyakova V.N., Ryutin V.S., Semeykina N.M., Yamshikova I.V. [2].

There is a well-known claim that the price of real estate is determined mainly by three factors: location, location, location.Any city has price zones. The market price is set and cannot be changed administratively. On the other hand, the costs of AB construction are also given. The costs are minimal and cannot be changed for economy class housing. Hence, if projects "promise" the profitability required by developers, then they can be implemented.

In the current situation the cost of AB construction increases by several percent due to the need to pay interest on this loan when a home loan is required during the housing sale at the construction stage. However, this is where the difficulty arises. The fact is the loan rate varies depending on the ratio of the loan amount to the escrow accounts amounts. The second difficulty is that projects are described by dynamic financial models [5-8, 12], i.e. (in the end) cash flows that change over time.

4. Conclusions

Obviously, concerning the impact of escrow financing on construction volumes, we can not calculate financial models for all projects, and even for the sample ones. We must have a formula that would allow us to draw a conclusion (on the minimum data set), the minimum prices should be (for a given cost and financial requirements of the bank) so that the implementation of AB project was possible. So, we can draw the opposite conclusions on maximum cost of construction should be in order to make AB project be possible.

In order to turn a dynamic financial model into a function, we must, firstly, set two dynamic parameters: the price index of the sold space and the schedule for the space's sale. Secondly, we must describe the rules of lending. Banks have established similar rules for making decisions on financing or on refusal to finance development projects. In general, two rules have the greatest impact: investing at least 15% of the OPR by the developer and the compliance of the project with the Loan Life Coverage Ratio (*LLCR*) requirement for debt coverage ratio.

$$LLCR = \frac{SCC}{\ell + BI}$$

where *SCC* is income according to participation contracts in shared housing construction, rubles;
L is loan amount is issued by the bank to the developer, rub.;
BI is bank interest accrued on a loan to a developer, rub.

The author of this work performed relevant studies. A formula is derived for determining the threshold value of the price (for a given cost, price indices for the sold space, schedule for the sale of space, the share of the minimum contribution of the builder's own funds to AB construction, loan rates, LLCR) from the LLCR formula, by a series of equivalent mathematical transformations. Calculations for Novosibirsk show that, the threshold price is 53.3 thousand rubles. per sq.m. at a cost of 41.6 thousand rubles per sq.m, the developer does not make sense to implement the project, and the bank does not make sense to finance projects if the prize is lower. This means that, AB cannot be constructed in Novosibirsk in those areas where prices are lower than the specified value. An exception is the AB with the construction cost which is less than that one above. So, it is possible to forecast the volume of AB commissioning in any city by combining these financial restrictions with a map of land plots, according which building permits have been issued.

References

[1] Churakov R S 2007 Escrow account in Russian law *Law* **8** pp 27-34
[2] Examination of investment and construction projects and programs: theory and practice: monograph Ed by V V Peshkov2017 Irkutsk National Research Technical University. Irkutsk, IRNITU P 370
[3] Information materials on the National Project "Housing and Urban Environment" URL: http://static.government.ru/media/files/fPs8qW8PjyNAAKRWDTk7Ac9KdUhBcETX.pdf.
[4] Kirsanov A R 2018 Property relations in the Russian Federation **2 (197)** pp 64-75
[5] Kogan A 2014 The criticism of net present value and equivalent annual cost *Journal of Advanced Research in Law and Economics* **5** pp 15-22
[6] Kogan A B 2015 Cost benefit analysis of investment projects, based on inter-industry models *Journal of Advanced Research in Law and Economics* **1** pp 112-120
[7] Kogan A B 2017 A method to determine the best way to finance development projects *Financial analytics: problems and solutions* **11 (341)** pp 1257-1272
[8] Kogan A B, Chaetskii A A 2019 Analyzing the efficiency of development projects funded with the use of escrow accounts *Economic analysis: theory and practice* **8 (491)** pp 1464-1477
[9] Konyakhin N Ya 2019 Difference of an escrow account from a letter of credit. *Accounting*. **No 1** pp 134-138.
[10] Kubasova T I 2017 Baikal Research Journal **4** DOI: 10.17150 / 2411-6262.8 (4) .23.
[11] Shestakov M 2019 the Minister gave a forecast for housing commissioning. URL: https://nsk.rbc.ru/nsk/28/11/2019/5ddfa20f9a7947b702e5a0f1?from=regional_newsfeed.
[12] Trynov A V 2016 Public-Private Investment Partnerships: Efficiency Estimation Methods *Economy of Region* 12 **(2)** pp 602-612.
[13] Unified Housing Information System. URL: https://xn--80az8a.xn--d1aqf.xn--p1ai/
[14] Yaskova N Yu 2018 Construction in the context of new views on the goals of economic development. *Proceedings of Universities. Investment. Construction. Real estate* **8** pp 133-142 DOI: 10.21285/2227-2917- 2018-1-133-142
[15] Peshkov A V 2019 Remote mortgage lending as a resource for housing development *Investments. Construction. Real Estate: New Technologies and Targeted Development Priorities IOP Conference Series: Materials Science and Engineering* **667** 012073

IOP Conf. Series: Materials Science and Engineering **880** (2020) 012111 doi:10.1088/1757-899X/880/1/012111

Development of a risk assessment methodology for the implementation of investment projects of a construction organization

G Z Nizamova[1], M M Gayfullina[1], D R Musina[1], A Yu Tumanova[1] and A A Battalova[1]

[1]Ufa State Petroleum Technological University, 450064, Kosmonavtov str. 1,Ufa, Russia

E-mail: Gulya182004@list.ru

Abstract. The article considers the risks of implementing an investment project in the construction sector. The purpose of the study is to develop a methodological approach to assessing the risks of implementing investment projects of a construction organization. The object of the study is an investment project of a construction company. The subject of the study is the risks associated with the implementation of the investment project of a construction organization. Research methodology-methods of logical analysis, grouping and comparison, generalization were used. Using the method of expert assessment, an integral indicator of risk assessment is proposed. The main results and conclusions of the study: a risk management algorithm is proposed, a system of external and internal risks for the implementation of investment projects in the construction sector is formed within the algorithm, which allows to identify market and investment risks of the project. A method for assessing the risks of investment projects in the construction sector, based on expert assessment, is proposed. It allows to quantify and rank the impact of risks on the implementation of an investment project of a construction organization, as well as to calculate the integral level of risk for the implementation of investment projects of a construction organization.

1. Introduction

Implementation of investment construction projects is a business with a high level of a risk. In these conditions, the problem of assessing the risks of investment projects and their reduction is enough for construction organizations. The reliability of construction companies is easily disrupted by external and internal factors. Therefore, taking into account the risk factors of a construction organization in the economic assessment of an investment project is an urgent task.

Problems of the theory and practice of assessment of an investment projects are considered in the works of foreign scientists (G. Birman, G. Schmidt, J. Brigham, L. Gapenski, R. Braille, J. Friedman, Nick. Oruduay) and Russian scientists (I.A. Blank, S.A. Vladimirov, N.V. Matveev, E.M. Chetyrkin)

In Russia, methodological and theoretical issues of optimizing the risks of investment projects received considerable attention in the works of P.L. Vilenskiy, M.V. Gracheva, A.G. Gryaznova, O.A. Zemlyanskiy, V.N. Livshitsa, M.A. Limitovskiy, N.A. Lysova, A.M. Margolin, I.A. Nikonova, E.R. Orlova, V.M. Rutgaizer S.A. Sirotin, S.A. Smolyaka, T.V. Teplova, M.A. Fedotova, Prodanova N A, Plaskova N S and Khamkhoeva F A [13], Vincent P and Schaaf T [17].

These works mainly deal with risk assessment issues and their impact on indicators of economic efficiency of investment projects.

Methods of risk analysis and management are presented in the works of Gharedaghi G and Omidvari M [8], Vilela M, Oluyemi G and Petrovski A [16], Kowalski J and Polonski M [11], Zelenakova M, Labant S, Zvijakova L [19], Tengyuan C, Xiaopeng D and Bon-Gang H [14], Jelena A M, Jiayuan W, Patrick Z. [10].

Valuation of investment projects and the impact of risks on cost are considered in the works of foreign scientists A. Damodaran, A. Gregory, T. Koller, T. Copeland, B. Myers, K. Mercer, D. Murrini, S. Pratt, P. Samuelson, L. Thomas, D. Fishman, W. Sharp, Philippe V, Schaaf T., Kroemer S. [12].

Among domestic researchers who study the pressing problems of risk assessment we can mention the works of such authors as Berval A V and Romanova A I [1], Barikaev E N and Eriashvili N D [2], Korovin E, Smirnova I and Sultygova and etc. [9]. The works of this author cover the problems of accounting and managing production risks in detail and deeply.

The specifics of the risks of investment projects in the construction sector also did not remain without attention of researchers. For example, in work [18] studies on the risks of modular integrated construction are systematized. The authors analyzed research in this area from 1992 to 2019. Analysis reveals that the research publications on risks of modular integrated construction witnessed a steady growth, with considerable progress occurring in the last decade. Existing empirical studies have focused heavily on perceived implementation risks, supply chain risks, schedule risks, investment risks, structural risks, ergonomic risks, and modular integrated construction risk management strategies, which indicate that modular integrated construction is associated with a host of risk events. Research gaps in existing studies are highlighted in this research, and areas for further studies are then proposed.

Dixit S, Sharma K and Singh S. [5] analyzed the key risk factors of construction projects.

In the work of Yu W, Chang H and Cheng S [15] a new method to tackle the duration risks of a construction project is proposed. The authors propose a quantitative model considering the seven risk-levels of activity duration is proposed. Авторский подход к оценке рисков в строительной сфере представлен также в работе Chang, H., Yu, W., Cheng, S. et al. [4].

Despite the existing developments of domestic and foreign scientists, existing approaches to assessing the risks of implementing investment projects do not take into account to a large extent the industry features of the functioning of construction enterprises, which is unacceptable from the point of view of studying the nature and essence of risks. Another significant drawback of existing approaches to assessing the risks of implementing investment projects is that most risk assessment methods do not take into account the dynamically changing economic conditions of the modern economy. Moreover, an increase in risk factors requires continuous improvement of the assessment methodology and their management mechanisms.

2. Methods

To assess the risks of implementing investment projects of the company, a number of steps will be required to ensure the scientific validity and methodological sequence of actions and decisions. We have proposed a method for assessing the risks of implementing investment projects of a construction organization, based on calculating the integral level of risk. An expert method of calculation is used by scoring the probability of risk occurrence and the expected value of losses in the implementation of this risk. The main stages of the risk management algorithm for the implementation of investment projects of a construction organization are as follows.

1 Analysis of the external and internal environment. The environment of an investment construction project consists of numerous variables: market conditions, various organizations, global risks, political and legal environment, competitors, suppliers, consumers. Also, during this stage, the data of the project itself are studied and analyzed in detail, including the contract agreement, design and estimate documentation; construction organization project; work execution plan; data on the experience of implementing previous projects, etc.

2 Investment analysis of the project, the calculation of the main indicators of the effectiveness of the investment project. The main performance indicators of the investment project are: Net Present Value (NPV), Internal Rate of Return (IRR), Profitability index (PI), Discounted Payback Period (DPP).

3 Identification of the risks of implementing an investment project. The purpose of this stage is to obtain a description of the risks of implementing an investment construction project.

The risks of construction are considered in detail in the works of researchers. In particular, Dixit S, Sharma K and Singh S. [5] analyzed the key risk factors in construction projects.

We propose to use the classification of risks by the sphere of occurrence. Taking this into account, the following risk groups can be identified:

1 External (market) risks in relation to the project:

- risk of economic deterioration (R1);
- risk of depreciation of the national currency (R2);
- risk of rising inflation (R3);
- risk of reduced availability and increased cost of borrowed funds to finance investment projects (R4);
- risk of reducing the solvency of demand for construction products (R5);
- risk of strengthening legislation in the construction sector (R6).

2 Internal (investment) risks are the risks of the project itself:

- risk of unscrupulous contractors (R7);
- risks of location of the construction site (R8);
- risk of reducing the planned profit for the project (R9);
- risk of failure to meet construction deadlines (R10);
- risk of increasing the volume of work and exceeding the project financing budget (R11);
- risk of lack of necessary technical resources (including construction equipment) (R12);
- risk of non-compliance with construction decisions (R13);
- risk of reducing the quality of construction (R14);
- -- the risk of violation of construction safety (R15);
- risk of lack of qualified personnel for the implementation of the construction project (R16).

4 Risk assessment of the implementation of the investment project. To assess the risks of implementing investment projects of a construction organization, we propose to calculate the integral level of risk. The authors' developments reflected in the results of previous studies are taken as the basis [3; 6; 7].

It is proposed to evaluate the integral risk level using an expert calculation method by scoring the probability of occurrence of risk and the expected value of losses in the implementation of this risk.

The sequence of calculation of the integral level of risk for the implementation of an investment project of a construction organization is as follows.

Estimates of the probability of occurrence of the i-th risk are made. Proposed scale for assessing the probability of occurrence of the i-th risk is given in table 1.

Table 1. The scale for assessing the probability of occurrence of the i-th risk.

Type of risk	Probability of occurrence of risk	Characteristics of probability	Number of points (Pi)
Weakly probable	$0 < P \leq 0,1$	Event can occur in exceptional cases.	Pi=1
Low probable	$0,1 < P \leq 0,4$	A rare event, but, as known, has already taken place.	Pi=2

Probable	0,4 < P ≤ 0,6	Sufficient evidence to suggest the possibility of an event.	Pi=3
Highly probable	0,6 < P ≤ 0,9	Event can occur.	Pi=4
Almost possible	0,9 < P < 1,0	Event is expected to happen.	Pi=5

Estimates of the amount of losses when the i-th risk occurs are made:
Wi=1 – if possible losses are estimated as minimal;
Wi=2 – low;
Wi=3 – medium;
Wi=4 – high;
Wi=5 – maximum.
Then the i-th risk index is calculated. The index of the i-th risk:

$$R_i = P_i * W_i, \qquad (1)$$

where Pi -the probability of occurrence of the i-th risk;
Wi – the amount of losses when the i-th risk occurs.

Next, the risk level is determined and the risks are ranked according to the degree of their acceptability for the construction organization. The risk level characterizes the degree of its acceptance for the construction organization performing the contract, and shows the place of each risk in the queue for processing. Based on the value of the risk index, we can determine the level of the i-th risk. Three levels of risk are suggested:
1 ≤Ri≤ 4 – acceptable risk;
5 ≤Ri≤ 10 –- justified risk;
12 ≤Ri≤ 25 – unacceptable risk.

Далее рассчитывается интегральный уровень риска при реализации инвестиционного проекта:

$$IR = \frac{\sum_{i=1}^{n} R_i}{n*25}, \qquad (2)$$

where Ri - the i-th risk index;
n -the number of risks;
25 - maximum risk index score on a 5-point scale, taking into account the probability of loss (5*5=25).

Next, the risk level is determined in accordance with the rating scale. The proposed scale for assessing the integral level of risk in the implementation of an investment project of a construction organization is shown in table 2.

Table 2. Scale for assessing the amount of loss when the i-th risk occurs.

IR	Risk	Risk level
0,04 ≤IR≤ 0,19	Ignored risk	Acceptable risk
0,20 ≤IR≤ 0,39	Insignificant risk	Justified risk
0,40 ≤IR≤ 0,59	Moderate risk	Justified risk
0,60 ≤IR≤ 0,79	Significant risk	Unacceptable risk
0,80 ≤IR≤ 1,00	Critical risk	Unacceptable risk

The integral level of risk, expressed as a percentage, is the possible amount of loss when risks occur. Thus, the losses of the construction organization during the manifestation of risks will amount to:

$$П = IR * S, \quad (3)$$

where S- the cost of construction work.

Based on the results of calculations, management decisions on risks are developed and made.

In accordance with the proposed risk management algorithm for the implementation of investment projects of a construction organization.

5 Processing of investment project implementation risks. The goal of this stage is to reduce the risk impact to an acceptable level.

Currently, in the practice of risk management there are four main ways of their processing: 1) risk mitigation; 2) risk acceptance; 3) risk deviation; and 4) risk transfer. It is recommended to choose the method of processing and managing risk in the following order: acceptance, mitigation, transfer, deviation, for justified risk-mitigation, transfer, acceptance and deviation, for unacceptable risk-mitigation, deviation, transfer, acceptance. To assess the effectiveness of the measures taken, the indicators of the relevant risks are compared before and after the implementation of the risk management program. In this case, the costs of implementing the proposed solutions should be taken into account. An effective solution is considered if the cost of implementing it does not exceed the possible damage.

6 Control and monitoring of the risks of implementing the investment project.

To take into account the impact of risks on indicators of economic efficiency of an investment project of a construction organization, it is proposed to introduce risk into the discount rate, i.e. use the method of adjusting the discount rate. The method for adjusting for the risk of the discount coefficient is as follows:

- the risk-free rate of return Emin is set-the minimum real rate of return (risk-free rate, this is the coupon-free yield of government securities);
- the forecast inflation rate is set;
- the risk premium is determined.

The b-method (beta method) was used to move from the integral risk level to the risk premium. The general standards for the b-coefficient value are shown in table 3.

Table 3. General standards for b - coefficient.

b-coefficient	Security dynamics
0	The income from the security does not depend on the risk (risk-free security – Treasury bill)
0,5	A security reacts at half its value to changes in the risk of securities
1,0	A security fully reacts to a change in the securities market (medium risk)
2,0	A security reacts doubly to changes in the securities market

The formula for moving from the integral risk level to the risk premium included in the discount rate calculation is as follows:

$$r = 2*IR, \quad (3)$$

where IR -the integral level of risk, fraction of units;

2 - maximum risk level (beta coefficient).

Discount rate E:

$$E = Emin + I + r, \quad (4)$$

Where E is the discount rate;

Emin - the minimum real rate of return (risk-free rate, this is the coupon-free yield of government securities);

I - the rate of inflation;

r-the risk premium.

3. Results

The proposed approach was tested on the example of an investment project for the construction of a multifunctional hotel complex. An expert assessment of the probability of risks and the amount of losses is given in table 4.

Table 4. Expert assessment of the significance of the risks of implementing the construction of a multifunctional hotel complex

Risk	Pi	Wi	Ri	Risk	Pi	Wi	Ri
R1	3	3	9	**R9**	3	5	15
R2	3	1	3	**R10**	4	3	12
R3	3	3	9	**R11**	4	5	20
R4	3	3	9	**R12**	2	3	6
R5	4	4	16	**R13**	2	5	10
R6	4	4	16	**R14**	3	3	9
R7	3	3	9	**R15**	3	3	9
R8	2	2	4	**R16**	3	3	9

The integral risk level of the project is equal to:

$$IR = \frac{\sum_{i=1}^{n} R_i}{n*25} = \frac{165}{16*25} = \frac{165}{400} = 0,4125 \quad (5)$$

In accordance with the risk assessment scale, the project is characterized by moderate risk. The risk premium included in the calculation of the discount rate for this risk value will be:: $r = 2*0,4125 = 0,825$.

The discount rate will be: $E = Emin + I + r =$ =6,55+3,4+0,825=10,78%.

Risks fall within the acceptable risk zone: R2, R8.

Risks fall into the zone of justified risk: R1, R3, R4, R7, R12, R14, R15, R16.

Risks fall into the zone of unacceptable risk: R5, R6, R9, R10, R1.

To transfer risks to the acceptable zone, it is necessary to minimize them. Risk minimization will be achieved by performing a number of measures for each risk separately. The risk index after these measures will be estimated at 84 points. The integral risk level of the project after the events will be 0.21. The risk premium included in the discount rate calculation for this risk value will be 0.42. .

The discount rate will be:

$E = Emin + I + r =$ 6,55+3,4+0,42=10,37%.

A comparison of how the project indicators will change without taking into account and taking into account the minimization of risks is given in table 5.

Table 5. Comparison of project indicators

Indicator	Excluding risk management measures	Taking into account risk management measures
IR	0,4125	0,2100
r, %	0,825	0,420
E, %	10,78	10,370
NPV, thousand rubles	317 578	332 305
IRR, %	14,2	14,7
PI	2,30	2,36
DPP, years	5,34	5,29

Thus, risk assessment and accounting can significantly improve the indicators of efficiency of construction projects. Using the proposed approach to assessing and accounting for risks in construction projects can encourage investors to take measures to reduce them.

4. Conclusion

1 Risk management algorithm has been developed that can be used to manage the risks of a real investment project, both at the stage of its design, and to serve as a basis for further development of management decisions made during the implementation of the project.

2 The classification of risks for the implementation of investment projects of a construction organization by the sphere of occurrence has been developed. With this in mind, the following risk groups are identified: external (market) risks in relation to the project and internal (investment) risks, i.e. the risks of the project itself.

3 A methodology has been developed for assessing the risks of implementing investment projects of a construction organization, based on the calculation of the integral risk level using an expert calculation method by scoring the probability of occurrence of risk and the expected value of losses in the realization of this risk. It is recommended to choose the method of processing and managing risk in the framework of the proposed method in the following order: acceptance, mitigation, transfer, deviation; for justified risk-mitigation, transfer, acceptance and deviation; for unacceptable risk-mitigation, deviation, transfer, acceptance.

4 A method of correction to the risk of the discount coefficient which is taken into account when calculating the effectiveness of an investment project is proposed. At the same time, it is proposed to use the b-method (beta method) to move from the integral risk level to the risk premium in investment projects included in the discount rate calculation.

5 Expert assessment of the probability of risk occurrence and magnitudes of losses have shown that the greatest influence on the effectiveness of the investment project construction of a multifunctional hotel complex have the following risks: the risk of increasing the scope of work and an excess of the project financing budget, risk of reduction of solvency of demand for construction products; the risk of increased legislation in the sphere of construction (related to escrow accounts); the risk of reduction in the planned profit on the project; the risk of a delay in construction; the risk of non-compliance with construction decisions.

6 Measures for transferring risks to an acceptable risk zone were proposed. By implementing the proposed measures, it is possible to prevent or reduce the probability of a risky event occurring as a result of the implementation of the investment project for the construction of a multifunctional hotel complex. According to the results of the calculations, the project risks will be reduced by almost 50%, the net present value of the project will increase by 14.7 million rubles or 4.6%, the internal rate of return will increase by 0.42 percentage points, amounting to 14.7%.

References

[1] Barikaev E N and Eriashvili N D 2015 Risk Management in the System of Economic Security: Theoretical Aspect (Moscow, Unity)

[2] Berval A V and Romanova A I 2015 Risk Management of the Main Participants in the Share Housing Market (Moscow, Infra-M)

[3] Burenina I V, Evtushenko E V, Kotov D V, Battalova A A, Gaifullina M M and Gamilova D A 2018 Integral Assessment of the Development of Russia's Chemical Industry *Journal of Environmental Management and Tourism* **8 (5)** pp 1075-108

[4] Chang H, Yu W, Cheng S. et al. 2019 The Use of a Multiple Risk Level Model to Tackle the Duration of Risk for Construction Activity. KSCE J Civ Eng **23** pp 2397–2408

[5] Dixit S, Sharma K and Singh S. 2020 Identifying and analysing key factors associated with risks in construction projects. In: Babu K., Rao H., Amarnath Y. (eds) *Emerging Trends in Civil Engineering. Lecture Notes in Civil Engineering* **61** Springer, Singapore pp 25-32

[6] Gaifullina M M, Nizamova G Z, Musina D R and Alexandrova O A 2017 Formation of strategy of effective management of fixed production assets of oil company *Advances in Economics, Business and Management Research* **38** pp 185-190
[7] Gaifullina M M, Nizamova G Z and Makov V M 2018 Forming a strategy for effective management of human resources of an oil company *Oil industry* **4** pp 8-11
[8] Gharedaghi G and Omidvari M 2019 A pattern of contractor selection for oil and gas industries in a safety approach using ANP-DEMATEL in a Grey environment *2019 International journal of occupational safety and ergonomics* **25 (4)** pp 510-523
[9] Korovin E, Smirnova I and Sultygova M 2020 Method development to support the enterprise economic security by using mathematical modeling tools 2020 *International transaction journal of engineering management & applied sciences & technologies* **11 (2)**
[10] Jelena A M, Jiayuan W and Patrick Z 2019 Fuzzy Logic-Based Method for Risk Assessment of Belt and Road Infrastructure Projects 2019 *Journal of construction engineering and management* **145 (12)**
[11] Kowalski J and Polonski M 2019 The method of planning the implementation of railway investments including risk *Archives of civil engineering* **65 (3)** pp 45-59
[12] Kroemer S 2019 Model risk regarding monthly wind energy production for the valuation of a wind farm investment *International journal of energy sector management V* 13 n 4 pp 862-884
[13] Prodanova N A, Plaskova N S and Khamkhoeva F A 2019 Modern tools for assessing the investment attractiveness of a commercial organization *International journal of energy sector management* **8 (24)** pp 145-151
[14] Tengyuan C, Xiaopeng D and Bon-Gang H 2019 Improving Quantitative Assessment of Political Risk in International Construction Projects: The Case of Chinese Construction Companies *Journal of construction engineering and management* **145 (12)**
[15] Yu W, Chang H and Cheng S 2018 A new method to tackle the duration risks of a construction project *8th International Conference on Engineering, Project, and Product Management (EPPM 2017). EPPM 2017. Lecture Notes in Mechanical Engineering*. pp 255-264
[16] Vilela M, Oluyemi G and Petrovski A 2018 Fuzzy logic applied to value of information assessment in oil and gas projects *Petroleum Science* **16 (5)** pp 1208-1220
[17] Vincent P and Schaaf T 2019 Reservoir and Economic-Uncertainties Assessment for Recovery-Strategy Selection Using Stochastic Decision Trees *SPE Europec Conference at the 80th EAGE Conference and Exhibition* **22 (4)** pp 1575-1592
[18] Wuni I Y, Shen G Q and Hwang B 2019 Risks of modular integrated construction: A review and future research directions *Front. Eng. Manag*
[19] Zelenakova M, Labant S and Zvijakova L Methodology for environmental assessment of proposed activity using risk analysis 2020 *Environmental Impact Assessment review* **80**

IOP Conf. Series: Materials Science and Engineering **880** (2020) 012112 doi:10.1088/1757-899X/880/1/012112

The model of ecologically oriented development of managerial technologies in construction sphere

I P Nuzhina[1,2] and L A Kaverzina[3]

[1]Tomsk State University of Architecture and Building, 634003, Solyanaya sq., 2, Tomsk, Russia
[2]Tomsk State University of Control Systems and Radioelectronics, 634050, Lenina pr., 40, Tomsk, Russia
[3]Bratsk State University, 665709, Makarenko str., 40, Bratsk, Russia

E-mail: irinanuzhina@yandex.ru

Abstract. The purpose of research is creation of concept model of ecologically oriented development of managerial technologies in construction sphere based on greening theory. The meaning of greening was shown as modern development of economics. In addition, principles of greening management in construction sphere were offered and discussed. Development directions of managerial technologies were reviewed through selecting of organization forms of interaction of project participants, methods of realization, regulators of building activity, instruments of ecological and economic efficiency of projects and building manufacture and also development of ecological competence of specialists. It was shown that adaptation of current system of building management to conditions of ecological oriented development is possible on base of institutionalization of social and economic responsibility of building business.

1. Introduction

The priority of modern Russia is providing of economic and social conditions of human's potential development. The most significant factors are: satisfaction of ecological needs of society and formation of comfortable environment [1]. The ecological conditions are directly related to level of greening. According to Yale University's research in the field of the state of ecology and natural resources management, Switzerland was recognized as the world leader in environmental efficiency in 2018. The list of 10 leaders include: France, Denmark, Malta, Sweden, the UK, Luxembourg, Austria, Ireland. Finland moved from 1st place in 2016 to 10th in 2018. Rating of Russia dropped from 32nd in 2016 to 52nd from 180 countries in 2018 [2].

Meanwhile in 90s of the xx century, Russia was actively involved in global process of solving environmental problems. From this period to present time, along with ratification of international protocols, acceptance of strategic documents for ecological improvement, significant additions have been made to federal laws, programs for environmental development have been developed and implemented just like regulatory acts and national standards which are governing to the transition to the principles of the best available technologies and also technologies of "green building" are actively developing. Since 2010 in Russia, key documents confirming the imperative of providing the energy efficiency and quality of environment during the building and exploitation processes. It is necessary to admit that increase of ecological efficiency of economic development can be provided only with greening of managerial technologies. In the context of management, greening means actualization of ecological orientation of the whole complex of methods and instruments, which are used to regulation of building activity and development of organization forms of participants of investment and construction activities and stakeholders to approve economic interests with social and ecological priorities of development.

The concept of "greening" has firmly entered the scientific vocabulary at the end of XX century and has meant a more systematic approach to the objective world and " bigger awareness of the role of nature in human's life" (Ramers, 1990). Works of Bobylev (2006, 2018), Matveev (2000), Ilyicheva (2006), Feraru(2007), Litovko, Pavlov, Fedorov (1998) studied the nature and content of the greening of economic development. [3-9]. The problems of development of building activity with ecological aspects are considered in works of Plotnikova (2008), Krygina (2015), Telichenko, Benzh (2016) [10-12]. Greening of economics has become an important direction of development. However, at the same time, the processes of managerial greening in urban development are not systematized and the system of relations between subjects of investment and building activities is not related to modern tendencies of the development of organizational forms of interaction between stakeholders of the project with social and ecological responsibility of business.

2. Materials and research methods

The theoretical and methodological foundation of research is based on states of the theory of economic development greening [3-9], institutional theory [13-14], works of native and foreign scientists and specialists in the field of problems of investment and construction activity regulation taking into account the ecological factors, environmental and economic support to building activities. [10-12, 15-17].

The modern progress of urban development in Russia is carried out under the conditions of dominant influence of the projects aimed at increase of housing security, safety and comfort of humans' environment. In the period from 2018 to 2024 the national project named "Housing and urban environment" is being implemented. Aims of the project are closely interconnected with ecological conditions of life. Nowadays, customers of construction production are increasingly focused on environmental conditions, presence of social and infrastructure facilities. The results of WCIOM polls evidence to this [18]. Therefore, 84% of respondents consider playgrounds to be a mandatory part of modern house's infrastructure, 71% - parking lots, 66% - grocery stores. Potential customers of residential properties are interested not only in economic and operational characteristics, but also in socio-psychological and socio-economic conditions of territory development.

The greening of economics has become an imperative of modern society development. The necessity of development of ecologically oriented model for the development of all kinds of economic activity, including the building one is caused by following this imperative. The analysis of institutional environment has showed that a lot of attention is payed to the development and realization of activities related to the technological part of ecologically oriented evolution of construction activities. Primarily, these are technologies of "green" building and architectural and urban planning of functional and spatial structure of construction object; resource-saving, energy-efficient and ecological technologies of building process, environmentally friendly systems of engineering support, utilization technologies. The priority of technological events for greening is well founded. Really, about 35% of energetic resources consumed in Russia are spent on heating the industrial, residential and public buildings. From 60 to 85 % of energetic balance is filled with costs of providing the comfortable conditions for people in room – heating and hot water. The effective development of building is impossible without adaptation the construction managing system to conditions of ecologically oriented model. Greening should not be limited to the system of events for greening the technological processes and equipment. Certainly, these funds can reduce the level of environment pollution in particular factory, but at the same time, it is necessary to green up managerial technologies in construction sphere to provide stable territory development. Greening management is a «type of management fundamentally oriented on formation and development of ecological manufacture, culture and human activity » [8. P. 32].

The methodological principles of forming the model of ecologically oriented development of managerial technologies in construction sphere are as follows:

1. Ecosystem approach. Making management decisions in the sphere of construction business must meet the requirements of biosphere compatibility, be based on a qualitative and quantitative assessment of relationship of construction activities, conditions of human living and environment generally. The practical implementation of this principle means rating of possible realization effects of planned construction projects on environmental and social conditions. Consequences manifest in a change in the state of environment, the natural resource base of the territory, socio-economic socio-psychological and cultural-historical environment.

2. Organization of ecological support for construction activity taking into account the ecological life cycle of building production. The boundaries of responsibility of project participants for the ecological consequences of the building activity must cover «all points of contact» with the environment arising in the process and connection with the creation and exploitation of construction facilities. Therefore, the objects of ecologically oriented management should be business processes from mining and processing of natural resources for production the building materials, products and structures, construction and installation works, to the disposal of the building materials and structures during reconstruction and demolition of facilities. The management system on every stage will have their own features associated with uncertainty of duration of the ecological life cycle. This duration is the uncertainty in time in space and can be multivariate. In some cases, it may coincide with traditional life cycle of building products (when there is no residual damage after liquidation of construction facility), and in other cases may go far beyond it (if there are no measures were taken for disposal and elimination of ecological consequences in form of residual damage with the liquidation of construction site).

3. Concordance of institutional environment of regulation of the building activity to the needs of ecologically oriented development of economics. The institutional environment integrates institutions and originations, methods, tools, organization forms of interaction between participants of the building activity. Institutional transformations are necessary for implementation of the ecological priorities in management. In substance, we are talking about formation the mechanism of the ecological and economic regulation.

4. The imperative of socio-ecological responsibility in construction business is an initiative activity of building companies aimed at adoption of additional requirements to ongoing projects, sites and construction works as well as compliance with mandatory ecological standards [19]. Building organizations go beyond the regulatory framework and initiate acceptance of increased social and ecological requirements, which must be met by the production activity of enterprises and its results.

5. Coordination of interests of authorities, business and society in the process of justification and adoption of investment decisions. The main aim of the coordination is an overcoming contradiction between environmental as well as social requirements and commercial interests of customers and investors. At the present time, various methods of citizen participation in decision-making on implementation of investment projects are actively developing. The technologies of organization of the participation may be different, but all of them mean availability of information that reflects the influence of building facility on environment and society.

6. Greening of needs is the primary factor, which determines the development dynamics of managerial technologies and also means demand of ecological building products by the market. Customers must be ready to pay for engineering decisions that provide comfort, environmental friendliness and resource saving just like for comfortable arrangement of external architectural and urban living environment. Only in this case, the initiatives will be supported not only with image, but with financial results of activity too. Society opinion polls (WCIOM) revealed that 53% of Russians identified the ecological state of the region as determining factor in deciding on the purchase of housing. The results of research [20] have showed that 62% of respondents are ready to choose more expensive ecological housing.

3. Results and discussion

The architecture of the development model of managerial technologies in urban planning reflects interconnection between ways and forms of its manifestation according to the principles of greening management (Figure 1). Success in solving of strategic objectives is largely determined by choosing the effective form of organization of project stakeholders' interconnection. The integration of stakeholders' actions in urban planning will allow minimizing risks of investors and customers, that are caused by discrepancy of the result with expectations of society. The forms of integration are different: platforms, associations, movements. Technologies of involving residents, public organizations, authorities and local communities in the decision-making process for urban development as a part of national projects are becoming increasingly popular in Russia nowadays. These processes provide collaborative planning, which technologies got huge development abroad. [21]. The collaborative planning may be the real constructive technology of management decision-making process, based on building the communication between stakeholders. Moreover, the communication building supposed to be on early stages of pre-planning researches, and the decision-making process is accompanied by distribution of responsibility and resources in general interests. The defining condition of implementing the managerial technology transformation is consolidation of socio-ecological responsibility as a paradigm of modern development of urban business that will require the transformation of the institutional environment of urban regulation activity.

4. Conclusion

Among the most important factors determining trends in the development of urban planning in Russia, it should be attributed: social priorities of the territory development; changing conditions and orientation on the efficiency criteria of resource consumption, greening the technical and technological development of construction processes and systems; greening the needs of society and relevance of ecological conditions to improve life quality; improvement of integration processes (intersectoral, intercountry, interdisciplinary). Changes in the system of building management at different levels of economics are necessary to solve the problems of ecologically oriented development of urban planning. Institutionalization of socio-ecological responsibility of construction business is a driver, which provides adaptation of management system to conditions of environmentally oriented model. In context of this research, institutionalization regarded as a formalization process of interconnection between urban planning subjects. There are socially and ecologically oriented actions become ordered into a system of sustainable structures and relations. Moreover, recognition of social and ecological priorities in the process of development and implementation of projects of construction buildings, facilities, development of the territories as an absolute imperative.

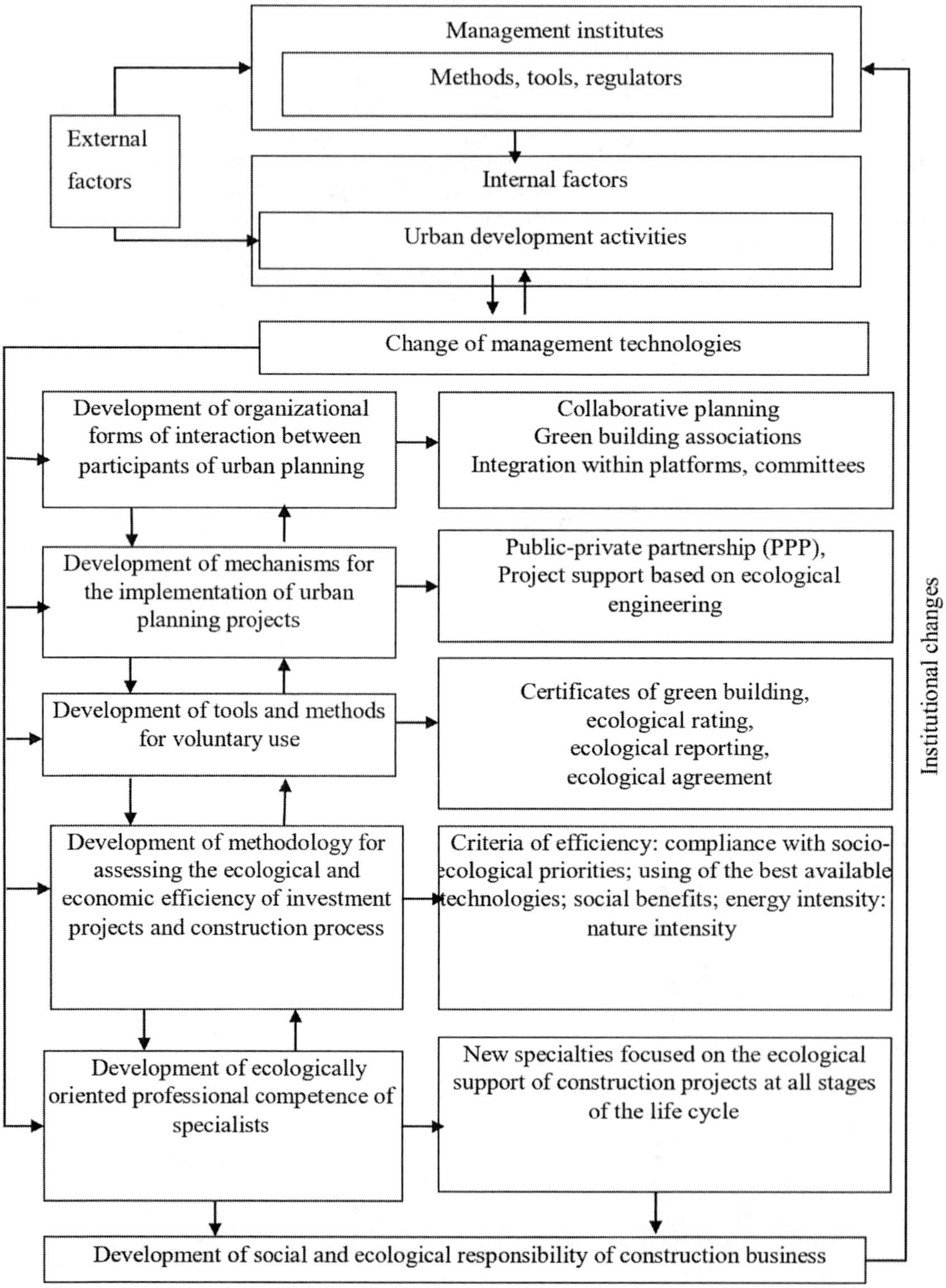

Figure 1. Model of ecologically oriented development of managerial technologies in construction sphere.

IOP Conf. Series: Materials Science and Engineering **880** (2020) 012112 doi:10.1088/1757-899X/880/1/012112

References

[1] Buzyrev V V, Nuzhina I P and Zolotareva M V 2018 Social-ecological priorities of town-planning activities on the urbanized territories ESCI 2018 Ho Chi Minh City 01007

[2] The Environmental Performance Index 2018 available at: https://gtmarket.ru/ratings/environmental-performance-index/info

[3] Reymers N F 1990 *Nature management: Manual-dictionary (*Moscow: Mysl) p 637

[4] Bobylev S N 2006 Russian economic reforms and the formation of "anti-sustainable" development *Nature management economics* **4** pp 31-41

[5] Bobylev S N, Revich B A 2018 Environmentalization of Economics and Public Health *Environmental Law* 5 pp 46-51

[6] Matveev M V 2000 Ecology, biotechnology, profit *Nature management economics* **4** pp 2-94

[7] Iljicheva E V 2006 *Ecologization and improvement of the accounting process in the conditions of transition to sustainable development* (Stary Oskol, TNT) 440

[8] Feraru G S 2007 Constructing of a strategic, ecologically oriented enterprise management *Nature management economics* **3** pp 31-50

[9] Litovka O P, Pavlov K V and Fedorov M. M. 1998 *Theory and practice questions of Nature management economics ecologization* (Saint-Petersburg — Izhevsk: Publishing house of the Economics and Management Institute of UdmSU) p.136

[10] Plotnikova L V 2008 *Environmental management of urban environment quality in highly urbanized areas, Ecology of a megacity* (Moscow ACV) p 239

[11] Krygina A M 2015 Resource, energy saving and environmental friendliness of construction as a basis for innovative sustainable development of housing real estate *Housing construction* **6** pp 57 -59

[12] Telichenko V I, Benuzh A A 2016 Current State and Development of Technical Regulation of the Green Technologies *Academia. Architecture and Building* **1** pp 118-121

[13] North D 1997 Институциональные изменения: рамки анализа *Questions of Economics* **3** pp 6-17

[14] Asaul N A 2005 *Institucional'noe vzaimodejstvie subjektov investicionno-stroitel'nogo kompleksa* Institutional interaction of the subjects of the investment and construction complex (Gumanistika Saint Petersburg) p 280

[15] Ilyichev V, Kolchunov V, Emelyanov S and Bakaeva N 2015 About the Dynamic Model Formation of the Urban Livelihood System Compatible with the Biosphere *Applied Mechanics* **V 725-726** pp 1224-1230

[16] Tetior A N 2008 Sustainable construction, sustainable architecture, creation of eco-cities are the most important tasks for Russia *Environmental engineering* **2** pp. 57-67

[17] Nuzhina I P, Kaverzina L A and Shershova L V 2019 Regulation of urban planning activities as a condition of socially and environmentally oriented development of territories *Issues of Social - Economic Development of Siberia* **4(38)** pp 45-53

[18] *Home, sweet home: dreams and reality 2017* available at: https://wciom.ru/index.php?id=236&uid=3457 3439

[19] Nuzhina I P, Zolotareva M V and Vasil'eva J.V. 2017 Institutionalization of social and ecological responsibility of construction business *Modern Management Technology* **11(83)** pp 1-10

[20] Nuzhina I P, Zolotareva M V and Vasil'eva J.V. 2016 Features of ecologization of needs of the population for construction products in the urbanized territories *Fundamental research* **12-2** pp 447-452

[21] Henry Sanoff 2015 *Souchastvuyushhee proektirovanie* Practices of public participation in shaping the environment of large and small cities(Vologda) p 170

IOP Conf. Series: Materials Science and Engineering **880** (2020) 012113 doi:10.1088/1757-899X/880/1/012113

Construction company administration of intangible resources in the face of uncertainty

A V Peshkov

[1] Irkutsk National Research Technical University, 83 Lermontov str., Irkutsk, 664074, Russia

E-mail: expertiza@istu.edu

Abstract. Management of intangible resources is an important component of a construction company's business processes. The rapid development of information technology forces construction companies to change their usual activities and rebuild business processes in accordance with the challenges of the real market. It is digital technologies that are the object of the company's intangible assets that can be used much more effectively to improve the current processes and information flows in the construction company. The advantages of the construction company's intangible assets, such as Smart spaces, are that they provide increased efficiency and lower production costs, achieved through rational actions, the use of faster and more secure communication methods that guarantee the necessary flow of information. They also improve efficiency through rational and timely actions, with an emphasis on creating value and using new technologies. The economic effect of using information technologies, namely the introduction of Smart spaces, can be either direct, material in the form of achieved cost reduction and increased productivity of operations, or non-material, in the form of implemented patents and licenses, and improving the company's image, as well as the potential of information technologies that can be used to make more effective decisions of a strategic nature.

At the present stage of business development in a knowledge-based economy, the share of intangible resources in the market capitalization of enterprises, and, consequently, their role in the production of goods and services, is steadily growing. In the knowledge economy along with financial resources are an important source of competitiveness of the organization are such "intangible" resources, such as: the ability of staff to successfully perform tasks, clever use of information environment of the enterprise, as well as the presence of organizational culture. The transition of the Russian economy to market-based forms of management required construction companies to radically revise the basic provisions of the concept of managing intangible resources. First of all, this is due to the fact that elements of unpredictability of changes in the economic environment of enterprises that occur at a rapid pace both at the microeconomic and macro-economic level of the developing market have appeared and are becoming more and more pronounced. In this regard, there is an objective need to develop effective methods of managing intangible resources of a construction company in the conditions of instability and uncertainty caused by poorly predictable changes in the modern market.

Real-time management of intangible resources is a complex scientific and technical problem, the practical solution of which will allow companies to manage the efficiency of business processes by quickly responding to events and eliminating the shortage or downtime of resources. Various software packages are used by construction companies to solve problems of managing intangible resources.

However, with the increasing complexity and dimension of these tasks, as well as due to the growing uncertainty in the forecasts of changes in demand and supply in the market, the use of classical methods and tools in practice meets significant difficulties, largely due to the situational nature of decisions on resource management and the requirement to take into account the individual characteristics of construction companies.

In these situations, it is necessary to develop standard models and methods for solving various problems of situational management of intangible resources of a construction company, for which no unified approaches have been created so far, as well as to build a software platform and technology that could be applied to create a wide range of intelligent resource management systems in real time and would reduce the labor intensity, time and cost of construction projects.

A construction company is a complex socio-economic object that closely interacts with a dynamic market environment during its operation. The modern Russian market, which determines the nature of the functioning and development of a construction company, can be characterized by instability, high risks associated with a priori unpredictable changes in the market environment, for example, due to unfair competition in the construction industry. Various factors of the market environment, changing, significantly affect the efficiency of the construction company, affecting its internal economic environment. Therefore, to effectively manage the production and economic activities of a construction company, it is necessary to know the causes of risk factors, the nature of their manifestation and the nature of their impact on production processes. If there are limited intangible resources, it leads to a constant shortage of information about the behavior of the market environment and reduces the effectiveness of the management object. The second reason why the management system is forced to function in conditions of uncertainty is the dynamism and a priori unpredictability of changes occurring in the market.

Implementation of methods for managing the construction company's intangible resources, based on extrapolation of the construction company's behavior in conditions of uncertainty, represents the formation of an effective mechanism for "proactive regulation" of the company's business processes. In this case, we are talking about planning and conducting proactive management measures in accordance with the forecast estimates of changes in the situations of the external and internal environment of the construction company, obtained through the use of the company's intangible resources.

Intangible resources are one of the most important components of the activities of construction companies in an unstable economic environment. The definition of a strategy for the use of intangible resources in a construction company depends fundamentally on the specific situation in which it is located and the planned prospects for its development. However, when designing a strategy for the use of intangible resources, it is necessary to proceed from the fact that there are common approaches to their formulation and some common frameworks that fit different strategies.

A significant influence on the choice of methodological support for the management of intangible resources of a construction company has a factor such as the dynamics of the surrounding economic environment. Depending on the speed of processes occurring in the external environment and the nature of their changes, environments can be slow and fast changing situations. For environments with a slow change of situations, the strategy of using the company's intangible resources in the external environment can be quite stable and adjusted if necessary in the event of changes in the internal environment of its functioning. With a rapid change in the external environment, the construction company faces uncertainty in the conditions of its behavior, on the basis of this, the management of intangible resources should be adaptive, that is, change in accordance with changes in the economic environment. At the same time, the following elements should be provided and implemented to form an adaptive strategy for managing the construction company's intangible resources:

- formation of NMR characteristics required for the implementation of strategic goals and key business processes of the enterprise
- assessment of the current state of the NMR as the degree of compliance of the actual characteristics with the required

- identification of possible solutions for managing NMR
- quantitative assessment of the effectiveness of control actions and implementation of the most effective ones
- verification of the effectiveness of decisions taken to manage NMR

In order to assess the current state of the enterprise's NMR, it is necessary to determine the strategic readiness of intangible resources. Strategic readiness indicates the degree of compliance of the actual state of intangible resources with the target characteristics and ensure the following:

- when allocating available or incoming funds, the construction firm should leave bumper stocks in case of unforeseen situations
- the main indicators that determine the technical and economic condition of a construction company should be set by interval estimates, whose boundary values are determined by the minimum allowed and maximum possible values.
- the organization of the structure of the construction company should be based on a matrix type, allowing the formation and dissolution, as necessary, of problem-oriented functional divisions
- availability of a powerful environmental monitoring system that regularly collects information about changes in the environment and determines trends in its development on this basis
- integration of feedbacks into various contours of the construction company's strategic management
- ensuring the possibility of changing the tactics of behavior in the market in the event of unforeseen short-term spontaneous changes at the expense of existing reserves.

All the above measures for managing intangible resources should be effectively organized using Internet technologies. Currently, the technological development of construction companies is accompanied by a sharp increase in the demand for information and the expansion of the use of information technologies. This is facilitated by the rapid development of information systems and new areas of information production, such as "Big Data", "Internet of things", "cloud technologies", "digital fields" (DMS) and other advanced technologies. Together, they can create a serious potential for improving the performance of construction companies.

Construction organizations differ significantly in the degree of integration of Internet technologies in their production and management practices. Some organizations have revised their approaches to the implementation of only certain management functions, such as marketing and personnel management, based on the active use of Internet technologies, while other construction companies have redesigned the main and supporting business processes based on Internet platforms.

Experts predict that the greatest success is possible when using real-time management methods and technologies, scenario modeling tools, and decision support in conditions of risk and uncertainty. In this context, Smart technologies are becoming a powerful tool for improving efficiency, since they allow you to design a single space for all holders of business processes of an investment and construction project, which allows you to provide remote control in the "online" mode, and also improves the quality of management decisions, which significantly affects the reduction of production costs.

Based on the analysis of the Smart technologies market, the key methods and areas of application of information technologies in the processes of evaluating the effectiveness of business processes of a construction company were identified (table 1).

Table 1. Methods and areas of application of information technologies in the processes of evaluating the effectiveness of business processes of a construction company

Investment-oriented Smart technologies	Cost-oriented Smart technologies	Smart technologies reflecting the factors of outer space	Smart technologies for quantification
1. Costs-benefits 2. Uncertainty of the future	1. Economic value added 2. Source of	1. Method balanced scorecard 2. Consumer index	1. Real Options Valuations 2. App

3. Diagnostics of business risks, management risks, and technological risks 4. Best practice 5. The total economic contribution 6. The definition of flexibility of costs and profits	economic value 3. Total costs of own production	3. Information economy 4. The performance of IT	3. Applied Information Economics

The effectiveness of the above methods can be more fully realized when numerous factors that determine the logic of the construction company's business processes are coordinated. In an ideal environment, a Smart space should be created in which all participants in the construction industry will realize the sale of goods and services, control over the fulfillment of obligations (Table 2).

Table 2. The basic characteristics of the Smart-space

Characteristic	Definition
1) Operational management of contractors	Relevant information of work process allows to correct rapidly the contractors' actions on the construction side
2) Simplified interaction with the regulator	a single trusted information environment allows banks and the government to get relevant data of work with a project without visits
3) Increasing of shareholder's loyalty	Access to verified information of dynamic of construction simplifies interaction with cost-sharing arrangement participants for shareholders
4) Reduction of administrative costs	Analytical reports and interactive panel of projects provide monitoring of building processes and reduce time gaps between execution and payment

The main benefits derived from the use of a single Smart-space in a construction company can be classified according to the categories presented in Table 3.

Table 3. Benefits of Smart-space

Category	Definiton	Example
Cost minimization	Complete rejection of a specific activity or process	Reallocation of resources with less added cost for more than effective activities
Avoidance of costs	Detection and correction errors that do not cause losses provided for in the budget, and which can lead to growth that is irreversible costs	Early correction of errors and the design flaws
Increased productivity and efficiency	Increasing productivity, that improves the use of existing resources	Evaluation processes, limiting the rational use of production capacity, or process adjustment during the production process
Benefits of an intangible nature	Improving business operations and the ability to control, protect, and expand the company's assets.	Implementation of solutions for managing the company's knowledge and intangible assets

The results shown in the table show that the greatest benefits from the use of Smart technologies are reflected in terms of economy and technology particularly increased productivity, improved process efficiency and increased profitability. Increasing productivity by using of Smart technologies will allow you to standardize processes and systems used throughout the enterprise, which optimizes the work of both management personnel and production. In turn, saving time by automating processes such as data collection, processing, and reproduction optimizes decision-making processes by obtaining more reliable information in a timely manner.

Based on the conducted research and analysis of the literature it can be concluded that the introduction of Smart technologies in the business processes of construction companies will reduce transaction costs by improving efficiency, increase revenues, by increasing the pace of construction, reduce overhead costs and provide for asset management and business processes of investment and construction project in real-time, based on the use of the latest information allowing them to make the most accurate, optimal and cost-effective solutions.

References

[1] Matveeva M 2018 Synchronization of the activity of participants during the implementation of housing construction projects *Proceedings of Universities. Investment. Construction. Real estate* **8** pp 31-41 DOI: 10.21285/2227-2917-2018-2-31-41.

[2] Matveeva M V, Doroshenko T G and Sukhanova I A 2019 On the issue of cost and timing of contract tenders for industrial and civil construction *Investments. Construction. Real Estate: New Technologies and Targeted Development Priorities IOP Conference Series: Materials Science and Engineering* **667** 012063

[3] Kalyuzhnova N, Violin S and Matveeva M 2019 Smart Specialization of Russian Regions: Prospects and Limitations *Investments. Construction. Real Estate: New Technologies and Targeted Development Priorities IOP Conference Series: Materials Science and Engineering* **667** 012036

[4] Skolubovich A Yu and Matveeva M V 2019 Development of a methodology for assessing the readiness of enterprises for public-private partnership in the field of public water supply *Proceedings of Universities. Investment. Construction. Real estate* **9** pp 274–283 DOI: 10.21285/2227-2917-2019-2-274-283

[5] Chupin V R and Maizel D I 2020 Aggregation and decomposition in routing oil and gas pipeline systems *Proceedings of Universities. Investment. Construction. Real estate* **10(1)** pp 108–115https://doi.org/10.21285/2227-2917-2020-1-108-115

[6] Biglova A A 2018 State of the public procurement system in Russia *Science, education, innovation: approbation of research results Materials of the International (correspondence) scientific and practical conference* pp 153-156

[7] Gaskova Yu.I. "Integrative approach to the formation of a mechanism for the coordinated development of industry and infrastructure in the region" // Proceedings of the II International scientific and practical conference "Economics and modern management: theory, methodology and practice". - Penza: 2018. - p.358-362.

[8] Komarov A K, Ivanov I A and Lundenbazar B 2019 Theory and practice of the use of gabions for forming protective structures *Proceedings of Universities. Investment. Construction. Real estate* **9(1)** pp 78–89 DOI: 10.21285/2227-2917-2019-1-78-89

[9] Peshkov V V and Gamayunova E V 2018 Specific features of investment support for the repair and reconstruction of buildings and facilities *Proceedings of Universities. Investment. Construction. Real estate* **8** pp 56-72 DOI: 10.21285/2227-2917-2018-2-56-72

[10] Yaskova N Yu and Sarchenko V I 2018Development of strategic approaches to the creation of aqua-territorial industrial complexes in the Arctic zone *Proceedings of Universities. Investment. Construction. Real estate* **8** pp. 84-93 DOI: 10.21285/2227-2917-2018-2-84-93

[11] Kudryavtseva V A, Zaslavsky A M and Larina O P 2018 Multicriterial comparison at the

inclusion of investment projects of manufacture of building materials into the state program of the region (at the example of Irkutsk region) *Proceedings of Universities. Investment. Construction. Real estate* **8** pp 60–72 DOI: 10.21285/2227-2917-2018-1-60-72
[12] Gertsekovich D, Gorbachevskaya L, Grigorova L and Peshkov V 2019 Return on investment in REIT real estate funds *Investments. Construction. Real Estate: New Technologies and Targeted Development Priorities IOP Conference Series: Materials Science and Engineering* **667** 012063
[13] Matveeva M, Kalyuzhnova N, Skripka S 2019 Intellectual specialization of Russian regions: prospects and limitations *IOP Conf. Series: materials Science and engineering* **667**
[14] Sayfutdinova R V 2016 *Modern Fundamental and Applied Researchers* **1(20)** pp 167-173
[15] Vovseenko E A 2018 *Financial Instruments and Hedging: New Rules of IFRS* European Proceedings of Social and Behavioural Sciences **L** DOI: https://dx.doi.org /10.15405/epsbs.2018.12.167
[16] Khokhlova G., Kretova N., Burov V. 2019 *The problems of investment activity of entrepreneurship and methodological aspects of credit risks assessment,* ICRE IOP Conference series: materials science and engineering **667** DOI:10.1088/1757-899X/667/1/012038

IOP Conf. Series: Materials Science and Engineering **880** (2020) 012114 doi:10.1088/1757-899X/880/1/012114

Inefficiency of the current system of forming the initial maximum price of the contract of the state construction order

T O Pinchuk[1] and I I Kazantseva[2]

[1] Irkutsk National Research Technical University, 83 Lermontov str., Irkutsk, 664074, Russia
[1] Baikal State University, 11 Lenin St., Irkutsk, 664003, Russia

E-mail: tanechka.pinchuk@list.ru

Abstract. Management of the investment cycle of construction requires the development of adequate measures to the modern challenges of innovative forms and methods of state regulation in the field of construction. The existing mechanism for the placement and implementation of the state construction order allows us to conclude that the public procurement management system is insufficient and requires further improvement, based on the development priorities of the innovation environment and the investment attractiveness of construction investment projects. In particular, one of the main components for placing a state construction order is to determine the initial maximum contract price, which affects the conditions for the implementation of a state construction order. The existing approach to the formation of the initial maximum price does not meet modern realities. The methodological foundations of the formation of the initial maximum price of the contract by dividing it into basic and managed components are considered. The separation of controlled factors into qualitative and quantitative indicators is proposed. Based on the study, it follows that through controlled factors, it is possible to influence the initial maximum price of the contract by changing the quality factors in particular.

The state construction order is understood as an organizational process for the acquisition of property rights for construction and installation works and services that meet state needs. Thus, the state construction order is implemented in the format of interaction of the state customer with participants of investment and construction activities (investors, developers, designers, construction contractors, manufacturers of construction materials, parts, structures, consulting and evaluation companies, etc.) on the basis of contractual relations. The implementation of the functional blocks of the state construction order assumes the following algorithm of actions (picture 1)

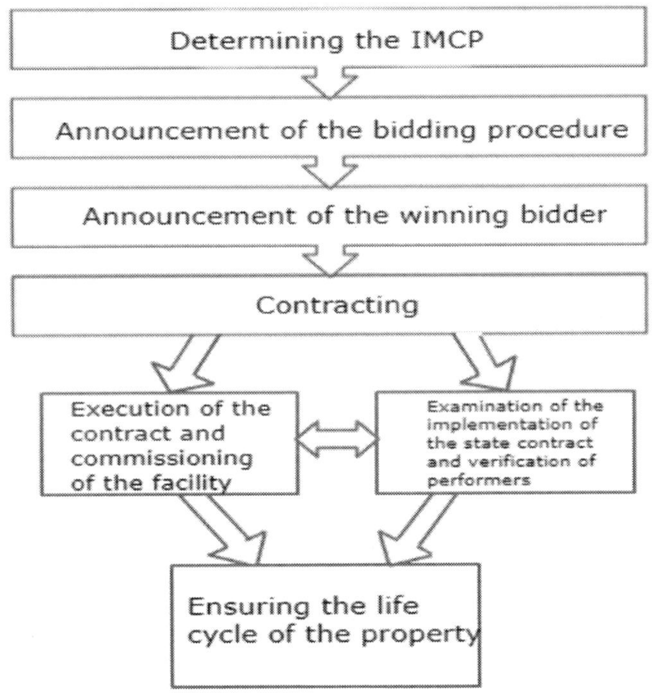

Picture 1. Algorithm for implementing the state order

In relation to investment and construction activities, bearing in mind the innovative long-term nature of the use of construction products, the state tools should acquire more predictive properties that allow you to anticipate the development of long-term sustainable trends in the innovative development of the national and world economy.

The analysis of the degree of development of the problem of public procurement of construction products as the main function of the federal contract system in the building, showed on the one hand, the fragmentation of views about the effectiveness of the federal contract system, the effectiveness of which, or reduced to a budget savings of resources by reducing the contract price or is considered from the standpoint of efficiency of investment construction project, as an object of state construction order, what requires a new approach to the formation of the initial maximum price of the contract of the state construction order in the conditions of innovative development of the economy and investment attractiveness.

It is proposed to allocate basic and conditionally managed components as part of the initial maximum contract price.

The structure of the initial maximum price contract for state construction order includes the so-called core costs, which include the cost of preparation of construction site; expenses for the basic construction facilities (cost of building materials, operating costs of construction machines and mechanisms, wages of workers-builders); costs for the objects of energy and transport services; costs of engineering infrastructure, improvement and landscaping; additional, limited, and unexpected costs, as well as construction control and supervision costs. Construction practice shows that when implementing an investment and construction project, contractors are forced to reduce the main costs, due to the quality and quantity of construction materials used, not qualified labor, not effective and not modern technical means, and as a result of the low quality of the work performed. This leads to negative consequences in the construction industry and a decrease in the pace and quality of

IOP Conf. Series: Materials Science and Engineering **880** (2020) 012114 doi:10.1088/1757-899X/880/1/012114

investment and construction projects.

In determining the initial maximum price of the contract (starting price), you should distinguish two groups of components: basic and managed. The basic components (factors) form a sedentary component in the initial maximum contract price, which is calculated on the basis of standard costs by the design and estimate method of calculating the cost. Managed components (factors) form a dynamic component that has greater variability due to qualitative and corresponding quantitative components

The result is shown graphically in picture 2.

The list of costs related to managed components in the estimated cost of the state construction order when forming the initial maximum contract price is determined, which differs from the existing system for determining costs as part of the contract price

As a result of the research, a list of costs that can be classified as manageable is proposed. This approach to the formation of the initial maximum contract price is a completely new approach, since previously the list of costs was determined by the design and estimate approach, and the costs themselves were classified only by cost elements. This division will allow you to manipulate the managed part as part of the initial maximum contract price and thus optimize costs.

The result is shown graphically in picture 3.

It is revealed that the initial maximum contract price contains basic and conditionally controlled components that change the contract price without negatively affecting other factors of the state construction order implementation, in contrast to the existing system for determining the initial maximum contract price based on the design and estimate method.

The list of costs that are included in the estimated cost of the state construction order to the basic and managed components in the formation of the initial maximum contract price, which differs from the existing system for determining costs in the contract price.

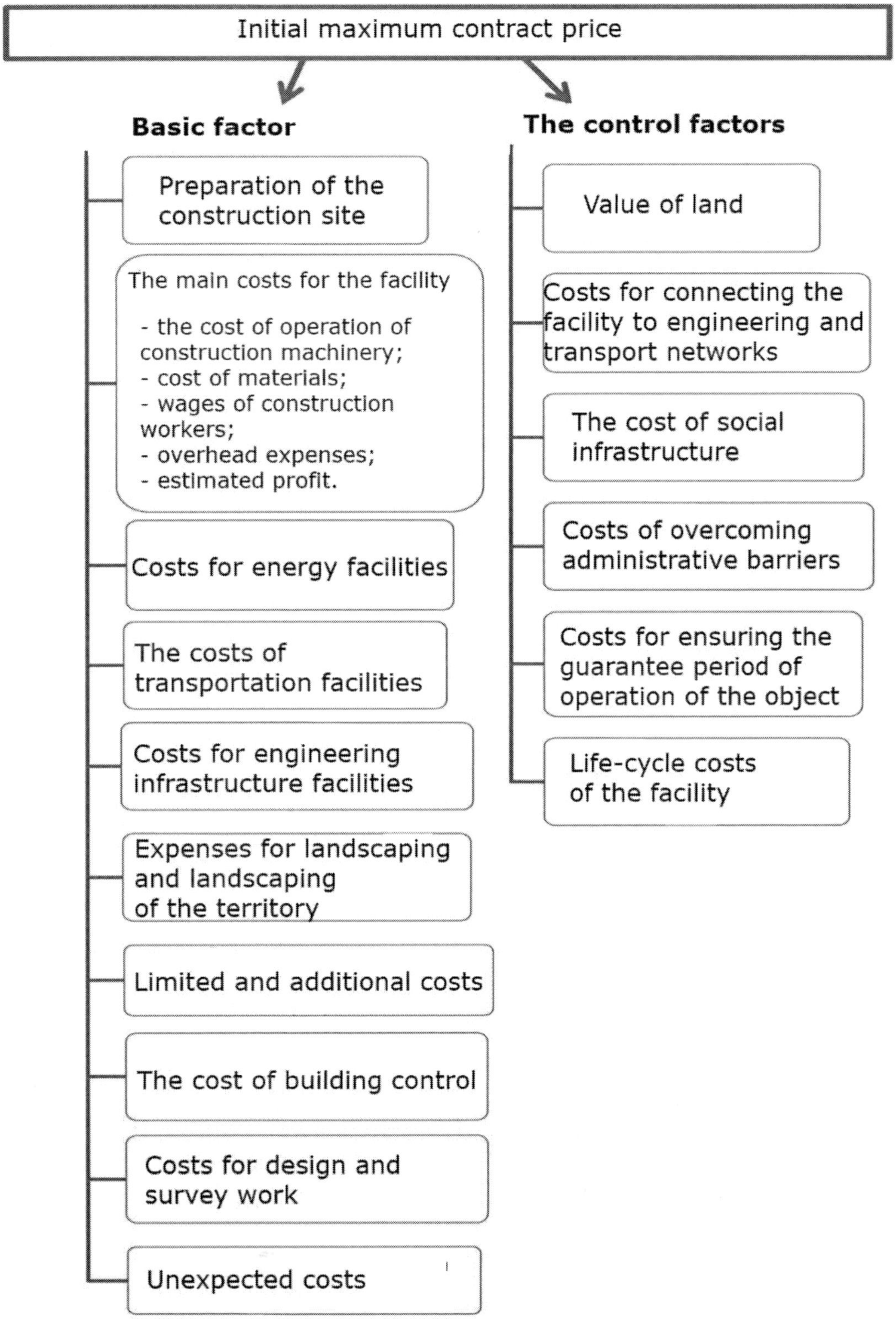

Picture 2. System of factors as part of the initial maximum contract price

IOP Conf. Series: Materials Science and Engineering **880** (2020) 012114 doi:10.1088/1757-899X/880/1/012114

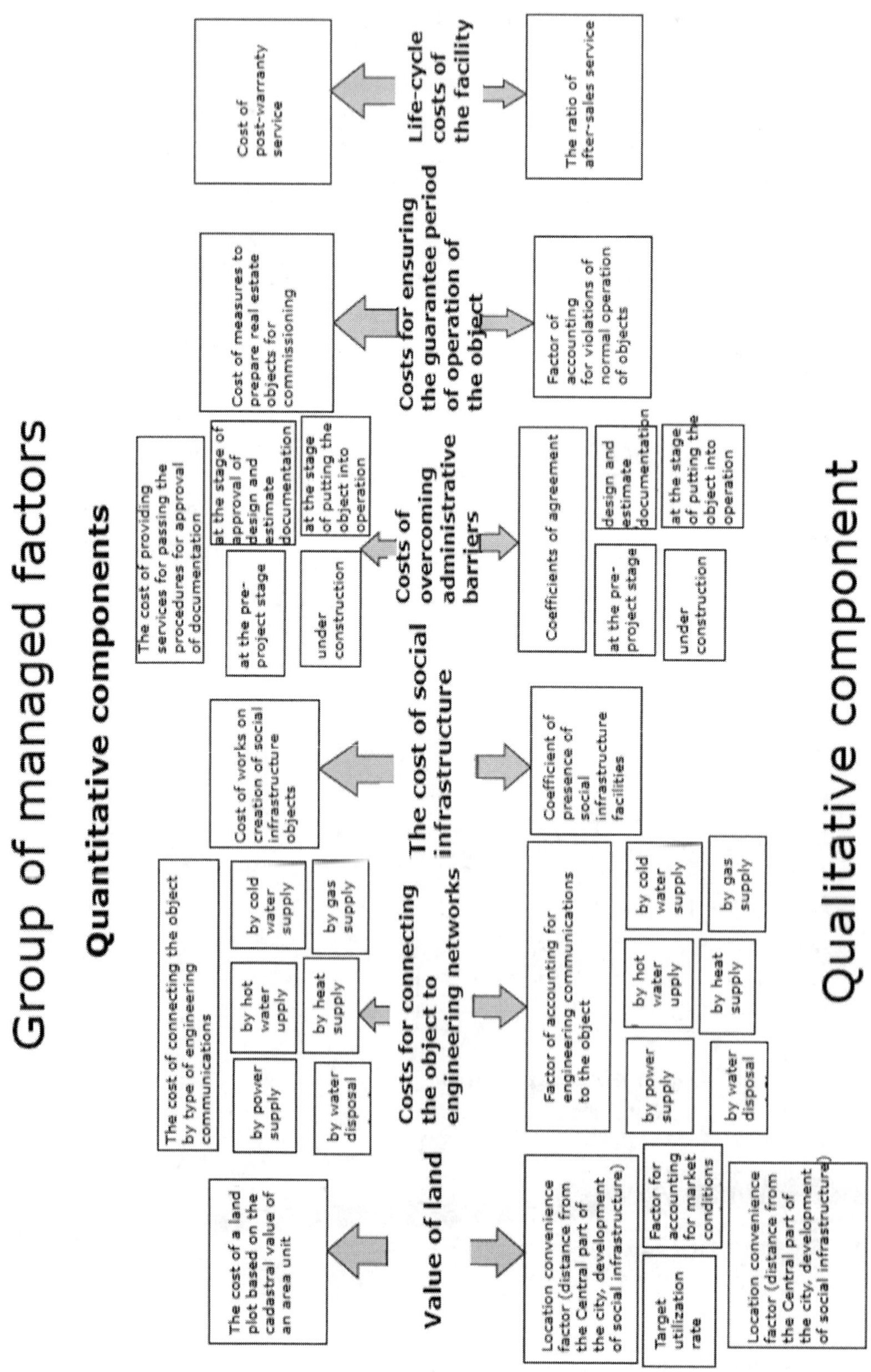

Picture 3. Composition and structure of quantitative and qualitative components of the group of managed factors

References

[1] Knyaginin V, Kuzmina A and Lipetskaya M 2018 *Priority Areas for the Implementation of Smart City Technologies in Russian Cities* (Moscow: Center for Strategic Research "North-West") p 178
[2] Vasilenko I, Lyulko A and Vasilenko E 2018 *Smart City of the XXI Century: Opportunities and Risks of Smart Technologies in City Rebranding* (Moscow: International Relations) p 256
[3] Volkov A, Sedov A and Chelyshkov P 2015 *The concept of "Smart City"* (Moscow: NRU MGSU) p 93
[4] Watson R 2014 *Smart Cities. Future. 50 Ideas You Need to Know About* (Moscow: Phantom Press) p 525
[5] Giffinger R, Fertner C and Kramar Hans 2007 *Smart Cities Ranking of European Medium-sized Cities* (Vienn: Centre of Regional Science) p 28
[6] Adcock J *Smart City* 2017 (Vancouver) p 42
[7] Shamlikashvili Ts A 2017 Mediation is a way to make our world a little better and safer *The Bulletin of the Federal Institute of Mediation* **1** pp 10-29
[8] Prorokov A N 2015 *Small business in Russia: socio-economic, legal and organizational aspect* monograph p 185
[9] Zaitseva L I 2015 Alternative methods for disputes settlement in the context of interstate relations and economic justice in the Russian Federation (Moscow: Publishing House "Original-Maket") p 216
[10] Skolubovich A Yu and Matveeva M V 2018 On the question of public-private partnerships in the field of municipal water *supply Proceedings of Universities. Investment. Construction. Real estate* **8** pp 45–53 DOI: 10.21285/2227- 2917-2018-3-45-53
[11] Peshkov A V 2019 Remote mortgage lending as a resource for housing development *Investments. Construction. Real Estate: New Technologies and Targeted Development Priorities IOP Conference Series: Materials Science and Engineering* **667** 012073
[12] Kalyuzhnova N, Violin S and Matveeva M 2019 Smart Specialization of Russian Regions: Prospects and Limitations *Investments. Construction. Real Estate: New Technologies and Targeted Development Priorities IOP Conference Series: Materials Science and Engineering* 667 012036
[13] Matveev M V 2000 Ecology, biotechnology, profit *Nature management economics* **4** pp 2-94
[14] Iljicheva E V 2006 *Ecologization and improvement of the accounting process in the conditions of transition to sustainable development* (Stary Oskol, TNT) 440
[15] Feraru G S 2007 Constructing of a strategic, ecologically oriented enterprise management *Nature management economics* **3** pp 31-50
[16] Litovka O P, Pavlov K V and Fedorov M. M. 1998 *Theory and practice questions of Nature management economics ecologization* (Saint-Petersburg — Izhevsk: Publishing house of the Economics and Management Institute of UdmSU) p.136

IOP Conf. Series: Materials Science and Engineering **880** (2020) 012115 doi:10.1088/1757-899X/880/1/012115

Concept of logistics processes in the activity of a construction company in the implementation of innovative construction projects

T O Pinchuk[1] and N V Chuvashova[1]

[1] Irkutsk National Research Technical University, 83 Lermontov str., Irkutsk, 664074, Russia

E-mail tanechka.pinchuk@list.ru

Abstract. The main objective of the study is the study and analysis of logistics processes in the activities of construction companies. During the study, methods were used to analyze existing approaches to the management of the logistics system in construction, as well as methods for the effective management of material and technical support. It was revealed that, based on the prevailing conditions, the mission of the logistics center is the introduction of an effective logistics system to ensure construction, which supports the growth of construction volumes by uninterrupted supply of construction materials to Group A. The main goal is to reduce construction time by implementing MRP2 and just-in-time concepts and a cost-effectiveness strategy. It can be concluded that logistics considers the movement of material flows in a company as a single and interconnected process, regardless of which unit controls these flows at the moment. Moreover, logistics technology involves close interaction with other departments at various stages of work with external contractors.

1. Introduction

Many companies have established special departments or logistics centers to achieve competitive advantages in the market. After collecting and analyzing the information, the employees of the department (logisticians) are engaged in planning the company's activities in different directions, followed by the organization of the relevant processes. Logisticians give recommendations to all services and departments of the construction company, based on the optimization of costs in the professional production of construction products and their sale to the customer. Control also concerns the fulfillment of the company's obligations on the basis of concluded contracts, including those related to the supply of construction materials. By acting correctly in all areas of logistics activities, they achieve the fundamental goal of increasing profits and gaining competitive advantages.

Thus, in order to optimize logistics at a construction company, it is necessary to apply a set of measures that reduce costs not only in a separate element of the supply chain, but in the aggregate in the entire chain of construction production.

To do this, it is necessary to implement a logistics management methodology. Logistics management is a set of methods and methods for managing the product distribution process based on determining the optimal parameters (the optimal criterion can be the minimum overall logistics costs, the level of service provided to customers, and the level of risks in the supply chains).

2. Materials and methods

Many construction companies that have switched to logistics management have become more efficient in organizing the entire production cycle [1]. Thanks to this concept, they were able to make more sense: use loans for the purchase of material resources, purchase raw materials, select suppliers, organize the processes of both production and distribution of finished products, as well as related information processes that accompany all stages of production organization [6].

New principles of organization and management, based on conceptual approaches and methods of thinking, United by the general concept of "logistics", are increasingly and successfully applied in practice by the most effectively functioning construction companies [3]. One of these competitive enterprises is «HCC "VostSibStroy"».

The company «HCC "VostSibStroy"» uses one of the levels of logistics management, i.e. a universal plan for optimizing logistics costs. However, based on the optimization methods presented below, you can make an individual plan to improve the efficiency of material flows at the construction company.

First of all, it is necessary to plan the supply of construction materials and related operations, which allows you to optimize purchases and inventory levels. It is necessary to analyze the entire range of purchased products and determine the part of it that brings the maximum effect, because, according to Pareto's law, 20% of efforts give 80% of the result, and the remaining 80% of efforts – only 20% of the result.

To do this, the *ABC*-analysis method is used: dividing the entire range into groups based on the demand for products-from the most popular, fast-turning products to the most low–consumption and unprofitable ones. In turn, *XYZ*-analysis allows you to evaluate the regularity of delivery of a particular group of materials. Based on a combination of these methods, the product range is optimized, which leads to a reduction in purchasing costs and helps to avoid the accumulation of excess products.

Second, suppliers of materials are determined after the rating is compiled. For this purpose, the main criteria for selecting a supplier are set, for example: reliability of delivery, price, quality of goods, payment terms, the possibility of unscheduled deliveries, and the financial condition of the supplier. The specific weight of each criterion is determined. For example, if you need to order construction material that is not in short supply, then the reliability of delivery will have the largest share. Then suppliers are evaluated according to the criteria on a point system from 0 to 10. Further, the rating of each is calculated as the sum of the products of the score of the criterion value by the specific weight of the criterion. The supplier with the highest rating, and is the best for the company «HCC "VostSibStroy"».

After optimizing the procurement procedure and inventory rationing, the most efficient warehouse logistics model is selected. When solving this issue, the financial capabilities of the enterprise, the volume of turnover, and others are taken into account. Your own warehouse allows you to change your storage strategy. But the use of such a warehouse is economically justified with a consistently high demand for construction products and profitability of the enterprise.

When optimizing logistics, we should not forget that the human factor plays a significant role in this process. It is the warehouse employees who accept the goods, place them on the shelves and complete them when ordering, as well as send materials to construction sites.

Optimization of transport logistics allows you to reduce the cost of transporting construction materials by increasing the number of transported goods, delivery points, reducing the cost of wages (by reducing the number of drivers), and the number of vehicles. This is achieved by using the company's own transport, which does not depend on tariffs, which makes it possible to plan transportation more accurately and improve the quality of services.

When using its own transport, the company «HCC "VostSibStroy"» forms the optimal routes for the delivery of materials. For this, first of all selected optimization criteria – the length of the route or time of delivery, and also takes into account parameters such as the area of delivery, schedule, delivery of the goods, the number of flights, the introduction of monitoring of mileage and consumption of fuel and lubricants (FAL). In addition to the standard fuel consumption accounting system, which includes mileage, fuel receipts submitted by drivers, and the amount of fuel consumed, modern fuel consumption monitoring systems based on *GPS*-monitoring are used, which can reduce fuel costs by up to 40%.

Specialists of the logistics center are responsible for purchasing products in accordance with requests received from internal consumers, which are the objects of construction. At this stage of implementation of the logistics system for construction, the logistics center is responsible for purchasing and delivering materials that cost 70-80% of the total cost of all materials (except for subcontracting works) - these are group *A* materials [4].

Currently, the list of group *A* materials includes the following:

– long and shaped products (fittings and other metal products);
– concrete;
– brick facing and effective;
– plywood;
– concrete goods;
– insulation material;
– elevator equipment;
– translucent structures (windows and stained glass);
– metal doors;
– ventilated curtain wall.

The mission of the logistics center «HCC "VostSibStroy"» is to implement an effective logistics system for construction support, which supports the growth of construction volumes by ensuring uninterrupted supply of construction objects with group a materials [5]. The main goal is to reduce construction time by implementing the *MRP2* and *just-in-time* concepts and the quality-cost ratio strategy [6].

It is necessary to identify the reasons that affect supply and demand in construction, improve the execution of operations, after which the stocks will disappear [2]. Managing staff in a construction company creates current, insurance and optimal reserves by purchasing additional capacity and installing backup equipment. However, these actions only hide the causes of problems. Effective management is about identifying the real problems and solving them. The *JIT* concept is an effective management system [4] that involves working in the following areas:

- *Stocks of construction materials*. HCC seeks to acquire the optimal batch of inventory [7].
- *Quality of construction materials*. Elimination of defects and improvement of quality of supplies of materials and performance of works [8].
- *Suppliers of building materials*. HCC is committed to long-term contracts with reliable suppliers of construction materials.
- *Lead time for delivery of construction materials*. Optimization of processes associated with deliveries and reduction of delivery times [10].
- *Reliability of supplies*. Continuity of delivery, without failures and downtime [9].
- *Workers*. Need a spirit of cooperation between workers, as well as between managers and workers, because the well-being of each of them depends on their overall success in work, and all employees should be treated equally and fairly. Any creative initiative expressed by any employee about possible improvements in the work is encouraged [8].
- *Information support* in the company must be perfect and allow rapid exchange of information in order to effectively manage the company [4].

The logistics center «HCC "VostSibStroy"» coordinates the processes of delivery of construction materials. Supply is made on the basis of requests. Delivery planning is carried out for a period of 1 year: the list is compiled in September and updated in December [10]. The task of the line staff of each object in the HCC structure is to submit a plan for the supply of construction materials for the year, broken down by months. Data for this plan is formed based on the estimated documentation [5]. Planning includes the following tasks:

1. Development and coordination of supply plans.

2. Choosing the best ways to supply construction.
3. Use of an effective inventory management system.
4. Implementation of information technologies in logistics.

The organization of deliveries provides:

1. Research of the current situation on the market of construction materials.
2. Evaluation and selection of suppliers.
3. Efficient inventory management.
4. Organization of storage facilities.
5. Control of material, money and information flows.

The implementation provides:

1. Purchase of construction products.
2. Delivering it to construction sites.

Implementation of the control system and analysis assume:

1. Contract enforcement.
2. Analysis and monitoring of the effectiveness of construction materials deliveries.

3. Results

The selection and evaluation of suppliers of building materials is carried out subjectively. It is necessary to further develop criteria and tighten the procedure for evaluating suppliers. In 2018, the logistics center «HCC "VostSibStroy"» formed and completed the Schedule of competitive selections for 2018. 23 competitive selections were held. 112 contracts have been concluded, with about 250 specifications attached to them.

Members of the competition committee are guided by the following criteria when evaluating submitted bids:

– quality;
– ability to fulfill obligations under the contract;
– delivery time;
– price;
– payment term;
– production capacity;
– availability of warehouse and transport infrastructure;
– warranty obligation;
– compliance of applicants proposals with the application conditions, technical requirements, russian standards, security requirements, and other criteria;
– business reputation and other criteria put forward by members of the competition commission [3].

After evaluating and selecting suppliers, the actual suppliers of building materials are determined.

The next stage is the implementation of the "management and document management concept":

1. Communication with the line staff of construction projects.

– collection of requests for the supply of construction materials;
– calculation of delivery schedules and models for acceptance of materials for the construction site;
– coordination of deliveries of construction materials.

2. Conducting analytical work:

– summarizing the results of the department's activities at the end of the reporting period, analyzing the effectiveness of its group of materials;
– creating, monitoring, and analyzing inventory;
– regulatory work of the delivery business process [9].

3. Document management support.

The logistics approach in the company contributes to the growth of construction volumes by ensuring uninterrupted supply of group *A* materials, despite the fact that:

- growth in the supply of materials (by 67% compared to 2017);
- late submission of applications by construction projects (870 applications were completed, of which about 20% were unscheduled);
- failure to meet deadlines for entering the site;
- "non-payment season" in August-September (the logistics center's budget was not fulfilled by 50% during this period). In connection with the implementation of the logistics approach in the company «HCC "VostSibStroy"» the following events were held:

1) Successfully implemented the project "Winter procurement 2018", which allowed to avoid a shortage of materials and get savings on the cost of fittings; organized responsible storage of purchased materials in the warehouses of suppliers, construction/retrofitting of the warehouse economy.

Project efficiency: at the time of completion – 21%, at the end of the year (due to falling prices on the metal market) - 17%.

2) The practice of planning competitive selection has been applied:

- a Schedule of competitive selections for 2019 has been formed and completed;
- 23 competitive selections were held; 112 contracts were concluded, with about 250 specifications attached.

3) 630 concrete fills were organized and coordinated, and 60 thousand cubic meters were delivered (an increase of 1.7 times compared to 2017). If there is a shortage of cement and non-payments in August-September, the volume declared by the objects is generally fulfilled with the minimum possible transfers. In relation to the main suppliers of concrete, a new system of price changes has been introduced: the cost of concrete changes in proportion to the change in prices for materials that make up the main cost of concrete. Thanks to this we managed to contain speculative price growth:

- the increase in the price of concrete on the market was 27%;
- increase in the purchase price-only 20%;

4) Organized storage of materials (rebar, brick, cellular blocks) in the warehouses of suppliers. Work was carried out with 7 warehouses of responsible storage. By working with suppliers and organizing storage of materials in their warehouses, significant cost savings were obtained (relative to the average market prices for storage of inventory items).

5) The centralization of material delivery continued: the delivery of 93% of rebar and 25% of piles was coordinated by the logistics center. The system of payment for transport services per flight was introduced, which allowed to perform 1 or 3 flights per day (in 2017, it was possible to make 2 flights per day).

6) Work was carried out within the framework of regulation of the supply process:

- the procedure for submitting applications based on group *A* materials has been put into effect, and training on application deadlines has been organized;
- in order to streamline the flow of documents for the supply of materials, together with the accounting department, the procedure for the movement of documents for the supply of materials of group *A* has been developed and put into effect.

7) The system of operational accounting of goods in the company's warehouses in natural units has been introduced. This made it possible to check in real time the balances of the central warehouse.

4. Conclusion

Based on the above description, as well as the analysis of approaches and methods, the following conclusions should be drawn: logistics considers the movement of material flows in the company as a single and interrelated process, regardless of which division controls these flows at the moment. Moreover, logistics technologies involve close interaction with other departments at various stages of working with external contractors.

For example, a contract with suppliers has two parts: commercial and legal. These parts are quite closely related to each other. The commercial part includes the choice of assortment, purchase price, substitute products, etc. the logistics part includes delivery conditions: their frequency, batch size, speed of order processing and preparation, distance from the supplier's warehouses, availability of goods in stock during the service period, etc.

When choosing a supplier and entering into a contract with it, it is necessary to work out not only the commercial part, but also the above-mentioned logistics parameters. This is what logisticians working in the logistics center «HCC "VostSibStroy"» do. The logistics center plans to support the overall strategy of this company. This year, we set goals for the introduction of additional information products, an effective system for managing the supply of construction materials, expanding the geography of purchases, finding new partners and concluding supply agreements. Due to the difficult economic situation in the country and the region, the logistics center is tasked with reducing production costs.

References

[1] Ching Francis D K, Winkel S R 2016 Building Codes Illustrated: A Guide to Understanding the 2015 International Building Code p 448

[2] Gang Chen 2012 Building Construction: Project Management, Construction Administration, Drawings, Specs, Detailing Tips, Schedules, Checklists and Secrets Others Don't Tell You p 196

[3] Kizim A A 2013 Efficiency of warehouse logistics based on WMS-systems *Economics of sustainable development* № **13** p 134-142

[4] Kizim A A and Kozenko V V 2013 Virtual logistics: problems and prospects *Economics of sustainable development* № **14** p 89-99

[5] Peshkov V V and Yas'kova N Yu 2015 The myths and reality of modern administration *Economic journal-XXI* № **3-4 (1)** p 32-36

[6] Pinchuk T O 2013 Innovative development of the state construction order system in the conditions of modernization of the construction industry *Bulletin of the Irkutsk state technical University* № **10 (81)** p 335-360

[7] Peshkov A V 2013 Logistics processes in the implementation of innovative energy-saving projects and their economic efficiency *Bulletin of the Irkutsk state technical University* № **11** (82) p 374-380

[8] World Summit on Sustainable Development (Johannesburg) [Electronic resource]. URL: http://www.johannesburgsummit.org (05.01.2020)

[9] Kazakbaev V, Prakht V, Dmitrievskii V, Ibrahim M, Oshurbekov S and Sarapulov S 2019 Efficiency Analysis of Low Electric Power Drives Employing Induction and Synchronous Reluctance Motors in Pump Applications *Energies* **12** doi: 10.3390/en12061144

[10] Topchiy D V, Yurgaitis A Yu, Kravchuk A S and Shevchuk D A 2019 Energy Audit of Capital Construction, Reconstruction and Reprofiling Facilities before Commissioning *University News. Investments. Construction. Real Estate* **9** pp 134–143

[11] Yanovsky A B 2019 Results of structural reorganization and technological re-equipment of the coal industry of the Russian federation and objectives for prospective development *Russian Coal Journal* **8** pp 8-16

[12] Volokitina V M and Gedich T G 2017 Investment Component in the Current Situation and Development of the Coal Industry in Russia *Economic Analysis: Theory and Practice* **16** p 1262

IOP Conf. Series: Materials Science and Engineering **880** (2020) 012116 doi:10.1088/1757-899X/880/1/012116

Alternatives of judicial economic disputes settlement in investment and construction sphere

N Yu Yaskova[1] and L I Zaitseva[2]

[1] Department of Investment and Construction Business Russian Presidental Academy of National Economy and Public Administration (RANEPA), Prospect Vernadskogo, 84, bldg 2, Moscow, 119571 Russian Federation
[2] 10th State arbitration appellate court, Sadovnicheskaya St, 68/70, bld. 1, Moscow, Russian Federation 117997

E-mail: mcua3@yandex.ru

Abstract. Modern state of construction industry signifies the necessity of coping with technological backwardness, accrued misbalances, low productiveness of construction business and also its persistent unprofitability and high level of conflict character of construction participants. Actual rise of economic interests contradictions caused by a more complicated character of projects, increase of social burden on construction business induced by inconsistent and even fragmentary measures of public regulation has led to enormous number of economic disputes. Bankruptcy caseload of investment and construction companies is at the highest possible level nowadays. Not only small and middle-sized business but also large companies cannot withstand modern challenges pressure. In this connection there is a huge field for elaboration of dispute resolution methods. Shifting towards project financing strengthens inner contradictions in the industry and also causes conflict of interests with banks. Decline of assets quality, increase of unfinished construction volume, loss of partners, withdrawal of turnover capital, waste of staff, enlargement of shadow economy segment – these are just the part of consequences which should be overcome immediately. Systemic possible solutions deal with the use of the whole specter of disputes resolution methods. Foreign experience, for example Belarusian one, proves that non-judicial methods are more flexible, rational, diverse and acceptable for parties in conflict. The most important thing is that these methods of settling differences have negotiating nature and let the parties find decisions with a high potential of practical fulfillment (not just winning the case in court). Experts expressed different opinions on the issue of conciliation procedure organizational form. The main requirement for arising disputes in investment and construction projects is an uninterrupted process and prevention of bankruptcies: debtors, contractors, subcontractors and construction shareholders. Participants of conference came to the conclusion that certain pilot projects promoting diverse disputes settlement procedures in investment and construction should be launched

One of the systemic features of modern construction business growth is the snowballing growth of disputes among its participants. Legal actions in their settlement have actually led to the crisis in shared-equity construction. Developers' bankruptcy, growth of the volumes of unfinished construction, hoodwinked investors - all these are just visible aspects of the issue. Shadow sector labour force replenishment, loss of sources of development and economic independence for enterprises in the construction industry, a drop in business activity and a decrease in the volume of commissioned housing still remain out of focus in public discussion. There is a number of reasons for that. First of all, this is an increased influence of uncertainty factors under the conditions of continuous crisis, significant changes in formats and conditions of activity, the complexity of development projects, the ongoing concentration of disproportions and contradictions in investment and construction sphere.

Current situation analyses in settlement of economic disputes in investment and construction sphere has demonstrated:

On the one hand, over past decades, business owners gained much more trust in judicial system and started to delegate the decision-making in disputes to it. It inevitably resulted in overloading the courts with cases and the forced formalism of judicial decisions;

On the other hand, the most intense crisis and call for global transformational processes over the last half of a century were reflected in growth of contradictions in investment and construction sphere participants.

Thirdly, economic interests contradiction enhancement, and sometimes even their polarization, as, e.g. in the field of shared-equity construction, led to the snowballing growth of bankruptcies and reaching the critical limit for business activities in construction.

The way out of existing situation, as the retrospective analysis of judicial practises in the Russian Empire and overseas suggests, can be found in the field of expanding the use of out-of-court methods for economic disputes settlement. Is out-of-court economic disputes settlement possible in Russia in general and in construction sphere in particular? Who and in which way is supposed to institutionalize alternative ways of disputes settlement, including mediation? Which competences will be in demand? The participants of the III International Scientific and Practical Conference "Out-of-court economic disputes settlement in investment and construction sphere" organized by the Department of Investment and Construction Business of the Russian Presidential Academy of National Economy and Public Administration (RANEPA) were trying to find the answers to those questions. For the first time the Academy served as s platform for gathering scientists and practitioners, representing business community interests, major developers and small enterprises, judges, possessing invaluable experience not only in bankruptcy cases, widely covered by media, but also in settlement rooms operating at Moscow City Arbitration Court.

All participants were unanimous, regarding the need for out-of-court methods for disputes settlement in construction. Under the conditions of market crisis and pressure on business, prompt actions of disputes settlement are of great demand than ever, pointed out the presidium member of "OPORA RUSSIA" Pozhidaev Ye.V. "Our experience shows that for small and medium-sized businesses, it is more important to take proactive measures to prevent crisis situations", he said. "The issue of economic disputes settlement between large and small construction organizations is no less acute, when courts turn into rigged trials with inevitable bankruptcies of either one, or the other, or both sides".

The representative of the Russian Guild of Property Managers and Developers (GMD) Dykov A.V. described characteristic features of three most common disputes settlement methods. The first, most common is a trial proceeding. Its obvious disadvantage is in overloading the courts with cases that often result in formal nature of judicial decisions that do not include real factors that lead the business to a dangerous line of activity termination. The second alternative mechanism is dispute settlement in arbitration. However, for many reasons, arbitration courts have lost their former popularity and business confidence in them (with the exception of large commercial arbitrations). In any case the dispute brought to court (either commercial arbitration or state) is the last point that can lead to final limit – the bankruptcy. Out-of-court disputes settlement involves the use of alternative methods designed to settle disagreements more quickly and efficiently. Developers' representative opinion was supported by the Associate Professor of the Department of Investment and Construction Business of RANEPA Folomeev Ye.V., who illustrated the critical situation having emerged in medium-size and small business segment by the department's academic research findings [1, 2].

Conference moderator, the Head of the Department of Investment and Construction Business of RANEPA Yaskova N.Yu., continuing the analysis of the current situation, noted that, on the one hand, over the past year in Russian judicial system in general proceeded 31 million cases, by the arbitration

IOP Conf. Series: Materials Science and Engineering **880** (2020) 012116 doi:10.1088/1757-899X/880/1/012116

courts - more than 1 million 900 thousand cases, among which only 1.5% of the cases ended in settlement agreements. Alongside with that, according to official data execution of the court ruling makes up 50 %, but actually it is higher than 30 %. That is, the judicial system effectiveness with respect to ensuring constructive cooperation between the parties to the disputes remains low, and most importantly, it adversely affects bona fide participants in entrepreneurial activity, which could have avoided adverse financial consequences, including bankruptcy. On the other hand, Yaskova N.Yu. provided the audience with some examples of finding the way out off the judicial system overloading from the historical perspective [3].

In the real estate sector, the aggregation of controversial issues regarding its mortgage security reached its tremendous scale in the second half of the XIX century. In 1859 the Emperor Alexander II had to issue the decree of the Senate to stop crediting from public credit organizations, thus putting an end to the first wave of mortgage lending. The new rules, symmetrical for both lenders and borrowers, and the suspension of legal proceedings gave Russian real estate the sustainable development nature, and the newly formed banks and societies became launch pads for the accumulation of capital in order to renew and rebuild urban real estate.

At the same time, the focus on the accelerated development of the real estate sector was the main point of legislative amendments. In fact, by the beginning of the 20th century it was completely updated: in central Russia, and in the Caucasus, and in the Baltic, and in the Polish provinces. Thus, in Russia as early as the 19th century, a preventive approach was used to eliminate the collapse of mortgages [4].

In modern conditions, especially in such a highly dynamic industry as construction, contradictions in economic interests and disagreements of participants are inevitable. In fact, all the speakers noted that there is no alternative to the development of efforts at conciliation designed to settle disagreements much more promptly and efficiently, including mediation. The latter present a special type of conciliation efforts, based on third arm's length party participation in the conflict. A mediator helps dispute parties have better understanding of their own interests, taking into account further cooperation prospectives, find the way out of the conflict on their own, reach mutually beneficial agreement in the process of dispute settlement and work out solutions, which both party are ready to fulfil.

As the President of the NPO “National Organization for Mediators” Shamlikashvili Ts. A. pointed out, unlike the arbitrator, the mediator does not express his or her opinion on specific methods and prospects for the dispute settlement. He or she also does not decide on the merits, as the arbitrator or judge does. Even in the arbitration court, the parties, in fact, waive the right to independently settle the disagreement, shifting this burden to the arbitrator. Mediation, in turn, gives them the opportunity to maintain control over the content of the agreements and be responsible for decisions taken [5]. That is why more than 85% of the agreements reached as a result of mediation are voluntarily implemented by the parties.

Another advantage of mediation lies in its confidentiality, as noted the judge of Moscow City Arbitration Court Ageeva L.N. “As an acting judge of the first instance, I can say that disputes in construction activities have the broadest potential for a peaceful settlement. All of the court rulings are subjects to mandatory publication, therefore, conciliation terms will be available to unknown persons. In case of mediation, nobody, including the court, will ever find out what the terms of peaceful settlement have been”. Mediator’s task is to make sure that the agreement is valid, acceptable for both parties and does not go beyond the framework of the current legislation” [6].

At the same time the Head of the Department of Provision of Proceedings of the Tenth Arbitration Appeal Court Zaitseva L.I. highlighted the fact that it would have been impossible to eliminate the court proceedings as such: the courts’ takes, beside everything else, includes judicial practice analysis and generalization, based on which it is possible to establish the specifics of the application of a particular rule of law. At the same time, it is unacceptable to assign to the court those functions that should be implemented by the parties themselves, for example, determining the total cost of work if the parties themselves agreed on a flexible price. Another example is a corporate conflict, in which 5-7 Chief Executive Officers participate and file the opposing claims. Under such conditions ensuring of an adequate and complete dispute settlement is quite complicated for the court [7, 8].

The practice of disagreements settlement in an alternative way should be so familiar for legal entities that they will use it automatically. Trying to analyse the courts activity, Ageeva L.N. in a metaphorical way compared it to a surgeon, who will not be asked to amputate the hand in case, for instance, of splinter in the finger. In addition, restrictions have been noted for judges in the framework of the procedural law, in particular, the fact they cannot go beyond the scope of the evidence and the stated requirements. The mediator is that general practitioner, who offers the party to discuss the matter and agree with each other. And only if those attempts cannot bring any positive results, then it is time for court proceedings.

The conference participants also discussed the extremely interesting experience of Moscow City Arbitration Court in organizing the Settlement Room. They first appeared in capital city in 2013 on the initiative of the Chamber of Commerce and Industry of the Russian Federation, supported by Moscow City Arbitration Court. The first agreement was reached the same year. Since that time their popularity has grown ten times. If at the beginning of their operations there were only 90 claims per year, in 2018 this number grew to up to 1,020.

While holding the discussions, special attention was paid to the experience of foreign states, as well as international organizations. In her speech the arbitrator of the International Commercial Arbitration Court at the Chamber of Commerce and Industry of the Russian Federation Gaidaenko Sher N.I. described the activitiesof the International Federation of Consulting Engineers (Federation Internationale des Ingenieurs-Conseils, FIDIC), in the framework of which the relevant provisions on dispute settlement methods have been developed, including in the investment and construction sector. Moreover, in 2017, the FIDIC Yellow Book introduced a special preventive function for Dispute Settlement Councils (commissions) (this is a permanent body being established from the start of the contract), in other words, the prevention of disagreements became their main goal [9]. The arbitrator noted that the International Chamber of Commerce (ICC) also made its provision on specified Councils more modern. The recommendations of such a body may be binding. Moreover, an appeal to the court is possible only after the project the completion, when "the ribbon is cut and all wishes are expressed". Until that moment there is a moratorium in force on disputing all decisions. At the same time ICC can suggest another option – an unbiased specialist who is constantly present at the construction site. A particular uniform is provided for him: a yellow vest and the lettering "mediator" on the back. "If the construction site happens to be in the central part of a small town, and a local resident, irritated by the noise and the dust, files a compliant, a mediator explains him what is going on and offers him an apology. It is enough to settle the disagreement and to release the tension", emphasised Gaidaenko Sher N.I. [10].

To respond to conference moderator's question concerning which country managed to find an optimal combination for court and alternative disputes settlement mechanisms, the international arbitrator came up with an example of Singapore. However, this is explained more by historical, cultural features and mentality, when maintaining a company's reputation is an absolute priority in doing business. And at the same time, market participants respect the state authorities, including the courts. As a result, the Ministry of Justice and judges of the Supreme Court become agents of mediation at the local level and in the courts. In particular, there is an institution of judges who, in the process of considering a dispute, have the right to act as a mediator, thereby helping to settle the dispute and finish the proceedings.

Conference participants also discussed the practise of conciliation efforts within the courts in Switzerland. The procedure provides for the possibility of replacing the conciliator with a mediator in the presence of the interaction difficulties. But the parties are obliged to use at least one of these pre-trial procedures. Only in case of the absence of a positive result, "big guns" come in and a court decision is made.

Getting back to the mediation usage in Russia, First Deputy General Director of the Etalon group of companies Viktorov M.Yu. expressed confidence that such an institution has great potential in the light of the ongoing changes in the investment and construction market. According to his data, more than a third of all construction companies are already unprofitable, and the industry is in a state of sharp intensification of all disputes between business entities. And this process will only intensify with the transition of the construction industry to project financing [11]. For instance, on February 1st, 2019, only 700 escrow accounts were opened in 12 Russian banks. The number of applications for project financing from the companies also did not exceed several hundreds. For comparison: in Moscow alone there are more than 800 construction sites, and even more in the Moscow Region. The imperfection of the project financing mechanism and the complexity of the transition to the new development standards, emphasized Yaskova N.Yu., led to a noticeable increase in the number of new arbitration cases on corporate bankruptcies. If in 2016, 7.2 thousand of claims were filed to declare the debtor insolvent; by the end of 2018 this number was twice as much. In 75% of cases creditors act as complainants in companies' bankruptcy. The share of claims from the state increased from 9% to 15% (in most cases, the complainant is the tax inspectorate).

Regarding the use of mediation, the results of which could be the conclusion of an amicable agreement in bankruptcy cases, the judge of the Arbitration Court of the Moscow Region Ponomarev D.A. noted that "such procedures are beneficial for participants in shared construction, as they allow to complete the facility, take into account all concerned parties' interests, including the debtor's employees, and pay off the claims in the foreseeable future when monitoring the execution on the creditors' behalf". At the same time, a significant number of persons severely complicate the process of concluding a settlement agreement. For instance, company "SU-155" had 25 thousand of shared construction participants and for Urban Group Company there were 10 thousand. Peaceful settlement in such situations is almost unachievable on the scale of the whole case, but nevertheless it is possible within certain isolated disputes. Such examples are available in the Moscow Region. As part of one of the cases of 2018, a settlement agreement was approved with respect to the construction company, which undertook to complete and obtain commissioning certificate for the residential complex located in the Lyubertsy District of the Moscow Region before May 1st, 2019. In another case (the settlement agreement was approved by the Arbitration Court of the Moscow Region in February 2019), the developer undertook the liability to pay off third-priority claim within 12 months and during the following year - the requirement of the 4th priority claim regarding payment of fines, penalties and forfeits, which were reduced by 20%.

The conference participants noted that within the framework of the disputes settlement alternative methods development, the construction and technical expertise (both out-of-court and legal) did not lose its importance. The Head of the Laboratory for Legal Construction and Technical Expertise of the Russian Federal Center of Legal Expertise under the Ministry of Justice of the Russian Federation Butyrin A.Yu. Shared the information on the latest methodological approach to expertise conducting, also having supported the initiative of experts professional training in out-of-court methods for economic disputes settlement. Doctor of Economics, Prof. Grabovoy K. P. presented his report to the conference participants. The author convincingly proved that the need for better control of each construction stage, starting with examinations of project documentation, stage-by-stage construction control and ending with the accepting of a completed construction project, significantly increased in modern conditions. In this connection construction and technical expertise should form a concept for settlement disputes between participants and if it is possible to settle the dispute out of court, so the independent expert's opinion can clearly demonstrate the prospects for reconciliation. From K.P. Grabovoy's point of view, both legal and alternative ways of settlement are impossible without the application of some particular construction and technical knowledge. It should be combined with procedural knowledge, and educational institutions should jointly begin to train qualified construction experts with the skill to settle disputes.

A certified mediator, a professional negotiator, a mediation and negotiation trainer Ye. E. Kiselev (Saint Petersburg) continued the discussion of the positive experience of out-of-court disputes settlement methods in Russia. He spoke about the initiative on creation M2B (Mediation to Business) team that is the group of mediators, operating in different cities and using the same standards. In particular, there are certain standards for case selection and working with the Commercial Case File. According to the mediator's data, one out of 10 cases is recognized as "mediable"one. Then a mediator finds the parties' contact details and initiates mediation himself/herself. The idea behind this is that neither of the parties finds itself in weaker position in case if it makes a settlement offer.

Shamlikashvili Ts.A., being a strong supporter for alternative methods non-government status, said: "If we rely on state institutions, we will come to the conclusion that we will create a quasi-state organization that is unlikely to be beneficial. Entrepreneurs should be aware of their own pragmatic interest and responsibility regarding citizens". As an option, the institution for mediators can be created as self-regulatory organization (SRO): "The most important thing is that SRO should actually, but not declaratively, include the tools of alternative disputes settlement in their global landscape. For instance, SROs can make it obligatory for their members to include mediation clauses, that is, to deal with risks using the mediation approach at the very beginning of the project". Representatives of the judicial bodies, based on their own experience, on the contrary, found it necessary to use legislative norms, obliging the parties to use options offered by alternative methods at the pre-trial stage. The opinion of the Department of Investment and Construction Business of RANEPA was less radical. For the integration of out-of-court disputes settlement methods into the real estate sector, and in particular, in construction, not only new standards and rules for conducting investment and construction activities should be developed, but also the industry Centre for out-of-court economic disputes settlement should be created. Its status still has to be determined. The main thing is that it should have the necessary functions, including dispute settlement activities scientific and methodological support, powers ensuring an equal dialogue with the authorities and large developers, as well as resources. "Only systemic activity, integrated into strategic priorities for the development of construction, operating within the mode of taking responsibility for the result under modern conditions is capable of mitigating the situation with court proceedings, if not completely resolving it. The Department's nearest plans for educational activity are the training of specialists with full-fledged management competencies and negotiation skills for resolving contradictions in the investment and construction field" concluded the Head of the Department Yaskova N.Yu.

At the final stage of the Conference, its participants became "debating parties" at a workshop on the negotiation skills development. The active discussions once again convinced the participants of the need to search for forms of alternative methods for economic disputes settlement that would have been adequate to the peculiarities of investment and construction activities.

References

[1] Yaskova N Yu 2019 *Spatial restructuring. New meanings and rules of investment and construction activities* ed N Yu Yaskova (Moscow: Publishing House "Delo" RANEPA) p 456

[2] Folomeev Ye V 2013 A process approach to the management of small businesses in construction as a way to increase the efficiency of their activities *University Bulletin* **5** (Moscow: State University of Management) pp 93-99

[3] Yaskova N Yu 2017 Methodological aspects of the cyclicality principle registration in the economic phenomena study of the construction industry *University Bulletin* **12** (Moscow: Moscow State University of Civil Engineering) pp 680-686

[4] Yaskova N Yu 2019 Non-profit forms of housing construction *Economics and Entrepreneurship* **1 (102)** pp 899-901

[5] Shamlikashvili Ts A 2017 Mediation is a way to make our world a little better and safer *The Bulletin of the Federal Institute of Mediation* **1** pp 10-29

[6] Prorokov A N 2015 *Small business in Russia: socio-economic, legal and organizational aspect* monograph p 185

[7] Zaitseva L I 2015 Alternative methods for disputes settlement in the context of interstate relations and economic justice in the Russian Federation (Moscow: Publishing House "Original-Maket") p 216

[8] Zaitseva L I and Yaskova N Yu Application of alternative dispute resolution in the field of construction projects *IOP Conference Series: Earth and Environmental Science* http://iopscience.iop.org/article/10.1088/1755-1315/90/1/012182
[9] Richbell D 2011 *Mediation of Construction Disputes* (New York: John Wiley & Sons) p 256
[10] Gaidaenko Sher N I 2018 *Agreement forms for disputes settlement arising in connection with business activities (experience of Russia and foreign countries)* (Moscow: Publishing House "Yustitsinform") p 398
[11] Viktorov M Yu and Larionov A N 2018 Modern organizational and economic problems of the transition of developers in the Moscow region from shared to project housing construction financing *Economics of Construction* **6 (54)** pp 12-21

Low-rise housing – a new tool for the development of the rental real estate market

E Yu Gorbachevskaya[1]

[1] Department of expertize and management of real estate, Irkutsk National Research Technical University, Irkutsk, 664074, Russia

E-mail: eugorbachevskaya@mail.ru

Abstract. The article deals with the development of such a segment of the residential real estate market as low-rise construction. The study analyzes existing approaches and concepts for implementing investment and construction projects in the low-rise housing segment. According to the results of the analysis, it is concluded that this segment is developing in the direction of "business" and "elite" in most of the Russian territories. The "comfort" segment is allocated only 8% of the total volume of implemented projects. According to the authors, the increase in the volume of low-rise construction in the "comfort" and "economy" class segment can become one of the main tools for solving the housing problem in the country and achieving the benchmarks indicated in the corresponding strategies. The study also considers possible options for building suburban areas by density, and it is determined that high-density low-rise housing can become relevant for "comfort" and "economy" class, and the main method of construction is large-panel housing construction. As a result of studying the collected analytical and technical information, methodological recommendations were developed for organizing mass low-rise construction of high-density low-rise housing in the comfort and economy segments. The potential advantages of the housing under consideration allow us to overcome the factors that hinder the development of rental housing, from which it is concluded that the developed recommendations are applicable not only for projects intended for sale to property, but also for renting to middle-income families.

1. Introduction

Starting around 2011, a new period of active program-targeted support for the low-rise housing market began. In accordance with the concept of long-term socio-economic development of the Russian Federation for the period up to 2020, one of the tasks of which is to stimulate the increase in low-rise development, as well as to ensure the comfort of the housing stock, the formation of a comfortable urban environment and the environment of rural settlements.

However, the ministry of construction has developed a strategy of development of housing sphere of the Russian Federation for the period till 2025, one of the important goals of which are to improve housing and increase the comfort of the urban environment and sustainable development of cities and territories. Despite the fact that the program does not mention the development of such areas as low-rise construction, however, modern concepts of low-rise buildings meet the objectives set in this program. In particular, the development of low-rise construction will make it possible to make the transition from building square meters to creating comfortable housing in the current urban environment, as well as to develop, improve and implement standards for integrated development of territories taking into account current trends. Special attention is currently paid to stimulating the use of new technologies in housing construction, which can reduce construction costs and increase its energy efficiency.

Another priority for the development of the residential real estate market is the institutionalization and legalization of the rental housing market. This issue has been raised regularly over the past decade. Pilot projects have been launched in some regions. However, this segment is still difficult to characterize as effectively functioning and solving the socio-economic tasks assigned to it. The importance, expediency and benefits of developing this segment of the housing market are obvious, but there are many barriers that prevent the institutionalization of the rental housing market in the Russian Federation, the main ones being economic and administrative.

In general, the segment of rental housing for investors from the point of view of attractiveness can be described as low-risk, since significant changes in demand and supply can only cause significant demographic changes that can be predicted for the future of 15-20 years. Another important factor is the development of the region's production sector and investment attractiveness, which can also be predicted for medium-sized and large cities, based on the results of the city-forming industries and retrospective information on the outflow and inflow of population.

The yield on such investment and construction projects as rental housing is comparable to the average rate on bank deposits, but the interest is an asset-a real estate object, the value of which, as shown by the analysis of recent decades, will only increase, and the impact of inflation processes will be minimal. This type of commercial activity may be of interest to investors in the "portfolio" approach to investment, when funds are diversified depending on the level of risk of the project.

Among the main constraints to the development of this segment of rental housing, two main factors can be identified: limited land resources of urban areas (high cost of resources, possible restrictions on use and development, requires the implementation of investment and construction projects with the corresponding expected economic results, which are not met by rental housing) and a long payback period for the project.

Despite the significant divergence of the strategy for the development of the Russian housing sector, the measures being developed and implemented should take into account all its main components as much as possible at the same time. In this regard, the rapidly developing market of low-rise construction is of particular interest. Increasing the share of the low-rise housing segment will, in our opinion, increase the availability of housing per inhabitant, make it more affordable, improve the comfort of living conditions, solve the problems of urban planning and regulation, and even create conditions for the development of the rental housing segment. Let's consider the main characteristics of this segment of the residential real estate market.

2. Analysis of the current situation in the low-rise housing segment

The existing approach to developing the concept of modern low-rise cottage settlements is a complex development that includes, in addition to the construction of low-rise houses, the construction of various commercial and social facilities. The growing market of materials encourages developers to use new materials and construction technologies in their projects, which allows reducing the cost of construction without reducing the quality of buildings. And new technologies make it possible to build more energy-efficient buildings, which subsequently reduce operating costs.

Interest in low-rise construction in Russia is growing every year. According to the analytical center under the government of the Russian Federation [1], the total area of low-rise housing in Russia as a whole was 32.9 mln m^2 in 2017; 32.5 mln m^2 in 2018; 7.7 mln m^2 in the first quarter of 2019. As a percentage, the share of individual housing construction in the total area of completed housing construction in Russia as a whole in 2017 - 41.6%; in 2018 - 43.1%; in the 1-3 quarter of 2019 - 51.7%.

If low-rise construction in Russia is only at the beginning of its development, the countries of Europe and the United States already have a wealth of experience in this area. This is reflected in the volume of construction, the structure of supply and demand, the rate of commissioning and energy efficiency of such construction, etc. In this regard, the us experience is indicative: in 2003, more than 250 million square meters of housing were built, of which 230 million are individual cottage – type houses. [12] in

IOP Conf. Series: Materials Science and Engineering **880** (2020) 012117 doi:10.1088/1757-899X/880/1/012117

Germany, Canada, the United States, and Finland, housing construction is more focused on low-rise housing, so more than 50% of the population lives in such homes. [10]

If we consider the parameter "availability of square meters per capita", we can say that in European countries and in the United States formed a large area of housing per person. [14] (picture 1)

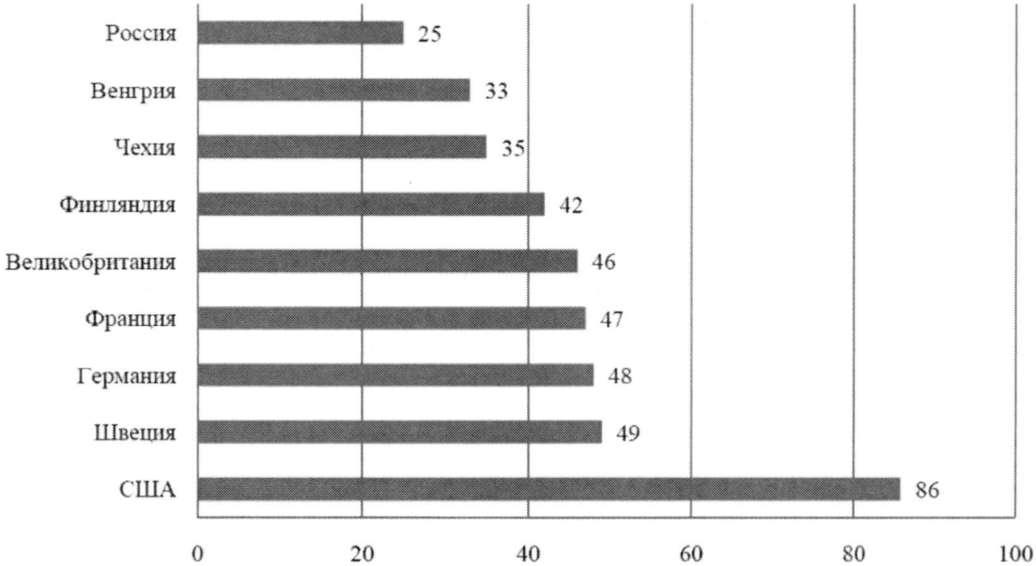

Picture 1. Number of square meters of living space per capita in Europe and Asia

The formation of a larger area of housing per person is a consequence of the popular in North America, Central and Northern Europe technology of prefabricated individual low-rise houses, at a price per square meter lower than urban. Fast-built low-rise individual houses based on modern technologies in Europe and America include:

- modular;
- panel (large element);
- frame-panel;
- glued beam;
- round timber.

The main construction technologies for low - rise buildings in North America, Central and Northern Europe are frame-panel and modular, as the fastest to build, with the lowest construction and installation costs. Also in Europe and the United States today, an approach to the design and construction of low-rise housing, called MMC - Modern Methods of Construction, (a variant in the United States-factory build housing, prefab housing), based on the design and modular assembly of the main elements of future homes [13]. The leading position in the sector of modular housing construction is occupied by Swedish housing construction companies (today in Sweden about 90 % of houses are built using prefab technology). Mass construction of low-rise residential buildings using prefabricated technology is also underway in Canada and Japan [6].

Analyzing international and domestic experience of realization of projects on construction of low-rise housing [15], it becomes obvious that large-scale construction of low-rise country houses and development of surrounding territories is the next stage of development of the real estate market [7].

State support in the form of programs to improve housing conditions and improve the comfort of the environment, as well as recently the interest of people to live in their homes attracted the attention of development companies, which began to massively implement projects of low-rise cottage settlements. To increase competitiveness in the housing market, developers are forced to rely on complex

IOP Conf. Series: Materials Science and Engineering **880** (2020) 012117 doi:10.1088/1757-899X/880/1/012117

construction, which includes the construction of various commercial and social infrastructure facilities in addition to the construction of low-rise buildings.

In view of the high competition that exists in the modern suburban real estate market, the creation of a unified concept of a cottage settlement becomes an important task for developers. The concept of development of complex low-rise residential development includes, as a rule:

- analysis of the area,
- marketing analysis,
- development of the architectural and construction concept,
- financial schemes for project implementation.

The concept is based on the analysis of the market-the history of the segment development, the existing image of the district and the price level, the current competition, the potential for the development of the district and the market. During the development of the marketing concept of the village, the target audience of the project is determined, which directly affects the positioning of the village, the development of a marketing and advertising strategy.

Architectural concept responsible for insights on the optimal, most popular sizes houses the town's future, the stylistic choice of project made on the basis of the prevailing market situation. Within the framework of the architectural and construction concept, it is necessary to make a decision on improvement (installation of fences, road construction, landscape, etc.), as well as consideration of various options for maintaining security in the village.

The set of infrastructure objects is determined both by the architectural and construction concept, and based on the conclusions of the marketing concept and the concept of development of the district.

The financial and economic part of the settlement concept clearly shows the financial results of the project, the costs associated with the implementation, and the movement of financial flows. It is on the basis of this item that the investor will make a decision on whether to implement the project.

The objects of a low-rise residential settlement may include several types of houses that differ in different volume and planning characteristics (number of floors, area, layout, etc.), as well as various construction technologies and types of houses (cottages, townhouses, etc.). Also, the object of a low-rise residential settlement can be separated from the surrounding territories and include various social and cultural objects (schools, parks, shopping complexes, etc.) [11] Usually, any modern cottage settlement is referred to a particular class of housing, which is determined already during the development of the marketing concept. Depending on the class, a low-rise village is characterized by its own characteristics according to the main requirements, such as: [11]

- Availability of failed communications. The minimum necessary for a comfortable stay of people should include central water supply, electricity, gas supply and sewerage.
- Availability of infrastructure. It involves placing pharmacies, shopping centers, schools, kindergartens, cafes, restaurants, recreation areas, and service centers in the immediate vicinity.
- Perimeter fencing, checkpoints and video surveillance.
- The presence of a management company. The management company must take over the arrangement and cleaning of the territory.
- Transport accessibility (taxis, trains).
- Concept of village architecture. On the market, you can distinguish three types of settlements: single-style settlements, settlements with housing estates and mixed settlements.
- Architectural and planning solutions for houses. These include the type of houses, the materials of which the house is built, the area of the house and the plot.
- Availability of unique structures or special natural areas. These may be non-typical structures (marinas, golf clubs, etc.), as well as special natural areas (nature reserves, sea coast, etc.)

According to the analysis, on the example of Irkutsk and its surrounding territories, approximately the same number of conceptual cottage settlements of elite class and comfort is currently being implemented - 34% and 36%, respectively. Business and economy classes account for 8% each, and premium classes account for 14%. The distribution of residential space in low-rise cottage settlements, depending on the class, is as follows: elite class accounts for 39% of the built residential area; 24% is comfort class; 25% - premium and business classes and 12% - economy class.

According to the results, it is obvious that little attention is paid to the "economy" segment, while increasing the volume of low-rise construction in this segment can become one of the main tools for solving the housing problem in the country and achieving the targets described at the beginning of this study.

In low-rise projects, it is more profitable to build houses consisting of several apartments than cottages designed for one family, since the effective use of the surrounding area is ensured, and the cost of connecting communications is reduced. In this case, the low-rise typology allows you to achieve a density comparable to a higher-rise building. [4] (picture 2)

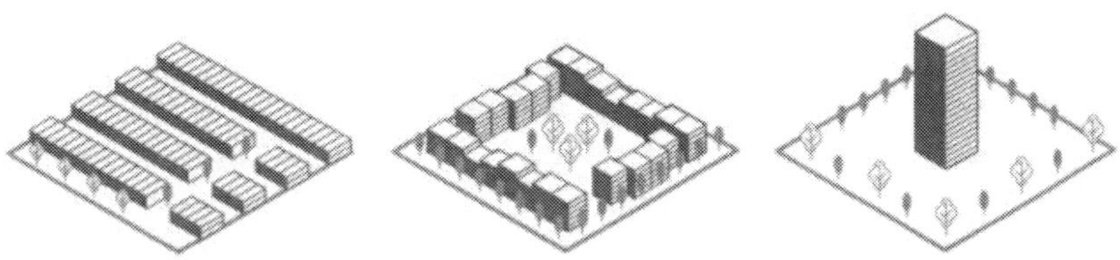

Picture 2. Plots with the same density of residential units that have different building typologies

High-density low-rise housing is recognized as the missing format in the modern market, which allows to restrain the spread of boundless suburbs, while combining the advantages of suburban and urban, high-rise and low-rise. The shortage of such homes, for example, has already been recognized nationally by the Canadian Association of Home Builders (CAHB), calling them the "missing middle" (picture 3). [2]

Picture 3. Missing segment of low-rise high-density housing [9]

In August 2016, canadian experts developed recommendations for the construction of townhouses and multi-family low-rise residential buildings, which include an expanded classification of low-rise housing. According to these recommendations, low-rise buildings are considered to be no higher than four floors, which, in addition to individual (one or two families) and apartment buildings, also include four types of townhouses (picture 4).

1 2 3 4 5 6 7

Picture 4. Canadian classification of low-rise residential buildings
(1-an individual house for one family; 2-an individual house for two families; 3-a townhouse; 4-a townhouse with several apartments located above each other and having a separate entrance; 5-a townhouse with several apartments located above each other, having a separate entrance and adjacent

side and rear walls; 6-low-rise buildings that have both apartments with a separate entrance and multi-apartment floors with a common entrance; 7-a multi-apartment low-rise house)

3. Low-rise high-density housing as one of the directions of development of the rental real estate market

The development of construction of low-rise apartment buildings in the economy segment is also possible within the framework of the institutionalization of the rental housing market. For this type of commercial housing, low cost is relevant (due to more affordable land resources and building construction technology). The following building concept will meet the main requirements.

In the economy class cottage settlement, it is proposed to provide for an increase in the density of development by adopting more dense typologies-blocked houses (duplexes, townhouses, etc.) and sectional houses. The number of floors of buildings in low-rise cottage settlements is accepted according to the urban planning code of the Russian Federation, and should not exceed three floors. Based on the analysis of cottage settlements in the suburbs of Irkutsk, as well as the recommendations of the set of principles [4], the number of floors of an ordinary building should be 2-3 floors - for a sectional house.

Table 1. "Recommended characteristics of a sectional house"

Name of the indicator	Value of indicator
Approximate floor area	350–450 m^2
Number of apartments per floor	2-6
Approximate minimum area of one apartment	40 m^2

Architectural and planning decisions of houses in the cottage village of economy class are accepted as standard. It is also recommended that some blocked and sectional houses take into account the free layout of the premises, thereby increasing the competitiveness of the project being implemented. It is proposed to adopt three-layer reinforced concrete wall panels as the wall material for blocked and sectional economy class houses.

The best option for providing settlements with an engineering infrastructure of economy class will be central water supply, drainage and electricity. Heat supply can be provided by individual electric boilers.

One of the technologies that involves reducing the cost of construction is the technology of panel housing construction. The main advantage of using this technology over others is that ready-made construction parts are delivered to the construction site, which the builders can only mount. As a result, labor productivity is very high, since long and labor-intensive processes such as installing rebar, concreting, and laying a brick wall are completely excluded.

If we consider Western countries, precast concrete is the second most popular building material after wood, as well as SIP panels. [5]

One of the main advantages of such buildings is the high speed of construction. It is also important that at the initial stage, you can accurately determine the final cost of the house. Three factors accelerate the construction process: [5]

- high installation readiness of structures (factory-made reinforced concrete structures);
- large panel area (reducing the number of mounted elements);
- effective use of insulation (insulation laid between the skin, several times lightens the weight of the wall and gives it excellent thermal insulation properties).

A special feature of this design is the high quality responsibility of panel joints.

4. Conclusion

Today, when the rental housing market is "shadow", which significantly reduces the volume of tax revenues to the budget, prevents increased mobility of labor resources both domestically and internationally, and, in the end, does not make it possible to increase the security of the tenant of housing, the construction of low-rise residential buildings in the complex development of suburban areas using pre-fabricated technology (large-panel housing construction) can become one of the main tools in the search for solutions to this issue.

The concept of mass development of districts with low-rise housing described in the article makes it possible to eliminate the main restrictions that hinder the development of the rental real estate market:

1. Land resources in remote areas of the city and suburban areas are more accessible, and the cost of their development is comparable to the cost of developing urban areas.
2. Reduced construction costs due to the use of pre-fabricated technology for large-panel housing construction and reduced construction time, as a result, higher economic results in general from the implementation of the investment and construction project.

References

[1] Analytical Center under the Government of the Russian Federation URL: http://ac.gov.ru (26.02.2019 г.)
[2] Canadian Home Builders' Association URL: http://www.chba.ca (01.04.2019г.)
[3] CasaSicurabyRapidcasa. URL: http://www.rapidcasa.com/ 1.06.2019 г.).
[4] Code of principles for the integrated development of territories. URL: https://дом.рф/wp-content/uploads/2018/04/Svod-printsipov kompleksnogorazvitiya-gorodskih-territorij.pdf (01.04.2019)
[5] Dmitrienko E A 2017 Innovative structural solutions for prefabricated precast buildings *Bulletin of the Donbass National Academy of Civil Engineering and Architecture* **4** p 108-115
[6] Gifford R 2007 The Consequences of Living in High-Rise Buildings *Architectural Science Review* **50**
[7] Gorbachevskaya E Yu 2019 Gentrification: causes and consequences for the regional housing market in Russia *IOP Conf. Ser.: Mater. Sci. Eng.* https://iopscience.iop.org/issue/1757-899X/667/1
[8] Madi H 2016 Townhouse and Low-rise Apartment Guidelines *Madi H., Bir-chall D.* URL: https://www.toronto.ca/wpcontent/uploads/2018/01/964c-townhouse-low-rise-guidelines-02-2017.pdf (09.04.2019г.)
[9] Missing Middle: Responding to the Demand for Walkable Irban Living URL: http://missingmiddlehousing.com (05.04.2019 г.)
[10] Shapochkin R A 2016 Analysis of the development of the foreign market of low-rise construction *Investments, construction, real estate as the material basis for modernization and innovative development of the economy VI International Scientific and Practical Conference* **2**
[11] Shapochkin R A 2017 Management of the development of low-rise residential settlements (on the example of the suburban zone of the city of Krasnoyarsk *Master's dissertation* p 62-97
[12] Shapochkin R A Analysis of the development of the foreign low-rise market Development of the infrastructure of low-rise construction in the city of Krasnoyarsk URL:https://core.ac.uk/download/pdf/84934011.pdf (28.04.2019 г.)
[13] Stadnik E B 2012 Design experience of modern low-rise housing based on the unification of building structures *Architecton - University News* **38**
[14] Trushin G 2019 Low-rise housing: pros and cons. URL: http://realty.vesti.ru/zhile/maloetazhnoe-zhile-plyusy-i-minusy (15.04.2019 г.)
[15] Yaroshevich I Yu and Gorbachevskaya E Yu 2019 Investment attractiveness of modern low-rise cottage villages in Russia *ISTU Youth Herald* **3** pp 80-85

IOP Conf. Series: Materials Science and Engineering **880** (2020) 012118 doi:10.1088/1757-899X/880/1/012118

Specifics of construction in Eastern Siberia in 2015-2019

N V Kaigorodova[1] and L G Rudykh[1]

[1] Irkutsk National Research Technical University, 83 Lermontov str., Irkutsk, 664074, Russia

E-mail: kaygorodova@mail.ru

Abstract. The article is devoted to the specifics of Eastern Siberia construction industry development in 2015-2019. Questions about a place of a building branch in structure of economy of region are actualized, data on priority directions of building in the investigated period are resulted, influence of bank sphere, mortgage crediting is specified, reasons of recession of the market of working trades are described, a degree of development of building in region is revealed. The article describes the current state and trends in the construction industry of Eastern Siberia for the last five years. The comparison of large-panel and reinforced concrete house-building is given, the advantages and disadvantages of each type of residential building construction are analyzed. The specificity of resource base of construction materials is noted, the reason of high cost of construction objects in the region, starting from complicated climatic conditions and finishing with economically expensive transport logistics is specified. The prospects for the development of other construction directions than civil construction are presented. The conclusions of the article reflect the specific character of the construction complex in Eastern Siberia in 2015-2019, which is expressed by the high demand of the population for the purchase of housing, the availability of a sufficient number of enterprises on the housing construction market, a wide mineral base for the production of construction materials.

1. Introduction

The construction industry has always played a leading role in the development of economic development in any region, as it creates conditions for the development of other sectors of the economy. The construction industry in Eastern Siberia takes one of the first places. According to statistics, the industry employs 6% of the region's workforce and 26% of the region's workforce. In recent years, there has been a high construction activity in the construction industry, and the region is ahead of the market average in terms of the actual volume of work index and ranks sixth in Russia.

In 1970-1980 the Irkutsk region had one of the most powerful construction complexes in the country. Such organizations as Glavvostoksibstroy, Angarsk Construction Department, Bratskgesstroy were able to erect the largest objects of power engineering, machine building, non-ferrous and ferrous metallurgy, timber industry, chemistry and whole cities in a few years. The construction industry supplied its structures to Surgut, Buryatia, Chita, Mongolia, Sakhalin, etc.

However, public funding for industrial and housing construction began to decline in the 1990s and then stopped. The advantages of the construction complex in Eastern Siberia - industriality, assembly, large-panel house-building, automated control systems - turned out to be its disadvantages [1]. There

IOP Conf. Series: Materials Science and Engineering **880** (2020) 012118 doi:10.1088/1757-899X/880/1/012118

were no funds for modernization of heavy technological lines and transition to another type of products, and only a part of enterprises managed to adapt to new market requirements, to master new types of materials, structures and new technologies [2]. Since the 2010s, there has been a certain growth in housing construction.

The modern construction industry of Eastern Siberia is represented by organizations performing design and expert works, general construction works, specialized types of work, electrical installation works, road works, production of basic building materials, structures, products. Eastern Siberia construction is characterized by the following sectors: 1) industrial construction objects, 2) civil construction objects, 3) agricultural construction objects, 4) transport construction objects, 5) military objects, 6) hydraulic engineering objects, 8) hydromeliorative objects.

2. Research questions

The development of the Eastern Siberia construction industry in 2015-2019 has a number of specific characteristics [3]. The construction of residential real estate has the largest specific weight in our region. It accounts for 7% of the gross regional product of the region and ranks fifth among all other industries. On the first place on introduction of habitation among cities of the region the regional centre takes the first place - in Irkutsk 332 thousand sq. m. of inhabited premises have been introduced (according to statistics for 2018), on the second place Angarsk - 26 thousand sq. m., on the third Bratsk - 16 thousand sq. m.

In Eastern Siberia there are more than 7000 large and medium construction organizations of various forms of ownership engaged in industrial and civil construction. The large organizations are based in the industrial centers of the region. Today in the territory of the region the following largest construction organizations work in various branches of construction: MUE "UKS of Irkutsk", "FSK Novy Gorod", GC "Vostok Center", GC "VostSibStroy", CJSC "Stroykompleks", "FSK DomStroy", "Holding SERS", "SK Maxstroy", JSC "SiberiaEnergoTrade", LLC "Vysota", etc. [4].

The development of the construction industry is supported by investment banking activities. Since 2018, the Russian Federation has introduced escrow accounts, which represent a deposit of funds for construction financing [5]. This makes it possible to impose more stringent requirements on the developer. So far, this practice is just beginning in Eastern Siberia.

Mortgage lending plays an important role in the development of the construction industry. The mortgage on sale of residential premises makes up 50%, and on some objects of residential development reaches 75-80%. Credit interest on mortgages in the region is 11-12% [6].

The development of the construction industry in Eastern Siberia is constrained by a limited market for workers - masons, welders and fitters. In 2019, the modal salary was 35-60 thousand rubles. However, the social prestige of the working professions in Russia is low, and construction organizations need to make efforts to raise the prestige of the working professions in construction.

Growth in housing construction is one of the top priorities of the construction complex, the most important task of social development in Eastern Siberia. At the same time, there are a number of negative factors affecting the rise in construction costs and volumes in the region, such as: 1) unfavorable climatic conditions (presence of permafrost); 2) high seismic activity; 3) remoteness from the main producers of construction materials; 4) high transportation costs; 5) inclusion of a part of the Irkutsk region territory into the Baikal natural area, where it is necessary to pass the state ecological expertise of project documentation.

The state of the construction industry and systemic problems in it in recent years have become one of the most powerful constraints for the development of the entire construction complex. Insufficient development of the given segment of the market in the region became the reason that we use a lot of imported materials that reduces economy and increases term of construction, complicates industrial logistics, puts builders in dependence on transport systems [7, 8]. At the same time, many materials produced in our company are inferior to our imported analogues.

Eastern Siberia is provided with certain capacities for the production of non-metallic materials, wall blocks including autoclave hardening, wood products, gypsum board, metal structures, reinforced concrete structures, bricks, insulators, paintwork materials, thermal insulation, there is a stock of capacity for cement production. It is necessary to deliver finishing materials, fiberglass fittings, carbon fabric, composite and other materials.

In the Irkutsk region there are 6 main enterprises producing bricks. Available capacities of brick producers are not fully loaded. The decrease in production is caused by several factors: 1) the quality of the products does not meet modern requirements (size, color, strength, etc.), 2) consumption is seasonal, and 3) the use of wall blocks in enclosing structures is increasing [9, 10].

Large-panel house-building (hereinafter - KPI) was widely used in Soviet times in the construction of buildings for various purposes, up to 60% of residential and social buildings were erected in the Irkutsk region from KPI mass series 1-335. In spite of the fact that buildings of efficiency have a number of advantages such as: all-season construction, controlled quality, shorter construction time, high factory availability. At present the buildings are mainly designed on a monolithic frame with wall fillings: from blocks, bricks with insulation, curtain facade systems, which does not always ensure energy efficiency of wall fillings, construction time is increased, the proper quality of work is not provided [11]. Wider application of prefabricated reinforced concrete products will allow to provide construction sites with quality products, to carry out installation works at any time of the year, to reduce the terms of construction of facilities, while carrying out modernization of existing facilities, it is possible to reduce costs (electricity, steam, etc.) for the production of reinforced concrete products, which will reduce the cost of the final product.

The main supplier of cement to the East Siberian market is Angarskcement. According to the data for 2015 this enterprise met 86% of cement needs in the region, being in fact a monopolist on the market. The current production of JSC "Angarskcement" produces cement grades 400, 500 without additives grade 400, which does not always meet the requirements for the production of responsible structures [12]. Some consumers import cement from "Achinsk Cement Plant", which has better quality characteristics and lower price.

Thermal insulating products are widely used for thermal insulation of enclosing structures, basement and attic floors, foundations of buildings and structures, pipelines. The projects mainly provide for insulation: such as mineral wool slabs and mats, basalt-based minplits, foamed polystyrene boards, foamed plastic, foamed polyurethane shells.

One of the most acute issues for the construction industry in the region is the issue of providing with thermal insulating materials (TIM). Delivery of qualitative building thermal protection from glass fibre (companies Ursa, Isover, Knauf, etc.) and mineral wool (Rockwool, Parok, etc.) is made mainly from the European part of Russia, in this connection its sharp increase in price at the expense of transportation costs occurs. Mineral wool insulation is mainly 60% imported from outside the region from the Khabarovsk and Krasnoyarsk Territories, and UPSA and Isover type insulation from the Novosibirsk Region and other regions of the country [13]. The available capacity in the Irkutsk Region is insufficient and the products are not always of a certain quality level.

The strategy for the development of the construction materials industry in the Irkutsk region in the future until 2030 offers: 1) modernization and technological development of production base; 2) creation of long-term conditions for sustainable development of building materials enterprises and increase of their competitiveness; 3) overcoming the technological gap of regional enterprises of building materials from the leading domestic and foreign companies; 4) reduction of energy consumption at the enterprises of building materials; 5) maximum use of household and industrial waste in the production of building materials [14].

3. Materials and methods

There are more than 30 producers of marketable concrete and reinforced concrete structures in the region, the production of marketable concrete is developed in almost all municipalities, but not everywhere is proper control over the quality of products. On available capacities it is possible to produce most types of reinforced concrete such as blocks, floor slabs, beams, trusses, supports, wall panels, stairs, rings, etc. Annually, the production of reinforced concrete slabs is reduced by 5-8% due to their mass application in monolithic projects. The equipment of most manufacturers of reinforced concrete structures was launched during the Soviet period of mass construction and has a wear and tear of up to 40%, is energy-intensive and requires a lot of labor during the production of structures. In this connection, the cost of produced structures is often higher than that of monolithic designs.

The article uses general scientific methods: induction, deduction, rise from the abstract to the concrete, analysis, synthesis, explanation, analogy, comparison.

4. Results

It should be noted that construction in Eastern Siberia has its own characteristics associated with the harsh Siberian climate, which implies an increased level of thermal insulation that allows you to save on heating.

The specific nature of the resource base has a number of issues that need to be addressed, including the high cost of construction and cost-effective transport logistics.

5. Discussion

Activation of building branch in other directions (building of objects of industrial, transport, hydraulic engineering appointment) can be connected with the project of formation of the North Siberian industrial belt developed recently.

The Irkutsk Region is one of the five most industrially developed regions in the Siberian Federal District and belongs to the group of regions where the role of the mining complex is lower than that of the processing complex. This is due to the presence of two large fuel and industrial complexes (hereinafter referred to as the Fuel and Processing Complex), established as early as 1940-1970: the Irkutsk-Cheremkhovsky and Sredneangarsky. These complexes specialized in chemistry, petrochemistry, non-ferrous metallurgy, power engineering and timber processing, and to a lesser extent in defense engineering [15]. The base for the project of the North-Siberian industrial belt is the Sredneangarsk TIC (MO of Bratsk city, Ust-Ilimsk city, MO of Bodaibo city and district, Bratsk district, Kazachinsk-Lensky district, Katanga district, Kirensky district, Mamsk-Chui district, Nizhneilimsky district, Nizhneudinsky district, Taishetsky district, Ust-Ilimsky district, Ust-Kutsky district, Chunsky district).

The project of forming the North-Siberian industrial belt is based on the following factors:

1. Significant natural and resource potential of the Irkutsk region in the northern regions of the region. The zone of concentration of productive forces and infrastructure in the north is the highway territory of the Baikal-Amur Mainline (BAM) zone, which covers the cities of Irkutsk and Kursk. Bratsk, Taishet, Ust-Ilimsk, Chuna, Bratsk, Kirensky, Kazachinsko-Lensky, Nizhneilimsky, Ust-Kutsky, Ust-Ilimsky districts. The BAM zone also includes Katanga, Mamsko-Chui and Bodaiba districts.
2. The Baikal-Amur Mainline is the 1st category road with the design carrying capacity of 18 million tons of cargo per year on the western section, 9 million tons on the eastern section, electrified on alternating current, double-track from Taishet to Lena (704 km) and single-track from Lena to Taximo (725 km) [16].

Implementation of such a project in the region will allow the construction industry to increase its capacity practically in all construction directions.

6. Conclusion

The specifics of the construction industry in Eastern Siberia in 2015-2019 is the high potential for the development of the construction industry with: 1) high demand of the population associated with rural-urban migration of the population, 2) availability of a sufficient number of enterprises in the housing construction market, 3) a wide mineral base for the production of construction materials. Nevertheless, the technological backwardness of a number of industries related to the construction industry, high mortgage rates for the population, and the low prestige of working professions in construction reduce the opportunities for the construction industry. Also specific to the construction industry in Eastern

Siberia in 2015-2019 is the prevalence of civil (primarily residential) construction in the general spectrum of the industry.

References

[1] Timoshenko A I 2018 Economic Development of Siberia as the Result of State Strategies Aimed at Industrial Modernization of the USSR *Historical-economic studies* **19** pp 91–118
[2] Tulokhonov A K, Batomunkuev V S and Boldanov T A 2018 The Role of the Transport and Communication Infrastructure in the Strategy for Spatial Development of Asian Russia: History, Problems, Risks and Prospects *New Economic Issues* **3** pp 4–13
[3] Rudykh L and Shilova O 2018 Analysis of the socio-economic indicators of the Irkutsk region, Buryatia, and the Far East in 2016-2017: investments and prospects *International Scientific Conference "Investment, Construction, Real Estate: New Technologies and Special-Purpose Development Priorities" (ICRE 2018)* **212** DOI: https://doi.org/10.1051/matecconf/201821208014
[4] The amount of work performed by type of "Activity" in 2019 in the Irkutsk region 2019 Statistics *Federal state statistics service Statistics for the Irkutsk region* URL: http://irkutskstat.old.gks.ru/wps/wcm/connect/rosstat_ts/irkutskstat/ru (28.01.2020)
[5] Construction in Priangarie: results and prospects 2019 *Sibdom.ru* URL: https://irk.sibdom.ru/journal/1849/ (28.01.2020)
[6] Shelomenceva N N 2018 To the problem of project financing in housing construction *Building economic* **6** pp 22–31
[7] Berliseva A E 2017 Construction industry analysis in the Irkutsk region *Science and education* **6** pp 73–77
[8] Ignatova O A 2018 Demographics of the construction industry in Irkutsk region *Current trends in spatial development and priorities of public geography. Barnaul*, 12-19 September
[9] Ryazantseva A A 2019 Analysis of development of the construction industry of the Irkutsk region in the current economic conditions *Informatization and virtualization of economic and social life* Irkutsk, May 14
[10] Saneev B G, Korneev A G and Agafonov G V 2016 Assessment of regional demand for electricity in the areas of the construction of new hydroelectric power plant in eastern siberia and the far east *Energy policy* **1** pp 37-49.
[11] Zubareva G I 2017 Perspectives of Straw-bale building as environmentally safe construction *Ecology and Industry of Russia* **21(6)** pp 10–14.
[12] Investment and construction in the Irkutsk region: statistical Bulletin, 2019.
[13] Fokin S 2018 Import substitution in the construction industry as an example from the Irkutsk region *Contract construction* **11** pp 66-76
[14] The draft Strategy for the development of the construction materials industry in the Irkutsk region was approved at a meeting of the regional government *Regional socio-political newspaper* URL: http://www.ogirk.ru/2017/02/20/proekt-strategii-razvitiya-promyshlennosti-stroitelnyh-materialov-irkutskoj-oblasti-odobrili-na-zasedanii-pravitelstva-regiona/ (28.01.2020)
[15] Zatsepin A 2019 Change proposal in the development strategy of electric power of the South of Siberia *National priorities of Russia* **4** pp 75-78

IOP Conf. Series: Materials Science and Engineering **880** (2020) 012119 doi:10.1088/1757-899X/880/1/012119

Impact of escrow accounts on construction rates

M V Matveeva

[1] Irkutsk National Research Technical University, 83 Lermontov str., Irkutsk, 664074, Russia

E-mail: expertiza@istu.edu

Abstract. Equity construction is actively developing in the Russian Federation. At the beginning of the second half of 2018, there were fundamental changes in the Russian real estate market. Federal Act No. 1 of 2018 July 175-ФЗ was adopted, amending the Act of 2004 on the procedure for citizens 'participation in equity construction. In particular, the system of subsidization of objects under construction has been completely revised - the equity participation was replaced by the mechanism of project financing. Federal Law No. 175-ФЗ changes the legal and regulatory regulation of housing construction, defining project financing as the only source of attraction of funds to equity construction. In this regard, interest in investments in real estate is growing, and there is an increase in interest of private capital in the housing industry. Thus, today the real estate market is changing, construction companies adapt businesses processes to new requirements. In order to assess the impact of escrow accounts on the real estate market, a comprehensive analysis of market development should be carried out and trends in development should be monitored.

The construction complex of the country is a significant segment of the economy. Activity in the sphere of construction is the most accurate indicator reflecting the state and level of development of economic spheres of the country. The extent to which the processes of the construction industry function effectively depends on the provision of demand from the state, business and citizens with new capacities, houses, high-quality objects [1].

In order to ensure the continuous development of the construction industry, the Government of the Russian Federation is carrying out a number of reforms aimed at optimizing the construction industry, providing affordable housing for citizens and improving the comfort of the urban environment. The most important and all covering reform has been taking place in Russia since July 1, 2017. The purpose of the reform is to protect the funds of shareholders and simplify the work of construction organizations with control bodies at all stages of construction. From July 1, 2019, the state launched a new reform mechanism, approving new rules of equity construction [2]. Since 2019, the transition to project financing of housing construction has been carried out - construction of apartment buildings with priority attraction of funds of the bank, in particular using escrow accounts. The purpose of this reform is to minimize the risks of buyers of apartments under construction, to provide certain guarantees for the safety of funds paid by persons purchasing housing at the stage of construction, to increase the requirements for developers and to control transactions involving funds of shareholders.

Escrow accounts are special accounts that are held by banks called escrow agents. Settlements using the escrow account are based on a three-way contract between the buyer deposit, the seller beneficiary and the escrow agent. The transaction includes several stages of Picture 1 (Picture 1) [3,4].

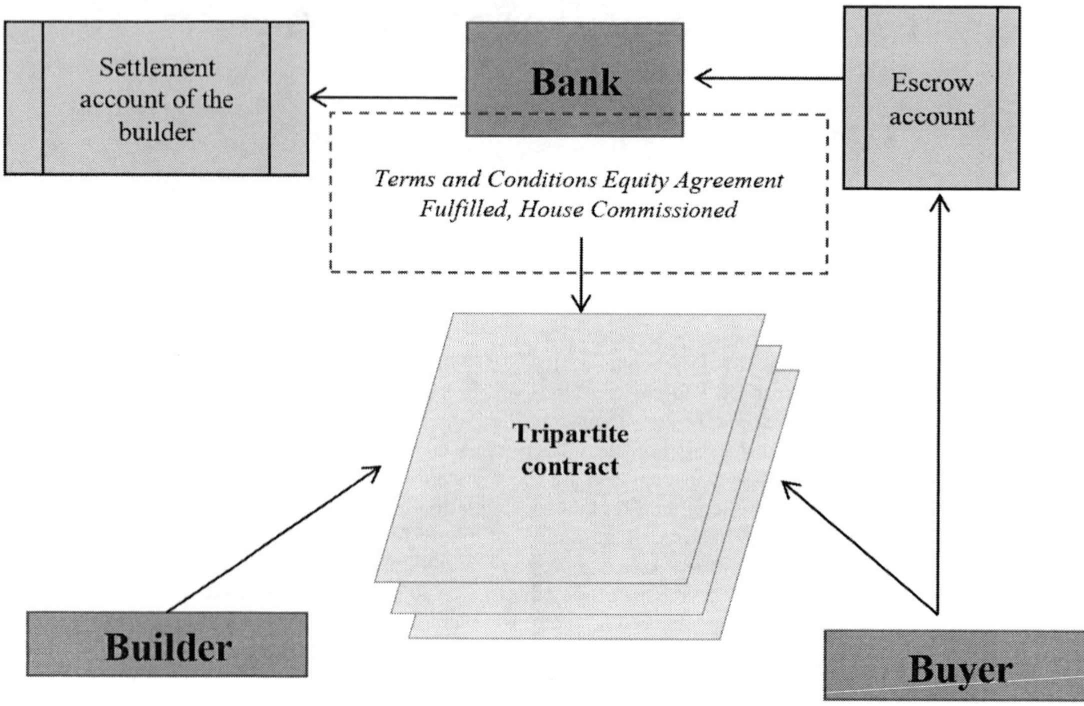

Picture 1. Scheme of transactions with new buildings using escrow accounts

Let 's look at each step in detail:

1. The parties conclude a contract: the seller undertakes to deliver the goods or provide the service, and the buyer pays for it. The bank in turn guarantees that the money will be transferred only after delivery.

2. The buyer opens an escrow account with the bank and contributes the required amount.

3. The bank is blocking money in your account.

4. The seller supplies the goods or service and presents evidence to the bank of this, such as the act of acceptance and transfer or the certificate of the new owner of the property, such as an extract from the Unified State Register of Taxpayers.

5. The bank releases the account and transfers the money to the seller.

This scheme of carrying out transactions with new buildings using escrow accounts caused a violent reaction in the professional community and posed a number of questions to the market participants. We will determine the advantages and disadvantages of innovations both for the developer and for the buyer (Table 1) [5].

Table 1. Advantages and disadvantages of using escrow accounts

Advantages	Disadvantages
1. The developer is guaranteed to receive money if he meets his obligations and the buyer is protected from fraud.	1. Not all banks in Russia work with escrow accounts [7].
2. If one of the parties decides to terminate the	2. Before the transfer to the developer, the money in the account is blocked, and the

transaction, the account is closed automatically and the money is returned to the buyer [6].	buyer can return it only if the contract is terminated.
3. The parties can choose which documents will be proof of the transaction for the bank.	3. As long as the money lies in the escrow account, it does not bring profit - unlike the deposit interest on it does not accrue.
4. If it is a matter of equity construction of housing, the content of the escrow account is free of charge, and amounts up to 10 million rubles are insured in the Deposit Insurance Agency.	4. The deponent must pay for the opening of the account a commission - usually 0.7 - 1% of the amount, in transactions on equity construction for the maintenance of the account is paid by the beneficiary - the developer company [8].
5. The escrow account cannot be seized even if the buyer has debts on taxes, fines, alimony, etc.	

The launch of the mechanism of purchase of housing at the stage of construction through escrow accounts first of all allows to remove from the market the situation when on the one side of the transaction on the not yet built object there is a shareholder (individual), and on the other side - a developer (legal entity) [19]. This imposed legal and information asymmetry: in practice, it is impossible for an ordinary citizen to assess the reliability of a developer, and in case of bankruptcy, shareholders had to co-operate in any way to protect their interests. In turn, with the launch of project financing, the interests of all shareholders are protected by the bank, which accredited the developer.

This takes all relationships to a fundamentally different level. Firstly, the bank has professional competencies in evaluating the developer and an individual object, which already seriously reduces the risks. Secondly, in the event of bankruptcy of the developer, the fate of the property will depend on the bank: either search and attraction of a new developer, or payment to equity holders. In any case, it is not profitable for the bank itself to keep an unfinished object on its balance sheet, thus it actually has the same interests as shareholders.

Shortcomings in the application of escrow accounts in the real estate market result in the most efficient construction companies absorbing business, and volumes that inefficient companies cannot conduct [10]. The consequence is a slowdown in construction growth. As the graph shows, in the Irkutsk region in 2019 there was a sharp decrease in the number of issued building permits (Picture 2).

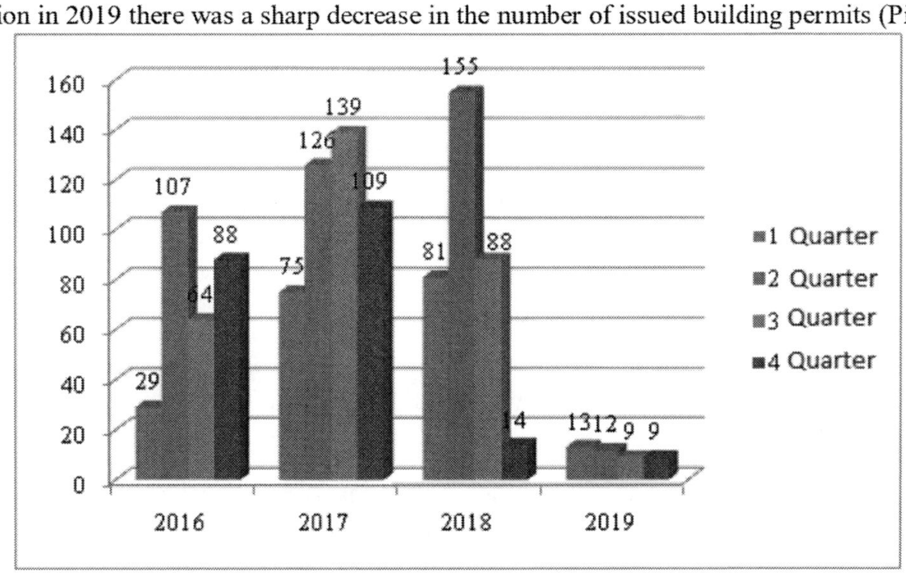

Picture 2. Dynamics of issued building permits in the Irkutsk region from 2016-2019

The number of building permits issued in 2019 decreased by more than 7 times. The decrease in the rate of construction permits issued is due to a decrease in the activity of developers of apartment buildings (Picture 3) [11].

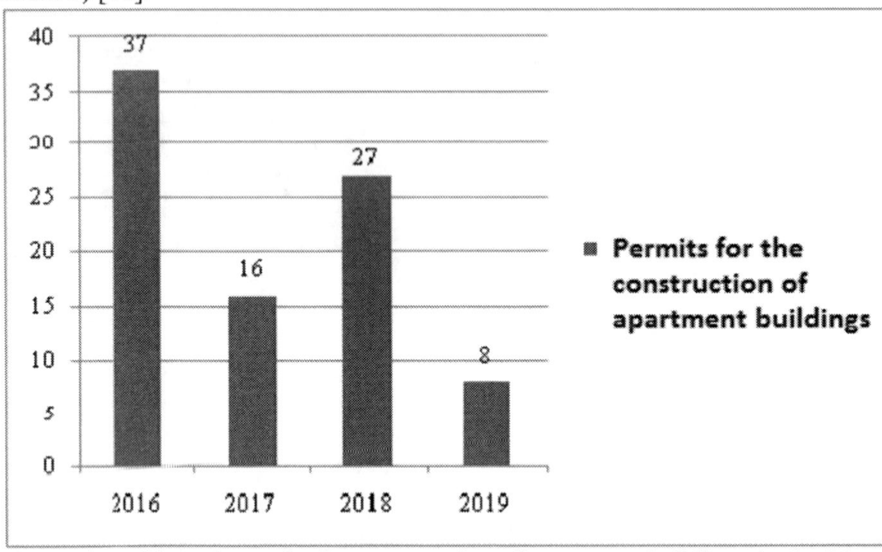

Picture 3. Dynamics of issued permits for construction of apartment buildings in the Irkutsk region since 2016-2019

Eight building permits were issued in 2019 four times the average for the past three years. Practice shows that the amendments made to Federal Law No. 214-ФЗ have had a negative impact on the activity of the construction sector. The application of escrow accounts in construction is still a challenge for developers [12,13]. Many of them are not ready to work on the market in accordance with the new legislation, as it has low profitability, which indicates that the necessary material resources are insufficient. The application of escrow accounts leads to a longer period of credit burden, as it is not possible to use funds from the sale of shareholders to pay for construction. At the same time, the floating lending rate depends on the security of the loan with funds from the actual sales received by the bank on escrow accounts. Accordingly, in the new conditions, developers will be forced to provide a smaller discount to apartment prices at the initial stages of construction relative to the prices of finished apartments. This will allow developers to compensate for the decline in profitability resulting from the increase in cost.

The sharp decline in construction permits is due to the fact that the implementation of equity construction reform is in an active stage, when the construction industry is only at the beginning of the path of forming the relevant competences that are necessary for the application of project financing using escrow accounts [14,15]. The purpose of escrow accounts realization is strict control over expenditure of funds by the developer. And it becomes a problem for those who are used to managing cash flows, taking money from one project, investing in another, taking first profits and then paying costs.

Accordingly, in order to build quality modern housing at affordable prices and in large volumes in a short time, developers will have to be able to attract financing with minimal costs and expenses. While developers and property developers bear a significant share of the costs at the pre-project stage before obtaining a building permit and starting sales.

Despite the difficulty of organizing an investment and construction project using escrow accounts, their number is open for settlements under agreements of participation in equity construction, And the amount of funds deposited by the participants of the equity construction on escrow accounts, In two months more than 2 times - with 10,1 thousand accounts and 30.2 billion rubles as of August 26 to 23,2 thousand accounts and 73.8 billion rubles as of October 28, 2019 (Picture 4). In 2019, five credit agreements totaling 3,372.7million rubles were concluded in the Irkutsk region.

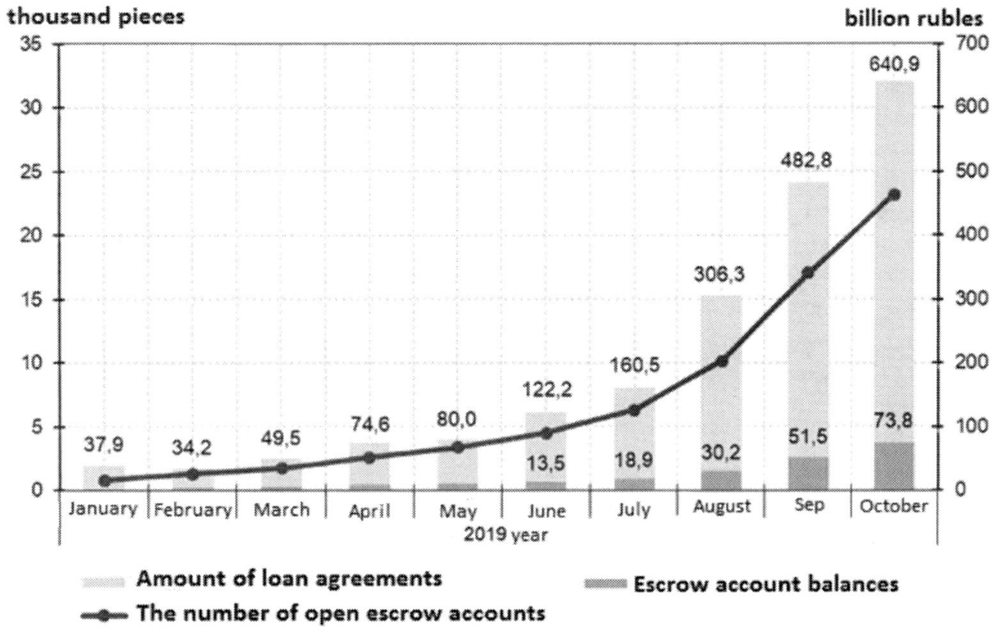

Picture 4. Dynamics of opening escrow accounts in 2019

Thus, the geography of project financing is expanding. Equity construction projects using escrow accounts are under way in most regions. Ultimately, the transition to project finance will create conditions for quality housing growth and increase public confidence in housing acquisition during the construction phase.

References

[1] Komarov A K, Ivanov I A and Lundenbazar B 2019 Theory and practice of the use of gabions for forming protective structures *Proceedings of Universities. Investment. Construction. Real estate* **9(1)** pp 78–89 DOI: 10.21285/2227-2917-2019-1-78-89

[2] Kazimirov I A and Peshkov V V 2019 Determination of price behaviour in the secondary residential real estate market using a multidimensional regression model *Proceedings of Universities. Investment. Construction. Real estate* **9(3)** pp 476–487 DOI: 10.21285/2227-2917-2019-3-476-487

[3] Larina O P and Khalabuda E I 2019 Accident prevention during construction and installation works at height *Proceedings of Universities. Investment. Construction. Real estate* **9(4)** pp 734–741 https://doi.org/10.21285/2227-2917-2019-4-734-741

[4] Yaskova N Yu 2018 Construction in the context of new views on the goals of economic development. *Proceedings of Universities. Investment. Construction. Real estate* **8** pp 133-142 DOI: 10.21285/2227-2917- 2018-1-133-142

[5] Baranova A A and Bobrova A A 2019 Dispersed reinforcement of cellular and fine-grained concrete based on silica fume *Proceedings of Universities. Investment. Construction. Real estate* **9(4)** pp 694–703.

[6] Shamlikashvili Ts A 2017 *Mediation is a way to make our world a little better and safer* (Moscow:The Bulletin of the Federal Institute of Mediation) **1** pp 10-29

[7] Zaitseva L I 2015 *Alternative methods for disputes settlement in the context of interstate relations and economic justice in the Russian Federation* (Moscow: Publishing House "Original-Maket") p 216

[8] Zaitseva L I and Yaskova N Yu *Application of alternative dispute resolution in the field of*

construction projects (IOP Conference Series: Earth and Environmental Science) http://iopscience.iop.org/article/10.1088/1755-1315/90/1/012182

[9] Richbell D 2011 *Mediation of Construction Disputes* (New York: John Wiley & Sons) p 256

[10] Gaidaenko Sher N I 2018 *Agreement forms for disputes settlement arising in connection with businessactivities (experience of Russia and foreign countries)* (Moscow: Publishing House "Yustitsinform") p 398

[11] Viktorov M Yu and Larionov A N 2018 *Modern organizational and economic problems of the transition of developers in the Moscow region from shared to project housing construction financing* (Moscow: Economics of Construction) No 6 (54) p 12-21

[12] Peshkov V, Gertsekovich A and Gorbachevskaya L 2019 Dilapidated and dilapidated housing in the aspect of the Federal project "Ensuring sustainable reduction of uninhabitable housing" *Investments. Construction. Real Estate: New Technologies and Targeted Development Priorities IOP Conference Series: Materials Science and Engineering* **667** 012075

[13] Gertsekovich D, Gorbachevskaya L, Grigorova L and Peshkov V 2019 Return on investment in REIT real estate funds *Investments. Construction. Real Estate: New Technologies and Targeted Development Priorities IOP Conference Series: Materials Science and Engineering* **667** 012063

[14] Peshkov A V 2019 Remote mortgage lending as a resource for housing development *Investments. Construction. Real Estate: New Technologies and Targeted Development Priorities IOP Conference Series: Materials Science and Engineering* **667** 012073

[15] Yaskova N Yu and Sarchenko V I 2018Development of strategic approaches to the creation of aqua-territorial industrial complexes in the Arctic zone *Proceedings of Universities. Investment. Construction. Real estate* **8** pp. 84-93 DOI: 10.21285/2227-2917-2018-2-84-93

IOP Conf. Series: Materials Science and Engineering **880** (2020) 012120 doi:10.1088/1757-899X/880/1/012120

Specifics of construction in China in 2019-2020

L G Rudykh[1]

[1] Irkutsk National Research Technical University, 83 Lermontov str., Irkutsk, 664074, Russia

E-mail: lili.rudih2011@yandex.ru

Abstract. China is currently building entire cities, and new settlements are rapidly being built. It is important to note that the ideology of construction has undergone significant changes. if earlier the high rate of construction was due to a large number of cheap labor, and the quality suffered, today the emphasis is on the quality of construction work and materials. For this purpose, China has trained specialists-professionals who have received the best education and passed training in construction companies in different countries of the world, so today Chinese young engineers can teach their foreign colleagues how to innovate in the field of construction. The successful development of the construction industry is facilitated by strict state control, and the fact that almost all construction equipment of modern samples is produced by China itself and does not depend on foreign investment. This, in turn, contributes to an increase in sales of construction vehicles to other countries. Today, after many years of development of construction industrialization, BIM technologies have become widespread, which has determined the acceleration of the growth of modular and industrial development. It is no secret that the Chinese economy is developing rapidly. Of course, effective development of the economy, industry and agriculture is impossible without the development of construction, services and infrastructure.

Keywords: information modeling concept, modular construction, large-scale construction, state control, high-rise buildings, planned economy, high-speed construction

1. Introduction

Due to China's strong interest in modular construction, the integration of BIM technology is seen as a strategy for expanding the market.

Applying the concept of building information modeling (BIM) provides many advantages for virtual design, including the use of 3D laser scanners for data collection, as well as the use of a robotic total station. BIM is a good way to improve construction efficiency by optimizing the construction schedule [1].

Modular construction with BIM support is based on the modularity of building components and is applied more often to high-rise buildings [2]. After the architectural project is ready, the specialists start creating a structural model by selecting the appropriate modules that will be analyzed using the software.

Today, this modular technology is widely used for the arrangement of large areas, for the placement of warehouses, retail premises, hospitals, and hotels. The technology is based on the use of light metal structures and sandwich panels. An important task is to prepare a platform for the Foundation, where a light metal frame and sandwich panels are being built, to which the engineering infrastructure is brought, taking into account the safety requirements [3].

The nearest practice in Wuhan has shown that innovations in modular construction using BIM give high quality in a short time of construction.

China demonstrated modern achievements in construction back in 2003, when the "XIAOTANGSHAN" hospital was built in Beijing in the shortest possible time during the fight against the SARS virus.

In connection with the coronavirus epidemic in China, thanks to the rapid construction in Wuhan, a modern medical center "HOSHENSHAN" for 1000 places, with an area of 34 thousand m^2, was built in 10 days [4]. The center is built of two-story modular container-type blocks using standard projects and modular structures that can be expanded to increase space if necessary. This method allows you to quickly, effectively, and most importantly inexpensive to solve problems of high-speed construction.

Despite the virus epidemic, construction in China continues. In February 2020, the construction of the metro in Beijing resumed, and a permanent orbital station is under construction, which is expected to be completed in 2023. It will consist of several modules and will have a volume of 160 m^3 [5].

China's membership in the world trade organization gave an impetus to the modernization of its own equipment and the development of new technologies. Many models of cranes and excavators were developed from scratch and put into mass production. China also shows good results in the field of export and import of engineering products.

2. Research questions

The main sector of the construction industry in China is: cement production; production of flat glass; production of bricks; mining and processing of limestone and gypsum; production of ceramics [6].

More recently, the us was the leader in the construction industry all over the world, and now China is the first in this industry [7].

The growth rate of cement production in China as the main building material is indicative (Fig. 1).

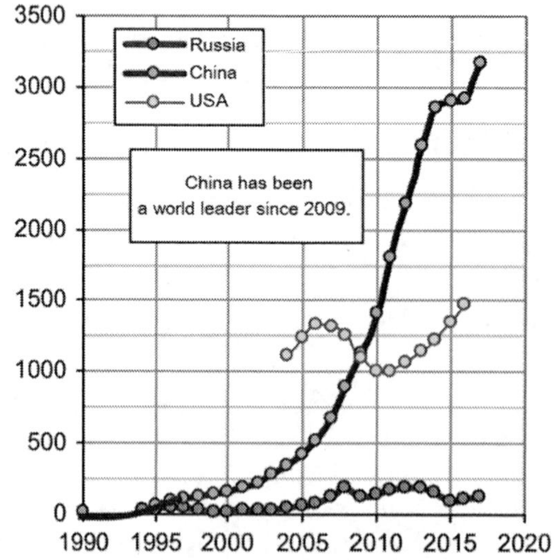

Figure 1. Growth rates of cement production

Today, China is a world leader not only in high-speed construction, but also in the construction of high-rise buildings. In 2019 appeared in China 15 skyscrapers above 200 m. This is 11.9 percent of the total global quantity. An example is The Tianjin CTF financial center, which was built in 2019. its height is 530 m (Tianjin CTF), and the Shanghai tower is 634 m high and has an area of 380 m2 [8]. All high-rise buildings are built near transport hubs [9].

Figure 2. Shanghai Center Tower

In recent years, China has surprised the world with its large-scale construction projects, high skyscrapers and endless bridges, for example, the Qingdao bridge is 42.5 km, the Tianxingzhou bridge over the Yangtze river is the longest screw bridge with road and rail traffic [10].

The state and local regional authorities exercise total control over construction, including road construction [11]. The contractor invests only its own funds in the construction, and investors pay the contractor only after the object is put into operation. This often motivates construction companies to perform work in the shortest possible time and at a high quality level.

China is a confident leader in the production of building materials, housing, roads, and power plants [12].

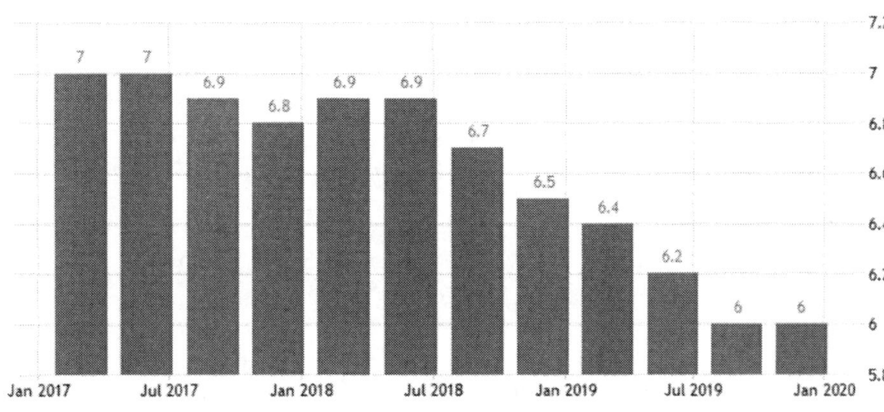

Figure 3. China's GDP growth rate, % [13]

3. Materials and methods

In recent years, China has overtaken all countries in the world in terms of housing construction [14].

The state strategy for 2014-2020 assumes maintaining high rates of urbanization of the population. The main issues of housing policy are the problem of empty housing in villages and the shortage of housing in cities. The urbanization program for 2014-2020 focuses on the construction of affordable housing at the expense of the Federal budget [15]. As part of this national program, construction is progressing rapidly, and many apartments are being rented out ahead of schedule.

The housing market in China is characterized by an active role of the state in its regulation. More than 80% of the construction market is occupied by private construction companies. State-owned companies are mainly engaged in infrastructure construction. Almost 40 % of construction is carried out through the reconstruction of built-up areas.

The work used General scientific methods of cognition: induction and deduction, progress from the abstract to the concrete, analysis and synthesis, explanation, analogy, comparison.

4. Results

The fact that construction in China has its own characteristics associated with the planned and market economy is determined. The important role of state control in the construction industry is noted.

5. Discussion

China's construction industry is developing at a tremendous pace. Analysis of the success of construction in China allows us to note the specifics of economic development, which is based on the planned economy.

6. Conclusion

The incredible speed of construction in China is already perceived by some experts in the field of construction as the norm. In 2018-2019 alone, more than half of the world's skyscrapers were built in China. It is necessary to name the reasons that contribute to the rapid development of the construction industry, including: close attention and control by the state; own production of construction equipment (up to the most complex complexes for the construction of high-rise buildings); high motivation of developers themselves and strict discipline; the emergence of Chinese highly professional specialists who received quality education abroad; the absence of corruption and planned high-speed construction.

Perhaps the secret of success in the construction industry of the PRC is not just the number of people and equipment, but rather specific approach, which are not only quantitative but also qualitative indicators.

References

[1] Domozhirova Y A, Stepanova J S and Vinidiktova M E 2019 Benefits of BIM technology on the example of the Chinese experience *Engineering journal of Don* **3(54)** *pp 45*

[2] Gerasimov A I, Nikonova E V 2019 Promising areas of application of BIM technologies in the operation of buildings and structures. *Engineering journal of Don* **1**

[3] Ten V 2020 Race against the virus. Construction newspaper **5**

[4] More than half of all those infected with coronavirus in China have already recovered. *Ria.ru*. URL: https://ria.ru/20200301/1565765766.html (accessed 02.03.2020)

[5] Next decade's space station: China. *Habr*. URL: https://habr.com/ru/post/404661/ (accessed 02.03.2020)

[6] Rudykh L G 2015 Specific character of Chinese economic strategy in South-East Asia. Proceedings of Irkutsk State Technical University **6** pp 258–260

[7] Churbanov A E, Shamara Yu A 2018 The impact of information modelling technology on the development of investment-construction process *Vestnik MGSU* **13** pp 824–835. DOI: 10.22227/1997-0935.2018.7.824-835

[8] Shanghai tower-a symbol of China's prosperity. *World attractions*. URL: https://cattur.ru/asia/china/shankhayskaya-bashnya.html (accessed 02.03.2020)

[9] Junjie L and Hui Y 2017 A Research on Development of Construction Industrialization Based on BIM Technology under the Background of Industry 4.0 *MATEC Web of Conferences* **100** pp 1–8. https://doi.org/10.1051/matecconf/2017100020469.

[10] Nosov V D 2017 China in the 2010s: The Economy of Overheating *Econolicheskaya*

Politika **12** pp 24–41 DOI: 10.18288/1994-5124-2017-6-02
[11] Sazonov S L and Chen Xiao 2017 Chinese high-speed railways have overtaken the whole world. Results of the 12th five-year plan (2011–2015) and prospects for the development of the Chinese economy until 2020 p 343
[12] Ryabtseva V V and Kravchenko A V 2015 Trends in the construction market in the conditions of international economic crisis *International journal of applied and fundamental research*, **8** pp 587-591
[13] Gui Y, Yuhe W, Yuxin Zh, Liming W, Houli X, Yuan F and Jian Z 2019 Impact of Migrant Workers on Total Factor Productivity in Chinese Construction Industry *Sustainability* **11** pp 926–297 https://doi.org/10.3390/su11030926
[14] People's daily (China): why are there no slums in China? *People's daily.* URL: https://inosmi.ru/social/20191124/246270424.html (accessed 02.03.2020)

AUTHOR INDEX